BOTANY

FOR DEGREE STUDENTS

ALGAE

B.R. VASHISHTA

Formerly Reader, Biology Study Group
Department of Botany
Punjab University, Chandigarh

Head of the Department of Botany
Multanimal Modi College, Modi Nagar
and
Punjab University College, Hoshiarpur

Revised by

Dr. A.K. SINHA
M.Sc., Ph.D., FPSI, FSB
Formerly Reader & Head, Department of Botany
Feroze Gandhi Postgraduate College
(Kanpur University), Raebareli, U.P.

Dr. V.P. SINGH
M.Sc., Ph.D., FSB
Reader, Department of Botany
Feroze Gandhi Postgraduate College
(Kanpur University), Raebareli, U.P.

S Chand And Company Limited

(ISO 9001 Certified Company)

S Chand And Company Limited

(ISO 9001 Certified Company)

Head Office: D-92, Sector–2, Noida – 201301, U.P. (India), Ph. 91-120-4682700

Registered Office: A-27, 2nd Floor, Mohan Co-operative Industrial Estate, New Delhi – 110 044, Phone: 011-49731800

www.**schandpublishing.com;** e-mail: **info@schandpublishing.com**

Marketing Offices:

Chennai	:	Ph: 23632120; chennai@schandpublishing.com
Guwahati	:	Ph: 2738811, 2735640; guwahati@schandpublishing.com
Hyderabad	:	Ph: 40186018; hyderabad@schandpublishing.com
Jalandhar	:	Ph: 4645630; jalandhar@schandpublishing.com
Kolkata	:	Ph: 23357458, 23353914; kolkata@schandpublishing.com
Lucknow	:	Ph: 4003633; lucknow@schandpublishing.com
Mumbai	:	Ph: 25000297; mumbai@schandpublishing.com
Patna	:	Ph: 2260011; patna@schandpublishing.com

First Edition 1960

Subsequent Editions and Reprints 1965, 67, 68, 70, 71, 72, 73, 74, 76, 77, 78, 79, 81, 83, 84, 85, 86, 87, 90, 91, 92, 93, 95, 96, 97, 99, 2002, 2004, 2005, 2007, 2008 (Twice)

First Multicolour Illustrative Revised Edition 2010; Reprints 2012, 2013, 2015, 2016 (Twice), 2018, 2019, 2020 (Twice), 2021 (Twice), 2022 (Twice)

Reprint 2025

ISBN: 978-81-219-3521-0 **Product Code:** H6BAL68BOTN10ENZX10O

PRINTED IN INDIA

By Vikas Publishing House Private Limited, Plot 20/4, Site-IV, Industrial Area Sahibabad, Ghaziabad – 201 010 and Published by S Chand And Company Limited, A-27, 2nd Floor, Mohan Co-operative Industrial Estate, New Delhi – 110 044.

PREFACE TO THE REVISED EDITION

We have the pleasure to present this thoroughly revised, updated, improved and multicoloured edition of the book ***Botany for Degree Students – Algae*** for the benefit of undergraduate and postgraduate students. In the present edition many new topics have been added to make the book more useful for those appearing at competitive examinations, in addition to postgraduate students. Topics and important genera added in the present edition of the book include Protochlorophyta, Toxic algae, colourless Algae, Biofertilisers, Cyanophages, Bioluminisense and genera Microcystis and Navicula. All the diagrams have been made of several colours making these more attractive. Questions of recent examinations have also been added.

During the course of revision of this book we have been guided and assisted by the suggestions of a large number of students and professors of colleges and universities to whome we express our sincere thanks.

We also wish to express our deep sense of gratitude to Mrs. Nirmala Gupta, CMD, Mr. Navin Joshi, Vice-President (Publishing), Mr. R.S. Saxena, Advisor, Mr. Shishir Bhatnagar, Manager (Pre-press), Mr. Ashutosh Mani Tripathi, Asstt. Editor and Mr. Dhan Singh Karki, Graphic Designer of S.Chand & Company Ltd., New Delhi for their help and cooperation in bringing out this book in the present form. We hope that the present multicoloured edition of the book will prove to be of much more use to all its users.

Suggestions and criticisms for the improvement of the book will be gratefully accepted and acknowledged.

Dr. A.K. SINHA
Dr. V.P. SINGH

PREFACE

The chief aim of writing this book is to meet the requirements of the botany students of degree classes of the various Indian Universities. It is the outcome of considerable experience of the author in teaching the subject to the college classes. It is proposed to bring out the book in five parts.

Chapter 1 deals briefly with the important system of classification of the plant kingdom and a general account of the thallophytes. The succeeding chapters are concerned with the detailed descriptions of the life histories of the important representative of each class of algae. Suitable and properly labelled diagrams have been given to illustrate the various stages in the life histories of the representative types. A list of diagnostic features of important types and graphic representation of their life histories have been given at the end of each chapter to enable the readers to have a clear cut idea of the type described. With the same object in view a list of revision questions has been added at the end of each chapter. The questions will help the readers to grasp the subject matter fully. The concluding chapter includes a general account of the various aspects of the life cycles of Algae. The subject matter has been arranged in what seems to the author a logical order. Care has also been taken to bring the subject matter up-to-date keeping in view the latest researches on the subject. Every attempt has been made to write in a simple and lucid style. Difficult phrases and lengthy sentences have been, so far as possible, avoided.

The author lays no claim to originality in the preparation of the book. It is a compilation work done in a manner in which it would prove most useful to the reader. All suggestions to improve the book will be thankfully accepted.

I am thankful to my publishers for providing me all the facilities in the publication of this book.

B.R. VASHISHTA

Modinagar
15th *July, 1960*

CONTENTS

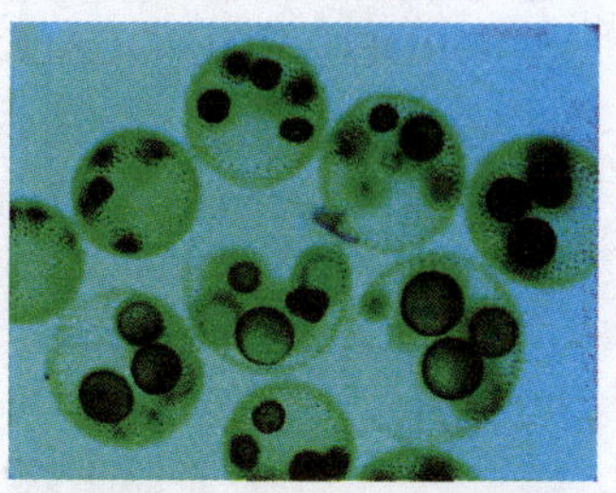

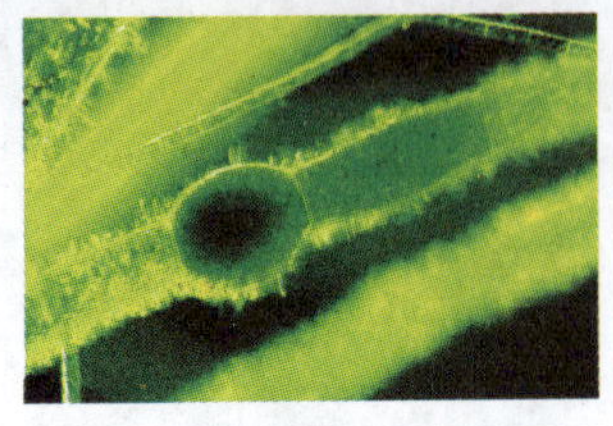

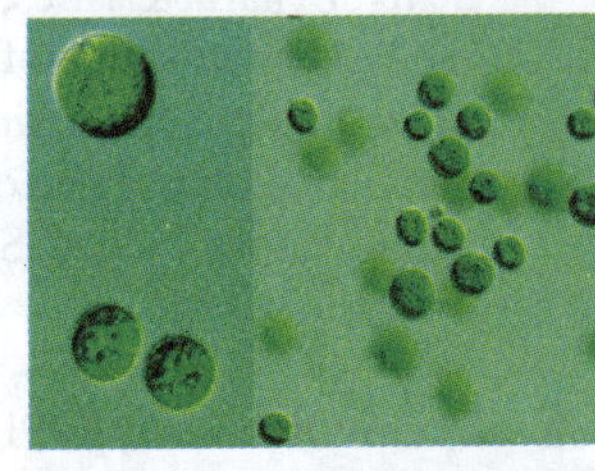

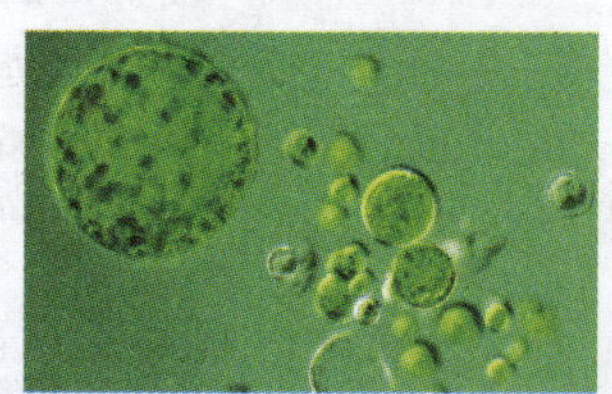

An Alga *Caulerpa sertularioides*—totally engulfing a Costa rican reef.

CHAPTER 1

Introduction

GENERAL CHARACTERISTICS

The algae, as a group, have now no official existence. It is a collective term for all those chlorophyll-bearing organisms which are thalloid. Pond scums, stoneworts, sea weeds and the like are collectively known as the **algae**. Many of them are small, and unattractive. However, they are important members of the plant world and several of them are significant to man in many ways. All are chlorophyll-bearing plants with a plant body showing no differentiation into true tissues. It never forms true roots, stems and leaves and is thus called a **thallus**. This term is used even if the plant is a unicellular. The thallus is non-vascular and thus has no elements for the transport of fluids. The algae can afford this simplicity because with only a few exceptions they are water dwellers. Even they are active only in very moist places or in times of moisture. They have non-jacketed, either unicellular sex organs or multicellular in which every cell produces a gamete. Most of them are among the simplest in the plant kingdom. In all there are about 30,000 species. On the basis of (*a*) thallus like non-vascular plant body, (*b*) simple, unicellular non-jacketed sex organs and (*c*) no embryo development after gametic union, the algae and fungi have long been grouped together in Thallophyta. Some algologists and mycologists hold that the above-mentioned common features between the algae and fungi are the result of parallel development and do not indicate any phylogenetic relationship.

DEFINITION OF ALGAE

It is very difficult to define Algae. According to Fritsch (1935) algae must include all holophytic organisms that fail to reach the higher level of differentiation characteristic of higher plants. G.M. Smith defines algae as simple plants with autotrophic mode of nutrition. Algae, according to R.N. Singh, are by and large simple plants displaying a wide range of photosynthetic pigments and evolve oxygen during photosynthesis. Sharma (1987) defines algae as an assemblage of chlorophyll bearing autotrophic thallophytes bounded by a cell wall made up of pure or mixed carbohydrates.

HISTORY OF PHYCOLOGY

The Study of Algae is called *Phycology* (*Phycos* = Algae, *logos* = Study of/Discourse of) and its history is quite old. *Phycos* is a Greek word which means sea weed and the references to algae are available in the early Chinese, Roman and Greek literatures. Romans named it *Fucus,* whereas Chinese called it *Tsao*. The ancient *Hawanians* used algae as food and called them *Limu*. The algae were used as manure on the north coast of France as early as the 12th century.

Though the algae have been used for various purposes since long but the orderly and systematized knowledge of algae started with the invention of microscope in the middle of 17th century. Linnaeus (1754) coined the term Algae for the first time, while giving the classification of Plant Kingdom. Later on A.L. de Jussieu (1789) classified the plants and delimited the algae as known to us at present. Even in the last quarter of 18th century much was not known about algae and only four genera or groups *i.e.*, *Fucus*, *Ulva*, *Conferva* and *Corallina* could be named. All others were placed under one of these four groups while *Chara* was grouped with *Equisetum* (horsetails).

In the beginning of 19th century when microscope developed into a workable tool many European biologists started taking interest in the study of algae. Roth (1797-1805) discovered and described *Hydrodictyon, Batrachospermum* and *Rivularia*. Turner (1802) described fertilisation in *Fucus*.

H.E. Link (1820-33) worked out the algae of Germany and described *Tertraspora, Oedogonium* and *Spirogyra*. Professor C.A. Agardh (1824) established two genera *Mougeotia* and *Zygnema* and established six orders of algae, namely Diatomaceae, Nostochineae, Confervoideae, Ulvaceae, Florideae and Fucoideae. At the same time Greville named and described *Polysiphonia* and *Rhodymenia*. J. Agardh worked out reproduction in *Fucus, Bryopsis, Griffithsia* and *Conferva*. Equally significant were the contributions of Kuetzing during this period in Germany. He described more new genera than any other phycologist. Hasall (1842-45) also made an outstanding contribution to the knowledge of algae in Great Britain. Braun (1835-55) studied intensively the development and taxonomy of *Chara* and made a valuable contribution to the sexual reproduction of Charales. Thurret (1843-55) similarly made valuable observations on zoospore formation and reproduction in many algae. Pringsheim (1855-60) described sexual reproduction in *Vaucheria, Oedogonium* and *Coleochaete*. At the same time Cienkowsky discovered reproduction in *Sphaeroplea*. De Bary (1858) saw the male gamete in *Spirogyra* push into the conjugation canal. He (1860), in addition, worked out the morphology of the Conjugatae. Areschoug (1866-84) described many new genera and species. He made valuable observations on zoospore and gamete formation in *Urospora* and *Cladophora*. In addition, he carried on morphological investigations in *Laminaria* and *Macrocystis*.

In the later part of the 19th century genus *Debarya* was established by Wittrock (1872),

Mougeotiopsis by Palla (1894), *Pleurodiscus* by Langerheim (1895) and *Temnogametum* by Wests (1897). Strasburger (1895-98) described the complete life cycle of *Fucus*, Williams (1897-98) of *Dictyota*, and Sauvageau (1899) of *Cutleria*. In the year 1897 Borzi segregated yellow green algae from the Chlorophyceae. O. Borge (1894-1936) worked out the fresh water algae of Sweden.

The beginning of 20th century witnessed Oltman's work (1905) on morphology of algae. The studies of fresh water algae of Britain by G.S. West and W. West are classical. G.S. West's book on 'Algae' gave an excellent account of structure and reproduction of algae. F.E. Fritsch and Rich (1907-37) investigated the fresh water algae of South Africa. Yamanouchi (1904) studied the life cycle of *Polysiphonia violacea*. In the year 1915 was established the laminarian life cycle. Pia (1910) published the result of his investigations of a fossil alga in Europe. He was followed by Walcott (1914) in U.S.A. Cotton (1912) worked out the marine algal ecology of Clare island. The investigations of V. Czurda (1922-39) on the morphology, physiology and cytology of Zygnemaceae are of great importance. Physiological and biochemical studies in algae made considerable advance with the works of Kylin, Kniep and Harder. Transleau's work on the Zygnemaceae and of Tiffany on Oedogoniales are of particular significance.

With the introduction of new techniques and the invention of electron microscope our knowledge of algal-cell, cell-wall, cell sap, nuclear division and structure of flagella and eye spot has greatly increased since 1930.

History of algal studies in India. The macroscopic forms were the first to attract attention. Lebeck in 1798 described a new variety of *Chara polyphylla* from Ceylon (now Sri Lanka). Later in 1806, *C. zeylenica* was reported and described by Wildenow from Ceylon and Tranquebar. Montague (1849) described *Calothrix indica* from India. Griffith (1847) described fertilisation in *Eudorina elgans* which he collected from the pools in Bombay. An army officer Wallich (1861) studied and published a paper on the desmids of lower Bengal and Dickie (1882) described some algae from the Himalayas. A memoir of the East India fresh water algae was published by Turner (1892-93). W. West and G.S. West (1897) reported and described 45 species of desmids from Singapore. Later in 1902, Wests published descriptions of 7 species of Red algae, 49 species of diatoms, 33 species of blue-green algae, 246 species of desmids and 84 species of green algae from Ceylon and again in 1907 reported 84 species of diatoms and 148 species of desmids. F.F. Fritsch (1907) published an account of the sub-aerial and fresh water algae of Ceylon. Boergeson (1930) laid the basic foundations of the systematic study of Indian marine algae. He published a number of papers.

Among the Indian phycologists Ghose (1919-32) was the pioneer worker. He studied the blue-green algae of Burma and Punjab. M.O.P. Iyengar who is regarded the 'Father of modern algology of India' started his work on algae in 1920. Assisted by his students Balakrishnan, Desikachary, Kanthamma, Ramanathan and Subrahmaniam, he described a number of new species and genera and worked out life histories of many Indian algae. Along with Subrahmaniam he investigated meiosis in auxospore formation of *Cyclotella meneghiniana*. His discovery of *Fritschiella tuberosa*, a terrestrial alga, is of particular significance.

The algae of Eastern India attracted the attention of Bruhl and Biswas (1922-26). A mention may be made here of the algal flora of Bengal and Assam by Biswas (1922-50). Bhardwaj (1928-36) established a school of algology at Banares Hindu University. He made a significant contribution to our knowledge of the Cyanophyceae in U.P.

R.N. Singh (1938-68) contributed a number of papers on the blue-green algae, Zygnemaceae, Oedogoniaceae and Chaetophorales of U.P. He investigated the life histories of *Fritschiella tuberosa* and *Draparnaldiopsis indica*. His work on the possibilities of the reclamation of saline usar land in India by the cultivation of blue-green algae is of singular interest, but his monograph on the "Role of blue-green algae in nitrogen economy of Indian agriculture" is his masterpiece. C.B. Rao (1935-1938) added to our knowledge of the blue-green algae and Zygnemaceae of U.P. and Madras by publishing a number of papers. V.P. Singh (1941) worked on some green algae from Chamba and U.P.

M.S. Randhawa (1932-59) worked on Zygnemaceae, Oedogoniales and Vaucheriaceae of Punjab and U.P. He contributed a number of papers recording 70 species of Zygnemaceae which included one new species of *Debarya*, 2 of *Mougeotia*, 2 of *Zygogonium* and 11 of *Spirogyra*. His monograph on the Zygnemaceae is monumental.

Desikacharya's contributions to the blue-green algae, diatoms and red algae of India deserve a special mention. His monograph on "Cyanophyta" (1950) is regarded as an important recent work on the taxonomy of the blue-green algae.

S.R. Narayana Rao (1941-49) worked on the fossil algae of India. R. Subrahmaniam (1954) described reproduction in some species of diatoms from Southern India. The Charophytes of India and Burma have been worked out by Allen, Pal, Dixit, Kundu and Sundaralingam. Sundaralingam's painstaking work on the "Developmental morphology of six Indian species of *Nitella*" deserves a special mention.

J.C. Dixit (1937) investigated 5 species of *Spirogyra* and some of *Zygnema* which include two new species. S.C. Dixit (1937) described one species of *Sirogonium* and two of *Spirogyra*. Useful contributions have been made to the algal flora of Bombay State by Gonzalves and her students. Ramanathan (1939-46) worked out the life history of *Enteromorpha compressa* and described sexual reproduction in *Carteria* and *Dictyosphaerium indicum*. In the year 1964, K.R. Ramanathan published a monograph on the 'Ulotrichales'. Y.S.R.K. Sharma (1956) made a significant contribution to the cytology of green algae. A.K. Mitra and his students investigated the blue-green algal flora of Allahabad and environs. Pandey and Mitra (1959) worked on the production of heterocysts in the blue-green algae. K.P. Singh made a contribution to our knowledge of the green algae of Naini Tal. He described some new species and their life histories.

G.S. Venkataraman made a valuable contribution in the field of algology, both systematics and physiology of nitrogen fixation. His monographs on Vaucheriaceae (1964) and Charophyta (1962) (the latter in collaboration with B.P. Paul, Kundu and Sundaralingam) are of special value. P.C. Vasisht (1956) worked on the blue-green algal flora of Punjab. It includes 17 new species. He also worked out the life histories of *Rivularia* and *Anabaena*. His work on the thermal Cyanophyceae of India is noteworthy (under publication).

R.S. Rattan (1960-) has made a significant contribution to the Zygnemaceae of Punjab.

H.D. Kumar (1970) has done important research work in algal physiology. His discovery of genetic recombination in blue-green algae is a significant contribution in the field of Cyanophyceae genetics.

J.N. Misra's monograph (1966) on the 'Phaeophyceae in India', is another valuable addition to the study of algal flora in India. It has been followed by a very useful monograph on the Chlorococcales from the Indian region comprising India, Sikkim, Pakistan, Afghanistan, Nepal, Burma and Ceylon by Philipose (1967).

B.N. Prasad (1970-1983) carried out extensive studies on diatoms, Zygnemaceae etc., and the algae of Andaman and Nicobar Islands. He also conducted studies on algae in relation to pollution.

G.L. Tewari (1990) has made significant contribution on blue-green algae and its use as biofertilisers.

HABIT AND HABITAT

The algae are predominantly aquatic and are found in fresh or salt waters. Fresh water forms occur abundantly in ponds, lakes, slow flowing streams and water reservoirs. In habit they may be free swimming, free floating or attached to the bottom in the shallow water. Some are terrestrial and grow in wet situations, such as, on damp soil, damp shaded sides of trees and walls or even rocks and thus have adapted themselves to a life in the air. They may live as epiphytes as well as epizoics. A few occur in association with fungi. Some algae are endophytic whereas a few live in symbiotic relationship with the higher plants. Most of the marine forms are *Sea Weeds*. They inhabit vast area of the ocean. Some of the marine forms are free floating. A few are attached. The free floating and free swimming minute algae together with similar other organisations constitute the *plankton* of the ocean and lakes. The algae which grow attached to the bottom in shallow water along the edges of seas and lakes form *benthos*.

According to the habitat, the algae may be classified as follows :

1. Aquatic algae
2. Terrestrial algae
3. Aerophytes
4. Cryophytes
5. Thermophytes
6. Algae of unusual habitats

A brief account of the above mentioned is given below :

1. Aquatic algae: Majority of the algal genera are aquatic and are found either completely submerged or free floating on the surface of water. Aquatic algae usually occur in ponds, pools, tanks, ditches, streams or in slow running rivers and are called fresh water forms. Marine algae are found in sea and macroscopic large thalli of brown algae are commonly known as "sea weeds". Fresh water algal forms like *Chlamydomonas, Volvox, Hydrodictyon* are found in stagnant waters, whereas *Cladophora, Oedogonium, Ulothrix* and few species of *Vaucheria* occur in slow running water bodies. Most of the members of Phaeophyceae and Rhodophyceae are found in sea either floating on the surface of sea water or attached with rocks or any other substratum. The free floating and free swimming microscopic algal forms together with other similar organisms constitute the *Planktons* of water bodies. Plankton forming algae may either be free floating from very beginning and are never attached (Euplanktons) *e.g., Microcystis, Chlamydomonas, Scenedesmus k, Cosmarium* or in the beginning may be attached but later on they get detached and become free floating (Tychoplanktons) *e.g., Zygnema, Oedogonium, Cladophora, Cylindrospermum, Rivularia* etc.

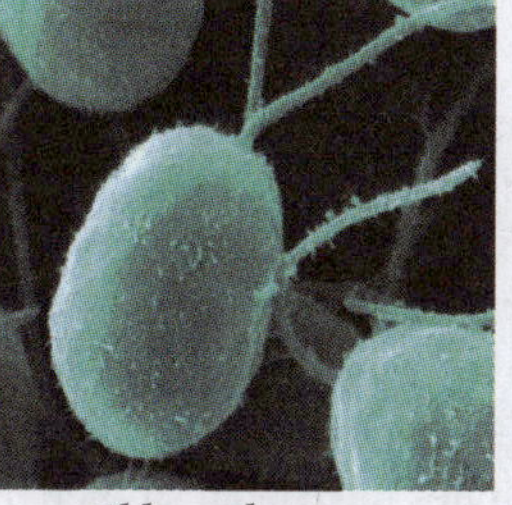

Chlamydomonas

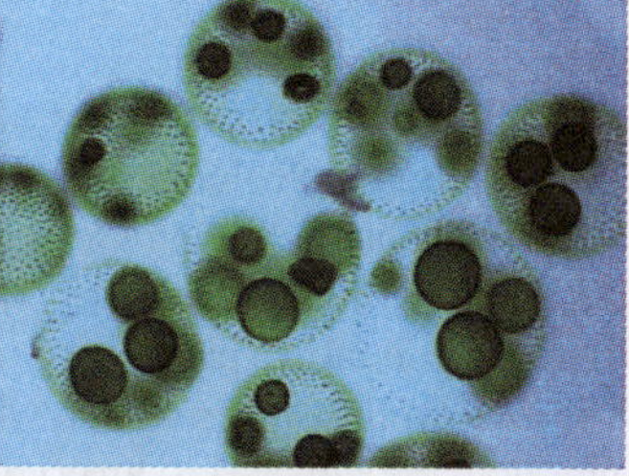

Volvox

AQUATIC ALGAE

2. Terrestrial algae: Many algal genera are found on or beneath the moist soil surface and are called terrestrial algae. The algal forms occurring on the surface of soil *e.g.,* few species of *Vaucheria, Botrydium, Fritschiella* and *Oedocladium* are called *Sapophytes* while algal genera having subterranean habit *e.g.,* few species of *Nostoc, Anabaena* and *Euglena* are known as *Cryophytes*.

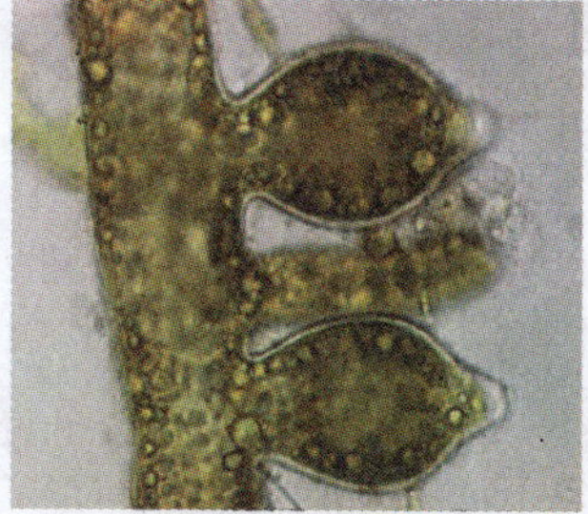

Vaucheria

Botrydium

TERRESTRIAL ALGAE

3. Aerophytes: Such algal forms as are adapted for aerial mode of life and occur on the tree trunks, moist walls, flower pots, rocks, fencing wires and get their water and carbon dioxide requirements completely directly from atmosphere are called Aerophytes. *Trentepohlia* is found on the bark of trees in moist and humid climatic conditions while *Phermidium, Scytonema* and *Hapalosiphon* have been observed to grow on bark of trees alongwith Bryophytes.

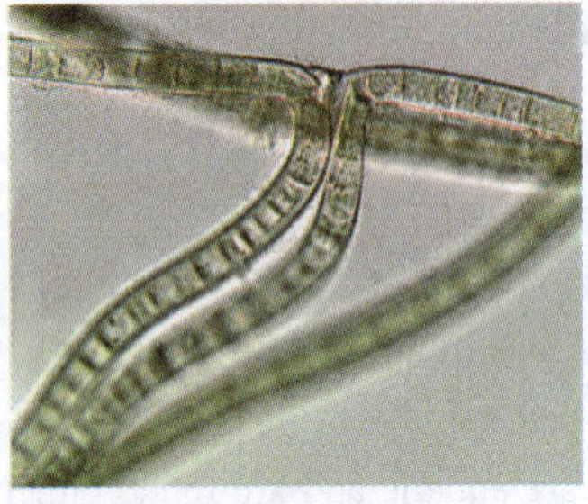

Scytonema *Trentepohlia*

THERMOPHYTE ALGAE

4. Cryophytes: These algae are found on the mountain peaks covered with snow and impart attractive colours to the mountains. *Haematococcus nivalis* gives red colour to Arctic and Alpine regions while *Chlamyodomonas yellowstonensis* (Kol, 1941) alongwith some species of *Ankistrodesmus* and *Mesotaenium* is responsible for the green colour of the snow of the mountains of European countries particularly in Arctic region. Certain species of *Protoderma* and *Scotiella* cause yellow or yellow green colour, whereas *Ancyclonema nordenskioldii* renders brownish to purple tinge colour to the snow. Alaskan cryophytes have been studied by Kol (1941) and following groups have been recognised:

(*a*) Those algae which are found on snow and not on ice *e.g.*, some species of *Raphidonema, Chlamydomonas* and *Scotiella*.

(*b*) Those algae which can grow only on ice and result in "ice bloom" *e.g.*, *Ancyclonema, Mesotaenium*.

(*c*) Those algae which can grow on snow and ice both *e.g.*, *Trachiscia* and *Cylmdrocystis*

(*d*) Those algae which are not true cryophytes and have their temporary growth on ice or snow *e.g.*, *Phormidium* and *Gloeocapsa*.

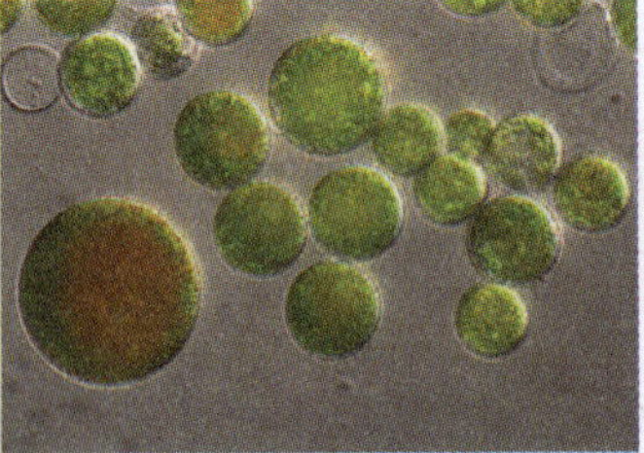

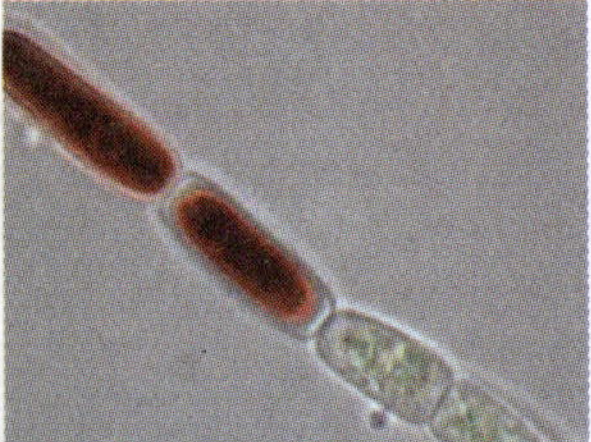

Protoderma *Haematococcus* *Ancylonema*

CRYOPHYTIC ALGAE

5. Thermophytes: The algal genera occurring in hot springs at quite high temperature are called *thermophytes*. There are certain algae which are known to tolerate the temperature upto 85°C *e.g.*, few genera belonging to family Chroococcaceae and Oscillatoriaceae. *Oscillatoria brevis, Synechococcus elongatus* and *Haplosiphon lignosum* are some common examples of thermophytes which can survive upto a temperature of 70°C at which generally plant life is not possible.

Oscilllatoria

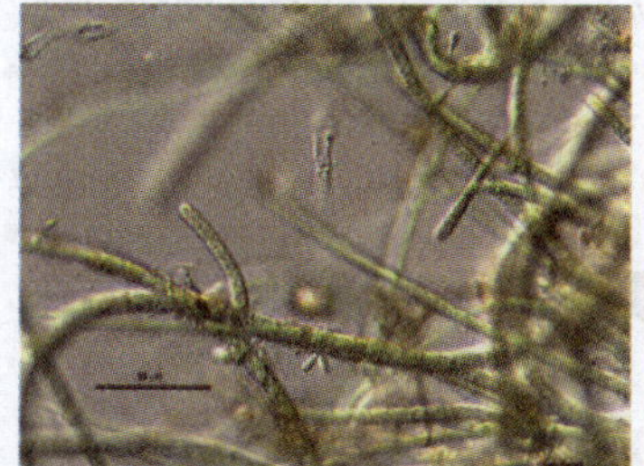

Hapalosiphon

THERMOPHYTE ALGAE

6. Algae of unusual habitats: Many algae are found at various interesting places and according to their habitats may be of following types:

(*a*) *Halophytic algae*. These algae are found in saline water containing high percentage of salts *e.g., Dunaliella, Stephanoptera* and *Chlamydomonas chrenbergii.*

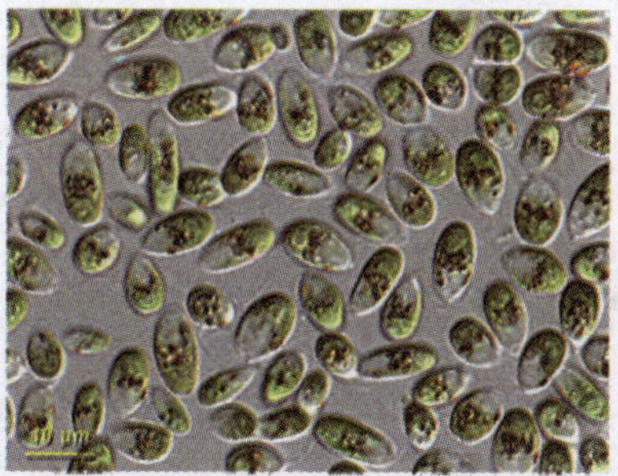

Dunaliella **(HALOPHYTIC ALGAE)**

(*b*) *Lithophytic algae*. Usually the members of Cyanophyceae grow on moist rocks, wet walls and other rocky surfaces. Blue-green algae *Rivularia* and *Gloreocapsa* occur on exposed rocks, whereas *Nostoc* is found growing in damp shady habitats. Several marine algae belonging to Rhodophyceae and Phaeophyceae are lithophytic in habit and grow on the submerged rocks or rocky surface *e.g., Ectocarpus, Polysiphonia* etc.

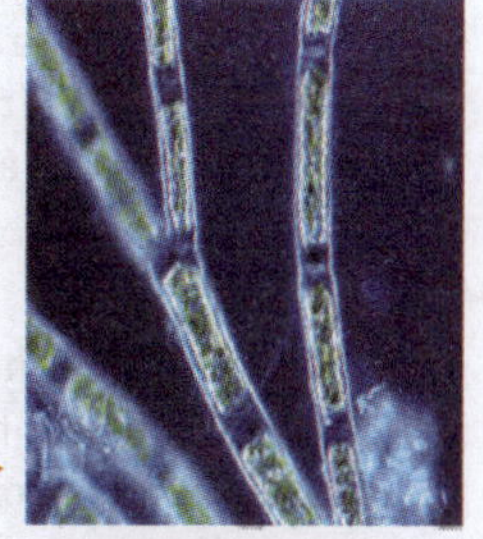

Ectocarpus **(LITHOPHYTIC ALGAE)**

(*c*) *Epiphytic algae*. Such algal forms which grow on the other aquatic plants are called *Epiphytic algae*. Examples are *Oedogonium, Bulbochaete, Aphanochaete* etc. Green algae *Coleochacte nitellarum* occurs on *Chara* and *Nitella* as an epiphyte, while *Chaetonema* has been found growing on *Tetraspora* and *Batrachospermum*. The colonies of *Rivularia* are observed to grow on Angiospermic plant *Scirpus,* whereas *Chaetophora* is found on the leaves of *Nelumbo* and *Vallisneria*.

Coelochaete
(EPIPHYTIC ALGAE)

(*d*) *Epizoic algae*. Many algae grow on the shells of molluscs, turtles and fins of fishes and are known was *epizoic algae. Cladophora* is found on snails and shells of bivalves, while *Protoderma* and *Basicladia* have been observed growing on the back of turtles.

(*e*) *Endozoic algae.* Contrary to epizoic algal forms endozoic algae are found inside the body of aquatic animals *e.g., Zoochlorella* is found inside *Hydra viridis* while *Zooxanthella* is known to occur inside the fresh water sponges. According to Langeron (1923) about 15 species of Blue-green algae belonging to family Oscillatoriaceae are known to occur in the digestive and respiratory tracts of vertebrates.

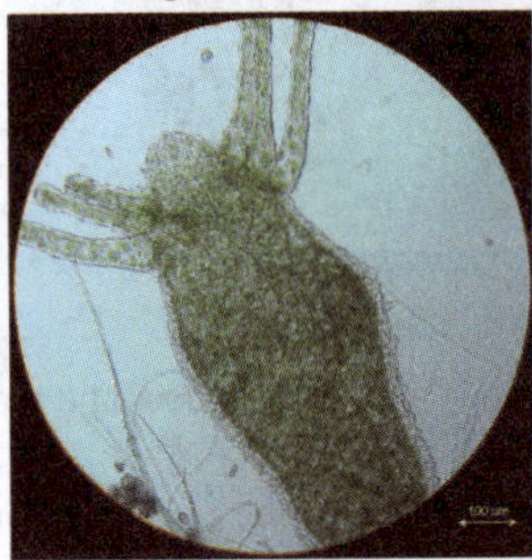

Zoochlorella

Zooxanthella

(ENDOZOIC ALGAE)

(*f*) *Parasitic algae. Cephaleuros virescens* which causes 'red rust of tea' is a striking example of parasitic algae and causes heavy damage to tea foliage in Assam and nearby states. *Chlorochytrium* and *Phyllosiphon* are other examples of parasitic algae. *Polysiphonia festigata* a member of Rhodophyceae is reported as semiparasite on *Ascophyllum nodosum.*

(*g*) *Symbiotic algae.* Several members of Cyanophyceae grow in association with other plants and lichens exhibit good example of it. Almost all the plant groups are known to have symbiotic association with Blue-green algae *e.g., Nostoc* is found within the thalli of *Anthoceros* and *Notothylas*, *Anabaena cycadeae* is reported in the coralloid roots of Cycas, *Anabaena azollae* occurs in *Azolla* etc. The association of *Chlorella* with a nitrogen fixing bacterium *Azotobactor chrooccocum*, and with that of certain species of *Ceratophyllum* and mosses are other examples of symbiotic algae.

Organisation of the thallus

The algae exhibit a great diversity in the organisation of the plant body. The simplest forms are motile or non-motile unicells (*Chlamydomonas* and *Chlorella)* In many species, the cells are grouped into aggregations called **colonies** *e.g., Volvox, Pediastrum*. These assume various forms and may be a hollow sphere, a flat plate or a filament. The filamentous types are usually multicellular and the filament may be simple *e.g., Ulothrix, Zygnema* or branched or an aggregation of filaments or a highly organised thallus of a large size. In some multicellular forms the cells may perform both functions, vegetative and reproductive as in *Oedogonium,* while in others special reproductive cells or organs may be developed. *e.g., Chara, Sargassum* etc. The most highly differentiated marine algae exhibit external differentiation and considerable size. *Macrocystis, Sargassum* and *Laminaria* are good examples of it. Some of them possess a plant body consisting of parts that bear a superficial resemblance to the roots, leaves and stems of higher plants. The length of the main axis equals or surpasses the height of the tallest tree *e.g.*, the thalli of *Macrocystis pyrifera* may attain a length of more than 60 meters. Why do algae with massive thalli are marine only? Obviously the marine water can provide minerals at the rate at which a huge thallus demands whereas the inland or fresh water cannot. Secondly, marine water is relatively more permanent in geological time as compared with inland water.

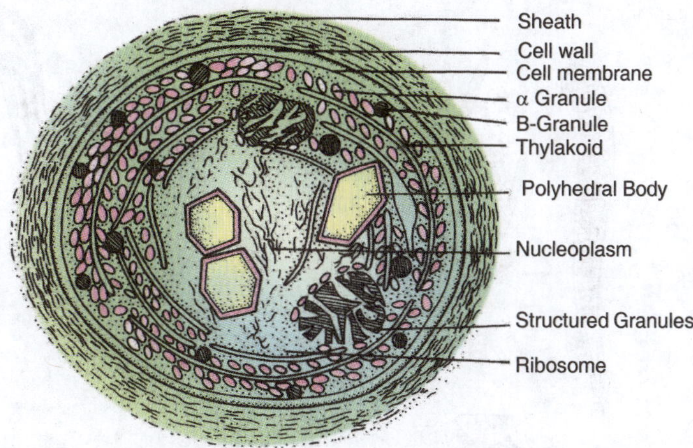

Fig. 1.1. Algae. Fine structure of a prokaryotic cell of a blue-green alga.

Structure of Algal Cell

The cells constituting the algal thalli are basically of two kinds, **prokaryotic** and **eukaryotic**. The prokaryotic cells (Fig. 1.1) which constitute thalli of cyanophytes (blue-green algae) have a cellwall which contains a specific strengthening component not found in the cell walls of other algae. It is **mucopeptide**. The **DNA** material representing the nuclear body consists of fibrils which may extend throughout the cell or are concentrated in the central part. The mitotic figures are also lacking. The chlorophyll pigment is bound to photosynthetic lamellae or thylakoids which may be arranged in parallel layers in the periphery of the cytoplasm or form a network extending throughout the cell cytoplasm. They are not organised into grana. The chloroplasts are thus absent and so are the mitochondria, golgi body and endoplasmic reticulum. The ribosomes are, however, present. The nuclear division does not take place by mitosis and no cell plate is formed. Instead there is a ring like

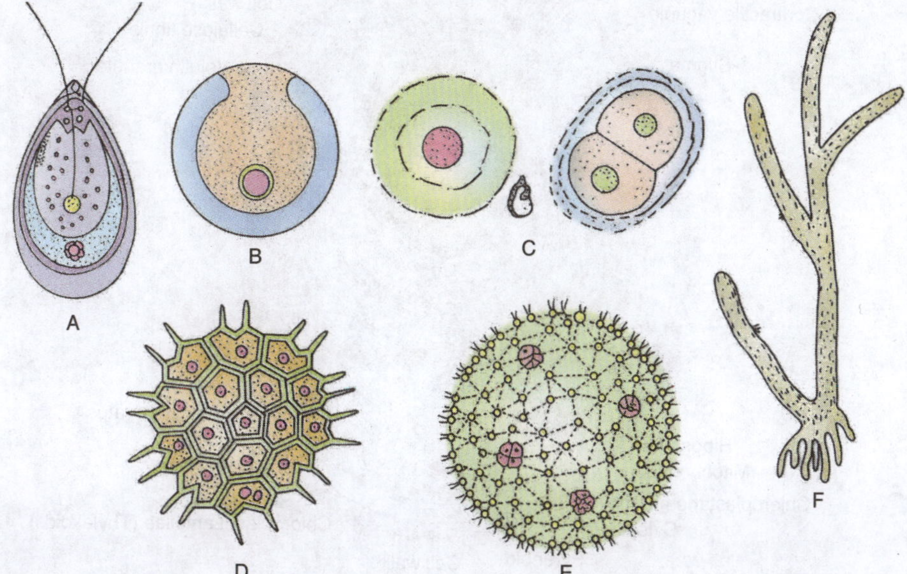

Fig. 1.2 (A). Range of thallus organisation in Algae. (A). *Chlamydomonas,* (B). *Chlorella,* (C). *Gloeocapsa,* (D). *Pediastrum* (E). *Volvox* (F).

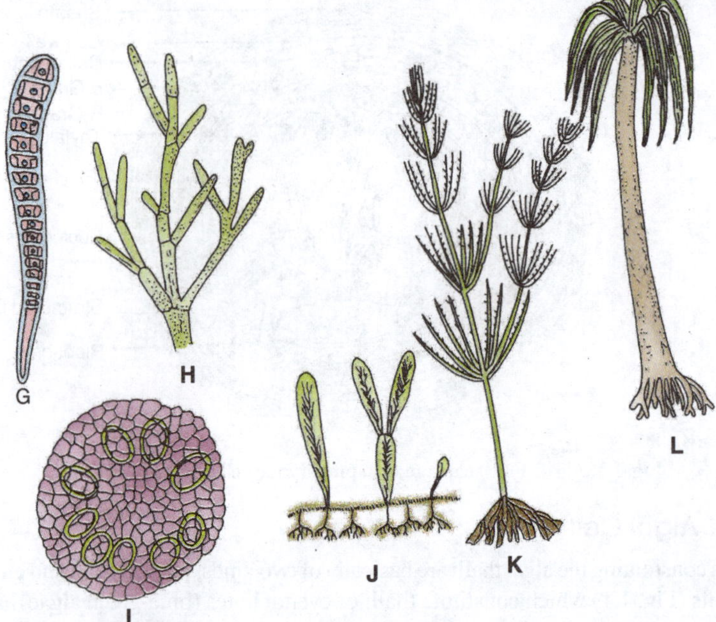

Fig. 1.2 (B). Range of thallus organisation in Algae. (G). *Ulothrix*, (H). *Cladophora*, (I). *Coleochaete*, (J). *Caulerpa* (K). *Chara* (L). *Postelsia*.

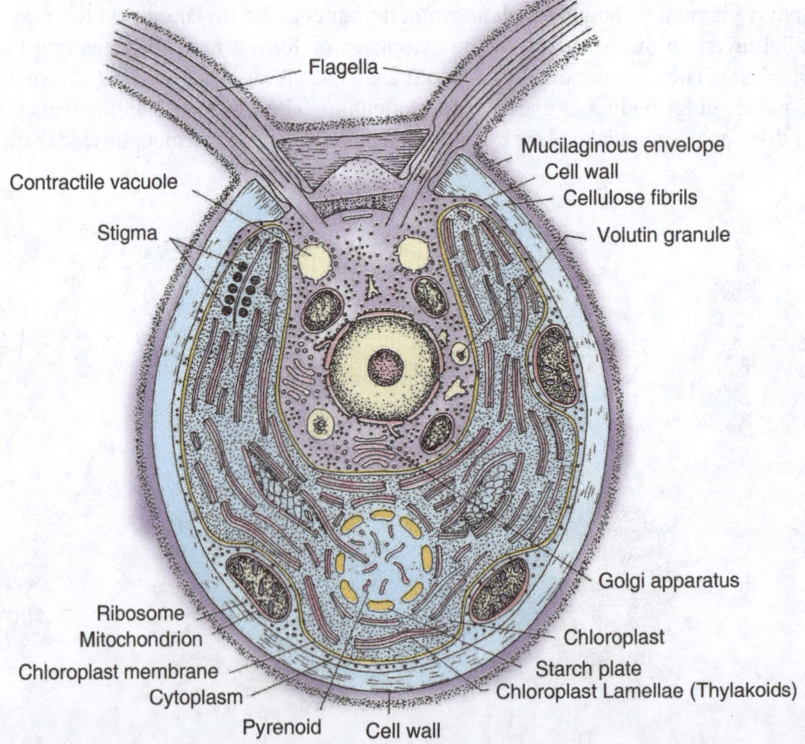

Fig. 1.3. Algae. Fine structure of eukaryotic cell of a unicellular green alga *Chlamydomonas* (Based on K. Vickerman and Cox).

extension of the cell wall. It extends inwards like a diaphragm with decreasing aperture and divides the cell into daughter cells. Such simple cells of blue-green algae (and bacteria) which lack a nuclear membrane, mitochondria, plastids and do not divide by mitosis are called **prokaryotic**.

The cells constituting the thalli of all other algae excepting the blue-greens are called **eukaryotic**. The eukaryotic cell (Fig. 1.3) has the same structure as is typical of the higher plants.

The nucleus is separated from the cytoplasm by a distinct nuclear membrane. It has mitotic figures and divides by mitosis. The cytoplasm contains membrane bound chloroplasts or chromatophores, mitochondria, golgi bodies and endoplasmic reticulum.

Algal Pigments. The colour of the algal thallus which varies in different classes of algae is due to the presence of definite chemical compounds in their cells. These are called the **pigments**. Each pigment has its own characteristic colour. The particular colour that a thallus has is due to the predominance of one pigment in a combination of several others. For example brown algae have predominance of Fucoxanthin and phycophein while red and blue green algae have excess of phycoerythrin and phycocyanin pigments respectively. Each group of algae has its own particular combination of pigments and a characteristic colour which is not found in the other algal groups. In fact the various algal groups show striking differences of colour. Thus the nature of the pigments present in the algal cells forms a quick guide to the primary classification of Algae into divisions. The photosynthetic pigments in algae are of three kinds, namely, **chlorophylls, carotenoids** and **phycobilins or biliproteins**.

The algal chlorophylls are characterized by green colour and in solution they show the phenomenon of fluorescence and emit red light. Chlorophyll pigments are fat soluble compounds and are of five different types *viz.* chlorophyll *a*, *b*, *c*, *d*, and *e*. Out of these chlorophyll *a* is universally present in all the groups of algae whereas chlorophyll *b*, *c*, *d* and *e* have restricted distribution. Carotenoids are fat soluble yellow coloured pigments and are subdivided into carotene, xanthophylls and carotenoid acids. Carotenes are unsaturated hydrocarbons which are fat soluble and yellow in colour whereas xanthophylls are oxygen derivatives of carotenes and have similar properties like carotenes. Carotenoid acid resembles very much with carotenes and xanthophyll, and are hydrocarbons consisting a chain of carbon atoms. Phycobilins are water soluble blue (phycocyanin) and red (phycoerythrin) coloured pigments and are present in the members of Cyanophyceae and Rhodophyceae.

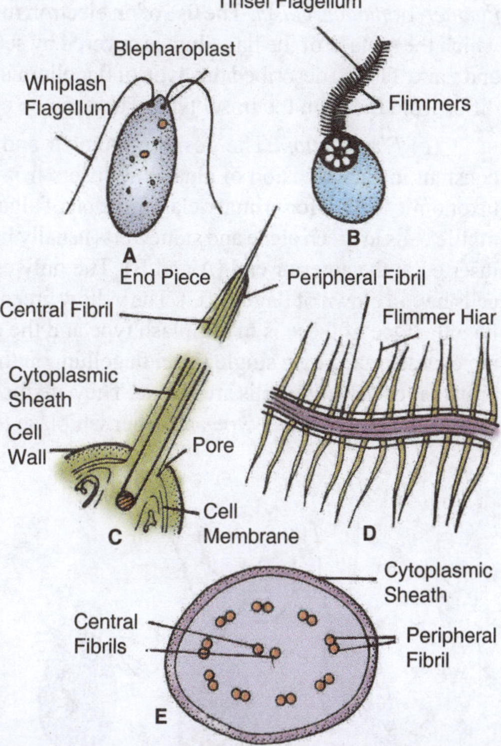

Fig. 1.4 (A-E). Algae. Kinds of Flagella and their structure. (A), Whiplash; (B). Tinsel; (C), L.S. Whiplash Flagellum; (D), L.S. Tinsel Flagellum; (E), C.S. Eukaryotic Flagellum.

Algal Flagella

The motile cells of algae are provided with fine, protoplasmic, whiplike threads, the **flagella**(A). They are extremely fine and hyaline emergences of the cytoplasm. In cells possessing firm cell walls, the flagella are connected with the inner cytoplasm through small pores in the cell wall (C). There is either a single anterior flagellum (rarely posterior) or the flagella occur in pairs (A), rarely in great numbers on the cell. The flagella on the cell may be equal (**isokont**) or unequal (**heterokont**) in length. When the flagella are inserted laterally one is directed forwards in motion and the other backwards. They function as the locomotory or propelling structures of the cell. Usually there is a single granule at the base of each flagellum. It is known as the **blepharoplast**.

(*a*) *Structure of the Flagellum.* Forming the core of the flagellum is an axial or central filament called the **axoneme**. The latter is surrounded by a cytoplasmic membrane or sheath which terminates short of the apex (C). The naked, terminal portion of the axoneme is called the **end piece**. The tip of the end piece may be blunt and rounded or pointed. In cross-section (E) the flagellum consists of two inner central simple fibrils forming an elastic axial thread. It is surrounded by nine united, peripheral contractile, thicker protein double fibrils. All are enclosed by sheath which is an extension of the plasma membrane. Each peripheral fibril is composed of two thin fibrils. The two central fibrils are single. They lie side by side and are sometimes enclosed by a sheath of their own. The fibrils are hollow and extend along the entire length of the flagellum. The nine peripheral fibrils join the basal granule (C) but the two central fibrils stop short of the granule. This '9+2' pattern of component fibrils is the basic structure of the flagellum of all organisms except the bacteria.

(*b*) *Kinds of Flagella*. They are of two main types, **whiplash** (A) and **tinsel (B)**. The whiplash flagellum has a smooth surface. The tinsel flagellum bears longitudinal rows of fine, minute flimmer hairs arranged along the axis almost to the tip of the flagellum. There may be a single row of hairs as in the Euglenophyta and Pyrrophyta or two as in Chrysophyceae and Phaeophyceae. The hairs arise from the margins of the peripheral fibrils. The whiplash or smooth flagella are also known by other names such as *acronematic* or *peitchgeisel.* The other names for the tinsel flagella are *pantonematic, flimmer* or *flimmergeisel.* The use of an electron microscope has revealed a third kind of flagellum in which the surface of the flagellum is covered by scales (*Chara*) and minute, short, stiff hairs. Manton and Parke (1960) described this type of flagellum in *Micromonas pusilla* (Prasinophyceae). The hairs differ from those on the tinsel type. They can be easily detached.

(*c*) *Flagellation.* The position, number and kind of flagella on the motile cells is strikingly constant in each division of algae but differs from division to division. Thus it forms an important taxonomic feature for primary classification of algae. The blue-greens and red algae lack flagella. The motile cells in green algae and stoneworts usually have two, rarely four equal flagella of whiplash type inserted at the anterior end (A and B). The only exception is the Oedogoniales in which the motile cells have a crown of flagella (C). The yellow green algae (Xanthophyceae) have two unequal anterior flagella. One of these is of whiplash type and the other tinsel (D). The diatoms (Bacillariophyceae) are characterized by a single tinsel flagellum on the male cell at the anterior end(E). In brown algae only the reproductive cells are motile. They are furnished with two laterally inserted unequal flagella. One of these is of tinsel type and other whiplash (F).

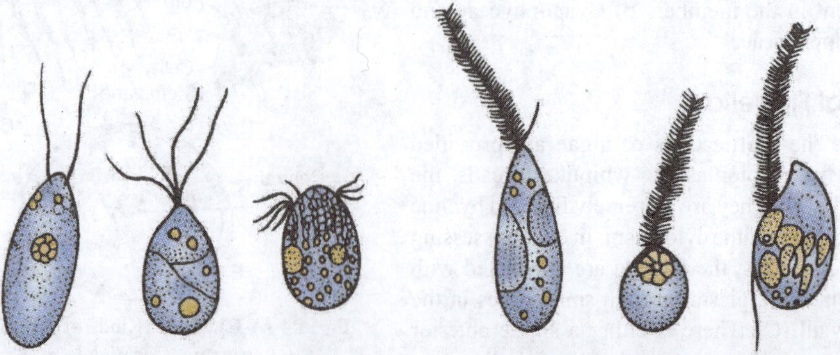

Fig. 1.5 (A-F). *Algae.* Flagellation., Chlorophyceae (A-C); Xanthophyceae (D); Bacillariophyceae (E) and Phaeophyceae (F).

Nutrition

With respect to their nutrition the algae are autotrophic. All or most of the cells of the thallus normally contain chlorophyll. The green cells can manufacture their own carbohydrate food from carbon dioxide and water through the agency of sunlight. The aquatic forms obtain water and carbon dioxide by osmosis and diffusion respectively from the medium in which they grow. The aerial forms obtain water from the damp substratum and carbon dioxide from the air. The algae are also able to synthesize oil and proteins from the carbohydrates which they manufacture and soluble forms of nitrogen and other minerals available in solution in the water in which they grow.

Food Reserves

Since the early steps in photosynthesis in all the algal groups are practically the same it is but natural that the primary products of this process must also be the same. The food materials which accumulate as food reserves in the form of polysaccharides, however, vary from group to group and thus provide useful data for preliminary classification of algae. True starch is typical of only two algal divisions namely, Chlorophyta and Charophyta. The two other kinds of characteristic starches are the **cyanophycean starch** and **floridean starch**. The former is characteristic of division Cyanophyta and the latter of division Rhodophyta. The three other important polysaccharides which accumulate as reserve food are **laminarin** found in the brown algae, **paramylon** characteristic of Euglenoids and **leucosin** peculiar to the Xanthophyta, Bacillariophyta and Chrysophyta. Besides, a proteinaceous compound **cyanophycin** is found only in the cells of blue-green algae. Mannitol which was formerly considered to be unique to the brown algae has recently been reported to occur in a few red algae. Fats occur as reserve food in appreciable amounts in the cells of Xanthophyta, Bacillariophyta and Chrysophyta. The environmental factors which favour growth of the algae are favourable temperature, suitable light, and proper supply of oxygen, carbon dioxide and essential elements.

Reproduction

In their methods of reproduction, the algae are as diverse as they are in the nature of the thallus. The primitive algae reproduce only by vegetative methods but in the higher forms both asexual and sexual reproduction are of common occurrence.

The common methods of vegetative reproduction are by simple cell division, fragmentation, Hormogone formation, primary and secondary protonema, tubers, bulbils, amylum starch or by formation of specialised adventitious branches or thalli called propagula. Asexual reproduction in algae takes place by different types of spores formed in favourable and unfavourable conditions by division of the protoplast. Zoospores are usually produced in favourable conditions whereas thick walled hypnospores are formed during adverse conditions. Other asexual spores which are formed in different groups of algae are aplanospores, autospores, endospores, neutral spores and monospores. Formation of daughter colonies in *Volvox,* palmella stage in *Chlamydomonas* and gongrosira stage in *Vaucheria* are some other notable specialised methods of asexual reproduction in algae in unfavourable conditions. Higher forms reproduce sexually which is an advanced method of reproduction in which fusion of two specialised cells known as sex-cells or gametes takes place. Sexual reproduction may be isogamous in which two identical gametes fuse to form zygote or heterogamous in which the gametic union takes place between two dissimilar gametes having different size and behaviour. Heterogamy is further divided into Anisogamy and Oogamy.

DISTINCTIVE CHARACTERS OF ALGAE

1. The algae are chlorophyll-bearing organisms with a thallus-like plant body.
2. The thallus shows little differentiation of true tissues.
3. Even the complex thalli lack vascular tissue and epidermis with stomata.
4. The sex organs are one-celled, when multicellular, each cell is fertile and there is no jacket of sterile cells.
5. There is no embryo formation after gametic union.
6. Both the generations when represented in the life cycle are independent. There are no algae with a sporophyte parasitic on the gametophyte plant.
7. Excepting a few all the algae are aquatic.
8. Under favourable conditions the gametophyte multiplies repeatedly by means of asexual spores called the mitospores.

CLASSIFICATION

The committee on the **International Code of Botanical Nomenclature** has recommended certain suffixes for use in the classification of **Algae.** These are **phyta** for *division,* **phyceae** for *class,* **phycideae** for *subclass*, **ales** for *order*, **inales** for *sub-order,* **aceae** for *family*, **oideae** for *sub-family*, Greek name for *genus* and Latin name for a *species.*

Algal characteristics basic to primary classification. The primary classification of algae is based on certain morphological and physiological features. The chief among these are: (*a*) pigment constitution of the cell, (*b*) chemical nature of stored food materials, (*c*) kind, number, point of insertion and relative length of the flagella on the motile cell, (*d*) chemical composition of cell wall and (*e*) presence or absence of a definitely organised nucleus in the cell or any other significant detail of cell structure. The importance of these features in the primary classification of algae has been discussed in the preceeding pages. The details of vegetative structure and reproduction are useful for algal classification only at the level of families, genera and species.

History of Classification of Algae

Although Carolus Linnaeus (1754) included Algae alongwith Lichens in his 25th class Cryptogamia, he did not elaborate further on the classification of Algae.

Vaucher (1803) was perhaps the first to propose a system of classification of Algae and he recognised three groups, *Conferves, Ulves* and *Tremelles*. While Link (1820) classified Algae on the basis of the colour of the pigment and structure, Harvey (1836) proposed a system of classification on the basis of the habitat and the pigment. J.G. Agardh (1849-1898) divided Algae into six orders: Diatomaceae, Nostochineae, Confervoideae, Ulvaceae, Florideae and Fucoideae.

Around 1880, *Algae* along with *Fungi* were grouped under THALLOPHYTA, a division created by Eichler (1836). Encouraged by this, Engler and Prantle (1912) proposed a revised scheme of classification of algae and included fungi in algae as they were of opinion that fungi have been derived from algae. The scheme proposed by Engler and Prantle is summarised as follows.

1. *Schizophyta*
2. *Phytosarcodina*
3. *Flagellata*
4. *Dinoflagellata*
5. *Bacillariophyta*
6. *Conjugatae*
7. *Chlorophyceae*
8. *Charophyta*
9. *Phaeophyceae*
10. *Rhodophyceae*
11. *Eumycetes* (Fungi)

West (1916), on the basis of the works of earlier phycologists divided algae into five categories as follows :

1. **Chlorophyceae**
2. **Isokontae**
3. **Akontae**
4. **Stephanokontae**
5. **Heterokontae**

Pascher (1914-1931) proposed altogether a new scheme of classification on the basis of phylogeny and interrelationships of the various groups and elevated the classes to the rank of divisions. He divided the algae into eight divisions which were subdivided into classes. The scheme proposed by him is summarised below.

Division 1. **Chrysophyta**
(*a*) *Chrysophyceae*
(*b*) *Diatomeae*
(*c*) *Heterokontae*

Division 2. **Phaeophyta**
(*a*) *Phaeophyceae*

Division 3. **Pyrrophyta**
(*a*) *Cryptophyceae*
(*b*) *Desmokontae*
(*c*) *Dinophyceae*

Division 4. **Euglenophyta**
(*a*) *Euglenophyceae*

Division 5. **Chlorophyta**
(*a*) *Chlorophyceae*
(*b*) *Conjugatae*

Division 6. **Charophyta**
(*a*) *Characeae*

Division 7. **Rhodophyta**
(*a*) *Bangineae*
(*b*) *Floridineae*

Division 8. **Cyanophyta**
(*a*) *Myxophyceae*

Tilden (1933) recognised the importance of reserve food materials, flagellation and pigments in the plastids in algal classification and divided algae as follows:

1. **Chlorophyceae**
2. **Myxophyceae**
3. **Rhodophyceae**
4. **Phaeophyceae**
5. **Chrysophyceae**

The most authentic and comprehensive account of the classification of algae was proposed by renowned phycologist of the time. F.E. Fritsch (1935) who did monumental work on Algae and published his voluminous work in the form of book entitled "*Structure and reproduction of the Algae*" in two volumes. The criteria selected for the classification of algae by Fritsch were : pigments in the plastids, chemical nature of reserve food material, kind, number and point of insertion of flagella of motile cells and presence or absence of organised nucleus in the cell. He classified algae into 11 classes, the names of which are given below:

Class 1. **Chlorophyceae**
2. **Xanthophyceae**
3. **Chrysophyceae**
4. **Bacillariophyceae**
5. **Cryptophyceae**
6. **Dinophyceae**
7. **Chloromonadineae**
8. **Euglenophyceae**
9. **Phaeophyceae**
10. **Rhodophyceae**
11. **Myxophyceae**

Nematophyceae a fossil group of Algae with doubtful affinity was also created by him.

G.M. Smith (1955) modified the classification of Algae proposed by Pascher (1914-1931) and proposed his own system of classification with minor modifications. He divided Algae into Seven divisions, which were further subdivided into classes. The names of Divisions and classes are given below:

Division 1 : *Chlorophyta* includes about 5700 forms out of which 90% are freshwater and the remaining 10% are marine. Dominant pigments are Chlorophyll a and b, the reserve food starch. Divided into two classes: (1) *Chlorophyceae* (green algae) e.g., *Volvox*, *Ulothrix*; (2) *Charophyceae* (stoneworts) e.g., *Chara*.

Division 2 : *Euglenophyta* includes 450 fresh water or terrestrial forms. Dominant pigments are chlorophyll and β-carotene and reserve foods are paramylum and fats. Has been divided into a single class *Euglenophyceae* (the euglenoids) e.g., *Euglena*.

Division 3 : *Pyrrophyta* include 1000 species mainly unicellular rarely colonial. Pigments are chlorophyll a and c, carotene and xanthophyll. Reserve foods is starch/oil. Divided into two classes: (1) *Desmophyceae* (dinophysids) e.g., *Desmarestia*; (2) *Dinophyceae* (Dinoflagelloids) e.g., *Dinophysis*.

Division 4 : *Chrysophyta* represented by 6000 species of which 75% are freshwater and the remaining 25% marine. Dominant pigments are carotene and xanthophyll and reserve food is leucosin and oil. Divided into three classes: (1) *Chrysophyceae* (golden brown algae) *e.g.*, *Chromulina*; (2) *Xanthophyceae* (Yellow green algae) *e.g.*, *Botrydium*; (3) *Bacillariophyceae* (diatoms) *e.g.*, *Pinnularia*.

Division 5 : *Phaeophyceae* (Brown algae) represented by 1000 mostly marine forms, dominant pigments are phycophyein and fucoxanthin and reserve foods are *laminarin* and *mannitol*. Divided into three classes: (1) *Isogeneratae e.g.*, *Ectocarpus* (2) *Hetero generatae e.g.*, *Myrionema* and (3) *Cyclosporae* e.g., *Sargassum*.

Division 6 : *Cyanophyta* (Blue-green algae). Represented by 1500 mostly fresh water species. Pigments are chlorophyll a & b, C-phycocyanin and C-phycoerythrin and the reserve food is cyanophycean starch. Motile cells absent. Divided into a single class *Myxophyceae* e.g., *Nostoc, Anabaena*.

Division 7 : *Rhodophyta* (Red Algae). Includes 2500 species mostly marine. Predominant pigments are r-*phycoerythrin*. Reserve food is floridean *starch*. Division contains only one class *Rhodophyceae e.g., Polysiphonia, Gracilaria, Batrachospermum* (fresh water).

G.E. Papenfuss (1955) proposed a classification of algae based on phylogenetic relationships. He recognised 8 Phyla and 12 classes as listed below:

PHYLUM I Chlorophycophyta
Class (*i*) Chlorophyceae
PHYLUM II Charophycophyta
Class (*i*) Charophyceae
PHYLUM III Euglenophycophyta
Class (*i*) Euglenophyceae
PHYLUM IV Chrysophycophyta
Classes (*i*) Xanthophyceae
(*ii*) Chrysophyceae
(*iii*) Bacillariophyceae
PHYLUM V Pyrrophycophyta
Classes (*i*) Dinophyceae
(*ii*) Cryptophyceae
(*iii*) Chloromonadophyceae
PHYLUM VI Phaeophycophyta
Class (*i*) Phaeophyceae
PHYLUM VII Schizophycophyta
Class (*i*) Schizophyceae
PHYLUM VIII Rhodophycophyta
Class (*i*) Rhodophyceae

On the basis of pigments in the plastids, morphological characters and biochemical differences, Chapman (1962) divided algae into four phyla which were further subdivided into classes as follows:

PHYLUM I Euphycophyta
Classes (*i*) Charophyceae
(*ii*) Chlorophyceae
(*iii*) Phaeophyceae
(iv) Rhodophyceae
PHYLUM II Myxophycophyta
Class Myxophyceae
PHYLUM III Chrysophycophyta
Classes (i) Chrysophyceae
(ii) Xanthophyceae
(iii) Bacillariophyceae
PHYLUM IV Pyrrophycophyta
Classes (i) Cryptophyceae
(ii) Dinophyceae

Christensen (1964) proposed a new scheme of primary classification of algae into *Procaryota* and *Eucaryota* on the basis of difference between the procaryotic and eucaryotic cells. It is briefly given below:

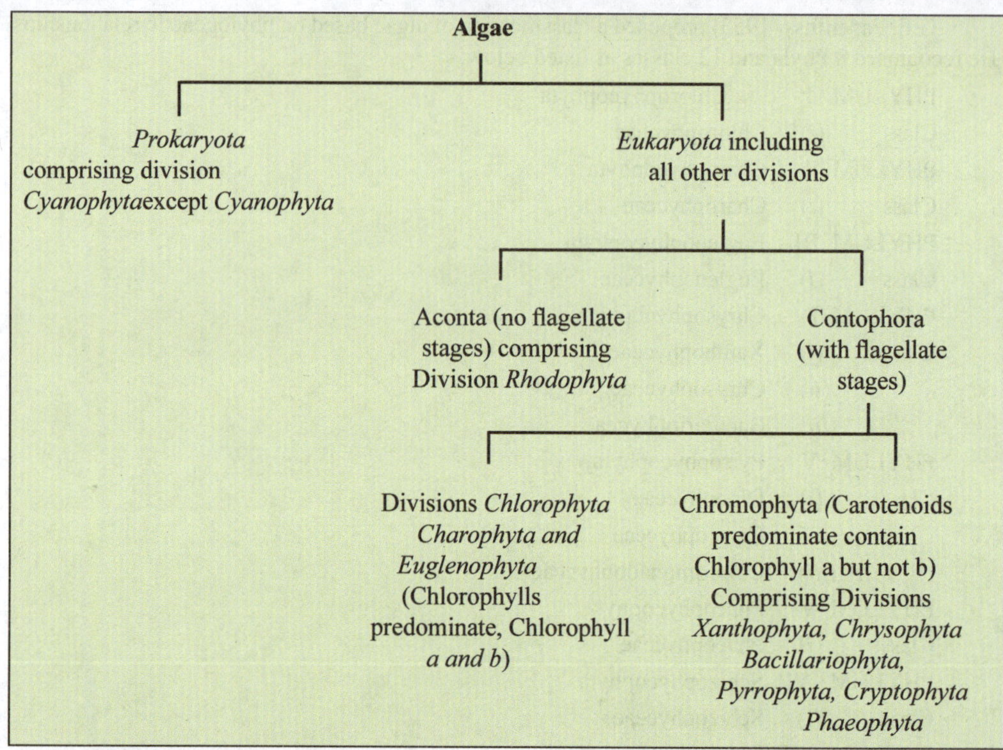

G.W. Prescott (1969) emphasized the presence or absence of true nucleus in the algal cells for their classification alongwith other characters viz. pigmentation, biochemical nature of cell wall and reserve food material and divided algae into nine phyla and fourteen classes. The scheme proposed by him is as follows:

PHYLUM I		Chlorophyta
Classes	(*i*)	Chlorophyceae
	(*ii*)	Charophyceae
PHYLUM II		Euglenophyta
PHYLUM III		Chrysophyta
Classes	(*i*)	Chrysophyceae
	(*ii*)	Bacillariophyceae
	(*iii*)	Heterokontae (Xanthophyceae)
PHYLUM IV		Pyrrophyta
Classes	(*i*)	Desmokontae (Desmophyceae)
	(*ii*)	Dinokontae (Dinophyceae)
PHYLUM V		Phaeophyta
Classes	(*i*)	Isogeneratae
	(*ii*)	Heterogeneratae
	(*iii*)	Cyclosporeae
PHYLUM VI		Rhodophyta
Classes	(*i*)	Bangioideae
	(*ii*)	Florideae

PHYLUM VII Cyanophyta
Classes (*i*) Coccogoneae
(*ii*) Hormogoneae
PHYLUM VIII Cryptophyta
PHYLUM IX Chloromonadophyta

F.E. Round (1973) also recognised the importance of presence or absence of well organised nucleus in algal cells in the classification of algae alongwith their phylogenetic relationships and other characteristics. He made certain modifications in the scheme proposed by Fritsch and raised the rank of classes of Fritsch to phyla. Besides this, Round placed blue-green algae in a major group Prokaryota which was put at the top whereas Fritsch has placed them near Rhodophyceae. The classification proposed by him is summarised as follows:

PROKARYOTA

PHYLUM I Cyanophyta

EUKARYOTA

PHYLUM II Chrysophyta
Classes (*i*) Xanthophyceae
(*ii*) Chryophyceae
(*iii*) Haptophyceae
(*iv*) Bacillariophyceae
PHYLUM III Chlorophyta
Classes (*i*) Chlorophyceae
(*ii*) Oedogoniophyceae
(*iii*) Bryopsidophyceae
(*iv*) Conjugatophyceae
(*v*) Charophyceae
(*vi*) Prasinophyceae
PHYLUM IV Euglenophyta
PHYLUM V Pyrrophyta
Classes (*i*) Desmophyceae
(*ii*) Dinophyceae
PHYLUM VI Cryptophyta
PHYLUM VII Phaeophyta
Class Phaeophyceae
Cohorts (*a*) Isogeneratae
(*b*) Heterogeneratae
(*c*) Cyclosporae
PHYLUM VIII Rhodophyta
Class Rhodophyceae
Sub classes (*a*) Bangiophycidae
(*b*) Florideophycidae

V.J. Chapman and D.J. Chapman (1973) divided algae into two major groups viz. Prokaryota and Eukaryota which were further subdivided into divisions and classes.

PROKARYOTA

Division	:	Cyanophyta
Class	:	Cyanophyceae

EUKARYOTA

Division	:	Rhodophyta
Class	:	Rhodophyceae
Division	:	Chlorophyta
Class	:	(*i*) Chlorophyceae
		(*ii*) Prasimophyceae
		(*iii*) Charophyceae
Division	:	Euglenophyta
Class	:	Euglenophyceae
Division	:	Chloromonadophyta
Class	:	Chloromonadophyceae
Division	:	Xanthophyta
Class	:	Xanthophyceae
Division	:	Bacillariophyta
Class	:	Bacillariophyceae
Division	:	Chrysophyta
Class	:	(*i*) Chrysophyceae
		(*ii*) Haptophyceae
Division	:	Phaeophyta
Class	:	Phaeophyceae
Division	:	Pyrrophyta
Class	:	(*i*) Dinophyceae
		(*ii*) Desmophyceae
Division	:	Cryptophyta
Class	:	Cryptophyceae

It would be amply clear from the above account that the older phycologists placed all the chlorophyll bearing thalloid organisms in a single sub-division called **Algae**. Such a grouping implies that these series of organisms are closely related. The modern phycologists, however, do not recognise any close relationship between the members of different classes of algae. During the second and third decades of the last century a number of fundamental differences among the several series of algae were found. These convinced the modern phycologists that the different series of algae, in fact, denote a number of parallel lines of development. This viewpoint resulted in the dismemberment of the former division Thallophyta and sub-division Algae and the elevation of its classes to the rank of a division. The modern algologists, therefore, classify these plants of algal organisation into 11 divisions, namely, Cyanophyta, Chlorophyta, Charophyta, Xanthophyta, Chrysophyta, Bacillariophyta, Pyrrophyta, Cryptophyta, Euglenophyta, Phaeophyta and Rhodophyta. The relevant portion of this classification with slight modifications is given below:

Division 1. **CYANOPHYTA** (Blue-green algae) including a single class.

Class 1. **Cyanophyceae** comprising 7 orders:

Order	1. **Chorococcales**
Family	1. Chroococcaceae : Type *Chroococcus*
Order	2. **Chamaesiphonales** comprising 3 families.
Order	3. **Oscillatoriales** with 1 family:
Family	1. Oscillatoriaceae : Type *Oscillatoria*
Order	4. **Nostocales** with a single family :
Family	1. Nostocaceae : Type *Nostoc* or *Anabaena*
Order	5. **Scytonematales** including 1 family :
Family	1. Scytonemataceae : Type *Scytonema*
Order	6. **Stigonematales** with a single family:
Family	1. Stigonemataceae : Type *Stigonema*
Order	7. **Rivulariales** comprising one family :
Family	1. Rivulariaceae : Type *Rivularia*
Division	2. **CHLOROPHYTA** with 1 class :
Class	1. **Chlorophyceae** comprising 9 orders :
Order	1. **Volvocales** with 3 families :
Family	1. Chlamydomonadaceae : Type *Chlamydomonas*
"	2. Volvocaceae : Type *Volvox*
"	3. Tetrasporaceae : Type *Tetraspora*
Order	2. **Chlorococcales**
Family	1. Chlorellaceae : Type *Chlorella*
Family	2. Hydrodictyaceae : Types *Hydrodictyon* and *Pediastrum*
Family	3. Protosiphonaceae : Type *Protosiphon*
Order	3. **Ulotrichales**
Family	1. Ulotrichaceae : Type *Ulothrix*
Family	2. Microsporaceae : Type *Microspora*
Family	3. Cylindrocapsaceae : Type *Cylindrocapsa*
Order	4. **Ulvales**
Family	1. Ulvaceae : Types *Ulva* and *Enteromorpha*
Order	5. **Oedogoniales**
Family	1. Oedogoniaceae : Type *Oedogonium*
Order	6. **Cladophorales**
Family	1. Cladophoraceae : Type *Cladophora* and *Pithophora*
Order	7. **Chaetophorales**
Family	1. Chaetophoraceae : Types *Stigeoclonium, Draparnaldia* and *Draparnaldiopsis*
Family	2. Trentepohliaceae : Type *Trentepohlia*
Family	3. Coleochaetaceae : Type *Coleochaete*
Family	4. Pleurococcaceae : Type *Pleurococcus*
Order	8. **Zygnemales**
Family	1. Zygnemaceae : Types *Zygnema* and *Spirogyra*
Family	2. Desmediaceae : Types *Closterium* and *Cosmarium*
Order	9. **Siphonales** or **Caulerpales**
Family	1. Caulerpaceae : Type *Caulerpa*

Division 3. **CHAROPHYTA** (Stoneworts) with one class :
Class 1. Charophyceae comprising one order :
Order 1. **Charales** with a single family;
Family 1. Characeae : Types *Chara* and *Nitella*
Division 4. **EUGLENOPHYTA** with one order :
Order 1. **Euglenales** : Type *Euglena*
Division 5. **PYRROPHYTA**
Division 6. **XANTHOPHYTA** with a single class :
Class 1. **Xanthophyceae** (Yellow-green algae)
Order 1. **Heterosiphonales**
Family 1. Botrydiaceae : Type *Botrydium*
Family 2. Vaucheriaceae : Type *Vaucheria*
Division 7. **BACILLARIOPHYTA** with one class :
Class 1. **Bacillariophyceae**; Type *diatoms*
Division 8. **CHRYSOPHYTA**
" 9. **CRYPTOPHYTA**
" 10. **PHAEOPHYTA** with three classes :
Class 1. **Isogeneratae**
Order 1. **Ectocarpales**
Family 1. Ectocarpaceae : Type *Ectocarpus*
Order 2. **Dictyotales**
Family 1. Dictyotaceae : Type *Dictyota*
Class 2. **Heterogeneratae**
Order 1. Laminariales
Family 1. Laminariaceae : Type *Laminaria*
Class 3. **Cyclosporae**
Order 1. **Fucales**
Family 1. Fucaceae : Type *Fucus*
Family 2. Sargassaceae : Type *Sargassum*
Division 11. **RHODOPHYTA** with one class :
Class 1. **Rhodophyceae**
Sub-class 1. **Bangioideae**
Sub-class 2. Florideae
Order 1. **Nemalionales**
Family 1. Batrachospermaceae : Type *Batrachospermum*
Order 2. **Ceramiales**
Family 1. Rhodomelaceae : Type *Polysiphonia*

In 1977, Lewin created a new division **Prochlorophyta** between Cyanophyta and Chlorophyta.

Salient features of important classes of Algae

Class 1. **Chlorophyceae** (Green Algae). The colour of members of the class is green or grassgreen due to presence of chlorophyll *a* and *b*, xanthophylls and carotenoid, the cells have organised nucleus (eukaryotic), reserve food material is starch rarely oil e.g., in *Vaucheria,* pyrenoids usually present in chloroplast except in the members of order siphonales; motile cells have variable number of flagella or cilia which are equal in size, sexual reproduction ranges from isogamy to oogamy; mostly fresh water but few forms are marine.

Class 2. **Xanthophyceae.** The members are yellow green in colour due to excess of xanthophylls; majority of the members are fresh water, few are marine; pyrenoids absent in the chromatophores and oil is reserve food material; sexual reproduction is of rare occurrence but if present is isogamous and motile sex cells have flagella-two in number and unequal in length.

Class 3. **Chrysophyceae.** The members of the class are marine and fresh water both, usually found in cold water bodies; brown or orange in colour owing to the excess of an accessory pigment Phycochrysin; reserve food material is oil and leucosin which is protein like substance; sexual reproduction rare and if present is isogamous, the motile cells have one, two or rarely three flagella of equal or rarely of unequal size.

Class 4. **Bacillariophyceae.** Commonly known as **diatoms** the members are fresh water and marine both; members have diatomin as principal pigment and are yellow or golden brown in colour; cell wall partly silicified and partly pectose and is divided into two symmetrical halves; variously ornamented which is of taxonomic importance; Pyrenoids present in chromatophores and reserve food material is oil, volutin and leucosan; sexual reproduction is of unique type and the members are diploid unicellular organisms but sometimes may form colonies.

Class 5. **Cryptophyceae.** The members are fresh water and marine and are of diverse colour i.e. brown, red, olive green or sometimes bluish green; the cells have two large perietal chloroplasts with pyrenoids and starch is reserve food material; Mostly the forms are flagellates and have two flagella which are unequal in length. Cysts are common and endogenous and sexual reproduction is isogamous in one species only.

Class 6. **Dinophyceae.** Most of the members are unicellular and motile, biflagellate and have sculptured cell wall; The nucleus is prominent and large and chromatophores are discoid, dark yellow or brown in colour due to presence of red phycopyrin, dark red peridinin and yellow green chlorophyllin which are collectively termed Pyrophyll; reserve food material is starch and fat; sexual reproduction rare and is isogamous when present.

Class 7. **Chloromonadineae.** Members included in the class are bright green in colour owing to the excess of xanthophyll and cells have numerous discoid chromatophores; Reserve food material is fat and individuals reproduce by longitudinal divisions.

Class 8. **Euglenineae** (Euglenophyceae) unicellular and motile forms, with one or two flagella and are found in fresh water bodies or in saline habitats; animal like naked flagellates. Mostly the members are green, free swimming and solitary but some forms may form gelatinous colonies whereas few of them may also get attached; holophytic forms have several green chloroplasts while other forms lack it. Reproduction usually takes place by fission.

Class 9. **Phaeophyceae** (Brown Algae). The members are mostly marine, brown in colour and have complicated and organised thalli; commonly known as 'sea weeds', of great economic importance; the cells have several chromatophores with excess of fucoxanthin alongwith other pigments; Reserve food material is alcohol and polysaccharides; reproduction is vegetative and sexual both and the motile sex cells are biflagellate having two flagella laterally inserted and of unequal length; sexual reproduction ranges from isogamy to oogamy.

Class 10. **Rhodophyceae** (Red Algae). Majority of the forms are marine and only few are fresh water forms, usually red or dark red in colour due to presence of accessory pigments c-phycocyanin and r-phycoerythrin; Reserve food material is 'floridean starch', thallus is organised and exhibits complex differentiation; Plasmodesmata is present between the cells except in the members of Protoflorideae, sexual reproduction is specialised and oogamous; motile cells altogether absent in the class and even the male sex cells are non motile structures.

Class 11. **Myxophyceae** (Blue-green Algae). Mostly fresh water forms with simple thallus organisation, look bluish green in colour owing to the presence of accessory pigments phycocyanin and phycoerythrin; c-phycocyanin is the overmasking blue pigment while the red pigment phycoerythrin is also found in peripheral regions in dispersed condition; pigments are not localized in organised and definite chromatophores and cells lack well-defined nucleus (prokaryotes), Photosynthetic products are polysaccharide and glycogen; sexual reproduction is completely absent and motile cells are not produced at any stage of life cycle.

QUESTIONS

ESSAY TYPE

1. Give the classification of Algae as given by Smith. Also mention the characters of different classes. *(Rohilkhand, 1997)*
2. Describe in detail the classification proposed by F.E. Fritsch and G.M. Smith. *(Poorvanchal, 2001; Gorakhpur, 2003; Rohilkhand, 2001)*
3. Give an illustrated account of the range of vegetative structure in algae.
4. With the help of suitable diagrams, describe the cell structure of an eukaryotic alga.
5. Describe the structure of flagella in algae and mention the three types.
6. Give a brief outline of Fritsch's classification. Mention characteristic features and two examples each of different classes studied by you. *(Rohilkhand, 2004, Awadh, 1991, 1999; Meerut, 1998, 1999; Punjab, 1998)*
7. What are the principal criteria for the primary classification of Algae? Enumerate the chief characteristics of any four main classes.
8. Describe the distinguishing features of Myxophyceae, Chlorophyceae, Phaeophyceae and Rhodophyceae. *(Awadh, 1994; Kanpur, 1996, 1999; Meerut, 1997; Bundelkhand, 1998)*
9. Describe methods of asexual reproduction in algae giving suitable examples. *(Lucknow, 1992)*
10. Give a detailed account of algal pigments and their significance in Algal classification. *(Lucknow, 1993; Bundelkhand, 1996)*
11. Describe the range of vegetative structures in algae with suitable examples. *(Gorakhpur, 1997; Rohilkhand, 1993; Punjab, 1993; Bharathiar, 1995)*
12. Describe the classification of algae with the help of important characters and examples. *(Kanpur, 1996; 20001; Bundelkhand, 1999)*
13. Name the four major groups of algae and give an account of their characteristics. *(Kanpur, 1998; Rohilkhand, 1992)*
14. On what characters Chapman's Classification of Algae is based? With supporting points give the details of phylum Euphycophyta. *(Agra, 1995)*
15. In which class you will place the minute plant which lacks Mitochondria,

Chloroplasts, true nucleus but contain 70S ribosomes and photosynthetic pigments, chlorophyll a, carotene and xanthophyll with phycocyanin? Give reasons for your answer. *(Kanpur, 2006)*

16. Give an account of classification given by Fritsch. Write characters of classes also in brief. *(Kanpur, 2003)*

17. Describe the most commonly used classification of algae with the help of important characters and suitable examples. *(Lucknow, 2004)*

18. Write any classification proposed for algae you like most. Describe its merits and demerits. *(Kanpur M.Sc. 2006)*

19. Write a note on contributions of Indian algologists. *(Kanpur M.Sc. 2005)*

20. What are the recent trends in classification of Algae? How does the modern system differ from the older systems? *(Lucknow M.Sc. 2006, 2004)*

21. Enumerate the criteria used in the classification of algae. Describe any classification upto order level and highlight its merits. *(Lucknow M.Sc. 2005)*

22. Give an account of classification of algae proposed by F.E. Fritsch. How would you justify the inclusion of the Cyanophycea and Charophyceae among algae? *(Poorvanchal, 2003)*

23. On what characters classification of Algae by Fritsch is based? In what respect is it different from Smith's classification? *(Poorvanchal 2004)*

24. Describe different characters of classification of algae studied by you. *(Poorvanchal, 2005)*

25. Describe the important characteristic features of different classes of algae as proposed by F.E. Fritsch. *(Poorvanchal, 2007)*

SHORT ANSWER TYPE

26. Describe the following:

(*a*) Algal cell walls

(*b*) Algal pigments *(Allahabad, 1992, Awadh 1991, 1995, Gorakhpur, 1999)*

(*c*) Flagella in Algae *(Poorvanchal, 2004, Awadh, 1991, 1993, 1996, 1997; Kanpur 1995)*

(*d*) Plastids in Algae *(Awadh, 1997; Bundelkhand, 1999)*

(*e*) Shapes of chloroplasts in Algae *(Lucknow, 2006; Rohilkhand, 1994)*

(*f*) Neuromotor apparatus in Algae *(Gurunanak Dev., 1991)*

27. Write short notes on the following:

(*i*) Distinctive features of Phaeophyceae *(Lucknow, 1997, 1998; Allahabad 1993, 1995)*

(*ii*) Distinctive features of Cyanophyceae *(Allahabad, 1994)*

(*iii*) Importance of reserve food materials in the classification of algae *(Allahabad, 1994, 2001)*

(*iv*) Characteristic features of Rhodophyceae *(Allahabad, 1995, Lucknow, 1993)*

(*v*) Difference between Zygospores and Oospores *(Allahabad, 1995)*

(*vi*) Heterotrichous habit *(Allahabad, 1995; Awadh, 1999)*

(*vii*) Distinctive features of Chlorophyceae *(Allahabad, 1996; Lucknow, 1993)*

(*viii*) Distinctive features of Xanthophyceae *(Allahabad, 1996)*

(*ix*) Alternation of generation in Algae *(Awadh, 1997)*

(*x*) Vegetative reproduction in Algae *(Awadh, 1997)*

(*xi*) Classification of Algae as given by Smith (1955) *(Gorakhpur, 1993)*

(*xii*) Epiphytic and Cryophytic Algae *(Kumaon, 1999)*

(*xiii*) Ultrastructure of eukaryotic algal cell *(Lucknow, 2002)* (*xiv*) Types of chlorophylls found in Eukaryotic algae (*Lucknow, 2005*)

(*xv*) Prokaryotic algal cell (*Lucknow, 2005*)

(*xvi*) Pigments and algal classification (*Kanpur M.Sc. 2005*)

28. Name the three different ways Zygote or Zygospore behaves at the time of germination.

29. Mention names of important methods of asexual and sexual reproduction in algae.

30. Write the names of four major classes of algae.

31. Compare the distinguishing features of Myxophyceae, Xanthophyceae and Phaeophyceae. (*Lucknow, 1995*)

32. Distinguish between the following pairs :

(*a*) Phaeophyceae and Xanthophyceae

(*b*) Phaeophyceae and Rhodophyceae

(*c*) Rhodophyceae and Cyanophyceae (*Lucknow 1998, 2001; Gorakhpur, 1996*)

(*d*) Chlorophyceae and Charophyceae (*Rohilkhand, 1992*)

33. Name the reserve food materials found in Red algae. (*Gorakhpur, 1990*)

34. Mention two important characters on which Smith based his classification. (*Gorakhpur, 1993*)

35. Name the reserve foods found in Cyanophyta. (*Gorakhpur, 1995*)

36. What is the reserve food material in brown algae? (*Gorakhpur, 1996, 1998*)

37. What is the function of eye spot in algal cell?

38. Define thermal algae, phytoplankton, phytobenthos, cryophytes and epiphytes. (*Himachal Pradesh, 1993*)

39. Name the pigments in Phaeophyceae, Rhodophyceae and Chlorophyceae.

40. List the main characters of distinction between Rhodophyceae and Chlorophyceae.

41. Define phycology.

42. Name any Indian algologist and his contributions. (*Kanpur M.Sc. 2006*)

43. Name reserve food materials in Phaeophyceae, Cyanophyceae and Xanthophyceae. (*Allahabad 2005*)

44. Name the father of Phycology in India. (*Poorvanchal, 2003*)

45. Give the name of pigments that impart brown colour to the members of Phacophyceae. (*Poorvanchal, 2003*)

OBJECTIVE TYPE

46. Fill in the blanks :-

(*i*) The study of algae is known as ....

(*ii*) The reserve food material in Chlorophyceae is ....

(*iii*) The alga .... is an example of heterotrichous habit.

(*iv*) Water blooms are generally formed by ....

(*v*) The example of Phaeophyceae is ....

(*vi*) The reserve food is floridean starch in the members of class ....

(*vii*) Phycobilins present in Red algae are ....

(*viii*) The classification of algae is mainly based on ....

(*ix*) Sexual reproduction is absent in the members of class ....

(*x*) A fresh water Rhodophycean alga is ....

(*xi*) The algae with prokaryotic organisation belong to the class............ .

(*xii*) All classes except .... include unicellular forms.

(*xiii*) The principal pigment of Phaeophyceae importing distinctive brown colouration is ....

(*xiv*) Plastids are absent in ....

(*xv*) Chlorophyll a and b is present in ....

(*xvi*) Luxuriant growth of some algae in water often imparting colour to water is called ....

(*xvii*) Smith based his classification of algae on two characters which are .... and ....

(*xviii*) .... is the reserve food material in Cyanophyceae.

(*xix*) The two features in which blue green algae resemble bacteria are .... and ....

(*xx*) Pyrenoids are meant for .... synthesis in various algae.

(*xxi*) Development of gametophyte directly into sporophyte without fertilisation is called................. .

47. Select the correct answer :

(*i*) Fucoxanthin is found in

(*a*) Blue-green algae

(*b*) Green algae

(*c*) Brown algae

(*d*) Red algae

(*ii*) Phycology is the study of

(*a*) Algae

(*b*) Fungi

(*c*) Bryophyte

(*d*) Bacteria

(*iii*) The photosynthetic, genetic and respiratory apparatuses are not bound by membranes in

(*a*) Chlorophyceae

(*b*) Phaeophyceae

(*c*) Rhodophyceae

(*d*) Cyanophyceae

(*iv*) Which of the following is a marine alga?

(*a*) *Oedogonium*

(*b*) *Chlamydomonas*

(*c*) *Ectocarpus*

(*d*) *Batrachospermum*

(*v*) An alga has septate branched filamentous body without true nucleus. To which class should this alga be placed?

(*a*) Cyanophyceae

(*b*) Chlorophyceae

(*c*) Phaeophyceae

(*d*) Rhodophyecae

(*vi*) The reserve food material in Chlorophyceae is

(*a*) Starch (*b*) Oil

(*c*) Protein (*d*) Glucosides

(*vii*) Which of the following has floridean starch as reserve food material?

(*a*) *Nostoc* (*b*) *Volvox*

(*c*) *Sargassum* (*d*) *Polysiphonia*

(*viii*) Trichothallic growth is found in

(*a*) *Volvox* (*b*) *Oedogonium*

(*c*) *Ectocarpus* (*d*) *Oscillatoria*

(*ix*) Which of the following shows heterotrichous habit?

(*a*) *Chlamydomonas*

(*b*) *Volvox*

(*c*) *Oedogonium*

(*d*) *Coelochaete*

(*x*) Myxophyceae are called so because of their

(*a*) Incipient nucleus

(*b*) Phycocyanin pigment

(*c*) Mucilaginous sheath

(*d*) Lack of plastids.

(*xi*) Heterotrachy means having

(*a*) Prostrate and erect branches

(*b*) Rhizoids and photosynthetic branches

(*c*) Long and short branches

(*d*) Branches differentiated into nodes and internodes

(*xii*) Plastids are absent in

(*a*) Chlorophyceae

(b) Phaeophyceae
(c) Rhodophyceae
(d) Myxophyceae

(xiii) Motile cells are completely absent in
(a) *Chlamydomonas*
(b) *Volvox*
(c) *Ectocarpus*
(d) *Polysiphonia*

(xiv) Which of the following classes does not include unicellular forms?
(a) Chlorophyceae
(b) Xanthophyceae
(c) Rhodophyceae
(d) Myxophyceae

(xv) Sexual reproduction is absent in
(a) Chlorophyceae
(b) Myxophyceae
(c) Phaeophyceae
(d) Rhodophyceae

(xvi) Xanthophyll is the principal pigment in the members of the class
(a) Chlorophyceae
(b) Phaeophyceae
(c) Rhodophyceae
(d) Xanthophyceae

(xvii) Fusion between gametes of unequal sizes is called
(a) Isogamy
(b) Anisogamy
(c) Oogamy
(d) Dichogamy

(xviii) An alga growing on snails is called
(a) Epiphyte (b) Symbiotic
(c) Endophyte (d) Cryophyte

(xix) The Red sea gets its name because of the presence in its water red coloured algae belonging to the class
(a) Rhodophyceae
(b) Cyanophyceae
(c) Chlorophyceae
(d) Phaeophyceae

(xx) Which are the two pigments in Phaeophyceae?
(a) Chlorophyll and Xanthophyll
(b) Chlorophyll and Fucoxanthin
(c) Phycocyanin and Phycoerythin
(d) *Phycocyanin and Xanthophyll*

Blue green algal bloom
(Afflicting Taihu lahe in China)

CHAPTER 2

DIVISION : CYANOPHYTA (*Cyanophycophyta*) Class : *Cyanophyceae*

- General Characteristics
- Distribution
- Microcystis
- Tribe : Hormogoneae : Order : Oscillatoriales
- Tribe : Hormogoneae : Order 2 : Nostocales
- Order 3 : Scytonematales

GENERAL CHARACTERISTICS

This division includes the blue-green algae which are the only known oxygen producing prokaryotes. It is a small primitive group comprising of about 2,500 species placed under 150 genera. All of them are included in a single class **Cyanophyceae** or **Myxophyceae**. The members of this class are considered the simplest, living autotrophic plants. Individually all the blue-greens are microscopic. The diagnostic features of the division and the class are:

(*i*) The cells constituting the thallus are **prokaryotic;**

(*ii*) The flagella are entirely lacking (even the reproductive cells are non-flagellated);

(*iii*) Where locomotion occurs, it is of gliding, jerky type;

(*iv*) The phycobilin pigments unique to this class are blue *C-phycocyanin* and *C-phycoerythrin*; in addition to *chlorophyll-a, B-carotene* and unique, xanthophylls, namely, *myxoxanthin* and *mycoxanthophyll;*

(*v*) There are no membrane bound chromatophores;

(*vi*) The unique food-storage compounds are the *myxophycean starch* and a proteinaceous material *cyanophycin;*

(*vii*) Many filamentous blue-greens possess specialised cells of disputed function known as the **heterocysts**;

(*viii*) Sexual reproduction is completely absent. Recently indications of genetic recombination which fulfil the function of sex have been reported in a few cases.

DISTRIBUTION

The blue-green algae are widely spread in the aquatic environment. Some terrestrial species are also known. The aquatic forms mostly occur in fresh water, a few however, are marine. A notable example of the marine blue-greens is *Trichodesmium erythraeum* in which the red pigment predominates. This species flourishes in the Red Sea and is responsible for the red colour of its water. The freshwater blue-greens occur in clean or polluted water reservoirs, ponds, springs, open tanks and lakes, where planktonic (free floating) forms, as a group, generally exhibit characteristic cyclic growth. The drastic increase in their number due to rapid algal growth and their buoyancy result in the sudden appearance of large, dense algal populations in calm weather usually in the summer in the temperate regions and any time of the year in the tropics. This accumulation or rapid appearance of planktonic blue-green algae and other microorganisms at the surface of lakes, ponds and reservoirs of freshwater which attracts attention is termed as *water bloom*. Some species of blue-green algae occur as scums on dirty stagnant water full of decaying organic materials. A few species are sub-aerial and inhabit damp surfaces where they commonly appear as green slime on wet rocks, flower pots or damp soil. In the moist regions of the world they are quite common on the bark of trees. Some are found in the soil and are able to fix atmospheric nitrogen and utilize it in their own metabolism.

Algal bloom

There are a sufficient number of thermal Cyanophyceae. They grow on snow and also constitute the principal vegetation of hot springs. They are able to live in water whose temperature is as high as 85°C. The Cyanophyta thus provide a good example of the adaptability of life to extremes of environment (high temperature of hot springs and low temperature of polar regions). Because of their gelatinous sheaths they can withstand long periods of desiccation. The compactness of protein molecules and their bonds in the protoplasm also helps the cells to face the extremes. Certain blue-green algae live intimately with other organisms as **symbionts**. Colonies of certain blue-green algae grow in the roots of *Cycas* and within the thallus of *Anthoceros. Anabaena* occurs in cavities in the fronds of *Azolla* (a water fern). Certain Cyanophyceae are associated with the fungi in the formation of Lichens. The ability of Cyanophyta to photosynthesize and fix atmospheric nitrogen may account for their symbiotic associations. Some blue-green algae which occur in soil up to a depth of several feet and in semi-aquatic environments (paddy fields) fix atmospheric nitrogen. The factors which contribute to their wide distribution are:

(*i*) ability to withstand prolonged drying and extremes of temperature,

(*ii*) capacity of some to assimilate atmospheric nitrogen,

(*iii*) labile metabolism,

(*iv*) ability of many of them to enter into symbiotic relationship with other plants, and

(*v*) their constancy of form along with their modes of reproduction.

Svmboisis: *Anabaena* in fronds of *Azolla.*

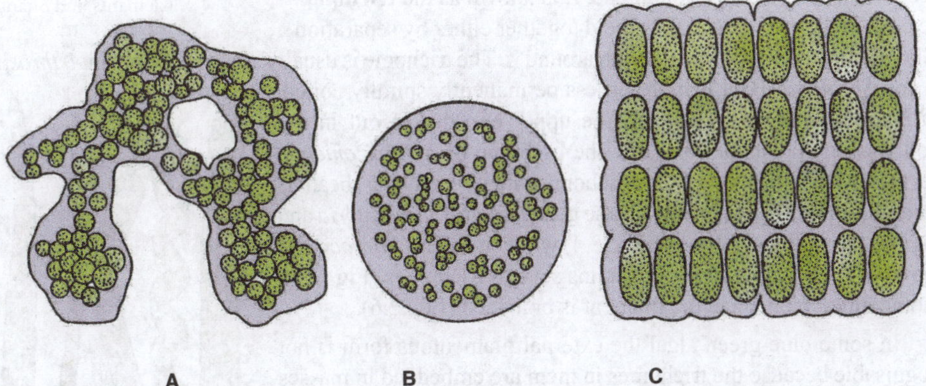

Fig. 2.1 (A-C). Non-filamentous colonial Blue-green Algae. (A) *Microcystis;* (B) *Aphanocapsa;* and (C) *Merismopedia.*

ORGANISATION OF THE THALLUS

The blue-green algae amongst the simplest photosynthetic plants living today. Just as do they vary in the choice of their habitats, so do they in the range of vegetative structure. Architecturally, the thallus may be a solitary cell or a colony.

1. Unicellular Forms. The thallus, in some species, is a unicellular which is usually spherical or oval (*Chroococcus, Synechococcus, Anacystis* and *Gloeocapsa*). There is immediate separation of the daughter cells from each other after cell division. Actual unicellular forms, however, are not many because the copious secretion of mucilage by the daughter cells results in the daughter cells remaining together after division.

2. Colonial Forms (Fig. 2.1). In most blue-greens, the cells after division remain attached by their walls or are held in a common gelatinous matrix to form a loose organisation of cells which is termed a **colony**. Of course the cells in the colonies are often aggregated into irregular, palmelloid forms of great variability. *Gloeothece* is an example of an aggregation of a few cells. *Aphanocapsa* (B) and *Aphanothece* are examples of aggregations of numerous cells. The colonies may either be **filamentous** or **non-filamentous**. Each colony is generally enclosed in a gelatinous sheath.

(*a*) *Non-filamentous colonies* (Fig.2.1). The *non-filamentous* colonies are of various forms.

They may be cubical, spherical, square or irregular depending on the planes and direction in which the cells divide. The formation of non-filamentous colonies is regarded to proceed along two lines:

(*i*) Cell divisions take place alternately in two planes at right angles to each other. The result is a flat plate of cells (C) or a hollow sphere in which the cells form a single layer near the periphery of the gelatinous matrix (B). Examples of hollow spherical colonies are *Coelosphaerum* and *Gomphophaera*.

(*ii*) Cell divisions proceed in three planes. If the sequence is regular as is the case in a few genera, the component cells become arranged into a colony having a definite shape. The cubical colony of *Eucapsis alpina* (Fig. 2.2) is the typical example. However, usually the sequence of division is irregular resulting in a colony with no regular arrangement of cells (*Microcystis,* A).

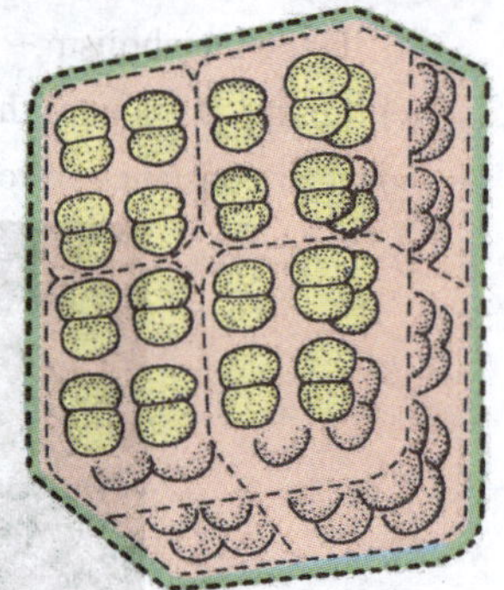

Fig. 2.2. Non-filamentous Cyanophyceae. *Eucapsis alpina* (After Clemants and Shantz).

(*b*) *Filamentous colonies* (Fig. 2.3). The *filamentous* colony is the result of repeated cell divisions in a single plane and in a single direction forming a chain or a thread. It is known as the **trichome**. The cells in the trichome may be held together either by separation walls or a common gelatinous sheath around it. The trichome is usually straight (A) but in *Arthrospira* (C) and *Spirulina* (B) it is more or less permanently spirally coiled. In *Rivularia* it is ship-like with the upper end drawn out into a colourless, multicellular hair (D). The trichome of *Aphanizomenon* tapers towards both ends (E). The trichome with its enclosing sheath is called a **filament**. The filament in some genera such as *Oscillatoria* and *Lyngbya*, has a single trichome (Fig. 2.4) while in others (*Microcoleus vaginatus* and *Hydrocoleus*) it contains several trichomes (Fig. 2.5). In some blue-green algae the filament is branched (Fig. 2.6).

In some blue-green algal the external filamentous form is not recognisable because the trichomes in them are embedded in masses of mucilage. For instance in *Phormidium* the trichomes are embedded in mucilage which looks like a gelatinous sheath whereas in *Nostoc* the contorted trichomes are embedded in mucilage which may form regular spheres (Fig. 2.21A).

Branching (Fig. 2.7). The trichomes may be branched (*Westiella* and *Hapalosiphon*) or unbranched (*Oscillatoria* and *Lyngbya*). On the whole very little branching occurs, in general, in the blue-greens. The trichomes show little differentiation and have no distinction into base and apex. The growth is diffuse except *Rivularia*. Forms like *Tolypothrix* (A) and *Scytonema* (B) exhibit false branching. In this case the hormogones or fragments germinate *in situ* with the ends piercing and growing out of the parent sheath in a different direction. A few genera of the blue-greens such as *Stigonema* (C) and *Hapalosiphon* (D) exhibit true branching. In this case the cells constituting the filament divide in two planes. The filament in *Stigonema* (C) is multiseriate and uniseriate in *Hapalosiphon* (D).

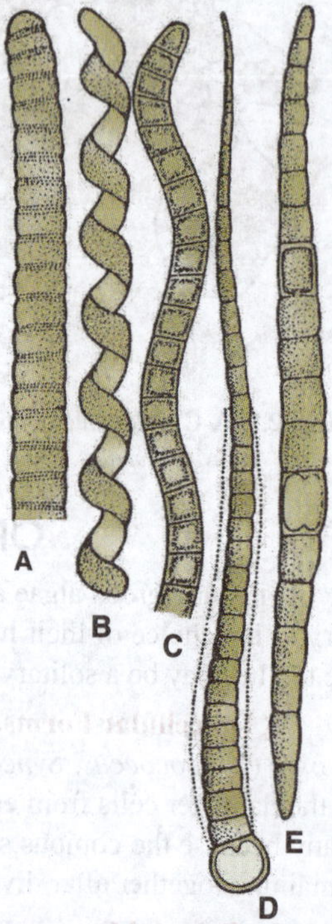

Fig. 2.3. Cyanophyceae. Different forms of trichomes. (A) *Oscillatoria*; (B) *Spirulina*; (C) *Arthrospira*; (D) *Rivularia*; (E) *Aphani-zomenon*.

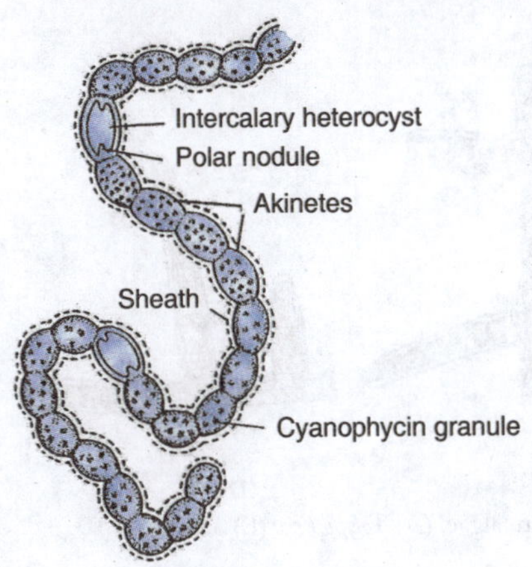

Fig. 2.4. *Nostoc* sp. A trichome with its own sheath.

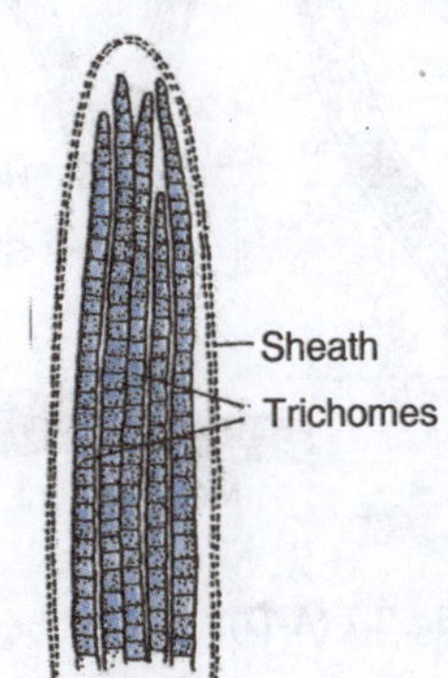

Fig. 2.5. *Microcoleus vaginatus.* A single filament consisting of several trichomes enclosed in the sheath (Diagrammatic)

The trichomes may be of uniform diameter throughout or taper from base to apex usually from a basal heterocyst (*Rivularia* or *Gloeotrichia*). The blue-greens with such trichomes usually exist in complex colonies. Each comprises many such trichomes aggregated within a common mucilage. Occasionally trichomes taper at both ends (Fig. 2.3 E). The enveloping sheath may often become very tough and enable the trichome or trichomes to withstand high temperature and considerable desiccation. The cells in the colonies are independent of one another. There is little coordination of activity between them and no division of labour. It is obvious therefore that the organisation of thallus in the Cyanophyta has not progressed very far. They can at the most form colonies of individuals rather than true multicellular organisms.

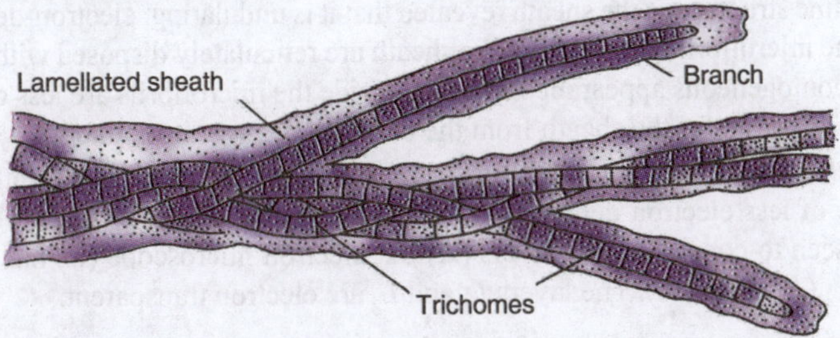

Fig. 2.6. *Schizothrix.* A branched filament with many trichomes.

Cell Structure (Fig.2.8.) The blue-green algal cell is an example of a typical prokaryotic cell. It lacks all the membrane bound cell organelles characteristic of a eukaryotic cell. Thus it has no discrete or well-defined nucleus, chromatophores, mitochondria, endoplasmic reticulum and dictyosomes. Nucleoli, pyrenoids and true vacuoles are also absent. It rarely exceeds 10 μ in diameter.

A. Gross Structure. Under light microscope the cell is seen to consist of a *cell envelope* enclosing the tiny *protoplast*.

Cell Envelope (Fig. 2.8.) Typically the cell envelope in the blue-green algae consists of two parts namely, the *sheath* and the *inner investment* commonly known as the cell wall (A).

(*a*) *Sheath*. It is a layer of extracellular mucilage external to the inner investment (cell wall). It is protective in function and irregular.

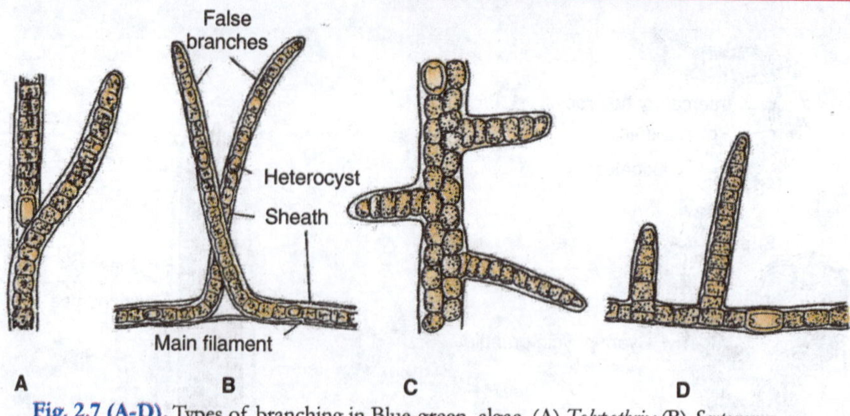

Fig. 2.7 (A-D). Types of branching in Blue-green algae. (A) *Tolypothrix* (B) *Scytonema*; (C) *Stigonema*; (D) *Hapalosiphon*.

Usually this mucilaginous covering of the cell is thick and dense and is called the *sheath*. In a few forms it is extremely delicate and diffluent (*Anacystis monatana*). The mucilage layer (sheath) is in the form of a cylinder around the entire trichome in the filamentous blue-green algae but in the unicellular forms it surrounds each cell. The secretion of copious mucilage (sheath) is one of the constant features of the blue-greens as a group and has earned for them the name Myxophyceae which means slime **algae**. Often the sheath is of considerable thickness. It may become lamellated or stratified and pigmented (yellowish or brownish). Usually it remains colourless. The sheath is useful in many ways. It serves to hold the cells in the colonies together. Its slimy nature endows the sheath with great water absorbing and water retaining capacity. Thus a firm pigmented sheath is an asset to the species growing under conditions of desiccation. It enables them to perennate. The pigmented sheath serves as a light screen. The sheath consists of pectic substances. The secretion of pectin by the protoplast is a primitive character. In some planktonic forms the sheath is of watery consistency and thus difficult to see. The study of fine structure of the sheath revealed that it is undulating, electron-dense and fibrillar in appearance. The microfibrils constituting the sheath are reticulately disposed within a amorphous matrix to give it homogeneous appearance. On the outside the microfibrils are less dense. A zone of less electron density separates the sheath from the cell wall or inner investment.

(*b*) *Cell Wall* (Inner investment). Internal to the extracellular mucilage (sheath) but separated from it by a zone of less electron density is a wall layer termed the *inner investment*. Under light microscope it is seen to consist of two layers (A) but electron microscope (B) has resolved it into four, L_1 through L_4 (Allen, 1968). The layers L_1 and L_3 are electron transparent.

Layer L_1. It is the innermost layer of the cell wall. It lies next to the plasma membrane and is visible as a clear electron-transparent space. It varies about 3 nm in thickness to about 10 nm and is covered by layer L_2.

Layer L_2. It is a thin, electron-dense mucopolymer layer which lies next to layer L_1. It contains mucopeptide and muramic acid. In addition there are glucosamine, alanine, glutamic and diaminopimelic acids. This layer shows remarkable similarity to the mucopolymer layer of the gram negative bacteria. The cell wall owes its shape and mechanical strength to this layer. The thickness of this layer ranges from 10 nm or less in *O. rubescens* and *Anacystis nindularia* to about 200 nm in *O. princeps*.

Layer L_3. It is external to L_2 and is again a clear electron-transparent space varying from about 3 nm to 10 nm in thickness. It separates L_2 from L_4.

Layer L_4. It is the outermost layer of the blue-green algal cell wall. It is thin, electron dense and has a wrinkled appearance. This layer is about the size of a unit membrane. Often it is undulating or even convoluted.

On the whole the cell wall of blue-greens shows granular structure. It lacks microfibrils. Its principal constituents are glucosamine, diaminopimelic acid, amino acids and muramic acid with some amino sugars. Cellulose is absent.

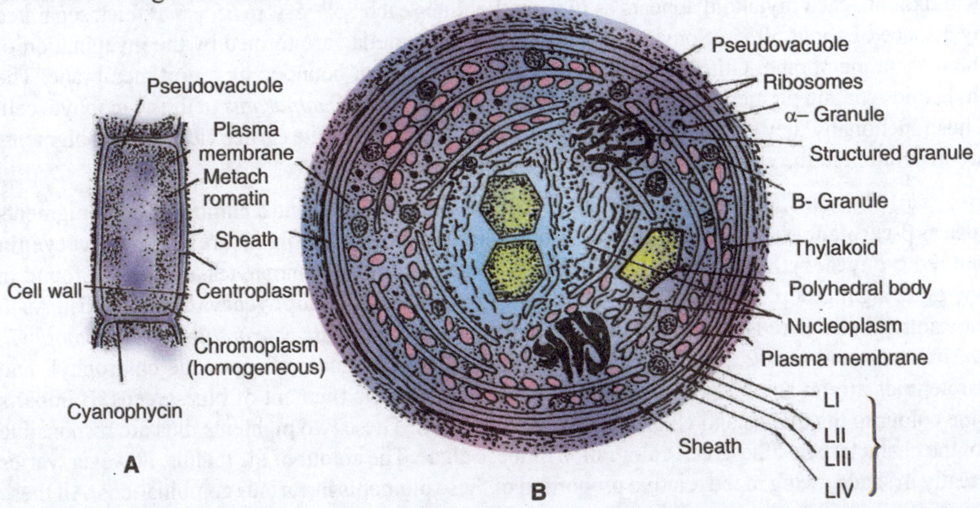

Fig. 2.8 (A-B). Cyanophyceae. Cell structure (A), Cell of a filamentous blue-green alga seen with light microscope; (B). Unicellular blue-green alga seen with electron microscope.

2. *Protoplast*. The protoplast of the blue-green algal cell shows elementary internal differentiation. The early investigators with light microscope found it divisible into two regions, (A), namely (*i*) a centrally located clear area forming the core and (*ii*) the peripheral denser region surrounding it. The former is termed the *centroplasm* (or nucleoplasm) and the latter *chromoplasm* (chromatoplasm). The plasma membrane which closely invests the chromoplasm is distinguishable. The centroplasm which forms the clear somewhat transparent central region of the protoplast contains chromatin or genetic material. The older cytologists considered this region to be equivalent to a primitive or incipient nucleus as it lacked both the nuclear membrane and nucleoli. The pigments under light-microscope appear to be in solution in the peripheral chromoplasm. Besides the latter contains a number of non-living inclusions. These are in the form of small, spherical or irregularly-shaped granules which may be of proteinaceous nature (*cyanophycin granules*) or carbohydrates stored in the form of a unique kind of starch akin to animal starch glycogen. It is known as the *cyanophycean starch*. In some species granules are arranged on either side of the transverse septa. These contain a large amount of proportion of *arginine*. In some planktonic species such as *Anabaena* and *Polycystis* chromoplasm contains clusters of gas containing cavities known as the *pseudovacuoles*. True vacuoles are absent.

B. Fine structure. The survey of the study of fine structure of the blue green algal cell by various investigators has revealed significant details about the structural features of the protoplast. These are :

(*i*) *Plasma membrane*. Internal to the cell wall, the protoplast at its periphery is differentiated into a distinct typical unit membrane known as the *plasma* or *cell membrane*. It shows small undulations and consists of two electron-opaque layers separated by a less opaque layer. Some workers have reported that the plasma membrane shows invaginations. The cytoplasmic matrix within the plasma membrane is distributed evenly throughout the cell. Thus the division of the cytoplasm into two regions, chromoplasm and centroplasm as seen under light microscope, is considered untenable.

(*ii*) *Photosynthetic apparatus*. Within the plasma membrane embedded in the peripheral region of the cytoplasm are elongated, flattened closed sac or disc-like structures called *thylakoids* or *lamellae*. Usually the thylakoids appear to be numerous and discrete within a cell. Most often they are seen closely appressed and organised in parallel arrays near the periphery of the cytoplasm but are not seen under light microscope. However they are not organized into a grana. In osmic acid fixation, each thylakoid appears as two parallel lines each 35 + 5A° to 65 + 5A° thick separated by a space of about 50A°. Some workers hold that the lamellae are formed by the invagination of the plasma membrane. Others hold that the thylakoids are not bounded by a unit membrane. The thylakoids contain pigments and thus constitute the *photosynthetic apparatus* of the cyanophyte cell. Thus functionally they appear to be analogous to the chloroplasts. The earlier view that the pigments are in solution in the chromoplasm is now considered erroneous.

(*iii*) *Pigments*. There is **chlorophyll**- a (but not chlorophyll-b) and a number of other pigments such as **β-carotene**, xanthophyll and phycobilins. The best known phycobilins are blue c-**phycocyanin** and red c-**phycoerythrin**. These two pigments are unique to the cyanophytes and are not found in any other algal group. Prescott (1969) added **phycocyanin**-r and **allophycocyanin** to the list. As to the xanthophylls there is a good number. Of these the best known are *zeaxanthin, myxoxanthophyll, myxoxanthin* and *oscilloxanthin*. The phycobilins are water soluble whereas the chlorophyll and carotenoids are fat soluble. Phycocyanin-c is the characteristic pigment of blue-greens. It imparts blue colour to the thallus and chlorophyll-a imparts green. These two pigments thus are responsible for the characteristic blue-green colouration of these algae. The colour of the thallus, however, varies greatly in shade owing to the relative proportion of these pigments in various combinations. All these pigments are neither present always nor are they in the same relative proportion. This accounts for the many hues of the blue-green algae from dark-green to blue-green, olive-green, grey-green, yellow, brown-purpled, orange or red. Phycocyanin fraction is considered to be responsible for more than one-half of light absorption in a cyanophyte cell.

(*iv*) *Gas Vacuoles* (or pseudovacuoles). The pseudo-vacuoles were discovered in certain planktonic blue-green algae (*Anabaena* and *Polycystis*). They encircle the photosynthetic apparatus in *Trichodesmium erythraeum*. Each pseudo-vacuole consists of a packed array of minute membranous cylindrical gas containing structures with closed conical ends. The membranous rigid wall of the gas vesicle is proteinaceous in nature. It is impermeable to water but freely permeable to gases (oxygen, nitrogen and argon). The gas vesicles are thus filled with metabolic gases. They are 0.2 to 1 m long and 70-75 nm in dia. Within the cell the gas-filled vesicles are subjected to turgor pressure. On increase in osmotic pressure within the cell, the gas vacuoles collapse with the conical end caps splitting away. The contained gas escapes by diffusion through the collapsing wall without forming a bubble. Fogl (1970) and Walsby (1980) suggested that the gas vesicles provide a buoyancy regulating mechanism enabling the planktonic procaryotes to poise at particular depths. At low light intensity, the gas vesicles are formed at a rapid rate making the thallus buoyant. Such cells rise to the surface when under high incident intensities the larger pressure may rise sufficiently to cause collapse of some of the weakest gas vesicles. Such cells lose their buoyancy and sink vertically to deeper layers of water. By determining buoyancy, gas vacuoles indirectly serve as a screen against intense light. Thus the gas vacuoles regulate, in addition to the light intensities and other environmental conditions to which the planktonic species of blue-green algae are exposed.

(*v*) *Granules*. Besides the thylakoids and pseudo-vacuoles, various kinds of granules are found in the protoplast. Of these, **ribosomes** which are tiny dense granules occur scattered between the thylakoids and throughout the cytoplasmic matrix but are less dense but longer in size than the ribosomes. They are usually elongated and are found in abundance between the thylakoids. They well surround the structured granules. Some think that the alpha granules are a reserve food, a glycogen-like substance. Pankratz and Bowen (1963) speculated that they are associated with metabolic functions related to either respiration or photosynthesis. **The structured granules** are large dense granules which usually occur near the cross walls. Frequently they are surrounded by the alpha-granules. Drews and Niklowitz consider them to be "mitochondrial equivalents". Fuhs (1958) described them to be granules containing phospholipids. Pankratz and Bowen (1963) considered them to be cyanophycin granules. The β-**granules** are spherical granules, less numerous than the ribosomes and alpha granules. Frequently they occur between the lamellae near the cross walls. Shatkin (1960) suggested that they are storage products probably lipoidal in nature. **Polyhedral bodies** are crystalline granules with a polygonal outline. They are consistently present in the central region (nucleoplasm) of the cell and are upto 0.5 μm in width.

(*vi*) *Nucleus*. The blue-green algal cell lacks a discrete nucleus. The nucleoplasm is largely central but partially peripheral. It is not separated from the surrounding cytoplasm by any membrane. Nucleoli are absent. Instead it is a region of low electron density than the surrounding cytoplasm and contains numerous fine randomly oriented fibrils of DNA. Histones are absent. RNA is diffusedly distributed in the nucleoplasm. **Genome** is the term suggested for this type of primitive nucleus. Cassel and Hutchinson (1954) were unable to find more than one disjoint mass of nuclear material per cell. Fuhs (1958) and Hagedoru (1961) reported that there may be one, two four or as many as eight "genomes" per cell. Associated with the nucleoplasm are the polyhedral bodies of unknown composition and function. There is concentration of ribosomes in the region of the nucleoplasm which may frequently be traversed by thylakoids.

NUTRITION

The Cyanophyceae contain chlorophyll-a, in addition to phycocyanin and other pigments. With the help of these pigments the blue-greens are able to synthesize their own carbohydrate food from carbon dioxide and water in the presence of sunlight. Thus the Cyanophyceae in general are **obligate photoautotrophs**. What role phycocyanin, the characteristic pigment of Cyanophyceae, plays in photosynthesis is still not very clear. It may facilitate the synthesis of food in these algae under the limited, light conditions in which they usually live. Its presence, however, enables the blue-greens to absorb and use more of green, red or orange and yellow wave-lengths of light than the green algae. Wolk (1973) reported that the light absorbed by phycocyanin appears to be about as active photosynthetically as chlorophyll. Duysens (1951) demonstrated energy transfer directly. Light energy absorbed by phycocyanin activates fluorescence of chlorophyll. In case the light energy absorbed by phycocyanin is transferred to chlorophyll the former must be located in proximity to the chlorophyll. Like other algae photosynthesis in blue-green algae involves both photosystem "II" and photosystem "I". The first evident products of photosynthesis are sugars and **glycogen.** They are converted into glycoproteins. Nitrogenous materials are also present in various forms such as nucleoproteins and albumins. Minute droplets of oil are often present which indicate the synthesis of fats.

Certain Cyanophyceae are able to fix the elementary nitrogen of the atmosphere. It is incorporated into their amino acids and proteins. Some diffuses from the healthy cells into their natural habitats. On death the nitrogenous materials of their dead bodies undergo decay and form ammonia. The latter is converted into nitrates by the nitrifying bacteria. The common nitrogen-fixing species are *Nostoc punctiforme, Anabaena variabilis, Tolypothrix tenuis, Cylindrospermum majus, Aulosira fertilissima, Calothrix parietina, and Mastigocladus*. Almost all of them are heterocystous forms. The heterocysts are the sites of N-fixation. These nitrogen-fixing species of the blue-greens are seeded in the rice fields to increase fertility in tropical countries.

As mentioned above the blue-greens in general are **obligate photoautotrophs**. They cannot grow in darkness even in the presence of organic nutrients in the substrate. However, experiments conducted by Moare, Hoare and Smith (1970) have shown that some blue-greens can assimilate organic compounds and incorporate them into specific cell contents. The capacity of blue-greens to assimilate and metabolise exogenous organic compounds is very limited and they cannot use organic compounds as a source of energy.

Differentiation of cells (Fig. 2.19.). The blue-green algae, in general show very little differentiation of cells. However, the trichomes of certain filamentous genera show differentiation of vegetative cells to either heterocysts or akinetes which possess special structural and biochemical properties. The *heterocysts* differ from the vegetative cells and occur between them along the length of trichome with characteristic cellular spacings.

Heterocysts. The heterocysts differ markedly from the vegetative cells and occur between them along the length of the trichome at some regular intervals. These are large, empty looking specialised cells found in the trichomes of certain filamentous blue-green algae. The occurrence of heterocysts is a feature unique to the Cyanophyceae. Thurret (1844) was perhaps the first to report their occurrence in the trichomes of blue-greens. The heterocysts can easily be distinguished from the vegetative cells of the trichome by their (*i*) larger size, (*ii*) thicker walls, (*iii*) homogeneous transparent, pale yellowish contents and (*iv*) a distinct pore either at one or at both ends.

Lateral Heterocyst
B
C
D
Germinating Heterocyst

Fig. 2.9 (A-D). Cyanophyceae. Distribution of heterocysts. (A) *Nostochopsis* (lateral): (B) (intercalary and single); (C) *Anabaenopsis* (intercalary in pairs): (D) *Anabaena ballansis* (germinating heterocyst) (After Bom and Fish).

Position of heterocysts in the trichomes (Fig. 2.9.). The heterocysts usually occur singly (B). In some genera they occur in pairs (*Anabaenopsis*, C) and rarely in chain. When they occur singly they are either terminal (*Gloeotrichia*) or intercalary in position (*Nostoc*, B). Rarely they occur on the sides of the filament. Lateral heterocysts are found in *Nostochopsis* (A). In shape they are identical to the vegetative cells. They are almost round in *Nostoc, Anabaena, Rivularia* and *Gloeotrichia* but somewhat rectangular in *Hapalosiphon, Aulosira* and *Scytonema*.

Fine Structure of Heterocysts (Fig. 2.10.). The envelope of the heterocyst is markedly thicker than that of the vegetative cell but like that of the latter it is divisible into two parts (*i*), the outer comparable to the sheath and (*ii*) inner to the cell wall or inner investment. The outer envelope is formed by the deposition of material secreted by the cell membrane external to the cell wall of the vegetative cell from which it has originated. It is differentiated into three layers. (*i*) outer loosely fibrous layer, (*ii*) middle homogeneous layer and (*iii*) inner laminated layer which borders on the inner envelope or cell wall of the heterocyst. The middle homogeneous layer and the inner laminated layer are of uniform thickness except at the poles where they are thicker still. All the layers of the outer envelope surround the heterocyst except at the junction of

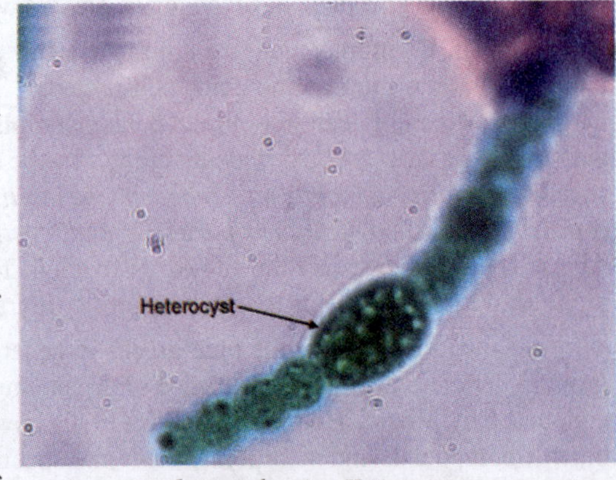

Anabaena showing Heterocyst.

the heterocyst with the vegetative cell. The inner envelope or cell wall of the heterocyst like that of the cell wall (inner investment) of vegetative cell consists of four layers L_1 - L_4 from within outwards. These are the electron transparent layer L_1 next to the plasma membrane. It is covered by electron dense mucopolymer layer L_2. External to the latter is again a clear electron-transparent space representing L_3. It is bordered by a wrinkled unit membrane constituting L_4. The inner envelope (heterocyst cell wall) is continuous with the inner investment cell wall of the adjacent vegetative cell, within the wall layers and surrounding the cytoplasm is the plasma membrane. The contents within the plasma membrane appear homogeneous, transparent and pale yellow in colour. Owing to the absence of pigments or their weak concentration the mature heterocysts appear empty and none or little carbon dioxide. However, they contain deoxyribonucleic acid and two lipids, *glycolipid* and *acyllipid* are not found in the vegetative cells. The ribosomes appear to be present but reduced in number. All other granular inclusions such as polyhedral bodies, polyphosphate and cyanophycin granules are absent. There are no gas vacuoles. The thylakoids are fewer in number. There is, however, concentration of contorted lamellae in the polar regions.

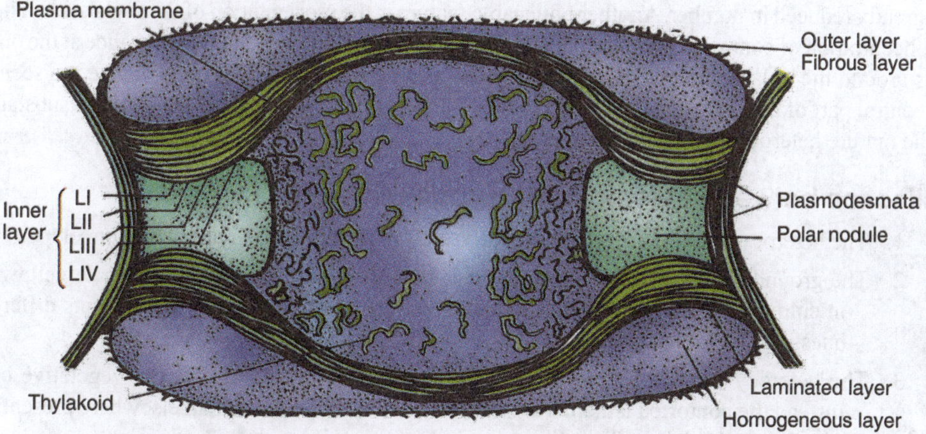

Fig. 2.10. Cyanophyceae. Diagrammatic representation of longitudinal section of heterocyst showing ultrastructure (Based on Long and Fay).

At the junction between the heterocyst and the vegetative cell there is a deep constriction. The constricted region has a pore for communication with the adjacent cells. The intercalary heterocyst thus has two pores (*Nostoc*), one at each pole whereas the terminal heterocyst has a single pore towards the trichome (*Gloeotrichia* and *Rivularia*). Iyengar and Desikachary (1957) and Venkataraman (1957) recorded these pores in the intercalary heterocyst of *Brachytrichia balani* and *Mastigocladus laminosus*. There is real communication between the developing heterocyst and the vegetative cell through the pore. Many workers have confirmed the existence of plasmodesmata connecting the plasma membranes of the heterocyst and the adjacent cells (Widlen and Mercer, 1963; Lang and Fany, 1971 and Wolk, 1968). In a mature heterocyst the pore channel is plugged with a *polar nodule* which is located between the plasma membrane and the inner envelope of the heterocyst. It is therefore cytoplasmic in nature. The plug material consists of highly refractive, electron-transparent, shiny granules called the **polar granules.**

Differentiation of Heterocysts. The heterocyst develops from an ordinary vegetative cell particularly the recently divided one. One of the daughter cells called the *proheterocyst* develops into a heterocyst and the other into a vegetative cell. Rarely both the daughter cells may develop into a pair of heterocysts. The main events associated with heterocyst differentiation are outlined under the following steps:

(*i*) *Synthesis of a thick, three-layered outer envelope*. Increase in size of the proheterocyst is the initial change. It is accompanied by laying down of an *outer fibrous layer* outside the inner investment of the enlarged vegetative cell (proheterocyst). A homogeneous intermediate layer is then synthesised below the outer fibrous layer and finally the laminate layer. The homogeneous and laminate layers become much thickened around the poles where the junction between the developing heterocyst and the vegetative cell is drawn out to form a neck-like channel which communicates with the vegetative cell through a pore. The existence of plasmodesmata in the pore region connecting the plasma membranes of the developing heterocyst and the vegetative cell has been well-established. However, the formation of pore channel and organisation of the polar nodule which fills the channel and thus plugs the pore are the structures to be completed last when the heterocyst reaches maturity.

(*ii*) *Changes in the ground substance*. During heterocyst differentiation there is gradual migration of the cellular material towards the poles. The granules such as the polyhedral bodies, phosphate and cyanophycin granules become reduced in the developing heterocysts but are totally absent in the mature ones. The ribosomes appear to be present in all stages of heterocyst development but greatly reduced in number. Another noticeable feature is the reorientation of the thylakoids. There is concentration of contorted lamellae presenting a lattice-like or honey comb appearance at the polar ends around the polar nodule. Consequently lamellae-free islands of ground substance are seen in the central part of the fully developed heterocyst. As a result of these changes, the ground substance of the mature heterocyst appears homogeneous.

Distinction between heterocyst and vegetative cell

1. The heterocyst differs from the vegetative cell in its larger size and thicker envelope.
2. The ground substance of the heterocyst is homogeneous, transparent and pale yellowish in colour whereas that of the vegetative cell is densely granular and presents different hues.
3. The heterocyst lacks the profuse thylakoid system characteristic of the vegetative cell. Instead the contorted lamellae are densely packed at the poles than elsewhere presenting a honeycomb-like configuration surrounding the polar nodule.
4. The presence of a polar nodule at one or both the poles of the heterocyst is another feature which distinguishes it from the vegetative cell which lacks them.
5. The several kinds of granules (cyanophycin, polyglucon and polyphosphate) which are typical of the lamellar system of the vegetative cell are absent from the heterocyst.
6. The pseudo-vacuoles may be present in the vegetative cell but they are always absent in the heterocyst.
7. The pigments characteristic of the vegetative cell are either absent or occur in weak concentrations in the heterocyst and thus the latter can fix no or little carbon dioxide.
8. The heterocysts of about 50 sp of active nitrogen fixers of blue-green algae are known to be the sites of fixing atmospheric nitrogen but the vegetative cells altogether lack this property.
9. There is no organisation of the nuclear material in the heterocyst whereas in the vegetative cell it occurs as aggregates of DNA fibrils in the nucleoplasm region of the vegetative cell.

Factors controlling heterocyst formation. (*i*) According to Fay, the production of heterocysts increases under conditions of low light intensity and increase in the amount of phosphate in the medium.

(*ii*) Tyagi (1973) found that heterocyst formation depends on the availability of carbon intermediaries and ATP. The former are supplied by photosynthesis and the latter by oxidative

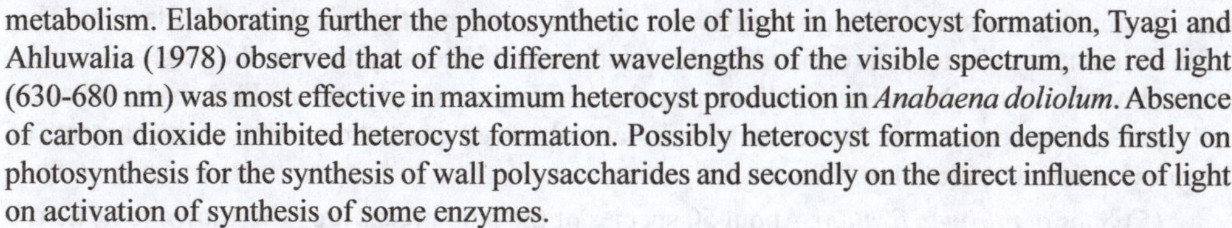

metabolism. Elaborating further the photosynthetic role of light in heterocyst formation, Tyagi and Ahluwalia (1978) observed that of the different wavelengths of the visible spectrum, the red light (630-680 nm) was most effective in maximum heterocyst production in *Anabaena doliolum*. Absence of carbon dioxide inhibited heterocyst formation. Possibly heterocyst formation depends firstly on photosynthesis for the synthesis of wall polysaccharides and secondly on the direct influence of light on activation of synthesis of some enzymes.

(*iii*) The concentration of nitrogen in the medium above a certain level results in complete inhibition of heterocyst production. Fog (1949) and Singh and Srivastava (1968) reported that heterocysts are formed in culture medium deficient in combined nitrogen. Mitra and his students (1970) found that the development of heterocyst depends on the amount of nitrogenous salts present in the medium. Grown in high concentration of nitrogen in the medium the heterocysts were absent in *Tolypothrix arenophila, Scytonema praegnans, Anabaena naviculoides*, *A. cycadae* and many others. The percentage of nitrogen necessary for inhibition of heterocyst formation, however, varies with the different species.

(*iv*) Kale (1970) found light as one of the essential factors which plays an important role in the differentiation of heterocysts in *Anabaena, ambigua* Rao, At least 6 hours exposure is necessary for initiation. The optimum light intensity necessary for an abundant crop of heterocysts is about 1,300 lux. Blue-green, light inhibits the formation of heterocysts. Red light promotes the differentiation process.

(*v*) Singh, *et. al.* (1970) found that heterocyst differentiation is inhibited in the presence of combined sources of nitrogen (nitrate and ammonium nitrogen) but is induced in the presence of nitrogen gas. However, they added that the differentiation of heterocysts, hormogones and spores in blue-green algae is genetically controlled. Their phenotypic expression is governed by growth conditions. Indumati (1970) also stated that in *Camptylonema lahorensis* Ghosh heterocyst differentiation is genetically controlled but its phenotypic expression is dependent on growth conditions in the medium.

Functions of Heterocysts. The subject is still controversial. Various suggestions put forth from time to time have been discussed beautifully by Tyagi (1973). These are:-

(1) *Vestigial reproductive structures.* Geitler (1921 *a*) held that originally heterocysts were developed for reproduction. Somehow, in the course of time, they lost this spore-like function and became functionless. Occasionally, however, they revert to their old function and germinate to produce new filaments (plants). Such degenerate organs which are of little or no use now but ancestrally well developed are called *archaic* or *vestigial*. This view is supported by the fact that in exceptional cases the heterocyst germinates to form a new plant. For instance heterocyst germination was recorded by Bornet and Thuret (1880) in *Nostoc ellipsosporium*. Subsequently it was reported by Steinecke (1932) in *Calothrix* and Desikachari (1946) in *Rivularia*. Spratt (1911) recorded endospore formation in the germinating heterocyst of *Anabaena cycadeae*. Each endospore, on germination, forms a new element. These observations support Geitler's view.

(2) *Storehouses of reserve food material* (Fritsch 1904 *a*) or *enzymic substances* (Canabaeus, 1929). Both the suggestions did not receive much attention because of experimental evidence.

(3) *Heterocysts produce substances which stimulate growth and cell division in adjacent vegetative cells* (Fritsch 1951). This revised hypothesis of Fritsch has been supported by recent researches. It has been found that heterocysts produce and secrete certain vital substances needed to keep the adjacent cells in active physiological state.

(4) *Play role in sporulation.* There is some convincing evidence in support of an inductive role of the heterocysts in sporulation. The akinetes, the so called spores of heterocystous blue-green algae, are formed only adjacent to the heterocysts. For instance in *Gloeotrichia* and *Cylindrosperm,* the single akinete always develops adjacent to the terminal heterocyst. In *Anabaena cylindrica*

spore formation starts near the heterocysts and proceeds towards the middle of the filament. This differentiation of akinetes is prevented by the detachment of heterocysts from the vegetative cells. The advocates of this hypothesis believe that the heterocysts supply metabolites to the adjacent cells undergoing sporulation. Some workers, in fact, observed gradual depletion of the contents of heterocysts during sporulation.

(5) *Sites of nitrogen fixation.* About 50 species of the blue-greens are now known to be active nitrogen fixers. With the exception of a few they are all heterocystous filamentous forms. Fay et. al. (1968) suggested that heterocysts are the sites where atmospheric nitrogen is fixed. The localization of acetylene reduction in the heterocysts supported this belief. A direct relationship between heterocysts and the ability to fix nitrogen has been provided by the fact that all the heterocystous species fix nitrogen. Inhibition of heterocyst formation by combined nitrogen in the medium lends further support to the hypothesis that heterocysts are the sites of nitrogen fixation. Because of their role in nitrogen fixation, the heterocysts are considered ecologically and agriculturally important in tropical rice fields.

(6) *Locus for filament breakage.* The junction between the heterocyst and the adjacent vegetative cell is fragile. It serves as the weakest link in the trichome and thus serves as a locus for breakage. Some workers, therefore, maintain that intercalary heterocysts serve to break the trichome into small sections of living cells called *hormogonia*.

(7) The definite relation between the position of a heterocyst and the point of origin of a true of false branch may be incidental.

Stewart (1970) concluded that the chief physiological functions of heterocysts are: (*i*) regulation of akinete formation, (*ii*) involvement in growth and differentiation, (*iii*) sites of nitrogen fixation, and (*iv*) archaic reproductive units.

Akinetes. These are specialized cells generally found in some heterocystous forms and often in old cultures under adverse circumstances. Since they function as perennating and reproductive structures, the akinetes are discussed under the heading spore formation (p. 46).

Sexual reproduction

Formation of sex organs, gamete differentiation, gametic union or formation of *zygotes* has not been observed in the Cyanophyceae. A parasexual phenomenon termed *genetic recombination* has been reported by Kumar (1962). It has been confirmed by Barin (1968) and others. In one experiment carried on by Singh and Sinha (1965) two strains of *Cylindrospermum majus* were cultured together in a medium which contained the antibiotics *streptomycin* and *penicillin*. One of these strains was resistant to the former and the other to the latter. All the cultures died except in two culture tubes. Since the surviving blue-greens in these two tubes continued to grow in the presence of both the antibiotics they were considered as genetic recombinants. Genetic recombination has been reported in two other genera, *Anacystis* and *Anabaena*. Shestakov and Khyen (1970) reported genetic recombination in *Anacystis nidulans*. In genetic recombination syngamy and meiosis are not involved. Absence of sexuality is generally considered a primitive feature. Stebbins (1960), however, considered that some form of genetic recombination which he called "meromixis" existed in the prokaryotic organisms since life first appeared. The loss of sexuality in simple algae is secondary.

Asexual Reproduction. It takes place by vegetative methods and by spore formation.

1. *Vegetative methods.* Vegetatively the cyanophytes multiply by (*a*) active cell division also known as *fission*, (*b*) *fragmentation* and (*c*) by the formation of *hormogones* or *hormogonia.*

(a) Cell division or *Fission* (Fig. 2.11.). In the unicellular blue-greens, cell division leads to multiplication of the species. It is preceded by changes in (or division of) the nuclear material. Some consider it to be a simple form of amitosis, others consider it to be a primitive type of mitosis and report the formation of a spindle. This is doubted by others. Anyway the division of the nuclear material

precedes cell division. This is followed by the division of cytoplasm and separation of the resultant two daughter protoplasts. Allen (1968) studied nuclear division in the coccoid form *Gleocapsa alpicola* at the level of fine structure using electron microscope. It comprises the following stages:

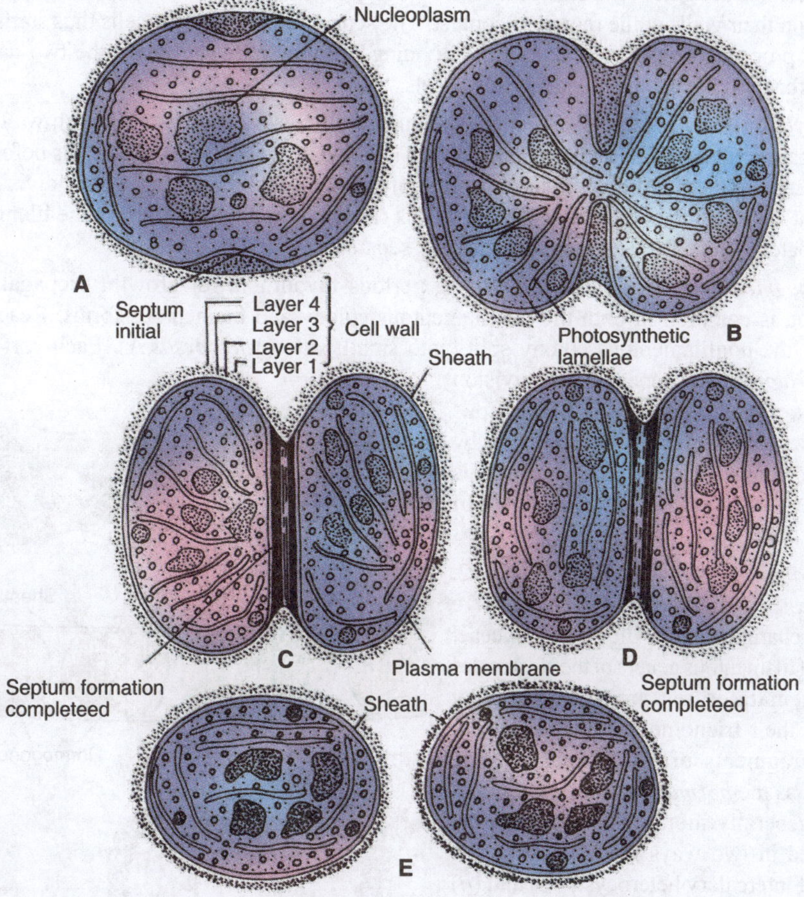

Fig. 2.11 (A-E). Cyanophyceae. Stages in cell division (Fission) in *Gloeocapsa alpicola* (After Allen).

(i) Septum formation (A-B). It is the normal prelude to cell division and is initiated by ring-like invagination and continuous ingrowth of the plasma membrane and inner two layers L_1 and L_2 of the 4-layered cell wall. One of these wall layers L_2 which is electron dense and consists of mucopeptide, is differentially synthesized to form the septum. The peripheral wall layers L_3 and L_4 which are not involved in septum formation simply pass over this region or show a small medial constriction at the early stages of cell division. (B). Septum when completely formed divides the parent protoplast into two daughter protoplasts (C). These are, however, still completely enclosed within the common wall layers L_3 and L_4 as a single structural unit.

(ii) Splitting of Middle Septum. The middle mucopeptide septum which is enclosed on either side by the electron transparent layer and plasma membranes, thickens. Eventually it splits centripetally into two layers. The two daughter cells, though completely separated are still enclosed within the common wall layers L_3 and L_4 as a single structural unit. At this stage, each daughter protoplast comes to possess a continuous plasma membrane, electron-transparent layer and electron-opaque mucopeptide layer.

(iii) Separation of daughter cells. The ends of the split septum are still continuous with the common peripheral wall layers L_3 and L_4. To achieve complete separation the small median constriction developed in these two layers at the early stage of division deepens at this stage and exposes the fission between the daughter cells externally. As the daughter cells grow they set up a turgor pressure which pulls on their walls at the region of contact. The separation of daughter cells thus starts at the periphery. It proceeds towards the axis of the dividing cell gradually separating the two daughter cells which function as two independent individuals.

(iv) Synthesis of wall layers L_3 and L_4. The outer wall layers L_3 and L_4 of the cell wall which do not take part in the division of the cell are formed *de novo* between the daughter cells before they actually separate. However, before final separation of the two daughter cells, a second division in a plane at right angles to the first in the coccoid forms (E) and parallel to the first in the filamentous forms is initiated before the daughter cells actually separate.

(b) Fragmentation (Fig. 2.12.A). During periods favourable for growth, propagation by fragmentation is common in both the nonfilamentous colony and filamentous forms. Reaching a certain size, the nonfilamentous colony splits into small parts (*Microcystis*, A). Each part which is called a *fragment* by repeated cell division grows into a new colony. In the filamentous forms, the trichome may break into fragments by mechanical means such as by the bite of animals or stress caused by water currents or death of certain cells weakening the trichome.

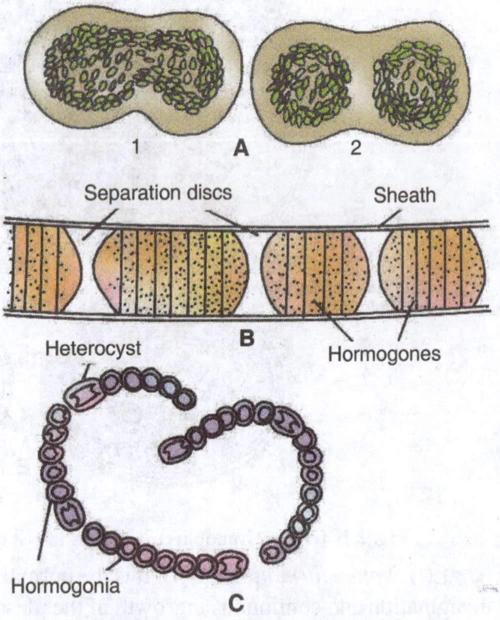

(c) Hormogonia formation (Fig. 2.12. B-C). It is a specialized process of vegetative propagation characteristic of the Cyanophyceae. Trichomes of filamentous genera of the Nostocales and Stigonematales regularly multiply by breaking of their trichomes within the sheath into short fragments of one to many living cells known as the *hormogones* or *hormogonia* which are generally motile. The hormogones are delimited in two ways namely, (*i*) by the formation of intercalary heterocysts (C) and (*ii*) by the development of intercalary biconcave separation discs or *necridia* (dead cells) at intervals along the trichome (B).

Fig. 2.12 (A-C). Cyanophyceae. (A) Fragmentation, (B-C) Hormogonia formation.

(i) By intercalary heterocysts. Many heterocystous genera of blue-green algal have intercalary heterocysts spaced quite a few cells apart from each other in the trichome (*Nostoc*). These intercalary heterocysts appear to be the sites of breaking up of the trichome into hormogones because of a weaker adhesion between a heterocyst and a vegetative cell than between the adjacent vegetative cells.

(ii) By the formation of separation discs. Genera with terminal or no heterocysts have hormogones delimited by the development of biconcave, gelatinous separation discs spaced quite a few cells apart from each other along the trichome. A living cell here and there in the trichome undergoes lysis perhapes by the secretion of muric acid. As a consequence the protoplast of the cell degenerates to form a viscous substance. These mucilage filled dead cells called *necridia*, lose turgidity. The mutual pressure on the end wall of adjacent cells having been released, the latter bulge and become convex. The dead cell (necridium) which is wedged in between the bulging walls of the two

cells (one on either side) appears biconcave. These biconcave, dead cells (necridia) are called the intercalary biconcave separation discs. The latter provide weak links at which the trichome breaks. The few-celled short segments that result are called hormogonia.

The hormogones may be two to several cells long. Typically any portion of the trichome may get abstracted as a hormogone (*Oscillatoria* and *Nostoc*). In some species they are formed from the apical portion of the branches. *Stigonema* forms hormogones on special branches. *Rivularia* abstricts hormogones from the intercalary meristematic zone after the hair has been shed. The hormogones are set free by a breakdown of the sheath at the end of the parent filament. In some species the hormogones are capable of locomotion and move away from the parent filament by creeping out of the gelatinous sheath. Finally they settle down and grow into new typical filaments by cell division. Hormogonia formation constitutes an efficient and quick method of vegetative propagation and dispersal of the species under conditions favourable for growth.

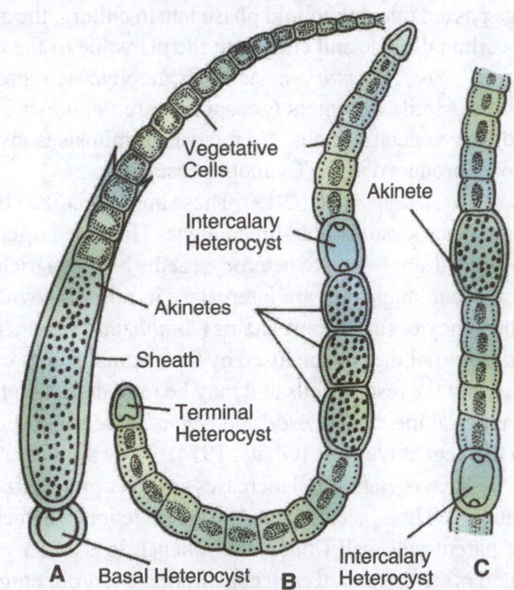

Fig. 2.13. Cyanophyceae. Distribution of akinetes.
(A) *Gloeotrichia*, akinete next to a basal heterocyst; (B) *Anabaena desikacharyiensis*, Series of akinetes next to an intercalary heterocyst. Note also two terminal heterocysts; (C) *Anabaena desikacharyiensis*, an akinete away from the heterocyst (After Vasishta).

The other less common methods of vegetative propagation in the Cyanophyceae are:

(d) Hormospores. Bozi (1914) and Fremi (1930) reported that short sections of living cells at the tips of the trichomes or short side branches of *Westiella lanosa* become invested by a thick, lamellated, pigmented sheath. Such multicellular spore-like structures function as perennating bodies. They are specially modified hormogones and are called the **Hormospores** or **hormocysts** (Fig. 2.16, D). Desikachary (1948) calls them **Pseudohormogonia**. They germinate *in situ*.

(e) Planococci. Borzi (1916) described the formation of single-celled hormogones in *Desmosiphon*. These exhibit slow creeping movements.

(f) Palmelloid Stage. Bausor and Agona (1973) reported the development of a palmelloid phase in a species of filamentous blue-green alga, *Anabaena*. First the cells separated from one another. This was accompanied by excessive production of mucilage. The individual cells were embedded and arranged at random in the copious mucilaginous matrix. Since the

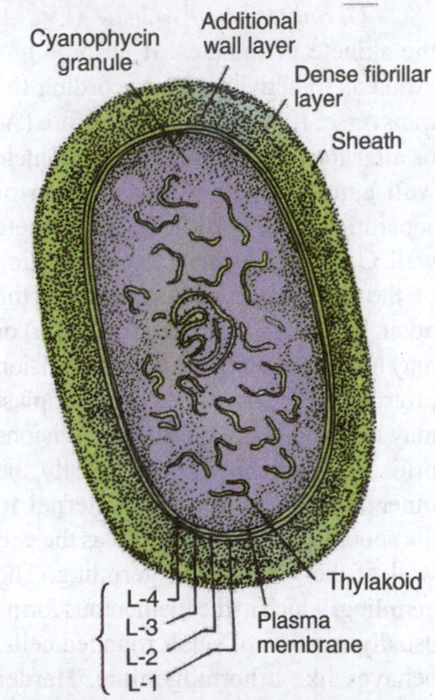

Fig. 2.14. *Cylindrospermum*. Diagrammatic representation of the structure of akinete as revealed under the electron microscope (Based on Miller and Lang).

alga passed into palmelloid phase late in culture, the authors suggest that autointoxication, depletion of carbon dioxide and change in the pH value of the medium may be the inducing factors.

2. *Spore formation.* Many Cyanophyceae reproduce by the formation of non-motile asexual spores. Motile swarmers (zoospores) are unknown in the blue-greens. The spores are produced in ordinary vegetative cells. No meiosis or mitosis is involved in their formation. The various kinds of spores produced in the Cyanophyceae are:

(*a*) *Akinetes* (Fig. 2.13). These are specialized thick-walled resistant cells formed generally in some heterocystous blue-green algae. They are larger than the vegetative cells. Frequently they are developed singly next to heterocyst at the base of a trichome (*Gloeotrichia,* A). In some Cyanophyceae they occur singly and are intercalary in positions without any relation to the heterocyst (C). In still others they occur in short chains (*Anabaena cylindrica,* B). These spore-like cells contain the entire *protoplast* of the cell enclosed by the original parent cell wall and the thick spore envelope. They are in reality the resting cells and may be regarded as highly modified resting one-celled fragments. For instance akinetes of *Nostoc* and *Cylindrospermum* survive heating to much higher temperature than do the vegetative cells (Glade, 1914). They are very resistant to water shortage.

The vegetative cell increases several times in size and accumulates food reserves (Cyanophycin granules). This is accompanied by the secretion of a thick multilayered spore wall envelope surrounding the parent cell wall (inner investment). It is often yellow or brown. Finally these modified cells round off and separate each constituting a reproducing unit and a perennating structure. The akinetes because of their survival value play a significant role in enabling the nitrogen fixing heterocystous species of blue-green algae to remain alive from year to year in rice fields. Phosphorus deficiency (Seenayya and Raju, 1970) and nitrogen deficiency (Singh and Srivastava, 1968) both resulted in the increased production of akinete bearing trichomes in *Anabaenopsis Raciborskie* and *Anabaena doliolum* respectively. The importance of other factors such as light intensity, temperature and density of culture medium has also been documented.

Germination of Akinete (Fig. 2.15). With the onset of conditions favourable for growth, the akinete germinates. It absorbs moisture. The thick, resistant wall softens and the protoplast awakens to activity (B). According to Bristol (1920) the first division of the protoplast is usually transverse. It may take place before (A) or after localized rupture of the akinete wall caused either by dissolution or separation of a portion of the akinete wall. Occasionally there is gelatinization of the inner layer and rupture of the outer. The protoplast, which may (A) or may not have undergone the first division protrudes (C). Sometimes the protoplast may undergo many transverse divisions prior to emergence (D). Generally the inner investment which is internal to the spore envelope is retained as the cell wall of the germinating germling. The germling, which in the filamentous forms usually consists of small rounded cells, behaves like a hormogonium. Harder (1918) stated that it moves forward and backward slowly in and out of the ruptured akinete wall. Finally, it settles and gives rise to a new plant (E).

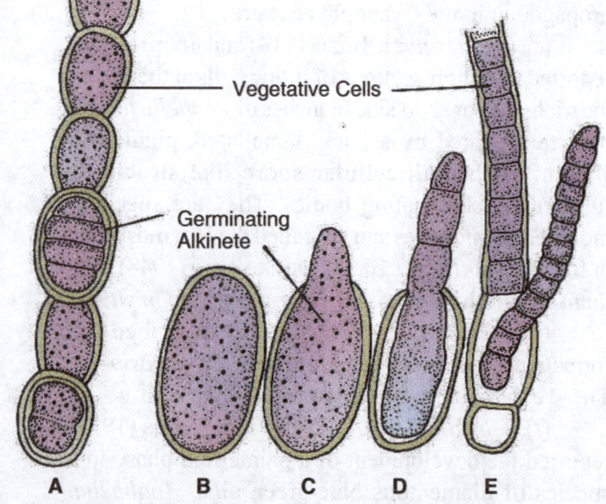

Fig. 2.15 (A-E). Germination of Akinete. (A) Germination of akinete in *Anabaena sphaerica,* (B-D) Akinete of *Anabaena* showing stages in germination; (E) *Calothrix fusca* germination of akinete in situ.

Akinete formation and its germination in *Cylinderospermum* was studied by Miller and Lang (1967) using the **electron microscope**. According to them cell enlargement is the initial step towards the differentiation of the vegetative cell into an akinete. This is accompanied by condensation of the mucilaginous sheath and formation of a dense fibrillar layer over the cell surface (Fig. 2.14). Simultaneously there is deposition of several additional spore wall layers between the fibrillar layer and inner investment and accumulation of cyanophycin granules in the periphery of the cytoplasm. The mature akinete retains cyanophycin granules and also the thylakoids, polyhedral bodies, lipid deposits and nucleoplasm regions but there are no alpha-granules. The latter remain absent during the resting period.

The akinete germinates in either of the following ways:

(*i*) Frequently the spore envelope remains intact but for a pore dissolved at one end. The germling emerges through the pore. Prior to emergence (*a*) it may have undergone one or two divisions or (*b*) the division begins after emergence.

(*ii*) At times there is dissolution of the entire spore envelope. It disintegrates into globular and rod-shaped fragments at the time of germination.

The alpha-granules reappear in the germling but the cyanophycin granules are almost absent. Probably cyanophycin is used up during the resting period of the akinete or at the time of germination.

(*b*) *Endospores* (Fig. 2.16 A-B). These are small spores formed endogenously within a vegetative cell of the unicellular or cushioned forms of Chamaesiphonales which do not form hormogonia (A). At the time of endospore formation the vegetative cell increases in size. The

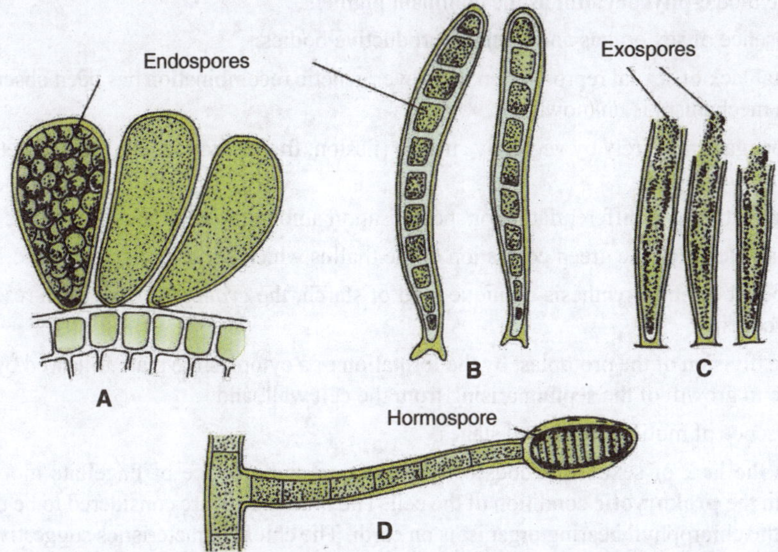

Fig. 2.16 (A-D). Cyanophyceae showing various types of spores; (A) *Dermocarpa sp.* with endospores (After Bornet and Thuret); (B) *Stichosiphon sensibaricus* with endospores; (C) Chamaesiphon sp. with exospores; (D) *Westilela lanosa* showing hormospore formation (After Fremy).

protoplast divides by the method of successive bipartition. A large number of small, uninucleated daughter protoplasts are formed. Each daughter protoplast secretes a wall around it to become an **endospore**. Usually the endospores are rounded in form (A) but sometimes **angular** (B). In some species the endospores are naked. The wall is secreted on liberation. The spore wall is quite distinct from the wall of the parent cell. The endospores are produced in *Dermocarpa pacifica* and

Stichosiphon. The liberated endospore germinates immediately without a resting period.

(c) Exospores (Fig. 2.16 C). In *Chamaesiphon,* the cell wall ruptures at the distal end of the vegetative cell. The spores are successively pinched off at the exposed end of the extruded protoplast. They are called the **exospores**. Each exospore is surrounded by a delicate membrane.

(d) Nannocytes. In some non-filamentous blue-green algae such as *Microcystis* the cell contents divide repeatedly without any cell enlargement. The successive divisions follow closely one upon the other. Numerous daughter cells are produced in each parent cell. These are called the **nannocytes**. The nannocytes are naked protoplasts. They differ from the vegetative cells in their extremely small size. They are closely packed. The nannocytes germinate *in situ* to give rise to new typical colonies.

Salient Features.

The members of the class Myxophyceae (Cyanophyceae) possess the following **important** features:

1. Extremely simple structure of the cyanophycean cell and low state of the thallus organisation,
2. More or less unstable condition of the cell wall,
3. Absence of a definitely organised nucleus, It lacks the nuclear membrane and nucleoli,
4. Simple protoplast without any plastids, endoplasmic reticulum, mitochondria, golgi apparatus and sap cavity,
5. Abundant secretion of pectin by the protoplast,
6. The blue **c-phycocyanin** as the dominant pigment,
7. Absence of sex organs and motile reproductive bodies,
8. Total lack of sexual reproduction. However genetic recombination has been observed but the mechanism is unknown,
9. Propagation entirely by vegetative means (fission, fragmentation, etc.) and sometimes by spores.
10. Very little or no differentiation or specialisation among the cells constituting the thallus,
11. Characteristic blue green coloration of the thallus which varies greatly in shade,
12. Product of photosynthesis a unique kind of starch, the *cyanophycean* starch resembling glycogen,
13. The division of the protoplast by the formation of a cytoplasmic plate followed by a ring-like in growth of the septum arising from the cell wall, and
14. Absence of motile, flagellated stages.

Perhaps the lack of sexual methods of reproduction and absence of flagellate motility are associated with the **prokaryotic** condition of the cell. The blue-greens are considered to be the most ancient of all the chlorophyll bearing organisms on earth. The chief characteristics suggestive of the Cyanophyceae being a **primitive** and **ancient group** of the plant kingdom are:

1. Primitive condition of the structural organisation of thallus as shown by:
   (*i*) Little or no differentiation of cells,
   (*ii*) The individual cells on separation from the parent plant are able to live as they do in the thallus. There is neither any differentiation of cells in the thallus nor division of labour. For this reason the thallus is considered a colony of unicellular plants rather than a multicellular individual,

2. The absence of plastids, mitochondria, golgi apparatus, a sap vacuole and a definitely organized nucleus in the cell are suggestive of its exceedingly primitive condition,
3. The incipient nucleus without a nuclear membrane and nucleoli in the cell illustrates a stage in the evolution of the nucleus of the higher plants,
4. More or less unstable cell wall,
5. Abundant secretion of pectin is also a primitive character,
6. Occurrence of pigments in the lamellae which are not organized into grana,
7. Absence of sex organs and motile reproductive bodies,
8. Lack of sexual reproduction,
9. Reproduction mainly by fission and other vegetative methods, and
10. Occurrence in the pre-paleozoic period. The imprints on the two billion years old rocks of the Archaeozoic era strongly resemble the present-day blue-green algae and thus bespeak of the great antiquity of these algae.

The above characteristics indicate that the Cyanophyceae is a **simple, primitive** and **ancient** group of the plant kingdom.

Classification

All taxonomists recognise only one class Cyanophyceae in this division. As to the number of orders included in this class the opinions vary. One system recognises only two orders, namely, *Chamaesiphonales* and *Hormogonales.* Smith makes out three, *Chroococcales*, *Chmaesiphonales* and *Hormogonales*. The only objection raised against this system is that many families, which should be placed in separate orders, have been put together in the order *Hormogonales.* To remove this objection Fritsch (1942) suggested that the *Chamaesiphonales* should be split into two series, namely, a group of uni-to multicellular epiphytes for which he retained the name *Chlamaesiphonales* and a group of heterotrichous epiphytes and lithophytes which he allotted to the order *Pleurocapsales*. Further he split the *Harmoganales* into *Nostocales* and *Stigonematales*. Fritsch thus divided the class into five orders, namely, *Chroococcales, Chamaesiphonales, Pleurocapsales, Nostocales* and *Stigonematales*. Another system divides the class into two tribes, *Coccogoneae* with two orders and *Hormogoneae* with five orders. The outline of this classification is given below:

A. Tribe **Coccogoneae** (unicellular and non-filamentous colonial blue-greens). Reproduce chiefly by fission and colonial forms, in addition, by fragmentation, cells spherical or oval in shape, sometimes rod-shaped. This tribe includes the following two orders :

1. Order **Chroococcales**. It includes two families. *Chroococcaceae* and *Entophysalidaeae*.

2. Order **Chamaesiphonales** with three families, namely, *Pleurocapsaceae, Dermaocarpaceae* and *Chamaesiphonaceae*.

B. Tribe **Hormogoneae** (Filamentous blue-greens). It includes cyanophytes in which cells are united into definite trichomes which may be simple or branched, straight or coiled. The branching may be false or true. Septa between the cells are frequently obscure. Heterocysts occur in some genera. Reproduction may be by hormogonia, akinetes, endospores and hormocysts. The tribe includes five orders:

3. Order **Oscillatoriales** with one family *Oscillatoriaceae.*
4. Order **Nostocales**. It includes a single family *Nostocaceae.*
5. Order **Scytonematales** with a single family *Scytonemataceae.*
6. Order **Stigonematales**. It comprises one family *Stigonemataceae.*
7. Order **Rivenlanales** which includes a single family *Rivenlanaceae.*

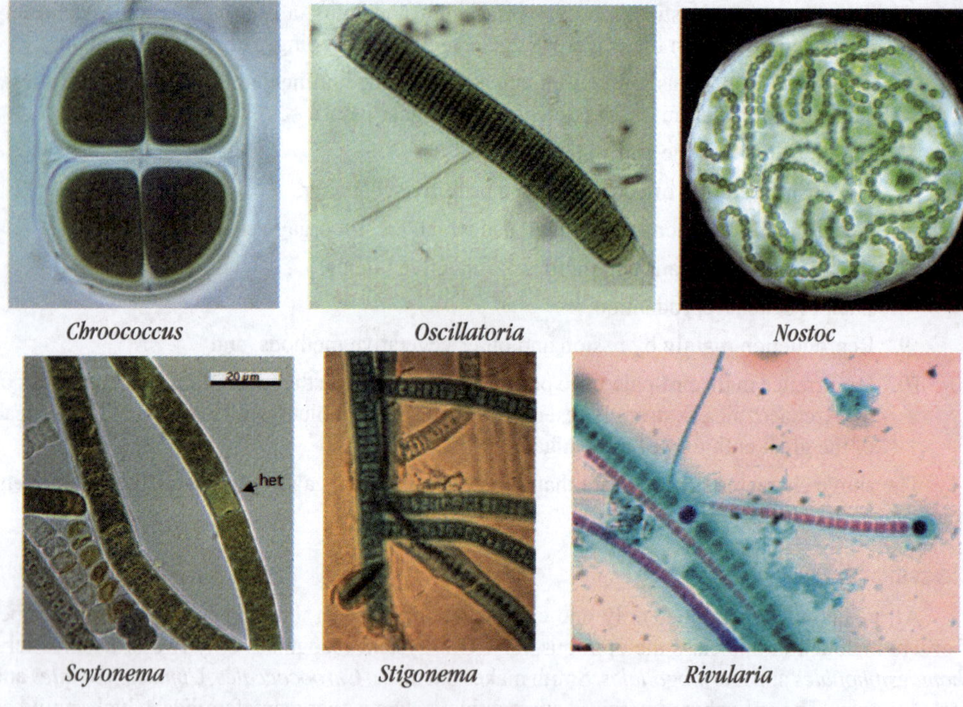

Chroococcus *Oscillatoria* *Nostoc*

Scytonema *Stigonema* *Rivularia*

SOME BLUE GREEN ALGAE

The latest system of classification outlined above has been followed in the present edition. Excepting order Stigonematales all the others are described here.

TRIBE COCCOGONEAE — ORDER : CHROOCOCCALES

It includes the most primitive members of the class Cyanophyceae. Almost all are fresh water. The order includes free living forms which either live singly or in non-filamentous colonies often with a regular and conspicuous symmetry. There is no trichome organisation. They reproduce chiefly by fission and fragmentation of the colony. The order comprises two families, *Chroococcaceae* and *Entophysalidaceae*. The former is discussed in this text.

Family **Chroococcaceae**. The members of this family are unicellular and free living. They occur either singly or in colonies. The cells are usually spherical, ellipsoidal or cylindrical. The enclosing membrane is thick, mucilaginous and often lamellated. Multiplication takes place by fission, fragmentation or sometimes by nannocysts. The important genera included in the family are *Chroococcus, Gloeocapsa, Synechocystis, Microcystis, Aplanocapsa, Merismepedia* and others. *Chroococcus* is discussed here in detail.

CHROOCOCCUS (NAEGELI, 1849)

Occurrence. It is one of the simplest members of the order. The genus is widespread. Some species are sub-aerial and are found in very moist habitats. Damp soils, damp cliffs, tree trunks and other sub-aerial habitats, where they are wet most of the time, are the favourite habitats. They occur intermingled with other Cyanophyceae and Chlorophyceae. A few grow free floating in permanent and semi-permanent ponds and pools.

Thallus (Fig. 2.17). Single cells are met with occasionally (A). More often *Chroococcus* occurs in colonies of 2.4(B), very seldom 8, 16 or more cells. They may be bluish green, olive-green or yellowish. The colony is invested by the distended, slightly gelatinous, colourless, not distinctly lamellated sheath of the original mother cell and may be spherical or flattened. On damp soil the sheath is often asymmetrical. Each cell in the colony also produces a gelatinous coat of its own which in a few instances is lamellate. The cells are generally spherical in shape but remain hemispherical for some time after division. The cells in the colony are independent of one another. Each is able to carry on all the essential life processes such as photosynthesis, respiration, reproduction, etc.

Cell Structure (Fig. 2.17 A). Each cell is a tiny, rounded or spheroidal, sometimes a flattened body surrounded by a sheath which is not very gelatinous. In some species it is quite thin. Within the sheath is the cell wall which surrounds a tiny mass of granular material of jelly-like consistency. It is the **cell protoplast**. The protoplast is differentiated into the central body and the surrounding cytoplasm. The central body is called an **incipient nucleus** as it lacks the nuclear membrane and the nucleoli. It consists of a network of colourless material. In the meshes are embedded the chromatin granules. Besides, the central body contains other granules. The cytoplasm, which is reticular, forms the outer portion of the cell protoplast. It is often called the **chromoplasm**. The network of the chromoplasm is continuous with the network of the central body. It contains pigments and inclusions. The pigments present are a mixture of chlorophyll a, carotene, xanthophyll and c-phycocyanin. The pigments are located in flattened sacs, the **lamellae** found in the chromoplasm. The inclusions in the cytoplasm are of the nature of food reserves. They occur in the form of granules and other inclusions. The carbohydrates are called the **cyanophycean starch** and the proteins **cyanophycin granules**. In addition there may be **oil drops.**

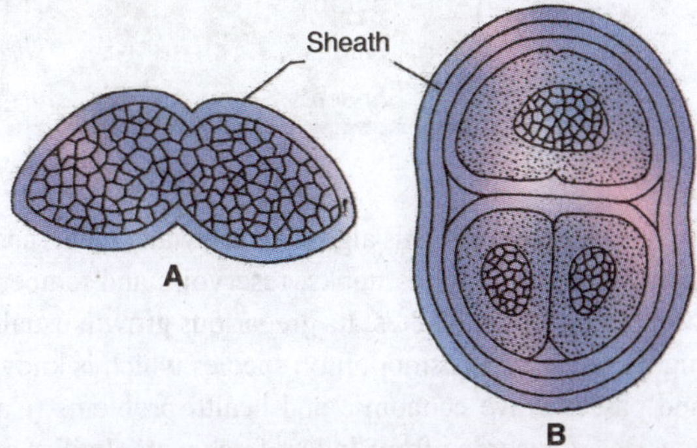

Fig. 2.17 (A-B). *Chroococcus* sp. (A) A single dividing cell; (B) 3-celled colony with the upper cell preparing to divide.

Reproduction. Multiplication takes place by two methods, **cell division** and **colony fragmentation.**

***(i)* Cell division or fission** (Fig. 2.11.). Single cell may be released from the colony by the disorganisation of the parent sheath. Each released cell grows into a new colony by cell division. For details refer to pages 42-43 and fig. 2.11. The mother cell divides into two. The process may be repeated. All the daughter cells are held together within the original sheath of the mother cell to form the colony.

***(ii)* Colony Fragmentation** (Fig. 2.12.A). Reaching a certain size the colony may break into fragments. Each fragment by cell division forms a new colony.

Gloeocapsa (Fig. 2.18. A-C). It is another member of the family Chroococcaceae closely related to *Chroococcus*. It is subaerial and occurs on rocks, or walls. The distinction between the two genera is slight and is often made upon the size of the palmelloid colony and nature of the sheath. There are numerous lamellate sheaths in *Gloeocapsa* and the cells occur in extensive gelatinous masses. The plant mass

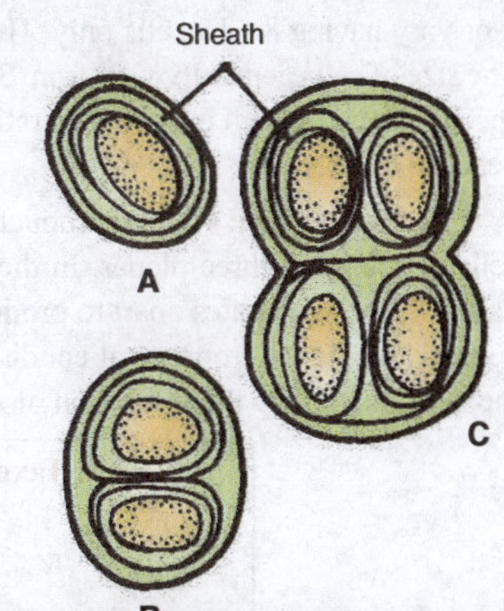

Fig. 2.18 (A C). *Gloeocapsa decorticans.*
(A) single cell enclosed in a lamellated sheath.
(B) Two-celled colony;
(C) 4-celled colony. (After Wille).

represents an association of many colonies. The cells in the mass are arranged in families. Each cell is more or less spherical in form and has its own individual conspicuous concentrically lamellate, gelatinous sheath and the daughter cells may be held together by the sheath of the mother cell. Further the sheaths of the two mother cells may be surrounded by the sheath of the grand-mother cell. The alga reproduces only by cell division in three planes at right angles to one another.

Taxonomic Position

Division	:	**Cyanophyta** (Cyanophycophyta)
Class	:	**Cyanophyceae**
Tribe	:	**Coccogoneae**
Order	:	**Chroococcales**
Family	:	**Chroococcaceae**
Genus	:	*Chroococcus*
Species	:	*turgidus (Kutz)*

MICROCYSTIS

Occurrence:- This alga is a fresh water genus and grows as dominant organism in the planktons of many tropical ponds, tanks, reservoirs and temperate lakes. It is also reported in stagnant rain water pools and salt lakes. Its gregarious growth usually causes water bloom and *M. aeruginosa* is a highly successful cosmopoliton species which is known to cause water bloom in fresh water bodies and passes grave economic and health problems (Callins, 1978; Raynolds etal, 1981). About 18 species are recorded from Indian fresh water bodies out of which *M. flos-aquae* and *M. aeruginosa* are of common occurrence.

Thallus: Cells are usually embedded in mucilage forming colonies.

Cell Structure: The cells are usually ovoid, capsule shaped or spherical end exhibit great variability in size and shape. They are embedded in *conspicuous* mucilage forming colonies. The shape and size of colonies varies greatly and it may be irregular - ovoid, elliptical, cylindrical, sub-spherical, strap-like or other similar type and quite big containing several thousands of cells within it or vary having 8 - 16 cells only. The colonies may be loose aggregates of cells or even compact. The size of cells generally vary from 3 - 10 μm and each cell has its own individual sheath. In several species the cell sheath is firm and stratified whereas in other species it may be colourless homogenous persistent or broken.

Reproduction: - The reproduction of individual cells takes place by means of fission and the cells divide in all three planes. In the elongated cells the plane of cell division is transverse and a part of colony separates apart to produce new colonies of various shapes and sizes which primarily depends on the environmental conditions of the water bodies. However, in *Microcystis flos-aquae* the reproduction by the formation of ***mannocytes*** is also reported.

Taxonomic Position

Division	:	**Cyanophyta**
Class	:	**Cyanophyceae**
Tribe	:	**Coccogoneae**
Order	:	**Chroococcales**
Family	:	**Chroococcaceae**
Genus	:	*Microcystis*

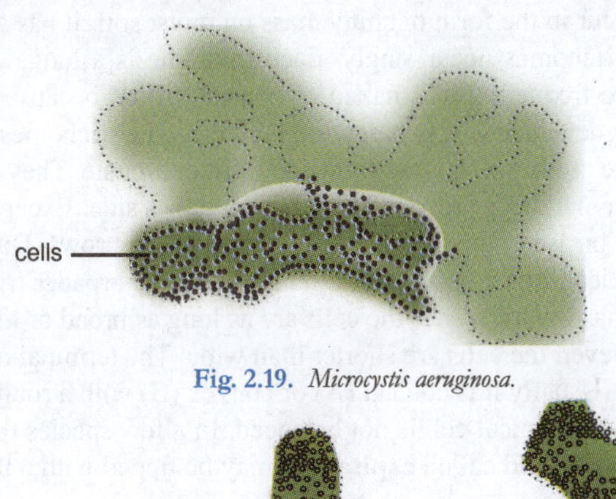

Fig. 2.19. *Microcystis aeruginosa.*

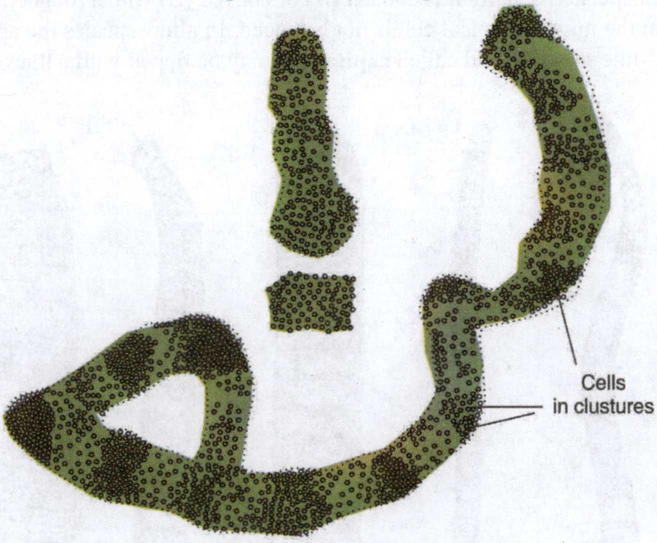

Fig. 2.20. *Microcystis pseudofilamentosa.*

TRIBE : HORMOGONEAE : ORDER : OSCILLATORIALES

The trichomes consist of undifferentiated cells and are unbranched and usually straight or trichome of more than one (*Microcoleus* Fig. 2.5). The order includes a single family Oscillatoriaceae.

Family **Oscillatoriaceae**. The trichomes consist of a single row of cells. The heterocysts and spores are absent. Hormogones develop along the longitudinal axis. The family includes several genera such as *Oscillatoria, Lyngbya, Spirulina, Arthrospira, Phormidium, Microcoleus, Trichodesmium* and others. *Oscillatoria* is taken here as an example.

OSCILLATORIA (VAUCHER, 1803)

Occurrence. It is an exceedingly common, fresh water, filamentous, dark, blue-green alga. It occurs in a wide variety of habitats. Usually it is found on damp soil, in temporary rain water pools and roadside ditches. Patches of entangled masses of filaments with the adherent mud are often found floating on the surface of fresh water channel. Bottoms of shallow temporary puddles and ditches, drains or sewers are usually covered with large patches of *Oscillatoria.* It is able to thrive in waters where other algae cannot survive. It often grows luxuriantly in stagnant water which abounds in decaying organic matter. A few species are marine.

Thallus (Fig. 2.21). The thallus consists of free-living trichomes which often form a compact, tangled floating mass or occur in the form of shiny mass on moist soil. It has a distinct filamentous texture. Occasionally the trichomes occur singly. Each trichome is a long, very fine, thread-like unbranched structure. More frequently it is naked or appears naked because the sheath around it is very delicate and poorly developed. It is barely perceptible. The trichomes are usually smooth, sometimes constricted at the cross wall (C), straight and rigid or arcuate. They are septate; the septa are faintly visible and often marked by rows of granules on either side. Except the end cell, all are alike and discoid. They are pressed flatly against one another in long rows. Due to mutual pressure the cells are shortly cylindrical. Some species of *Oscillatoria* have broader trichomes (C) than the others. In the species with narrow trichomes the cells are as long as broad or longer than broad (A). In most of the species, however, the cells are shorter than wide. The terminal or end cell is variable in form in different species. Usually it is conical (A) or convex (B) with a rounded free end. This is due to the fact that the turgor of apical cell is not balanced. In a few species the apical cell may be swollen into a cap-like structure and called **capitate** or may be tipped with a thickened membrane, the **calyptra** (D).

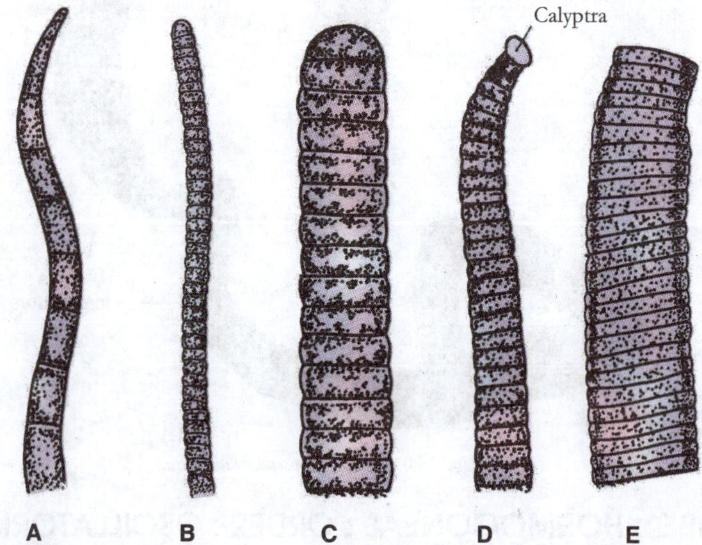

Fig. 2.21 (A-E). *Oscillatoria*, (A) *O. acuminata*, a trichome with acuminate apical cell; (B) *O. foreani*, a narrow trichome with constrictions. The apical cell is rounded; (C) *O. annae*, portion of trichome with hemispherical apical cell; (D) *O. proboscidea* with a proboscia-like terminal cell. Also note the calyptra; (E) O. *princeps*. a broad filament with discoid cells (after Vasishta).

Cell Structure (Fig. 2.22.). All the cells in the trichome are similar in structure. The cell wall being stable there is hardly any conspicuous mucilaginous sheath external to it. According to Halfen and Castenholtz (1970) the cell wall in *O. princeps* ultrastructurally consists mainly of a thick (2000 Å) structural layer external to the plasma membrane. Outside the structural layer is 160 Å thick another layer. There is a third 90 Å thick membrane-like layer loosely wrapped around the inner two. The structural layer has a series of 700 Å wide pores which terminate at the 160 Å layer. The 160 Å layer and the outer membrane are continuous over the trichome surface. They do not take part in the formation of cross walls. The cell protoplast shows distinction into an outer coloured cytoplasm, the **chromoplasm** which surrounds the colourless central area called the **centroplasm** or **central body**. The chromoplasm is invested by a two-layered plasma membrane. The protoplast lacks the plastids, mitochondria, endoplasmic reticulum and dictyosomes. The pigments are located in flattened sac-like structures called the **lamellae** or **thylakoids** forming an array or parallel, paired membranes embedded in the chromoplasm (Fig. 2.21.). The lamellae are not organised into grana. Besides the **cyanophycean starch** and cyanophycin granules, the contents of the cell, especially in the planktonic species, may include **pseudo-vacuoles**. For structures and function of pseudo-vacuoles refer to page 36. The central body represents the **incipient nucleus**. It consists only of a mass of scattered chromatin granules but lacks the *nuclear membrane* and the *nucleoli*. Under electron microscope the centroplasm is seen to consist of a faintly granular ground matrix in which are found embedded crystalline-like bodies and clear areas.

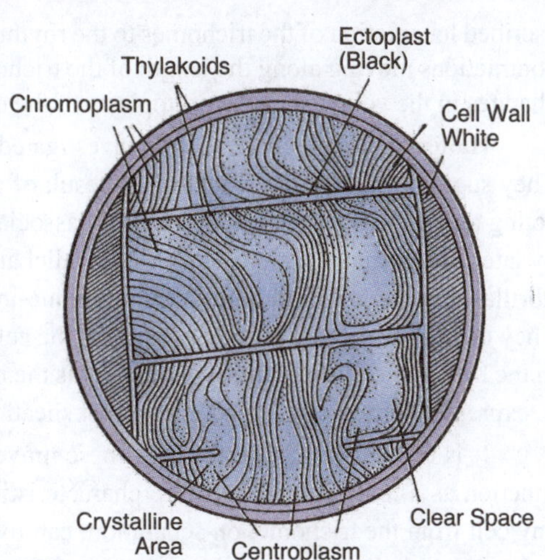

Fig. 2.22. *Oscillatoria chalybia* showing cell structure as revealed by electron microscope. (Based on Menke).

Growth. It is **intercalary**. The cells constituting the trichome divide in unison during growth. The division is at right angles to the longitudinal axis. This results in increase in the length of the trichome which is never more than one cell wide. Cell division is preceded by nuclear division which according to some is a primitive type of **mitosis** and according to others is **amitotic**.

Movement. The striking characteristic of *Oscillatoria* is the slow, rhythmic but active movement of its trichomes. Place a mass of living material on the wet surface of a petridish. The trichomes will soon spread out in all directions across the substrate and may slide up on to the sides. The movements are spontaneous. The frequency can, however, be increased by higher temperature or greater light intensities. Under the microscope the movements are seen to be of the following three types:

1. *Gliding or creeping movements*. Jarosch (1962) defined gliding as, "the active movement of an organism in contact with a solid substratum where there is neither a visible organ responsible for the movement, nor a distinct change in the shape of the organism". These rhythmic movements take place in the direction of the long axis of the trichome. They are called the **axial movements** and their general effect is that of propulsion. The trichome glides forwards and backwards. It moves back and forth (backwards and forwards) in water accompanied in some species, by clockwise or anticlockwise rotation of the forward end of the trichome.

2. *Oscillatory movements*. *Oscillatoria* also exhibits slow waving movements. These are somewhat jerky, pendulum-like oscillations of the front end, hence the generic name (*oscillare* to swing). Oscillations generally occur when there is no forward progression.

3. *Bending movements.* At the end of each oscillation there is generally a rapid bending of the extreme apex.

The actual **mechanism** responsible for these movements is not known definitely. A number of hypothesis have been advanced to explain these. One of these ascribes these movements to the presence of invisible cilia at the tip of the trichome. West and others ascribe it to the secretion of slime. They think that active and continual secretion of gelatinous material, along the sides of the trichomes, brings about locomotion. It is secreted through the minute pores in the cell walls. The arrangement of slime secreting pores in two crossing spiral series induces rotating movements. This causal relationship between localised secreting of slime and gliding is based on the indirect evidence from the fact that in *Oscillatoria*, gliding is restricted to the species which secrete slime. Some investigators attribute these movements to the growth of the trichomes or osmotic currents. Ullrich ascribed locomotion of the trichomes to the rhythmic longitudinal waves of alternate expansions and contractions moving along the length of the trichomes. These peristaltic contractions are induced by changes in the volumes of the protoplasts of the cells.

Halfen and Castenholtz (1970) investigated the mechanism of gliding motility in *O. princeps*. They suggest that the movement is the result of unidirectional travelling waves on the cell surface acting against solid substrate or a closely associated elastic sheath. These waves may be produced by lateral deformation of elements of a parallel array of fibrils over the surface of the trichome. The fibrillar array is wrapped around the trichome in a right-handed helix with a pitch of 60 degrees. They demonstrated that in the L_3 region of the cell wall is a layer of fibrils in parallel array pointing in the helical direction of motion. Motility is the result of torsion of these fibrils producing waves in L_3 expressed through L_4 and acting against sheath or solid substratum.

It is evident that in performing these movements the cells of the trichome co-operate. They function as a unit. This is a feature characteristic of truly multicellular plants. On the other hand any cell from the trichome, on separation, can live as well, as it does in the trichome. It lives as an independent individual. For this some scientists consider *Oscillatoria* as a colony of unicellular plants rather than a multicellular individual. This view is supported by the lack of any differentiation of cell in the trichome.

Reproduction. *Oscillatoria* reproduces vegetatively. The only known method is by the formation of **hormogones** (Fig. 2.23 A). The hormogones are short sections or lengths or living cells separated from the trichomes. The break takes place where a **dead cell (necridium)** is situated.

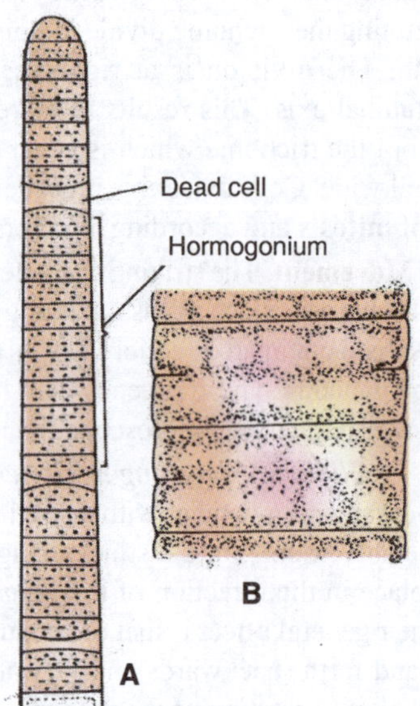

Fig. 2.23 (A-B). *Oscillatoria.* (A) Formation of hormogones by dead cells, (B) Cell structure.

During the growing season a cell here and there in the trichome will collapse (die). The protoplast of such cells changes into a transparent, viscous substance called the **mucilage**. The mucilage filled dead cells are called **necridia**. The dead cell loses its turgidity (Fig. 2.23 A) . The mutual pressure on the walls of the adjacent cells is released. They become convex so that the dead cell appears biconcave. These biconcave dead cells or separation discs provide weak links in the trichome and thus mark the points of disjunction of the trichome into **hormogones**. The mucilage swells and necridia break down releasing the hormogones. Each hormogone may consist of a few to several living cells. Sometimes a hormogone may consist of a few to several living cells. Sometimes hormone break off from the extremity of the trichome. The hormogones exhibit a greater capacity for a slow forward motion than the vegetative trichome. They move away from the latter. By repeated cell division (fission) in all the cells hormogone grows into a full-length, typical trichome.

Occasionally propagation by fragmentation also takes place. It may result from the bites of the animals feeding on the trichomes. Formation of spores, heterocysts and akinetes has not been observed in *Oscillatoria*.

Taxonomic Position:

Division	:	**Cyanophyta (Cyanophycophyta)**
Class	:	**Cyanophyceae**
Tribe	:	**Hormogoneae**
Order	:	**Oscillatoriales**
Family	:	**Oscillatoriaceae**
Genus	:	*Oscillatoria*
Species	:	*princeps*

TRIBE : HORMOGONEAE : ORDER 2 : NOSTOCALES

Family **Nostocaceae**. The trichomes are composed of a row of similar cells and are either free or embedded in a common matrix of mucilage. Certain cells in the trichome form heterocysts and some akinetes. The former are intercalary, rarely terminal in position and occur either singly or more than one together. Reproduction is by hormogonia and in some by spores in addition. The important genera included in the family are *Nostoc, Aulosira, Nodularia, Anabaena* and *Aphanizomenon*. Here *Nostoc* is taken as a type.

NOSTOC (VAUCHER, 1803)

Occurrence. It is filamentous form of both terrestrial and aquatic habitats. It does not occur in single filaments but grows in large colonies of closely packed trichomes embedded in a firm matrix of gelatinous material. Nostoc colony thus forms a mucilaginous lump or thallus which occurs floating or attached. The thalli are of various sizes and shapes and may be solid or hollow. They may be balls of a jelley-like substance (Fig. 2.24A) or may be irregularly shaped. In size they may be microscopic, pea-size, walnut sized or as large as a plum. Sometimes they may reach a diameter of eight centimetres. The colour may be olive-green or blue-green and the surface of the colony warty or smooth. Terrestrial species, *Nostoc commune* grows on damp soil and forms leathery or rubbery sheaths. It is common in the Arctic and alpine meadows. Aquatic species occur either as free floating thalli in the sunny pools, ponds and lakes or lying at the bottom attached to submerged vegetation. A few species favour running water, especially fast flowing mountains streams. There they are found attached to stones along the stream beds. *Nostoc* also occurs in symbiotic association with fungi to form lichens. A certain species of *Nostoc* occurs in the thalli of *Anthoceros.* One species (*N. punctiforme*) occurs in the underground stems of *Gunnera manicata* (Angiospermic marsh plant). Some species of *Nostoc* have been reported to fix atomspheric nitrogen and tend to maintain fertility of paddy fields in which these forms regularly occur. Large colonies of terrestrial species are eaten in the orient.

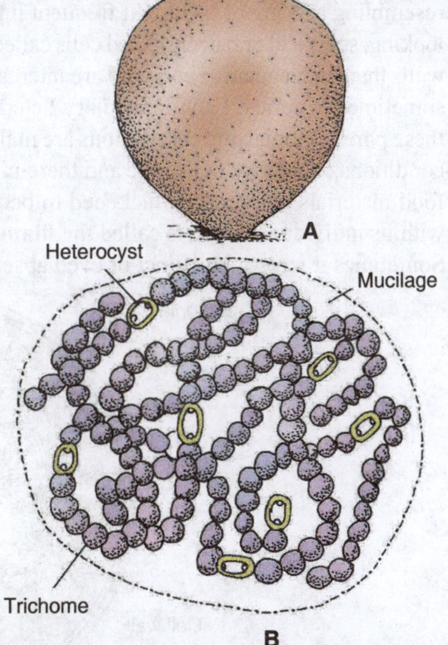

Fig. 2.24 (A-B). *Nostoc* sp. (A) Colony; (B) Portion of the colony as seen under low power of microscope.

Structure of the Colony (Fig. 2.24 B). Innumerable chains of bead-like cells (trichomes) of varying lengths are intricately tangled in a copious gelatinous matrix to form a colony (B). Each trichome (Fig. 2.25) is usually enclosed by its own mucilaginous sheath and is called a **filament**. The numerous filaments in the colony are held together by a soft mucilaginous envelope formed by the fusion of the individual trichome sheaths and also secreted copiously by the component cells. The mucilage lump is bounded externally by a firm, tough, pellicle-like bounding membrane to form a definite colony. Each colony appears like a bluish green or yellowish mass of jelly. Young colonies are small and microscopic. The mature ones are large. Enclosure of several colonies growing in the neighbourhood, within a common gelatinous mass results in the formation of a macroscopic, mucilaginous lump or thallus. It may be called a **compound colony**. Frequently the compound colony breaks open as it gets large and forms a flat expanse with lacerated margins.

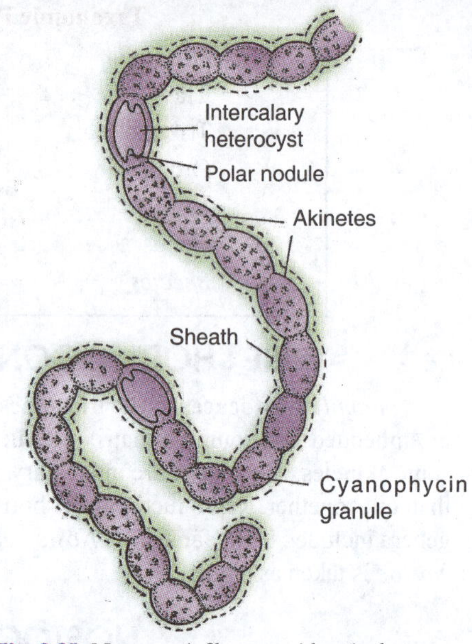

Fig. 2.25. *Nostoc* sp. A filament with a single trichome.

Structure of the Trichome (Fig. 2.25). The trichomes are much contorted, moniliform and intertwined. They wind about in every direction in the gelatinous matrix. They are more crowded towards the periphery of the colony to form a dense limiting layer. Each trichome is composed of numerous rounded or oval cells. The cells are joined loosely from end to end into a trichome somewhat resembling a string of beads. At frequent intervals along with trichome are found colourless, empty looking, spherical or barrel shaped cells called the **heterocysts**. They are slightly large and have thicker walls than the vegetative cells and are intercalary, sometimes terminal (*N. linckia*). They occur singly, sometimes in series. Each intercalary heterocyst, when first formed, has two polar pores. Through these pores cytoplasmic connections are maintained with the adjacent vegetative cells. Under certain conditions vegetative cells here and there in the trichome may become greatly enlarged, stored with food materials and greatly thickened to become resting bodies called the **akinetes**. Each trichome with its individual sheath is called the **filament** (Fig. 2.25). The sheath may be hyaline or coloured. Sometimes it is very indistinct or even absent.

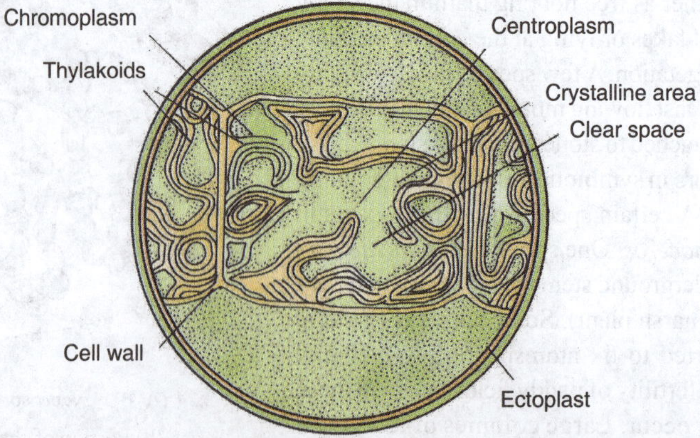

Fig. 2.26. *Nostoc muscorum.* Showing cell structure as revealed by electron microscope (Based on Menke).

Cell Structure (Fig. 2.26). The cell consists of the **cell wall** surrounding the **protoplast**. The protoplast, as usual in the cyanophycean cell, is vaguely defined into the outer pigmented cytoplasmic region called the **chromoplasm** and the inner colourless **centroplasm**. The chromoplasm has the usual pigments (phycocyanin, chlorophyll a, etc.) located in the lamellae which form a system of peripheral area of parallel paired membranes embedded in the granular matrix. There is no chromoplasm without lamellae. It also contains the colourless granules of myxophycean or cyanophycean starch and the cyanophycin granules of proteinaceous nature. The centroplasm is not separated from the chromoplasm by any kind of membrane.

Reproduction. *Nostoc* reproduces entirely vegetatively by the following methods:-

1. Colony Fragmentation. The *Nostoc* colony as it gets larger frequently breaks up into flat expanses as a result of storms and other disturbances. Each of these grows up to the size of the parent colony.

2. Hormogonia (Fig. 2.12 C). Hormogone formation is very common in *Nostoc*. The trichome ruptures at places where a heterocyst and the vegetative cell adjoin. This junction is the weakest link in the chain. In fact some algologists believe that heterocysts represent a mechanism for the fragmentation of the trichome. In this way, short segments of living cells called the **hormogonia** became isolated. The hormogonia slip out of the enclosing gelatinous matrix and establish new colonies by division (Fig. 2.27 A-B). The terminal cells of the hormogonia differentiate as heterocysts (B). The intercalary cells then divide in a plane parallel to the axis of the trichome (B) forming a packet of cells (C). This is the *aseriate stage*. Sometimes the hormogones develop into fresh trichomes without being liberated from the colonial sheath and increase the number of trichomes in the adult colony.

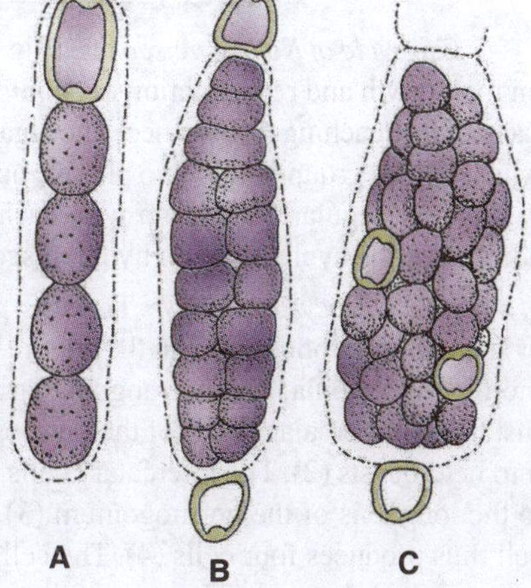

Fig. 2.27 (A-C). *Nostoc.* Stages in germination of a hormogone to form a new colony (C).

3. Resting spores or akinetes (Fig. 2.25). Under certain conditions any cell or some of the vegetative cells of the trichome become enlarged and each secretes a thick, highly resistant wall around it. They get gorged with reserve food materials. Such specially modified vegetative cells are called the **akinetes** or **resting spores**. These are well adapted to survive unfavourable conditions such as water shortage and unsuitable temperature. It is not unusual to find all the cells between two heterocysts and occasionally the entire trichome converted into akinetes. With the return of favourable season each akinete germinates to form a new filament of *Nostoc*. Recently Tewari (1979) discovered a new species of *Nostoc* described as *N. spinosa* which produces akinetes with spinose spore walls.

4. In exceptional cases (*N. commune*) the heterocyst may become functional (A). It germinates to form a new filament (Fig. 2.28). At first, a 2-celled (B) and later, a 4-celled germling is formed (C). It escapes either by the rupture of the heterocyst wall at the equatorial region or by gradual dissolution and widening of the pore.

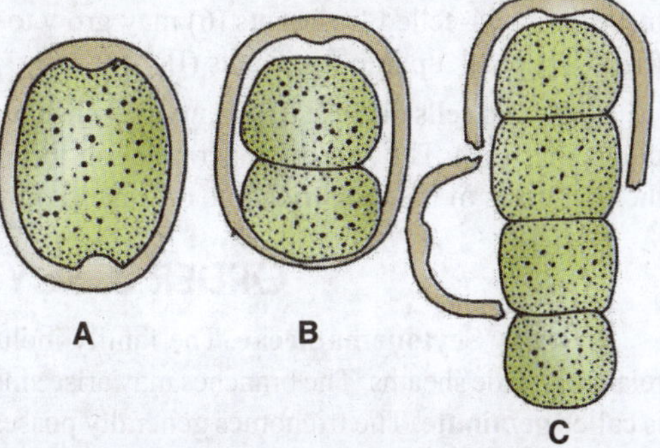

Fig. 2.28 (A-C). *Nostoc.* Stages in germination of heterocyst (After Geitler).

5. **Endospore formation**. Brand (1901) and Spratt (1911) reported that the contents of the heterocysts in *Nostoc commune* and *N. microscopicum* divide to produce **endospores**. The endospores, on liberation, give rise to new filaments.

Taxonomic Position:

Division	:	**Cyanophyta** (Cyanophycophyta)
Class	:	**Cyanophyceae**
Tribe	:	**Hormogoneae**
Order	:	**Nostocacales**
Family	:	**Nostocaceae**
Genus	:	*Nostoc*
Species	:	*muscorum*

Life cycle of Nostoc muscorum (Fig. 2.29). Observations of Lazaroff and Vishaniac (1961, 1962) on the growth and reproduction of *N. muscorum* under cultural conditions have revealed interesting facts of far-reaching importance. The alga passes through a regular sequence of development changes which can be grouped into two phases, namely, **heterocystous phase** and sporogenous phase. These two phases regularly alternate with each other in the developmental cycle. Following is the brief account of the cycle as given by the discoverers:

(*a*) *Heterocystous phase*. The alga passes through this phase in complete darkness with glucose or sucrose as carbon source in the culture medium. The trichomes break up at the heterocysts to form motile hormogonia. The hormogonia do not grow into normal heterocyst containing trichomes (1). Instead the intercalary cells of the hormogonia increase in size and the terminal cells metamorphose into heterocysts (2). The intercalary cells of the hormogonium separate and divide in a plane parallel to the long axis of the hormogonium (3). The division is suggested to be meiotic. Each intercalary cell thus produces four cells (4). The cells divide further so that a packet of several undifferentiated cells results from each. The cells in the packet may occur singly (5) or in chains of two or three (6). This is the *aseriate stage*. The developmental cycle is arrested at the aseriate stage in the dark. On exposure to light the cells of each packet of the aseriate stage form a single trichome. The 2 or 4-celled fragments anastomose. The products of anastomosis may grow in lengths by cell division and develop intercalary heterocysts (7). The liberated filaments break up into heterocysts and hormogonia. These seven growth stages constitute the heterocystous phase of the developmental cycle.

(*b*) *Sporogenous phase*. This phase in the cycle starts with two or four cell long chains of the aseriate stage (6). On exposure to light and with carbon dioxide as source of carbon in the culture medium 2 to 4-celled fragments (6) may grow to form long trichomes without heterocysts (I). The trichomes break up into fragments (II).

The end cells of these fragments may change into heterocysts (III). The intercalary cells develop into spores (IV). The spores on germination form 2 or 3-celled germlings (V) apparently similar to the fragments of the long trichome of stage (II).

ORDER 3. SCYTONEMATALES

Family **Scytonemataceae**. The family includes filamentous forms with **false branching** and relatively wide sheaths. The branches may arise singly or two together. In the latter case, the branching is called **geminate**. The trichomes generally possess **heterocysts** which may be either round, oval or quadrate in form. The vegetative cells are usually quadrate or slightly longer than broad, *Plectonema,* however, is an exception. It has no heterocysts. The filament usually consists of a single trichome but in *Hydrocoryne* the filament possesses more than one **trichome**. The heterocysts are generally

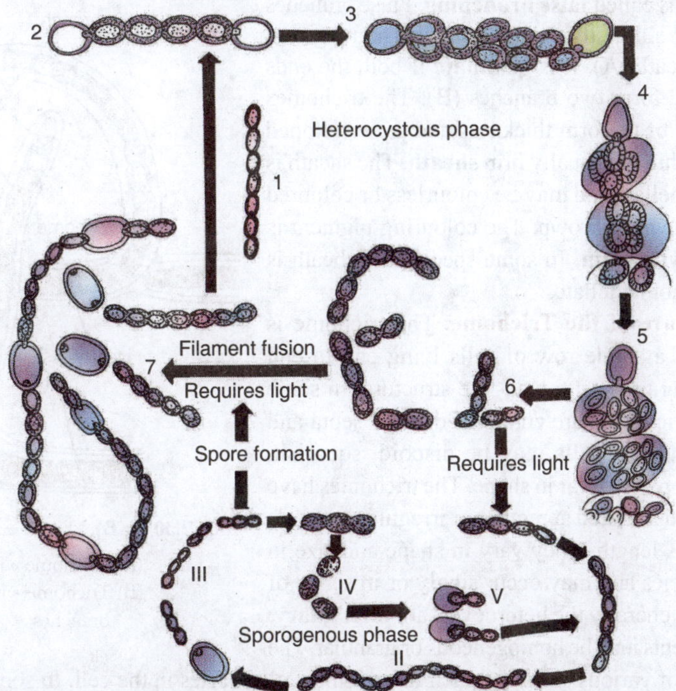

Fig. 2.29. *Nostoc muscorum*. Alternative scheme showing alternation of sporogenous and heterocystous generations in the developmental life cycle (After Lazaroff and Vishniac, 1962).

intercalary except in *Tolypothrix* where terminal heterocysts are also present. Reproduction is chiefly by means of **hormogonia**. Akinetes are absent in most genera, rare in others (*Scytonema*). In India this family is represented by eight genera. These are *Scytonema, Plectonema, Tolypothrix, Hydrocoryne, Camptylonemopsis, Scytonematopsis,* and *Pseudoscytonema*. The genus *Scytonema* is very common in India. Its life-history is considered here.

SCYTONEMA

Habit and Habitat. It is predominantly a sub-aerial and aquatic mostly freshwater genus. Forty species of this genus have been reported from India (Desikachari, 1959). Of these 31 occur in damp terrestrial situations. The remaining 9 are aquatic. The aquatic species form floccose, woolly, cespitose, tomentose and tufted thalli. They vary in colour from blue green, dirty blue green, blackish green, yellowish brown to brown. They may float freely in the water of ponds, lakes and tanks. Some may grow submerged along the walls of the tanks or as epiphytes on submerged plants. *Scytonema simplex* is an extremely common aquatic species. The terrestrial species grow on moist soil, long tree trunks, on the walls of houses and mountains. *Scytonema ocellatum* is the commonest terrestrial species. It forms cushion-like, blackish or greyish blue thalli along the walls of houses, tree trunks and walls of mountains. It sometimes covers extensive areas and gives a peculiar hue to the landscape. The terrestrial species are endowed with a remarkable capacity of absorbing moisture from the air.

Thallus. (Fig. 2.30). It is filamentous. The filaments are variously interwoven to form different types of thalli. Each filament consists of a single trichome. It may be sparingly or profusely branched. The branching in *Scytonema* is peculiar. It is not formed by the lateral protrusion of a cell and its subsequent division to form a full-fledged branch. On the other hand, the trichome breaks at one point and one or both the broken ends protrude out of the sheath and develop into branches. This type of branching is called **false branching**. False branches are said to be **single** if one end of the trichome grows out of the sheath (A). It is **geminate** if both the ends grow out and form two branches (B). The trichomes are generally of uniform thickness and are enveloped in a thick or thin but usually firm **sheath**. The sheath is generally lamellate and may be colourless or coloured yellowish brown or brown. The colouring pigment is known as **scytonemin.** In some species the sheath is smooth and not lamellate.

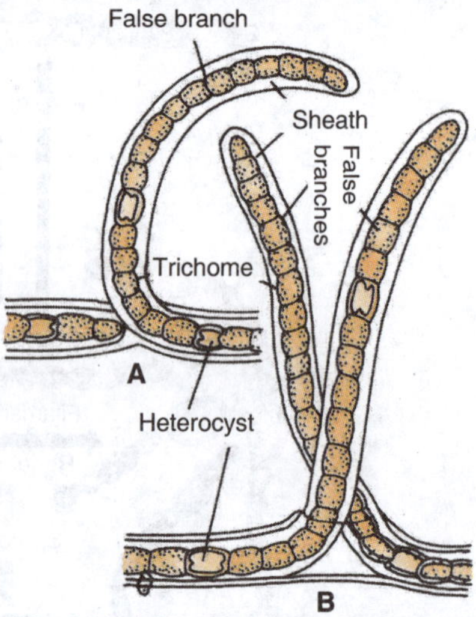

Fig. 2.30 (A-B). *Scytonema.* False branching. (A) Trichome with one false branch; (B) Trichome with two false branches.

Structure of the Trichome. The trichome is composed of a single row of cells lying end to end to form a long or short thread-like structure. In some species the trichomes are constricted at the septa and in others not. The cells may be discoid, squarish, cylindrical or rectangular in shape. The trichomes have **heterocysts** distributed at regular or irregular intervals throughout its length. They vary in shape and size in different species and may occur singly or in series of two to four. Generally the heterocysts are **intercalary**. The cell contents may be homogeneous or granular. The granules are of various sizes and their arrangement also varies in the cell. In some species they are present along the cross walls.

Growth (Fig. 2.31). The trichomes increase in size by the repeated division of its cells. The cell division is mainly restricted to the apical or sub-terminal cells. The apical cell is generally hemispherical and flattened. In *Scytonema leptobasis* it is spherical and much broader than the other cells of the trichome (Fig. 2.31). The apical cells are generally characterised by pale or light, blue-green colour and have dense cytoplasm. In the terrestrial species the apical cell may undergo disorganisation and be replaced by a new cell. The intercalary cells also divide and add to the length of the trichome.

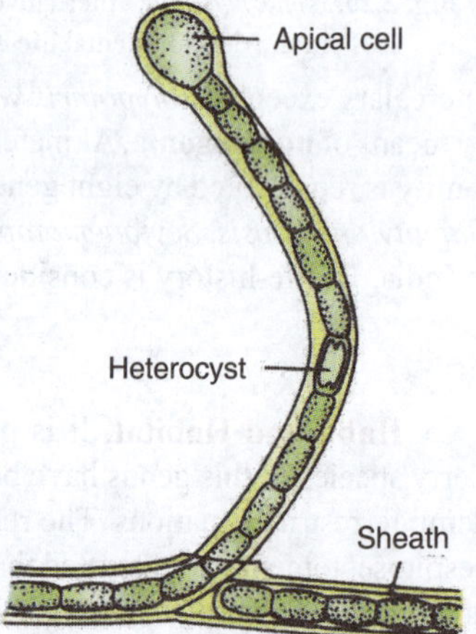

Fig. 2.31. *Scytonema leptobasis.* Note the spherical apical cell at the top of a false branch.

Branching of the Trichome (Fig. 2.32). *Scytonema,* as stated above, exhibits false branching. The branches may be single or geminate. This type of branching starts with the breaking of the trichome at a certain point. One or both the ends may then perforate the sheath and grow out as branches. In fact it is a kind of vegetative multiplication in which the fragments or hormogones grow *in situ*. The branching in *Scytonema* is affected by either of the following methods:

(*i*) By the degeneration of one or more intercalary cells (A). In this case one or more cells of the trichome lose their contents and become more or less triangular (A). Ultimately they degenerate leaving a space. One of both the ends of the trichome may then pierce through the sheath and grow out as lateral branches.

(*ii*) By the formation of separation discs. An intercalary cell of the trichome takes on a dark green appearance. This is due to the diffusion of the pigments throughout the protoplast of the cell.

The walls become thin. Finally the cell becomes a biconcave, disc-like structure (B). Ultimately the disc loses its contents and becomes colourless. It then disorganises to break the trichome. One or both the ends grow as false branches.

***(iii)* By breaking of the trichome near the intercalary heterocysts.** In case the trichome breaks on one side of the heterocyst a single branch is formed (C). If the trichome breaks on both the sides of the heterocyst, the branching is geminate.

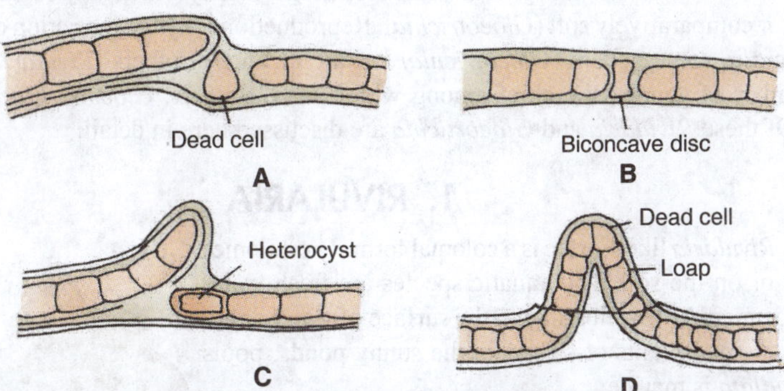

Fig. 2.32 (A-D). *Scytonema.* (A) Development of a branch by the formation of a dead cell; (B) Formation of a biconcave disc which initiates branching; (C) The trichome has broken at the heterocyst. In this case the branch is on one side; (D) Formation of a loop.

(iv) **By the formation of a loop.** Branching by loop formation was described by Geitler (1936). The trichome increases in length at a particular point. This leads to the formation of a loop (D). The loop becomes prominent and grows in size. A cell in the loop becomes distinct and undergoes division into two cells, both these cells eventually die and result in the formation of two branches (Bhardwaja, 1933).

Reproduction. Hormogone formation is the principal method of reproduction. The hormogones are delimited by the death of an intercalary cell, by the formation of biconcave discs or by breaking at the heterocyst. Usually the hormogonia are formed at the ends of the trichomes. They are of short lengths. The hormogones creep out by breaking open the sheath. The liberated hormogone germinates to give rise to a new filament of *Scytonema*. In most of the terrestrial species the hormogones during the dry period of the year secrete thick, lamellate sheaths. The cell contents become densely granular and pale yellow or yellowish brown. During the rainy season the sheathed hormogones become active and grow into new filaments.

Akinete formation. Burzi (1879) reported the formation of akinetes in some species of *Scytonema*.

Taxonomic position:

Division	:	**Cyanophyta** (Cyanaphycophyta)
Class	:	**Cyanophyceae**
Tribe	:	**Hormogoneae**
Order	:	**Scytonematales**
Family	:	**Scytonemataceae**
Genus	:	*Scytonema*
Species	:	*simplex*

ORDER 4. RIVULARIALES

Family **Rivulariaceae**. The trichomes which are unbranched or exhibit false branching are attenuated or terminate each in a colourless hair consisting of elongated cells. Heterocyst is usually present, sometimes absent (*Hormoethrix*). When present it occurs singly and is usually basal in position and in some with an adjacent akinete. Growth is trichothallic. Filaments may occur singly, but usually numerous within a mucilaginous matrix forming a colony which may be hard and rubbery as in *Rivularia* or comparatively soft (*Gloeotrichia)*. Reproduction is by the formation of hormogonia. Spores are found in some genera (*Gloeotrichia)* but are lacking in others (*Rivularia*). The family includes a number of genera, the chief among which are *Calothrix*, *Gloeotrichia, Rivularia* and Leptochaete. Of these *Rivularia* and *Gloeortichia* are discussed here in detail.

1. RIVULARIA

Habitat. *Rivularia* like *Nostoc* is a colonial form. The colonies occur in water or on the soil. The aquatic species are fresh water forms which occur either free floating on the surface of water or are attached to submerged plants or stones in the sunny ponds, pools and lakes. *R. bullata* is marine.

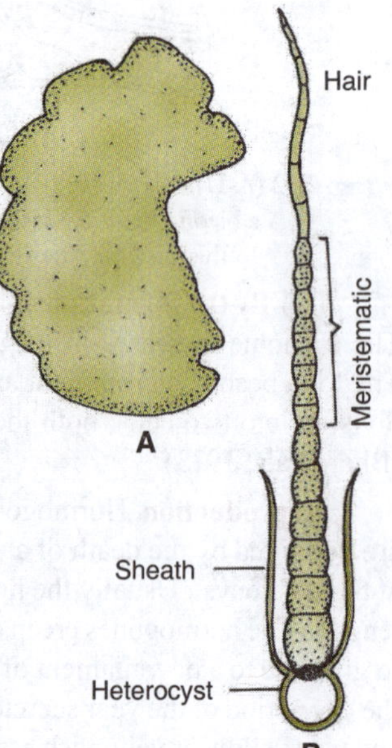

Fig. 2.33. *Rivularia* (A) Colony; (B) A filament.

Structure of the colony (Fig. 2.33 A). The colonies are macroscopic thalli of various sizes and shapes. They are yellowish brown in colour. The colony may be thick and discoid or a spherical, hemispherical or irregularly lobed gelatinous mass. It is soft, sometimes solid and rarely verrucose (*R. Joshi,* Vasishta). Each colony contains numerous radially arranged trichomes. The trichomes are loosely arranged in the central portion, but are more or less crowded towards the periphery of the colony. They are unbranched but sometimes exhibit false branching. The numerous trichomes constituting the colony are embedded in the soft, mucilaginous matrix secreted by the cells of the trichomes. The mucilage envelope has more or less a firm boundary to form a definite colony.

Structure of the trichome (Fig. 2.32 B). The trichomes are whip-like structures. They are straight or slightly curved. Each trichome is broad at the base but gradually tapers further up to terminate in a long, colourless, multicellular hair at its distal end. The hair is straight or slightly curved rarely irregularly curved and coiled as in *R. mehrai,* Vasishta. Situated below the hair is the meristematic region of the trichome. Each trichome has its own mucilage envelope or sheath which is colourless and non-lamellated. The trichome with the surrounding sheath is called the **filament.** The sheath is indistinct or distinguishable only near the base of the trichome and is diffuent further up. The younger trichomes are pale, blue-green in colour and show slight constrictions opposite the septa. The heterocysts occur singly and at the base of the trichome. The basal heterocyst is spherical, ellipsoidal or cylindrical and normally has a single pore. Two pored heterocysts have been recorded in *R. mehrai,* Vasishta (1960). The contents of the heterocyst are homogeneous. Absence of akinete (spore) next to the basal heterocyst is a noteworthy feature of *Rivularia.*

The trichomes sometimes exhibit false branching, each branch, in turn, ends in a colourless hair. Owing to the copious secretion of mucilage the branches become displaced and it becomes difficult to make out their origin.

Cell structure. The cells in the basal and meristematic region are usually barrel-shaped or subspherical. They are as long as broad, sometimes shorter than broad. In the region of the hair they are long and cylindrical.

The cell, as usual, consists of two parts, the *cell envelope* enclosing the *protoplast*. The cell envelope is divisible into two regions, the outer *sheath* and the inner *cell wall* proper. The latter is usually called the *inner investment*. It encloses the cell protoplast whereas the cell sheath envelopes the entire trichome. The septa between the cells are formed by the inner investment. Internal to the inner investment is the living **plasma membrane** which is differentially permeable. The cell protoplast has the same structure as is characteristic of a typical cyanophycean cell. For details refer to pages 33-37.

Growth. The growth of the trichome in *Rivularia* is **trichothallic**. It takes place by the activity of the meristematic cells of the **intercalary meristematic zone** located at the base of the terminal hair. The meristematic cells are small in size, have denser granular contents and pseudo-vacuoles.

Reproduction. *Rivularia* multiplies vegetatively by the formation **hormogonia** and rarely by the germination of the heterocysts. Desikachary (1946) reported the germination of an heterocyst in *R. manginii,* Asexual reproduction by spores and sexual reproduction are lacking.

Delimitation of hormogonia (Fig. 2.34). At the time of hormogone formation the trichomes usually shed the hair. The delimitation of hormogonia starts in the meristematic zone and gradually extends to the basal portion of the trichome. They are delimited singly or in long series (A) by the formation of **biconcave separation discs**. Each hormogonium consists of two to several living cells (B). On germination one of the two end cells of a hormogonium becomes modified to form the basal heterocyst (*c*). All others divide repeatedly in one plane. The daughter cells grow to form the trichome (D).

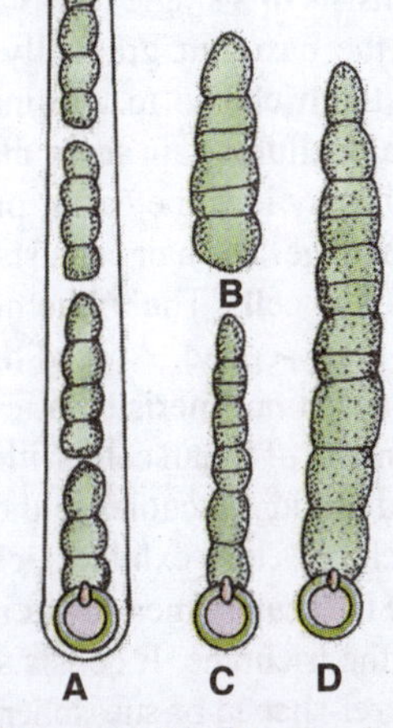

Fig. 2.34 (A-D). *Rivularia joshi*, Vasishta. (A) Portion of a trichome with hormogonia in a series; (B-D) stages in the germination of hormogones.

Taxonomic Position :

Division	:	**Cyanophyta** (Cyanophycophyta)
Class	:	**Cyanophyceae**
Tribe	:	**Hormogoneae**
Order	:	**Rivulariales**
Family	:	**Rivulariaceae**
Genus	:	*Rivularia*
Species	:	*mehrai*

2. GLOEOTRICHIA

Habitat. *Gloeotrichia* like *Nostoc* is a colonial form comprising about 10 species (Desikachary, 1959). The colonies occur in fresh water. Mostly they are found free floating on the water surface in sunny ponds, pools and paddy fields. A few species occur attached to the plants submerged in water.

Structure of the colony (2.35 A). With few exceptions the colonies are macroscopic thalli of mucilage with a firm boundary. The latter gives definite shape to the colony. The colonies vary in size, shape and are of various tents. Usually they are soft, spherical or hemispherical or ellipsoid to an irregularly lobed gelatinous mass. Hyaline when young but turn yellow or yellowish brown to dark brown or even olive-green towards maturity. The number of trichomes in the colony varies with the species from a few (40+5) to numerous. These are radially arranged within the colony (C) with a more or less parallel configuration. The soft mucilaginous matrix in which they are embedded is secreted by the cells constituting the trichomes.

Structure of the trichome (Fig. 2.35 B). The trichomes are straight or slightly curved. Each consists of a single row of cells. The cells are broad at the base but gradually taper in the upper part of the trichome to terminate in a long colourless, multicellular hair at its distal end. The tips of the hair may in some cases, project beyond the colony mucilage. The hair consists of elongated, cylindrical hyaline cells. The trichome below the hair is a few to many-celled. According to Schwendener there is a distinct meristematic zone below the hair. It consists of small cells with dense, granular contents and pseudo-vacuoles in the planktonic species. The trichomes thus exhibit **trichothallic growth**. Below the intercalary meristematic zone is the basal region of the trichome. It consists of comparatively larger barrel-shaped or sub-spherical cells usually as long as broad. The lowermost cell of the trichome is modified to form the basal one-pored heterocyst with the pore directed toward the trichome. It is spherical to ellipsoidal, sometimes conical in form with homogeneous contents. Rarely there is a pair of one-pored basal heterocysts. Intercalary heterocyst have also been recorded. At the base of the trichome next to heterocyst is generally a single large spore or akinete (D) sometimes there may be two or more.

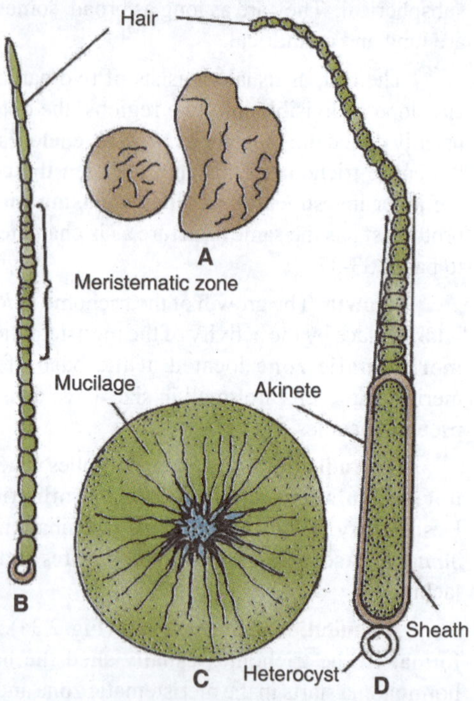

Fig. 2.35 (A-D). *Gloeotrichia* sp; (A) colony; (B) developing trichome with a basal heterocyst; (C) section of the colony showing gradual arrangement of filaments; (D) filament with a basal heterocyst and an akinete above it.

Each trichome in the colony has its own sheath. The trichome with its enclosing sheath is called the **filament**. The individual sheaths of the trichomes, being colourless, nonlamellated and confluent with one another, are not distinctly seen in the purely vegetative condition (prior to spore formation).

Cell Structure. The cell has the same structure as is characteristic of a typical cyanophyte cell. For details refer to pages 33-37.

Reproduction. Akinete formation is the only common method of asexual reproduction and perennation in *Gloeotrichia.*

Akinete Formation. In the vegetative state, the basal region of the trichome is the broadest. The cells constituting it do not divide. With the onset of akinete formation, the lowermost cell, which is next to the heterocyst, increases in size and divides. The process is repeated. Typically, the lowermost of the recently divided cells is destined to form the spore (akinete). It is next to the basal heterocyst. The cell marked out to become the akinete enlarges considerably in size, accumulates food reserves and elongates to many times the length. Eventually it becomes converted into an akinete (spore) by secreting a thick highly resistant *spore wall* or *envelope*, which is often yellow to brown in colour, external to the inner investment (parent cell wall). With the initiation of akinete development, a smooth or ornamented often pigmented thick sheath begins to be formed in the basal region of the filament around the developing akinete. When fully formed the sheath appears like a thick protective covering around the akinete. It is firm. It gelatinizes only at the surface. In colour it may be hyaline or yellow to yellow brown.

The fully developed akinete of *Gloeotrichia* is a large, cylindric to ovoid structure with a thick smooth sometimes ornamented wall. It is much more resistant to shortage of water and unfavourable temperature than are the vegetative cells. Generally a single akinete is formed next to the basal heterocyst. Sometimes two or more may be formed in succession.

Germination of Akinete. The akinete germinates on the onset of conditions favourable for growth. It absorbs moisture. The thick resistant wall softens and the protoplast awakens to activity. The latter divides by a transverse wall into a 2- celled germling which by further divisions in the same plane becomes many-celled. It is still enclosed within the old akinete wall. When the germling attains a certain length, it escapes through a localized rupture of the akinete wall caused either by dissolution or separation of a portion of the akinete wall. Generally the inner investment is retained as the cell wall of the germling. The liberated germling breaks up into a number of hormogones which become bundled together. In the course of time each hormogone becomes transformed into a typical filament of *Gloeotrichia.*

Taxonomic Position:

Division	:	**Cyanophyta** (Cyanophycophyta)
Class	:	**Cyanophyceae**
Tribe	:	**Hormogoneae**
Order	:	**Rivulariales**
Family	:	**Rivulariaceae**
Genus	:	*Gloeotrichia*
Species	:	*indica*

DISCUSSION AND CONCLUSION

Systematic Position of Blue-green Algae. The question of taxonomic position of blue-green algae in the plant kingdom is most interesting. They, like bacteria, have a simple structure which indicates their primitive (prokaryote) nature. Besides they include many colourless forms which bear a superficial resemblance to some bacteria. Both reproduce almost entirely by asexual means and largely by the method of **fission.** On the basis of these resemblances, Cohn (1871-72) suggested that the blue-greens (**schizophyceae**) and bacteria (**schizomycetes**) be considered as two classes of phylum, **Schizophyta** meaning **fission plants**. Later Stainier and van Niel (1941) supported Cohn's viewpoint on the basis of three common features which the two groups share between them. These are absence of a distinct nucleus, lack of plastids and absence of sexual reproduction. Pringsheim (1949), an opponent of this viewpoint, questioned the significance of negative features which two groups share between them. On the other hand, he pointed to the differences between the blue-greens and the bacteria.

Recent investigations with electron microscope have revealed the presence of several positive features which the two groups share between them in addition to the three negative features of Stainier and van Niel. Echlin and Morris (1965) who reviewed in detail the common features between the blue-greens and bacteria club them together and suggest their separation from all other cellular organisms. There are others who place them in a separate kingdom, the **Monera** with the blue-greens forming phylum Cyanophyta and bacteria phylum Schizophyta.

Majority of the botanists still do not contribute to the above views. They point out to the significant difference between the two. The blue-greens possess chlorophyll like other algae whereas bacteria lack chlorophyll. Oxygen is liberated in blue-green algal photosynthesis but not in that of bacteria. The blue-greens mostly have a thallus-like plant body which is either a filamentous or a non-filamentous colony whereas in the bacteria the thallus in predominantly a unicell. Because of these differences besides others they favour the inclusion of blue-greens among the algae as a separate class known as Cyanophyceae or Myxophyceae. The modern algologists, however, consider that there is no near or remote kinship between the blue-greens and the other algae. The evolutionary differences between them are so great that the blue-greens should be considered a separate phylum, the Cyanophyta comprising a single class Cyanophyceae.

Christensen (1962, 1964) proposed a scheme for primary classification of algae into **Prokaryota** (cyanophyta) and **Eukaryota** (all other algae except the blue-greens) on the basis of the difference between prokaryotic and eucaryotic cell. To Ian Morris (1968) the terms Prokaryota and Eukaryota appear excellent as basis for primary classification of all cellular organisms and not only algae.

Origin and Affinities of Blue-green Algae. The affinities of Cyanophyceae with other groups of plants are uncertain. Except for the presence of certain photosynthetic pigments and thallus, in many forms, being either a many-celled filament or a colony, they have little in common with any other divisions of algae. Some algologists suggest a relationship between the *blue-greens* and *red algae* on the basis of the following features which the two share between themselves:-

(*i*) Presence of chemically similar phycobilin pigments which are phycocyanin and phycoerythrin and absence of chlorophyll b in both.

(*ii*) Lack of motile cells. The absence of flagella places the blue-greens apart from every other algal group except the red algae.

(*iii*) Presence of inter-connecting wall pores in some genera of both groups.

(*iv*) Storage of starch variants as reserve food in both. In the cyanophytes it is cyanophycean starch and in red algae it is floridean starch.

(*v*) Occurrence of unicells and false branching in certain blue-green algae and primitive protoflorideae.

On the basis of the above-mentioned similarities, some algologists suggest a close relationship between the two groups. The suggestion is that either both these algal groups have originated from some non-flagellated common, unknown ancestor or the red algae evolved from some cyanophycean ancestor by way of primitive protoflorideae in which unicells and false branching occurred. The ardent supporters of this view suggest *Porphyridium* as a primitive red alga to have descended from a blue-green ancestor, from forms like *Porphyridium* evolved *Porphyra*. The higher red algae (Floridae) in turn evolved from the Bangiales.

This hypothesis of close relationship between blue-green and red algae is based on very inadequate grounds. The opponents of this view point out that there is no evidence of similarity in the cell structure of blue-greens and red algae. In addition the highly specialised thallus of red algae and their elaborate methods of sexual reproduction are features which markedly set them off from the blue-green algae. In the face of all these differences the probability of a direct affinity between these two groups of algae is very small. The possibility, however, is that both could have evolved from a common stock such as a non-flagellated ancestor in the remote past.

Many algologists consider Cyanophytes to be the simplest and the most primitive among the algae. They argue that they have no organized nucleus and membrane restricted plastids in the cell. The gene containing DNA-protein forms one or two clumps. These clumps do not contain chromo-somes of the type found in other organisms. The cytoplasm does not contain vacuoles and it undergoes no cyclosis and streaming movements characteristic of other cells. These negative features and absence of sexual methods of reproduction reflect the evolutionary antiquity of these simplest photosynthetic members of the plant kingdom. This viewpoint has been supported by the discovery of certain imprints in the oldest sedimentary rocks of the Precambrian, Archeozoic era. These imprints closely resemble some of the extinct forms of the Cyanophyta and lend support to the view that blue-greens are the oldest of all algae and appeared first. They had some representatives living on the planet some two to three billion years ago.

Similarities with Bacteria

The other equally simple and believed to be the ancient members of the plant kingdom are the bacteria with which the blue-greens have the following characteristics in common :-

(*a*) Both groups include members which are one-celled.

(*b*) Frequent development of sheaths in both.

(*c*) Absence of a membranes delimiting nuclear material and absence of membranes restricted organelles, endoplasmic reticulum, golgi bodies, mitochondria and true vesicles in the cytoplasm.

(*d*) Presence of diaminopimelic acid in the wall, substance lacking in all other algae but present in bacteria and blue-greens.

(*e*) The DNA rich chromatin in bodies either concentrated in the centre of the cell or scattered throughout the cell.

(*f*) Ability of some members of both the groups to withstand extreme desiccation and endure temperatures approaching boiling point of water.

(*g*) Ability of a few blue-green algae to live on dead organic matter like saprophytic bacteria in the absence of light.

(*h*) Ability of certain species to fix atmospheric nitrogen,

(*i*) Cell division not by the formation of a cell plate but by a ring-like (annular) inward extension of the cell membrane resembling a diaphragm with a decreasing aperture.

(*j*) Non-mitotic division of the nuclear material.

(*k*) Reproduction or multiplication by fission.

(*l*) Formation of resting cells or spores in both.

(*m*) Lack of motile spores.

The geological evidence also indicates that bacteria are as ancient a group of plants as the blue-greens, if not more. The geological evidence, the similarity of organisation (prokaryotic nature) of cell and other sufficiently large number of features in common with the photosynthetic bacteria support the view that the blue-greens are related to bacteria and the two had a common ancestry (Echlin and Morris, 1965). It is probable that blue-greens and the bacteria are possibly the descendants of ancient organisms in which elaboration and specialization of cellular structure had not progressed very far and they lacked chlorophyll. The evolutionary relationship of these two groups with each other is still not clear.

One group of phylogeneticists believes that the blue-greens and bacteria arose from a common prokaryotic ancestor. This hypothesis is supported by the long list of features common between them and is outlined as follows:

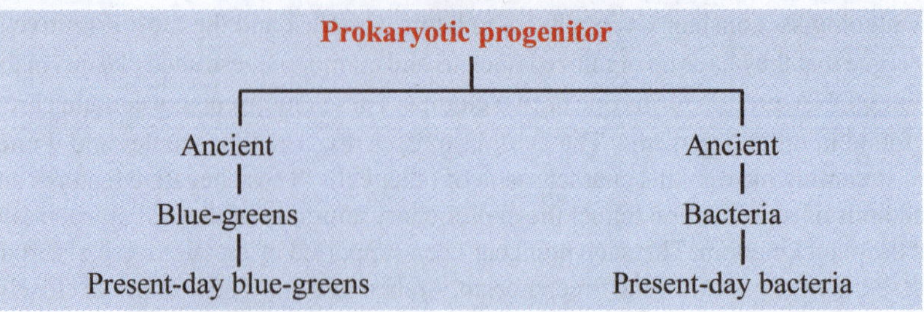

Another school of phylogeneticists advocates that the blue-greens and bacteria do not share a common ancestry. They have evolved independently of each other. The supporters of this view-point emphasize the great differences between the two groups which preclude any close affinities between them. These are:

1. The blue-greens are invariably a flagellate whereas many bacteria are flagellate.

2. All blue-greens are photosynthetic and use chlorophyll *a* to capture light energy. Only a few bacteria are photosynthetic but lack chlorophyll *a*.

3. In photosynthesis the blue-greens use hydroxyl ions derived from water electric donors and release oxygen as a by-product. In bacterial photosynthesis hydrogen sulphide is the source of hydrogen and by-product is sulphur and not oxygen.

4. Blue-greens are aerobic whereas many bacteria are anaerobic.

The advocates of the second hypothesis thus advance the following evolutionary scheme:

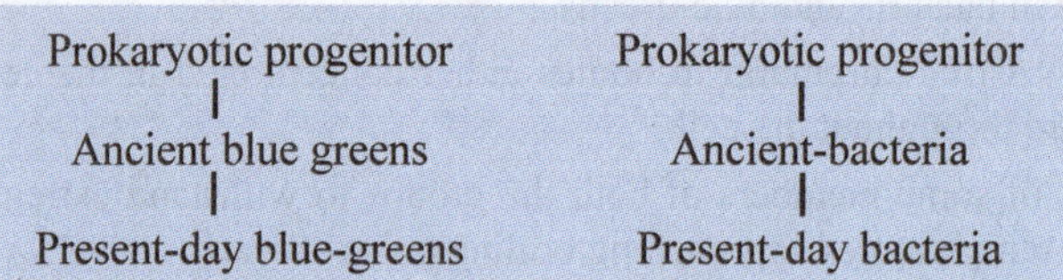

The third suggestion is that the blue-greens gave rise to bacteria by the loss of chlorophyll and adaptation to a saprophytic or parasitic mode of existence. The mechanism of photosynthesis and the photosynthetic pigments, however, suggest a relationship of cyanophytes with the eukaryotic algae.

Economic Importance

(1) The blue-green algae furnish food for fish and other aquatic animals. According to Chacko (1970), "*Oscillatoria* is the most favoured blue-green alga consumed by 56 species of fishes. Others in order of preference are *Spirulina, Anabaena, Microcystis, Lyngbya* and *Merismopedia*."

(2) The blue-greens add organic matter to the soil and increase fertility. Barren alkaline soils in India have been reclaimed and brought to a productive state by inducing a proper growth of certain Cyanophyceae. Decomposition of successive crops of these neutralize the alkalinity of the soil.

(3) Recent investigations have definitely proved that some of the blue-green algae increase the fertility of the soil by fixing atmospheric nitrogen. The important nitrogen-fixing blue-green algae are *Oscillatoria princeps, O. formosa* and some species of *Anabaena, Spirulina, Nostoc,* and *Cylindrospermum.* Some species of *Scytonema* and *Mastigocladus* also have the ability to fix atmospheric nitrogen. The list at present includes about 50 species which are known to the active nitrogen fixers (Stewart 1970 and Fogg et. al., 1973). Verma (1970) stated that the blue-greens form an important group of soil organisms which are of great agricultural importance because of the ability of some of them to synthesize organic substance as well as to fix atmospheric nitrogen. They thus

serve as an excellent source of nitrogen and organic matter. The practical application of these algae is the seeding of rice fields with the nitrogen-fixing species. This results in appreciable increase in yields of rice. Subrahmanyan (1970) writes, "In addition to supplying nitrogen, the nitrogen fixers improve the fertility status of the soil and the residual effect left after three crops was found enough to support ensuing crops".

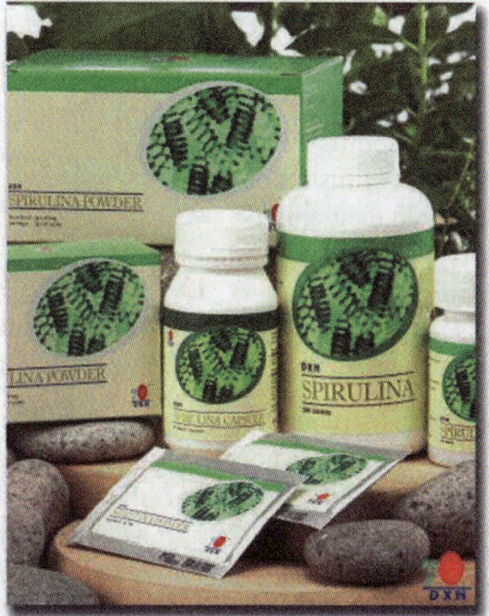

Spirulina Health Products.

(4) Marathe (1970) suggested that some blue-green algae bring about aggregation of loose, unbound soil particles and thus improve the physical texture of the soil which helps in checking soil erosion. The blue-green algae, therefore, play a role in soil conservation to certain extent.

The blue-green algae are of negative value as well. They contaminate water supplies by producing colour, odour and giving a fishy taste to the drinking water. Their growth can, however, be checked by addition of very small quantities of copper sulphate and dichlorophen to the water. Some of the blue-green algae are poisonous to fish and domesticated animals. Examples are *Microcystis, Anabaena; Aphanizomenon,* etc. The active principle in *Microcystis* has been reported to be **neurotoxin.** Chacko (1970) reported that thick blooms caused by *Microcystis, Nostoc, Anabaena, Oscillatoria*, and some other blue-green algae bring about depletion of oxygen resulting in large-scale death of fish and other aquatic animals.

Harmful *Microcystis* bloom.

QUESTIONS

ESSAY TYPE

1. Describe the range of thallus structure in Cyanophyceae.
2. Describe the different modes of reproduction met within Cyanophyceae.
3. Give an account of cell structure and reproduction in Cyanophyceae. Illustrate your answer with neat labelled diagrams.
4. Give an account of the habit and habitat of *Nostoc*. Explain how its form and structure is adapted to the environmental conditions.
5. Give an account of the thallus structure and life history (reproduction) of *Nostoc*.
   (*Poorvanchal, 2003; Punjab, 1992*)
6. List the features which indicate that the blue-green algae are primitive and ancient members of the plant Kingdom. In what respects do they resemble the bacteria?
7. "Heterocyst is popularly called a botanical enigma". Discuss the statement in the light of its structure and function.
   (*Kanpur M.Sc. 2006;Gorakhpur, 1991; Punjab 1991*)
8. Describe the distinguishing features of Cyanophyceae. Illustrate the structure of a cell of a cyanophycean algae under electron microscope. (*Gorakhpur, 1992; Himachal Pradesh, 1994*)
9. Discuss habitat, structure and reproduction of *Oscillatoria.*
   (*Kanpur, 1997; Allahabad, 1997; Bundelkhand, 1999*)
10. Give an account of the habitat, structure and reproduction of *Nostoc.*
    (*Kanpur, 1998; Himachal Pradesh, 1997*)
11. Identify the following giving their important characters with the help of diagrams:
    (*a*) *Oscillatoria*
    (*Kanpur, 1998, 1999: Awadh, 1998*)
    (*b*) *Nostoc*
    (*Allahabad, 1991; Bundelkhand, 1998; Rohilkhand, 1991, 1996; Bihar, 1991*)
    (*c*) *Rivularia*
    (*d*) *Gloeotrichia*
    (*e*) *Scytonema.*
    (*Awadh, 1997; Bihar, 1990*)
12. What are heterocysts? Describe their structure and function in detail.
13. Describe the structure of thallus and modes of reproduction in *Oscillatoria* or *Rivularia.*
14. What do you know about the distinguishing characters of Cyanophyceae? By means of labelled diagrams only show the structure of cyanophycean cell under light and Electron microscope. (*Lucknow, 1997*)
15. Give the distinguishing characters of Cyanophyceae and describe the structure of the cell and reproduction in any type which you have studied.
    (*Awadh, 1995; Allahabad, 1994; Rohilkhand, 1993*)
16. What is the cell type of Bacteria and blue-green algae called? Give an account of Morphological features, physiological functions and methods of reproduction in any of the type studied by you.
    (*Agra, 1995*)
17. Describe the various methods of reproduction in *Nostoc* and *Oscillatoria.*
    (*Punjab, 1993*)
18. Give an illustrated account of the ultrastructure of any prokaryotic algal cell. (*Bundelkhand, 1996*)
19. Write an essay on economic importance of Blue-green algae. (*MDS Univ., 1998*)
20. Describe the different modes of reproduction in Cyanophyceae.
    (*Bhagalpur, 1990*)

21. Give an account of the ultra structure of a cell of a blue green algae studied by you. Why are blue-green algae considered to be most primitive of all algae? *(Bhagalpur, 1996)*

22. In which class you will place the minute plant which lacks mitochondria, Chloroplasts, true nucleus 70 S ribosomes but contain chl. a as photosynthetic pigment.

23. Write general characters of blue-green algae and also justify the statement that they are close to Bacteria. *(Kanpur M.Sc., 2002)*

24. Are Cyanophyceae algae? Justify. *(Kanpur M.Sc., 2006)*

SHORT ANSWER TYPE

25. Write short notes on :

(*i*) Movements in *Oscillatoria* or oscillatory movement *(Kanpur M.Sc. 2002, Gorakhpur, 1991, 1994; Lucknow, 1995)*

(*ii*) Akinete *(Kanpur, 2007, Awadh, 1995; Allahabad, 1994, 1996; Bundelkhand, 1997)*

(*iii*) Branching in *Scytonema* *(Gorakhpur, 1993, 1997; Kanpur, 1995, 1997; Awadh, 1991; Magadh, 1993)*

(*iv*) Structure of a cell in *Oscillatoria*/ *Nostoc* *(Poorvanchal 2005; Gorakhpur, 1993; Himachal Pradesh, 1990, 1992; Meerut, 1998, 1999)*

(*v*) Structure of mature heterocyst *(Gorakhpur, 1995, 1998)*

(*vi*) Hormogonium of *Oscillatoria*/ Hormogones *(Allahabad 2005; Gorakhpur, 1998; Agra, 1992; Punjab, 1996, 1998; Bundelkhand, 1993)*

(*vii*) Reproduction in Cyanophyceae *(Gorakhpur, 1999; Awadh, 1993, 1997)*

(*viii*) Cell structure in Blue-green algae *(Poorvanchal, 2004; Kanpur, 1995, 1997; Allahabad, 1992)*

(*ix*) Economic importance of Blue-green algae/Cyanophyceae. *(Lucknow M.Sc., 2006; Kanpur, 1995; Awadh, 1991, 1997; Bundelkhand, 1999)*

(*x*) Heterocysts *(Lucknow, 1992, 1994, 1995,1998; Awadh, 1994, 1997, 1998; Allahabad, 1993, 1995, 1996, 1999, 2004, 2005; Agra, 1992, 1994; Bundelkhand, 1994, 1997, 1999; Rohilkhand, 1991, 1992, 1995; MDS Univ, 1998; Kumaon 1999; Magadh, 1993)*

(*xi*) Nuclear structure of *Nostoc* *(Kanpur, 2007, Poorvanchal, 2004, 2005, 2007)*

(*xii*) Ultrastructure of cyanophycean cell *(Poorvanchal, 2003, Kanpur M.Sc. 2005, Lucknow, 1993; Allahabad, 1998)*

(*xiii*) *Gloeotrichia* *(Lucknow, 1996)*

(*xiv*) Endophytic Blue-green algae *(Awadh, 1996, Kanpur, 2007)*

(*xv*) Nitrogen fixation by Blue-green algae *(Awadh, 1999; Kanpur, 2007; Kanpur M.Sc., 2007)*

(*xvi*) Use of Blue-green algae as fertilisers

(*xvii*) Akinete and heterocysts *(Lucknow, 2002)*

(*xviii*) Fertile filament of *Gloeotrichia* *(Lucknow, 2007)*

(*xix*) Pigments and food reserves of blue-green algae. *(Allahabad, 2002)*

(*xx*) Ultrastructure of *Nostoc cell.* *(Allahabad, 2006)*

26. Describe asexual reproduction in *Oscillatoria.* *(Kanpur, 2005)*

27. Write short notes on heterocyst and akinetes. *(Kanpur, 2005)*

28. Draw a portion of *Nostoc* trichome and explain why *Nostoc* balls are called fallen stars. *(Kanpur, 2003)*

29. What is an heterocyst? *(Lucknow, 2001)*

30. Draw neat and labelled diagrams of the following and mention their characters briefly.

(*a*) *Microcystis* *(Lucknow, 2004, 2007)*

(*b*) *Oscillatoria*

(*c*) *Nostoc*

31. Describe the following:

(*a*) Reproduction-blue-green algae. *(Lucknow, 2005)*

(*b*) Reasons for considering blue greens as algae or as Cyanobacteria *(Lucknow M.Sc., 2005)*

(*c*) Affinities of blue-green algae *(Allahabad, 2002)*

32. Name the alga of Cyanophyceae which show false dichotomy. *(Kanpur M.Sc., 2002)*

33. Name an alga which helps in nitrogen fixation. *(Allahabad, 2001)*

34. Why the members of blue-green algae are known as Cyanobacteria. *Poorvanchal, 2003)*

35. Name the sites where N_2 fixation takes place in blue greens. *(Poorvanchal, 2003)*

36. Draw labelled diagrams of the following:

(*i*) False branching in *Scytonema* *(Gorakhpur, 1994; Awadh, 1992)*

(*ii*) A fertile trichome of *Gloeotrichia* *(Gorakhpur, 1996, 1997)*

(*iii*) Ultrastructure of Myxophycean cell *(Poorvanchal, 2003; Lucknow, 1994, 1996, 1999; Awadh, 1992)*

(*iv*) Structure of *Oscillatoria* *(Awadh, 1994)*

(*v*) Cell structure in Cyanophyceae *(Bundelkhand, 1993)*

(*vi*) *Microcystis* *(Lucknow, 2004)*

37. In what respects do Cyanophyceae resemble bacteria?

38. In what respects the following are different from each other. Or Distinguish between:

(*i*) *Oscillatoria* and *Gloeotrichia*

(*ii*) *Nostoc* and *Anabaena* *(Lucknow M.Sc., 2003)*

(*iii*) Chlorophyceae and Cyanophyceae *(Kanpur, 2007)*

(*iv*) *Scytonema* and *Rivularia*

(*v*) Akinetes and Heterocysts *(Lucknow, 2006)*

(*vi*) Nostoc and *Oscillatoria*

39. Give reasons for the following:

(*i*) It is not necessary to add nitrogen fertilizers to paddy fields.

(*ii*) Although all the blue-green algae have a blue pigment, they appear as red or yellow brown.

(*iii*) Blue-green algae are considered primitive of all the algae.

(*iv*) Blue-green algae can thrive in relatively dry places.

40. (*a*) Explain hormogones and Heterocysts in *Nostoc*. *(Himachal Pradesh, 1992)*

(*b*) Reproduction in *Gloeotrichia* *(Lucknow, 2003)*

41. Blue-green algae can thrive in relatively dry places as compared to other algae. What structural peculiarities make this possible? *(Himachal Pradesh, 1996)*

42. Describe the general characteristics of Cyanophyceae. *(Rohilkhand, 1993, 1994)*

43. Describe the plant body of *Oscillatoria*. *(Bharathiar, 1995)*

44. Describe the reproduction in *Oscillatoria/ Nostoc.* *(Bharathian, 1995; Rohilkhand, 1996; Bhagalpur, 1995)*

45. Discuss briefly the thallus structure of *Scytonema.* *(Bhagalpur, 1991)*

OBJECTIVE TYPE

46. Fill in the Blanks:

(*i*) The commonest mode of Reproduction in *Oscillatoria* is by ....

(*ii*) In Cyanophyceae, the site of nitrogen fixation is ....

(*iii*) False branching is characteristic feature of ....

(*iv*) Water blooms are generally formed by ....

(*v*) Myxophyceae are so named because of ....

(*vi*) Plastids are absent in ....

(*vii*) *Nostoc* does not reproduce ....

(*viii*) In Algae, the prokaryote cell is present in the members of ....

(*ix*) The name of an alga which helps in nitrogen fixation in ....

(*x*) The prokaryotic alga having heterocyst is ....

(*xi*) The genetic apparatus is not bound by membranes in members of ....

(*xii*) The two principal components of cell wall of blue green algae are .... and ....

(*xiii*) The reserve food material in Cyanophyceae is ....

(*xiv*) The alga in which basal heterocyst and akinete are present is ....

47. Select the correct answer:

(*i*) Which of the following has a single polar nodule in the heterocysts?

(*a*) *Volvox*
(*b*) *Oscillatoria*
(*c*) *Gloeotrichia*
(*d*) *Scytonema*

(*ii*) The photosynthetic apparatus, the respiratory apparatus and the genetic apparatus is not bound by membranes in

(*a*) Chlorophyceae
(*b*) Myxophyceae
(*c*) Rhodophyceae
(*d*) Phaeophyceae

(*iii*) Which of the following has prokaryotic cells?

(*a*) *Chlamydomonas*
(*b*) *Volvox*
(*c*) *Polysiphonia*
(*d*) *Nostoc*

(*iv*) The alga in which basal heterocyst and akinete is present is ....

(*a*) *Oscillatoria*
(*b*) *Gloeotrichia*
(*c*) *Nostoc*
(*d*) *Chlamydomonas*

(*v*) False branching is characteristic of

(*a*) *Scytonema* (*b*) *Gloeotrichia*
(*c*) *Nostoc* (*d*) *Anabaena*

(*vi*) Vegetative reproduction through hormogonia is present in

(*a*) *Oscillatoria*
(*b*) *Scytonema*
(*c*) *Gloeotrichia*
(*d*) *Volvox*

(*vii*) Plastids are absent in the members of class

(*a*) Chlorophyceae
(*b*) Phaeophyceae
(*c*) Rhodophyceae
(*d*) Myxophyceae

(*viii*) Phycocyanin-c is the dominant pigment in

(*a*) *Volvox*
(*b*) *Ectocarpus*
(*c*) *Polysiphonia*
(*d*) *Oscillatoria*

(*ix*) The members of the class Myxophyceae are characterised by

(*a*) Presence of motile spores and gametes

(*b*) Sexual reproduction
(*c*) Fusion and meiosis
(*d*) Absence of motile reproductive bodies and incipient nucleus.

(*x*) The algologist who has done extensive research on cyanophycean algae is
(*a*) F.E. Fritsch
(*b*) R.N. Singh
(*c*) M.O.P. Iyengar
(*d*) R.S. Randhawa

(*xi*) The dominant pigment of blue-green algae is
(*a*) Phycoerythrin
(*b*) Xanthophyll
(*c*) Phycocyanin
(*d*) Fucoxanthin

(*xii*) *Scytonema* usually reproduces by
(*a*) Hormogonia and Endospores
(*b*) Hormogonia and Akinetes
(*c*) Heterocysts and Akinetes
(*d*) Heterocysts and Endospores

(*xiii*) The outer portion of the protoplast in a cyanophycean cell is called chromoplast because of the presence of
(*a*) Coloured plasma membrane
(*b*) Chromosomes
(*c*) Pigment
(*d*) Chromoplast

(*xiv*) Psuedovacuoles in Cyanophyceae help as organs to
(*a*) Regulate buoyancy
(*b*) Store food
(*c*) Store oxygen
(*d*) Store CO_2

(*xv*) In *Rivularia* growth of the filament is due to the activity of
(*a*) An apical cell at the apex of the hair
(*b*) Intercalary meristem at the base of the hair
(*c*) All cells of the trichomes
(*d*) Basal heterocysts

CHAPTER 2i

DIVISION : PROCHLOROPHYTA (*Pro* = *Primitive*)

- Distinguishing features

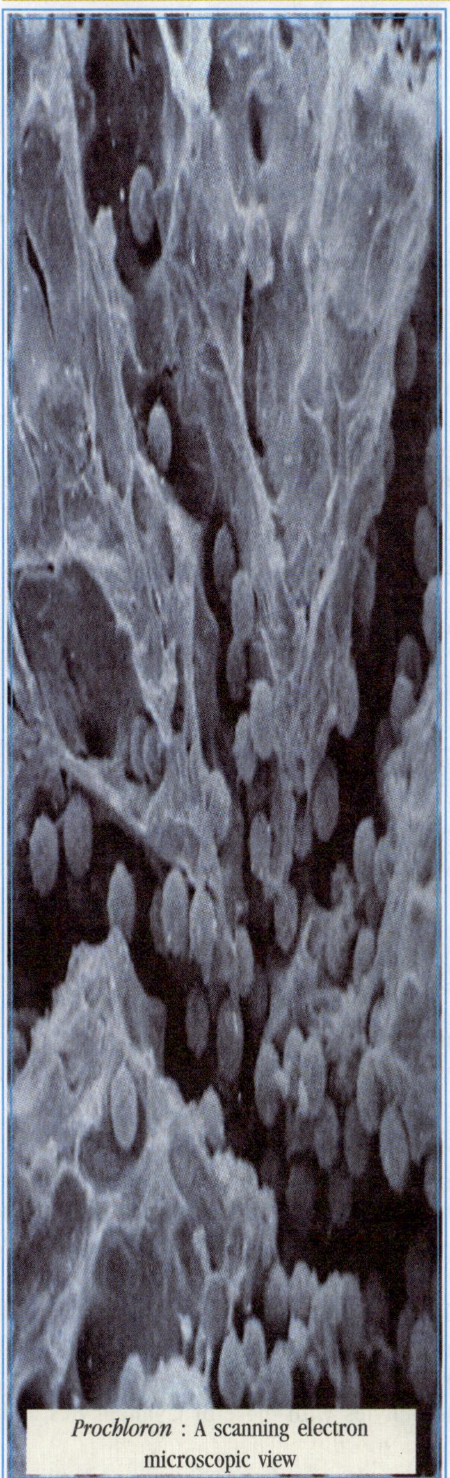

Prochloron : A scanning electron microscopic view

During 1976-77, Lewin while studying unicellular green prokaryotic symbiont of didemnid ascidians from tropical and subtropical marine seashores and in coral reef areas on Mexican coasts, discovered an alga which neither has all the characters of Cyanophyta nor the Chlorophyta. These algal symbionts were given a new generic name *Prochloron* and to include this alga he created a new division between Cyanophyta and Chlorophyta and named it as **Prochlorophyta** with a single class **Prochlorophyceae**. The class is presently represented by two genera ***Prochloron*** and ***Prochlorothrix***.

DISTINGUISHING FEATURES

(*a*) Thallus single celled or filamentous

(*b*) Cells prokaryotic

(*c*) Plastids are absent

(*d*) Thylakoids present, paired

(*e*) Chlorophyll-b present

(*f*) Phycobilinosomes are absent

(*g*) Biliproteins are absent

(*h*) Reproduction not yet reported

Prochloron didemni is a typical Prochlorophyte and resembles blue-green algae (Cyanobacteria) in its general cell structure but differs in pigmentation and lack phycobilinosomes and phycobiliproteins. Besides, it is reported to contain both chlorophyll *a* and *b* along with β-carotene and resembles with the eukaryotic green algae.

Prochloron is not found in free living state in nature but can survive only when associated with larvae of symbiotic didemnid **[Lewin & Cheng (1989) & Lewin (1994)]**. The cells are not endosymbiotic but are epizoic and appear to be obligate symbionts where they are intimately integrated with their host tissues. It is one of the largest known unicellular prokaryotes having spherical cells measuring upto 30 μm in diameter. Cells have stacked thylakoids, one or more vacuoles which occupy the major parts of cells. Studies by **Lewin (1993)** revealed that prochlorophytes may have developed chlorophyll 'b' and stacked thylakoids independently of the chlorophyta and other land plants. The cells of *Prochloron* in living state are highly turgid and refractile but on death they shrink, become translucent and release pinkish exudate which has refractive index greater than that of sea water **(Lewin, 1994)**, and have absorption spectrum arround 334 nm, a wavelength absorbed by certain UV-B absorbing compounds *e.g.,* polythenol, polythine and mycosporin. **Lewin (1994)** is of opinion that the transfer of glucose or maltose present in the vacuolar sap of *Prochloron* cells may be one of the reasons to sustain the symbiotic association with the didemnid ascidians.

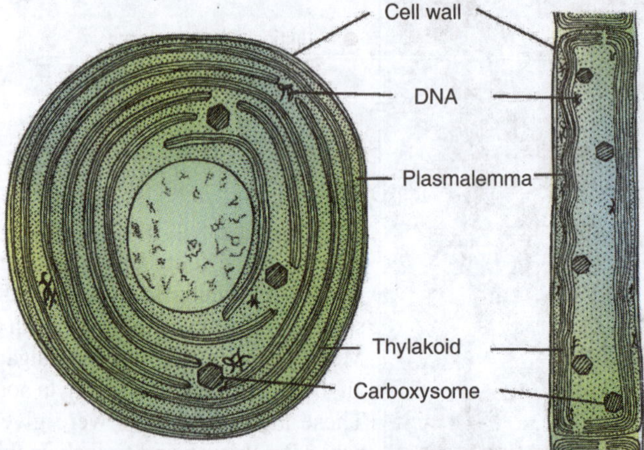

Fig. 2.1(*i*). Structure of Prochlorophytes: A, *Prochloron* B, *Prochlorothrix*, a filamentous Prochlorophyte.

Another Prochlorophyte *Prochlorothrix* was reported by **Burger-Wiersman (1986)** which is a free floating phytoplanktonic filamentous form in a fresh water lake in the Netherlands. It also has chlorophyll 'a' and 'b' and lack phycobilin. **Pinevich et. al., (1999)** also reported it from a lake in Norway and cultured it.

The current phycologists are of the the opinion that the prochlorophytes may not be a natural assemblage. Their members may be scattered throughout the cyanobacteria. They might have developed chlorophyll *b* during the course of evolution and lost phycobilinosomes and biliproteins.

A comparison of general characters of Prochlorophytes with those of Cyanophytes and chlorophytes has been given below:

	Characters	Cyanophyta	Prochlorophyta	Chlorophyta
1.	Cell type	Prokaryotic	Prokaryotic	Eukaryotic
2.	Plastids	Absent	Absent	Present
3.	Thylakoids	Single	Paired	Paired
4.	Chlorophyll - b	Absent	Present	Present
5.	Phycobilinosomes	Present	Absent	Absent
6.	Biliproteins	Present	Absent	Absent

QUESTIONS

ESSAY TYPE

1. Describe the details of the division Prochlorophyta.
2. Why was Prochlorophyta separated from Cyanophyta and Chlorophyta? Give reasons.
3. Compare fine structure of cells found in Cyanophyceae and Prochlorophyceae. *(Lucknow M.Sc., 2002)*

SHORT ANSWER TYPE

4. **Write short notes on**
   (*a*) Prochlorophyta
   (*b*) *Prochloron* *(Lucknow, 2002)*.
   (*c*) *Prochlorothrix*
   (*d*) Prochlorophyceae. *(Lucknow M.Sc., 2003)*
5. **Differentiate between**
   (*i*) Chlorophyta and Prochlorophyta
   (*ii*) Prochlorophyta and Cyanophyta
   (*iii*) Prochlorophyceae and Cyanophyceae. *(Lucknow, 2005)*

OBJECTIVE TYPE

6. (*i*) Algae which do not resemble with either blue-green and green algae but have some characters of both have been placed under division
   (*a*) Rhodophyta
   (*b*) Cyanophyta
   (*c*) Prochlorophyta
   (*d*) Phaeophyta

   (*ii*) *Prochloron* and *Prochlorothrix* have been placed in the Division
   (*a*) Cyanophyta
   (*b*) Chlorophyta
   (*c*) Rhodophyta
   (*d*) Prochlorophyta

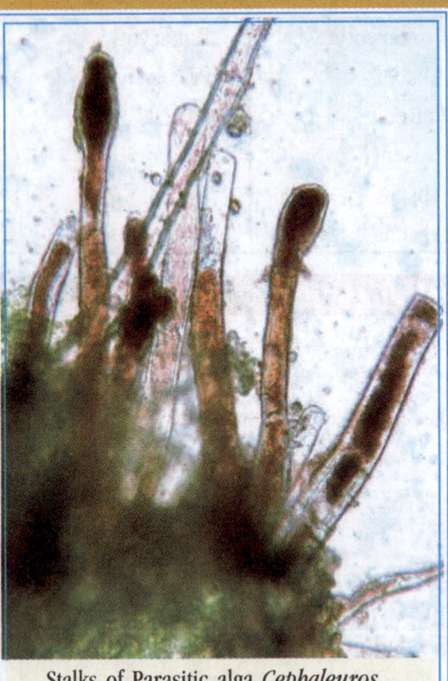
Stalks of Parasitic alga *Cephaleuros* Virescens

Cephaleuros Virescens on *Memecylon umbellatum*

CHAPTER 3

DIVISION : CHLOROPHYTA (Chlorophycophyta) Class : Chlorophyceae

- Occurrence
- Range of Vegetative structure or Organisation of Thallus
- Reproduction

The division Chlorophyta more appropriately called Chlorophycophyta includes a large number of species. Prescott makes as many as 20,000 species. These are included in a single class Chlorophyceae. The Chlorophyta and the Chlorophyceae have the same features. The cells constituting the thallus are eukaryotic and thus contain all the membrane bound cell organelles such as the definitely organised nucleus, plastids, mitochondria, dictyosomes, endoplasmic reticulum, and true vesicles. The thallus is typically green in colour due to the presence of a green pigment, the **chlorophyll**. It is contained in plastids called the **chloroplasts**. Embedded in the chloroplasts are rounded, proteinaceous bodies one or more in number, the **pyrenoids**. The pyrenoids are intimately associated with the elaboration of starch, which is the principal storage product. The cytoplasm contains **vacuoles**. Some of these may be contractile. The motile cell of the primitive forms contains an **eyespot** or **stigma**. The presence of membrane bound chloroplasts, a sap cavity and a definite nucleus in the cell are the chief characteristics besides others which distinguish the Chlorophyta from the Cyanophyta. The reserve carbohydrates are usually stored in the form of **starch**. The cell wall invariably contains cellulose. Unlike the blue-greens most of the green algae produce motile reproductive bodies generally furnished with two to four flagella. The flagella are of equal length and of whiplash type. They are inserted at the anterior end. Occurrence of sexual reproduction is another feature which distinguishes the Chlorophyceae from the Cyanophyceae.

OCCURRENCE

Most of the Chlorophycease are aquatic but some are subaerial. The subaerial forms are generally found in moist situations and form common constituents of the soil flora and also occur on moist rocks of cliffs and on damp wood work. A few species are epiphytic on the larger algae such as sea weeds or on the bark of shaded sides of trees (*Protococcus)*. Of the aquatic forms, majority occur in fresh cold water where they seem to have had their major evolution, if not origin. They occur both in standing and flowing waters and may be attached or planktonic (free-floating). They either form greenish scum on the surface of quiet or stagnant water or grow firmly attached to the submerged rocks, pieces of wood and other objects in the water. Several species of the orders Ulvales and Siphonales are marine. They generally grow attached to the rocks in the water along the shore. A few attached forms are found at great depths. Some strains of *Chlorella* have been reported to be thermophilic. Species of *Chlamydomonas* and some Chlorococcales occur in snow. A few of the green algae are endophytic (*Coleochaete nitellarum*). Very few parasitic forms are known to occur among the green algae. Only one genus of green algae has been reported to cause considerably economic loss to the crops of tea and coffee. This is *Cephaleuros* belonging to the order **Chaetophorales** and family **Trentepohliaceae**. *Cephaleuros virescens* is the cause of '*red rust of tea*' in North East India and Assam. It causes serious damage to the tea plantation of *Thea siaensis* and attacks young leaves and shoots. The thallus is composed of interwoven and branched filaments that grow through the host tissue and sub-aerial branches tipped with groups of sporangia which produce zoospores. C. *coffeae* occurs as a parasite on *Coffea liberica* and causes much damage to the plantations. *C. parasitica* is another species that attacks tea plantations and causes virulent tea disease in the form of organic rust. Recently Joshi *et. al.* (1978) reported "*brown rust disease*" of grapes (*Vitis vinefera*) caused by *C. parasitica. C. virescens* has also been reported to cause damage to the piper plantations in some parts of Eastern India. There are instances of green algae which live epizoically on or endozoically within the bodies of lower animals. For example, *Chlorella* (Zoochlorellae) is found in *Hydra*, *Characium* on the antennae of mosquito and *Chlorella* beneath the scales of fish. A few green algae in association with the fungi constitute **lichens**.

Algal leaf spot on avocado caused by *Cephaleuros virescens*.

RANGE OF VEGETATIVE STRUCTURE OR ORGANISATION OF THALLUS

The Chlorophyceae are a heterogeneous group of plants exhibiting a wide range of the body plan (architecture). The plant body may be single-celled or many-celled. In size it varies from minute unicells no more than a micron or two in diameter to a few feet long, strand-like structure. Morphologically it may consist of motile or non motile unicells, motile and non-motile colonies, simple and branched filaments, siphonaceous and coenocytic filaments, foliaceous thalli, cushion-like thalli and highly evolved heterotrichous filament. However, it is always of a simple construction. It shows no differentiation into true root, stem and leaves. For this reason the plant body of the algae is called a **thallus**.

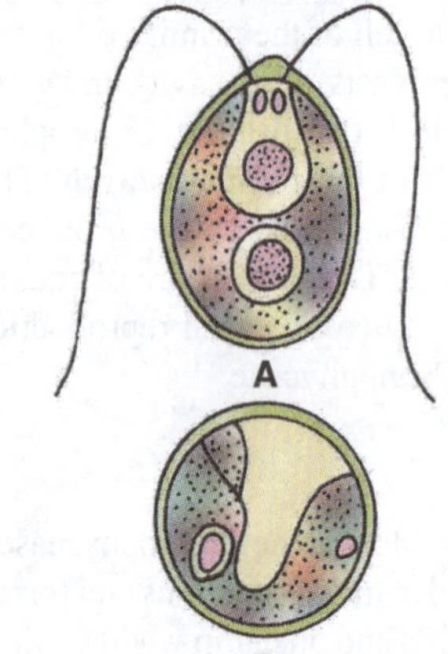

Fig. 3.1 (A-B). Chlorophyceae. Unicellular thallus. (A) motile unicell of *Chlamydomonas*; (B) non-motile unicell of *Chlorella*.

1. Unicellular thallus. In the simplest forms the thallus is a unicell. It is motile in some and non-motile or amoeboid in others.

(a) Motile Unicellular Thallus. Chlamydomonas (Fig. 3.1 A) is an example. The cell may be spherical, oval or pear-shaped. It is furnished with two flagella at its anterior end and has a single cup-shaped chloroplast situated in its posterior part. The single nucleus lies in the cavity of the cup surrounded by cytoplasm. There are two pulsating contractile vacuoles at its anterior end and a single red eye spot. *Chlamydomonas* has animal-like locomotion but is autophytic in its nutrition.

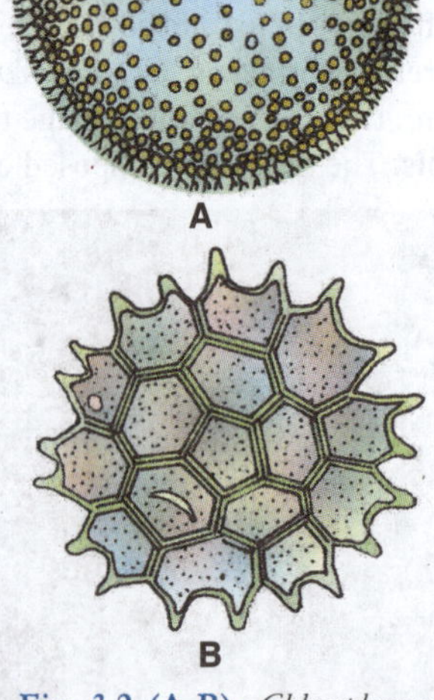

Fig. 3.2 (A-B). *Chlorophyceae.* Colonial thallus. (A) Motile coenobium of *Volvox;* (B), Non-motile coenobium of *Pediastrum.*

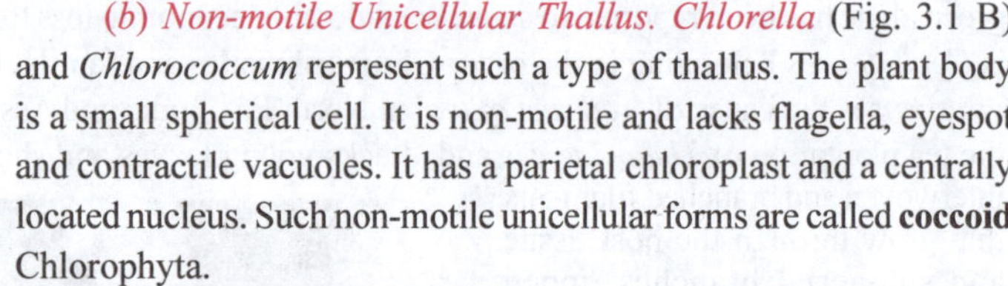

(b) Non-motile Unicellular Thallus. Chlorella (Fig. 3.1 B) and *Chlorococcum* represent such a type of thallus. The plant body is a small spherical cell. It is non-motile and lacks flagella, eyespot and contractile vacuoles. It has a parietal chloroplast and a centrally located nucleus. Such non-motile unicellular forms are called **coccoid** Chlorophyta.

2. Colonial thallus. Many of the green algae have a thallus consisting of a loose assemblage of cells mechanically held together generally in a gelatinous envelope. It is known as a **colony**. The cells in the colony have little or no dependence upon one another. The colonial thalli are of diverse forms. They may be plate-like or hollow spheres, motile or non-motile.

(a) Motile colonial thallus. It consists of a definite number of motile, unicellular *Chlamydomonas* or *Sphaerella* type cells held together in a common mucilaginous sheath. The individual cells are complete in themselves and independent of each other.

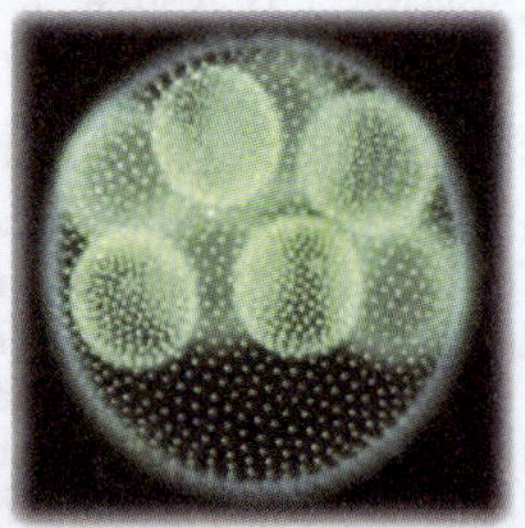

Colonial Thallus of *Volvox*.

They act together and bring about the movement of the entire colony. Examples are *Gonium* (Fig. 4.14 A-B), *Pandorina* (Fig. 4.17 A) and *Eudorina* (Fig. 4.20). Hollow spherical colony of *Volvox* (Fig. 3.2A) represents the most highly developed motile colonial thallus. It is made up of thousands of cells, which in some species are connected with one another by cytoplasmic strands.

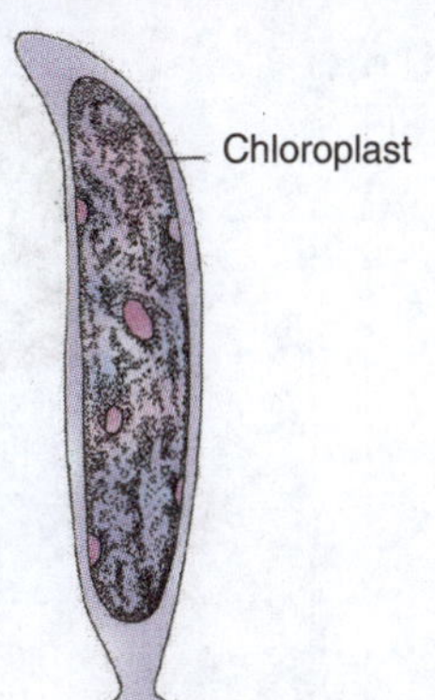

Fig, 3.3 (A). *Chlorophyceae.* Tubular uninucleate thallus of *Characium.*

(*b*) *Non-motile colonial thallus.* In certain members of the order Chlorococcales the small, non-motile cells are held together to form non-motile colonies with either a definite or an indefinite number of cells. They are free floating colonies. *Pediastrum* (Fig. 3.2 B) and *Hydrodictyon* (Fig. 5.4 A) are the important examples. *Hydrodictyon* represents the highest state of organisation of this type.

3. Siphonaceous or Coenocytic thallus. In this case the unicellular thallus is enlarged to form a non-septate multinucleate sac-like or tubular structure which is not divided into cells in the somatic phase. *Acetabularia* and *Characium* represent the early stage in the development of a siphonaceous body. *Characium* (Fig. 3.3A) and *Acetabularia* (Fig. 3.3 B) both are unicellular green algae. The former is tubular and the latter umbrella-shaped. *Acetabularia* consists of a stalk ending in an umbrella-like cap. The single nucleus lies at the base of the stalk. In both cases as the unicell reaches maturity and enters the reproductive phase, the nucleus undergoes division to form a number of nuclei. This tendency to become multinucleate for a short time becomes more permanent in *Protosiphon* (Fig. 3.4A) which has an unseptate, unbranched, tubular thallus containing numerous nuclei (A). It grows in damp situations exposed to air. In the submerged siphonaceous forms buoyed up by water, the coenocytic thallus is more elaborate. *Caulerpa* (Fig. 3.4B) is an example of such a type of thallus. It is a tropical marine genus in which the thallus extends to a foot or so in length. It mimics the creeping shoots of aerial plants superficially resembling a large moss or a small angiosperm or a fern. It is differentiated into a creeping structure resembling a rhizome. The rhizome gives rise to root-like holdfasts from its under face and erect leafy shoots from its upper face.

4. Multicellular filamentous thallus. In some species of green algae, the cells are arranged in linear rows called the **threads** or **filaments**. The filament is the result of repeated cell division of a non-motile cell in a single plane. The divisions are all in parallel planes resulting in the formation of a chain of cells. The daughter cells in the chain remain united and further keep on dividing in the same plane forming a simple filament. It is the forerunner of various branched types. The filamentous thallus is well-adapted for a great variety of fresh water and marine habitats.

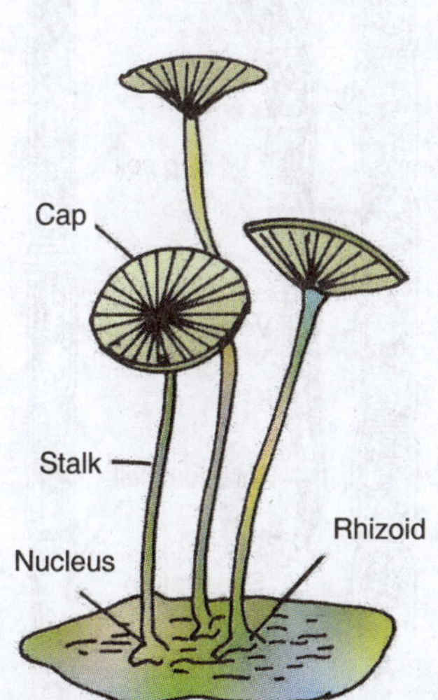

Fig. 3.3 (B). *Chlorophyceae.* Umbrella-shaped uninucleate thallus of *Acetabularia.*

(*a*) *Simple filament* (Fig. 3.5). The simple multicellular forms such as *Spirogyra* (A) have a plant body in which the cells are arranged from end to end in a single file and are held together firmly. It is called the **filament**. It is a long, threadlike, unbranched, elementary type of multicellular thallus. Like the colonial thallus, all the cells in the filament are alike, self-sufficient and independent of one another. It is because of this that the simple filament of *Spirogyra* is often regarded as an advanced type of colonial thallus in which the cells are arranged in a row. It will be called a *filamentous colony.*

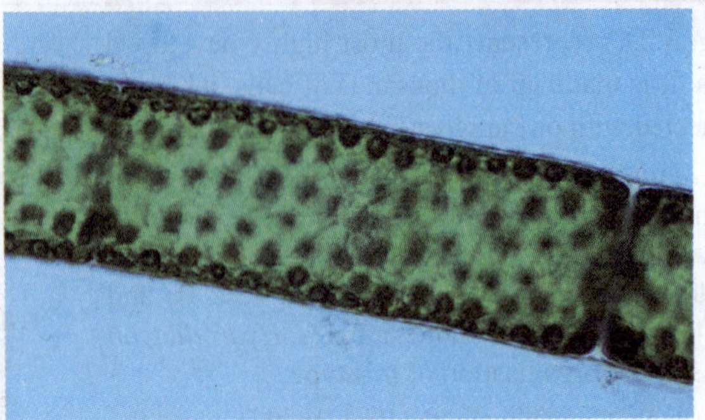

Filamentous Thallus of *Spirogyra*.

Ulothrix (B) illustrates a step further in the development of a multicellular filamentous thallus. The thallus is an unbranched filament. As in *Spirogyra* the simple filament of *Ulothrix* consists of similar cells but it is attached to the substratum at one end by a **rhizoidal cell** specially modified for this purpose. The simple filament of *Ulothrix*, thus shows the beginning of differentiation of cells accompanied by slight division of labour.

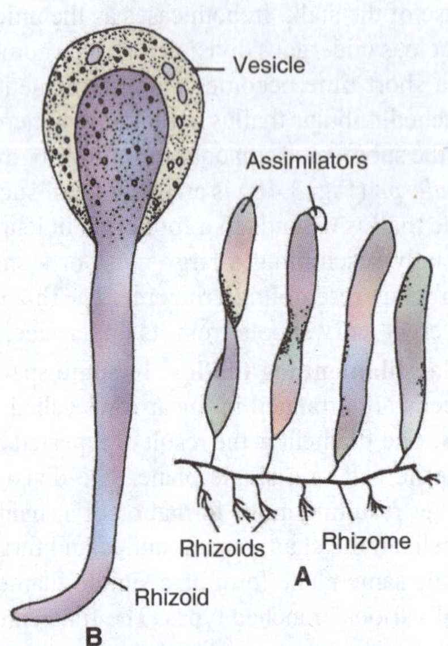

Fig. 3.4 (A-B). *Chlorophyceae.* Siphonaceous thallus; (A) *Protosiphon*; (B) *Caulerpa*.

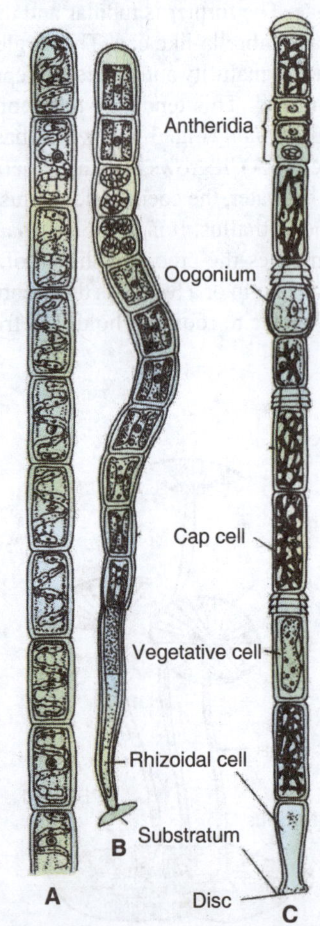

Fig. 3.5 (A-C). *Chlorophyceae.* Simple unbranched filamentous thalli of *Spirogyra* (A), *Ulothrix* (B), and *Oedogonium* (C).

The simple filament of *Oedogonium* (C) represents a step still further in the differentiation of cells accompanied by division of labour. It is an example of a simple multicellular filament with a branched or lobed rhizoidal cell. Besides the rhizoidal cell, the filament consists of green **vegetative cells** for nutrition, **cap cells** for cell division and zoospore formation and highly specialised reproductive cell which function as **sex organs** (antheridia and oogonia). *Ulothrix* may be taken as a transition stage between filamentous colonial forms like *Spirogyra* on the one hand and simple multicellular filamentous forms like *Oedogonium* on the other.

(b) Branched filament. In more advanced multicellular green algae such as *Bulbochaete* (Fig. 3.6) and *Chaetophora*, the thallus is a branched filament. It is formed when an occasional cell in the filament divides in a second plane. The branches thus arise as lateral outgrowths from the main filament. The branched filament of *Cladophora* forms a more tufted and bushy thallus.

(c) Foliaceous thallus. When cell divisions occur in two planes (transverse and longitudinal) in the cells of a simple filament, the result is a thin, flat, plate-like thallus resembling papery expansions (large-green sheets of paper). *Ulva* (Fig. 3.7) is an example possessing a foliaceous type of thallus which is a modification of a filamentous habit.

Foliaceous Thallus of *Ulva Lactuca*

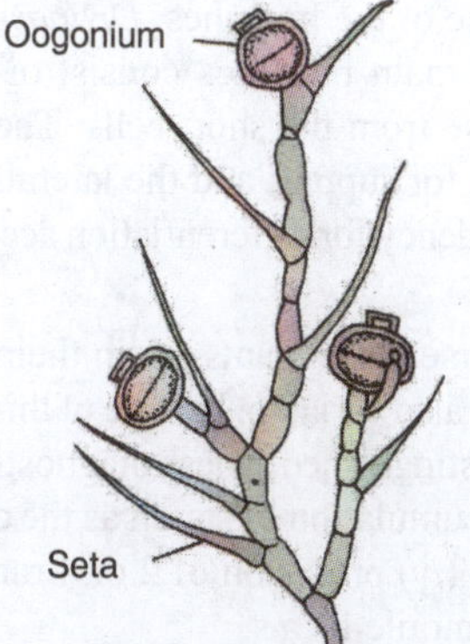

Fig. 3.6. *Chlorophyceae.* Branched filamentous thallus of *Bulbochaete.*

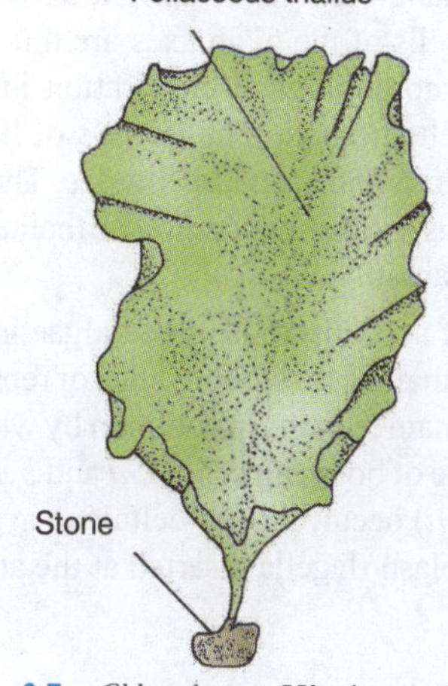

Fig. 3.7. *Chlorophyceae. Ulva lactuca.* Foliaceous thallus attached to a stone.

(d) Massive parenchymatous thallus. It is the result of divisions in a plane parallel to the surface in a foliaceous type of thallus. In this way a massive bulky thallus, two three or more layers of cells thick, is formed.

(e) Heterotrichous filament (Fig. 3.8) It is the most advanced and evolved type of thallus met with in the Chaetophorales among the green algae. In this case the basal portion of the thallus creeps along the substratum and is called the **prostrate system**. From it arise the upright filaments constituting the **erect system**. In *Stigeoclonium* (Fig. 3.8) both the erect and prostrate systems are equally well developed.

The prostrate system in *Fritschiella* (Fig. 3.9), which grows on the beds of drying pools on moist mud, consists of clusters of rounded cells. It is attached to the substratum by colourless septate rhizoids. From the prostrate system arise a number of upright, branched threads consisting of very short cells. These constitute the **primary projecting system**. Above the latter is the **secondary projecting system** comprising elongate cells bearing branches.

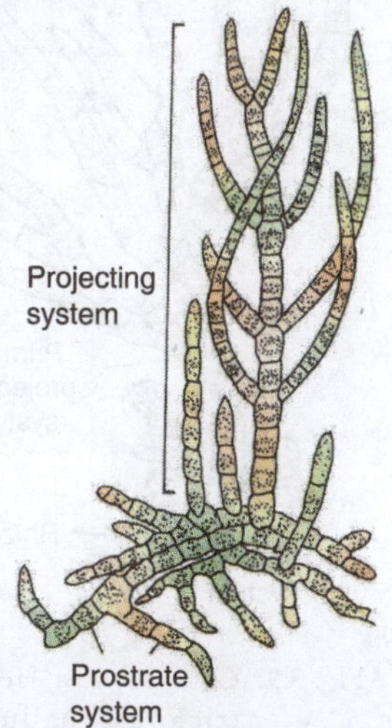

Fig. 3.8. *Chlorophyceae.* Heterotrichous thallus of *Stigeoclonium.*

The heterotrichous filament thus represents the highest differentiation of the thallus reached in green algae. It consists of two parts-the prostrate and erect projecting systems. The heterotrichous filament has undergone a variety of modifications by reduction or elimination of one or other of the systems. The disappearance of erect system has resulted in the discoid forms of thalli in different species of *Coleochaete* (Fig. 3.10A). The discoid thallus represents the prostrate system. In *Coleochaete soluta* the discoid thallus consists of loosely arranged branched filaments distinct from one another. They radiate from a common centre In *C. scutata* (Fig. 3.10A) and *C. orbicularis* the discoid thallus is more compact. It consists of branched filaments which are laterally united. In some species of *Coleochaete* such as *C. pulvinata* (Fig. 3.10B) the thallus is a typical heterotrichous filament. In it the upright threads combine to form a cushion-like thallus covered by mucilage.

Draparnaldia and *Draparnaldiopsis* are examples of green algae in which there is complete disappearance of the prostrate system and a corresponding elaborate development of the erect projecting system. The prostrate system is vestigial and represented by a holdfast. The erect system which is dominant, consists of upright axes from which arise whorls of branched laterals at short intervals. The cells of the main axes are much larger than those of the branches. *Draparnaldiopsis* shows further morphological elaboration in that the cells of main branches consist of long and short cells which alternate. The laterals of limited growth arise from the short cells. There is thus differentiation into nodes and internodes. The main axes serve for support and the laterals function in photosynthesis and reproduction. The thallus thus shows a tendency for differentiation accompanied by division of labour.

This brief account of the green algae indicates how diverse these plants are in their form and structure of the thallus. In their methods of reproduction they are also variable. In spite of this diversity they have four main features in common by which they can be distinguished. These diagnostic features are: (*i*) presence of both chlorophyll *a* and *b* in the cells, (*ii*) accumulation of starch as the customary reserve food, (*iii*) occurrence of cellulose in the cell walls and (*iv*) provision of 2 or 4 rarely a ring of isokont whiplash flagella inserted at the anterior end in the motile forms.

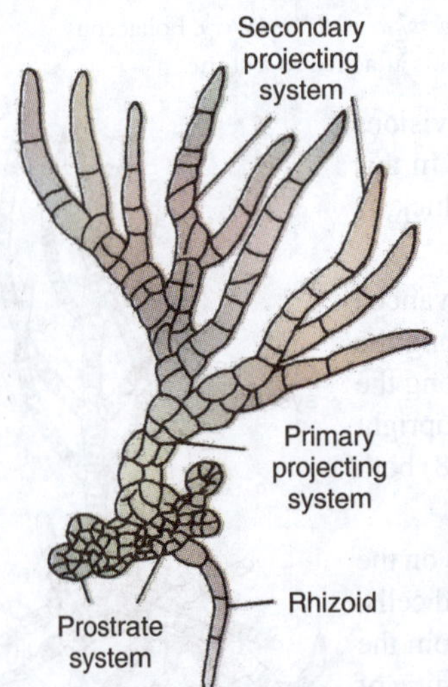

Fig. 3.9. *Chlorophyceae*. Heterotrichous thallus of *Fritschiella* with greater elaboration of projecting system.

CELL STRUCTURE

Typically the cell possesses a wall of nonliving material. It encloses a tiny mass of *protoplast*. Of course in a few primitive flagellate forms, the protoplast lacks the cell wall and is naked. The protoplast is bounded by a thin plasma membrane and shows an advance over the Cyanophyta in the presence of a definite nucleus, a distinct cytoplasm, one or more membrane limited chloroplasts, mitochondria, dictyosomes, endoplasmic reticulum and frequently a sap cavity. These algal cell organelles basically have the same structures as those of the higher plants.

1. *The Cell Wall.* The protoplasmic contents of the cell are bound by a definite cell wall which is typically stable in the green algae. It is a secretion product of the cell protoplast deposited in the form of two concentric layers external to it. The inner layer is invariably in the form of a firm membrane. It is mainly composed of **cellulose** which occurs in the form of microfibrillar units. The only exceptions are the Siphonales in which it is replaced by **callose** (Mirande, 1913) and pectin and not cellulose. The outer layer is made of **pectose**. It is probably secreted by the cell protoplast and filters through the microfibrils of the cellulose layer to form an outer layer of the cell wall.

In forms like *Spirogyra,* the outermost portion of pectose changes into **pectin**. The latter dissolves in water to form the so-called gelatinous sheath which is slimy. It envelopes the entire filament. Frequently an insoluble substance impregnates into the external portion of the pectose layer and prevents dissolving away of pectose (*Ulothrix, Oedogonium* and *Cladophora*). According to Tiffany (1924), the insoluble substance is of the chemical nature of chitin whereas Von Wettstin denied the presence of chitin. In some cases the cell wall is composed of hemi-celluloses (*Chlorogonium, Haematococcus*) and in still others occur lipids and proteins. The cell wall of some marine Caulerpales (old name Siphonales) of the tropics are impregnated with lime. Deposition of iron compounds has been reported on the walls of certain *Desmids*. In the order Volvocales a few genera have no cell wall. The plasma membrane serves as a protective pellicle.

More recent investigations have revealed that cellulose is present in the cell walls of some green algae (*Chaetomorpha*) but is absent in the some others *(Bryopsis* and *Chlamydomonas*). In the latter case the cell wall is largely composed of polymers of glucose or xylose with a covering layer of mannose. These polymers occur in the form of fine, thread-like structures called the **microfibrils**. The microfibrils are in a network at right angles to one another. These are embedded in a non-fibrillar matrix composed of hemi-celluloses.

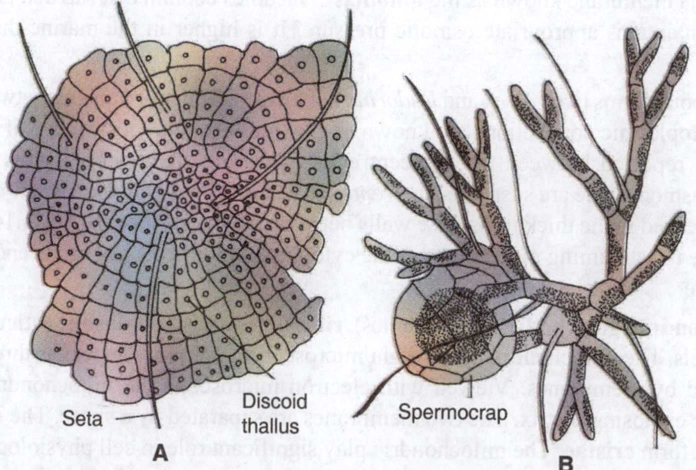

Fig. 3.10 (A-B). *Chlorophyceae.* (A) Discoid thallus of *Coleochaete scutata;* (B) Cushioned thallus of *C. pulvinata.*

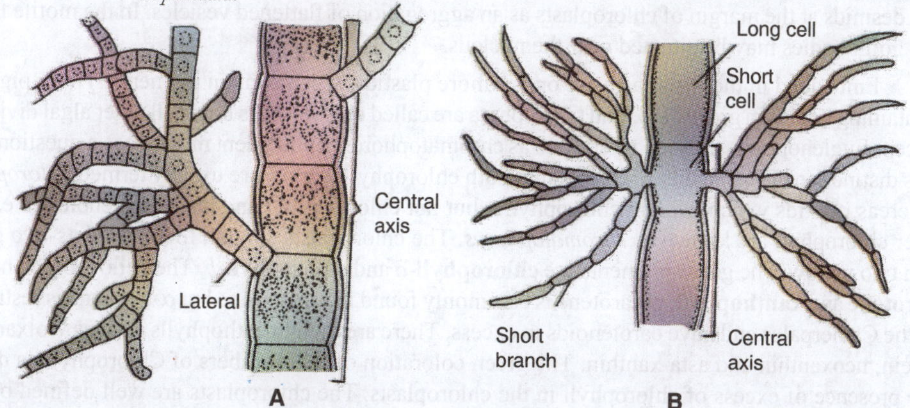

Fig. 3.11 (A-B). *Chlorophyceae.* Showing elaboration of projecting system in *Draparnaldia* (A) and *Draparnaldiopsis* (B).

2. *The Protoplast.* Internal to the cell wall is the living substance of the cell called the **protoplast**. It is bound by a living, differentially permeable, extremely thin **plasma membrane**. Electron microscope reveals that the plasma membrane is differentiated into three layers. The outer and inner are proteinaceous. The middle lipid layer is sandwiched between the protein layers. Within the plasma membrane is the cytoplasm. It tends to be viscous and granular and often exhibits streaming movements. Granules other than the plastids and reserve products are rare. In the advanced Chlorophyta, the cytoplasm surrounds a big central vacuole. The central vacuole is absent in the simpler motile forms of Volvocales and non-motile forms of Chlorococcales in which it is replaced by several smaller vacuoles. Some of these may be contractile. The contractile vacuoles are considered to be excretory in function. The latest interpretation is that they function as osmoregulatory organelles responsible for the removal of extra water of the cell. The contractile vacuoles are situated mostly in the flagellar or apical region. The same condition is prevalent in the motile swarmers of higher forms like *Ulothrix, Cladophora,* and *Oedogonium*. Dangeard considers it to be a reversion to the condition usual in the unicellular algae. The big central vacuole is usually traversed by a number of cytoplasmic strands that shoot out from the lining layer of the cytoplasm. Such a condition of the protoplast is often referred to as the **primordial utricle**. The large central vacuole is bound by a definite, plasma membrane known as the **tonoplast**. Vacuoles contain cell sap that is rich in tannins. The cell sap maintains appropriate osmotic pressure. It is higher in the marine than in the fresh-water forms.

The colonial forms like *Volvox* and *Eudorina* have cytoplasmic connections between the adjacent cells. Such cytoplasmic connections are known as **Plasmodesmata**. Occurrence of plasmodesmata has also been reported between the adjacent cells of certain filamentous forms of green algae (*Ulothrix*). Plasmodesmae are suspected in *Trentepohlia*. Pits filled with extensions of the protoplast have been observed in the thick transverse walls between the adjacent cells (Fig. 3.14 D). Unlike the Cyanophyceae the streaming movements of the cytoplasm are of frequent occurrence in the cells of Chlorophyceae.

Mitochondria, golgi bodies (dictyosomes), **ribosomes and endoplasmic reticulum** are present in the algal cells. The mitochondria under light microscope commonly appear as threads or granules and are bound by membranes. Viewed with electron microscope the mitochondrion is a double membrane sac enclosing matrix. The two membranes are separated by a space. The inner membrane invaginates to form **cristae**. The mitochondria play significant role in cell physiology as originators of enzyme. The latter oxidize carbohydrates and synthesize proteins. In fact the mitochondria are known to be the site of important metabolic reactions. The golgi bodies have been especially seen in the desmids at the margin of chloroplasts as an aggregation of flattened vesicles. In the motile forms the golgi bodies may be situated near the nucleus.

Embedded in the cytoplasm are one or more plastids which contain pigments. These pigment containing bodies in green algae and Charophyta are called **chloroplasts** but in all other algal divisions except Euglenophyta, they are referred to as chromatophores. The student might pose a question why this distinction. The plastids which contain both chlorophyll *a* and *b* are usually termed *chloroplasts* whereas plastids which contain chlorophyll *a* but not chlorophyll *b* and have carotenoids in excess over chlorophyll are known as *chromatophores*. The chloroplasts contain four pigments-two green and two yellow. The green pigments are **chlorophyll-*a*** and **chlorophyll-*b***. The yellow pigments are **carotene** and **xanthophyll**. β-carotene is commonly found in green algae but α-carotene is restricted to the Caulerpales and have carotenoids in excess. There are many xanthophylls such as violxanthin, lutein, neoxanthin and asta-xanthin. The green coloration of the members of Chlorophyta is due to the presence of excess of chlorophyll in the chloroplasts. The chloroplasts are well defined bodies met with in every cell of the members of this class. The number and shape of the chloroplasts varies in different orders of the class (Fig. 3.12).

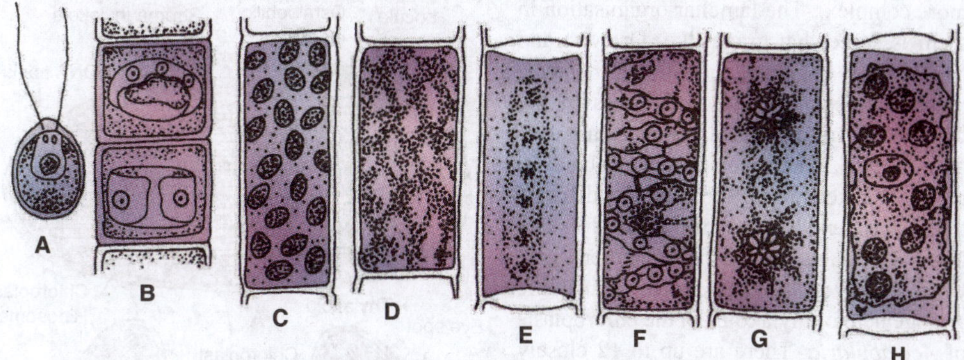

Fig. 3.12 (A-H). *Chlorophyceae.* Principal types of Chloroplasts.

In the Volvocales there is usually a single large cup-shaped chloroplast with one or a few pyrenoids. It occupies the posterior part of the cell (A). In the Chlorococcales and Chaetophorales too, there is a single parietal chloroplast. In the Chaetophorales, the chloroplast is parietal and girdle-shaped with incised margins (*Draparnaldia*). It is more or less reticulate in *Draparnaldiopsis* (D). Shape of the chloroplast in *Coleochaete* is irregular. In *Oedogonium* there is an elaborate, reticulate, cylindrical chloroplast with narrow sub-parallel meshes. In *Ulothrix* there is a single, collar-shaped chloroplast in each cell (B). Siphonales have a large number of discoid chloroplasts lying in the lining layer of the cytoplasm (C). Nature of the chloroplast in *Cladophora* is disputable. Some regard it to be a single reticulate structure whereas others consider it to be in the form of numerous discoid structures. The chloroplasts in the Conjugales are of four types, namely, an axile plate (H), two parietal plates with a nucleus in between, a pair of axial, stellate bodies (G) and one to several parietal ribbon-shaped, spirally wound structures (F). In *Debarya* it is band-like (E).

Fine structure of Chloroplast (Fig. 3.13). Under light microscope the chloroplast is seen to consist of a clear structureless matrix containing a number of small discs. Investigations with electron microscope revealed that the chloroplast is bounded, by a narrow double membrane (*Chloroplast membrane* or *envelope*) which encloses a colourless granular matrix known as the *stroma*. Traversing the matrix is a complex system of a completely closed network of membranes called the *lamellae*. Menke (1962) termed the lamellar units as *thylakoids* but Sagar and Palde (1957) preferred to call them discs. The thylakoids or discs probably contain all the chlorophyll pigment of the chloroplast and are made up of proteins and lipid substances. Each thylakoid is a membrane bound flattened sac, the opposite sides of which are compressed to come together. In a section it appears like a pair of parallel mem-branes joined at each end and enclosing a very narrow space termed the **loculus**. The thylakoids in algae are generally long and may run nearly the entire length of the chloroplast.

In many green algae the thylakoids occur in multiple layers called **stacks** or **bands**. The number of discs in a stack varies from 2-20 or even more. Usually there are 4-6 thylakoids in a stack. The thylakoids in the stack do not adhere and remain separated by a narrow space. The space between the two adjacent thylakoids is known as **interdisc** or **inter-thylakoid space**. In a number of other green algae the thylakoids do not form stacks. They come together and move apart in an irregular manner.

The matrix in the large cup-shaped chloroplast of *Chlamydomonas* (B) is traversed by thylakoids which are organised into regular stacks or bands. Commonly there are 2-6 discs per stack (*C. moewusii*) but in *C. reinhardii* the number goes up to 8. The discs run parallel to the cell wall. The structure of chloroplast of *Nitella* is peculiar. The matrix contains 40-100 discs. All of them may be associated together in a single stack. The structure of the chloroplast in some green algae is more complex. The lamellar organisation in them is somewhat grana-like. Drawert and Misc (1961) reported that in *Microsterias rotata* (a desmid), the thylakoids in the chloroplast are organised into regular well-defined stacks, each consisting of about 50 closely packed discs. Some of the discs in the stack extend beyond the edge of the stack and run through the stroma to another stack. Crawley (1964) also observed grana-like arrangement of thylakoids in the chloroplasts of *Acetabularia*. There are up to 12 closely packed thylakoids in a stack. Some of these are restricted to the stack area whereas others extend beyond and cross the stroma of another interstack region.

The granular chloroplast matrix (stroma) contains small **ribosomes** and a variable number of **osmiophilic globules**. A small amount of DNA in the form of minute wispy fibrils has also been reported in certain regions of the stroma. Presence of starch grains within the chloroplast is a feature in which the green algae, differ from all other algal groups.

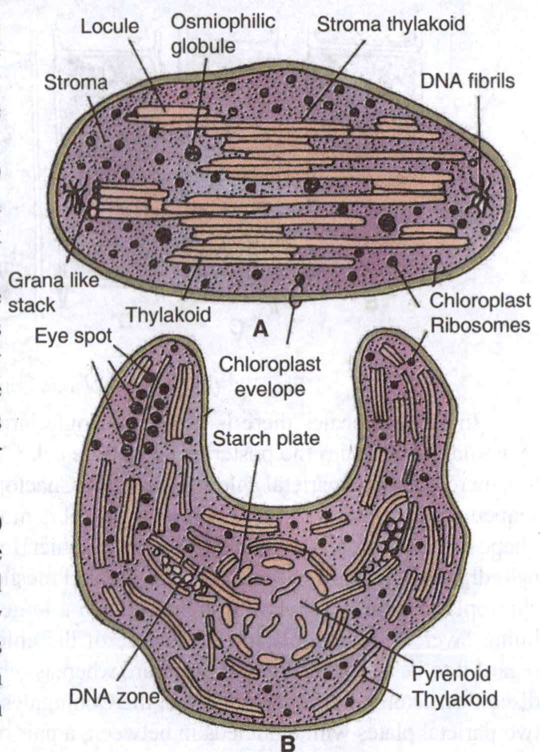

Fig. 3.13 (A-B). Chlorophyceae. Fine structure of Chloroplast. (A) Diagrammatic sketch of a chloroplast of a green alga showing grana-like arrangement of thylakoids; (B) cup-shaped chloroplast of *Chlamydomonas.*

The chloroplasts of green algae contain one or more distinct, rounded, proteinaceous bodies called the **pyrenoids**. Each pyrenoid consists of a central, viscous, granular core surrounded by tightly packed minute plates of starch. The pyrenoid core is not traversed by lamellae in some green algae (*Scenedesmus quadricanda*). In *Spirogyra* a number of thylakoids pass through it. The pyrenoid in *Chlamydomonas* is penetrated by a number of thylakoids but usually they extend into the pyrenoid for a short distance. There are two views regarding the nature of pyrenoids. According to Dangeard and others they are masses of reserve protein. The other school of thought regards them as special organs of the cell. The former view seems to be more prevalent. It has been noticed that pyrenoids diminish in size and ultimately disappear if the plant is under conditions of starvation. They reappear when the conditions become favourable.

Eye-spot or Stigma (Fig. 3.14). Certain mobile, unicellular green algae possess a structure which with the light microscope appears as a small bright reddish or brownish red spot or streak. It is known as the **Stigma** or **eye-spot**. In green algae the eye-spot is usually associated with the chloroplast. In *Chlamydomonas* it lies laterally at the anterior end of the unicell inside the chloroplast close to its rim and parallel to the chloroplast surface (A). Eye-spot is considered a photoreceptive organ. According to Mast (1927) it consists of a curved **pigment plate** or **pigmentosa** carrying the pigment and a biconvex hyaline *lens* in front (C). The latter lies in the concavity of pigment plate and external to it. The use of an electron microscope has revealed that in a vertical section the eye-spot consists of one to four rows of globules (A). Each row comprises closely packed globules containing carotenoid pigments. If the eye-spot is independent of the chloroplast it is enclosed in the same type of a membranous envelope as is found around the chloroplast. In a section parallel to the surface (face view), the eye-spot shows facet-like structures (B).

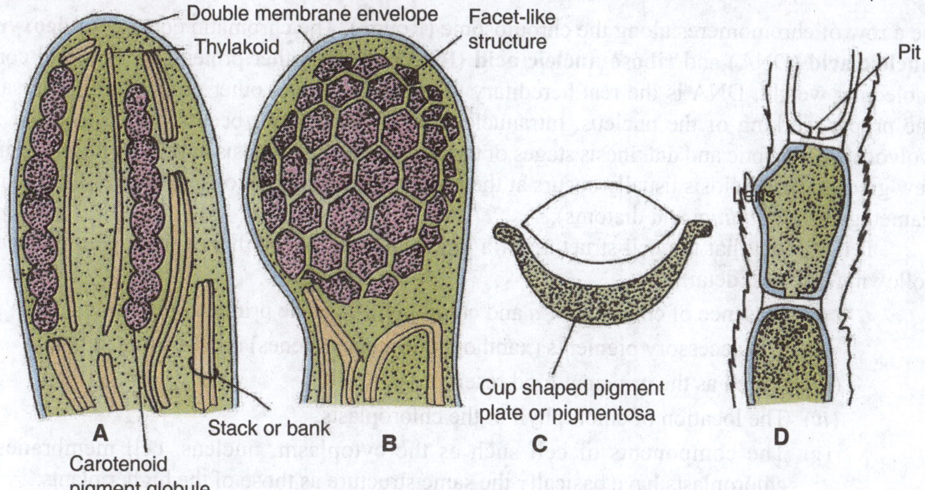

Fig. 3.14 (A-D). *Chlorophyceae.* Structure of eye-spot or stigma. (A) fine structure of stigma as seen in vertical section; (B) fine structure of stigma as seen in face view (in section parallel to the cell surface); C, structure of stigma as seen in a vertical section under light microscope; (D) *Trentepohlia* showing pits in the septa (C after Smith and D after Brand).

The cell has a definitely organized **nucleus** with a **nuclear membrane** enclosing **nucleoli, chromatin network** and **karyolymph**. In the electron microscope the nuclear membrane has pores. It is a double layered structure made up of proteins and lipids. There is a narrow space between the two layers. The number of nuclei per cell varies. Both uninucleate and multinucleate cells have been reported in the Chlorophyta. Majority of them are uninucleate but in forms like *Cladophora* the cells are multinucleate. The Siphonales have a thallus which is coenocytic. Older cells in *Pediastrum* have more than one nucleus. Pronounced nucleoli are often discernible. The number of nucleoli is variable. There may be one or more than one. Nuclei of desmids show realtered nucleoli. The nucleolus is made up of **RNA** and proteins. The chromatin network which is hereditary material of the cell becomes conspicuous during early stages of division. It resolves into chromosomes at the time of division which is mitotic and involves spindle formation. The chromosomes are of different shapes in the green algae. They are small and spherical in some desmids but are in the form of small rods or mere granules in others. Long thread-like chromosomes have been reported by Tuttle in *Oedogonium*. There appears to

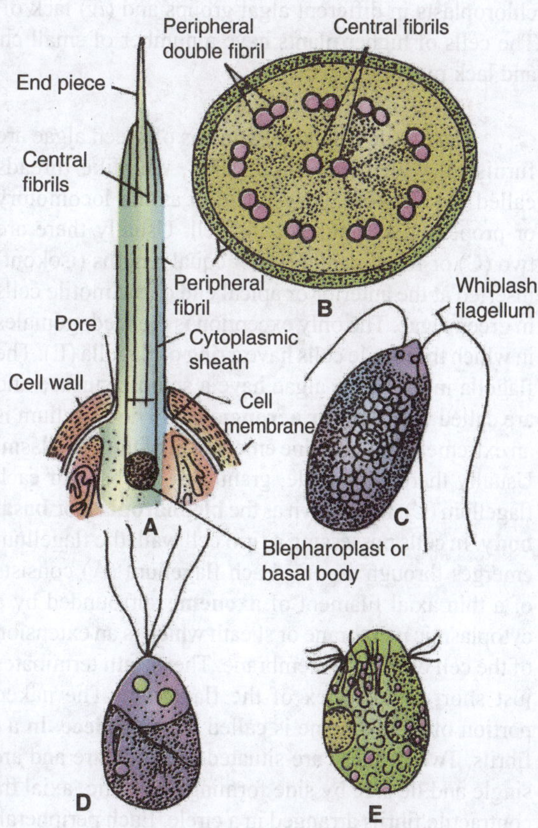

Fig. 3.15 (A-E). *Chlorophyceae* : Structure of flagellum (A-B.) and flagellation (C-E).

be a row of chromomeres along the chromosome (Round). The chromatin consists of **deoxyribose nucleic acid** (DNA) and **ribose nucleic acid (RNA)** besides other proteins of low and complex molecular weight. DNA is the real hereditary material. RNA and other proteins are necessary for the proper working of the nucleus. Intranuclear centrosomes have been reported to occur in the Volvocales. Synaptic and diakinesis stages of the nucleus during meiosis have been observed in very few green algae. Meiosis usually occurs at the time of differentiation of spores rarely at the time of gametogenesis (*Codium* and diatoms).

It is evident that the cell structure of a green alga resembles that of the higher plants in the following essential details:

(*i*) Presence of chlorophyll-*a* and chlorophyll-*b* as the principal pigments.

(*ii*) The accessory pigments (xanthophylls and carotenes) are the same in both.

(*iii*) Starch as the principal food reserve.

(*iv*) The location of chlorophyll in the chloroplasts.

(*v*) The components of cell such as the cytoplasm, nucleus, cell membranes, and chloroplasts have basically the same structure as those of the higher plants.

The above-mentioned similarities particularly the presence of identical pigments and starch as the end product of photosynthesis in both suggest that the green plants at a higher level have evolved from some green alga similar to the present-day green algae. The cell structure of the green algae, however differs from the higher plants in certain features namely, (*i*) presence of pyrenoids, (*ii*) chloroplasts few in number but larger in size, (*iii*) great diversity in shape and arrangement of chloroplasts in different algal groups and (*iv*) lack of organisation of cells into tissues and organs. The cells of higher plants have a number of small chloroplasts which have attained fixity of form and lack pyrenoids.

Flagella (Fig. 3.15)

The motile cells or organisms of green algae are furnished with fine protoplasmic, whiplike threads called the **flagella.** They function as the locomotory or propelling structure of the cell. Usually there are two (C) or four flagella (D) of equal lengths (isokont) inserted at the anterior or apical end of the motile cells in green algae. The only exception is the Oedogoniales in which the motile cells have a ring of flagella (E). The flagella in the green algae have a smooth surface and are called **whiplash or acronematic**. The flagellum is an extremely fine, hyaline emergence of the cytoplasm. Usually there is a single, granule at the base of each flagellum (C). It is known as the **blepharoplast or basal body**. In cells possessing a firm cell wall, the flagellum emerges through a pore. Each flagellum (A) consists of a thin axial filament of **axoneme** surrounded by a cytoplasmic membrane or sheath which is an extension of the cell or plasma membrane. The sheath terminates just short of the apex of the flagellum. The naked portion of the axoneme is called the **end piece.** In a cross section (B), the axoneme consists of 11 fibrils. Two of these are situated in the centre and are called the **axial** or **central fibrils**. They are single and lie side by side forming an elastic, axial thread which is surrounded by nine, peripheral, contractile fibrils arranged in a circle. Each peripheral fibril consists of two thin microfibrils. All the fibrils are hollow. External to the peripheral fibrils is the common cytoplasmic sheath.

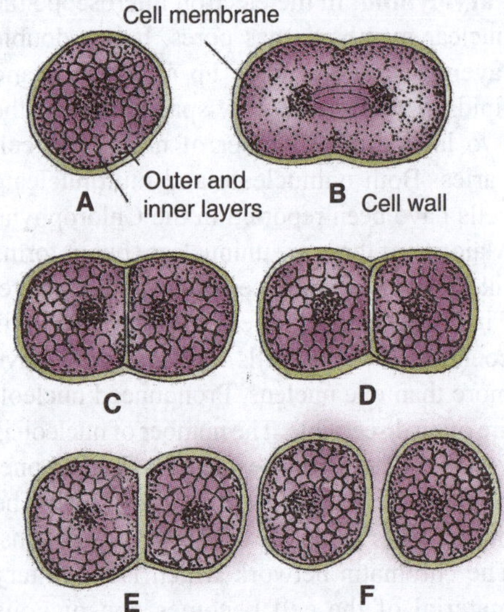

Fig. 3.16 (A-F). *Chlorophyceae.* Stages in cell division in a green alga *Pleurococcurs.*

REPRODUCTION

In green algae it takes place by all the three methods, namely, vegetative, asexual and sexual.

1. Vegetative reproduction. Propagation of the species by any method which uses only vegetative cells is known as **vegetative reproduction**. In this method the parent cell wall is retained. It may take place by **cell division, fragmentation** or **akinete formation**.

(i) Cell division (Fig. 3.16). The unicellular forms commonly rep-roduce by a simple process involving cell division. It is called **fission.** The division of the cell is preceded by the division of the nucleus which is mitotic. The nuclear division is followed by cleavage of the cytoplasm which begins by a median constriction at the periphery (B). Into the constriction introverts the inner layer of the cell wall which invaginates with the wall membrane. The inward growth of the latter accompanied by that of the inner layer of the parent cell wall finally cuts the cytoplasm into two halves, each containing a daughter nucleus. At this stage the transverse septum splits into two layers. The outer layer is formed *de novo* before the two daughter cells separate. Cell division in unicellular forms thus is a method of reproduction and is synonymous with vegetative reproduction. In the multicellular forms it leads to growth.

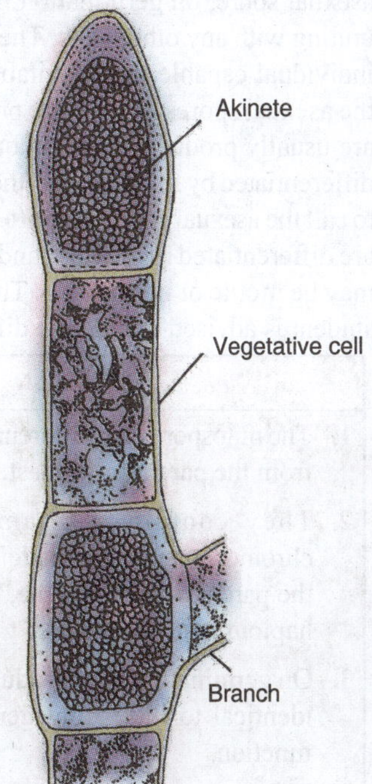

Fig. 3.17. *Chlorophyceae.* Portion of a filament of *Pithophora* forming akinetes.

(ii) Fragmentation. It involves the breaking up of the multicellular filamentous thallus (*Spirogyra*) into one or many-celled segments of living cells. These are called the **fragments**. The fragmentation of the filament may be the result of external forces or formation of mitospores (zoospores and aplanospores) or gametes in certain parts of the filament followed by the breaking up of empty cells. Each fragment functions as a reproductive unit. It is less than the whole thallus but more than a single living cell. The fragment by repeated cell division and subsequent growth of the component cells grows into a new filament. Fragmentation thus serves to increase the number of filaments in a mass. Cell division and fragmentation both constitute vegetative methods of reproduction which takes place under exceptionally favourably conditions.

Akinetes of Green Algae (*Pithophora*).

(iii) Akinetes (Fig. 3.17). These are one-celled specially modified resting vegetative structures. Some scientists consider them to be modified one-celled fragments. Under certain conditions cells here and there in the filament (*Pithophora*) lose water and contract to become rounded or oval in form. They become gorged with reserve food materials. The parent cell wall becomes thickened and highly resistant. These specially modified vegetative cells are called the **akinetes**. They occur either singly or in chains in the vegetative filament and enable the plant to tide over unfavourable conditions. All other vegetative cells of the filament may perish, the akinetes survive. With the onset of the favourable season, each akinete may germinate directly to form a new

individual or its protoplast divides to form asexual spores which on liberation germinate to form a new plant.

2. Asexual reproduction. It involves the multiplication of the species by the formation of highly specialised reproductive cells called the **asexual spores**. Reproduction by asexual spores is a very common method of propagation under normal conditions of life and is often called **sporulation**. Each asexual spore, on germination, is alone capable of directly giving rise to a new individual without uniting with any other cells. The asexual spore is thus a unit of multiplication. It is potentially a new individual capable under suitable conditions of taking on the form of the species. In green algae, the asexual spores are usually produced by the haploid plants and thus are genetically haploid. They are usually produced endogenously in more or less specialized cells called the **sporangia** and are differentiated by **mitosis** from the protoplast of the sporangium. It will, therefore, be more appropriate to call the asexual spores as **mitospores** in order to distinguish them from another kind of spores which are differentiated by **meiosis** and are called the **meiospores**. The mitospores are asexual spores. They may be motile or non-motile. The former are called the **zoospores** and the latter **aplanospores**. The student is advised to note the differences between the two kinds of spores. These are :

Mitospores (asexual spores)	*Meiospores (Sexual spores)*
1. The mitospores are differentiated by **mitosis** from the parent protoplast.	1. Meiospores are differentiated by **meiosis** from the diploid parent protoplast.
2. They contain the same number of chromosomes in the nuclei as the nuclei of the parent. In green algae, they are usually haploid plant only.	2. They contain half the number of chromosomes in their nuclei as compared with the nuclei of the parent cell. Thus they are always haploid.
3. On germination they produce an individual identical to the parent genetically and in function.	3. On germination they develop into an alternate individual in the life cycle which is genetically and functionally different from the parent.
4. They are purely vegetative in origin and asexual in character.	4. Their formation is contingent upon sexual reproduction and thus are reproductive in nature. Meiospore formation, in fact, is a stage in sexual reproduction.
5. They play a definite role in the phenomenon of alternation of generations.	5. They play no role in the phenomenon of alternation of generations.

(i) Zoospore formation (Fig. 3.18 A-C). The common and characteristic method by which asexual reproduction operates in the green algae is by the formation of motile, naked asexual spores called the **zoospores**. They may be bi-or quadri-flagellate (*Ulothrix*), occasionally with a ring of flagella and thus multiflagellate (*Oedogonium*). They are usually formed during night and develop either in any of the vegetative cells (A) or in modified and specialised cells called the **zoosporangia**. The protoplast of the cell may develop into a single zoospore (*Oedogonium*) or it may undergo divisions resulting in the formation of several zoospores (*Ulothrix*). They escape in the morning from the parent cell through a pore in the surrounding cell wall or by rupturing or gelatinisation of the cell wall (B). The liberated zoospore is tiny, naked mass of protoplast containing a single nucleus, an eye-spot and one or more chloroplasts (C). The flagella are of equal lengths (isokont) and inserted at the anterior end. They function as locomotory organs. Each zoospore on coming in contact with a suitable substratum and under favourable conditions develops into a new plant resembling the parent. Zoospore formation is the normal, most effective and rapid means of multiplication in the aquatic green algae under favourable environmental conditions.

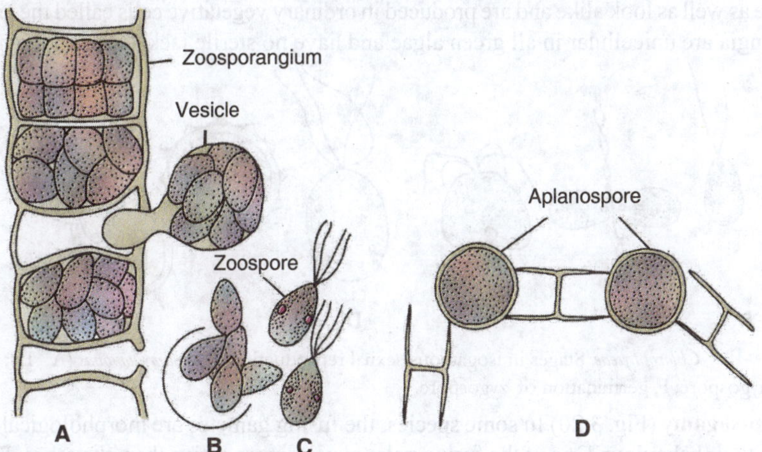

Fig. 3.18 (A-D). *Chlorophyceae.* Stages in zoospore formation and their liberation (A-C) in *Ulothrix* and discharge of aplanospores (D) in *Microspora.*

(*ii*) *Aplanospore formation* (Fig. 3.18 D). The non-motile asexual spores often called **aplanospores** are normally formed in the algae of subaerial habitats. In the aquatic forms they are produced under unsuitable conditions. The protoplast of the cell may form a single aplanospore (*Microspora*) or more than one. Each aplanospore secretes a wall around itself before liberation from the parent cell. It lacks flagella and thus is non-motile. It has no eye spot and is enclosed by a cell wall. The aplanospore germinates directly to give rise to a new individual resembling the parent. When the non-motile spores produced appear identical to the parent cell, they are referred to as **autospores** (*Chlorella*). Some scientists consider the aplanospores as abortive zoospores with the motile phase omitted. Under certain circumstances the aplanospores secrete thick walls around them and store abundant food reserves. These thick-walled aplanospores are called the **hypnospores.**

The green algae may reproduce repeatedly by sporulation and vegetative methods of reproduction during the growing season. This results in the reduplication of the parent individuals which produce them. Reproduction by *vegetative methods* and by *sporulation* is considered by some scientists as **asexual reproduction**. Asexual and vegetative reproduction serve to maintain the species through several generations without a change in the chromosome number.

3. Sexual reproduction. This kind of reproduction involves the fusion of two specialized reproductive cells or units called the **gametes.** The gametes alone cannot develop normally into a new individual. They must undergo the sexual process which comprises the fusion of the gametes. Plasmogamy is the fusion of the cytoplasm of the gametes and karyogamy or fertilisation is the fusion of the gamete nuclei. The resultant fusion cell is known as the zygote. It has a diploid nucleus.

In some algae the two fusing gametes may be produced by the same thallus and in others by two separate thalli. The former are called **monoecious** or **homothallic** species and the latter **dioecious** or **heterothallic** species. Sexual reproduction always entails three steps namely *formation of gametes, fusion of gametes* (syngamy or fertilization) and *germination of zygote.* It takes place only once during the growing season when the climax of vegetative activity has just been passed. It is of three kinds namely isogamy, anisogamy and oogamy.

(*i*) Isogamy (Fig. 3.19). The fusing gametes in the lower forms are alike in size, structure and behaviour (A). They are indistinguishable as to sex. Such gametes are called the **isogametes.** (A). Sexual reproduction which involves the fusion of isogametes (B and C) is termed **isogamous**. It is a primitive type of sexual reproduction and is common in some species of *Chlamydomonas*. It is typical of certain species of *Zygnema*. The isogametes are usually naked protoplasts and are often flagellated.

They behave as well as look alike and are produced in ordinary vegetative cells called the **gametangia**. The gametangia are unicellular in all green algae and have no sterile jacket around them.

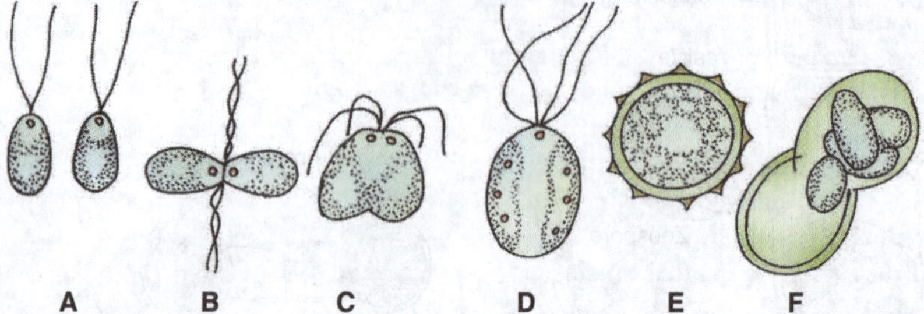

Fig. 3.19 (A-F). *Chlorophyceae.* Stages in isogamous sexual reproduction in *Chlamydomonas* (A - D); E, resting zygospore; F, germination of zygospore.

***(ii)* Anisogamy** (Fig. 3.20) In some species, the fusing gametes are morphologically alike but dissimilar in their behaviour. One of the fusing pairs may be more active than the other. The gametes of *Spirogyra* are distinguishable by their degree of motility. This difference in behaviour is the first step towards anisogamy. It is called **physiological anisogamy**. It is more marked when the fusing gametes are of different sizes (A). Such gametes are called **anisogametes** and the reproduction as **anisogamous**. In most green algae the anisogametes are of two different kinds so that they can be designated as **male** and **female**. The male gamete remains small and is active. The female gamete is large and is passive. Both may be shed and thus fuse externally (C). This is an advanced type of anisogamy. It illustrates a step further towards oogamy. The cells in which gametes are produced are called the **gametangia.**

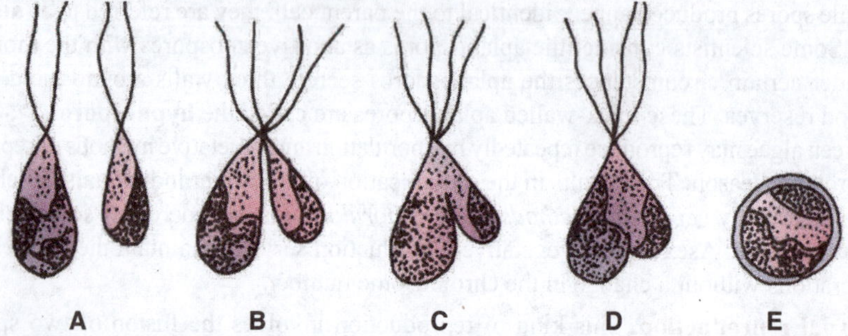

Fig. 3.20 (A-E). *Chlorophyceae.* Stages in anisogamous sexual reproduction in *Enteromorpha intestinalis* (A-D); E, resting zygospore.

***(iii)* Oogamy** (Fig. 3.21). The higher forms develop distinct sex organs which are specially modified vegetative cells. Male sex organ is called an **antheridium** (A) and the female **oogonium**

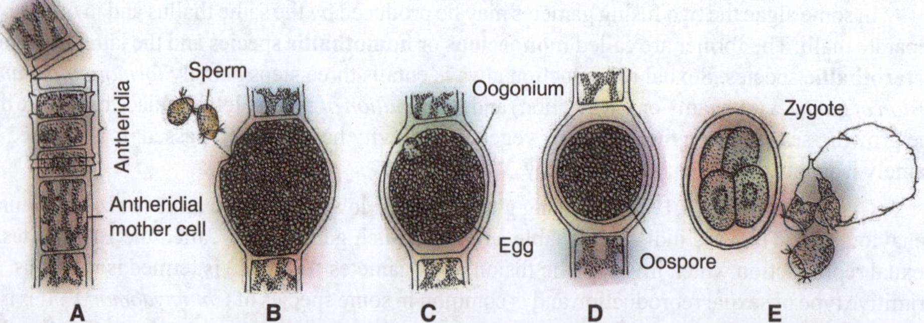

Fig. 3.21 (A-F). *Chlorophyceae.* Showing oogamy (A-D) and germination of oospore (E-F) in *Oedogonium.*

(B). The smaller, active flagellated gametes produced in the antheridia are called the **sperms**. They are developed in large numbers in an antheridium and on maturity are shed in the water. The female gametes are large and passive. They are produced singly. The female gamete is retained within the oogonium and is called an egg or **oosphere** (C). The male and the female gametes in oogmay thus differ from each other in every respect such as size, motility, behaviour and structure. Such gametes are called **heterogametes.** The sperm swims towards the egg and fuses with it (C). The structure resulting from this union is called the **zygote** (D). Oogamy is the most advanced type of sexual reproduction.

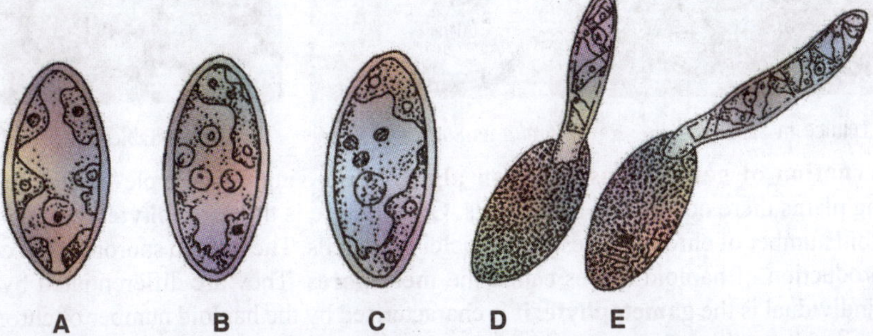

Fig. 3.22 (A-E). *Chlorophyceae.* Showing zygotic meiosis (A-C) and germination of zygospore (D-E) in *Spirogyra.*

Germination of zygospore (Fig. 3.22)

The zygote is the pioneer, diploid structure in the life cycle (A). In the green algae it always passes through the resting stage and is thus called a **zygospore**. Its subsequent behaviour prior to germination differs in different species of green algae. (*i*) In *Spirogyra* and *Zygnema* the diploid zygospore nucleus undergoes meiosis (B). It is known as **zygotic meiosis**. Of the four resultant haploid daughter nuclei three degenerate (C). The zygospore protoplast containing the surviving haploid nucleus directly grows into a new gametophyte filament (D-E.). (*ii*) There are other green algae (*Ulothrix* and *Oedogonium*) in which all the four haploid nuclei resulting from zygotic meiosis remain functional and are organised into **meiospores** (Fig. 3.21 E). The zygospore containing the meiospores illustrates a step further in the evolution of the diploid phase or sporophyte. It is considered the **incipient** or **primitive sporophyte**. Each meiospore, on liberation, gives rise to a haploid or gametophyte filament and not the parent zygospore. (*iii*) the third category of green algae are represented by *Cladophora*. In this alga the zygote nucleus undergoes no meiosis. Meiosis is delayed until some later stage of development in the life cycle. This kind of meiosis is called **intermediate meiosis**. The unreduced zygote in these green algae germinates to form the **diploid filament** or the **sporophyte** which produces meiospores in diploid zoosporangia. The diploid protoplast of the sporangium undergoes meiosis to form the motile meiospores. Each meiospore on liberation and germination gives rise to the haploid or gametophyte filament-an alternate or second plant in the life cycle. Forms like *Cladophora* exhibit distinct alternation of generations in the life cycle.

Economic Importance

Some of the green algae constitute an important source of food for fish and other aquatic animals. As result of considerable experimental work on a unicellular green alga, *Chlorella* it has been found that it can be used as an efficient source of food for man. Thacker and Babcoke, however, concluded that the production of *Chlorella* was not an economic proposition. Still extensive experiments to test the feasibility of artificial culturing of certain species of green algae as a means of producing such harvest of food for human consumption are under way. The 'sea lettuce'-a marine alga, *Spirogyra* and *Oedogonium* are eaten directly by certain people. *Enteromorpha* and *Ulva* are used in salads and in soups. *Chlorella* yields an antibiotic, **Chlorellin**. Some filamentous green algae are used in sugar purification. Investigations are in progress to discover certain green algae as a source of fuel gas. Soil-inhabiting green algae increase soil fertility by adding organic matter to the soil.

Sea Lettuce in Salad.

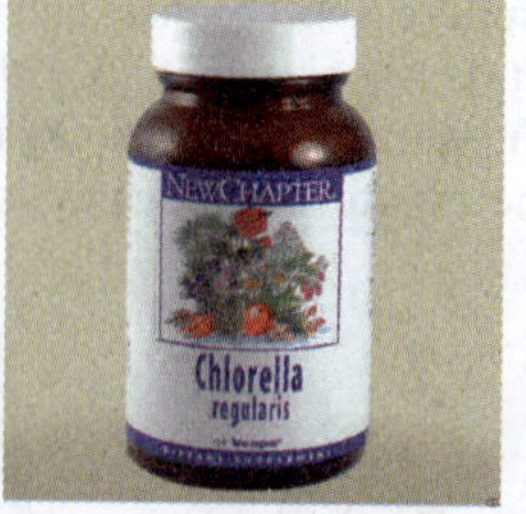

Chlorella as Source of Energy.

Chlorella Tablets as growth factor.

Alternation of generations in green algae. In the single life cycle of certain sexually reproducing plants there occurs two individuals. One of these is the **sporophyte**. It is characterised by the diploid number of chromosomes in the nuclei of its cells. The diploid sporophyte is concerned with the production of haploid spores called the **meiospores**. They are differentiated by meiosis. The other individual is the **gametophyte**. It is characterised by the haploid number of chromosomes in the nuclei of its cells. It is responsible for sexual reproduction. It bears the haploid gametes. These two individuals normally follow each other in a single life cycle. This is called *alternation of generations*. It means the alternation in the life cycle of two distinct individuals having not only different chromosome numbers but different functions as well.

(*a*) In the majority of the green algae such as *Spirogyra* (Fig. 3.23) and *Zygnema*, the thallus (filament) is haploid. It bears the gametes. The gametes fuse to form the zygospore. It is a diploid structure. The diploid zygospore nucleus, prior to germination, undergoes meiosis. As a result four haploid daughter nuclei are formed. Three of these disintegrate. The surviving functional daughter nucleus becomes the nucleus of the first cell of the new haploid or gametophyte filament. The *zygospore,* therefore, is the only diploid structure in the life cycle of these green algae. It represents a simple, one-celled sporophyte. There is no organised sporophyte phase with a different function. Such species of green algae indicate the origin of the diploid body in the life cycle. It is closely associated with the sexual process. In the life cycle of these forms there is alternation of a well developed free living gametophyte with a single-celled sporophyte.

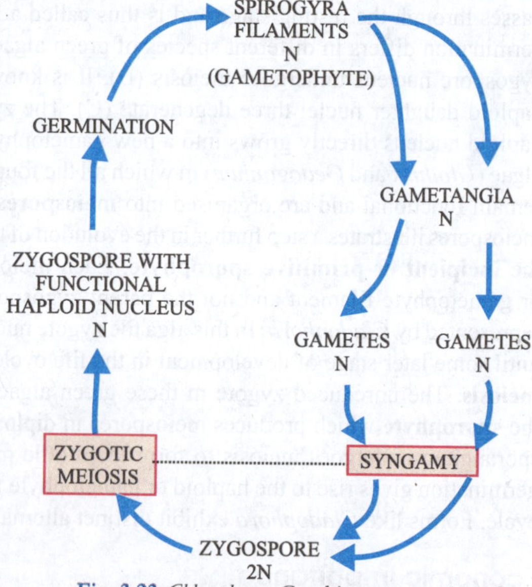

Fig. 3.23. *Chlorophyceae.* Graphic representation of the haplontic life cycle of *Spirogyra.*

This results in the alternation of chromosome numbers from haploid to diploid and back to the haploid stage. There is, however, no alternation of individuals with different functions. Meiosis in the zygote nucleus and fertilization form the cytological basis for alternation of chromosome numbers in the life cycle.

A life cycle characterized by a haploid thallus, and zygotic meiosis is called **haplontic** life cycle (Fig. 3.23). Svedelius (1931) termed it **Haplobiontic** because only a single type of free living individual is involved in the life cycle. Since the free living individual is haploid in its generic constitution he called such a life cycle *haplontic haploid*.

(*b*) A further step in the evolution of the sporophyte is illustrated in green algae *Ulothrix* and *Oedogonium* (Fig. 3.24). In these forms the protoplast of the zygospore, at the time of germination divides by meiosis and is organised into haploid spores. Each spore on germination, gives rise to a gametophyte filament. The diploid zygospore containing the haploid spores constitutes an **incipient sporophyte** in the life cycle of these green algae. The development of an organised or adult sporophyte phase with a different function is not possible in such forms, because of the meiotic division of the zygospore nucleus. It is called the **zygotic or initial meiosis.**

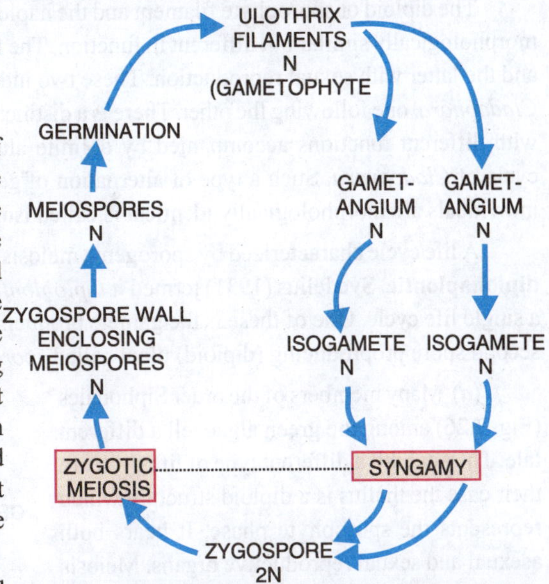

Fig. 3.24. *Chlorophyceae.* Word diagram of the haplontic life cycle of *Ulothrix.*

(*c*) *Cladophora* and *Ulva* among the green algae illustrate a still further stage in the evolution of the sporophyte. The free living sexual or haploid filaments of *Cladophora* represent the **gametophyte phase** (Fig. 3.25). They bear the gametes. The latter fuse to form a zygote. Meiosis is delayed in *Cladophora*. It does not occur at the time of germination of the zygote. The latter, on the other hand, by repeated equational mitosis forms a well-developed free living diploid filament. Certain cells in the diploid filament function as **zoosporangia.** The diploid protoplast of the zoosporangium undergoes **meiosis** to form the meiospores. Meiosis in *Cladophora* is thus delayed to later stage. This kind of meiosis is known as **intermediate meiosis.** It has resulted in the development of another well-developed, free living individual or filament (in the life cycle) which is diploid and concerned with the production of motile meiospores. The diploid filament represents the **sporophyte phase** in the life cycle. The meiospores, on germination, give rise to sexual or gametophyte filaments and not to sporophyte filaments on which they are borne. In *Cladophora*, therefore, the spores play a definite role in the alternation of generations.

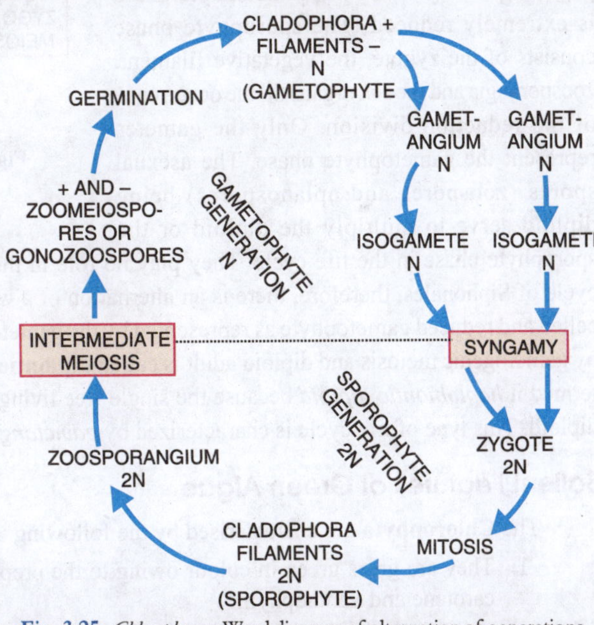

Fig. 3.25. *Chlorophyceae.* Word diagram of alternation of generations in *Cladophora.*

The diploid or sporophyte filament and the haploid or gametophyte filament of *Cladophora* are morphologically similar but different in function. The former is concerned with asexual reproduction and the latter with sexual reproduction. These two individuals regularly alternate in the life cycle of *Cladophora*, one following the other. There is a distinct alternation of different individuals (filaments) with different functions accompanied by definite alternation of chromosome numbers in the life cycle of *Cladopyora*. Such a type of alternation of generations in which the alternating free living individuals are morphologically identical is called **isomorphic** or **homologous.**

A life cycle characterized by sporogenic meiosis and distinct alternation of generation is called **diplohaplontic**. Svedelius (1931) termed it *diplobiontic* because two free living individuals occur in a single life cycle. One of these is the gamete producing (haploid) plant-called *gametophyte* and the second spore proproducing (diploid) plant called *sporophyte*. Both are free living.

(*d*) Many members of the order Siphonales (Fig. 3.26) among the green algae tell a different tale. They exhibit a different type of life cycle. In their case the thallus is a diploid structure which represents the sporophyte phase. It bears both asexual and sexual reproductive organs. Meiosis occurs only at the time of differentiation of gametes. The latter fuse to form a zygote, in which the diploid condition, characteristic of sporophyte phase is reestablished. The zygote by repeated equational mitosis develops into a diploid filament. In this type of life cycle the sporophyte phase is prolonged and the gametophyte phase is extremely reduced. The sporophyte phase consists of the zygote, the vegetative filament, zoosporangia and the sex organs till the occurrence of the reduction division. Only the gametes represent the gametophyte phase. The asexual spores (zoospores and aplanospores) being diploid serve to multiply the diploid or the sporophyte phase in the life cycle. They play no role in the alternation of generations. In the life cycle of Siphonales, therefore, there is an alternation of a well developed sporophyte with a single celled and reduced gametophyte as represented by the gametes. The life cycle which is characterized by gametogenic meiosis and diploid adult is called **diplontic life cycle** (Fig. 3.26). Svedelius (1931) termed it *haplobiontic diploid* because the single free living individual involved in the life cycle is diploid. This type of life cycle is characterized by *gametangial meiosis*.

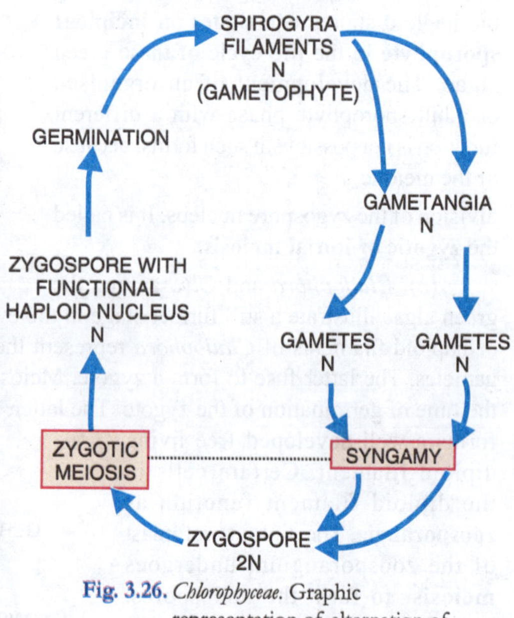

Fig. 3.26. *Chlorophyceae.* Graphic representation of alternation of generations in *Caulerpa.*

Salient Features of Green Algae

The **Chlorophyta** are characterised by the following :

1. They are grass green in colour owing to the preponderance of chlorophyll *a* and *b* over carotene and xanthophyll.
2. The pigments are localised in the green plastids known as chloroplasts.
3. The reserve carbohydrate food is stored as starch.

4. The chloroplasts normally contain the pyrenoids.
5. The cell has a well-defined nucleus and in the higher forms a central sap cavity in addition.
6. The cell wall is stable and invariably contains cellulose.
7. The majority produce motile reproductive cells which may be bi-or quadriflagellate rarely with a ring of flagella as in Oedogoniales. The flagella are of equal length and of whiplash type inserted at the anterior end.
8. Sexual reproduction ranges from isogamy to oogamy. The zygote nearly always enters upon the resting period.
9. The sex organs are always unicellular.
10. Zygote generally is the only diploid structure in the life cycle.
11. The plant body is unicellular or multicellular. The former may be solitary or colonial. The multicellular form show a considerable range of variation in the form and structure of the plant body. Commonly they are filamentous, sometimes massive. However it is always simple in construction and shows relatively little cellular organisation.

Classification of Green Algae

Fritsch (1935) placed all the green algae in the class **Chlorophyceae**. Pascher, however, considered the stone worts (*Charales*) to be markedly different from the other green algae. He therefore, placed the stoneworts in a separate division, **Charophyta** (Charophycophyta) and the other green algae in the division **Chlorophyta** (= Chlorophycophyta). Smith (1958) followed the *via media*. He included all the green algae in the division **Chlorophyta** and divided it into two classes **Chlorophyceae** and **Charophyceae**. The latter includes the stoneworts.

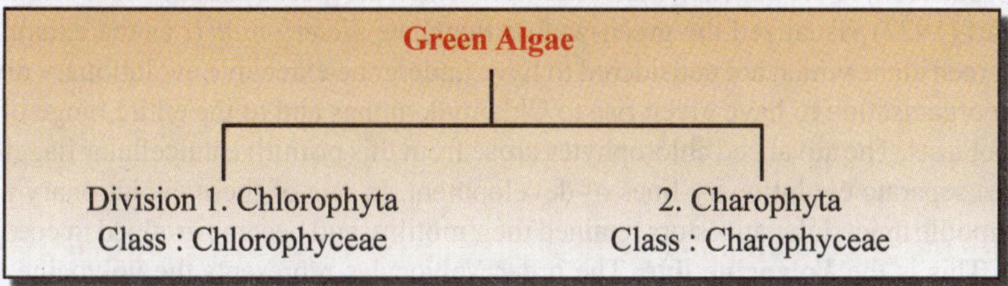

The overwhelming majority of modern phycologists, on the basis of the equisetoid habit and presence of enveloping sheaths around the sex organs, favour an independent status of a division to the stoneworts as Charophyta (Charophycophyta). This view has been followed in this text. Fritsch divided the class Charophyceae into 9 orders namely, Volvocales, Chlorococcales, Ulotrichales, Oedogoniales, Cladophorales, Chaetophorales, Conjugales, Siphonales and Charales. Smith makes a departure. He recognises 12 orders which are Volvocales, Tetrasporales, Ultotrichales, Ulvales, Schizogoniales, Cladophorales, Oedogoniales, Zygnematales, Chlorococcales, Siphonales, Siphonocladales and Dasycladales. In this text are discussed 9 orders which are Volvocales, Chlorococales, Ulotrichales, Ulvales, Oedogoniales, Cladophorales, Chaetophorales, Zygnematales and Caulerpales (= Siphonales).

Stewart and Mattox (1975) and Pickett-Heaps (1975, 76) proposed a new system of classification of the green algae on the basis of nuclear and cell division, organisation of zoospores or gametes and their comparative biochemistry. They include all the green algae in the division *Chlorophyta* (Chlorophycophyta) and subdivide it into the following two classes :-

1. Chlorophyceae. Telophase nuclei in close proximity and spindle absent, *phycoplast* (consisting of microtubules perpendicular to the axis of the spindle) organised at the site of cytokinesis which may take place by furrowing or cell plate motile cells with 4 cruciately arranged flagellar roots at the flagellar bases, flagella apical, do not produce glycolate oxidase. The class includes orders Volvocales, Microsporales, Ulvales, Chaetophorales and Oedogoniales.

2. Charophyceae. Telophase nuclei relatively far apart, spindle persistent unlike the completion of telophase; cell plate organised from a **phragmoplast** (like land plants) as cytokinesis by furrowing proceeds, motile cells with flat band of microtubules at the flagellar bases, flagellar insertion slightly lateral, produce glycolate oxidase. The class includes orders Klebsormidiales, Zygnematales, Coleochetles and Charales.

Origin of Chlorophyta

There is a suggestion that the green algae have evolved from a group of photosynthetic bacteria as have the blue-green algae. The progenitors which constituted the procolgal stock possessed chlorophyll *a* and B-carotene as the principal photosynthetic pigments. They used water as a source of hydrogen and released oxygen as a by-product in photosynthesis. The Chlorophyceae and Cyanophyceae originated from this pro-algal stock along two parallel lines of descent. All the other algal groups except the Rhodophyceae and Cyanophyceae are considered to have evolved from the Chlorphyceae as is the entire sub-kingdom Emleryphyta. The majority of the phycologists, however, hold that the green algae have evolved from simple, unicellular, flagellated ancestors possibly similar to some of the present unicellular Volvocales. This view point was first put forth by two English botanists, Blackman and West (1900). It is based chiefly on three observations namely, (*i*) occurrence of unicellular forms in the Chlorophyta, (*ii*) frequent production of a unicellular stage (spore or gamete) by the multicellular green algae, and (*iii*) similarity of this stage in the different species of green algae and its resemblance to unicellular Volvocales. Among the Unicellular Volvocales, *Chlamydomonas* is considered very much similar in all respects except the presence of a cell wall and sexual reproduction) to the flagellated ancestor from which the greens possibly have originated. Mattox and Stewart (1977) visualized the green-walled flagellate *Hetero-matrix* as the extant remnants of the early green algae which are considered to have undergone extensive evolutionary diversification in cellular organisation to have given rise to Chlamydomonas and to the entire range of green algae with phycoblasts. The advanced chlorophytes arose from this primitive unicellular flagellate ancestor along three separate evolutionary lines of development. In one of these evolutionary pathways the primitive motile unicellular ancestors retained their motility and became grouped in coenobia (motile colonies). This is the **Volvocine line.** The order Volvocales represents the volvocine trend. In the second series the flagellate ancestors lost both motility and capacity to divide into somatic cells resembling the parental cells. It is the **Chlorococcine line** illustrated by the Chlorococcales and the Coenocytic orders Caulerpales, Dasydales and Siphonocladales. The third evolutionary trend known as the *tetrasporine line* involves the loss of motility and retention of capacity of these cells to divide directly into somatic cells resembling the parent cells. This evolutionary trend resulted in the present day order Tetrasporales.

Origin of filamentous habit (multicellularity) in green algae

Three different hypotheses, which attempt to explain this debatable point, have been put forth. The first suggests an origin of multicellular condition in green algae from the motile unicellular ancestor and the second from the palmelloid forms (Tetrasporales). The third envisages origin from the colonial coccoid forms (Chlorococcales). Of course most of the algologists agree that both the Tetrasporales and Chlorococcales have directly evolved from the motile green unicellular ancestor like *Chlamydomonas*.

1. *Origin of filamentous forms from a motile unicellular ancestor* (Fig. 3.27). The advocates of this hypothesis maintain that the multicellular condition in green algae arose through the adherence

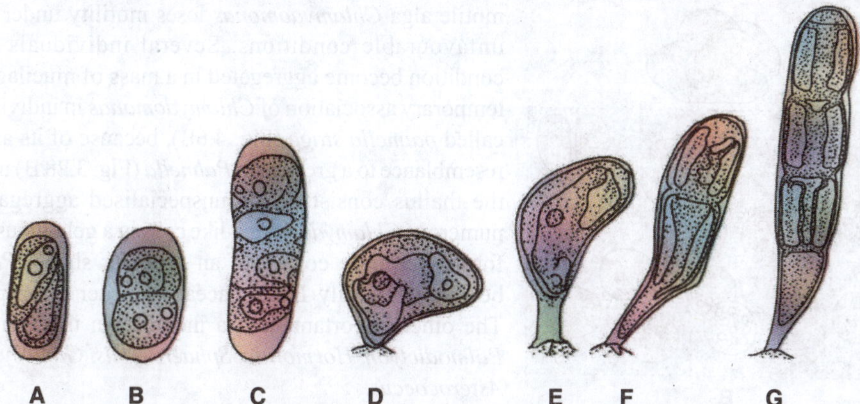

Fig. 3.27 (A-G). *Chlorophyceae.* Origin of filamentous habit. (A-C) derivation of transverse division from longitudinal division of flagellates via *Chlamydomonas seriata* (After Pascher); (D-G) stages in germination of zoospore of *Ulothrix* to form a filament.

of daughter cells after cell division. The first step is the loss of motility of the ancestral unicell in the vegetative condition with retention of ability to divide freely. The divisions occurred in a single plane (transverse) and the daughter cells remained united. The result was a row of cells called a **simple filament** (*Ulothrix*). The transverse division of cells is possibly derived from longitudinal division of Flagellates via forms like *Chlamydomonas seriata*. The first division in this alga is somewhat oblique (A). The two daughter protoplasts after division rotate so that the division comes to be transverse with respect to the parent cell as a whole (B). The two daughter cells again divide longitudinally producing a row of 4 daughter cells, (C). This arrangement has a significant bearing in understanding the origin of filamentous habit in algae from the unicellular ancestor. This hypothesis was supported by Fritisch (1935) who cited as evidence the early growth of many filamentous green algae (Fig. 3.27 D-G). For instance in *Ulothrix, Oedogonium* and many others, the unicellular motile reproductive cells such as the zoospore comes to rest on its base after a short period of activity. It retracts its flagella and attaches itself to the substratum. In the quiescent condition it secretes a thin cellulose wall around it (D). Its protoplast turns through 90° so that its axis coincides with the transverse axis of the filament which arises from it (E). The quiescent-zoospore protopla st then divides into two daughter cells which remain united (F). With further divisions in the same plane a filament of cells (multicellular condition) is produced (G). The supporters of this hypothesis believe that the behaviour of zoospore, at the time of its germination, reflects what happened during evolution of the filamentous green algae like *Ulothrix* from a unicellular ancestor. In other words, the behaviour of the zoospore at the time of germination recapitulates the evolution of the filament. As to whether the unicellular ancestor was motile (*Chlamydomonas*-like) or non-motile (*Chlorella* or *Chlorococcum*), the opinion is divided. Some phycologists believe that the filamentous form arose directly from the motile unicells. Another suggestion is that the filamentous forms evolved from the non-motile unicells which evolved from the motile forms. There is a third suggestion which holds the via media. According to this the filamentous forms with uninucleate cells such as Ulotrichales, Ulvales and Chaetophorales evolved from the motile, unicellular ancestor and the filamentous forms with multicelluleate cells such as the Cladophorales, Siphonocladales, Dasycladales and Siphonales originated from the non-motile unicellular Chlorococcales such as *Chlorococcum*. The Oedogoniales and Conjugales diverged early from the main stream of evolution (tetrasporine line) as represented by the Ulotrichales, Ulvales and Chaetophorales.

2. *Origin of filamentous (multicellularity) from palmelloid forms.* According to this hypothesis, filament evolved through the aggregation of separate cells into a colony. A number of species of green algae serve to illustrate as the likely stages in this evolutionary line. The unicellular

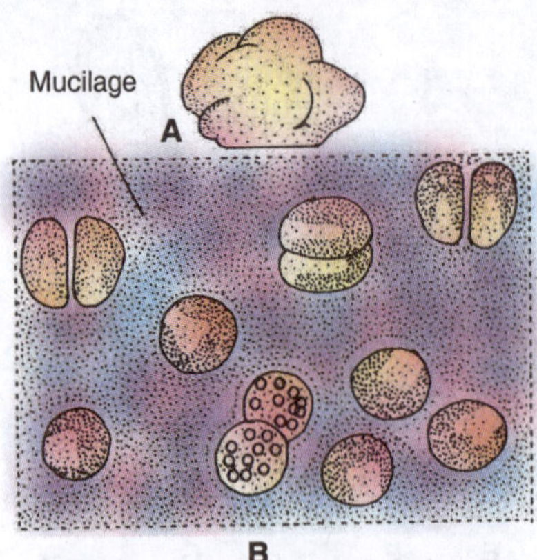

Fig. 3.28 (A-B). *Chlorophyceae. Palmella miniata.* (A) thallus; (B) portion of thallus magnified (After Smith).

motile alga *Chlamydomonas* loses motility under certain unfavourable conditions. Several individuals in this condition become aggregated in a mass of mucilage. This temporary association of *Chlamydomonas* in individuals is called *palmella stage* (Fig. 4.6B), because of its apparent resemblance to a green alga *Palmella* (Fig. 3.28 B) in which the thallus consists of an unspecialised aggregation of numerous *Chlamydomonas*-like cells in a gelatinous matrix forming a loose colony of an irregular shape. *Palmella* belongs to family Palmellaceae of order Tetrasporales. The other important genera included in this family are *Palmodictyon, Hormotilla, Sphaerocystis, Gloeocystis* and *Asterococcus*.

Blackman (1900) advocated the origin of filamentous forms in the Ulotrichales from the palmelloid forms (Fig. 3.29). As evidence he cited existence of palmelloid forms such as *Palmodictyon (P. varians)* in which the cells are sometimes arranged in linear series (A). He assumed that the restriction of cell division in such forms to a single plane (transverse) led to forms like *Radiofilum* and *Geminella interrupta* (B), members of order Ulotrichales. In these genera, the cells are arranged in a row to form a simple filament but they are separated from each other by gelatinous material. In *Geminella minor* the cells in the filament are sufficiently close for their ends to touch (C). Blackman considered these forms more primitive than *Ulothrix* in which the cells in the filament are attached to each other by a cementing material. Smith (1955) supported this hypothesis.

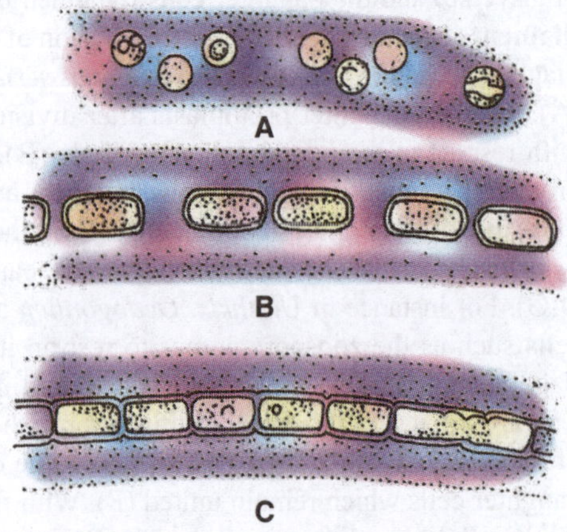

Fig. 3.29 (A-C). *Chlorophyceae.* Illustrating origin of filamentous habit in green algae. (A) *Palmodictyon varians*; (B) *Geminella nterrupta;* (C) *G. minor* (Fritsch).

3. Gupta and Nair (1962) suggested the evolution of filament in green algae from a colonial form *Gloeotaenium loitlesbergerianum* (Fig. 3.30) of family Oocystaceae order Chlorococcales. The mature thallus in this alga is a colony consisting of 2, 4 or 8 cells. Four-celled colonies with the cells

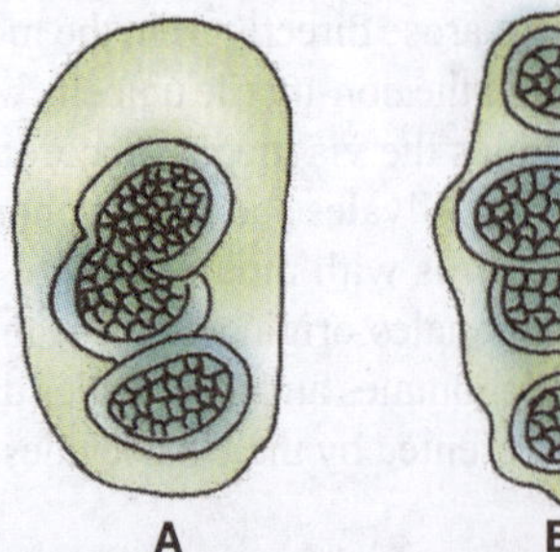

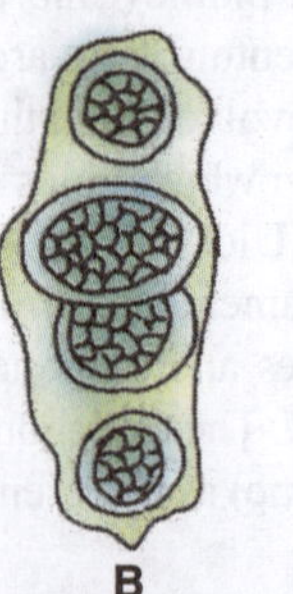

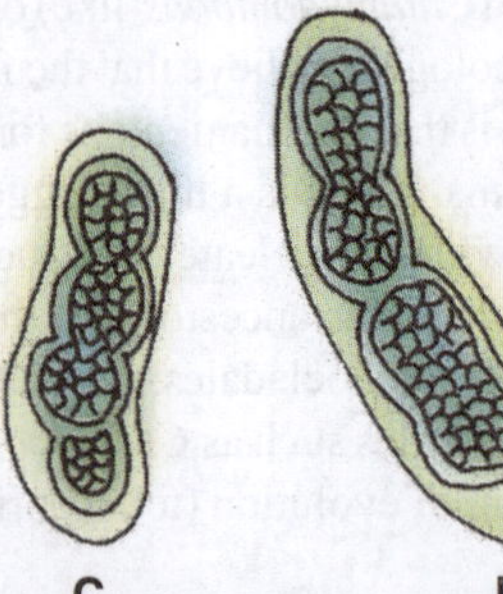

Fig. 3.30. *Chlorophyceae.* Four-celled colonies of *Gloeotoenium loilesber-gerianum.* (A) cells arranged in a tetra-chedral manner; (B-D) stages in the arrange-ment of cells in a single row (After Gupta and Nair).

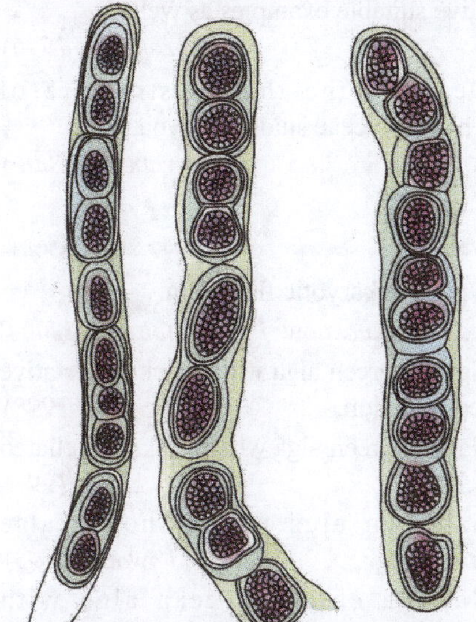

Fig. 3.31 (A-C). *Chlorophyceae.* Long filamentous colonies of *Gloeotaenium loitlesberg eranium* consisting of more than eight cells (After Gupta and Nair).

arranged in a tetrachedial manner or in the same plane are of frequent occurrence (A). They observed a tendency in the cells of the colonies of latter type to become rearranged in a row (B-D). This is followed by repeated cell division in the same plane (transverse) resulting in filamentous colonies (Fig. 3.31 A-C) apparently resembling the young filaments of *Geminella* in the order Ulotrichales. This behaviour of four-celled colonies of *Gloeotaenium loitlesbergerianum* strongly reflects what happened during evolution of some of the filamentous forms in the green algae.

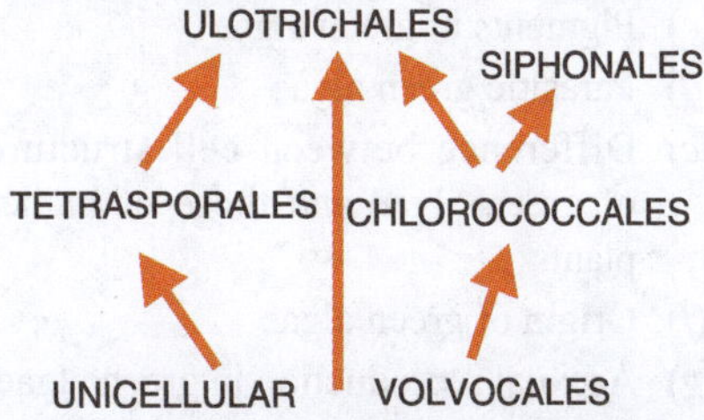

Fig. 3.32. Scheme for origin of filamentous habit in the Ulotrichales as proposed by Gupta and Nair.

From the conflicting opinions stated above, Morris (1968) suggested that it is probable that all possible methods actually happened during evolution of the filament. Gupta and Nair (1962) also suggested that the filamentous habit in the Ulotrichales may have evolved from Unicellular Volvocales, Tetrasporales or from such Chlorococcales as G. *loitlesbergerianum* and not exclusively from any one of these groups (Fig. 3.32).

Advantage of multicellularity (filamentous habit)

The student might pose a question. Why did some groups of algae adopt the multicellular state? The plausible explanation appears to be that multicellularity permits cell specialisation. Some cells in the thallus trap solar energy to use it in photosynthesis, some serve the purpose of reproduction. Some afford protection and still others anchor it to the soil. This division of labour accompanied by differentiation of cells increases the ability of the organism to exploit its environment fully and also increase its chances of survival in the struggle for existence.

QUESTIONS

ESSAY TYPE

1. Discuss the range of thallus structure in Chlorophyceae. Give suitable diagrams also. *(Kanpur MSc, 2005)*
2. Give an account of cell structure in green algae.
3. Describe in detail the various modes of reproduction in green algae. *(Bharathiar, 2000)*
4. Describe in detail the various types of chloroplasts found in different genera belonging to Chlorophyceae.
5. Give an account of different life cycles met within green algae.
6. Give an account of various types of chloroplast in green algae. *(Kumaon, 1995)*

7. Describe the various modes of perennation in green algae. (*Kanpur, 1995; Awadh, 1992*)
8. Describe the various modes of sexual reproduction in green algae studied by you. Give suitable examples as well. (*Gorakhpur, 1996*)
9. Describe the thallus structure of Chlorophyceae studied by you. (*Poorvanchal, 2006*)

SHORT ANSWER TYPE

10. **Write short notes on:**
    (*a*) Distribution of green algae
    (*b*) Chlorophyceae
    (*c*) Pigments in green algae
    (*d*) Parasitic green algae
    (*e*) Difference between cell structure of green algae and those of higher plants
    (*f*) Origin of green algae
    (*g*) Asexual reproduction in green algae
    (*h*) Sexual reproduction in green algae
    (*i*) Structure of chloroplast in green algae/Chlorophyta (*Allahabad, 2002*)
    (*j*) Alternation of generation in green algae
    (*k*) Economic importance of green algae
    (*l*) Shape of chloroplasts in chlorophyta (*Allahabad, 1991, 1993*)
    (*m*) Zoospores of chlorophyceae (*Allahabad, 1995*)
    (*n*) Chief characters of green algae (*Lucknow, 2005*)
    (*o*) Variation in chloroplast structure in Chlorophyceae (*Lucknow M.Sc., 2006*)
11. In what respects does the protoplast of the cell of green algae show an advance over that of the blue green algae?
12. Describe the salient features of Chlorophyceae and name the various orders placed in the class. (*Allahabad, 1996*)
13. Name an alga which is used as food. (*Allahabad, 1994*)
14. Name the reserve food material in the cells of green algae. (*Allahabad, 1991; Gorakhpur, 1991*)
15. Name an alga which has starch as reserve food. (*Allahabad, 1997*)
16. Draw an ultrastructure diagram of T.S. of a typical eukaryotic flagellum. (*Lucknow, 1993, Allahabad, 2004*)
17. Name a green alga which lacks vegetative reproduction. (*Lucknow, 1998*)
18. Name a green alga which lacks flagellated cells. (*Lucknow, 1998*)
19. Name an alga which shows false branching. (*Awadh, 1993*)
20. Name a chlorophycean alga with *heterotrichous* habit. (*Awadh, 1993*)
21. Name an alga which is rich in protein. (*Gorakhpur, 1992*)
22. Name the alga where the thallus consists of an erect and postrate systems. (*Gorakhpur, 1995*)
23. What is a coenobium? (*Gorakhpur, 1996*)
24. Distinguish between
    (*a*) Chlorophyceae and Charophyceae (*Rohilkhand, 1992*)
    (*b*) Chloroplast and Pyrenoid
25. Differentiate between isogamy, anisogamy and oogamy modes of reproduction. (*Gurunanak Deo, 1991*)
26. Describe the structural details of neuromotor apparatus.
27. Differentiate between
    (*a*) Coenobium and Colony (*Lucknow, 2001*)
    (*b*) Akinete and Zoospore (*Allahabad, 2005*)
28. Describe briefly
    (*a*) Range of thallus structure in chlorophyceae (*Lucknow, 2002*)
    (*b*) What is coenobium? Give two examples. (*Lucknow, 2004*)
    (*c*) Characters of eukaryotic algae (*Lucknow, 2007*)
29. Write a note on the structure of algal flagella with diagrams.

OBJECTIVE TYPE

30. Fill in the blanks:

(*a*) Fusion between gametes of unequal sizes is called ....

(*b*) A motile flagellated asexual cell is known as ....

(*c*) .... is fusion between gametes of equal sizes.

(*d*) .... is the common name of *Chara*.

(*e*) Unicellular non-motile spherical green alga is ....

(*f*) An alga growing on snails is called ....

(*g*) An alga in which cup shaped chloroplast is present is known by the name ....

(*h*) Chlorophyll *a* and *b* are present in the members of the class ....

(*i*) The reserve food material of green algae is ....

(*j*) Coenobium is found in ....

(*k*) Luxuriant growth of some algae in water often imparting colour to the water is called ....

(*l*) The green algae may be broadly divided into two classes namely .... and ....

(*m*) Pyrenoids are meant for .... synthesis in various algae.

(*n*) Zoospores are usually produced under .... conditions.

(*o*) Asexual spores which are motile and have two flagella are called ....

31. Select the correct answer:

(*i*) Heterotrichy means having
- (*a*) Prostrate and erect branches
- (*b*) Rhizoidal and photosynthetic branches
- (*c*) Long and short branches
- (*d*) Branches differentiated into nodes and internodes

(*ii*) Reserve food is starch in
- (*a*) Chlorophyceae
- (*b*) Myxophyceae
- (*c*) Phaeophyceae
- (*d*) Rhodophyceae

(*iii*) Asexual spores which are motile and have two flagella are called
- (*a*) Zoospores
- (*b*) Synzoospores
- (*c*) Zygospores
- (*d*) Chlamydospores

(*iv*) Which of the following has reserve food as starch?
- (*a*) *Volvox*
- (*b*) *Vaucheria*
- (*c*) *Ectocarpus*
- (*d*) *Batrachospromum.*

(*v*) Chlorophyll *a* and *b* are together present in
- (*a*) Chlorophyceae
- (*b*) Myxophyceae
- (*c*) Phaeophyceae
- (*d*) Rhodophyceae

(*vi*) Which of the following has heterotrichus habit?
- (*a*) *Ulothrix*
- (*b*) *Coliochaete*
- (*c*) *Oedogonium*
- (*d*) *Oscillatoria*

(*vii*) Which of the following has Cenobium?
- (*a*) *Volvox* (*b*) *Vaucheria*
- (*c*) *Ectocarpus* (*d*) *Ulothrix.*

(*viii*) Asexual zoospores are usually produced under
- (*a*) Unfavourable conditions
- (*b*) Favourable conditions
- (*c*) Both of the above
- (*d*) None of the above

(*ix*) Zygospores are produced as a result of
- (*a*) Isogamous fusion
- (*b*) Gametangial contact
- (*c*) Gametangial copulation
- (*d*) Oogamy

(*x*) Non motile thin walled spores of algae are called
- (*a*) Macrospore
- (*b*) Microspore
- (*c*) Aplanospore
- (*d*) Zygospore

(*xi*) Fusion between gametes of unequal sizes is called
- (*a*) Isogamy
- (*b*) Anisogamy
- (*c*) Planogametic fusion
- (*d*) Oogamy

Volvocales: *Pleodorina sp.*

CHAPTER 4

CLASS : CHLOROPHYCEAE

Order 1 : Volvocales

The Volvocales include chiefly the microscopic genera of the green algae in which the thallus is one-celled and generally motile throughout life. The unicell is furnished with 2, rarely 4 whiplash type flagella of equal length. This is the only order of the green algae in which the vegetative cell is actively motile. In some genera the motile cells are joined into groups to form an organisation called a **colony**. A colony is an aggregation of individuals mechanically held together generally in a gelatinous sheath. The individuals in the colony have little or no dependence upon others. The colonies included in this order have a definite number of cells arranged in a specific manner. This type of colony is called **coenobium**. The coenobia are motile. The order

Volvocales, therefore, includes both **unicellular** and **colonial** forms which occur widely in fresh water plankton. They are absent from the sea. Almost all the representatives are fresh water. Many of them prefer water rich in nitrogen and organic substances and thus occur in quiet pools. The order comprises at least six families (Chlamydomonadaceae, Volvocaceae, Polyblepharidaceae, Chlorodendraceae, Phacotaceae and Sphaerellaceae) with about 60 genera and about 500 species. Of these Chlamydomondaceae and Volvocaceae are described here. A mention, however, be made of another small family of the order. It is Polyblepharidaceae. All the members of this family are naked. Polyblepharides, a member of this family only divides asexually by longitudinal division into two.

Coenobia in *Volvox.*

FAMILY 1. CHLAMYDOMONADACEAE

The family includes unicellular motile forms with a distinct cell wall. It comprises about 20 genera. The cell is bi-or quadriflagellate (*Carteria*). The flagella are of equal length (**isokont**). Asexual reproduction takes place by zoospores. Sexual reproduction ranges from isogamy to oogamy through anisogamy. The zygote undergoes the resting period. The division of the zygote nucleus is meiotic. The important genera included in this family are *Chlamydomonas, Haematococcus (Sphaerella), Carteria, Chloromonas, Polytoma, Polytomella* and *Hyaliella.* Of these *Chlamydomonas* is usually taken as a type.

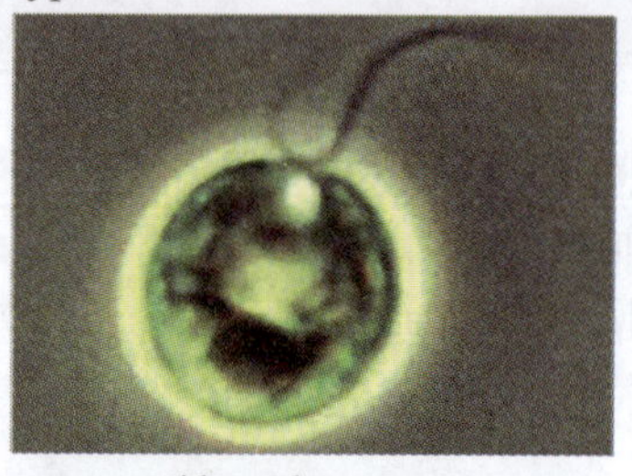

Chlamydomonas.

Carteria.

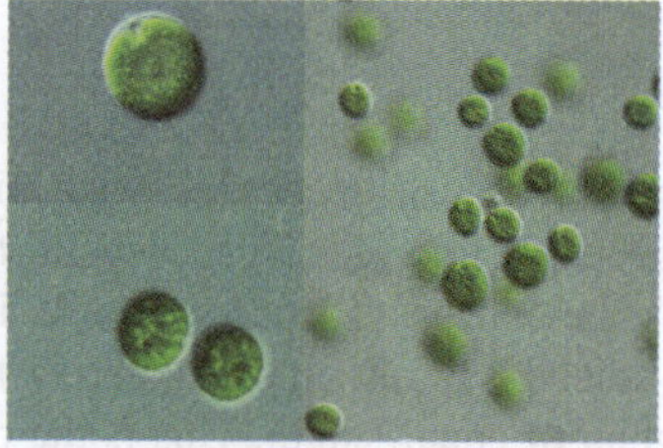

Chloromonas.

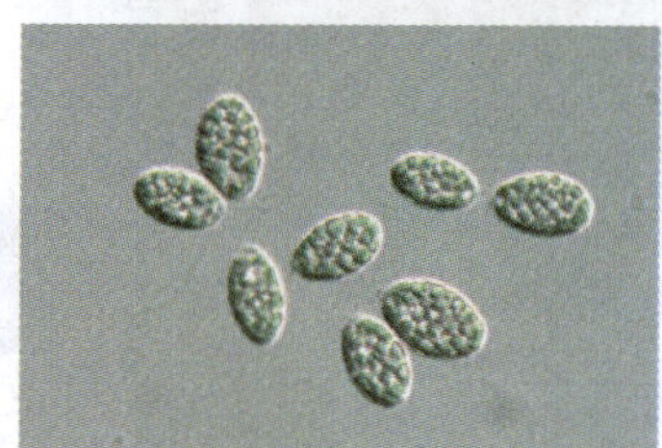

Polytoma.

MEMBERS OF FAMILY CHLAMYDOMONADACEAE

CHLAMYDOMONAS

Occurrence

Chlamydomonas is one of the most primitive, free swimming, nucleated (**eukaryotic**) organisms. The principal plant characteristics are cellulose cell wall typical of plants, presence of chloroplast and autotrophic nutrition. The chief animal characteristic is its motility in the vegetative phase. Some zoologists therefore consider it a plant-like flagellate. Botanists take it as a widely distributed fresh water, free swimming alga. Many species occur in waters having a high content of organic matter. Thus it is frequently found in the standing water or ponds, ditches, swimming pools, lakes, rain-water tanks and also on damp soil. Sometimes *Chlamydomonas* forms a green scum over the surface of small bodies of water such as barnyard, pools, and drinking troughs. Often it occurs in such great number as to colour the water green. It prefers water rich in ammonium compounds. *C. grandistigma* and *C. eugametos* have been recorded from Allahabad by Mitra (1957). In all about 500 species of *Chlamydomonas* have been described. Some occur in moist soil.

Structure of Thallus. (Fig. 4.1 A).

(a) Form. The plant body is a thallus which consists of a single biflagellate cell (A). In some species the unicell is about 20 microns in length. It rarely exceeds 30 in major diameter. The unicellular thallus is biflagellate and usually oval (egg-shaped) or rather oblong in form. Other forms such as cylindric, pear-shaped and spherical are by no means rare.

(b) Structure. With light microscope the unicell appears to consist of a tiny mass of protoplast enclosed in a distinct **cell wall**. The cell wall is thin and transparent. It is smooth and closely applied to and surrounds the protoplast. Early reports suggested that it is cellulose in nature but Fritsch reported that the cell wall contains hemicellulose and not true cellulose. The cellulose nature of the cell wall was reconfirmed by Lewis *et al* (1951) and Barnett and Priston (1969). Ultrastructural study of the cell wall revealed that it is finely striated with the parallel fibrils oriented circumferentially. The determination of chemical composition of the cell wall showed that it is principally hydroxyproline-rich glycoproteins, glycostylated with arabinose and *galactose sugars* (Muller *et al* 1975; Roberts, 1974 and Catt *et al*, 1976). In many species there is a pectose layer external to the cell wall. The pectose at the surface changes into pectin which dissolves in water to form a mucilaginous pectin layer surrounding the cell wall (Fig. 4.2). It is also called the *capsule layer*. It appears as a fibril felt with a frayed outer surface.

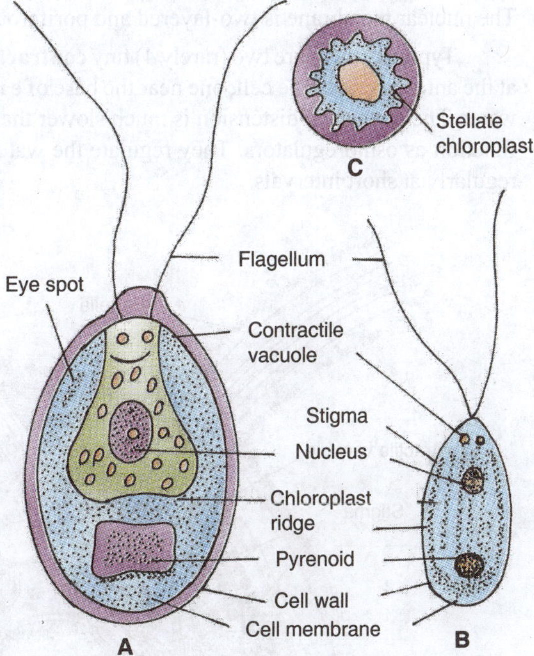

Fig. 4.1 (A-C). *Volvocales* (A) *Chlamydomonas angulosa*; (B) *C. steinu steinii*; (C) Optical T.S. of B (After Dill).

The protoplast is differentiated into a plasma membrane, cytoplasm, a single nucleus, a single chloroplast with one or more pyrenoids, two contractile vacuoles, a red eye spot and other cell organelles such as the mitochondria, dictyosomes, endoplasmic reticulum, ribosomes and the neuromotor apparatus. However, there is no central vacuole. Within the cell wall closely investing the cytoplasm is the plasma membrane which consists of two electron opaque layers. Included in the cytoplasm in *majority of the species* is a single massive roughly cup-shaped chloroplast. It almost fills the oval or pear-shaped body of the cell at its blunt hinder end. The chloroplast is located in the peripheral part of the cytoplasm within the plasma membrane. It is thus **parietal** in position. In some species the chloroplast has ridges. It is thus laminate or stellate (Fig. 4.1 B,C). In the narrow region of the cytoplasm external to the chloroplast are visible a few strands of granular endoplasmic reticulum, free ribosomes and mitochondria. The hollow of the cup is filled with colourless cytoplasm which contains mitochondria, volutin granules, a few strands of granular endoplasmic tubules, free ribosomes and golgi bodies besides the single nucleus. The mitochondria have plate-like cristae. The golgi bodies (dictyosomes) are paranuclear in arrangement. The volutin granules are perhaps reserves of phosphate.

The single **nucleus** is located centrally, sometimes nearer the anterior end. It is embedded in the colourless cytoplasm that fills the hollow of the chloroplast cup. The nucleus has a distinct nucleolus. The nuclear membrane is two-layered and poriferous.

Typically there are two (rarely 4) tiny **contractile vacuoles** situated in the colourless cytoplasm at the anterior end of the cell, one near the base of each flagellum. These contract alternately to expel water. The process of distension is much slower than contraction. Probably the contractile vacuoles function as osmoregulators. They regulate the water contents of cell by discharging their contents regularly at short intervals.

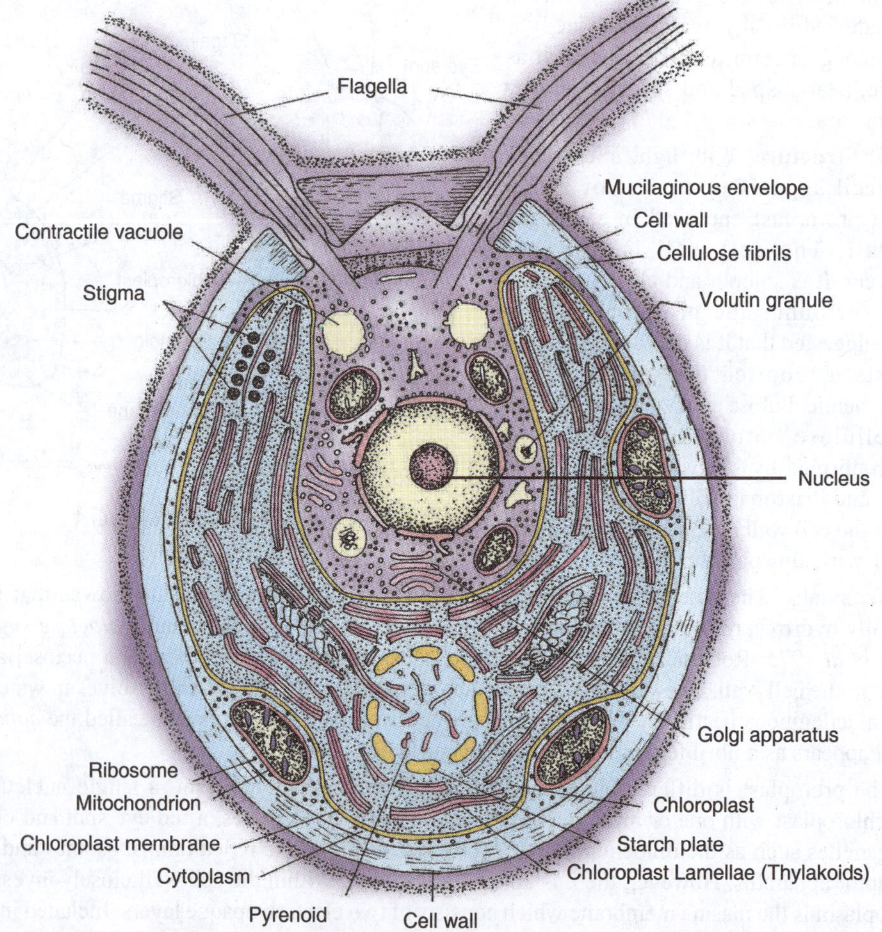

Fig. 4.2. *Chlamydomonas sp.* Cell structure as revealed by electron microscope (Based on Vickerman).

Ultrastructurally the parietal **chloroplast** is a double membraned, cup-shaped, sac-like structure enclosing granular matrix called the **stroma**. Traversing the stroma are stacks or bands of membranous tube-like structures called the discs or *thylakoids*. The latter contain the chlorophyll pigment. The thylakoids usually run parallel to the cell wall. Usually there are 2-6 thylakoids in a single stack but in *C. reinhardii* the number goes up to 8. The stacks, however, are not organized into grana-like structures. The stroma contains ribosomes smaller than those of the cytoplasm. They are considered to be the sites of chloroplast protein synthesis. Located medianly in the lower part of the chloroplast is the solitary medium sized usually spherical pyrenoid. The number of pyrenoids however varies. It may be two to several (*C. sphagnicola*). In a few species the pyrenoids may be absent. The pyrenoid has a central proteinaceous core surrounded by minute plates of starch. The pyrenoid lacks ribosomes. The pyrenoid core is not traversed by chloroplast lamellae. The pyrenoid functions as a centre around which starch is formed. It is also lamellated. The pyrenoid lamellae are thicker than those of the chloroplast. A tiny spot of an orange colour or reddish pigment (carotenoid) lies laterally at the anterior end of the cell inside the chloroplast close to its rim. It is called the **stigma** or the **eye spot** (Fig. 4.2). The eye spot thus is an integral part of the chloroplast. It is commonly found in flagellated motile cells only. It functions as a **photoreceptive organ**. According to Mast (1927), it consists of two parts: (Fig. 3.14 C) (i) a curved pigmented plate and (ii) a biconvex, hyaline photosensitive substance which functions as a lens. The latter lies in and external to the pigmented plate. The eye spot is sensitive to light and directs the movements of swimming cells. It is in fact the site of light reception and helps in bringing the plant into light of moderate intensity which plays an important part in its nutrition. Ultrastructurally the eyespot (Fig. 4.4) is composed of two sometimes three rows of globules concentrically arranged near the anterior end of the cell inside the chloroplast close to its rim between the outermost lamella and the chloroplast envelope. Each layer consists of equal sized closely packed globules containing carotenoid pigment. The globules are not membrane-bound either individually or in groups.

The cell is provided with two long, whip-like protoplasmic threads called the **flagella**. They are of whiplash type with a typical 9+2 arrangement of the component fibrils. They pass out through the cell wall in fine canals, at the narrow anterior end of the cell and lie close together on either side of a distinct protoplasmic papilla. In some species the papilla is absent. Immediately after emergence from the anterior end the flagella curve towards the back of the cell. Each flagellum arises from a basal granule, the **blepharoplast** (basal body) which controls the movement of its respective flagellum. The flagella are of equal length and longer than the length of the cell. They are the locomotory organs, the energy for locomotion is derived from ATP. To bring about locomotion they move in the same fashion as the arms of a man swimming with the breast stroke. Another but conflicting view is that a sine wave passing from the base of the flagella to the tip brings about the forward movement of the cell. Kater (1929) stated that the flagella are closely associated with the **neuromotor apparatus** of the cell. In fact the flagella owe their origin to this apparatus. The neuromotor of **flagellar** apparatus (Fig. 4.3) consists of :

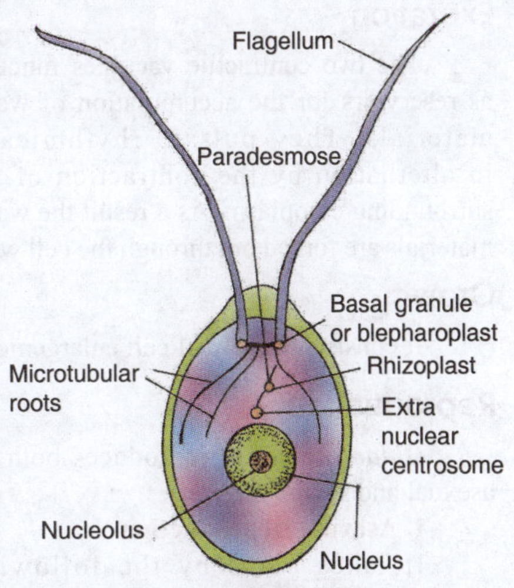

Fig. 4.3. *Chlamydomonas sp*. Showing neuromotor apparatus (diagrammatic).

(*i*) *Two* **blepharoplasts** (basal granules) one at the base of each flagellum.

(*ii*) A delicate transverse fibre known as the **paradesmose**. It connects the two blepharoplasts.

(*iii*) A descending delicate fibre called the **rhizoplast**. It runs down from one of the blepharoplasts and joins the **centrosome**, which, may be intranuclear or may lie just outside the nucleus (Elliot).

(*iv*) The centrosome is connected by a delicate fibril with the nucleolus.

(*v*) **Microtubular roots**. These originate from the paradesmose and are distributed to different parts of the body giving support.

The neuromotor apparatus constitutes the incipient nervous system. When the mother cell becomes quiescent the rhizoplast and basal granules disappear.

The motile forms like *Chlamydomonas* have an advantage over the non-motile blue green algae in that they swim toward better conditions of light and also move away from shallow water which becomes unbearably hot in summer months.

Nutrition

Chlamydomonas is an **autotroph**. The mode of nutrition like other green plants is holophytic. By virtue of its chlorophyll *Chlamydomonas* cell manufactures sugar in the presence of sunlight from water and carbon dioxide. The raw materials for the photosynthetic process are absorbed over its surface from the surrounding water. The sugar formed is used up as a source of food. The excess of sugar synthesised during photosynthesis is converted to starch which accumulates in the form of little grains or plates around the pyrenoids. All the species of *Chlamydomonas* thus are obligate phototrophs. The only exception is *C. dysosmos* which is a facultative heterotroph.

Respiration

Oxygen is absorbed from the surrounding water over the entire surface of the cell. It combines with the sugar and sets free the potential energy (absorbed from the sun during photosynthesis) in the form of kinetic energy which keeps the vital processes of the cell going.

Excretion

The two contractile vacuoles function as reservoirs for the accumulation of waste materials. They pulsate rhythmically in alternation by the contraction of the surrounding cytoplasm. As a result the waste materials are forced out through the cell wall.

Growth

It consists entirely of cell enlargement.

Reproduction

Chlamydomonas reproduces both by asexual and sexual methods.

1. Asexual Reproduction

It takes place by the following methods:

(a) *Zoospore Formation* (Fig. 4.4). *Chlamydomonas* multiplies asexually by **zoospores** in the growing season. The parent cell comes to rest. The flagella are resorbed, contractile vacuoles disappear and the protoplast withdraws from the cell wall (A). In this quiescent state the cytoplasm, chloroplast and the nucleus divide along a longitudinal plane into two daughter protoplasts (B).

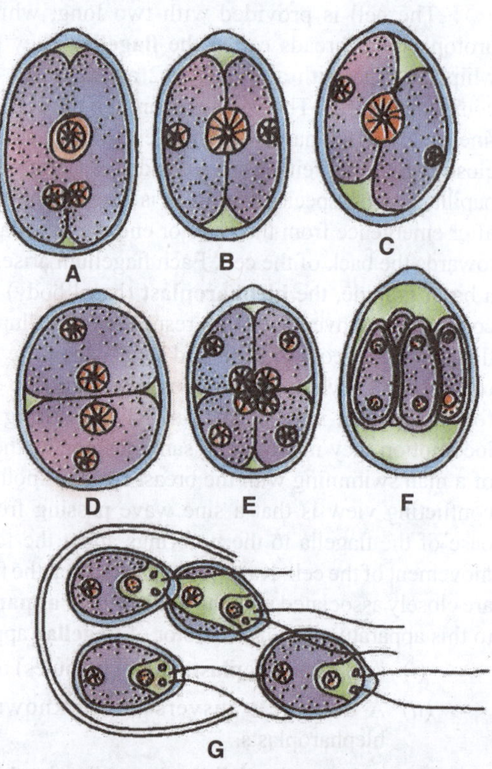

Fig. 4.4 (A-G). *Chlamydomonas sp.* Successive stages in zoospore formation.

The nuclear division is mitotic. A few species, however, remain motile during division. The division of the parent protoplast into two daughter protoplasts is brought about by a constriction which appears at the back and front ends and gradually deepens. In some species the protoplast during or after division rotates through a right angle (C and D). Consequently the division comes to be transverse with respect to the cell as a whole (D). It lies across the short diameter of the parent cell. However it is still longitudinal with respect to the protoplast. No rotation of the protoplast during division takes place in other species. The second division is at right angles to the first. It is again longitudinal to the axis of the protoplast. In this way usually four daughter protoplasts may be formed (E). The chloroplast is halved along with the pyrenoid at each successive division. One daughter cell receives the eye spot of the parent, the other forms it afresh. The aflagellate daughter protoplasts remain bound within the parent cell wall (F).

Occasionally when the conditions are exceptionally favourable mitotic cell division continues longer and eight daughter protoplasts or more may be formed by this method of *successive bipartition*. The divisions are always longitudinal and simultaneous. When the division ceases the daughter protoplasts usually become ranged parallel to one another within the distorted and distended parent cell wall (F). These daughter protoplasts within the original cell wall look like a colony (F). This association, however, is temporary. Each daughter protoplast finally acquires a new cell wall, develops flagella and a neuromotor apparatus of its own. The contractile vacuoles reappear. The flagellate daughter cells take on the shape of the parent. In fact they resemble in all respects the parent *Chlamydomonas* plant but are smaller in size. The parent cell wall then ruptures (G) or gelatinizes. The daughter cells, each of which is a miniature edition of the parent cell, are released. They are sometimes called the **zoospores** or more appropriately **mitozoospores**. However they differ from the zoospores in having a cell wall. The liberated daughter cells (mitozoospores) soon grow to the full size and repeat the process. Under favourable conditions the division is complete in a few hours and takes place daily. The mitozoospores are liberated towards evening.

(*b*) Wille (1908) reported that sometimes the *Chlamydomonas* cell comes to rest and the flagella are resorbed. The protoplast withdraws from the parent wall and rounds up. It secretes a thin wall around it to become an **aplanospore** (*C. caudata*). To survive severe conditions the single spore secretes a thick wall around it. Such a resting spore is called a **hypnospore** (*C. nivalis*). It often develops a strong red coloration.

(*c*) **Palmella stage** (Fig. 4.5) Under certain unsuitable conditions the motile cells come to rest and lose flagella. The protoplast of each divides repeatedly by the method of successive bipartition into two (A and B), four or eight daughter cells. The resultant daughter protoplasts fail to develop flagella (particularly in species growing on moist soil) and thus do not escape. They remain clustered together in the parent cell wall which subsequently becomes mucilaginous and swells up considerably. Numerous aflagellate daughter protoplasts thus become embedded within a common mucilaginous matrix formed by the gelatinisation of the parent cell walls (C). Each such daughter cell may in turn divide and the process of division may continue indefinitely. Repeated divisions of these daughter cells are accompanied by progressive gelatinisation of the walls of the successive generations. Finally a colony of considerable size containing numerous cells embedded in a common mucilaginous mass is formed. This assemblage of the cells is known as

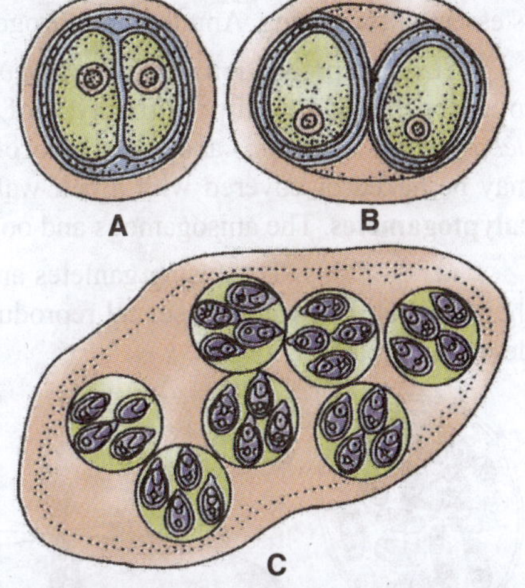

Fig. 4.5 (A-C). *Chlamydomonas sp.* Successive stages in the formation of palmella stage (After Goroschankin).

the **palmella stage** (C) because the older phycologists mistook them to be species of another alga of that name (*Palmella*). This stage is usually of brief duration. The individual cells, after some time when conditions for growth become favourable, become motile and escape from the mucilage envelope to produce the motile stage again. Palmella stage is an immobile reproductive phase. It helps *Chlamydomonas* to survive periods of partial desiccation or tide over spells of adversity when the ionic balance of the medium becomes unfavourable. Chapman, however, reported that in *C. kleinii* it is the dominant phase in the life cycle. Vischer (1926) reported that in *C. pulvinata* the palmella stage is favoured by liquid media and ions that induce swelling. Franck (1904) considered concentrated solutions to have favoured palmella stages in *C. Franki*.. Smith holds that *Chlamydomonas* forms palmella stages when it grows on damp soil or is cultured on agar. Sometimes the individuals in the palmella stage develop into red non-motile resting spores called the **hypnospore** by secreting thick walls around them. Their contents may frequently be coloured red, due to the presence of a red pigment called **haematochrome**. The red hypnospores of quite a number of species of *Chlamydomonas* (*C. nivalis*) are responsible for the phenomenon of 'red snow'.

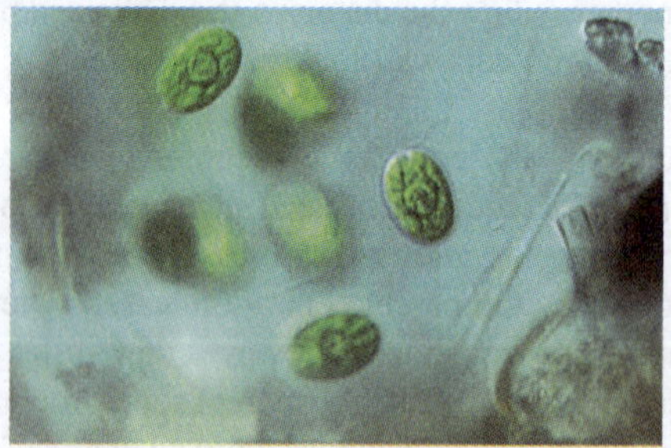

Palmella Stage in *Chlamydomonas.*

Red snow caused by *Chlamydomonas nivalis.*

2. Sexual Reproduction

Sexuality is controlled by certain environmental factors. The chief among these which favour sexual reproduction are (i) low supply of nitrogen, (ii) deficiency of nutritive materials, (iii) temperature and ion concentration, and (iv) bright sunlight and high CO_2 concentration. The presence of calcium is essential for mating. Ammonium nitrogen inhibits sexuality.

Sexual reproduction in *Chlamydomonas* varies through a wide range. It ranges from isogamy to anisogamy and even primitive type of oogamy. Some isogamous species are **homothallic** (*C. debaryanum, C. media*) and some **heterothallic** (*C. reinhardii*). The gametes are biflagellate and may be naked or covered with a cell wall. The former are called **gymnogametes** and the latter **calyptogametes**. The anisogamous and oogamous species produce only calyptogametes.

(*a*) Isogamy. The fusing gametes are similar in size, form and structure and thus are called the isogametes. Isogamous sexual reproduction in homothallic and heterothallic species are being described separately.

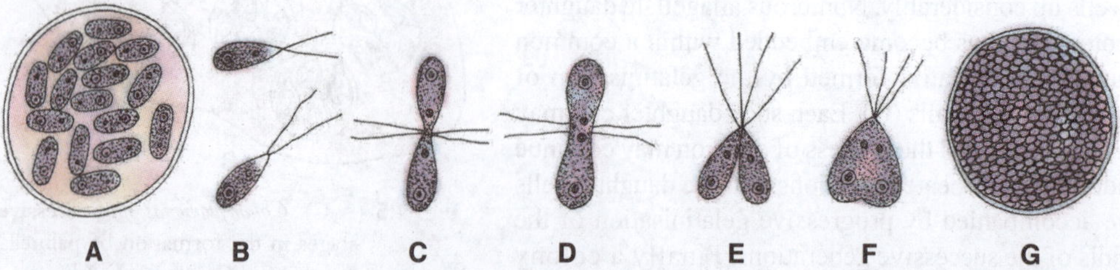

Fig. 4.6 (A-G). *Chlamydomonas sp.* Showing isogamous sexual reproduction in the homothallic species.

(*i*) *Homothallic species* (Fig. 4.6). *C. debaryanum* and *C. longistigina* are the typical examples of homothallic species which illustrate isogamous type of sexual reproduction. Sexual fusion takes place between biflagellate gametes coming from the same parent cell. The fusing gametes are motile, naked and similar in form, size, structure and behaviour (B). At the time of gamete formation *Chlamydomonas* cell withdraws its flagella and comes to rest. The protoplast of the cell then divides by the method of successive bipartition repeatedly to form 16, 32, or 64 daughter protoplasts (A). Each daughter protoplast furnishes itself with two flagella to become a gamete. It is pear-shaped in form and has no cell wall (gymnogamete). The gymnogametes are smaller than the zoospores. The gametes are released in the surrounding water by the rupture of the parent cell wall. The released isogametes swim about for a while. Finally they come near each other and fuse in pairs either end to end (C) or side by side (E). The cytoplasm mingles with the cytoplasm and nucleus fuses with the nucleus. The fusion cell or zygote is at first, a naked, motile quadriflagellate structure (F). It may continue to swim for a while. Soon it comes to rest, retracts its flagella, rounds up and secretes a thick wall around it to become a **zygospore**. *C. media* is also homothallic but it produces calyptoisogametes. At the time of fusion the protoplasts of fusing gametes emerge leaving behind their coats of cellulose.

(*ii*) *Heterothallic species* (Fig. 4.7). *C. reinhardii* and *C. moewusii (C. eugametos)* are examples of heterothallic species which illustrate isogamous type of sexual reproduction. In them sexual fusion occurs only between gametes from two different parent cells which belong to mating types (A). One of these is called the plus strain (+) and the other minus strain (–). The gametes are alike (isogametes) but often chemically differentiated. The chemicals result in mutual attraction between + and – strain gametes which have cell wall (calyptogametes). In *C. reinhardii*, the gametes are generally formed by

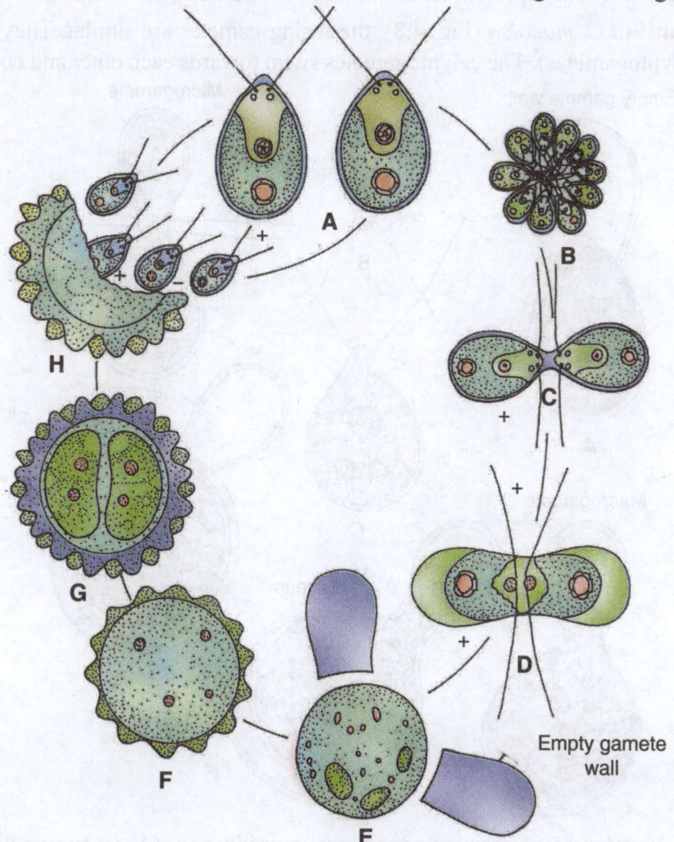

Fig. 4.7. *Chlamydomonas sp.* Showing isogamous sexual reproduction in the heterothallic species.

repeated division of the protoplasts of the mother cells as in *C. debaryanum.* But the adult individuals themselves of *C. moewussi* or the young vegetative cells of opposite mating types as released from the zygospore (H) function as calyptogametes. Before fusion the calyptoisogametes exhibit a marked tendency towards clumping (B) which involves clustering of + and – strain calyptoisogametes through entanglement of their flagellar tips. The affinity of the flagellar tips of opposite mating types for each other is attributed to the presence of complementary mating type substance of a glycoproteinaceous nature. These are known as *isoaglutinins*. The calyptoisogametes emerge from the clump in pairs (C), each pair consisting of + and – gametes still united at their tips (C). Soon after the walls between the two papillae dissolve and flagella become parallel (D). The pairs continue to swim with the help of only one pair of flagella which remain active and belong to the plus strain. The protoplasts of the conjugating individuals gradually slip out of their respective cell walls leaving the empty walls behind and unite completely (in the water), cytoplasm with cytoplasm and nucleus with nucleus (E). The naked protoplasts thus assume the function of gametes. The resultant fusion cell is called the **zygote** (E). It comes to rest, rounds off and secretes a resistant wall. According to Minami and Goodenough (1978) it remains wall-less for some time after mating (150 mts) in *C. reinhardii*. By 180 minutes after mating a distinct wall surrounds the entire cell. During wall formation discrete fibres accumulate around the cell surface until a very thick wall is formed. The heavy walled zygote enters upon a period of rest and is called the **zygospore** (F). This type of sexual reproduction which consists in the fusion of gametes similar in size, form and structure (**isogametes**) is called **isogamy**. The isogametes in this case though morphologically alike exhibit functional disparity. It means they are physiologically heterothallic as one belongs to the plus strain and the other to the minus strain. *C. eugametos* and the like thus represent an initial stage in sexual differentiation.

***(b)* Anisogamy.** In *C. monoica* (Fig. 4.8), the fusing gametes are similar. They are provided with cell walls (calyptogametes). The calyptogametes swim towards each other and come in contact

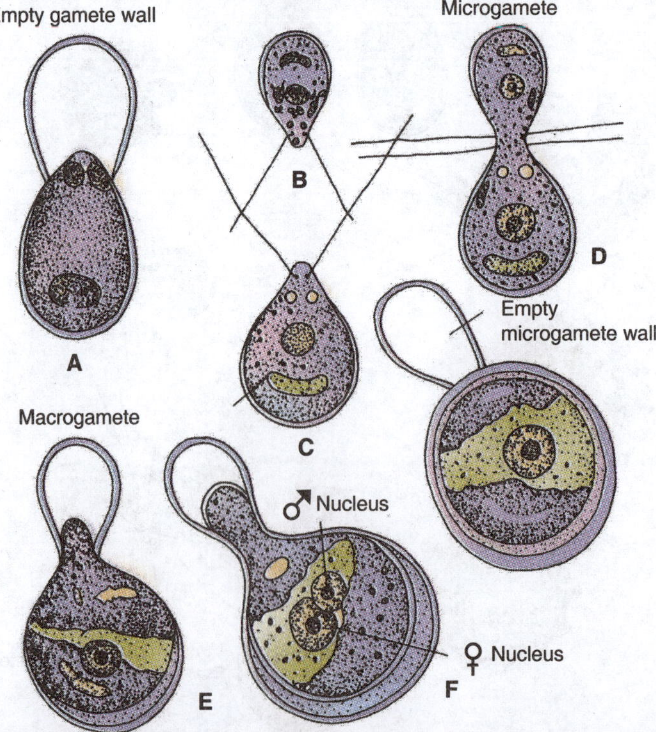

Fig. 4.8 (A-G). *Chlamydomonas sp.* Anisogamous sexual reproduction. (A) *C. monoica* illustrating the first step towards anisogamy (After Strehlow); (B-F) *C. braunii* illustrating three stages in anisogamous fusion; (G) Zygote with the empty microgamete wall. (After Goroschankin).

at their anterior ends (A). The intervening walls dissolve at the point of contact. The protoplast of one of the gametes slips out of its cell wall and flows entirely into the envelope of the other to fuse with it. It is the initial step towards anisogamy and is called **physiological anisogamy**. Forms like *C. monoica* thus serve to link isogamy with anisogamy which is a more advanced type of sexual reproduction.

C. braunii exhibits well marked **anisogamy** (Fig. 4.8 B-D). It shows distinction between the sexes. The gametes differ appreciably in size. The female mother cell (gametangium) produces two or four large **macrogametes.** They resemble the parent cell. The male parent cell gives rise to eight or sixteen, small **microgametes**. They are half the size of the macrogametes and relatively long and narrow. Both macro (C) and the microgametes (B) are calyptogametes. The macrogametes are less motile and soon come to rest without losing their flagella. The microgamete swims towards the macrogamete. The two become attached at their flagellated ends where the walls coalesce (D). The protoplast of the microgamete escapes from its cellulose cell wall (E) and flows into the envelope of the macrogamate (E). The flagella disappear as the two protoplasts fuse with each other. A zygote is formed in the cell envelope of the macrogamete (G). Another step towards oogamy is illustrated by *C. suboogama* in which the female cell produces three macrogametes and the male four microgametes. In both isogamy and anisogamy fusion between the gametes takes place in the surrounding water. The anisogamous species are all heterothallic.

(*c*) Oogamy (Fig.4.9). Oogamous state of sexual reproduction is reached in *C. coccifera.* The female mother cell withdraws its flagella (A). The protoplast rounds off to form a single large macrogamete. The globose macrogamete without flagella is considered an egg or ovum. It is immobile and is not shed. The protoplast of the male parent cell divides to form sixteen nearly spherical **microgametes** (B). Each microgamete has two flagella markedly longer than the body (C). Both kinds of gametes possess a cell wall. The microgametes are shed in the surrounding water (B). The active microgamete swims towards the immobile macrogamete and the two become attached at their anterior ends (D). The intervening walls between the two at this point dissolve. The protoplast of the microgamete separates from its cell wall and passes over to the macrogamete. The flagella of the microgamete are resorbed while fusion is going on. The two protoplasts fuse to form the zygote which in this case is non-motile. In oogamy the macrogametes are produced singly. They are large in size and non-motile. Fertilisation takes place *in situ* as the macrogamete is never shed. The sexual differentiation in this species amounts to a primitive type oogamy. We might speak of the immobile macrogamete as the ovum and the motile microgamete as the spermatozoid though the microgametes are not the typical spermatozoids.

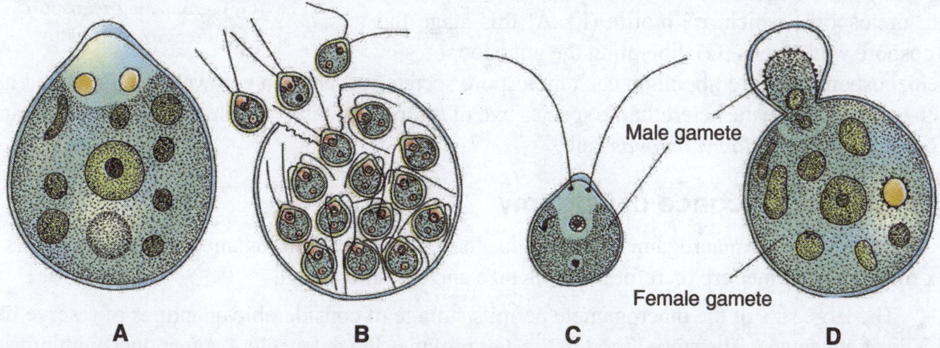

Fig. 4.9 (A-D). *Chlamydomonas coccifera* : Showing oogamous sexual reproduction. (A) Female gametangium with an egg or ovum; (B) Male gametangium with male gametes; (C) Liberated male gamete; (D) Fusion of the male gamete with the egg. (After Goroschankin).

Zygospore (Fig. 4.10, A-B)

The freshly formed zygote in isogamy is motile at first. It has four flagella, two chloroplasts and two eye spots (A). Soon the flagella are resorbed. The non-motile zygote secretes a thick wall around it to become a resting zygospore (B).

The resting period coincides with the period unfavourable for growth. During this period it undergoes ripening. The pigment spots, chloroplasts and pyrenoids cannot be distinguished. The ripened zygospore possesses a thick, smooth or stellate wall and is spherical in form (B). It is orange red in colour and contains fats and reserve food materials other than starch. It can resist drought and can also be distributed through air to colonize other ponds where it awaits the return of favourable season.

Ultrastructural investigations have revealed that the cytoplasmic fusion of two gametes is followed by the fusion of the gamete nuclei and chloroplasts (Brown *et al*, 1968; Cavalier Smith, 1970; Ladygin *et al.*, 1975, and Grobe and Arnold 1977). Their reports about the time sequence of chloroplast and nuclear fusions, however, vary. Brown *et. al.* (1968) found that in *C. moewusii* the chloroplasts fuse before the nuclei. Ladygin, *et. al.* (1975) reported that the chloroplasts begin to fuse when the nuclear fusion is complete in *C. reinhardii*. Blank *et. al.* (1978) confirmed the findings of Ladygin *et. al.* As regards the mitochondrial behaviour in the zygotes no evidence for their fusion could be found. Grobe and Arnold (1977) reported that the two reticular mitochondria contributed by the copulating gametes in *C. reinhardii* degenerate with the increasing age of the zygote into many small mitochondrial fragments.

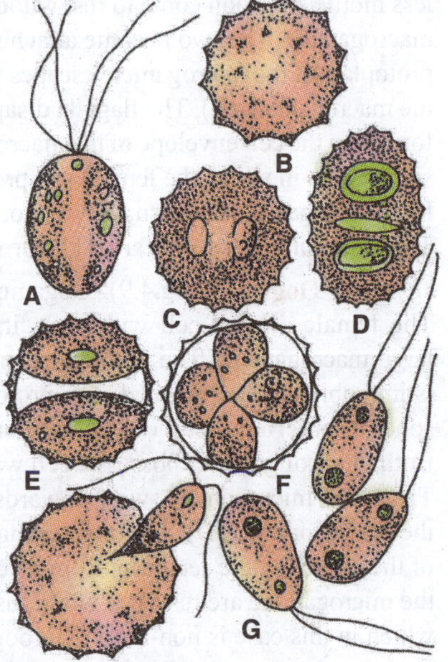

Fig. 4.10 (A-G). *Chlamydomonas sp.* (A) Freshly formed zygote; (B) resting zygospore; (C-F) Stages in meiosis; (G) Germinating zygospore. (C-F after Goroschankin).

Germination of Zygospore (Fig. 4.10 C-G). When once again in water it germinates. Prior to germination its red colour changes to green. The diploid nucleus undergoes **meiosis** to form four rarely eight (*C. reinhardii*) haploid nuclei (C-F). In the heterothallic species, segregation of nuclei of opposite mating types (plus and minus strains) takes place. The haploid nuclei are eventually incorporated into meiospores which are motile (F). At this stage the zygospore wall cracks (G), liberating the gonozoopores or meiozoospores. Before liberation each meiospore secretes around it a cell wall and furnished itself with two flagella. In the heterothallic species two of these are of either strain. The released meiospore grows to an adult *Chlamydomonas* cell.

Biological significance of Oogamy

In **Oogamy** the macrogametes are produced singly and the microgametes in large numbers per sex organ. The former are therefore large in size and the latter small.

The large size of the macrogamete permits storage of considerable quantities of reserve food. This is of advantage. The stored food in the egg provides the defenceless young ones with nutritive capital until they have had opportunity to establish themselves. The loss of motility is a disadvantage because motility favours contact and fusion. The small size of the microgamete, on the other hand, favours motility. Oogamy therefore combines both the advantages of large size with accumulation of food in the macrogamete or egg and small size with motility in the microgamete or sperm.

Evolutionary Trends in Chlamydomonadaceae

(*a*) Origin of sex. When the environmental conditions are favourable. *Chlamydomonas* repeatedly reproduces by the formation of motile spores differentiated from the cell protoplast by the method of **successive bipartition**. These motile, biflagellate structures may be called the swarmers. Swarmer is the general term commonly used for any motile reproductive cell irrespective of the fact whether it behaves as a gamete or as a zoospore. Each swarmer directly grows into a new *Chlamydomonas* individual resembling the parent. In this case there is no union of the elements of two sex cells. For this reason this method of multiplication is generally called **asexual reproduction**. At the close of the growing season when the environmental conditions become unfavourable *Chlamydomonas* resorts to sexual reproduction. It involves the fusion of two sex cells resulting in the formation of thick-walled resting **zygospore**. The resting period of the zygospore generally coincides with the period unfavourable for growth. The cells which fuse to form the zygospore are called the **gametes**. The origin of gametes means the **origin of sex**.

In the primitive form of sexual reproduction which is **isogamy** the fusing gametes are motile and similar in every respect. They are called the **isogametes**. There is striking resemblance between the **isogametes** and the zoospores in *Chlamydomonas*. In their form, structure, development and the mode of liberation, the two are exactly alike. They differ only in size and subsequent behaviour. The gametes are invariably smaller in size and incapable of growing into a new individual alone. The marked resemblance between the two suggests that the isogametes have originated by the subdivision of swarmers of zoospores into still smaller swarmers of zooids in response to the unfavourable conditions in the environment. These smaller swarmers or reduced zoospores had become too small and thus weak to grow into new plants by themselves because of an insufficiency of stored food or some other factors. What these factors are is not known. Most of them perished in their wanderings. A few met by chance and fused in pairs. This accidental pairing between the undersized zoospores turned to be a happy accident as it resulted in pooling their food reserves. It proved of advantage to the species in the following respects :

(*i*) Increased vitality and vigour to give rise to a new individual.

(*ii*) The ability of the resulting structure, the fusion cell or **zygospore** to enable the species to tide over conditions unfavourable for growth.

The accidental pairing which brought a benefit to a species was perpetuated by inheritance. The reduced zoospores became the **gametes** and their union, the **sexual fusion**. The sex thus originated by the accidental fusion of undersized zoospores in response to the unfavourable conditions in the environment. As it proved to be of advantage to the species the process was maintained. It finally became fixed and was known by the name of sexual reproduction. This theory of the origin of sex in plants is called the "hunger theory of sex". In *Polytoma uvella,* the most primitive form of family Chlamydomonadaceae we are close to the origin of sexuality. It produces swarmers which are similar to the vegetative cells. These swarmers can either fuse in random pairs or behave as zoospores and reproduce vegetatively by directly growing into new individuals.

(*b*) Evolution and differentiation of sex. The different species of *Chlamydomonas* illustrate all the stages in the differentiation of sex. They can be arranged in ascending series indicating the transitional reproductive stages from isogamy to oogamy through anisogamy. The series start with an allied genus *Polytoma uvella*. It produces swarmers similar to the vegetative cells. They either fuse in random pairs or develop directly into new individuals. *Polytoma uvella* thus provides an example very close to the origin of sexuality. *Chlamydomonas debaryanum, C. media and C. botryoides* are species which exhibit *isogamy*. The fusing gametes in them are of equal size and shape. They exhibit no morphological difference. *C. braunii* exhibits *anisogamy*. The fusing gametes in this species are unequal in size. The protoplast of the smaller gamete flows into the envelope of the larger gamete. The next step in the evolution of sex is furnished by *C. suboogamum* and *C. ooganum*. In the former, the flagella of the female gamete are functionless and in the latter, completely absent. The ovum is fertilised after liberation by a needle-shaped, biflagellate, naked spermatozoid. This is **oogamy**. *C. coccifera* provides the final stage in the evolution of sex. The cell loses its flagella and its entire protoplast functions as a female gamete which is not liberated. It is fertilised *in situ* by a spermatozoid which has a cell wall. Hence *C. coccifera* provides an example of most advanced level of sexual reproduction called **oogoniogamy** in which the ovum is not released so that fertilization is internal.

(*c*) Evolution of plastids. The most primitive species of *Chlamydomonas* (*C. arachne* and *C. eradians*) have single axile (central) stellate chloroplast. The majority, however, possess a single cup or basin-shaped parietal chloroplast (*C. angulosa*). In the higher forms (*C. reticulata*) it is reticulate. *C. alpina* has numerous, small discoid chloroplasts. The cup-shaped chloroplast in primitive forms thus breaks up into nets and discs in the advanced species.

(*d*) Origin of evolutionary lines of development of thallus. *Chlamydomonas* is generally considered to be one of the most primitive of the green algae and one that played an important role in the evolutionary considerations of this group. There is consensus of opinion among algologists that from a *Chlamydomonas*-like ancestor arose different evolutionary lines of development of thallus in green algae. The chief among these are the *Volvocine, Chlorococcine* and the *Tetrasporine*.

(*e*) It is suggested that the transverse division of cells, by which the filaments grow in length, has been derived from the longitudinal division of the flagellates *via* forms like *C. seriate* (Fig. 3.27 A-C). The first division of the protoplast in this species is oblique (A). The two daughter protoplasts rotate and take up a transverse position (B). Each daughter protoplast again divides longitudinally. A linear series of four protoplasts is formed in the mother cell (C). This orientation of cell division and the arrangement of daughter cells in a linear series is of significance in understanding the development of filamentous habit in algae. The continuous division in one plane accompanied by holding together of daughter protoplasts by the formation of a common cross wall between the adjacent cells results in the development of a filamentous habit from an ancestral unicell.

Relationships. *Chlamydomonas* and other closely related unicellular forms are considered primitive and ancient members of green algae. The features which indicate their primitiveness are, (1) presence of contractile vacuoles, (2) an eye spot, (3) a single basin-shaped chloroplast, (4) longitudinal division of the protoplast not by the formation of a cross wall but by the activity of the cell protoplast itself, and (5) sexual reproduction by the fusion of isogametes in the primitive forms. In the above-mentioned features, *Chlamydomonas* resembles a small family of green flagellates called Polyblepharidaceae, which includes the most primitive members of green algae. All the members of this family, in addition, are naked. The primitive members of this family (*Polyblepharidaceae*) reproduce only asexually be longitudinal cell division into two. More advanced members of this family exhibit sexual reproduction by the fusion of isogametes.

Some phycogenetists suggest that the green algae arose from some *Chlamydomonas*-like ancestor having a unicellular body showing a simple form of the differentiation of cell organelles. This assumption is based on the fact that all green algae including even the advanced forms possess zoospores and gametes similar to the permanently motile forms such as *Chlamydomonas*. Here the student might pose a question. Was the common ancestor of green algae *Chlamydomonas*? It is difficult to answer this question with any assurance. At the best one can say if the green algae are monophyletic in their origin, the probable progenitor is not *Chlamydomonas* but resembled somewhat *Chlamydomonas*. In its method of reproduction *Chlamydomonas* shows a higher degree of differentiation than is expected in an ideal hypothetical ancestor of green algae. The vegetative cells in this alga differentiate into gametes. This is a snag. One would expect the ancestral form of green algae to divide into two to form new individuals. Mottox and Stewart (1977) visualized the green walled flagellates, such as Heteromatrix as the extant remnants of the early green algae which are considered to have undergone extensive evolutionary diversification to have given rise to *Chlamydomonas* and the entire range of green algae.

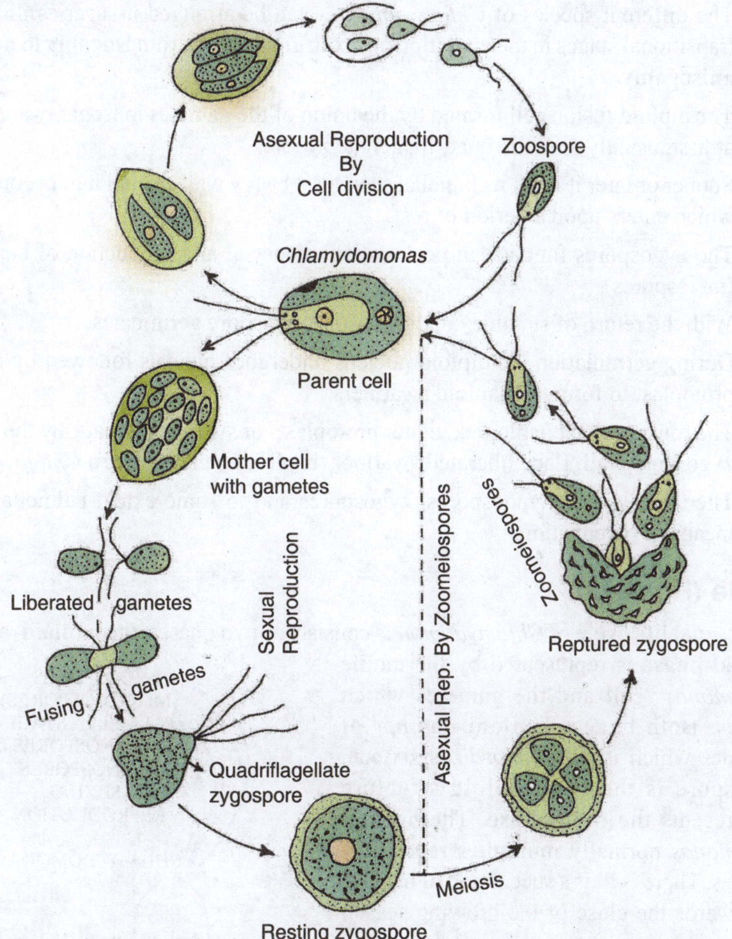

Fig, 4.11. *Chlamydomonas sp.* Pictorial sketch of asexual and sexual life cycles.

Salient Features

1. The plant body of *Chlamydomonas* is a unicellular, motile, biflagellate usually egg-shaped **thallus**.
2. It consists of a tiny protoplast surrounded by a thin delicate but firm cell wall.
3. The cell protoplast is differentiated into a **plasma membrane, cytoplasm, neuromotor apparatus**, a single **nucleus**, a green, cup-shaped **chloroplast** with usually a single **pyrenoid, a red eye spot**, two **contractile vacuoles**, ribosomes, perinuclear dictyosomes, mitochondria and endoplasmic reticulum.
4. The *Chlamydomonas* cell is haploid. It reproduces, repeatedly during the growing season by the asexual method of zoospore formation. There is thus a succession of haploid phases.
5. In addition there is production of aplanospores or hypnospores, each of which on liberation behaves like a *Chlamydomonas* cell.
6. Sexual reproduction takes place only once during the season and usually towards the close.

7. The different species of *Chlamydomonas* can be arranged in a series illustrating all the transitional stages in the evolution of sex in green algae from **isogamy** to **oogamy** through **anisogamy.**
8. The diploid fusion cell formed by the union of the gametes in isogamy and anisogamy is, at first, usually a motile quadriflagellate zygote.
9. Sooner or later it loses its flagella, secretes a heavy wall around it to become a **zygospore** which enters upon a period of rest.
10. The zygospores function in perennation, dispersal and production of haploid swarmers (meiospores).
11. With the return of suitable conditions, the zygospore germinates.
12. During germination the diploid nucleus undergoes meiosis followed by cleavage of the protoplast to form the haploid swarmers.
13. The four resultant haploid daughter protoplasts or swarmers escape by the cracking of the zygospore wall. Each liberated swarmer (haploid) behaves like a *Chlamydomonas* cell.
14. The thick-walled hypnospores, zygospores and to some extent palmella stage serve as means of perennation.

Life Cycle (Fig. 4.12).

The sexual life cycle of *Chlamydomonas* consists of two phases, the **haploid** and the **diploid**. The **haploid-phase** is represented by the motile *Chlamydomonas* cell and the gametes which it produces. Both have a haploid number of chromosomes which in *C. reinhardii* is sixteen. The zygospore is the only diploid structure which represents the **diplophase**. The haploid *Chlamydomonas* normally multiplies repeatedly by zoospores. There is thus a succession of haploid phases. Towards the close of the growing season *Chlamydomonas* resorts to sexual reproduction. It takes place only once in the growing season. During sexual reproduction there is fusion between the two motile or one motile and the other non motile gametes. With the fusion of the gametes or fertilization the haplophase ends. The resultant fusion cell is diploid and is called the **zygote.**

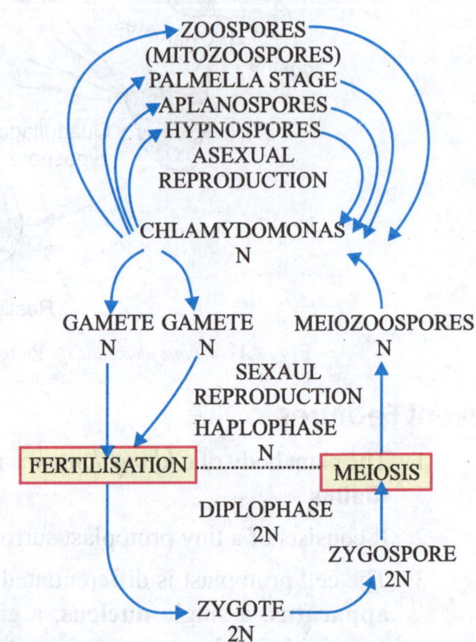

Fig. 4.12. *Chlamydomonas sp.* Word diagram of the life cycle.

The zygote secretes a wall around it to become a resting zygospore. On the onset of suitable conditions the zygospore germinates. The diploid nucleus undergoes zygotic meiosis followed by cleavage of the protoplast. The four resultant haploid uninucleate daughter protoplasts escape as swarmers with the cracking of the zygospore wall. Each liberated swarmer (meiozoospore) behaves as a haploid *Chlamydomonas* cell. There is no succession of diploid phase in the life cycle. In the single life cycle of *Chlamydomonas* there is not regular alternation of the haploid and diploid generations. The haploid phase is reduplicated and thus lengthened by mitospore (zoospore) formation.

Systematic Position:

Division	:	**Chlorophyta** (Chlorophycophyta)
Class	:	**Chlorophyceae**
Order	:	**Volvocales**
Family	:	**Chlamydomonadaceae**
Genus	:	*Chlamydomonas*
Species	:	*eugametos*

FAMILY 2. VOLVOCACEAE

It includes all the forms in which the flagellated cells do not live singly but are held together and are surrounded by a common mucilaginous envelope to form a colony. The colonies included in this family may be flat, disc-like or spherical. Each colony is composed of a definite number of cells with the flagella protruding outwards through the mucilage. The cells in the colony are in a multiple of two and their number in the colony is determined during the embryonic stage. There is no increase in the number of cells in the colony after this stage until a new reproductive phase sets in. The cells in the colony are arranged in a specific manner. Such a colony of a definite number of cells arranged in a specific manner and forming an integrated whole is called a **coenobium**. The coenobial colony is of constant size and shape for any given species. The component cells of the colony show vegetative division. The coenobia are of two kind, motile and non-motile. All the motile coenbobial forms are placed in the family **Volvocaceae**. They are all fresh water forms and are of singular interest. The common and the important motile coenobial genera are *Gonium, Pandorina, Eudorina* and *Volvox*. Asexual reproduction takes place by the formation of daughter colonies by the vegetative division of all or certain cells (*gonidia*) in the colony. All the cells in the colony in the Volvocales invariably arise from a single cell. Sexual reproduction ranges from isogamy to oogamy through anisogamy.

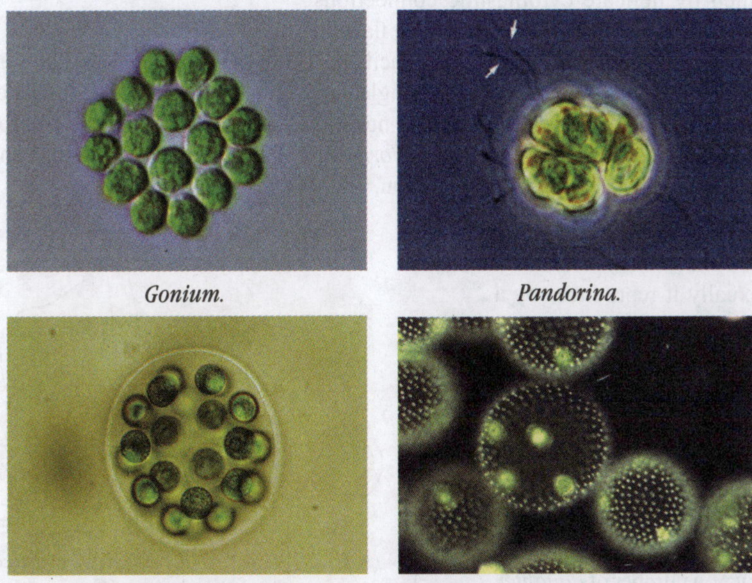

Gonium. *Pandorina.* *Eudorina.* *Volvox.*

REPRESENTATIVE ALGAE OF FAMILY VOLVOLACEAE.

GONIUM (Fig. 4.13)

Occurrence

Gonium is the simplest representative of the motile coenobial forms included in the family Volvocaceae. It is a fresh water planktonic form found in ponds, pools and ditchets. The two common species of this genus are *G. sociale* with four-celled coenobia (A) and *G. pectorale* (C, D) with 16 celled coenobia. Another species *G. octanarium* has 8-celled coenobia.

Coenobia

The cells in the curious little coenobia are arranged in a single plane but in two dimensions. The coenobium thus has the form of a thin plate which is rhomboidal or squarish in surface view (A). The plate is slightly curved. The colony size ranges from 24-100 μ. Each cell in the colony is connected to its neighbouring cells by short protuberances of the cell wall forming oval or quadrangular interstices. In the 16-celled species (C) there are four central and 12 peripheral cells (3 on each side). The cells in the colony are embedded in a common gelatinous matrix with a firm bounding layer. The matrix is formed by each cell but is united into a common sheath enclosing the entire colony. The centre of the coenobium is occupied by a square area containing mucus (A,C).

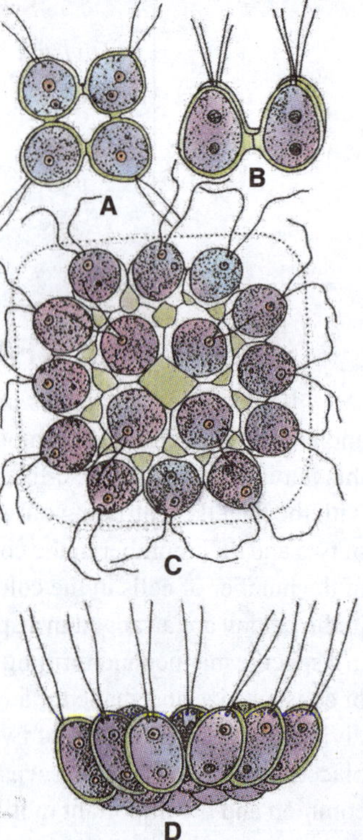

Fig. 4.13 (A-D). *Gonium.* (A) *G. sociale* top view; (B) Side view; (C) Top view of *G. pectorale*; (D) Side view of C.

Cell Structure

All of the cells in the colony are alike and each is of *Chlamydomonas* type. The adjacent cells in the colony are reported to be joined by delicate cytoplasmic connections (A,B). Each cell is furnished with two whiplash type flagella at its anterior end protruding outwards through the mucilage. There are contractile vacuoles and an eye spot at the anterior end. In addition it contains a single, large cup-shaped chloroplast with one or two pyrenoids located near the hinder end. The single nucleus lies in the hollow of the chloroplast. The cells are ovoid in *G. pectorale* and pyriform in *G. formosum*. The colony has no distinct anterior and posterior ends. It moves by a somersaulting motion.

Reproduction

Gonium reproduces both asexually and sexually. Asexually it reproduces by a vegetative method. No spores are formed.

Vegetative Reproduction. (Fig. 4.14). It takes place by the formation of a daughter colony within each component cell of the mother colony. All the cells become involved simultaneously in the asexual process. The protoplast of each of the cells of the parent colony divides by longitudinal fissions within its cell wall. In this way, a number of *daughter colonies* or the *autocoenobia* are formed. The divisions are longitudinal. After division the

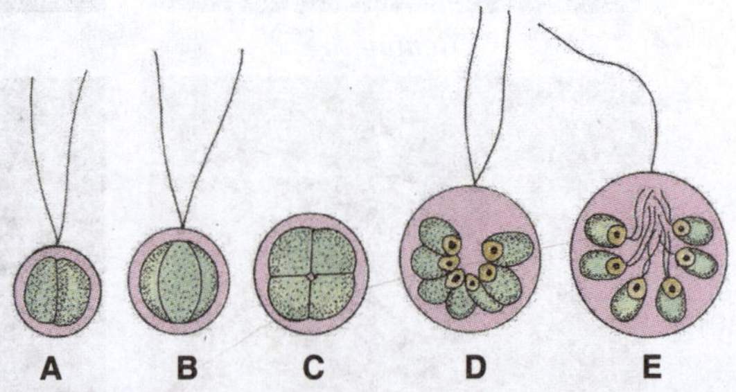

Fig. 4.14 (A-E). *Gonium pectorale.* Side views of successive stages in the development of a new colony by division of one of the cells in the old colony (After Hartmann). (C) Top view of (B-E) L.S. mature colony.

daughter protoplasts lie in the mother cell in the form of a bowl or curved plate, the **plakea** (D). The plakea is some-what hemispherical. By a modified form of inversion, the bowl-shaped daughter colony bends back so that flagellar ends of cells are directed outwardly and cells are arranged in the form of a flat plate characteristic of the parent colony. Prior to liberation by gelatinisation of the envelope of the parent colony, each daughter colony has an envelope of its own. The liberated colonies increase only in size. There is no further cell division until a new reproductive phase sets in.

Sexual Reproduction. (Fig. 4.15) *G. pectorale* is heterothallic. Fusion between gametes (C and D) takes place when they come from two coenobia (A and B) of opposite mating types called plus and minus strain. There is thus physiological differentiation of sex. Morphologically the gametes are alike. Each cell in the freshly formed coenobium functions as a gamete. The sexual coenobia are thus comparatively smaller in size. The naked gametes escape from the colony into the surrounding medium. During swimming the plus and the minus strain gametes pair by means of their flagella. They come in contact at their anterior ends but the fusion between the gametes is lateral (E). It starts at the anterior end. The cytoplasm fuses with the cytoplasm and the chloroplast with the chloroplast. Soon after, the nuclei do the same.

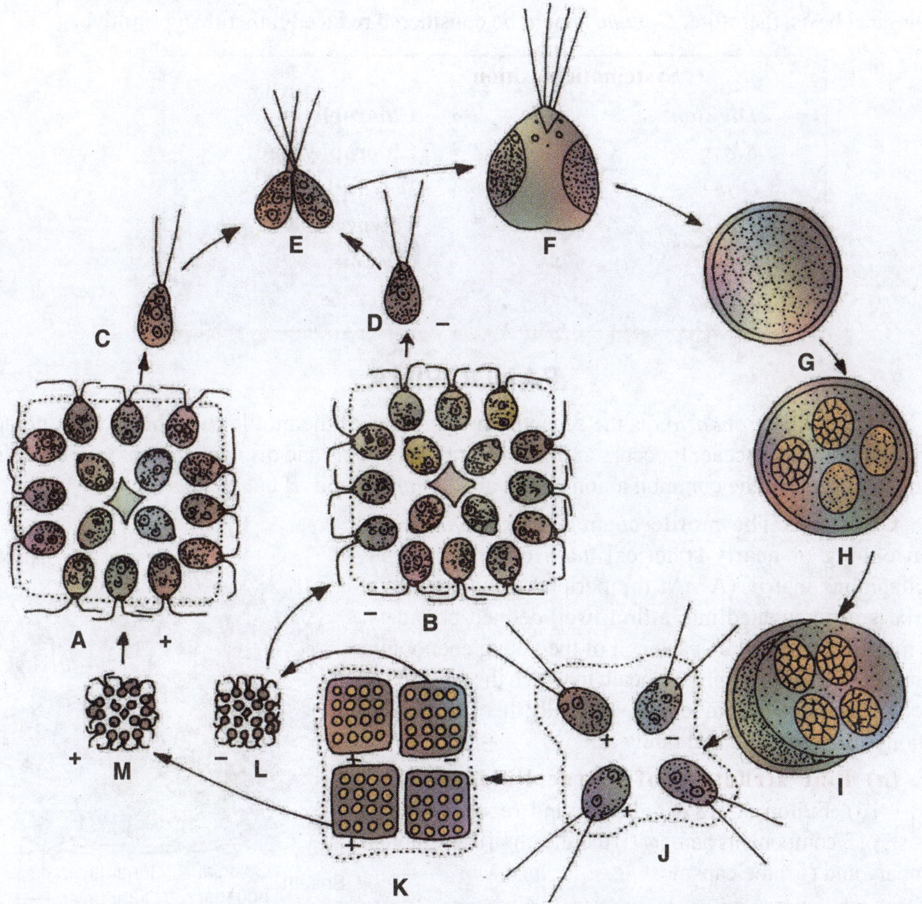

Fig. 4.15 (A-M). *Gonium sp.* Sexual life cycle. Explanation in the text.

Zygote. The zygote, at first is, quadriflagellate (F). It soon comes to rest, withdraws its flagella and rounds off. It then secretes a wall around it which is smooth in *G. pectorale* but shows reticulations in *G. sociale* (Starr 1955). The thick-walled zygote or zygospore enters upon a period of rest (G). In the resting stage it remains viable for weeks or even months.

Germination of zygote. With the onset of conditions favourable for growth (rewetting) it germinates. Prior to germination the resting zygote or zygospore nucleus undergoes meiosis (Stein, 1958). Segregation of strains takes place. Two of the resultant nuclei are of plus strain and two of minus strain. After meiosis the zygospore protoplast cleaves into four daughter cells termed the **gones** (H-I). Each cell or gone develops two flagella (J). the biflagellate cells remain united to form a tiny four celled germ colony (J). The zygospore ruptures and the germ colony swims away leaving the zygospore wall behind. Each cell of the germ colony divides four times by mitosis. In this way four tiny daughter colonies, each consisting of 16 cells are formed within the mucilage envelope of the germ colony (K). The new four daughter colonies then separate and each grows in size to form a typical 16- celled mature colony characteristic of the species (L-M).

Is Gonium a primitive or a reduced form? There is no definite answer to this question. Till recently it has been held that *Gonium* is the most primitive genus of family Volvocaceae. The morphological and reproductive features support this view point. The cytological investigations concerning the nuclear condition of Volvocaceae by Cave and Pocock (1951) point the other way. *Gonium* has the largest number of chromosomes (n=17) whereas species of *Volvox* have less (n=14 or 15). On the cytological basis, therefore, *Gonium* should be considered reduced rather than primitive.

Systematic Position

Division	:	**Chlorophyta**
Class	:	**Chlorophyceae**
Order	:	**Volvocales**
Family	:	**Volvocaceae**
Genus	:	*Gonium*
Species	:	*pectorale*

PANDORINA

Occurrence. *Pandorina* is the second representative of the motile coenobial forms included in the family Volvocaceae. It occurs in fresh water ponds, pools and also in the plankton. The genus includes 3 species. The common among these are *P. morum* and *P. charkowensis*.

Coenobia. The motile coenobium of *Pandorina* is an oblong or nearly spherical mass of extracellular, mucilaginous matrix (A). At the periphery the coenobial matrix is differentiated into a firm, well-defined boundary. The matrix is secreted by each cell of the young coenobium but it is united into a common sheath in which the cells of the developing colony are embedded. Typically there are cells in a colony sometimes 32 and rarely 8.

***(a)* Fine structure of extracellular matrix.** (Fig. 4.16). Fulton (1978) studied it and reported that it consists of 3 components namely (i) the sheath (ii) the colonial boundary and (iii) the capsule.

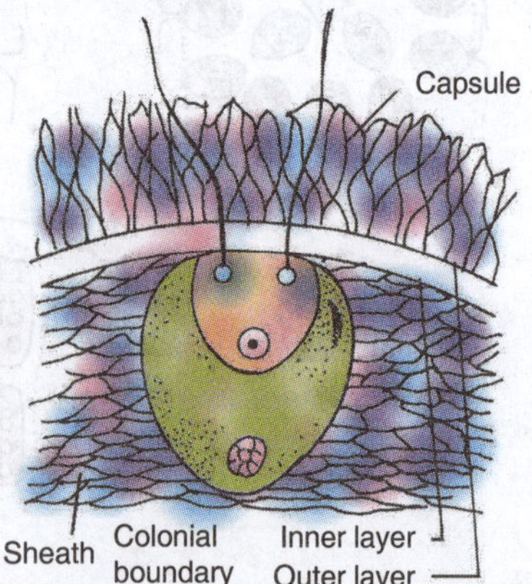

Fig. 4.16. *Pandorina.* Showing diagrammatic representation of the fine structure of colonial extracellular matrix (based on Fulton).

(*i*) *Sheath.* It is a broad inner layer of refractile matrix varying in thickness between 1 and 4 mm. The cells of coenobium are embedded, closely packed in the sheath. Ultrastructurally it consists of circumferentially oriented fibrils of hydroxyproline rich protein and polysaccharide.

(*ii*) *Colonial boundary*. The sheath on the outside ends in a well-defined boundary consisting of a thin layer of uniform thickness. It surrounds the coenobium as a whole and not penetrating between the cells. It consists of two layer, outer and inner separated by a elecrolucent space of narrow constant width.

(*iii*) *Capsule*. It is a fragile layer, measuring between 2 and 3 m thick lying outside the colonial boundary. It consists of radically oriented fibrils containing carbohydrates and sulphate.

Cell structure. All the cells in the colony are alike and each resembles *Chlamydomonas* in its structure. It is biflagellate, somewhat wedge-shaped in form with the pointed end directed inwards and broader outwards (B). The nucleus is centrally located. From the middle of the broader end arise the two equal whiplash flagella which extend through the mucilaginous envelope to the exterior. The chloroplast is cup-shaped and has a pyrenoid near the posterior end. There are two contractile vacuoles above at the broader anterior end with an eye spot to the right. The eye spot tends to be less prominent in the posterior cells. The cells are embedded in a homogeneous gelatinous matrix. Cellular connections between the adjacent cells in the mature colony of *Pandorina* are absent. The cells are closed-packed in a hollow sphere. The central cavity is small. The vegetative cells in the developing colony increase in size but undergo no division during the somatic phase. There is thus no increase in the number of cells in the colony. The coenobium of *Pandorina* shows a further advance over *Gonium* in that it is spherical. Besides, it shows some degree of polarity in its progression and decrease in the size of eye spot towards its posterior end.

Reproduction. *Pandorina* colonies reproduce vegetatively and sexually.

Vegetative Reproduction (Fig. 4.17 C). During the reproductive phase, cells of the colony withdraw flagella and the coenobium becomes immobile. It sinks to the bottom of the pool. The mucilage envelope swells and becomes more watery. Each cell then undergoes four mitosis in a short period to form daughter coenobium. All the cells of the colony become involved simultaneously, the divisions are all longitudinal and usually a sixteen-celled curved plate, the **plakea** is formed in each cell. The curved plate undergoes inversion. It turns itself inside out. At first it becomes flat. Finally it curls in the opposite direction (backwards) with the ends meeting to form a hollow sphere. With this orientation each cell develops a pair of flagella. Formation of extracellular matrix begins after inversion, the colonial boundary is formed first. Capsule and sheath are then formed on the

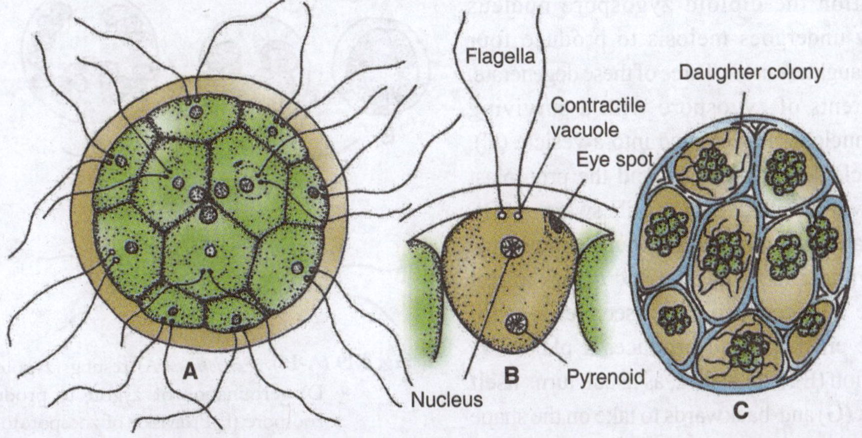

Fig. 4.17 (A-C). *Pandorina.* (A) Top-view of vegetative colony; (B) Portion of the same magnified to show cell structure; (C) Parent colony with daughter colonies.

outer and inner faces of the colonial boundary. The sixteen daughter colonies then swim away from the watery mucilage envelope of the parent coenobium and grow into adults.

Sexual Reproduction. (Fig. 4.18) *Pandorina* is heterothallic and exhibits a slight tendency to anisogamy. Deficiency of sulphur and nitrogen stimulate sexual reproduction (Raybwen and Starr, 1974). All the cells of the colony divide simultaneously to form gametes. Each vegetative cell in the colony produces as numerous gametes as the cells in the colony. The groups of gametes separate. The coenobia containing gametes are *Eudorina*-like in appearance and have a watery gelatinous envelope (A). Shortly after the individual gametes escape from the colonial matrix and swim about in water. They are biflagellate. Gametes from deferent coenobia only fuse. Fusion occurs in water and may be terminal (D) or lateral (E). One of the fusing gametes is slightly larger and less active (C) than the other (B) thus showing an advance over *Gonium* towards anisogamy. The quadriflagellate zygote remains motile for a while (F). Finally it comes to rest, retracts its flagella and secretes a wall around it. The old zygospores have a smooth thick wall and have contents coloured red. After a period of rest the zygospore germinates.

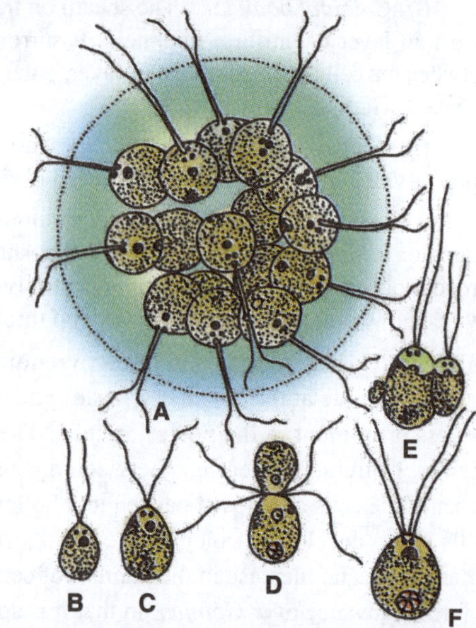

Fig. 4.18. (A-F). *Pandorina.* Sexual reproduction. (A) Female colony containing gametes; (B) male gamete; (C) female gamete; (D-E) fusion of gametes; (F) quadri-flagellate zygote.

Germination of Zygote (Fig. 4.19). On germination the diploid zygospore nucleus probably undergoes meiosis to produce four haploid daughter nuclei. Three of these degenerate. The contents of zygospore with a surviving haploid nucleus are extruded into a vesicle (C). The vesicle soon disappears and the protoplast swims away as biflagellate, motile swarmer (D). The liberated swarmer swims about for sometime. It then withdraws its flagella and comes to rest (E). The quiescent protoplast secretes a broad mucilage envelope and produces a plakea by cell division (F). The plakea, as usual, turns itself inside out (G) and backwards to take on the shape of a typical colony.

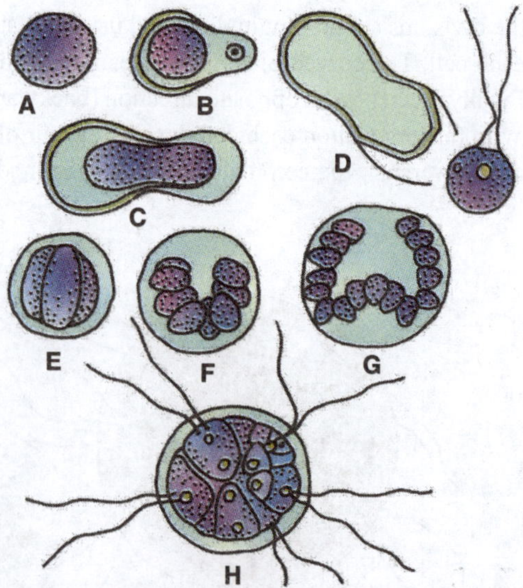

Fig. 4.19 (A-H). *Pandorina.* (A) resting zygote; (B-D) germination of zygote to produce a zoospore; (E-F) division of zoospore to form a plakea stage; (G) inversion of new colony; (H) new colony (Except E and G all after Pringsheim).

Systematic Position:

Division	:	**Chlorophyta**
Class	:	**Chlorophyceae**
Order	:	**Volvocales**
Family	:	**Volvocaceae**
Genus	:	*Pandorina*
Species	:	*morum*

EUDORINA

Occurrence

Eudorina is found in the fresh water of ponds, pools and ditches in the summer especially during the rainy season in India. It also occurs in the plankton in the soft water of lakes. It is represented by 5 or 6 species. The commonly known among these are *E. elegans, E. indica* and *E. illinoisensis*.

Coenobia (Fig. 4.20A)

The vegetative colony is spherical to ellipsoidal in shape. The number of cells in the colony are 16, 32, 64 or even 128. However 32 is the usual number. The coenobia are larger than those of *iandorina*. In *E. indica* the posterior end of the colony shows mamillate projections. The globose cells of the colony are embedded in an extremely viscous hydroxyproline rich glycoprotein mucilaginous matrix which is bound by a flexible colony wall (Tautvydas, 1978). They are arranged in a single peripheral layer (usually in tiers) away from one another and are not close-packed as in *Pandorina*. The coenobuim of *Eudorina* is thus a hollow sphere with the interior of the colony showing no cellular organisation. Instead it is filled with colourless mucilage.

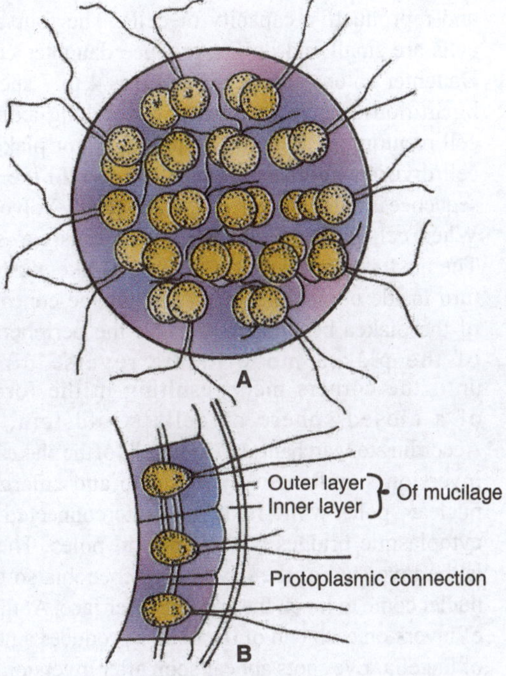

Fig. 4.20 (A-B). *Eudorina elegans.* (A) Vegetative colony; (B) T.S. showing cell structure and protoplasmic connections. (After Fritsch).

Frequently the cells are arranged in transverse tiers across the short axis of the ellipsoidal colony. In a 320-celled colony there is an anterior and posterior tier of 4 cells each and 3 medium tiers of 8 cells each (A). There are fine cytoplasmic connections between the adjacent cells (B). In some species all the cells in the colony are equal or nearly equal in size and each cell possesses equal reproductive potential (*E. elegans* and *E. unicocca*). But in *E. illinoisensis* and *E. californica* the cells of the anterior one tier (*E. illinoisensis*) or two tiers (*E. californica*) are reduced in size. They remain sterile in the short axis of the ellipsoidal colony. In a 32-celled colony there is an anterior and a posterior tier of four cells each and three median tiers of eight cells each. Apparently the cells in the colony appear separated from one another but there are fine protoplasmic connections between them (B). The cells are of *Chlamydomonas* like in structure. It possesses a pair of apically inserted flagella which project into the surrounding water. The single cup-shaped chloroplast may have one or more pyrenoids according to the species. There are two contractile vacuoles at the base of the flagella. There is progressive decrease in the size of the eye spot in the cells from the front to the rear end of the colony. The eye spots may even be absent in the posterior cells of the coenobium. The distinction in the anterior and posterior end of the colony is recognisable not only by the decrease in size of the eye spots but also in the size of cells and their reproductive capacity. For instance in *E. illinoisensis* the anterior four cells are much smaller and are either purely vegetative in function or take part only in asexual reproduction. The *Eudorina* colony exhibits distinct polarity. The anterior end is always directed forward while swimming.

Asexual Reproduction (Fig. 4.21). It is similar to that in *Pandorina*. Each cell in the colony forms an autocolony by cell division. All the cells in the colony may become involved at the same time but some times certain cells in the colony may remain sterile. A few species *(E. illinoisensis*) exhibit polarity in size and reproductive capacity of cells. The four anterior cells are small and do not produce daughter colonies. Daughter colony formation involves 4 to 7 successive bipartitions occurring parallel to the long axis of the cell resulting in a curved plate of cells or plakea. The cell division to form daughter colonies in the plakeal sequence is thus characteristic of the family Volvocaceae. When cell division ceases the plakea is cup-shaped. The next step is inversion. The bowl-like cluster cells turn inside out and backwards. Thus the concave side of the plakea becomes convex as the peripheral cells of the plakea move in the reverse direction until the corners meet resulting in the formation of a closed sphere of cells (Goldstein, 1967) According to Marchant (1977) all cells of the plakea during inversion simultaneously elongate and enlarge their nuclear poles while remaining interconnected by the cytoplasmic bridges at their plastid poles. The result is the inversion of the developing coenobia so that the nuclei come to lie on the convex outer face. At the onset of inversion each cell of the plakea produces a new pair of flagella. Eye spots appear soon after inversion. When daughter colony formation is complete the parental (gonideal) cell walls break down and the newly formed colonies swim away through the gelatinised envelope of the parent colony.

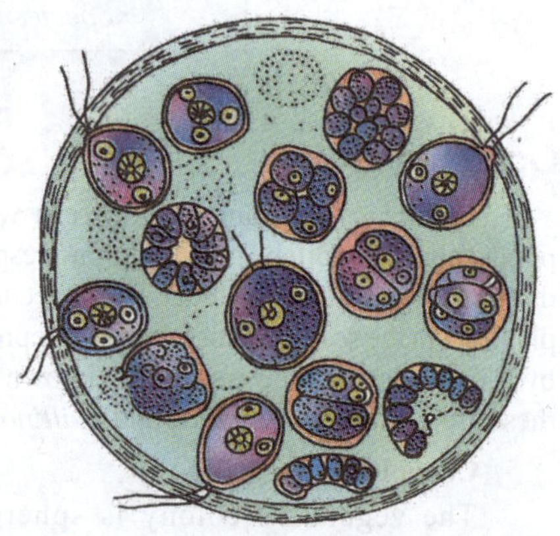

Fig. 4.21. *Eudorina elegans.* Asexual reproduction. Vegetative cells dividing to form daughter colonies (After Goebel).

Sexual Reproduction (Fig. 4.22). There is distinct advance in sexual reproduction over Pandorina. It is anisogamous and shows an approach to oogamy. The coenobia are commonly heterothallic (E. elegans) but some species are homothallic. In the heterothallic species the male and the female gametes are formed in different colonies so that the colonies in the heterothallic species *Eudorina* are either male or female. In the monoecious forms the four cells in the anterior tier divide to form the male gametes and the remaining cells form eggs

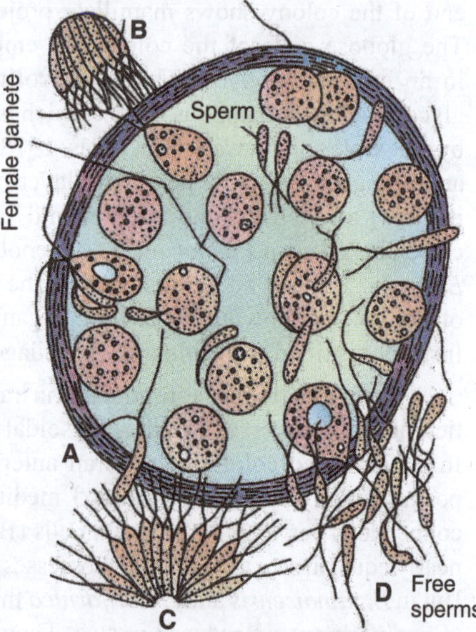

Fig. 4.22. *Eudorina elegans.* Sexual reproduction. Female colony (A) with two groups of sperms (B and C) and numerous free sperms (D) surrounding the colony. Some of these have penetrated the colony to fuse with the female gametes. (After Goebel).

without division. The two kinds of gametes are dissimilar in size and shape. The female gamete is large and immobile. The male gametes or sperms are biflagellate and spindle-shaped. They are shed into the surrounding water. The liberated sperms swim to the cells containing the female gamete by penetrating through the gelatinous matrix. Finally a single sperm enters the egg or ovum and unites with it to accomplish fertilization.

In the heterothallic species the vegetative cells of the female colony slightly enlarge by the accumulation of food and come to the surface of the matrix and function directly as female gametes or ova without division. The biflagellate globose female gametes remain *in situ*. They are never shed. Each vegetative cell of the male coenobium by successive divisions forms a flat or curved plate of 64 elongated spindle-shaped biflagellate spermatozoids. The spermatozoids are yellowish in colour (Fig. 4.22 B). They are liberated as a group or packet of 64 cells which is called the male packet or platelet. It swims about as an individual unit through water separating into individual sperms near the female colony. The sperms swim directly into the gelatinous matrix of the female coenobium. There they fuse with the ova or oospheres. Szostak *et. al.* (1973) reported that increasing light intensity speeds up the formation of sperm packets.

Zygote (Fig. 4.23 A). Each zygote soon retracts flagella and secretes a smooth wall. Finally the zygotes are liberated by the decay of the coenobial envelope. The liberated zygote has a thick smooth wall and red contents (A). It undergoes the resting period.

Germination of Zygote (Fig. 4.23). After the resting period it germinates. Prior to germination it probably undergoes meiosis. The zygote wall swells and a sac-like vesicle is formed on one side (B). The vesicle contains a single biflagellate zoospore and two or three degenerating small bodies (C). The functional zoospore is eventually liberated (D). It comes to rest, divides and redivides to form a new coenobium in the usual manner (E-F). Mainx (1929) recorded parthenospores in *E. elegans*.

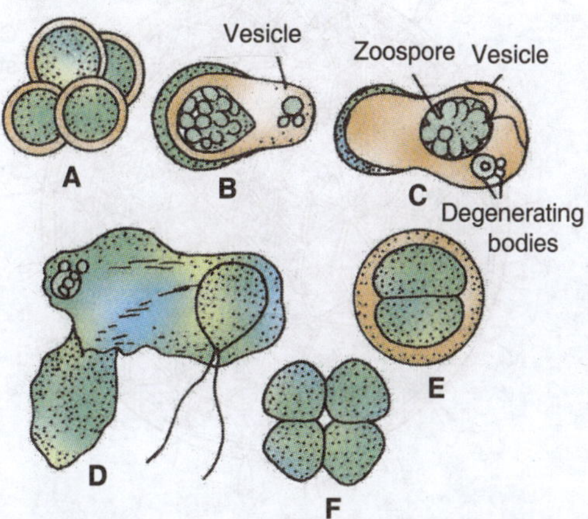

Fig. 4.23 (A-F). *Eudorina elegans.* (A) zygotes; (B-D) stages in the germination of zygote; (E-F) two earlier stages in the division of zoospore to form a new colony (After Schreiber).

Systematic Position:

Division	:	**Chlorophyta**
Class	:	**Chlorophyceae**
Order	:	**Volvocales**
Family	:	**Volvocaceae**
Genus	:	*Eudorina*
Species	:	*elegans*

VOLVOX

Occurrence

Volvox is a green, flagellate colonial alga of worldwide distribution comprising twenty species. It occurs both in temporary and permanent fresh water of ponds, pools and ditches. It is also found in the soft water of lakes. This motile coenobial genus grows in spring and summer but is abundant in the rainy reason all over India. It passes winter in the resting condition as a zygospore or oospore. It is one of the planktonic organisms which helps to colour the water of the pond green.

Size, Form and Structure of Coenobium

The *Volvox* coenobium is the largest and most highly differentiated of all the free swimming colonial forms. The coenobia look like minute green balls just the size of a small pinhead varying from 0.5 to 2.0 mm in diameter.

In form, the coenobium is a sphere of extracellular mucilage with a firm bounding layer (Fig. 4.24 A). The numerous cells of the colony are embedded in the gelatinous matrix and are arranged in a single peripheral layer. The *Volvox* coenobium is thus a hollow sphere with the interior showing no cellular organization. The matrix and the colony wall both contain large amounts of hydroxyproline-rich glycoproteins. The major sugars are galactose and arabinose. The former is present in higher amounts than the latter. According to Lamport (1979), *Volvox* unlike *Eudorina* has not clearly defined colony wall separable from the mucilage.

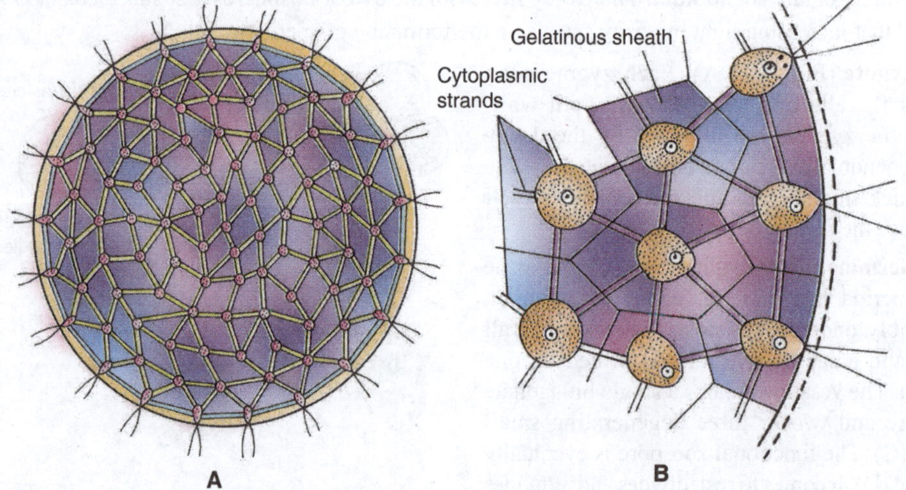

Fig. 4.24 (A-B). *Volvox* sp. (A) Vegetative colony; (B) Portion in detail.

The *Volvox* colony is of constant size and shape for any given species. A colony consisting of a definite number of cells arranged in a specific manner and forming an integrated whole is called a **coenobium**.

Depending upon the species the number of cells in the colony varies between 500 or 1000 and 50,000. Each cell has its own gelatinous sheath or mucilaginous envelope and thus is separated from its neighbours by a considerable expanse of the gelatinous material. This shows that *Volvox* is not an individual but an association of a number of independent cells. The mucilaginous envelopes in most species are not confluent with those of adjacent cells. In majority of the species each cell is connected to its neighbours by a series of broad protoplasmic strands (B). The latter form a fine, continuous, spherical network. The protoplasmic fibrils connect the cells through the mucilage which extends in a wedge-shaped mass towards the interior. The coenobium has a central cavity. It is filled with a watery mucilage. In *V. mononae* and *V. tertius* the interconnecting cytoplasmic strands are, however, absent. All the cells of the young colony are green and vegetative in function. They are the **somatic** cells. The coenobia of *Volvox* are motile.

The somatic cells of a mature coenobium of *Volvox* are oriented in two ways, Huskey, (1979): (*i*) Each cell is oriented along a radial line with the flagella to the exterior and chloroplast to the interior; (*ii*) The cells are also oriented rotationally around the radial line. The latter orientation is crucial for the phototactic motion.

Cell Structure

In *V. carteri* and *V. aureus,* each cell is remarkably similar to *Chlamydomonas* (B). In *V. aureus* the number of cells ranges from 500 to 1,000. Each cell is provided with two flagella at its anterior end. The flagella arise from the basal bodies and project beyond the surface of the colony into the water. There are two contractile vacuoles near the base of the flagella. The green chloroplast is cup-shaped with usually one pyrenoid. The single nucleus with its prominent centric nucleolus is located in the cytoplasm which fills the concavity of the chloroplast. The reddish brown eye spot is located in the anterior region of the chloroplast immediately beneath the double chloroplast envelope. *Volvox* has thus carried the *Chlamydomonas* cell into a highly complex colony in which the *Chlamydomonas* type of cells are arranged in the form of a hollow sphere.

The coenobia of *V. globater* and *V. rousseletii* are composed of cells of *Sphaerella type*. The number of cells in the colony of former varies from 15,000 to 20,000 whereas in the latter it ranges from 20,000 to 60,000. In this case each biflagellate cell is spherical to ovoid. It has a thick cell wall differentiated into an inner mucilaginous portion bound by an outer firm layer. The protoplast is stellate with a number of coarse unbranched processes traversing the mucilaginous portion of the cell wall. There are several contractile **vacuoles** (2 to 6) distributed irregularly near the surface of the protoplast. The chloroplast is a curved plate with one or more pyrenoids.

Volvox obviously is an assemblage of similar and independent cells. Each cell functions like an individual carrying out its own nutrition, respiration and excretion. There is no co-operation between the cells in these functions. All the cells are alike in the vegetative stage, yet the whole assemblage moves like a single organism in a definite direction. It moves through water rotating slowly with one end of the sphere always leading. Further, distinction between the anterior and posterior ends (polarity) of the colony is shown by the gradual decrease in the size of the eye-spot from the front to the rear and reproductive cells being confined to only the hinder part of the *Volvox* colony. The colony does not roll forward like a ball but rotates about its axis of movements with one end of the colony always leading in progression. This shows that the beat of flagella of all the cells in the colony is co-ordinated. Obviously *Volvox* is a true colony or a community of cells, in which all the members co-operate, and work together to certain extent. They have some form of interconnection. There is, however, no co-operation in the functions of nutrition, respiration and excretion.

Reproduction

Individual cells of simple coenobial forms such as *Gonium, Pandorina* and *Eudorina* have equal reproductive potentialities and show little differentiation other than a decrease in the size of the eyespot in the posterior cells of the colony. In the young *Volvox* colonies as well, all the cells are similar and vegetative in function. These somatic cells age and die. A certain number, especially in the posterior half of the colony, become considerably enlarged and function either as **asexual** or **sexual** reproductive cells. The reproductive cells are recognisable by their large size, definite large nuclei and more denser granular protoplasmic contents. They divide and form new colonies. The adult *Volvox* colony thus exhibits a degree of polarity and differentiation of cells accompanied by division of labour. It is indicated by the position of vegetative and reproductive cells. A coenobium has all its reproductive cells either entirely asexually when the season is favourable for growth but sexually towards the end of the growing season.

1. Asexual Reproduction (Fig. 4.25). A few cells (2 to about 50) in the posterior half of the coenobium take part in asexual reproduction. The rest are vegetative. These special reproductive cells gradually push back into the colony, withdraw their flagella, increase ten or more times the size of the vegetative cells and become more or less rounded in form (B). They are recognisable by their well-defined nuclei and dense granular cytoplasm. Such reproductive cells are called the **parthenogonidia** or **gonidia** (B). The number of gonidia in the colony varies from species to species and also within the species. The protoplast of each gonidium by successive longitudinal divisions forms a daughter coenobium (autocolony) within the parent cell wall. It is a tightly packed ball of cells (H).

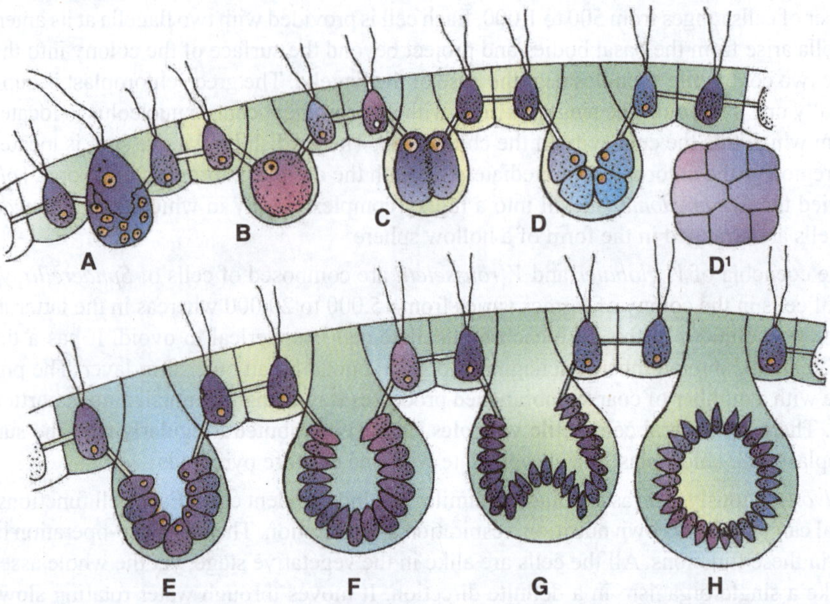

Fig. 4.25 (A-H). *Volvox.* Asexual reproduction. (A) vegetative cell enlarges to form the gonidium; (B) gonidium; (C-E) stages in the development of daughter colony; (F) daughter colony just before inversion; (G) during inversion; (H) after inversion with phialopore closed.

Development of the Daughter Colony (Fig. 4.25 C-H). The first division of the protoplast of the gonidium is longitudinal with respect to the parent coenobium as a whole (C). It is in the anterior-posterior plane of the coenobium. The second longitudinal division is at right angles to the first. After this each of the daughter cells divides again lengthwise so that an eight-celled plate is formed. It becomes more or less incurved with concave surface outwards. This curved eight-celled plate is the **plakea stage** (D). Simultaneous longitudinal divisions of daughter cells continue for several cell generations. At the 16-celled stage the cells in the young daughter colonies become arranged in the form of a hemisphere. Finally a hollow sphere with a small aperture is formed (F). This aperture is called the **phialopore**. It is directed towards the exterior of the mother coenobium. The divisions of cells continues until a specific number of cells in the colony characteristic of the species is attained. At the end of division stage all the cells are naked and in contact with one another. The anterior flagellar end of each cell is directed inwards towards or the centre of the sphere.

For the colony to be motile the flagella must be situated externally. This reversed orientation is corrected by a fascinating process in the life cycle of *Volvox* called *inversion*. In this process the young daughter colony turns inside out through the phialopore to bring flagella from its interior to its outer surface. Extension of flagella commences in the daughter colonies at about the inversion stage, Kocher, (1968). Prior to inversion the cells of young coenobia are pear-shaped or oval.

Mechanism of inversion. According to Pickett-Heaps (1970) during inversion, the pear-shaped cells become markedly elongated at their chloroplast ends where they are interconnected by protoplasmic connections. After inversion the cells gradually resume their normal form. According to him, this change in cell shape (cell elongation) is involved in the mechanism of inversion, the peripheral cytoskeleton of microtubules which runs longitudinally around the cells of the daughter coenobia close to the plasma membrane is involved in cellular elongation (change in cell shape).

The intercellular connections play a vital role functioning as hinges in the highly coordinated movements which turn the young daughter colony inside out.

After inversion the ends of the philapore finally meet. The flagella elongate rapidly. The cells of the daughter colony which uptil now have been in contact with each other begin to separate from one another by the secretion of matrix. The marginal cells of the plakea become the rear and central cells the front of the new colony. The daughter colony is still retained within the parent (gonidial) cell wall. The latter eventually develops into a mucilaginous membrane surrounding the young volvox colony. Several daughter colonies may be developed simultaneously in the parent colony (Fig. 4.26). At first, they hang down but later fall into the soft mucilage filling the hollow (central cavity) of the parent colony where further growth takes place. Finally the young auto colonies escape either by the disintegration of the parent colony (Fig. 4.27) or through a pore at the position of the original gonidium. In the former case the parent colony immediately perishes. This is in sharp contrast to the primitive coenobial form (*Gonium* and *Pandorina*). In them the entire living substance of the parent coenobium is perpetuated in the daughter coenobia. The total number of cells in the colony is reached before release. Increase in size of the colony later is partly due to the cell enlargement but mainly to the swelling of the mucilaginous cell walls. Sometimes one may find coenobia containing daughter coenobia which in turn may contain grand daughter coenobia.

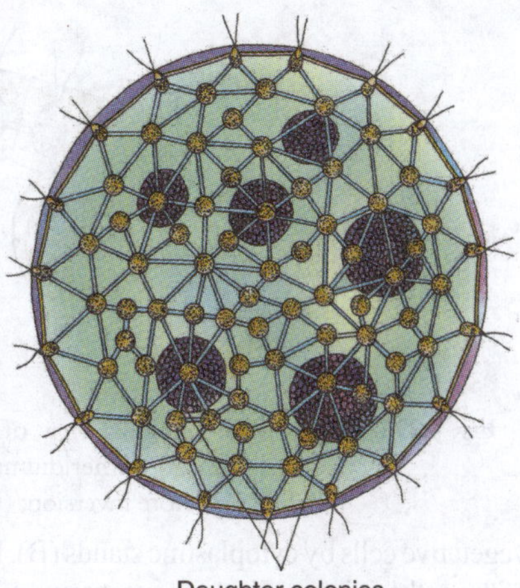

Fig. 4.26. *Volvox sp.* Parent colony with a number of daughter colonies.

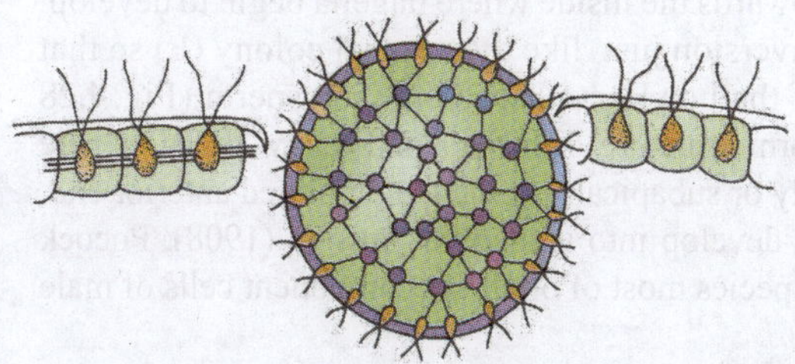

Fig. 4.27. *Volvox sp.* Daughter colony escaping after the rupture of the parent colony.

2. Sexual Reproduction. It is of advanced type with well-developed **oogamy**. Some species are **dioecious** (*V. aureus*) and others **monoecious** (*V. globater*, Fig. 4.28). Monoecius coenobia are protandrous. Certain special cells in the posterior region of the mature colony enlarge, retract the flagella to become **gametangia**. Their number is usually small. The gametangia are large, round cells with many pyrenoids but no flagella. They are connected with the adjacent vegetative cells by fine, protoplasmic threads (Fig. 4.28). The male gametangia are called the **antheridia** or **androgonidia** and the female **oogonia**.

Antheridium (Fig. 4.29 A-E). The biflagellate cell destined to form the antheridium (androgonidium) enlarges, retracts its flagella and pushes back into the colony but keeps its connection with the adjacent

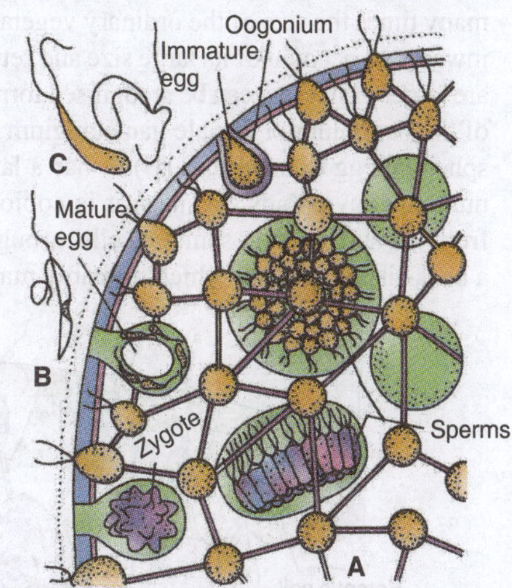

Fig. 4.28 (A-C). *Volvox globator.* Sexual reproduction. (A) portion of monoecious colony; (B) sperm; (C) sperms of *V. aureus* (A, slightly diagrammatic after Colin; B, after Strassburger; C, after Fritsch).

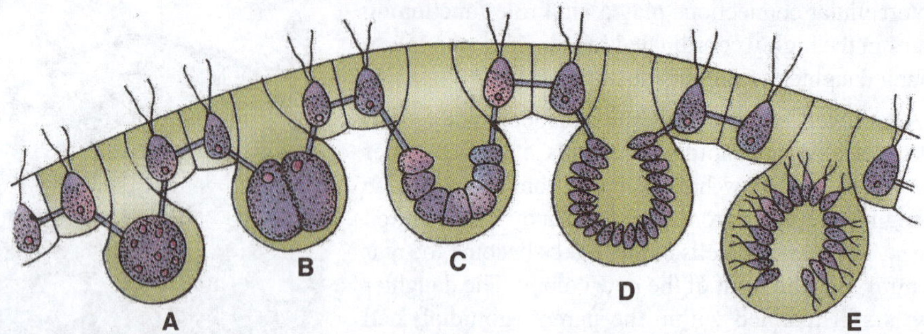

Fig. 4.29 (A-E). *Volvox.* Sectional view of a portion of the colony showing stages in the development of sperms. (A) antheridium; (B-C) stages in the development of sperms; D, antherozoidal bundle before inversion; (E) after inversion.

vegetative cells by cytoplasmic stands (B). Plasma membrane forms over the basal bodies. The protoplast of the antheridium undergoes successive mitotic divisions. In this way usually 64-128 small conical cells are formed. They are grouped as a bowl-shaped plate (C). In some species (*V. aureus*) the divisions continue beyond 128-celled stage and a cluster of as many as 512 cells is formed. The cells in the cluster are arranged in a hollow sphere with a pore, the **phialopore**. The sperm cells both in the bowl or hollow sphere have their anterior ends directed towards the inside where flagella begin to develop. As the bowls or spheres mature they undergo inversion just like the asexual colony (E) so that the flagella are on the exterior (convex) side of the bowl or sphere (E). Each sperm (Fig. 4.28 B,C) is a biflagellate elongated, conical or fusiform structure. It has a small yellow green or pale green chloroplast. The flagella are inserted apically or subapically at the long, pointed anterior end.

Usually only a few cells in the coenobium develop into antheridia. Powers (1908), Pocock (1937) and Smith (1944) reported that in certain species most of or all the component cells of male coenobium may function as antheridia.

Oogonium (Fig. 4.30 A-D). The cell predestined to form the oogonium (A) enlarges considerably, many times the size of the ordinary vegetative cell. It becomes rounded or flask shaped and projects inwards on account of its large size and retracts its flagella (A). The oogonia look like the gonidia but are larger in size and can be recognised form the latter by the absence of division. The entire protoplast of the oogonium or female gametangium gets metamorphosed into a single, non-flagellated, green spherical egg or **oosphere** (B). It has a large central nucleus, and a parietal chloroplast containing numerous pyrenoids. Besides, it is copiously stored with reserve substances which are absorbed from the surrounding somatic cells through cytoplasmic connections. The oosphere often develops a beak-like protrusion which probably marks the point of entrance of the sperm (C).

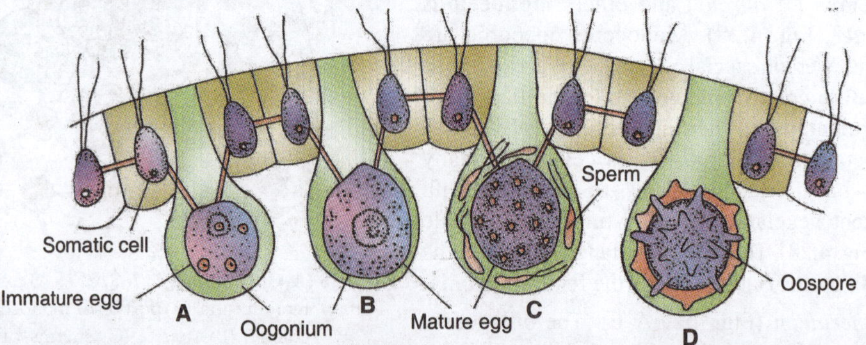

Fig. 4.30 (A-D). *Volvox.* Sectional view of a portion of the colony showing development of oogonium. (A) oogonium with an immature egg; (B) oogonium with a mature egg; (C) oogonium showing fertilization; (D) oogonium containing an oospore.

Fertilization (Fig. 4.30 C)

When mature, the oosphere is fertilized by the sperm. The entire mass of 64 or 128 sperms from an antheridium usually swims as a colonial unit until it approaches the vicinity of female cells of another colony in the dioecious species. There the bowl-shaped unit breaks up into individual spermatozoids. In the second case when the sperms are arranged in the form of a little sphere they are liberated singly and gradually. In either case a single sperm swims slowly through the gelatinous oogonial wall, enters the egg probably through the beak-like receptive spot and fuses with the egg nucleus to form the zygote. The fertilisation is thus internal which means that the ovum of egg is fertilized without becoming detached from the parent coenobium.

Zygote (Fig. 4.31)

Soon after fertilisation the zygote secretes a thick, three-layered smooth (A) or spiny wall (B) around it to become an oospore. The thick-walled oospores remain in the parent coenobium for some time and develop sufficient haematochrome or red pigment to colour their protoplast orange red. Mainx (1929) reported parthenogenesis in *V. aureus*. Eventually the parent colony decays and the oospores fall to the bottom of the pool. They are resistant to adverse circumstances. *Volvox* thus provides the first example of the regular formation of a corpse in the motile coenobial forms. All the cells of the colony not involved in reproduction die. In the fact that *Volvox* leaves behind a corpse, has interconnections in the forms of fine protoplasmic strands between the neighbouring cells and exhibits some specialisation of cells accompanied by division of labour, it approaches the multicellular condition of organisation.

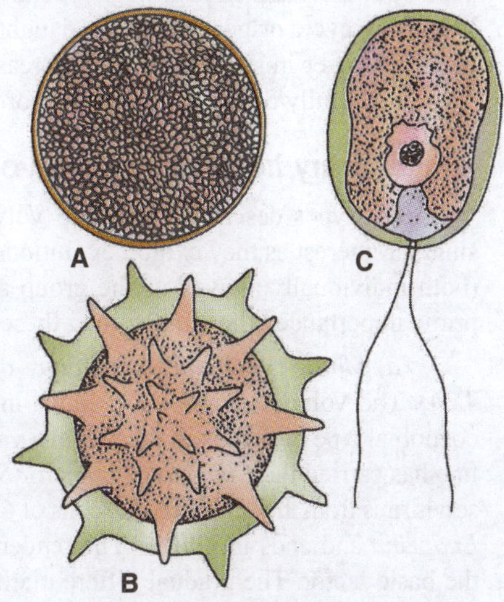

Fig. 4.31. *Volvox*. A, smoothwalled zygote; B, zygote with a spiny wall; C, meiozoospore of *V. rousseletii* formed within the oospore (After Pocock).

Germination of the oospore

With the return of circumstances favourable to growth the oospore germinates. Prior to germination the zygote nucleus undergoes meiosis to form four haploid daughter nuclei. In one species (*V. rousseletii*) only one of these survives. The outer two layers of the zygospore wall (exospore and mesospore) gelatinize or split. The innermost layer extrudes in the form of a small, thin vesicle. The haploid protoplast containing the functional daughter nucleus migrates into the vesicle. There it develops into a single large biflagellate meiospore (Fig. 4.31 C). The meiozoospore or gonozoospore rarely escapes from the vesicle. It divides to form a

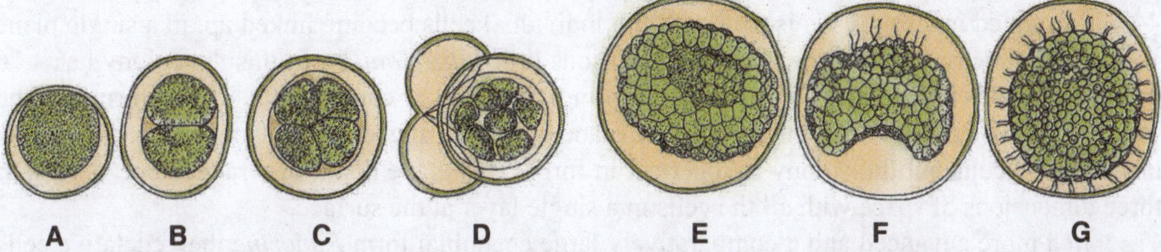

Fig. 4.32 (A-G). *Volvox minor.* Direct germination of zygote to form a new colony. (A) zygote; (B) zygote protoplast divides to form two cells; (C) 4-celled stage; (D) Outer thick layer bursts open to allow multicellular young colony still surrounded by the inner layer to escape; (E) later stage with the cells arranged in a hollow sphere; (F) young colony inverting through the phialopore; (G) young colony surrounded by the inner layer of oospore (All except F after Kirchner, F diagrammatic).

hollow sphere of cells. The sphere after usual inversion, develops into a new colony or coenobium. In *V. aureus* and *V. minor* (Fig. 4.32) the two thick outer layers of the zygote wall spilt. The haploid protoplast surrounded by the inner layer escapes. By repeated division it directly develops into a colony in the same manner as in asexual reproduction. The coenobia formed directly from the zygote are small and have a lesser number of component cells. The maximum size constant for the species is attained after it has passed through several asexual cycles. With each cycle or generation the daughter coenobium becomes larger in size and has an increased number of cells until a fully developed colony is formed.

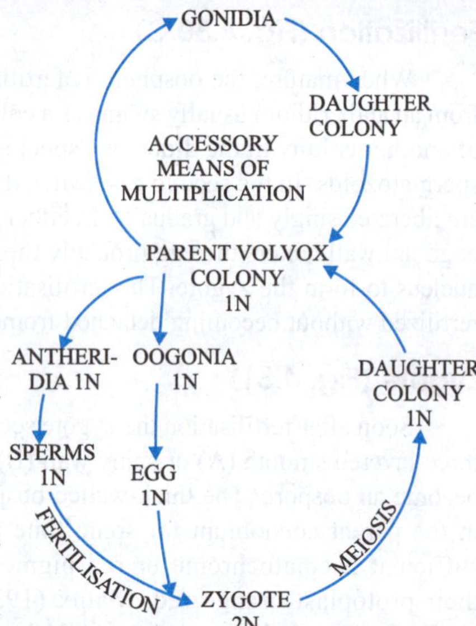

Fig. 4.33. Word diagram of the life cycle of *Volvox.*

Evolutionary trends in the Volvocales

The types described from the Volvocales are of singular interest as they exhibit evolutionary tendencies (both individually as well as the group as a whole) of prime importance. The chief among these are :-

(a) Motile coenobial architecture of thallus (Fig. 4.34). The Volvocales can be arranged in a series showing progressive development of the motile coenobial type of thallus. It is called the **volvocine line** of development of the thallus. The volvocine line has carried the *Chlamydomonas* (or *Sphaerella*) type of cell into a highly complex colony. The series runs from the simple motile cells of *Chlamydomonas* through *Oltmansiella, Gonium, Pandorina, Eudorina* and ends in *Volvox*. The tendency for the daughter cells to remain united appears to be the basic factor. The gradual differentiation and increasing complexity of the thallus in this series represents some evolutionary lines of development. These are :-

(i) Increase in the number of cells in the colony leading to increase in size and change in the shape of the coenobium.

(ii) Morphological differentiation of the cells.

(iii) Specialisation of reproductive evolution of sex from isogamy to oogamy through anisogamy.

The primitive unicellular motile *Chlamydomonas* cell performs all the vital functions, both vegetative and reproductive. There is no division of labour. The simple, primitive or less specialised coenobial forms of *Oltmansiella* (B), *Gonium* and *Pandorina* are only aggregations of similar *Chlamydomonas*-like cells with little differentiation and no division of labour. They consists of cells all of which are alike and able to reproduce. These cells are independent. Any cell separated from the colony is able to develop into a new colony by cell division. The *Oltmansiella* colony consists of 4-cells arranged in a row. This is formed if the individual cells become linked up, in a single plane and in one dimension only. Next comes the curious little alga *Gonium*. In this the colony has 4,16 or rarely 32 cells (depending on the species) arranged as a flat or curved plate. This is formed if the individual cells become linked up in a single plane but are arranged in two dimensions. *Pandorina* has 16 or 32 cells but the colony is spherical in form. This is the result of arrangement of cells in three dimensions of space with all the cells in a single layer at the surface.

In a more advanced and a comparatively large coenobial form *Eudorina* the vegetative cells in the hollow spherical colony may be 16, 32 or 64 in number. They are connected by delicate protoplasmic strands. This results in the formation of a more definite plant body with a tendency towards co-ordination of cells. This high organisation of the thallus in *Eudorina* is also accompanied by division of labour and differentiation of cells. According to Harmann, the four anterior cells in

V. elegans (F) do not take any part in the formation of sexual elements. In *Pleodorina illinoiensis* they remain much smaller and are either purely vegetative in function or take part only in asexual reproduction (G). This increasing differentiation between the anterior and posterior cells is seen in a more pronounced form in *Pleodorina californica* in which half of the cells of the anterior region of the coenobium are smaller and purely vegetative in function (H). The number of cells in the *Pleodorina* colony varies from 32-128. The stigma tends to become less prominent in the posterior cells of the colony in *Pandorina* and *Eudorina* and is absent in the posterior cells of the colony in *Pleodorina* and *Volvox.*

The culmination in increase in the number of cells accompanied by increase in size and complexity of the motile coenobial forms is reached in the hollow spherical coenobia of the genus *Volvox.* The individual *Volvox* colony is much larger in size. The number of cells in it reaches up to thousands (500-50,000 cells). The cytoplasmic connections between the cells are more pronounced. The cells in the coenobium are definitely differentiated. Only a few, more specialised cells in the posterior half of the colony take part in asexual or sexual reproduction. The chief weakness of the *Volvox* series is the retention of motility by the whole coenobium. This feature presents mechanical limitations to further progress in the plant body beyond the *Volvox* stage. Thus in the opinion of botanists *Volvox* is the end of the **volvocine line**. The series could not develop further.

(b) Evolution and differentiation of sex. Another beauty of the volvocine series is the progressive and gradual evolution of sexual reproduction. In *Chlamydomonas* sexual reproduction is commonly isogamous. Among the more primitive coenobial forms such as *Gonium* the sexual process is simple. The gametes are naked and fusion is isogamous but there is some functional disparity as some species of *Gonium* are heterothallic. *Pandorina* is also heterothallic. The sexual fusion in some species is essentially isogamous but a certain species of *Pandorina* exhibits a tendency for gametes of different strains to be unequal in size. The smaller male gamete fuses with larger more passive female gamete. In other words the different strains to be unequal in size. The smaller male gamete fuses with larger more passive female gamete. In other words the different species of *Pandorina* fluctuate between isogamy and anisogamy. This tendency towards increasing anisogamy becomes fixed in *Eudorina*. Some of the cells in the colony divide to form clusters of spindle-shaped sperms. These may be as many as 64 sperms in the cluster. The other cells in the colony enlarge slightly by the accumulation of food and function directly as female gametes. Both the male and the female gametes are biflagellate. The cluster of sperms is liberated as a single unit which swims in water. Near the vegetative cells which directly function as

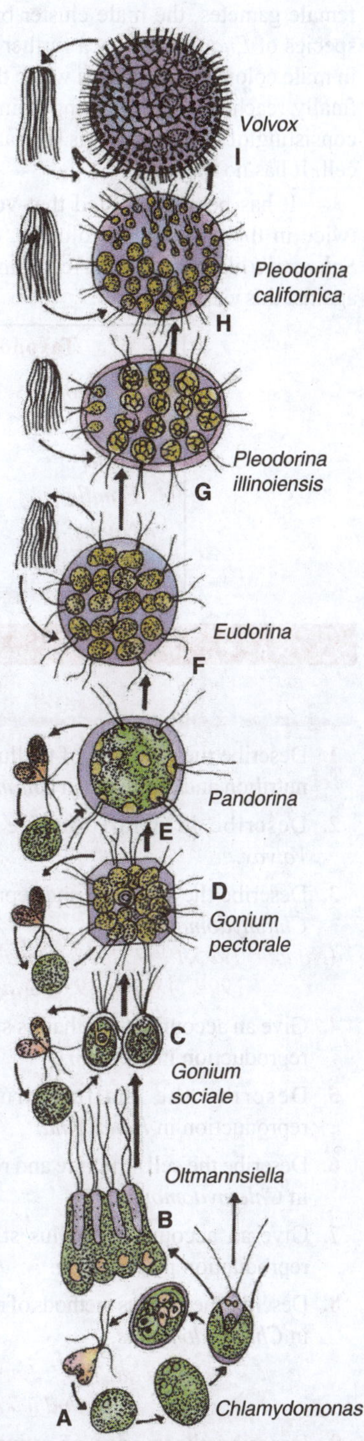

Fig. 4.34 (A-I). *Volvocales* Evolutionary line of development of thallus in green algae.

female gametes, the male cluster breaks up into separate sperms which fuse with the eggs. Some species of *Eudorina* show a further degree of sexual differentiation in being heterothallic. All cells in male colony form sperms while those in the female colony function as eggs. Sexual reproduction finally reaches its culminating point in the well-developed oogamy of *Volvox*. The unit of sperms consisting of 64–512 cells is liberated as in *Eudorina* but the female gamete is a large, non-motile cell. It has no flagella.

It has been suggested that volvocine line is diphyletic in its origin. It means it has arisen twice in the course of evolution, once from a *Chlamydomonas*-like ancestor and once from a *Sphaerella* like progenitor. According to Fritsch (1945) great uniformity in sexual reproduction goes against this view.

Taxonomic Position:

Division	:	**Chlorophyta** (Chlorophycophyta)
Class	:	**Chlorophyceae**
Order	:	**Volvocales**
Family	:	**Volvocaceae**
Genus	:	*Volvox*
Species	:	*globater*

QUESTIONS

ESSAY TYPE

1. Describe the structure of thallus, modes of nutrition and excretion in *Chlamydomonas*.
2. Describe in detail the life history of *Volvox*.
3. Describe the structure and reproduction in *Chlamydomonas*.
   (*Kanpur, 1995,1999, 2007; Gorakhpur, 1993, 1999; Awadh, 1994; Magadh, 1990*)
4. Give an account of the thallus structure and reproduction in *Gonium*.
5. Describe the thallus structure and reproduction in *Pandorina*.
6. Describe the cell structure and reproduction in *Chlamydomonas*.
7. Give an account of thallus structure and reproduction in *Eudorina*.
8. Describe the various methods of reproduction in *Chlamydomonas*.
   (*Allahabad, 1993; Lucknow 1993; Bundhelkhand, 1999*)
9. Describe the modes of reproduction in *Volvox*. (*Allahabad, 1994*)
10. Describe sexual reproduction in *Chlamydomonas*. What light does it throw on the origin and evolution of sex in Algae?
    (*Agra, 1991; Himachal Pradesh, 1992*)
11. With the help of suitable diagrams, describe the asexual and sexual reproduction in *Volvox*/life cycle in *Volvox*.
    (*Agra, 1996; Gorakhpur, 1990, 1994; Bihar, 1991; Rohilkhand, 1991*)
12. Give an illustrated account of asexual/sexual reproduction in *Volvox*.
    (*Lucknow, 1995; Bundelkhand, 1996*)
13. (*a*) Give an account of sexual reproduction in unicellular alga studied by you.
    (*b*) Discuss how *Volvox* is more advanced than *Chlamydomonas*.
    (*Himachal Pradesh, 1997; Poorvanchal, 2007*)
14. Describe asexual reproduction in *Volvox*. Explain the term *coenobium*.
    (*Punjab, 1999*)

15. Describe the structure and method of reproduction in *Volvox*. *(Bundelkhand, 1993; Bihar, 1991, 1992; Bhagalpur, 1999; Rohilkhand, 1996)*

16. Describe the life history in *Chlamydomonas*. *(Bundelkhand, 1995)*

SHORT ANSWER TYPE

17. Write short notes on:

(*i*) Coenobium in *Volvox* *(Allahabad, 1992, 1993, 1996, 1999, 2002; Kanpur, 1995; Gorakhpur, 1991; Lucknow, 1993; Himachal Pradesh, 1997; Rohilkhand, 1994, 1995)*

(*ii*) Palmella stage *(Allahabad, 1994, 2001, 2005; Agra, 1996; Kanpur, 1996, 2006; Awadh, 1997, 1999; Himachal Pradesh, 1990; Bundelkhand, 1993, 1999)*

(*iii*) Asexual reproduction in *Volvox* *(Allahabad, 2005; Agra, 1993; Kumaon, 2000; Awadh, 1991; Trichy, 1995)*

(*iv*) Neuromotor apparatus in *Chlamydomonas* *(Poorvanchal, 2003; Agra, 1993, 1996; Punjab, 1993; Himachal Pradesh, 1993; Rohilkhand, 1991, 1995)*

(*v*) Asexual reproduction in *Chlamydomonas* *(Kumaon, 1999)*

(*vi*) Sexual reproduction in *Chlamydomonas* *(Gorakhpur, 1994; Lucknow, 1997, 1999; Trichy, 1995)*

(*vii*) Gonidium *(Lucknow, 2003; Himachal Pradesh, 1993)*

(*viii*) Daughter colony *(Meerut, 1998, 1999)*

(*ix*) Structure and germination of zygote in *Chlamydomonas* *(Rohilkhand, 1993)*

(*x*) Monoecious coenobium of *Volvox* *(Lucknow, 2002)*

(*xi*) Pyrenoid. *(Lucknow, 2003, Lucknow M. Sc., 2006)*

(*xii*) Nanandria *(Lucknow, 2003)*

(*xiii*) Hydrodictyon *(Allahabad, 2003)*

18. Draw neat and well-labelled diagrams of the following:

(*a*) Cell structure of *Chlamydomonas*

(*b*) Monoecious coenobia of *Volvox*

(*c*) Section through *Volvox* colony

(*d*) Male coenobium in *Volvox*

(*e*) Colony of *Volvox* showing antheridia and oogonia

(*f*) Cell structure of *Chlamydomonas* as seen in electron microscope. *(Kanpur, 2006)*

19. Define coenobium. What is coenobium? *(Allahabad, 2006)*

20. What is anisogamy? *(MDS Univ., 1998)*

21. Describe the structure of chloroplast in *Chlamydomonas*. *(Awadh, 1993)*

22. Give the name of the alga in which neuromotor apparatus is found. *(Awadh, 1994)*

23. Discuss the plakea stage in *Volvox*. *(Punjab/Himachal Pradesh, 1990)*

24. Describe Gonidia in *Volvox*. *(Himachal Pradesh, 1991)*

25. What is palmella stage? *(Bharatidasan, 1995)*

26. (*a*) Describe the structrual details of neuromotor apparatus. *(MDS Univ., 1998)*

(*b*) Differentiate between isogamy, anisogamy and oogamy modes of reproduction with examples. *(MDS Univ., 1998)*

(*c*) Describe zygospore germination in *Chlamydomonas*. *(Gurnanak Deo, Univ. 1991)*

27. Name the species of *Volvox* in whch cytoplasmic strands are lacking. *(Rohilkhand, 1994)*

28. Where does reduction division occur in *Volvox*? *(Allahabad, 1999)*

29. (*a*) With the help of diagrams give the details of *Chlamydomonas* cell/thallus as seen under electron microscope. *(Agra, 1995; Himachal Pradesh, 1996)*

(*b*) Give a brief account of sexual reproduction in *Volvox*. *(Agra, 1995)*

30. Classify the following:

(*i*) *Chlamydomonas*

(*ii*) *Volvox* *(Kumaon, 1995)*

(*iii*) *Gonium*

(*iv*) *Pandorina*

(*v*) *Eudorina.*

31. Identify the following by giving one important character :

(*a*) *Chlamydomonas* *(Kanpur, 1996; Awadh, 1992, 1998)*

(*b*) *Volvox* *(Kanpur, 1996, 1998; Awadh, 1992, 1996)*

32. Describe the sexual reproduction in *Volvox* with the diagrams only. *(Kanpur, 2006)*

33. Explain

(*a*) Palmella stage in *Algae* *(Lucknow, 2006)*

(*b*) Male coenobium in *Volvox* *(Lucknow, 2006)*

(*c*) Inversion in *Volvox* *(Lucknow, 2006)*

(*d*) Formation of daughter colonies in *Volvox* *(Lucknow, 2005)*

(*e*) Thallus organisation in volvocales *(Kanpur M. Sc., 2006)*

(*f*) Cell structure of *Chlamydomonas.* *(Poorvanchal, 2005, 2006)*

34. Name an alga which is single-celled and eukaryotic. *(Allahabad, 2002)*

35. Briefly give description of sexual reproduction in *Chlamydomonas*. *(Kanpur, 2005)*

36. Describe the structure and function of neuromotor apparatus of *Chlamydomonas.* *(Kanpur, 2003)*

37. Draw neat and labelled diagrams of the following and mention their characters briefly.

(*a*) *Volvox* *(Lucknow, 2004)*

(*b*) *Pediastrum* (*c*) *Gonidium*

(*d*) *Pandorina* (*e*) *Eudorina.*

OBJECTIVE TYPE

38. Fill in the blanks :

(*i*) Thick walled resting spores formed after isogamy or anisogamy is called ....

(*ii*) Coenobium is found in ....

(*iii*) Spores produced by *Chlamydomonas nivalis* are coloured red by ....

(*iv*) Number of chloroplasts in *Chlamydomonas* is ....

(*v*) Neuromotor apparatus is present in ....

(*vi*) Sexual reproduction ranges from isogamy to oogamy in ....

(*vii*) Specialised cells meant for asexual reproduction in young coenobia are called ....

(*viii*) The number of cells found in coenobium in *Volvox* ranges from .... to ....

(*ix*) The species of *Chlamydomonas* which forms the red snow is ....

(*x*) *Chlamydomonas* belongs to the family ....

(*xi*) *Volvox* belongs to the family ....

(*xii*) The type genus of the family Volvocaceae is ....

(*xiii*) The reduction division in *Chlamydomonas* occurs at .... stage ....

(*xiv*) The number of contractile vacuoles in *Chlamydomonas* is ....

(*xv*) Asexual reproduction in *Chlamydomonas* takes place by ....

39. Select the correct answer:

(*i*) Coenobium is the characteristic feature of
(*a*) *Chlamydomonas*
(*b*) *Volvox*
(*c*) *Spirogyra*
(*d*) *Ectocarpus*

(*ii*) Unicellular motile thallus is present in
(*a*) *Chamydomonas*
(*b*) *Volvox*
(*c*) *Spirogyra*
(*d*) *Ectocarpus*

(*iii*) The chloroplast in *Chlamydomonas* is
(*a*) Parietal (*b*) Spiral
(*c*) Cup-shaped (*d*) Reticulate

(*iv*) Neuromotor apparatus is present in
(*a*) *Chlamydomonas*
(*b*) *Volvox*
(*c*) *Spirogyra*
(*d*) *Ectocarpus*

(*v*) The number of blepharoplasts in *Chlamydomonas* is
(*a*) One (*b*) Two
(*c*) Three (*d*) Four

(*vi*) In *Volvox* the number of cells in coenobium ranges between
(*a*) 100–200 (*b*) 200–300
(*c*) 300–500 (*d*) 500–1000

(*vii*) The type of genus of the family Volvocaceae is
(*a*) *Chlamydomonas*
(*b*) *Volvox*
(*c*) *Spirogyra*
(*d*) *Ectocarpus*

(*viii*) In *Chlamydomonas* the reduction division occurs at
(*a*) Zoospore stage
(*b*) Thallus stage
(*c*) Gamete stage
(*d*) Zygospore stage

(*ix*) The asexual reproduction in *Chlamydomonas* takes place by
(*a*) Zoospore
(*b*) Gamete
(*c*) Zygospore
(*d*) Cell

(*x*) The vegetative reproduction by daughter colonies takes place in
(*a*) *Volvox*
(*b*) *Chlamydomonas*
(*c*) *Spirogyra*
(*d*) *Ectocarpus*

(*xi*) The gametes in *Chlamydomonas* are
(*a*) Motile and uniflagellate
(*b*) Motile and biflagellate
(*c*) Non-motile and uniflagellate
(*d*) Non-motile and biflagellate

(*xii*) Palmella stage is characteristic of
(*a*) *Chlamydomonas*
(*b*) *Volvox*
(*c*) *Gonicium*
(*d*) *Ectocarpus*

(*xiii*) The cytoplasmic strands are absent in
(*a*) *Volvox aureus*
(*b*) *Volvox globator*
(*c*) *V. mononae*
(*d*) *Volvox carteri*

(*xiv*) Asexual reproduction in *Volvox* takes place by
(*a*) Zoospores
(*b*) Aplanospores
(*c*) Mitospores
(*d*) Daughter colonies

(*xv*) Sexual reproduction ranging from isogamy to oogamy takes place in
(*a*) *Chlamydomonas*
(*b*) *Volvox*
(*c*) *Eudorina*
(*d*) *Pandorina*

(*xvi*) In *Volvox* sexual reproduction is of advanced type known as

(*a*) Isogamy
(*b*) Anisogamy
(*c*) Herkogamy
(*d*) Oogamy

(*xvii*) The reduction division takes place in *Volvox*

(*a*) In daughter colonies
(*b*) During formation of gametes
(*c*) In zygote nucleus prior to germination
(*d*) During formation of zoospores

(*xviii*) The mode of nutrition in *Chlamydomonas* and *Volvox* is

(*a*) Symbiotic
(*b*) Heterotrophic
(*c*) Autotrophic
(*d*) Symbiotic

(*xix*) In *Volvox* the excess photosynthicites are stored in the form of

(*a*) Glycogen
(*b*) Oil droplets
(*c*) Glycogen bodies
(*d*) *Starch*

CHAPTER 

Chlorococcales : *Eutetramorus sp.*

CLASS : CHLOROPHYCEAE

Order 2 : Chlorococcales

- General Characteristics
- Family 1. Chlorellaceae
- Chlorella
- Family 2. Hydrodictyaceae
- Pediastrium
- Family 3. Protosiphonaceae
- Family 4. Coelestraceae

GENERAL CHARACTERISTICS

The order includes unicellular, coenocytic and colonial, **non-motile** green algae. The colonial forms consist of a definite number of non-motile cells arranged in a specific manner. Such colonial forms are called **coenobia**. This coenobia included in this order are non-motile. Motility is confined to the gametes and zoospores only. The thallus in non-motile. Vegetative division of the cell is absent. Division takes place only at the time of reproduction. Unlike the Volvocales, the nuclear division in the reproductive cells is not immediately followed by cleavage of cytoplasm. There is thus a tendency for the cells in the Chlorococcales to become multinucleate for a short while. Even the unicellular forms do not multiply by binary fission. The vegetative cell in most species resembles *Chlamydomonas* in having a well-defined cell wall, a single nucleus and a single, massive parietal chloroplast with a single pyrenoid but differs in the absence of flagella, eyespot and contractile vacuoles. Some species reproduce by the formation of biflagellate zoospores but some are azoosporic and reproduce by aplanospores. The order comprises about 173 genera and 1,079 species. Of these 208 species have been reported from India. *Chlorococcum, Chlorella, Scenedesmus, Pediastrum, Hydrodictyon, Characium* and *Protosiphon* are the most important genera. Most of the members occur as fresh water plankton. Only a few species of *Chlorella, Characium* and *Oocystis* are marine. Some live in the moist soil, on walls and bark of trees. Some live in symbiotic relationship with fungi to form lichens and some in the lower animals.

The order is divided into eight families by Fritsch (1935) and Smith (1955). The chief among these are Chlorellaceae, Selenestraceae, Dictyosphaeriaceae, Hydrodictyaceae, Coelastraceae, Protosiphonaceae. Here Chlorellaceae, Hydrodictyaceae, Protosiphonaceae and Coelastraceae are discussed in some detail.

FAMILY I. CHLORELLACEAE

The family Chlorellaceae includes forms which are usually solitary, free living or symbiotic and a few colonial members (*Radiococcus*). The cells are usually spherical or ellipsoid. Reproduction is by autospores. The autospores are non-motile spores resembling the parent in form and structure. The important unicellular member of this family is *Chlorella*. We study it in detail.

Chlorella

CHLORELLA

Occurrence

The genus comprises a number of species. Of these 4 have been reported from the Indian region. These are *C. vulgaris, C. gonglomerata*, C. *conducterix* and *C. parasitica*. All are small, spherical, green unicells. They are generally found in fresh water of ponds and ditches, in moist soil or other damp situations such as the surface of tree trunks, water pots and damp walls. The common examples of the free living species are *Chlorella vulgaris* and *C. variegata*. The former is the commonest of the free living species. Some form symbiotic association in lichens and in certain invertebrates such as *Hydra, Paramecium* and *Sponges*. The species living as symbionts in these animals are called *C. zoochlorella*. A few species are reported to occur as parasites *(C. parasitica)*.

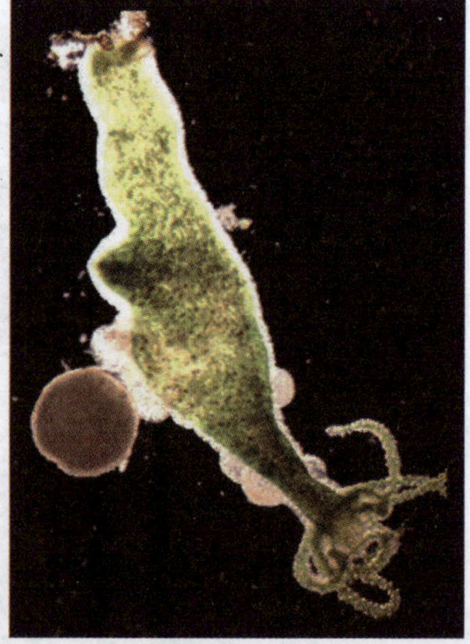

Chlorella zooxanthellae
(Lives Symbiotically inside *Hydra*.

Structure

Chlorella (Fig. 5.1) is known for its extreme simplicity. The plant is a unicell which at the most may grow to 10 μ in diameter but usually it is much smaller. Most of the cells of *C. variegata* are approximately 7 to 7.5 μ in diameter (Silberverge and Sawa, 1974). The small cells are non-motile, round or oval, usually found solitary, some times in groups. The **cell protoplast** is enclosed in a **membrane** which is selective in what it will allow to enter the cell. External to the cell membrane is the thick **cell wall**. There is a single thin, usually cup-shaped or bell-shaped **chloroplast** which is parietal in position. Sometimes the chloroplast is a curved band or of irregular shape usually flattened against the cell wall. In the cavity of the chloroplast is the colourless cytoplasm in which lies the single central **nucleus**. The pyrenoids are usually absent. The stigma and contractile vacuoles are lacking.

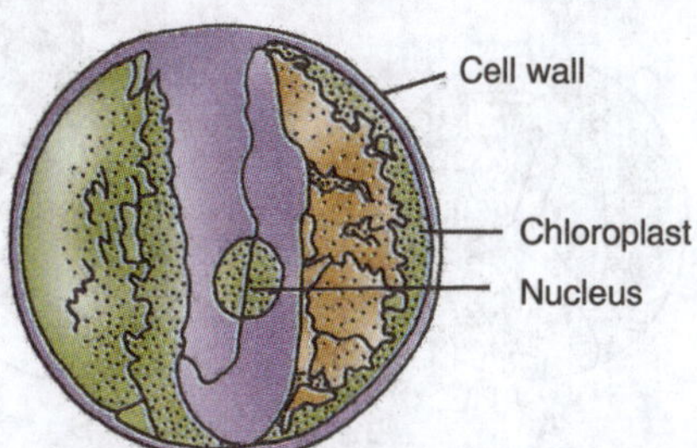

Fig. 5.1. *Chlorella.* A cell under light microscope.

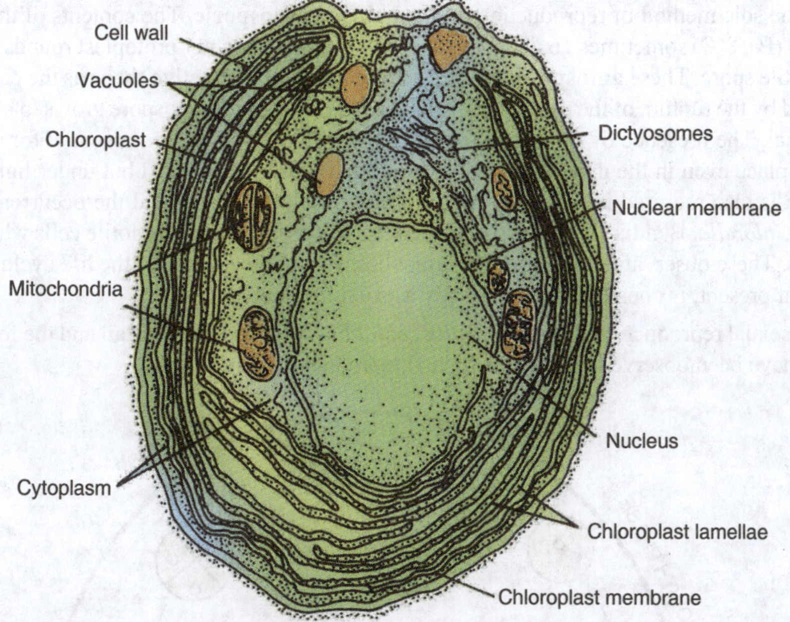

Fig. 5.2. Median section of *Chlorella* unicell showing cell structure as revealed by the electron microscope. (After Bilasputra).

By the use of an electron-microscope which magnifies the *Chlorella* unicell about 25,000 times further details have been discovered. The large lobed chloroplast which occupies most of the cell consists of an array of photosynthetic lamellae displaying C-shaped outline (Fig. 5.2). The lamellae or thylakoids in *C. pyrenoidosa* are organised into granal stacks, each consisting of three thylakoids and interconnecting tracts. Sometimes there are as many as seven but they are not organised into a grana. The chloroplast is enclosed in a double membrane which separates it from the hyaloplasm. The chloroplast has a centrally located pyrenoid which is specialised region of the chloroplast matrix. It consists of a denser matrix material and is traversed by a single central disc of the band. The pyrenoid is bordered by numerous starch plates. The centrally located nucleus has a nuclear membrane which is two-layered and has pores. Mitochondria are present in the cytoplasm within the chloroplast cup. In addition, there are dictyosomes (golgi apparatus), a few vacuoles, endoplasmic reticulum and cytoplasmic lipid bodies (Silverberge and Sawa, 1974).

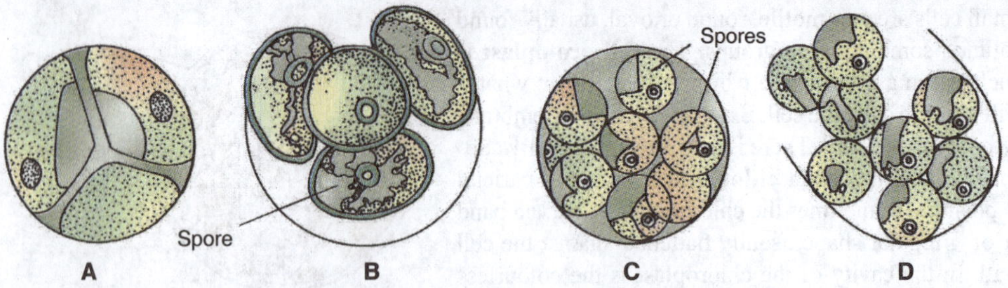

Fig. 5.3 (A-D). *Chlorella.* Asexual reproduction.

Reproduction (Fig. 5.3, A-D)

The sole method of reproduction is **asexual** and **azoosporic**. The contents of the cell divide into 2, 4 (B), 8(C) sometimes 16 daughter protoplasts. Each daughter protoplast rounds off to form a non-motile spore. These **autospores** (spores having the same distinctive shape as the parent cell) are liberated by the rupture of the parent cell wall (D). On release each autospore grows to become a new individual. The presence of sulphur in the culture medium is considered essential for cell division. It takes place even in the dark with sulphur alone as the source material but under light conditions nitrogen is also required in addition. Pearsal and Loose (1937) reported the occurrence of motile cells in *Chlorella*. Bendix (1964) also observed that *Chlorella* produces motile cells which might be gametes. These observations have an important bearing on the concept of the life cycle of *Chlorella* which, at present, is considered to be strictly asexual in character.

Asexual reproduction in *Chlorella ellipsoides* has been studied in detail and the following four phases have been observed during the asexual reproduction (Fig. 5.4).

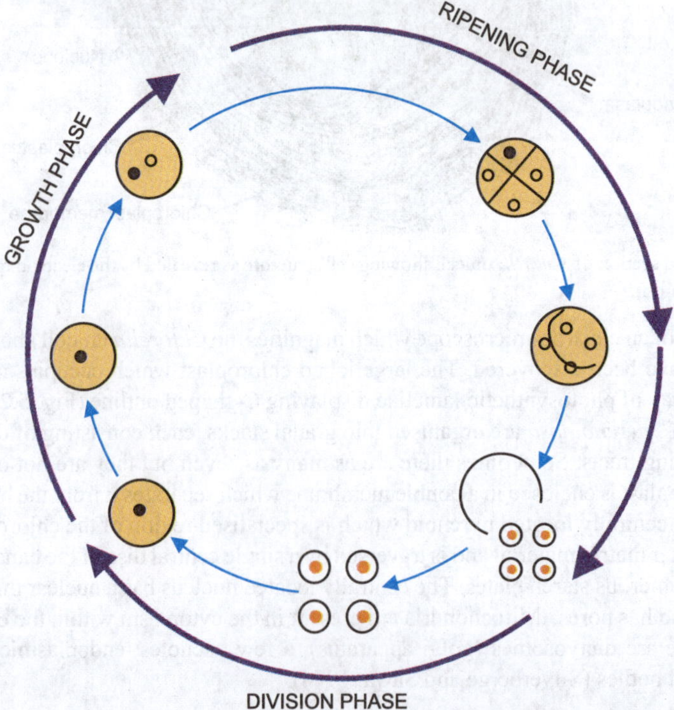

Fig. 5.4. Phases in the life cycle of *C. ellipsoides.*

(*i*) *Growth phase*. During this phase the cells grow in size by utilising the photosynthetic products.

(*ii*) *Ripening phase*. In this phase the cells mature and prepare themselves for division.

(*iii*) *Post ripening phase*. During this phase, each mature cell divides twice either in dark or in light. The cells formed in dark are known as *dark nascent cells* which in turn give rise to photosynthetically active cells. During transition from dark to light phase, cells again grow in size.

(*iv*) *Division phase*. During this phase the parent cell wall ruptures and unicells are released.

Physiology of Chlorella

The alga is autophytic with its photosynthetic system biochemically similar to an angiosperm. For this reason it has become a classic material for studies in physiological research on chlorophyll, photosynthesis and respiration because it can be grown with ease and speed under controlled conditions. It was a German scientist Otto Warburg who in 1919 for the first time employed this alga in experiments on photosynthesis. Since then other scientists avoiding the complicated photosynthetic apparatus of the leaf favoured the use of this alga in experiments on photosynthesis for the following reasons :

1. It can easily grow in varied environments,
2. The chlorophyll pigments it contains and the end products of photosynthesis are similar to those of the other green plants,
3. In young cultures the protein content is high but as the cultures grow older fat and carbohydrate percentage increases,

The nutritional role of *Chlorella* depends on two factors, namely its high protein content and the essential amino acids being in the same proportion as in other green plants. For these reasons *Chlorella's* role as a potential source of food for mankind is being studied seriously.

Some species occur as saprophytes (occasionally with loss of chlorophyll) in sewage and sap exuded from the trees. Many species grow readily on artificial cultures. Chodat (1913) recorded luxuriant growth on agar containing glucose. Muenscher (1923) reported protein synthesis in *Chlorella* in the absence of light at the expense of inorganic nitrogen. Beijerinck (1904) found that *C. variegata* grown on agar containing glucose exhibited decolorisation. The decolorised cells retained their colourless character for about 30 years in successive cultures. The normal green colour, however, reappeared when the decolorised cells were transferred to agar containing no glucose. Chapman reported form variations in *Chlorella* on pure cultures when grown in the light and in the absence of light. The cells produced in the dark were small and dark in colour. They were more active photosynthetically. The dark cells multiplied by division to produce dark cells. These dark cells grown under light conditions (day light) produced 'light' cells which were larger in size but poorer in chlorophyll content. Photosynthetically the light cells were less active. The 'light' cells gave rise to 'dark' cells in the absence of light.

Evolutionary Importance. The non-motile unicells of *Chlorella* illustrate the first step in the **tetrasporine line** of the evolution of thallus in algae. This evolutionary trend involves the loss of motility in the motile unicell of the ancestral type like *Chlamydomonas* through the disappearance of flagella. This results in the origin of non-motile unicells as in *Chlorella.* Repeated division of the non-motile cell and the daughter cells in single plane resulted in the formation of a chain of cells which is called a **filament**. The simple filamentous thallus is considered the forerunner of various branched types. Chodat (1929) reported the origin of new elementary species by micromutations in *Chlorella rubecene* grown on agar culture containing glucose.

Economic Importance

Chlorella contains high percentage of proteins, fats and contains most of the known vitamins (Carotene, Riboflavin, vitamin B_{12}, Choline, etc.). As regards carbohydrates both starch and sucrose have been isolated from *Chlorella.* It multiplies rapidly in the presence of carbon dioxide, sunlight and mineral nutrients. All essential amino-acids are present in *Chlorella* and the amino-acid index is about 62. It is more or less the same as in white flour or pea-nut meal. For these reasons the possibility of *Chlorella* for human consumption as food is being explored all the world over. In fact in countries like Japan, Israel, Germany, Holland and America, "farms" of this alga on a factory scale, with a high efficiency of photosynthesis, have been established. To maintain these farms much water is not required. If such experiments succeed, *Chlorella* will be employed to augment food supplied in arid regions.

Chlorella Farming.

The biologists have as well given a serious thought if *Chlorella* could be utilised to keep the air in space vehicles pure and supply food in space stations and prolonged space flight trips. In fact it has been reported that an apparatus and process for the purpose have been developed. The stale air rich in carbon dioxide is fed into a floodlight container containing water, mineral nutrients and *Chlorella*. The alga restores oxygen by photosynthesis. The additional alga services as food. The space travellers could feed on *Chlorella* soup. It is nourishing but not appetizing food. *Chlorella* also yields an antibiotic called **Chlorellin**. It is considered to be a useful drug in the control of bacterial diseases.

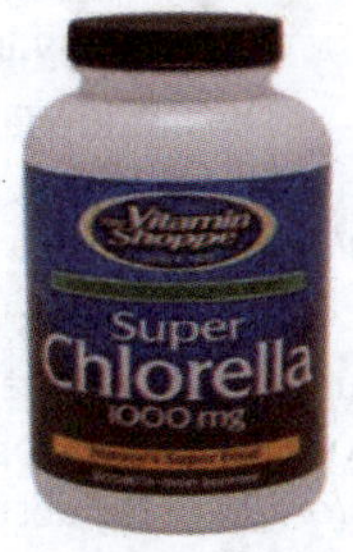

Chlorella nutritional supplement.

Chlorella helps to aerate water by removing carbon dioxide and restoring oxygen in the process of photosynthesis. The growth of this alga is therefore encouraged in the sewage disposal plants where it crowds out by its rapid rate of multiplication and suppresses harmful bacteria. It takes up nitrates and phosphates from the shallow tanks of effluent for its metabolism liberating oxygen in photosynthesis. It helps the aerobic bacteria to decompose raw sewage. This is a very cheap and effective means of purifying sewage.

Taxonomic Position

Division	:	**Chlorophyta** (Chlorophycophyta)
Class	:	**Chlorophyceae**
Order	:	**Chlorococcales**
Family	:	**Chlorellaceae**
Genus	:	*Chlorella*
Species	:	*vulgaris*

FAMILY 2. HYDRODICTYACEAE

The family includes the non-motile, coenobial forms. Sexual reproduction is isogamous and never oogamous. The new colonies are formed by the apposition of swarmers at the end of the swarming period without being discharged resulting in the formation of a miniature colony within the parent cell wall or mucilaginous vesicle. The latter thus liberates a small plant with the complete number of cells. No further division takes place in the young colony until reproduction occurs. *Hydrodictyon* and *Pediastrum* are the two common genera of this family. Both are non-motile, coenobial, free floating forms.

Pediastrum

HYDRODICTYON

Occurrence. *Hydrodictyon* or the common water net, as it is popularly called, is beautiful alga. It is of a large, macroscopic size and occurs in great masses in still, fresh water of ponds, lakes and slow flowing streams. *Hydrodictyon* is of worldwide distribution and is found in various localities in the Punjab and other states of India. The two common Indian species are *H. reticulatum* and *H. indicum. H. patenaeforme* and *H. africanum* are natives of Africa.

Thallus (Fig. 5.5A). The plant body is a non-motile, coenobium of macroscopic size. In form it is a free-floating, hollow, cylindrical network closed at either end. The colony may be 60 cm. in length when mature. Often small fish may be trapped in the meshes of the net where it may die and decay. There is every possibility that *Hydrodictyon* may absorb some nitrogenous compounds released by the decaying body of the fish. The meshes of the net are each composed of five or six cells. The cells are thus arranged in pentagons or hexagons (A) to form the net. Three cells meet at each angle of the mesh. The colony consists of several hundred to many thousand cells. As it grows it breaks up and perfect nets are therefore rare.

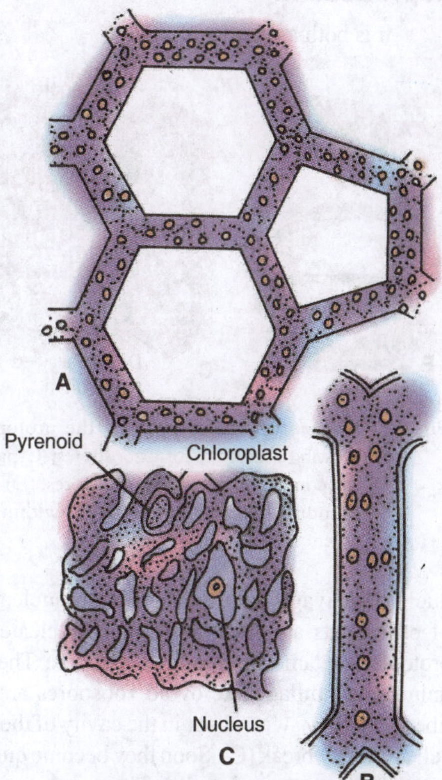

Fig. 5.5 (A-C). *Hydrodictyon.* (A) portion of a coenobium; (B) single coenocyte; (C) surface view of a portion of the reticulate chloroplast. (After Klebs).

Cell Structure (Fig. 5.5 B, Fig. 5.6). The individual cells are elongate, cylindrical and somewhat ovoid in form. Each is joined terminally with two others to form a net-like structure (A). The cell is uninucleate when young and has a simple parietal, band-shaped chloroplast with a single pyrenoid. The cell wall is made up of cellulose. As the cell grows in size the simple chloroplast becomes spiral and finally reticulate. The single nucleus also divides repeatedly but the nuclear divisions are not accompanied by wall formation. The older cells thus become coenocytic and multinucleate with a parietal, reticulate chloroplast containing many pyrenoids. The cytoplasm forms a lining layer and encloses a large central vacuole. Iyengar (1925) discovered a new species *H. indicum* in Madras. It is a larger form with more elongated cells and bigger meshes. The cells have thick, lamellated walls.

According to Harper (1908), the grouping of cells in the colony is very economical and judicious. It allows the largest number of cells in the smallest area permitting the formation of a network with pentagonal or hexagonal meshes. The latter afford greatest elasticity and strength. According to Mcreynolds (1961) the cells grow when in contact with others. They do not grow when not touching the adjacent cells.

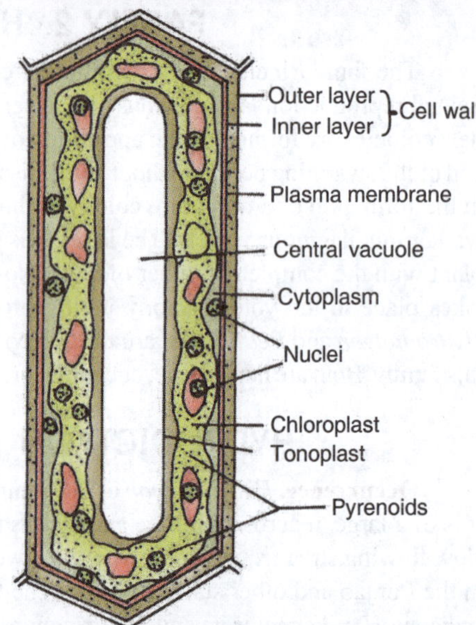

Fig. 5.6. *Hydrodictyon.* A vegetative cell (coenocycte) in section showing structure.

Reproduction

It is both asexual and sexual.

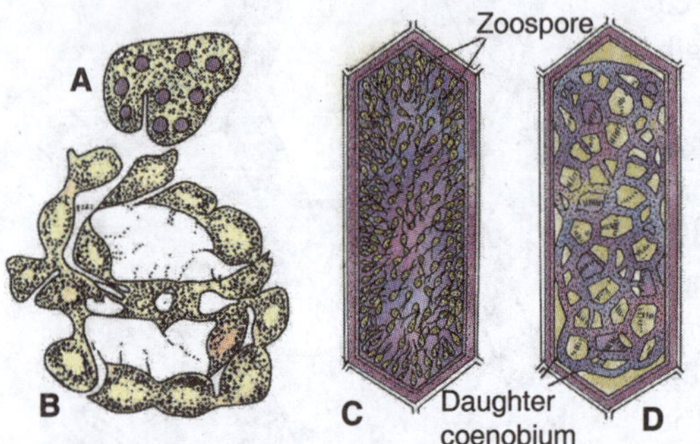

Fig. 5.7. *Hydrodictyon.* (A-B) parts of the protoplast showing cleavage to form zoospores; (C) parent cell containing numerous zoospores; (D) new cylindrical colony (autocolony) within the parent cell (A,B,D after Klebs).

(*a*) Asexual reproduction (Fig. 5.7 A, B). It takes place by the formation of zoospores. They are produced in thousands within any cell of the coenobium by the repeated cleavagc of its protoplast. During cleavage the multinucleate protoplast of the cell slightly recedes from wall and cleaves to form a number of multinucleate segments. The latter undergo further cleavage into smaller and smaller segments with fewer nuclei. Cleavage continues until uninucleate bits of protoplasts are formed. Each uninucleate bit of protoplast metamorphoses into a zoospore. The small, uninucleate, biflagellate ovoid zoospores are never liberated. They swim about in the cavity of the parent cell about day break (C). Soon they become quiescent and withdraw their flagella. They come together, secrete thin walls around them and get apposed in three directions to form the daughter coenobium or net. The latter remains within the parent cell and its component cells elongate slightly (D). The daughter

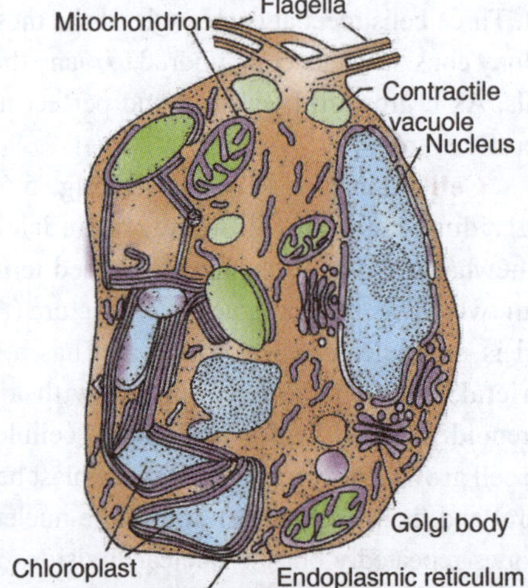

Fig. 5.8. *Hydrodictyon.* Diagrammatic represen-tation of the ultra structure of the zoospore.

net is finally liberated into the surrounding water by the gelatinisation of the parent cell wall. There it grows to the adult size by the elongation and growth of its component cells. There is not further increase in the number of cells by division.

The most characteristic phenomenon of the family is that the zoospores arrange themselves in the manner of the parent colony. There is no vegetative division of the zoospores. *Hydrodictyon* may, therefore, be regarded as an assemblage of individual plants. It differs in its construction from *Volvox* which is formed from a single gonidium or zygote by repeated division. *Volvox* is a single plant made up of many cells.

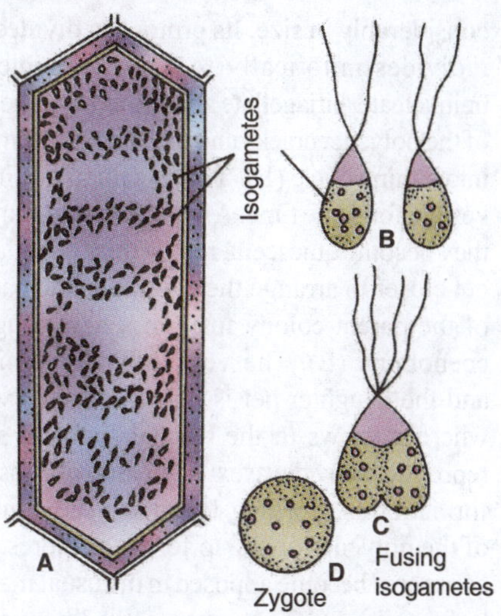

Fig. 5.9 (A-D). *Hydrodictyon.* (A) cell containing gametes; (B) liberated isogametes; (C) isogametes fusing; (D) resting zygote (B-D after Klebs).

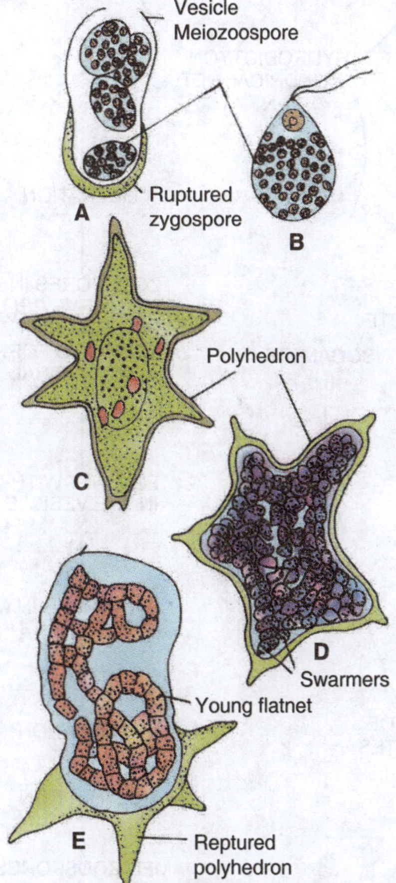

Fig. 5.10 (A-E). *Hydrodictyon.* Post fertilization changes. (A) germinating zygospore with 4 meiozoospores escaping into a vesicle, (B) discharged meiozoospore, (C) polyeder or polyhedron formed by a meiozoospore; (D) swarmer formation in a polyhedron; (E) young flat net escaping from a polyhedron into a vesicle (A-E after Pringsheim).

(*b*) Sexual Reproduction (Fig. 5.9 A-D). *Hydro-dictyon* is **homothallic** (mono- ecious) and sexual reproduction is **isogamous**. The isogametes are tiny, uninucleate, biflagellate structures produced in large numbers by the division and redivision of the protoplast of any mature cell of the coenobium. According to Marchant and Picketi Heaps (1971) the mature gametes are smaller than the zoospores. Each gamete has a small anterior cap. The gametes escape into the surrounding water through a pore in the parent cell wall. The liberated gametes (B) swim about for some time and then fuse in pairs (C) to form a **zygote**. Even gametes from the same coenocyte may fuse. The flagella are withdrawn after fusion. The green zygote becomes spherical and secretes a thin wall around it (D).

Germination of Zygote (Fig. 5.10 A-E). At the time of germination the zygote enlarges. Its diploid protoplast undergoes **zygotic meiosis**. Four uninucleate, haploid daughter protoplasts are formed (A). Each daughter protoplast furnishes itself with two flagella to become **a gonozoospore meiozoospore**. The zygote wall bursts and the four motile meiospores enclosed in a thin vesicle emerge (A). The vesicle soon disappears and the meiosperes are liberated into the surrounding water. Each liberated biflagellate meiozoospore (B) has many discoid chloroplasts. The pyrenoids are absent. The meiozoospore comes to rest. It retracts its flagella. The quiescent protoplast enlarges, secretes a thick wall and becomes angular or polyhedral in shape (D). These thick-walled angular cells are called the **polyeders** or **polyhedra** (Fritsch). Each polyeder or polyhedron increases considerably in size. Its protoplast divides and redivides mitotically to produce numerous uninucleate, biflagellate zoospores (C). The wall of the polyeder cracks and the zoospores emerge into a thin vesicle (E). They swim about in the vesicle for some time. After a period of motility they become quiescent, retract their flagella and get closer to arrange themselves in the pattern of the parent colony to form a new daughter coenobium (E). The vesicle finally vanishes and the daughter net is liberated in the water where it grows to the normal size and starts reproducing by the asexual method of zoospore formation. According to Fritsch the contents of the polyeder divide to form zoospores. The zoospores become apposed in the usual manner to form the net. The young net is liberated by the rupture of the outer layers of the polyhedral cell. Of course the young colony remains surrounded by the dilated inner layers for some time. Some algologists hold that the nuclear division at the time or zygote germination is not meiotic and 2-8 diploid zoospores are produced in the zygote. The vegetative colony is thus considered to be diploid (Morris, 1968).

Net Formation in Hydrodictyon. The nets of *Hydrodictyon* are of two kinds namely, (i) flat nets and (ii) a single layered network of cells arranged in a hollow cylinder closed at either end (Pocock, 1960 and Marchant and Pickett-Heaps, 1971, 1972, a,c). The flat nets arise from the germination of zygotes. The cylindrical nets are formed by the aggregation of zoospores derived from multinucleate parent cells. Each vegetative cell in the colony is thus capable of asexually reproducing a new net or coenobium. Asexual reproduction involves cleavage of the parent protoplast of the cell into 2 zoospores. The cell number in the net is determined by the number of mitosis in a parent cell. No further cell division occurs in the development of the net.

The zoospores in the parent cell remain confined between the cell membrane and the tonoplast. They move freely in a single layer within this space about the longitudinal axis. Eventually their flagella are resorbed in the cytoplasm. The quiescent cells become joined in groups.

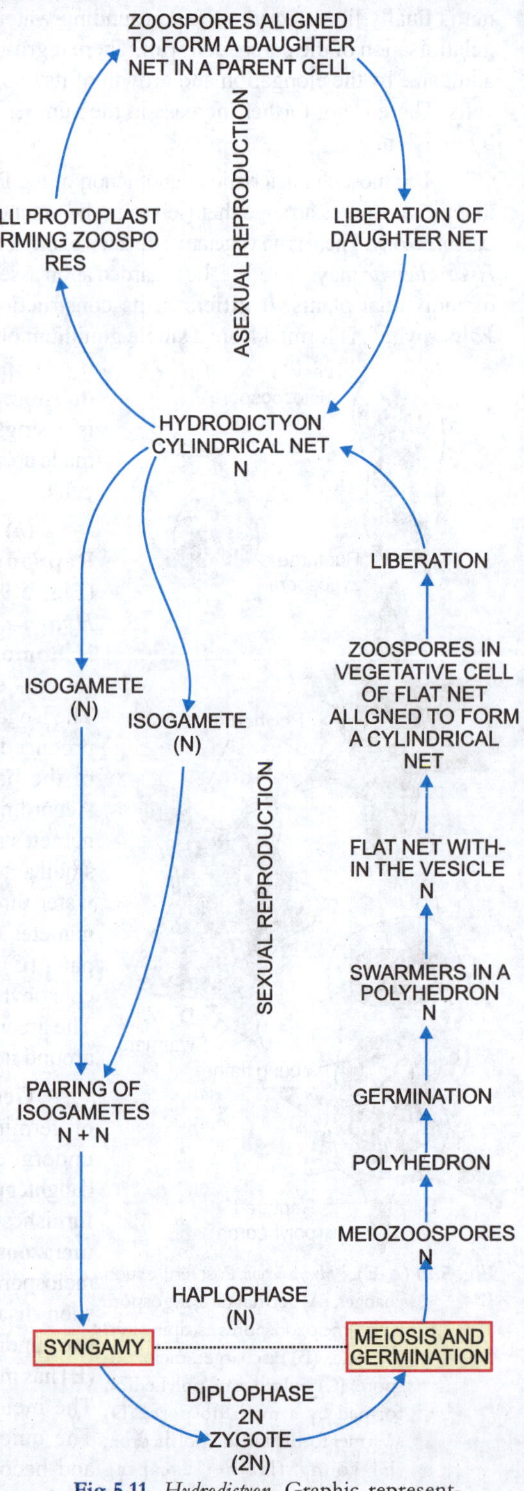

Fig. 5.11. *Hydrodictyon.* Graphic representation of the life cycle.

Origin of Cylindrical Form of the Net

There are different views expressed to explain the origin of the cylindrical net in *Hydrodictyon*.

Klebs (1891) suggested that the zoospores are connected by cytoplasmic processes during the motile phase but Pocock (1960) denied this. Recent observations by electron microscope confirmed Pocock's view.

Pocock (1960) suggested that the form of the space between the plasma membrane and tonoplast affects the overall shape of the net but not the process by which individual zoospores unite in the net. Thus the central vacuole, tonoplast and cell wall together act as a mold in which the young cylindrical net is organised.

Hawkins and Leedale (1971) are of the opinion that planar orientation is a feature of colony-forming zoospores (both in *Hydrodictyon* and *Pediastrum*). The closed cylindrical net of *H. reticulatum* is imposed upon the primary plannar system by the shape of the space available between the plasma membrane and persistent tonoplast.

Mellington and Gawlik (1970) and Marchant and Pickett-Heaps (1972 b and 1974) demonstrated the presence of bands of microtubules beneath the plasma membrane at the initial sites of contact of the aggregating zoospores irrespective of shape of the net (flat or cylindrical) they are about to form. These are oriented in the plane of the developing net. According to them, the microtubules have a cytoskeletal function and play an important role in determining the arrangement of cells in the developing net by affecting change in the shape of the zoospores before and during aggregation. Amorphous material appears at the site of contact (between the appeared surfaces). At this stage cell wall formation follows.

Gupta and Srivastava (1963) reported that *H. reticulatum* produces an antibacterial substance.

Taxonomic Position

Division	:	**Chlorophyta** (Chlorophycophyta)
Class	:	**Chlorophyceae**
Order	:	**Chlorococcales**
Family	:	**Hydrodictyaceae**
Genus	:	*Hydrodictyon*
Species	:	*indicum*

PEDIASTRUM

Occurrence. It is a non-motile, coenobial green alga met with in ponds, ditches, and plankton of fresh water lakes. It prefers still water and avoids flowing or running water of streams and ravines. It rarely occurs in abundance. The coenobia are free-floating. They usually float on the surface of water alone or with other colonial forms and water plants. They sink to the bottom of the pond when water level is low. The genus includes 30 species. Eleven of these are known from the Indian region. Of the Indian species *P. biradiatum, P. boryanum* (with 3 varieties), *P. constrictum, P. duplex* (with 5 varieties), *P. integrum, P. Kawraiski,* and *P. Nortatum, P. Simpex* (with one variety) and *P. tetras* (with 3 varieties) have been reported by Patel (1970) from Baroda and Kaira districts (Gujarat) alone. All are fresh water.

Plant body (Fig. 5.12). The word *astrum* means star, and refers to the more or less radiate form of the colony. The colonies are disc-shaped to stellate in form. The cells in the colony are in

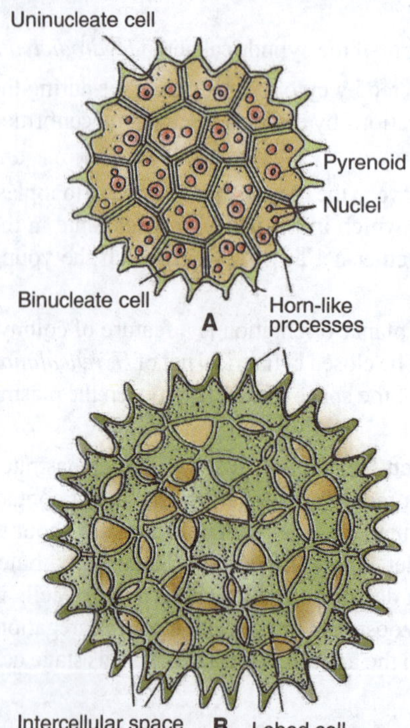

Fig. 5.12. (A-B). *Pediastrum.* (A) young colony of *P. boryanum* consisting of 16 cells forming a flat disc; (B) Colony of *P. duplex* showing species of variable size between the lobed cells (After Smith).

some multiples of two and are usually arranged in a single layer (A) forming a flat circular plate. Rarely the large coenobia have cells arranged in two layers in the centre (*P. integrum*). The number of cells in each colony varies from 4 to 128 depending on the species. They are grouped together into a non-motile colony in a regular sequence. For example in *Pediastrum boryanum* which is a common alga in the plankton of lakes and ponds, the cells in the discoid coenobia are arranged in concentric circles around a core of 1, 2 or rarely three individuals (Ingold, 1973). Fritsch (1935) reported that in 8-celled colonies (Fig. 5.15 E) there is usually a single cell in the centre and the rest 7 are arranged around (1 + 7). In the 16-celled colony there is one cell in the centre surrounded by two concentric rings of 5 and 10 cells each (A). The 32-celled colony has 1 + 5 + 10 +16. According to Ingold (1973) the pattern in the 8-celled colony was 2+6, in the 16-celled coenobia, 1 + 5 + 10 is the usual arrangement, in 32-celled coenobia 1 + 6 + 10 + 15 is the commonest. In the 64-celled colony he reported 2 (or 3) cells with a pattern 2 + 8 + 13 + 19 + 22. The number of cells in *P. boryanum* may go up 128 in one coenobium. Departures from the above arrangement are also met with. The 8, 16 and 32-celled coenobia of *P. simplex* are irregular in form in the sense that the cells are rarely arranged in concentric rings. *P. tetras* has 4 to 8-celled coenobia whereas *P. duplex* (B) may have 32-celled coenobia.

There are two kinds of coenobia in *Pediastrum* (Fig. 5.12). In one kind represented by *P. boryanum* (A), the individual cells in the coenobium fit tightly together, there being no intercellular spaces between the component cells. Such compact coenobia are called **unfenestrated**. The second kind of coenobia are called **fenestrated**. *P. duplex* provides an example of fenestrated species (B). The constituent cells of the coenobium being lobed, species of varied sizes are left between them.

Cell Form

The cells are generally polyhedral in shape. The marginal cells of the colony often differ from the rest. They bear one (*P. simplex*), two (*P. boryanum*) or four (*P. tetras*) hornlike projections or processes often called the **prongs**. *P. bradiatum* has bifid horns. The disc-shaped pronged colony of *Pediastrum* presents the appearance of a cog or gear wheel. Rarely the prongs are absent. In *P. clathratum.* Peterson reported the presence of groups of stiff, gelatinous bristless (Fig. 5.13) arising from the

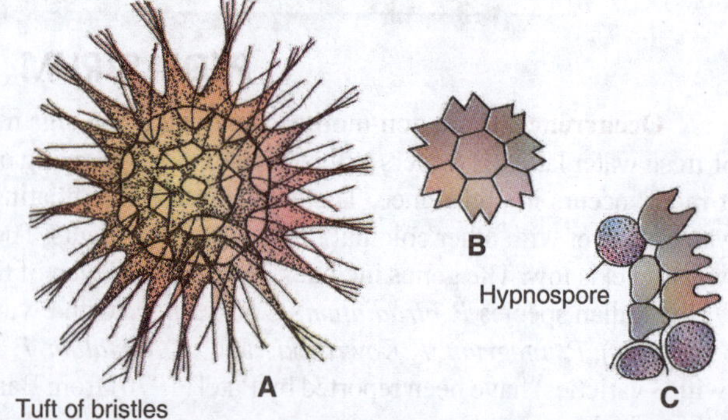

Fig. 5.13. (A-C). *Pediastrum.* (A), coenobium of *P. clathratum* showing groups of stiff, gelatinous bristles arising from the apeces of peripheral cells and surface of some of the central cells; (B), eight-celled colony of *P. tetras*; (C), part of a colony of *P. duplex* with cells containing Hypnospores (after Peterson, B-C after West).

apices of the processes of the peripheral cells. Mellington and Gawlik (1975) suggested likelihood of cell shape being under genetic control and the number of horns predetermined.

Cell Structure (Fig. 5.14)

(a) *Cell wall*. The cell consists of the cell wall and the protoplast. The cell wall which is composed of cellulose derivatives is resistant to decay. It is usually smooth or granulate but may also show reticulations in some cases. It is differentiated into two layers, outer and inner. The inner layer is cellulose and the outer is of a loose mesh construction (Moner, 1955). According to Mellington and Gawlik (1975) the outer wall layer is thin but uniformly thick. It is traversed by strands of electron dense material evident as fine dots in surface view. The inner cellulose layer is thicker and is granular or fibrillar in appearance. The cementing material between the two layers is seen as a fine granular line.

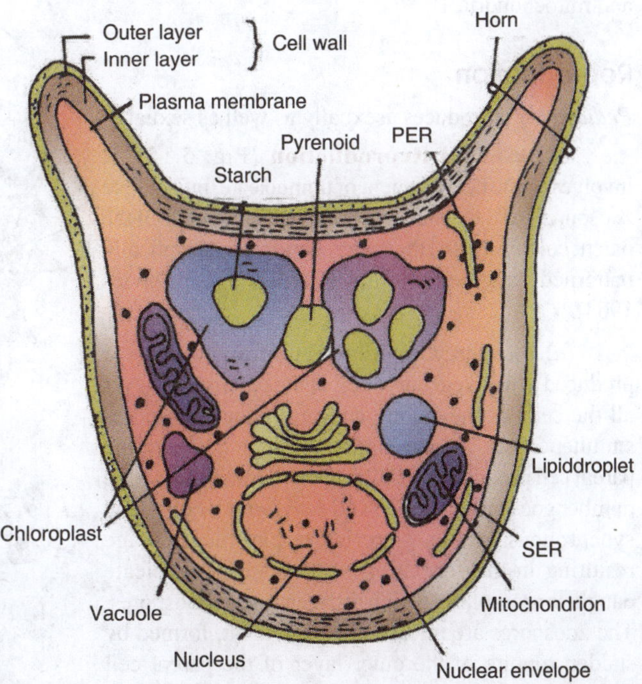

Fig. 5.14. *Pediastrum.* Diagrammatic representa-tion of fine structure of cell.

(*b*) *Protoplast*. It is differentiated into the plasma membrane, cytoplasm and the usual cell organelles (nucleus, chloroplast or S. mitochondria, dictyosomes, endoplasmic, reticulum, ribosomes, and membrane bound vesicles). The lipid droplets and vacuoles are sometimes common. Young cells are uninucleate. The prominent nucleus is peripherally situated usually on the inner side of the cell. The nuclear envelope is double-membraned. The outer membrane contributes blabs along its inner face to the forming face of the single active perinuclear dictyosome in *P. boryanum*. It is associated with the flattened ribosome free region of the nuclear envelope. The mature cells in the colony prior to asexual reproduction may become multinucleate by repeated mitosis which will produce 2 nuclei (where n = 1–7). The dictyosome in *P. boryanum* consists of seven cisternae. Davis (1964) reported the presence of two equal-sized chloroplasts in the outer half of the cell especially in the peripheral cells of the colony. Marchant (1974) stated whether there are two equal chloroplasts or a single bilobed chloroplast as reported in *P. duplex* by Hawkins and Leedle (1971) could not be determined. Thylakoids within the double-membraned chloroplast are randomly arranged in groups of 2,3 and 4 as well as in larger stacks (Rogalski *et al*, 1977). The horns develop on the chloroplast side of the cells (Harper, 1918, Davis 1964, Hawkins and Leedale, 1971). Marchant and Pickett-Heaps (1974) concluded that all the cells in the colony have the potential to develop horns but this potentiality remains suppressed by contact of the sides with other cells. The cytoplasm is rich in free ribosomes. Smooth endoplasmic reticulum (SER) underlie the plasma membrane. Rough endoplasmic reticulum (RER) are found scattered in the cytoplasm. Lipid bodies and large vacuoles filled with electron-dense material are also evident. Single membraned bodies containing fine granular matrix and resembling the microbodies in dia. are reported by Rogalski *et al* (1977) in the mature cells of *P. tetras*. These were found associated with the RER, chloroplasts and mitochondria.

Reproduction

Pediastrum reproduces asexually as well as sexually.

1. Asexual Reproduction (Fig. 5.15). It involves (*i*) the development of uninucleate, biflagellate zoospores from the multinucleate vegetative cells of the parent colony and (ii) their aggregation into specifically patterned daughter colonies (Harper, 1918; Davies 1964).

(*a*) *Zoospore formation*. The zoospores may be produced in any vegetative cell of the colony. However all the cells of the colony do not produce zoospores simultaneously. By repeated mitosis the nuclei of the parent cell produce 8, 16, 32 nuclei. In some species the number goes up to 64 or 128 (*P. boryanum*). Following synchronous mitosis there is cytoplasmic cleavage resulting in the formation of as many uninucleate daughter protoplasts which differentiate as zoospores. The zoospores are released through a slit, formed by sudden rupture of the outer layer of the parent cell wall within a vesicle (A_5). The origin of the vesicle is uncertain. Usually it is considered as a part of the inner layer of the parent cell wall. Marchant (1979) suggested that it could as well be a secretion of the zoospores prior to their release. The slit in the parent cell which permits zoospore release is perhaps caused by the swelling of the inner pectic layer of the parent cell wall. Rogalski *et al* (1977) attributed it to the rapid degeneration of the parent cell wall induced by the occurrence of RER along the periphery of the developing zoospores.

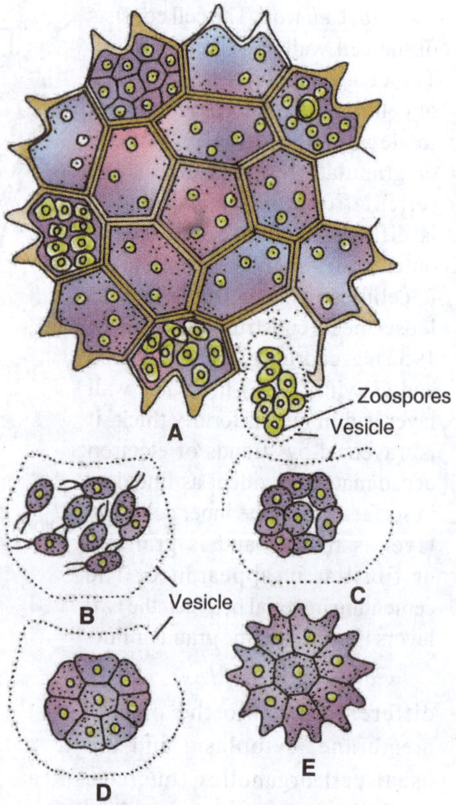

Fig. 5.15 (A-E). *Pediastrum*. Asexual reproduction. (A), *P. boryanum* coenobium with the cells showing stages in the development of zoospores (A1 - A4) and their escape into a thin vesicle (A5); (B) Swarmers jostling around inside the vesicle; (C-D) Successive stages in the retraction of flagella and alignment of zoospores into a flat, plate-like colony one cell thick inside the vesicle; (E) A young coenobium just liberated as the vesicle vanished (A based on Braun and the rest based on Smith).

(*b*) *Zoospore structure* (Fig. 5.16). The biflagellate zoospores lack a cell wall. When discharged into the vesicle they are oval or spherical in outline. The two flagella emerge from the anterior end. The basal bodies of the flagella are interconnected by a bridge. The bilobed chloroplast and irregularly-shaped nucleus is placed asymetrically along the periphery of the zoospore with the chloroplast occupying much of the volume of the cell. Embedded in the zoospore cytoplasm, in addition, are the perinuclear dictyosome, mitochondria, endoplasmic reticulum and contractile vacuoles. Marchant (1979) described 3 systems of microtubules (Fig. 5.16) in the zoospores of *Pediastrum* namely, (*i*) 4 bands of *rootlet microtubules* which arise from the microtubule organizing centre (MTOC) associated with the bridge between the basal bodies; (*ii*) An array of 12-20 curved microtubule underneath the plasma membrane encircling the zoospore; and (*iii*) microtubules emanating from the microtubule organizing centre (MTOC) attached to the tips of the developing horns. The microtubule girdle running below the plasma membrane functions as a cytoskeleton. It is involved in the change of shape of the individual zoospores at the time of their aggregation. The extensive cytoskeleton of microtubules thus plays a significant role in the patterned arrangement of cells in colony formation. The microtubules emanating from the tips of the developing horns are involved in their extension.

*(c) **Colony formation.*** (Fig. 5.15 C-E). At the time of their release from the paternal cell into the vesicle, the zoospores are oval or spherical in form (B). In the vesicle they swarm for about 10 minutes. During the swarming period they become somewhat rectangular. At the end of swarming period they become passive and retract their flagella. The aflagellate zoospore cells, which are devoid of cell walls link up with one another to form a characteristically patterned multicellular daughter colony with all the cells disposed more plane. Aggregation takes place along regions of the plasma membrane proximal to the nucleus (Rogalski *et al,* 1977). In this cell to cell contact interaction between zoospore membranes as well as disposition of prewall cementing material at these specific adhesive sites on their surface could be implicated (Schraudolf and Frauenkron, 1979). Linking up initiates further change in the shape of the zoospores. The cells on the periphery of the colony develop horn-like cytoplasmic projections on the side of the bilobed chloroplast (E). After zoospore aggregation and extension of horns, cell wall deposition occurs within one hour. Shortly after wall deposition on the daughter cells, the swarming vesicle disintegrates to release the daughter colony. The number of cells in the *Pediastrum* colony is determined by the number of mitosis in the parent cell at the time of zoospore formation. No further nuclear divisions occur during colony development. The number of cells in the *Pediastrum* colony is in multiples of two. This number is determined during the juvenile (embryonic) stage. The cells in the colony are arranged in a specific manner. Such a colony is called a **Coenobium**. The component cells of the *Pediastrum* coenobium are aflagellate. The genus thus provides an excellent example of non-motile coenobium.

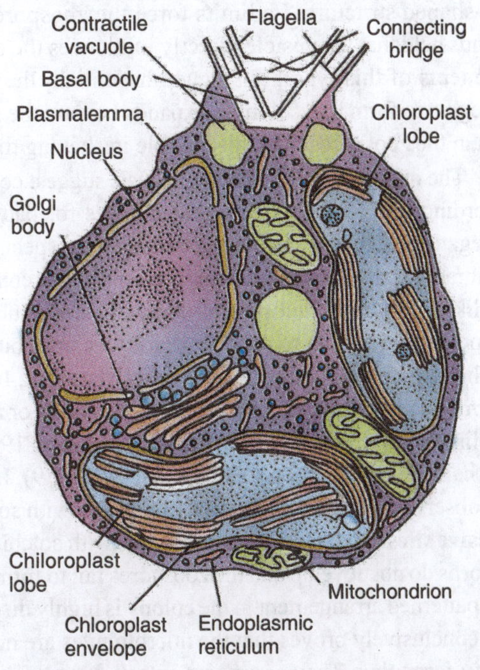

Fig. 5.16. *Pediastrum.* Diagrammatic representation of the ultrastructure of a zoospore.

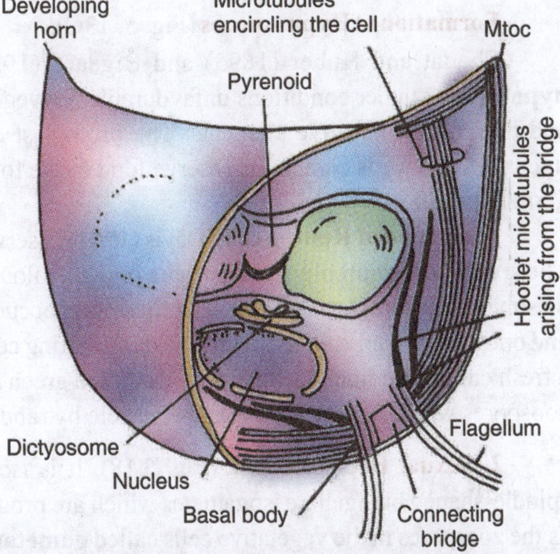

Fig. 5.17. *Pediastrum.* Diagrammatic representation of the shape and fine structure of the zoospore at time of aggregation to form the colony (Based on Marchant).

Patterned arrangement of cells in the colony. The question arises how the zoospores which swarm in a disorderly mass in the vesicle, link up with one another in a patterned manner to form a multicellular colony consisting of a flat plate of cells. Some ingenious theories have been put forth to explain this. Davis (1964) suggested that the enclosing zoospore vesicle is relatively a rigid, lens-shaped structure. Its limits force the zoospores to become arranged in one-cell thick layer. He thus held that the vesicle directly influences the cellular arrangement in colony formation. The opponents of this hypothesis argued that initially the vesicle is close about the zoospores but as they aggregate to form the colony it expands beyond the swarming cells. The vesicle thus only serves to contain the zoospores but plays no role in shaping the circumperiphery of *Pediastrum* colony.

The opponents of Davis' hypothesis suggest correltaion of cell contact with pattern formation. According to them one of the prerequisites for pattern formation is the specific cell shape of the aggregating cells and second the existence of specific adhesive sites on the surface of the plasma membrane of the zoospores which on coming in contact initiate colony formation. Mellington and Gawlik (1970) observed parallel arrays of microtubules underlying the plasma membrane of the zoospores aggregating to form the colony. Microtubules were also observed in the developing horns which protrude from the peripheral cells (Marchant, 1979). The possible role of these microtubules in determining the shape of cells and their role direct or indirect in colony formation has been discussed (Mellington and Gowlik 1970, 1975; Marchant 1974, a, c.; Marchant and Pickett-Heaps 1974; Marchant 1979; Schraudolf and Frauenkron 1979). In addition to the peripheral microtubule arrays they observed amorphous material associated with some regions of the plasma membrane called the **adhesive sites.** On treatment of zoospores with colchicine, the microtubules disappear. Consequently the horns do not develop and the zoospores fail to form normal colonies. The zoospores do link up but their patterned arrangement in the colony is highly disturbed. They fail to aggregate in ordered arrays. This conclusively proves that the microtubules are not involved in the adhesion of the zoospores in colony formation. They serve as a cytoskeleton in determining the specific shape of the individual zoospores at the time of their adhesion and thus play a significant role in the patterned arrangement of cells in the colony (Marchant and Pickett Heaps, 1974 and Marchant, 1979).

Formation of Hypnospores (Fig. 5. 13C).

Chodat and Huber (1895) and Bigeard (1936) reported the formation of thick-walled **hypnospores** under conditions unfavourable for vegetative multiplication such as low temperature and decrease in nutritive elements. The protoplast of the cell may form one or two spores with thick resistant walls containing reserve food in the form of dispersed red oil. The hypnospores have germinated after 12 years.

Formation of Resting cells. Davis (1964) observed the formation of **resting cells** in old cultures. These cells lost chlorophyll and became orange-coloured due to the formation of carotenoid pigment dissolved in fat droplets. No cell wall thickening occurred. The resting cell contained abundant starch and one or two orange fat droplets. The dried resting cells remained viable for 4 years. When placed in a fresh culture medium, the resting cells turned green and reproduced like vegetative cells producing zoospores which are released into the vesicle by random rupture of the outer layer of cell wall.

2. Sexual Reproduction. (Fig. 5.18). It is isogamous and takes place by the formation of spindle-shaped biflagellate isogametes which are produced in large numbers and in the same manner as the zoospores in the vegetative cells called **gametangia**. They are small in size than the zoospores and are discharged into a vesicle through a hole in the gametangial wall. The vesicle perishes and the liberated gametes (A) fuse in pairs (B) in the surrounding medium. It is not definitely known whether *Pediastrum* is homothallic or heterothallic. The quadriflagellate zygospore (C,D) soon retracts its flagella and secretes a thin wall around it to become a zygospore (E).

Germination of Zygospore. (F) The zygospore enlarges and germinates immediately. Sometimes it may perennate. The diploid nucleus of the germinating zygospore undergoes repeated division. Most probably the first two divisions are meiotic. The numerous haploid nuclei in the zygospore are then apportioned into uninucleate, biflagellate, haploid swarmers known as the gonozoospores or meiozoospores. Eventually the zysospore wall ruptures and the haploid swarmers escape into the water (F). After a brief period of activity they become passive, retract their flagella and come to rest. Each of the resting swarmers becomes angular in shape and secretes a thick wall around it. These thick-walled non-motile angular structures are called the **polyeders** or **polyhedrons** (G).

Germination of polyeders (Fig. 5.18 H-K). The polyeders are resting structures. With the onset of season favourable for growth, the polyeder or polyhedron grows in size and its protoplast divides mitotically (H) to produce a number of biflagellate zoospores (I) which are discharged into a vesicle through an aperture in the polyeder wall (I). After a brief period of activity in the vesicle the zoospores withdraw flagella, come to rest and arrange themselves appropriately in one plane (K) to form a daughter colony (Askenasy 1888). Prescott (1969), however suggested that the zoospores aggregate appropriately inside the polyeder to form the young colony which is then released into a vesicle. Shortly after wall segregation of the cells in the colony, the vesicle disintegrates to liberate the colony.

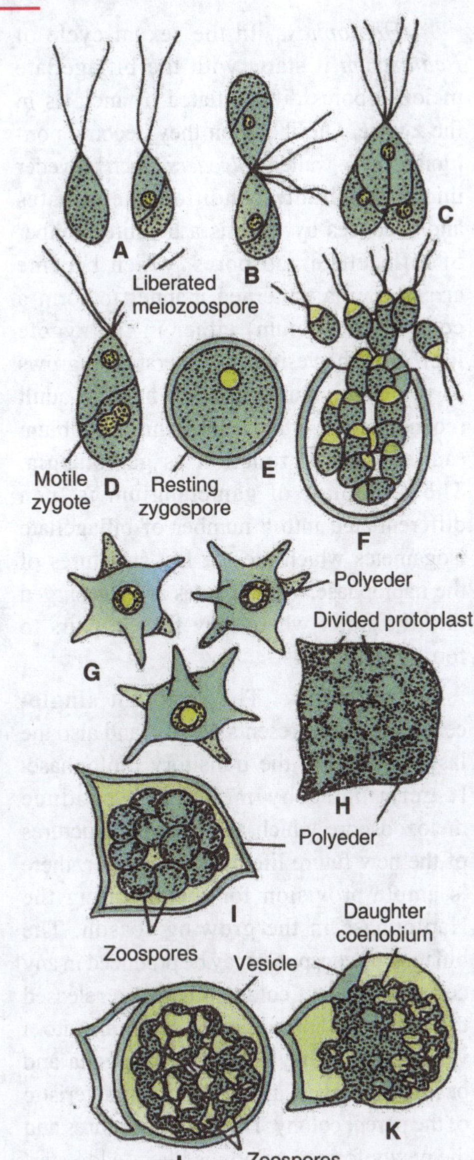

Fig. 5.18. (A-K). *Pediastrum.* Sexual reproduction and post fertilisation changes. (A), isogametes; (B-C), stages in the fusion of gametes; (D), young quadriflagellate zygote; (E), resting zygospore; (F), germinated zygospore discharging the meiozoospores; (G), resting meiospores having developed into thick-walled angular polyeders or polyhedrons; (H-I), stages in the germination of a ployeder or polyhedron; (J), liberation of zoospores into the vesicle; (K), zoospores with retracted flagella aligned to form the daughter coenobium within the vesicle (G-K after Askenasy).

Life cycle (Fig. 5.19)

Pediastrum boryanum, like other green algae, in general, has a prolonged vegetative phase representing the *haplophase* or *gametophyte*. Single-celled zygote is the only diploid structure which represents the *diplophase* or *incipient sporophyte*. Meiosis takes place in the zygote. Such a life cycle is called *haplontic*. It is characterized by *zygotic meiosis* and dominant, *haploid adult* (coenobium in this case).

Haplophase. In the sexual cycle of *Pediastrum* it starts with the biflagellate meiozoospores differentiated by meiosis in the zygote. On liberation they become non-motile, thick-walled *polyeders*. Each polyeder under favourable conditions germinates and produces by mitosis a definite number of biflagellate zoospores, which become arranged in a patterned manner to form a colony (coenobium) either in the zygote itself or in the vesicle. On liberation it grows in size and matures into the haploid adult (coenobium). Certain cells of this coenobium (adult haploid) function as gametangia. The protoplast of gametangium is then differentiated into a number of biflagellate isogametes which are the last structures of the haplophase. The gametes are discharged into the water where they fuse in pairs to form zygotes.

Diplophase. The resultant single-celled zygote represents the first and also the last structure of the transitory diplophase. It germinates by meiosis to produce meiozoospore which are pioneer structures of the new future life cycle. However, there is ample provision for reduplicating the Haplophase in the growing season. The biflagellate zoospores may be produced in any cell of the parent colony. These are released through a slit in the parent cell wall into a vesicle where they retract their flagella and become arranged in the form characteristic of the parent colony. The vesicle ruptures and the newly formed colony is liberated to start the life cycle afresh.

It is suggested that certain sexually reproducing forms of Chlorococcales are diploid. The parent plant in them thus is a sporophyte. Example of this type is *Chlorochytrium* belonging to the family **Chlorococcaceae.** This tendency met with in some of the Chlorococcales is suggestive of the fact that the Siphonales might have originated from this group.

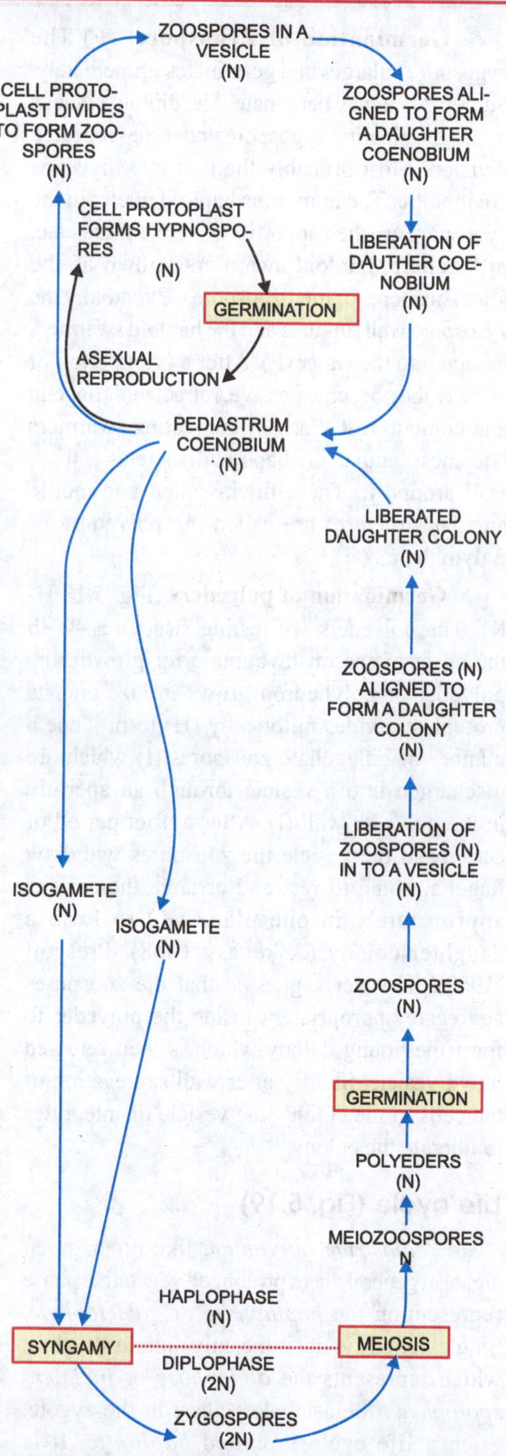

Fig. 5.19. *Pediastrum.* Graphic representation of the life cycle.

Systematic Position

Division	:	**Chlorophyta** (Chlorophycophyta)
Class	:	**Chlorophyceae**
Order	:	**Chlorococcales**
Family	:	**Hydrodictyaceae**
Genus	:	*Pediastrum*
Species	:	*duplex*

FAMILY 3. PROTOSIPHONACEAE

The family includes chlorococcaceous forms which are solitary, permanently multinucleate and spherical to tubular in form. They are attached to the substratum by a narrow, colourless rhizoid. Asexual reproduction takes place by means of biflagellate zoospores and sexual by morphologically similar uninucleate biflagellate gametes. Both are differentiated by successive cytoplasmic contents. The family includes a single genus *Protosiphon* (*Proto* meaning first and *siphon* meaning tube).

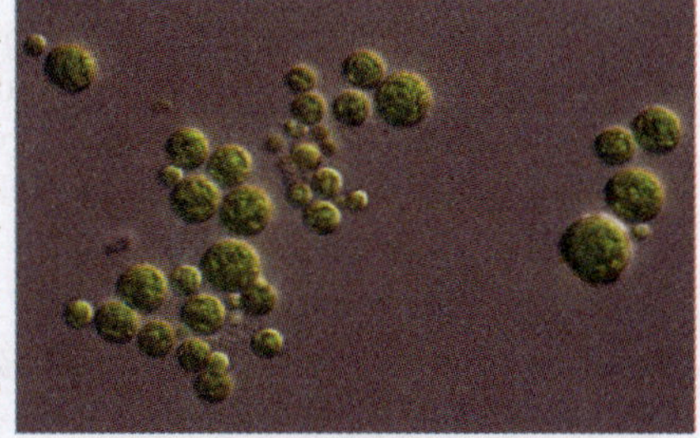

Protosiphon bortryoides.

PROTOSIPHON (KLEBS, 1896)

Occurrence. This terrestrial green alga occurs intermingled with *Botrydium* on damp mud at the edges of ponds and lakes. Often its green saccate cells form extensive patches. It has also been reported to occur on damp walls. *Protosiphon* can, however, be distinguished from *Botrydium* by the starch test. It contains starch whereas *Botrydium* does not. The genus is represented by a single species *P. botryoides* (Kutz) Klebs. It is widely distributed.

Thallus (Fig. 5.20). The multinucleated (coenocytic) thallus of this alga is differentiated into a green, overground, more or less spherical (bladder-like) to tubular portion tapering into a narrow, colourless, simple unbranched, rarely branched subterranean portion (A). The saccate aerial part is called the vesicle and the underground part the **rhizoid**. The former may grow up to 0.3 mm in diameter and the latter may reach a length of 1 mm. The colourless rhizoid anchors the vesicle to the substratum.

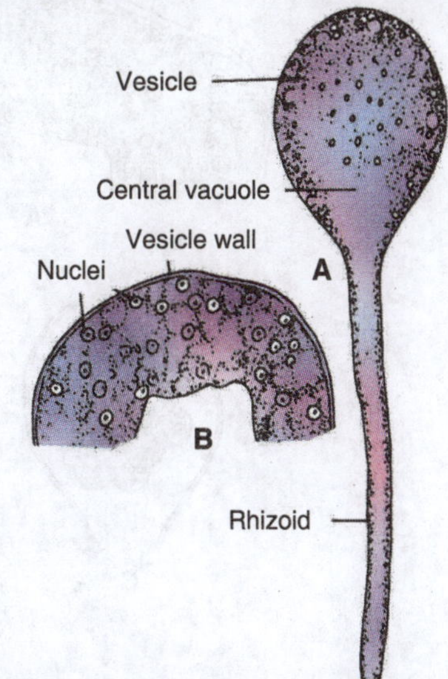

Fig. 5.20 (A-B). *Protosiphon botryoides.* (A), the plant showing habit; (B), Portion of vesicle magnified to show details of structure (A, aftr Fott; B, after Bold).

Structurally the thallus (B) consists of a thin, cellulose cell wall enclosing the protoplast which is continuous in both the parts. Within the cell wall the cytoplasm forms a lining layer. The peripheral cytoplasm harbours numerous nuclei. Such a multinucleate structure is called a *coenocyte.* It results from repeated free nuclear divisions but no cytokineses. There is a single reticulate or perforate parietal chloroplast which is restricted to the vesicle region only. The chloroplast contains a number of pyrenoids each surrounded by starch plates. The central region of both the parts of the thallus is occupied by a continuous central vacuole filled with cell sap. The vacuole lies internal to the chloroplast in the vesicle region. Iyengar (1933) reported that the thalli of *Protosiphon* growing on the drier parts of wall in India show progressive abbreviation of the rhizoid portion until the vesicle could scarcely be distinguished from a *Chlorococcum* cell. In nutrient cultures, Klebs found that *Protosiphon* thallus consists of branched thread-like outgrowths. Chapman (1964) reported that in bright light and low moisture the old vesicles of *Protosiphon* turn brick red.

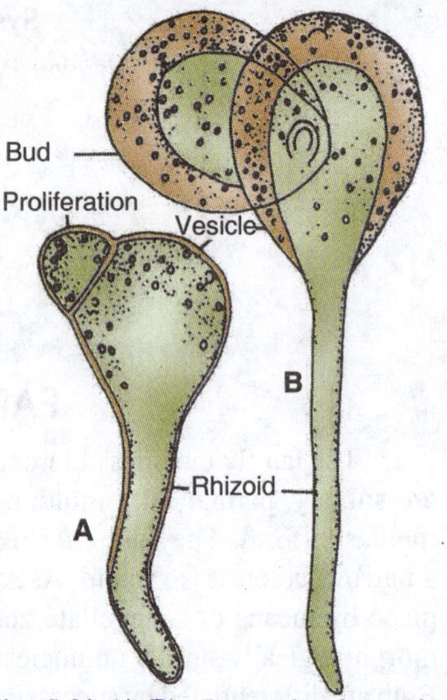

Reproduction. It takes place by, all the three methods, vegetative, asexual and sexual.

1. *Vegetative Reproduction* (Fig. 5.21). It takes place by budding. The bud arises as a lateral proliferous outgrowth from the vesicle (A). The proliferation subsequently becomes cut off by a cross wall as a bud (B) which separates to form a new plant.

Fig. 5.21 (A-B). *Protosiphon botryoides* Showing budding. (A) proliferation of young thallus; (B), separation of bud (A after) Smith and B after Fott).

2. *Asexual Reproduction* (Fig. 5.22). Under normal conditions the contents of the young vesicle undergo progressive cleavage (A-B) to form 4-16 rounded uninucleate bodies. Each of these secretes a membrane around it (C). After liberation each thin-walled structure or aplanospore germinates to form a new *Protosiphon* individual. Flooding of plants growing on damp soil or rain leads to the formation of uninucleate biflagellate zoospores by progressive cleavage of the contents of the vesicle.

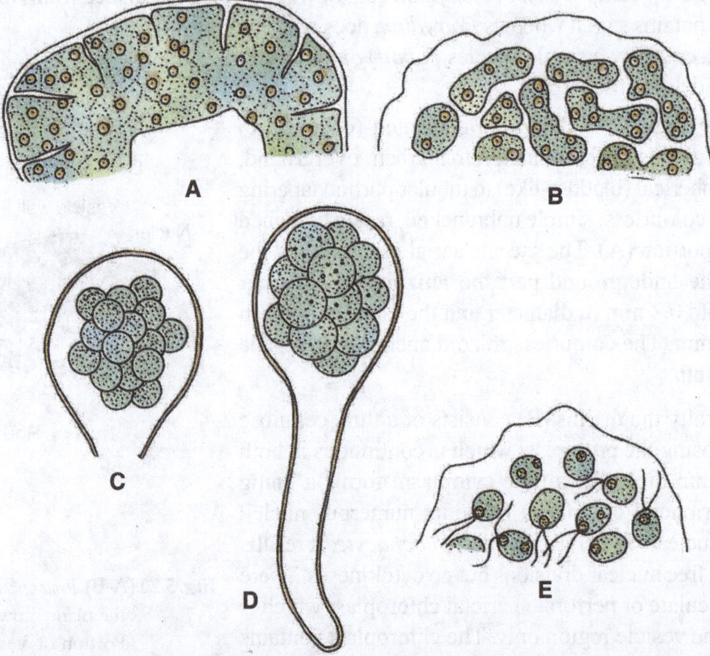

Fig. 5.22 (A-E). *Protosiphon botryoides.* (A), initiation of cleavage furrows; (B), later stage of A; (C), aplanospores; (D), coenocysts; (E), swarmers (b and E after Bold).

Scarcity of water in the substratum or exposure to strong insolation may cause the protoplast of the vesicle to divide by centripetally advancing furrows (A-B) into a number of daughter protoplasts (Fritsch, 1945) Each daughter protoplast develops a red pigment (*haematochrome*) and secretes a thick wall around it to become an *hypnospore* or a *coenocyst* (D). When the conditions return to normal the coenocysts germinate and either grow into new individuals directly or produce biflagellate swarmers (Fig. 5.23). The latter may behave either as zoospores (Chapman, 1964), or as gametes (Klebs, 1896). The liberated zoospore (B) comes to rest, retracts its flagella and secretes a wall around it (C). It then germinates to produce a new individual. The germination stages consist of free nuclear division and synthesis of more cytoplasm resulting in rapid increase in size to form a new saccate individual (D-F).

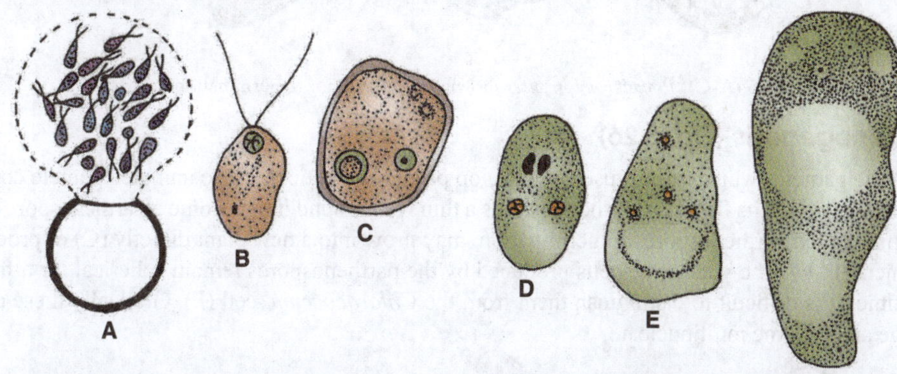

Fig. 5.23 (A-F). *Protosiphon botryoides.* (A), germination of coenocyst to produce zoospores; (B), liberated zoospore; (C-F), stages in the germination of zoospore (A, after Smith; B-F, after Bold).

3. *Sexual Reproduction* (Fig. 5.24). Plants growing in typical habitats along the edges of ponds and lakes under submerged conditions are stimulated to produce a large number of morphologically similar swarmers (Fig. 5.24, A-B). They are differentiated by progressive cleavage of the entire protoplast of the vesicle by an inward growth of furrows from the plasma and vacuolar membranes

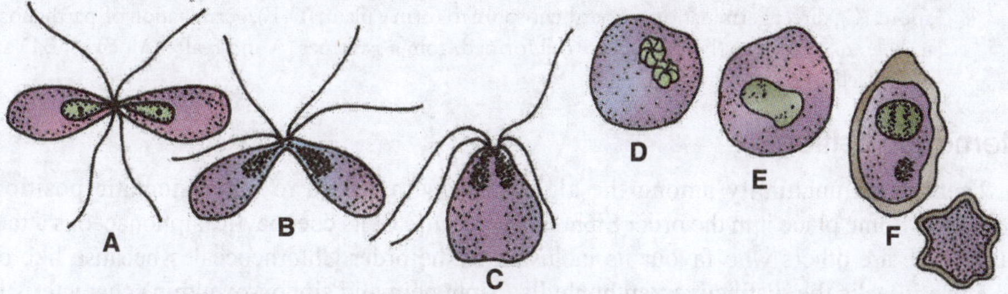

Fig. 5.24 (A-F). *Protosiphon botryoides.*Sexual reproduction. (A-B) Fusion of isogametes; (C) quadriflagellate zygote; (D-E), fusion of nuclei in the fusion cell; F, zygote (A-F, after Bold).

until uninucleate biflagellate swarmers result (Fig. 5.22 A, B and E). They escape from the vesicle through an apical aperture and often behave as gametes. The gametes from the same or different individuals (vesicles) may conjugate in pairs (Moewns, 1933). This denotes that some strains of *Protosiphon* are *homothallic* and others *heterothallic*. According to Bold (1933), the fusing gametes become opposed end to end and then fuse laterally (B-C). The quadriflagellate fusion cell or zygote (C) swims for a short while, comes to rest and retracts its flagella (D). It then secretes a wall of substellate shape (F) around it (Smith, 1955). The zygote either germinates immediately to give rise to a new plant directly or may remain dormant for some time (Chapman, 1964). Fritsch (1945) reported that the zygote is a thick-walled lobed resting spore (F). Smith (1955) also considered that the zygote enters upon a resting period before germination. Moewns (1933) stated that the zygote may germinate directly to give rise to a haploid individual (Fig. 5.25 A-C).

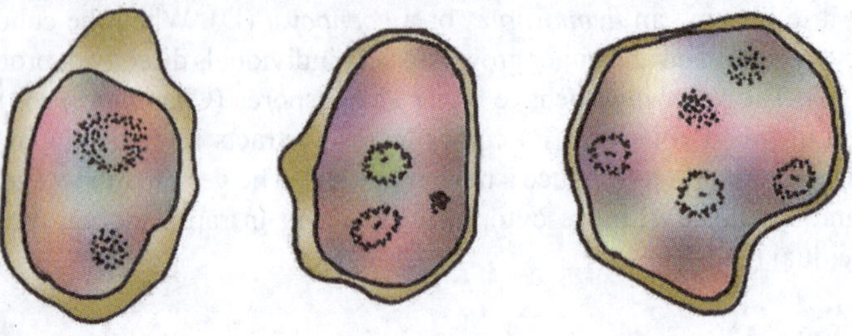

Fig. 5.25 (A-C). *Protosiphon botryoides* showing meiosis prior to germination (after Bold).

Parthenogenesis (Fig. 5.26)

The gametes which fail to fuse can develop parthenogenetically. The unmated gamete comes to rest (A), retracts its flagella and then secretes a thin wall around it to become a parthenospore (B). The thin-walled parthenospore, on germination, may grow into a new plant directly (C) or produce swarmers (D-E). The vegetative cells produced by the parthenospores remain spherical for a fairly long time. It is difficult to distinguish them from the *Chlorococcum* cell (F). Gradually these cells enlarge and become multinucleate.

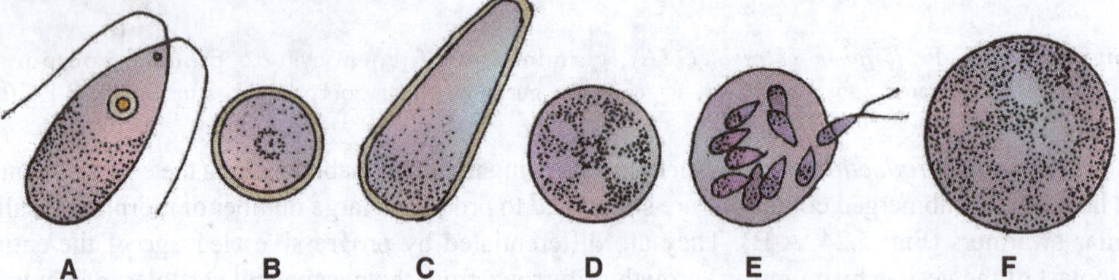

Fig. 5.26 (A-F). *Protosiphon botryoides* showing parthenogenesis. (A), gamete which has failed to fuse; (B),parthenospore, (C), direct germination of parthenospore to form a plant; (D-E), germination of parthenospore to liberate swarmers; (F), *Protosip-hon* cell formed from a swarmer (A and F after Moewns; B,D and E after Klebs; C after Smith).

Systematic Position

There is no unanimity among the algologists with regard to the systematic position of *Protosiphon*. Some place it in the order Siphonales because of its coenocytic, siphonaceous attached thallus. There are others who favour its inclusion in the order Chlorococcales because like other green algae it lacks the distinctive xanthophylls (Siphonein and siphon-oxanthin) characteristic of Siphonales (Strain, 1951) and has β-*carotene* with zygotic meiosis whereas Siphonales are diploid with gametogenic meiosis. Siphonales are marine whereas Protosiphon like the Chlorococcales is a fresh-water form. Furthermore *Protosiphon* has a single reticulate chloroplast instead of many discoid ones and approaches some of the chlorococcaceous forms like *Characium* in certain respects. Both are elongated cells with a tendency to become attached to the substratum. The uninucleate cell of *Characium* undergoes free nuclear division to become temporarily multinucleate (before zoospore formation), a tendency which is permanent in *Protosiphon*. In both the genera biflagellate zoospores and gametes are differentiated by progressive cytoplasmic cleavage and sexual reproduction is isogamous. In view of these facts most of the algologists have agreed to place *Protosiphon* in the order Chlorococcales. The taxonomic position of this alga thus, at present, is :

Division	:	**Chlorophyta** (Chlorophycophyta)
Class	:	**Chlorophyceae**
Order	:	**Chlorococcales**
Family	:	**Protosiphonaceae**
Genus	:	*Protosiphon*
Species	:	*botryoides*

FAMILY 4. COELESTRACEAE

The members of the family form two, four, eight or sixteen celled coenobia. The cells are uninucleate and of variable shape. The members reproduce only by autospores.

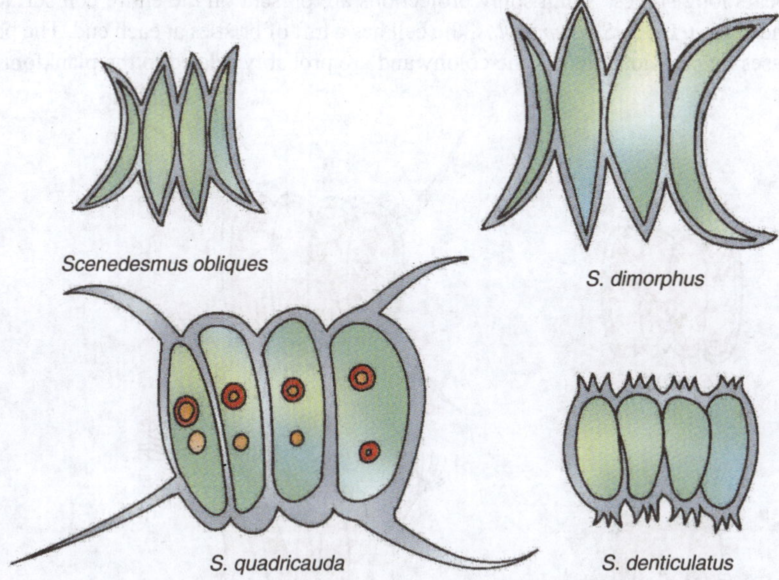

Fig. 5.27. Different shapes of the cells in *Scenedesmus.*

The most common and well-known genus of the family is *Scenedesmus* which is being described here in some detail. The genus was however, placed in a separate family Scenedesmaceae by Smith (1955).

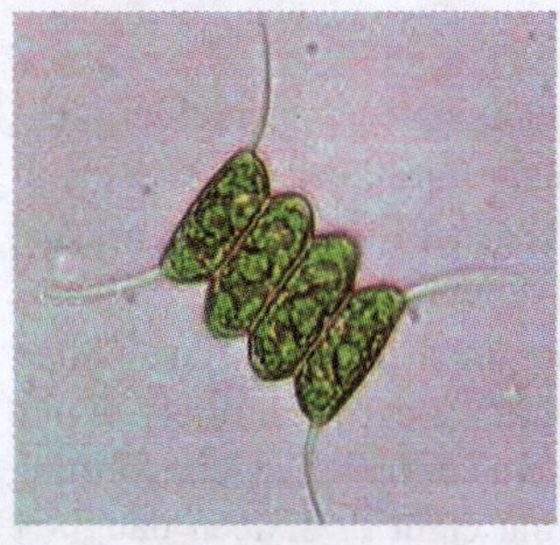

Scenedesmus obliquus.

Scenedesmus dimorphus.

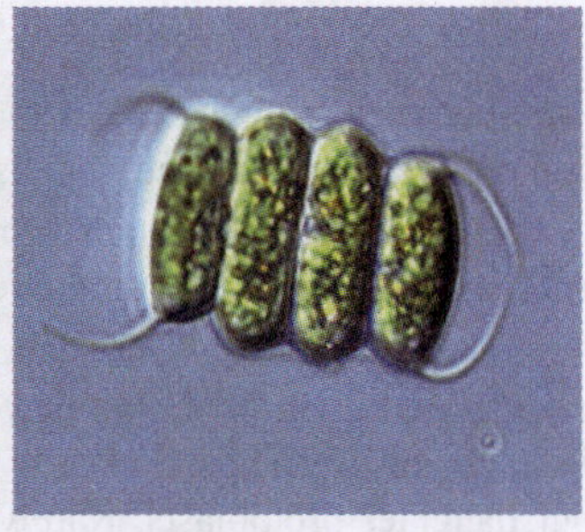

Scenedesmus quadricauda.

SCENEDESMUS

Occurrence. *Scenedesmus* is a common fresh water alga commonly found in standing water of lakes and ponds. The genus is represented by about 100 species and can be grown in laboratory in almost pure culture in aquaria and jars holding standing water for sometime. The common species of the genus *Scenedesmus* are: *S. obliquus, S. dimorphus, S. quadricanda* and *S. denticulatus.*

Structure. The thallus of *Scenedesmus* is a coenobium consisting of 4, 8 or 16 cells. The coenobium is a flat sometimes curved plate of fusiform to elliptic cells which are arranged in a single to double series with their long axes parallel to one another. The cells are fusiform in *S. obliquus* and *S. dimorphus* while in *S. quadricanda* and *S. denticulatus* the cells are ellipsoid. The cell wall may be smooth, corrugated, granulate or spicate with or without lateral or terminal spines. The elaborations of the cell envelope are of taxonomic importance while in *S. quadricanda* and *S. dermatus* the terminal cell bears long spines. Small spiny projections are present on the entire cell surface in *S. denticulatus* and *S. hystrix.* In *S. acuminatus*, the cell has a tuft of bristles at each end. The presence of these processes help in floatation of the colony and are probably related to the planktonic mode of life.

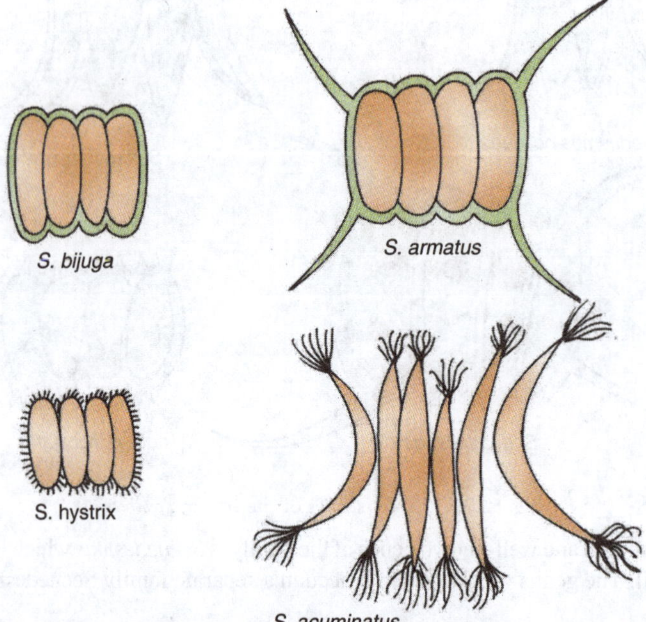

Fig. 5.28. Morphology of the Cells.

Each cell is uninucleate with a single haploid nucleus and possesses a single laminate chloroplast with one pyrenoid. In mature cells, the chloroplast generally occupies the entire cell cavity.

Reproduction. Though the reproduction in *Scenedesmus* may take place by vegetative, asexual and sexual methods, the commonest mode of reproduction is vegetative reproduction.

Vegetative reproduction takes place by means of daughter colonies. Each cell of the coenobium is capable of giving rise to a new daughter colony but all the cells of the coenobium rarely divide simultaneously. The protoplast of the cell divides through transverse and vertical divisions forming four daughter protoplasts. Rarely further divisions take place forming eight or sixteen daughter protoplasts. The last generation of protoplasts become autospores that remain laterally united to one another according to the arrangement of cells in the coenobium. Fully developed coenobium is liberated by longitudinal rupture of the parent cell wall.

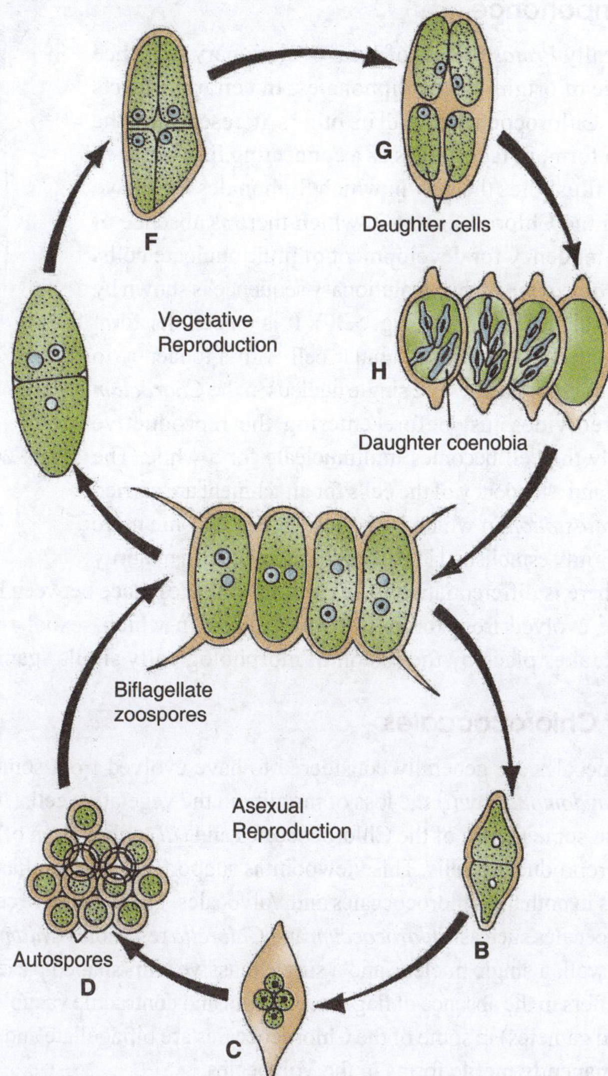

Fig. 5.29. *Scenedesmus.* Reproduction and life cycle.

Asexual reproduction takes place occasionally by means of motile biflagellate zoospores.

Sexual reproduction though rare takes place by fusion of biflagellate isogametes. The zygote thus formed divides meiotically to form four haploid daughter cells which arrange to form the coenobium.

Systematic Position:

Division	:	**Chlorophyta** (*Chlorophycophyta*)
Class	:	**Chlorophyceae**
Order	:	**Chlorococcales**
Family	:	**Coelestraceae**
Genus	:	*Scenedesmus*
Species	:	*armatus*

Evolutionary Importance

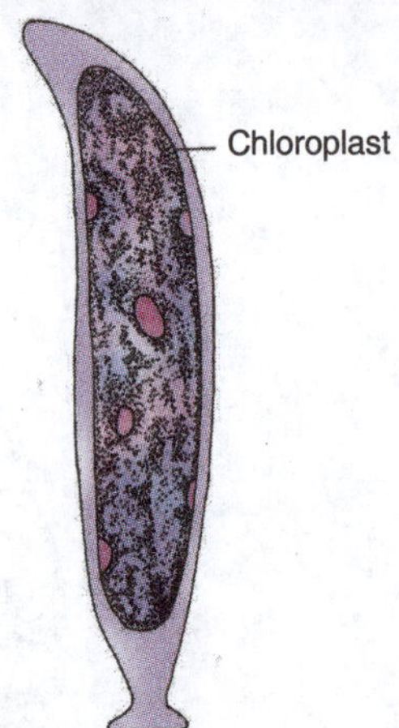

Fig. 5.30. *Chlorococcales.* Tubular uninucleate thallus of *Characium.*

Morphologically *Protosiphon* is of great evolutionary sequence as a possible source of origin for the Siphonales. In certain respects it approaches the Chlorococcales and in others it resembles the Siphonales. Such a form, in fact, serves as a connecting link between the two orders and illustrates the way in which Siphonales may have been evolved from the Chlorococcales in which there is absence of cell division and a tendency for development of multinucleate cells. A step earlier to *Protosiphon* in this evolutionary sequence is shown by a Chloracoccaceous alga *Characium* (Fig. 5.30). It is unicellular form consisting of an elongated, somewhat tubular cell with a tendency to become attached to the substratum. The single nucleus in the *Characium* cell divides and redivides just before entering the reproductive phase. Consequently the cell becomes multinucleate for a while. The multinucleate habit and a tendency of the cells for attachment are carried a step further in *Protosiphon* in which both these habits become more permanent and are firmly established in the order Siphonales. In majority of the Siphonales there is differentiation of sex and fusion takes place between heterogametes. This tendency may have evolved from forms like *Protosiphon* in which sexual reproduction is of a primitive type and takes place by the fusion of morphologically similar gametes.

Phylogency of Chlorococcales

The Chlorococcales are generally considered to have evolved from some motile unicellular ancestor like *Chlamydomonas* by *(i)* the loss of motility in the vegetative cell, (*ii*) its retention in the reproductive cells in some genera of the Chlorococcales and (*iii*) obliteration of cell division except in the formation of reproductive cells. This viewpoint is supported by the similarity in the vegetative organisation of cells in both the Chlorococcales and Volvocales. The vegetative cell in most species of unicellular Chlorococcales such as *Chlorococcum* and *Chlorella* resembles *Chlamydomonas* in having a well-defined cell wall, a single nucleus and a single massive, cup-shaped parietal chloroplast with one pyrenoid but differs in the absence of flagella, eye-spot and contractile vacuoles. The reproductive cells (zoospores and gametes) in some of the Chlorococcales are biflagellate and also similar in other respects to the permanently motile forms in the Volvocales.

Apart from the unicellular forms the order contains characteristically shaped no-motile colonies which, as in the Volvocales, can be arranged in an ascending series from the unicellular forms (*Chlorococcum*) through simple aggregates of 4 or 8 cells united transversely in a row (*Scenedesmus*) to a free floating, flat, plate-like *Pediastrum*. The series ultimately culminates in globular aggregates. The resemblance to the corresponding series in the Volvocales is however, that the position of each cell in the colony is fixed from the very beginning. In the Chlorococcales, on the other hand, the whole brood of young cells formed by the cleavage of one protoplast can freely intermingle and subsequently come together randomly.

The coenocytic condition of the chlorococcoid cell is likely to have arisen from the repeated mitosis of the nucleus of the volvocoid cell in which cytokinesis does not accompany mitosis. This tendency to become multinucleate is accompanied by an increase in the length of the cell and its attachment to the substratum. This has resulted in the evolution of a permanently multinucleate form like *Protosiphon*, a border line genus which illustrates the way in which the coenocytic Siphonales may have arisen from the Chlorococcales.

The Chlorococcales, however, are considered polyphyletic. They evolved along two lines, **azoosporic** and **zoosporic**. The former multiply by means of autospores (aplanospores similar in shape to the parent vegetative cell) and the latter by zoospores (*Chlorococcum*). The phylogenetic connection between the azoosporic and zoosporic forms is shown by genera which reproduce by both types of spores.

The group as a whole is diversified both in form and behaviour so that some algologists consider it as an artificial assemblage of one-celled and colonial algae which are non-motile in vegetative phase but otherwise indicate few interrelationships. Motility is restricted only to the reproductive cells (zoospores and gametes). The only common feature between them is their inability to undergo ordinary cell division (simple binary fission).

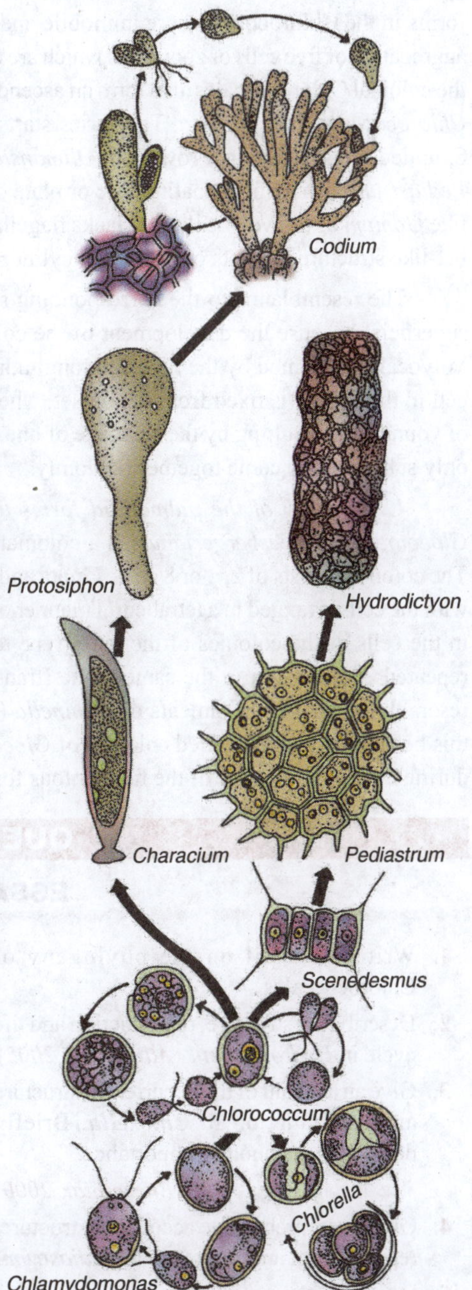

Fig. 5.31. Chlorococcine line of development of thallus in green algae.

Evolutionary Trends in the Chlorococcales. The Chlorococcales represent the third evolutionary line of development in the algae. It is called the **chlorococcine line**. The simplest forms in this order are non-motile and unicellular (*Chlorococcum* and *Chlorella*). The advance forms are non-motile colonies (*Scenedesmus* and *Pediastrum*). Non-motility in fact is the permanent feature of the chlorococcoid cell in the vegetative condition. Motility is found only in the reproductive cells. The other characteristic feature of this order is the absence of vegetative divisions. There may be division of the nucleus but no cytokinesis. The members of this order exhibit the following evolutionary tendencies :

1. *Tendency to become multinucleate*. The uninucleate chlorococcoid unicell has a tendency to become multinucleate, at least for a short while just before reproduction (*Chlorococcum*). This contrasts with the Volvocales in which each nuclear division is accompanied by wall formation. In *Characium* the young cells are uninucleate. The older ones become multinucleate or nuclear division occurs before zoospore formation. *Protosiphon*, however, is permanently multinucleate.

2. *Tendency to increase in length and become attached.* The multinucleate habit in some species is accompanied by increase in length of the chlorococcoid cell and a tendency to become attached to the substratum. In *Characium* the fusiform cell is affixed to the substratum by a small disc. The permanently multinucleate *Protosiphon* is attached

to the soil by a long rhizoid. According to some algologists the coenocytic condition seemed to have reached its climax in the production of complicated coenocytes of the Siphonales. They consider *Protosiphon* as a border line genus between the Chlorococcales and Siphonales. The Siphonocladiales are considered to have arisen by the separation of the siphonaceous coenocyte of the Siphonales.

3. *Tendency of the non-motile chlorococcoid cell to form non-motile colonies*. The colonial forms in the Chlorococcales are immobile and have a characteristic shape. They are formed by the aggregation of free cells or zoospores which are the product of a single parent cell. As in the Volvocales, the colonial Chlorococcales also form an ascending series starting with the spherical unicellular forms *Chlorococcum* or *Chlorella*. The series starts with *Scenedesmus* which consists of 4 cells rarely 8, united transversely in a row as in *Oltmansiella* of the colonial Volvocales. Next in order comes *Pediastrum* with its free-floating disc or plate of cells varying in number from 2-128. *Pediastrum* is like *Gonium* of the Volvocales but lacks flagella. *Hydrodictyon* which is an elongated, hollow, open, net-like structure consists of numerous cylindric cells joined terminally to two others.

The resemblance to the corresponding series in the Volvocales is however misleading. It is superficial because the development of the colony in the two cases is different. The colony in the Volvocales is formed by the repeated longitudinal division of the cells so that the position of every cell in the colony is fixed from the outset, whereas in the colonial Chlorococcales the whole brood of young cells resulting by the cleavage of one parent protoplast can freely intermingle and the cells only subsequently come together randomly.

4. *Tendency of the palmelloid forms to form filamentous colonies* (Fig. 3.30 and 3.31). *Gloeotoenium loitlesbergerianum* is a colonial form of family Oocystaceae order, Chlorococcales. The colony consists of 2, 4 or 8 cells. Gupta and Nair (1962) reported 4-celled colonies of the species with the cells arranged in a tetrahedral manner or in the same plane. They further observed a tendency in the cells of the colonies of the latter type to become re-arranged in a row. This is followed by repeated cell division in the same plane (transverse) resulting in filamentous colonies apparently resembling the young filaments of *geminella* (order Ultotrichales). Gupta and Nair suggested that this behaviour of the 4-celled colonies of *Gloeotaenium loitlesbergerianum* reflects what happened during evolution of some of the filamentous forms in green algae.

QUESTIONS

ESSAY TYPE

1. Write in detail on the phylogeny of Chlorococcales.
2. Describe the structure, reproduction and life cycle in *Hydrodictyon*. (*Allahabad, 2005*)
3. Give an account of the occurrence, structure and reproduction in *Chlorella*. Briefly describe its economic importance. (*Bharathiar, 2000*)
4. Give a comprehensive account of structure, reproduction and life cycle in *Pediastrum*.
5. Describe the occurrence, structure and reproduction in *Scenedesmus*.
6. Describe mode of reproduction in *Hydrodictyon*. (*Allahabad, 1998*)
7. Describe the cell structure and reproduction in *Chlorella*. (*Annamalai, 1995*)
8. Give a comprehensive account of reproduction/life history of *Pediastrum*/*Scenedesmus*.
9. Describe the structure and mode of reproduction in *Hydrodictyon*. (*Allahabad, 2001*)

SHORT ANSWER TYPE

10. Describe the characteristic features of the order Chlorococcales.

11. In what respects are the followng different from each other :

(*i*) *Volvox* and *Hydrodictyon*

(*ii*) *Chlamydomonas* and *Chlorella*

(*iii*) Coenobium of *Sceneedesmus* and *Volvox*

(*iv*) *Hydrodictyon* and *Vaucheria*

(*v*) *Chlorella* and *Scenedesmus.*

12. Draw well-labelled diagram of the following:

(*i*) Net of *Hydrodictyon* (*Lucknow, 1995*)

(*ii*) Cell structure of *Chlorella.*

13. Name a green alga which lacks flagellated cells.

14. Name a unicellular non-motile green alga.

15. Name an alga which is used as a source of food.

16. Name an alga popularly known as 'water net'.

17. Does vegetative reproduction occur in *Hydrodictyon*?

18. Write short notes on :

(*i*) *Hydrodictyon* (*Allahabad, 2003*)

(*ii*) Sexual reproduction in *Hydrodictyon* (*Allahabad, 1991*)

(*iii*) Cell structure in *Chlorella*

(*iv*) Economic importance of *Chlorella.*

(*v*) Water net

(*vi*) Reproduction in *Scenedesmus*

(*vii*) Germination of zygote in *Hydrodictyon*

(*viii*) Systematic position and evolution in Chlorococcales

(*ix*) *Chlorella.* (*Lucknow M.Sc., 2005*)

19. Write down the systematic position and draw well labelled diagrams alongwith description of the following

(*a*) *Hydrodictyon* (*Lucknow, 2005*)

(*b*) *Chlorella* (*Lucknow M.Sc., 2006*)

(*c*) *Protosiphon*

(*d*) Scenedesmus.

20. How many nuclei are present in the mature cells of Hydrodictyon? (*Allahabad, 2004*)

21. Compare asexual reproduction in *Volvox* and *Hydrodictyon.* (*Allahabad, 2004*)

OBJECTIVE TYPE

22. Fill in the blanks:

(*i*) Unicellular, non-motile, spherical green alga is ....

(*ii*) The alga which forms water net is ....

(*iii*) In *Chlorella,* the chloroplast is ....

(*iv*) The sole method of reproduction in *Chlorella* is .... and ....

(*v*) The alga which is used as a source of food is ....

(*vi*) The chloroplast in *Hydrodictyon* is ....

(*vii*) The zoospores in *Hydrodictyon* is .... , ... and ....

(*viii*) Sexual and Asexual reproduction is present in water net also known as ....

(*ix*) In *Pediastrum* the chloroplast is ....

(*x*) In *Protosiphon* the thallus is .... , ... and ....

(*xi*) The chloroplast in *Protosiphon* is reticulate or .... in position.

(*xii*) Vegetative reproduction takes place by budding in ....

(*xiii*) Chloroplast is laminate in ....

(*xiv*) In *Scenedesmus* vegetative reproduction takes place by ....

23. Select the correct answer:

(*i*) Non-motile unicellular alga is
(*a*) *Chlamydomonas*
(*b*) *Chlorella*
(*c*) *Pandorina*
(*d*) *Volvox*

(*ii*) The alga which may be used as a food in spaceship is
(*a*) *Chlorella*
(*b*) *Chlamydomonas*
(*c*) *Chlorococcum*
(*d*) *Haematococcus*

(*iii*) The alga which is popularly called 'water net' is
(*a*) *Pediastrum*
(*b*) *Hydrodictyon*
(*c*) *Cladophora*
(*d*) *Pithophora*

(*iv*) Which of the following does not reproduce sexually?
(*a*) *Chlamydomonas*
(*b*) *Chlorella*
(*c*) *Volvox*
(*d*) *Hydrodictyon*

(*v*) Reticulate or perforated pareital chloroplast is present in
(*a*) *Chlorella*
(*b*) *Hydrodictyon*
(*c*) *Pediastrum*
(*d*) *Protosiphon*

(*vi*) Simple pareital band-shaped chloroplast is present in
(*a*) *Chlorella*
(*b*) *Hydrodictyon*
(*c*) *Pediastrum*
(*d*) *Protosiphon*

(*vii*) Cup or bell shaped chloroplast is present in
(*a*) *Spirogyra*
(*b*) *Nostoc*
(*c*) *Chlorella*
(*d*) *Hydrodictyon*

(*viii*) Autospores are produced in
(*a*) *Spirogyra*
(*b*) *Nostoc*
(*c*) *Chlorella*
(*d*) *Hydrodictyon*

(*ix*) The alga which is autophytic with its photosynthetic process similar to angiosperm is
(*a*) *Spirogyra*
(*b*) *Nostoc*
(*c*) *Chlorella*
(*d*) *Hydrodictyon*

(*x*) In which of the following, the thallus is coenobium consisting of two, four or eight cells?
(*a*) *Chlorella*
(*b*) *Chlamydomonas*
(*c*) *Scenedesmus*
(*d*) *Volvox*

Ulotrichales: *Ulothrix zonata*

CHAPTER 6

ORDER 3. ULOTRICHALES

- General Characteristics
- Order : Ulotrichales
- Suborder : Ulotrichineae
- Family – Ulotrichaceae
- Ulothrix

GENERAL FEATURES

This order chiefly includes fresh water forms. A few are marine. The plant body typically consists of a simple unbranched filament, sometimes branched. *Uronema* is known to branch rarely. With the exception of one genus in which the cells are multinucleate, in all others they are uninucleate. The cell wall consists of two layers, the outer of pectic substances and inner of cellulose. The single parietal chloroplast has one or more pyrenoids. The accumulated food reserve is starch. Asexual reproduction usually takes place by the formation of bi-or quadriflagellate zoospores or sometimes by aplanospores or hypnospores. Sexual reproduction is generally isogamous and the zygote nucleus undergoes meiosis at the time of germination of zygote.

Classification

The Ulotrichales have been classified differently by different algologists. A review of the classification proposed by them is beyond the scope of this book. Suffice it to say that apart from the Oedogoniales, Conjugales and Siphonales, the earlier algologists often included all other filamentous forms, simple, branched or heterotrichous (Ulotrichales, Chaetophorales, Sphaeropleales, Acrosiphonales and Cladophorales) and thalloid forms (Ulvales and Prasiolales) with uninucleate cells under the order Ulotrichales. Fritsch (1935) considered all the forms with heterotrichous habit as distinct from the Ulotrichales and placed them under a separate order Chaeophorales. He further removed Cladophoraceae and raised it to the rank of an order Cladophorales. He divided the Ulotrichales into three suborders comprising six families. Smith (1938) objected to the separation of Chaetophorales from the Ulotrichales and inclusion of Ulvaceae in it. He elevated Ulvaceae to the rank of an order Ulvales and divided Ulotrichales into two suborders comprising eight families. The latest classification as proposed by Ramanathan (1964) recognises two suborders with four families. It is based on the nature of the thallus. The bare outline of this classification is as follows:

ORDER : ULOTRICHALES		
SUBORDER	I.	ULOTRICHINEAE
FAMILY	1.	ULOTRICHACEAE
FAMILY	2.	MICROSPORACEAE
FAMILY	3.	CYLINDROCAPSACEAE
SUBORDER	II.	SPHAEROPLEINEAE
FAMILY	4.	SPHAEROPLEACEAE

SUBORDER : ULOTRICHINEAE

This sub-order is characterized by the presence of a uniseriate thallus consisting of a simple unbranched filament. The cells are uninucleate. Of the three families included in this suborder only Ulotrichaceae is discussed here.

FAMILY — ULOTRICHACEAE

The thallus is an unbranched filament with little differentiation. The cells are uninucleate. Each has a single girdle - shaped, parietal chloroplast occupying only a portion of the cell. Sexual reproduction is isogamous and by the union of biflagellate gametes. The family includes about 16 genera. Almost all of them are fresh water forms. Of these *Ulothrix, Hormidium, Uronema* and *Stichococcus* are important. Here *Ulothrix* is considered in some detail.

Girdle Shaped Chloroplast in *Ulothrix*.

ULOTHRIX

Occurrence

It is a common, fresh water, thread-like alga found in rather cold, flowing water. It occurs in a variety of habitats. Usually it is found attached to the substratum such as rocks or stones or other solid objects. It commonly occurs in streams or under the taps where water is constantly renewed or in other places where water does not become stagnant. The best place to look for this algae is on the submerged rocks in cold, shallow, shaded streams. It forms bright green attached masses. *Ulothrix* comprises about 30 species. Some species are marine and occur in the intertidal zone. Of these

U. flacca is well known. Mostly the marine forms are epiphytic. *U. implexa* is a lithophyte in estuaries. *U. zonata* is the common, cold, fresh-water species found in flowing water in autumn and spring. It occurs attached to stones forming bright, green masses which disappear in summers. Of the species that are reported from India, the common species are : *Ulothrix zonata, U. acqualis, U. pectinalis* and *U. variabilis*.

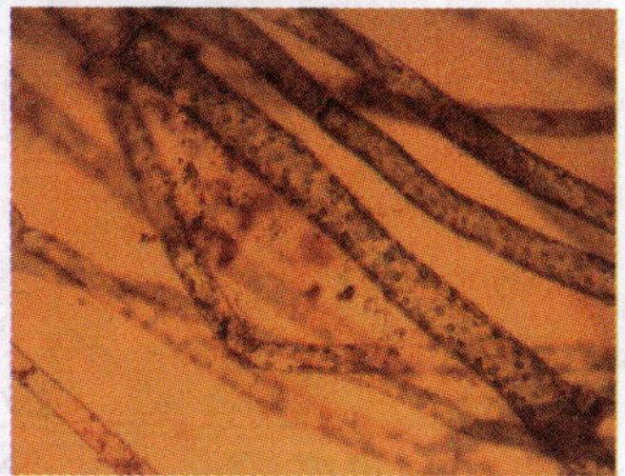

Ulothrix zonata

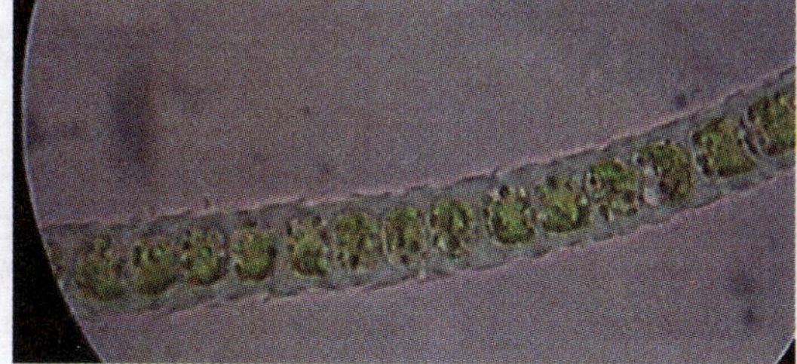

Ulothrix moniliformis

Plant body (Fig. 6.1)

It is a thallus consisting of an extremely fine, unbranched filament. The cells of the filament are usually short, cylindric or quadrate and often numerous. They are placed end to end in a single line or row. The filament is usually attached at one end. The plants, however, live equally well as detached free floating, bright, green masses on the surface of water. The cells in the filament are all alike excepting, the apical cell which is dome-shaped rather than cylindrical and a basal, anchoring one is the **rhizoidal cell**. The latter is longer and narrower than the others. At its lower end it conforms to the irregularities on the substratum to which it is attached. Thus it serves as an anchor or a *holdfast.* The holdfast lacks chlorophyll and is either brown or nearly colourless. It does not divide following the first division in the germine state. All the cells in the filament except the basal cell or holdfast are capable of cell division and are involved in reproduction. This slight differentiation of the basal cell from the rest of the cells in the filament is very important. It is a striking example of the beginning of the process of differentiation of cells accompanied by division of labour, so characteristic of the advanced multicellular forms.

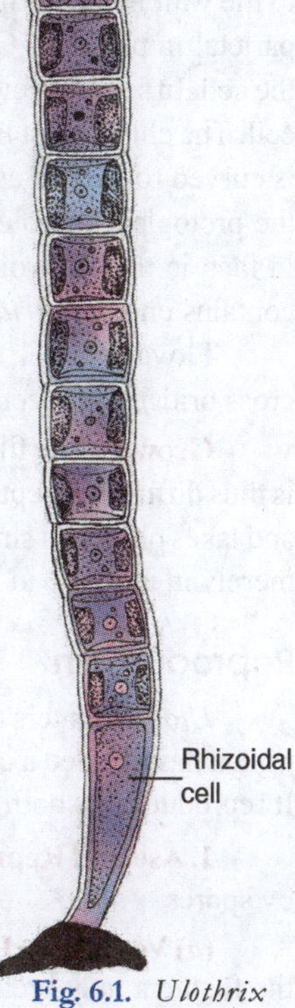

Fig. 6.1. *Ulothrix* filament.

Is Ulothrix a colony or a multicellular plant? All colonies of single-celled algae are spherical or disc-like or irregular. In many species successive cell divisions occur in parallel planes so that a chain of cells is formed. Such colonies are filamentous. The best example of this type of colony is *Spirogyra*. Each cell is potentially capable of independent life. In *Ulothrix* the basal cell may be colourless, somewhat modified in shape and attached to the substratum. It is different from the other cells in the filament. It simply anchors the plant to its substrate whereas the others are capable of photosynthesis, cell division and form either spores or gametes. Because of this slight differentiation between cells, a filament of *Ulothrix*

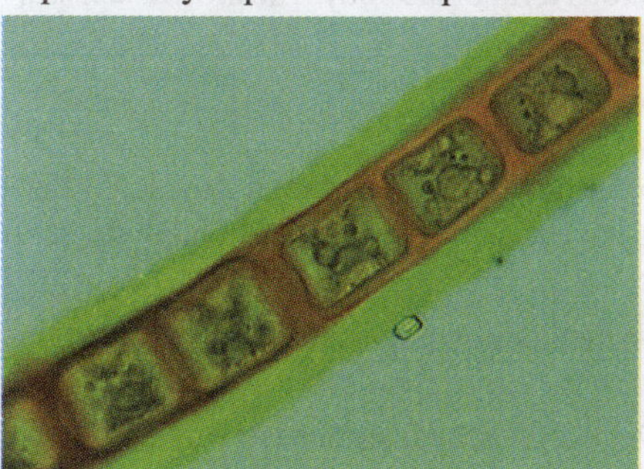

Extracellular matrix around the Filament of *Ulothrix*.

represents a transition between a filamentous colony of single-celled plants and a very simply constructed multicellular plant such as *Oedogonium*.

Cell structure (Fig. 6.2). The vegetative cells are often wider than long (*U. zonata*) or as long as broad (*U. moniliformis*). In *Ulothrix subconstricta* they are 2 to 6 times as long as broad. Each cell, as usual, consists of a cell wall enclosing the protoplast.

(*a*) Cell wall. It may be thick and lamellated (*U. moniliformis*) or thin and homogeneous (*U. oscillatorina*) according to the species. It consists of two concentric layers, inner and outer. The inner layer which is next to the protoplast, mainly consists of cellulose. The outer layer consists of pectic substances chiefly *protopectin* which is insoluble in water. For this reason the *Ulothrix* filaments feel like wet threads.

(*b*) Protoplast. Within the cell-wall is the protoplast. It is differentiated into a cell membrane, cytoplasm or cytoplast, a single nucleus, a chloroplast and a central vacuole. The cytoplast forms the lining layer or *primordial utricle* and is closely invested by the cell or plasma membrane. In it lies embedded the single nucleus. In the centre is the sap vacuole containing cell sap. The single, green girdle-shaped chloroplast lies next to the wall in the peripheral layer of cytoplasm. It is thus parietal in position. Generally it occupies the middle of the cell. In some species it extends the whole length of the cell. The chloroplast is in the form of a broad band which is curved round the circumference of the cell to encircle the protoplast completely (*U. zonata*) or partially. It is, in fact, in the form of a simple girdle round the cell and is thus commonly called girdle-shaped. It contains one (*U. rorida*) or more pyrenoids (*U. zonata*).

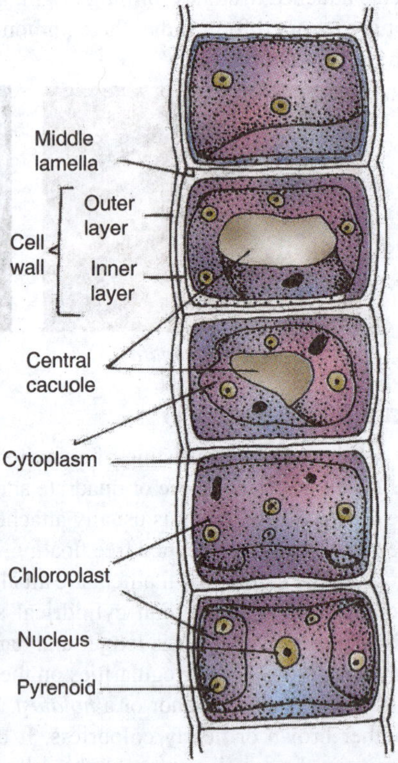

Fig. 6.2. *Ulothrix sp.* Segment of the filament to show cell structure and disposition of the chloroplast.

Floyd *et al.* (1971) reported the occurrence of *plasmodesmata* forming minute cytoplasmic cross bridges between the adjacent cells of the filament.

Growth. The filament grows throughout its length. It has no special growing point. The growth is thus diffuse. Excepting the basal rhizoidal cell, all others undergo division which thus is intercalary and takes place in a single plane. This maintains the filamentous form of the body. The growth consists merely in increase in length and number of cells in the filament.

Reproduction

Ulothrix enters the reproductive phase after a certain period of active vegetative growth. By this time it has reached a certain stage of maturity and the cells have accumulated reserve food materials. It reproduces by both the methods, asexual and sexual.

1. Asexual Reproduction. It includes both vegetative methods of propagation and reproduction by spores.

(*a*) Vegetative Reproduction. *Ulothrix* multiplies vegetatively by **fragmentation**. In this case the filament may break into small portions or segments. Each segment consists of a few to several living cells and is called a **fragment**. The filament may be broken into fragments by, (*i*) accidental breaking of the filament, (*ii*) dying out of some intervening portion or portions of the filament (*iii*) through the formation of zoospores or gametes here and there along the filament. Fragmentation is chiefly caused by accidental breakage. Simultaneous dissociation of the filament into fragments has rarely been observed. The fragment forms a new filament by repeated intercalary division and subsequent growth of daughter cells.

(*b*) Spore formation. It takes place by the formation of small, specialised cells. They are differentiated from the protoplast of the parent vegetative cell by **mitosis**. These specialised reproductive cells are called the mitospores. These are the true asexual spores which serve to multiply or reduplicate the thallus plant (filament) on which they are produced. The mitospores thus play no role in the phenomenon of **alternation of generations**. They are of two kinds, motile and non-motile. The former are called the **zoospores** and the latter **aplanospores.**

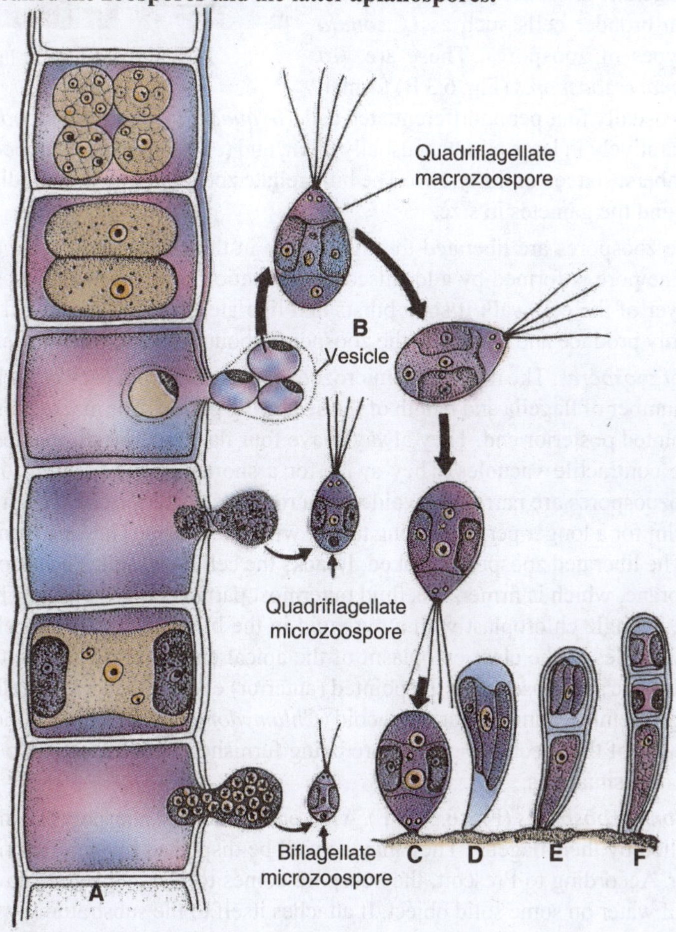

Fig. 6.3 (A-F). *Ulothrix sp.* (A), stages in zoospore formation and liberation of zoospores; (B), liberated zoospore; (C-F), stages in germination of zoospore.

(*i*) *Zoospore formation* (Fig. 6.3,A). It is the usual method of multiplication of the species when the conditions of life are favourable. Zoospore formation commences generally near the apex and progresses towards the base of the filament. All cells, except the holdfast, are capable of reproduction. The protoplast contracts slightly from the cell wall and becomes filled with food materials. The contracted protoplast divides by the method of *successive bipartition* until a required number of daughter protoplasts is produced. First the nucleus divides by mitosis. This is followed by the cleavage of the cytoplast in a plane at right angles to the long axis of the filament (*a*). Two daughter protoplasts are formed. The second division is at right angles to the first and so on. The daughter protoplasts develop into **zoospores**. Occasionally only a single zoospore may be produced. Species with narrow filaments produce 1, 2 or 4 zoospores per mother cell. Species with broad cells produce 2, 4, 8, 16 or rarely 32 zoospores.

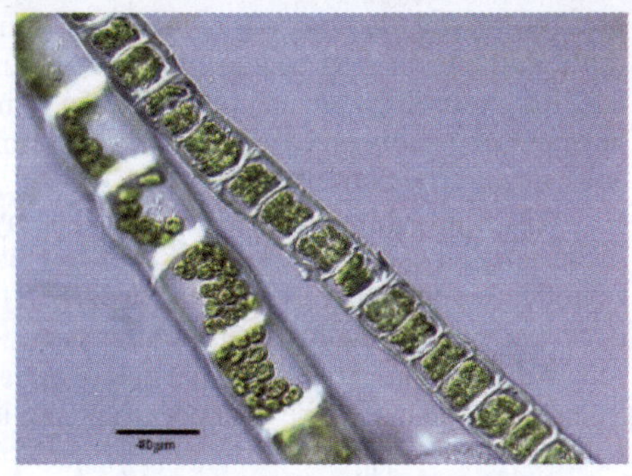

Zoospore Production in *Ulothrix*.

Kinds of zoospores. The narrow-celled species produce quadriflagellate zoospores of one kind only. The species with broader cells such as *U. zonata* produce three types of zoospores. These are, *(a) quadriflagellate macrozoospores* (Fig. 6.3 B) formed in small numbers usually four per undifferentiated cell, (*b*) *quadriflagellate microzoospores* (Fig. 6.3 A) formed comparatively in large numbers usually eight, and (*c*) *biflagellate microzoospores* formed still in larger numbers, sixteen or thirty-two. The biflagellate zoospores are intermediate between the macrozoospores and the gametes in size.

The mature zoospores are liberated through a pore in the lateral wall into a delicate vesicle (Fig. 6.3 A,C). The pore is formed by a localised gelatinisation of the wall. The vesicle represents the innermost layer of the cell wall. It soon bursts and liberates the zoospores. All the cells in the filament commonly produce and discharge the zoospores about the same time just after sunrise.

Structure of zoospores. The macro and microzoospores differ from each other in size, position of the eye spot, number of flagella and length of the swarming period. The macrozoospores are pear-shaped with a pointed posterior end. They always have four flagella. The stigma lies at the anterior end and also the contractile vacuoles. They swim for a shorter period coming to rest within 24 hours. The microzoospores are narrowly ovoid with a rounded posterior end. The stigma lies in the middle. They swim for a longer period coming to rest within 2-6 days. They are furnished with four or two flagella. The liberated zoospore is naked. It lacks the cellulose wall. The shape is maintained by the cell membrane, which is firmer, less fluid outermost part of the cytoplasm. The zoospore has a single nucleus, a single chloroplast with a pyrenoid in the broader part of the cell. It has one or two contractile vacuoles in the clear cytoplasm of the apical end. It also contains the red pigment spot or eye spot on one side. Inserted at the pointed (anterior) end are the long flagella. The released zoospores closely resemble a unicellular volvocoid (*Chlamydomonas,* like) cell without its cell wall. Besides propagation of the species, the zoospore being furnished with flagella help in the dispersal of the species to new situations.

Germination of zoospores (Fig. 6.3, C-F). After emergence the zoospores swim about actively for a time propelled by their flagella. They may as well be dispersed by being carried along in the currents of water. According to Prescott, the zoospore comes to rest and settles down anterior end first the bottom of water on some solid object. It attaches itself to the substratum by the secretion of mucilage which fits itself like a drop of jelly into its rough surface (Fig. 6.3,C). It retracts its flagella and germinates by secreting a cell wall around it (D-F). The clothed protoplast elongates and divides. The lower cell forms the holdfast and the upper by repeated divisions forms the filament. Some other investigators hold that the macrozoospore attaches itself by its posterior end to which the stigma has shifted (Fig. 6.4B). The flagella are then resorbed. The naked protoplast, if lodged in a suitable situation, secretes around it a cell wall. The clothed zoospore has a chloroplast. It makes food and broadens out laterally to form the holdfast on one side (posterior end) and green cells at the other (Interior). Soon it grows to a size which evokes cell division. The first division produces two cells. The cell at the posterior end functions as the holdfast. The cell at the other end by repeated divisions produces all other cells of the adult *Ulothrix* filament. The basal holdfast cell becomes colourless and

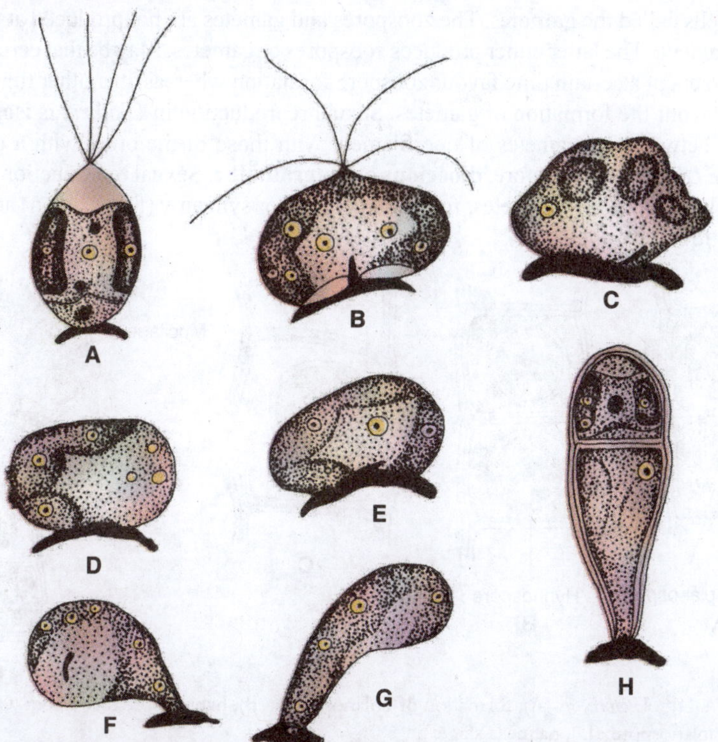

Fig. 6.4 (A-H). *Ulothrix sp.* Stages in the germination of zoospore (After Goss).

functions as the rhizoidal cell. Under favourable conditions *Ulothrix* may multiply by this method through a number of generations in a single growing season.

(*ii*) *Aplanospore formation* (Fig. 6.5 A). Occasionally the daughter protoplasts are not liberated as zoospores from the parent cell. These undischarged protoplasts fail to develop flagella. On the other hand each daughter protoplast secretes a thin wall around it to become an **aplanospore.** The aplanospores are eventually liberated by the disintegration of the parent cell wall (Fig. 6.5 C). Many may, however, germinate *in situ* before liberation (Fig. 6.6A)

Thick-walled aplanospores or hyponospores. Sometimes during period of drought the protoplast of a cell may round up to form a single, large, thick-walled aplanospore or **hypnospore** (Fig. 6.5). With the return of favourable conditions the hypnospore germinates to give rise to a new filament. The hypnospore and the aplanospore germinate in the same way as the zoospore. Sometimes the protoplast of the hyponospore may divide to form daughter protoplasts which may function as zoospores. The latter eventually escape from the cell wall and give rise to the new filaments.

(*iii*) Akinete formation has been reported in *U. idospora* by West (1903). In this case certain cells in the filament, under conditions of stress and strain, secrete thick resistant walls and become gorged with food materials. The thick-walled resting vegetative cell is called an **akinete** (Fig. 6.6B).

(*iv*) *Palmella stage*. (Fig. 6.5D). Occasionally the walls of the parent cells, which give rise to aplanospores, gelatinise. This is followed by the gelatinisation of the walls of the aplanospores.

In this way a number of green rounded bodies become embedded in a mass of mucilage. This is the *palmella stage*. Each green structure is finally liberated as a zoospore which grows into a new filament.

2. Sexual Reproduction. (Fig. 6.7, A-F). The conditions prevalent toward the end of the normal growing season stimulate the vegetative cells to a new activity. It is the production not of zoospores but of sex cells called the **gametes**. The zoospores and gametes are not produced at the same time in the same filament. The latter either produces zoospores or gametes. May be that certain conditions in the environment at a certain time favour zoospore formation whereas at another time environmental conditions favour the formation of gametes. Sexual reproduction in *Ulothrix* is **isogamous**. Fusion takes place between the gametes of one filament with those of the other with a different genetic constitution. *Ulothrix*, is therefore, **dioecious or heterothallic**. Sexual reproduction comprises three steps, namely, formation of gametes, fusion of gametes or syngamy (fertilization) and production of meiospores following meiosis.

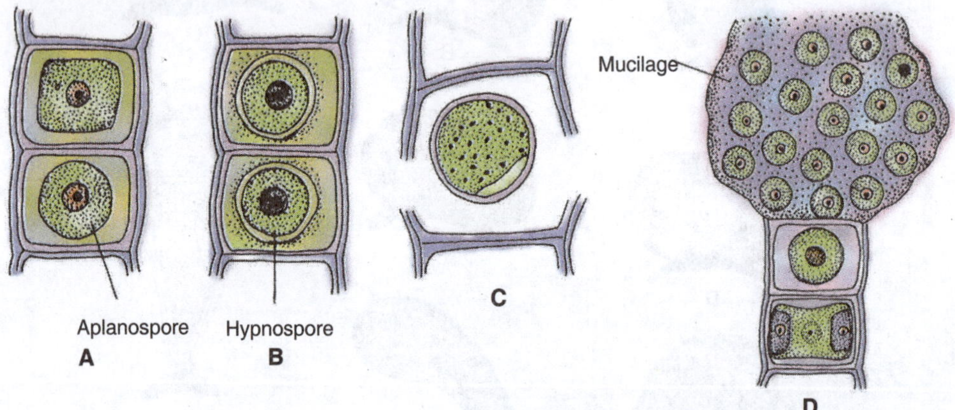

Fig. 6.5 (A-D). *Ulothrix sp.* (A), formation of aplanospores; (B), hypnospore formation; (C), liberation of aplanospore; (D), palmella stage.

***(i)* Gamete formation**. The gametes are produced in the same way as the zoospores. The protoplast of any vegetative cell of the filament (except the rhizoidal cell) may undergo repeated division by the method of successive bi-partition. The gametes may be formed on the same filament which has produced zoospores or on different filaments. The cells producing the gametes may be called the **gametangia**. The gametangium produces 8, 16, 32 or 64 gametes depending upon the species. They are all alike. There is no distinction in size, shape and activity between the fusing gametes. Such gametes are known as the **isogametes** (Fig. 6.7 C-D). The isogametes of *Ulothrix* though alike in every respect show functional disparity. It is for this reason that after liberation two gametes from filaments of different genetic constitution fuse to produce a zygote. The fusing gametes being morphologically identical, they cannot be called male and female. Instead they are denoted by the signs + and – strains.

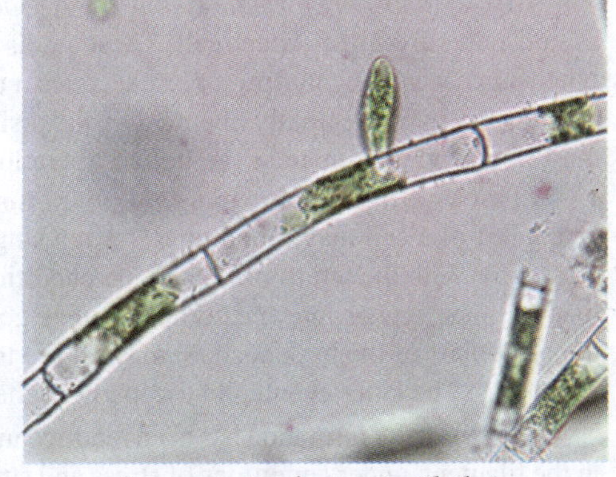

Isogamete Production in *Ulothrix*.

The gametes resemble the quadriflagellate macro-and microzoospores in structure, origin and development. Like the zoospores they are complete cells each with its nucleus, cytoplasm and a tiny chloroplast. They are formed and liberated in the same manner as the zoospores and swim about aimlessly for a time. They are, however, smaller in size and uniformly have two flagella instead of four. But this morphological difference in size and number of flagella has been bridged over by reported occurrence of biflagellate microzoospores. The main difference between the two, however, is that the gametes are incapable of producing new *Ulothrix* filaments alone.

***(ii)* Liberation and fusion of gametes.** The biflagellate gametes are liberated usually in the morning through a pore in the lateral wall of the gametangium (Fig. 6.7, B). On liberation they are surrounded by a common thin vesicle (Fig. 6.7, B) which soon disappears. The liberated gametes swim about in water at random. While swimming in the small area of their liberation, the two will occasionally collide. Many collisions come to naught. But when gametes from filaments of different genetic constitution happen to collide they conjugate in pairs laterally (Fig. 6.7 E). Fusion starts at the anterior end of the fusing gametes. This is the sexual act.

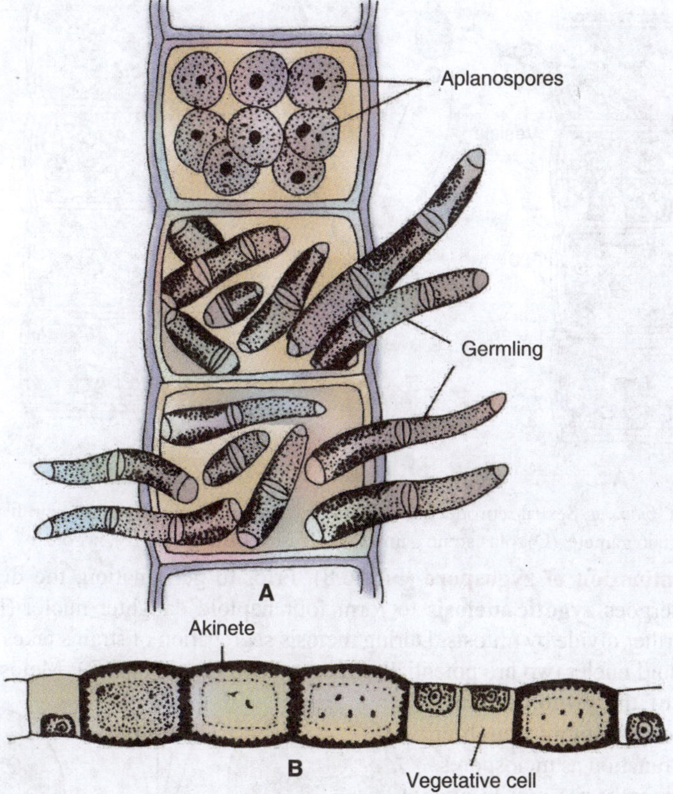

Fig. 6.6 (A-B). *Ulothrix* sp. (A), Germination of aplanospores *in situ*; (B), akinete formation in *U. idiospora.*

The liberated reproductive cells (zoospores and gametes) of *Ulothrix* closely resemble volvoccoid (*Chlamydomonas*-like) cells. From this one may infer that *Ulothrix* in its reproduction reverts to the condition of *Chlamydomonas*-like ancestor. The latter settled down, resorbed its flagella, secreted a wall around it and divided, repeatedly in a single plane and direction by mitosis to produce a chain of cells (*Ulothrix* filament). The individual cells of this filament still retain the capacity to produce *Chlamydomonas*-like cells at the time of reproduction.

Zygospore (Fig. 6.8). The combined protoplast which is the result of fusion of the two gametes is called the **zygospore**. The zygospore of *Ulothrix* is quadriflagellate and thus motile at first (Fig. 6.7 F). It is spindle-shaped. It has two eye spots and two chloroplasts. Its fusion nucleus has two sets of chromosomes-one set being contributed by each fusing gamete. Such a nucleus is known as **diploid**. The zygospore remains motile for a short period. It soon comes to rest, withdraws its flagella and rounds off. It then secretes a thick wall to become a spherical, heavy-walled structure (A). In this condition it enters upon a period of rest. The resting zygospores may be carried by wind along with dust to colonize other ponds. In rare cases the zygospore is reported to form a rhizoid-like attaching organ (Dodel, 1976). During the resting period there is considerable accumulation of reserve food.

The zygospore germinates after a period of rest by undergoing meiosis and generally produces four motile or non-motile **meiospores**. The diploid phase in the life cycle of *Ulothrix* is thus represented by a single-celled zygospore. It is short-lived and inconspicuous. The cells of this stage have a double set of chromosomes.

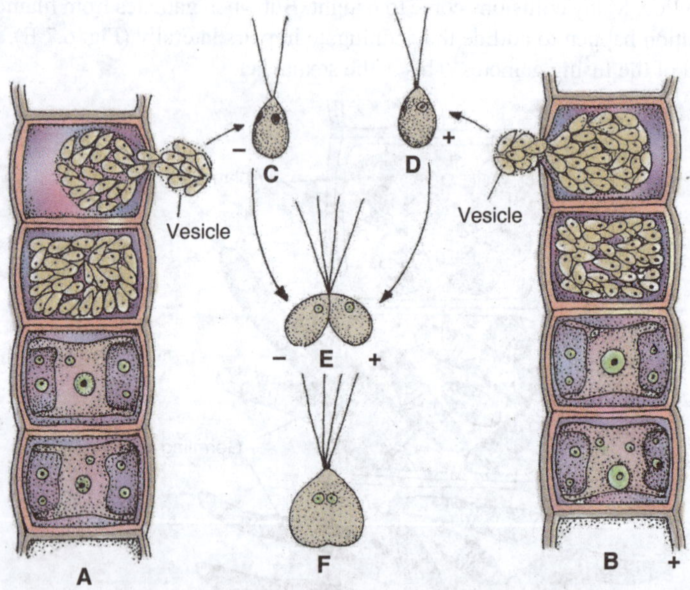

Fig. 6.7 (A-F). *Ulothrix sp.* Sexual reproduction. (A), minus strain filament; (B), plus strain filament; (C), minus strain gamete; (D), plus strain gamete; (E), fusion of gametes; (F), zygote.

(*iii*) Germination of zygospore (Fig. 6.8). Prior to germination, the diploid nucleus of zygospore undergoes **zygotic meiosis** to form four haploid daughter nuclei (B). The latter sometimes may further divide by mitosis. During meiosis segregation of strains takes place. Of the resultant four haploid nuclei two are potentially of + and two of – strain (D). Meiosis is followed by the cleavage of the protoplast into 4 to 16 uni-nucleate haploid daughter protoplasts which function as meiospores. The meiospores usually are motile and quadriflagellate. They are called the **meiozoospores** (E). In certain cases they develop thin walls but not flagella and are called the **meioaplanospores**. Some phycologists hold that the contents of the zygospores usually develop as non-motile meiospores. At this stage the zygospore wall ruptures (D) and the meiospores are liberated into the surrounding water. Each meiospore by repeated mitotic divisions develops into a new filament of *Ulothrix*. The cells of the filament have a single set of chromosomes.

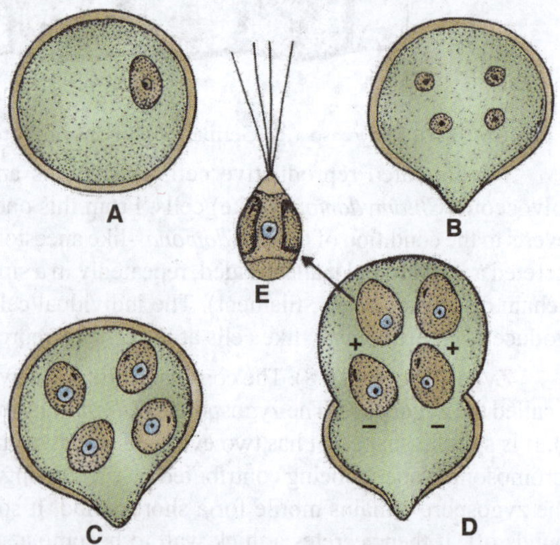

Fig. 6.8 (A-E). *Ulothrix* sp. (A), zygospore; (B-D), stages in the germination of zygospore; (E), liberated meiozoospore.

Reproduction by meiospores serves to multiply the benefits of a single act of fertilization.

Biological Importance of Ulothrix

Algae like *Ulothrix* are considered of biological importance as they throw light on certain important biological phenomenon.

1. Specialisation leading to Interdependence. The *Ulothrix* filament shows slight differentiation of cells. There is the basal, colourless **rhizoidal cell** which attaches the filament to the substratum. All others are green and similar. This differentiation of cells in the filament leads to **specialisation**. The rhizoidal cell neither takes part in nutrition nor in reproduction. All other green cells in the filament function in nutrition, reproduction, growth, cell division, etc. The differentiation of cells or specialisation is thus accompanied by division of labour. The latter leads to interdependence of cells in the filament upon one another. The rhizoidal cell is dependent for its nutrition on the green vegetative cells and the latter for attachment on the rhizoidal cell. *Ulothrix* thus provides a striking example of the beginning of specialisation accompanied by interdependence of cells. It is more marked in *Oedogonium.*

2. Origin of Sex. The gametes of *Ulothrix*, in structure, development and mode of liberation, are exactly similar to the zoospores. Structurally, they differ only in their small size and number of flagella which is two instead of four. Even this difference has been bridged over by the reported occurrence of biflagellate microzoospores in *U. zonata*. Now remains the difference in their sexual behaviour.

Under certain unfavourable environmental conditions the protoplast of the cell produces swarmers smaller than the zoospores. They alone had not the resources to produces new *Ulothrix* filaments. In their wandering they met and fused by accident to form a thick-walled fusion cell. This accidental fusion of swarmers or smaller zoospores in response to adverse conditions proved of advantage to the species. This advantage ensured the continuation of the process which gradually became fixed. The fusing swarmers came to be known as the **sex cells or gametes**, the act of fusion as **fertilization** and the fusion cell as **zygote**. The following facts support this view of the origin of sexuality in the green algae:

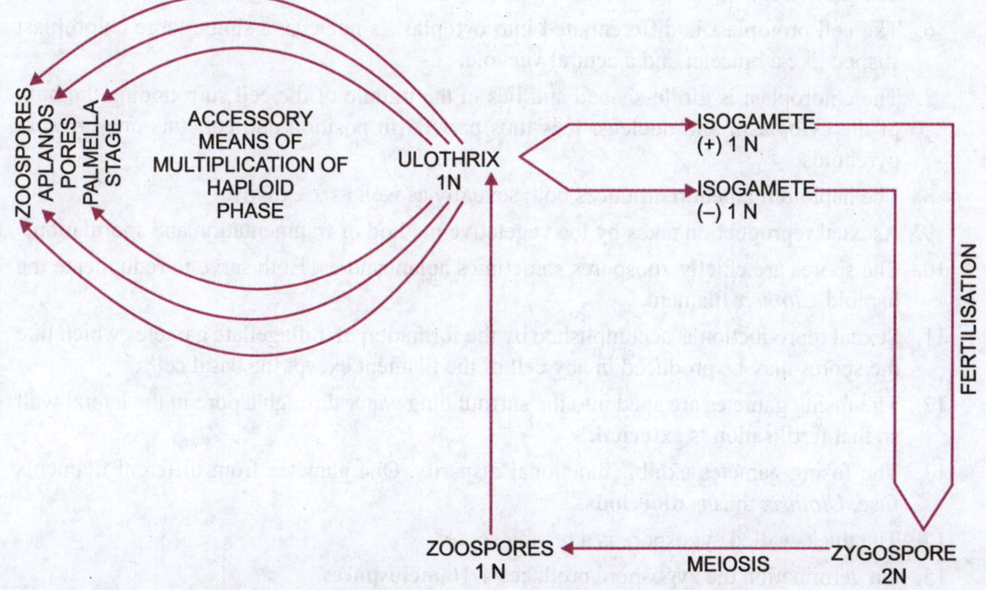

Fig. 6.9. *Ulothrix sp.* Graphic representation of the life cycle.

(*i*) Sexual reproduction in the green algae invariably takes place only once and towards the end of the growing season.

(*ii*) In certain green algae the gametes, which fail to fuse, function as **azygospores**. Each azygospore develops directly into a new filament.

Both these facts lend additional support to the view that sexual reproduction originated by the accidental fusion of zoospores in response to the conditions unfavourable for growth.

3. Origin of Sporophyte. Algae like *Ulothrix* illustrate the first step in the origin and organisation of the **diploid** or **sporophyte** generation in the life cycle of plants. The diploid zygospore, on germination, does not directly develop into a new filament. Instead its nucleus undergoes zygotic meiosis. The resultant four haploid daughter nuclei are organised into **meiospores** which may be motile (**meiozoospores**) or non-motile (**meioaplanospores**). The meiospores, on liberation, produce the haploid *Ulothrix* filaments. The zygospore with its contained meiospores is considered by some scientists as the forerunner of the sporophyte generation. They look upon it as an **incipient** or **primitive sporophyte**. In the life cycle of *Ulothrix,* therefore, there is an alternation of a multicellular, filamentous gametophyte generation with a one-celled, diploid (sporophyte) generation.

Salient Features

1. The thallus is an unbranched filament consisting of short, cylindric or quadrate uninucleate cells.
2. The filament shows distinction into base and apex.
3. The basal, colourless **rhizoidal cell** is longer and narrower than the others in the filament. It functions as an organ of attachment. All others are green, cylindric or quadrate and similar in all respects. The cell at the free end, however, is dome-shaped rather than cylindrical.
4. The filament illustrates the beginning of **specialisation** leading to **interdependence** of cells.
5. The green vegetative cell is made up of the **cell wall** surrounding the protoplast.
6. The cell protoplast is differentiated into cytoplast, a nucleus, a single large chloroplast shaped like a bracelet and a central vacuole.
7. The chloroplast is girdle-shaped and lies in the middle of the cell surrounding the bulk of the cytoplasm and nucleus. It is thus parietal in position and contains one or more pyrenoids.
8. The haploid filament reproduces both sexually as well as asexually.
9. Asexual reproduction takes by the vegetative method of fragmentation and sporulation.
10. The spores are chiefly zoospores sometimes aplanospores. Both serve to reduplicate the haploid *Ulothrix* filament.
11. Sexual reproduction is accomplished by the formation of biflagellate gametes which like the spores may be produced in any cell of the filament except the basal cell.
12. The fusing gametes are shed into the surrounding water through a pore in the lateral wall so that fertilisation is **external.**
13. The fusing gametes exhibit functional disparity. One gametes from different filaments fuse. *Ulothrix* thus is **dioecious**.
14. The thick-walled zygospore is a resting spore.
15. On germination the zygospore produces 4-16 **meiospores**.
16. On liberation each meiospore produces a haploid *Ulothrix* filament.
17. The zygospore with the contained meiospores may be regarded as the forerunner of the sporophyte generation and called the **incipient** or **primitive sporophyte**.

18. Zygospore thus has a three-fold function : *perennation, dispersal* and *production of meiospores.*
19. The simple life history of *Ulothrix* consisting of a haploid plant body (filament) with the zygote as the diploid stage is called **haplontic**. It is characterised by **zygotic meiosis.**

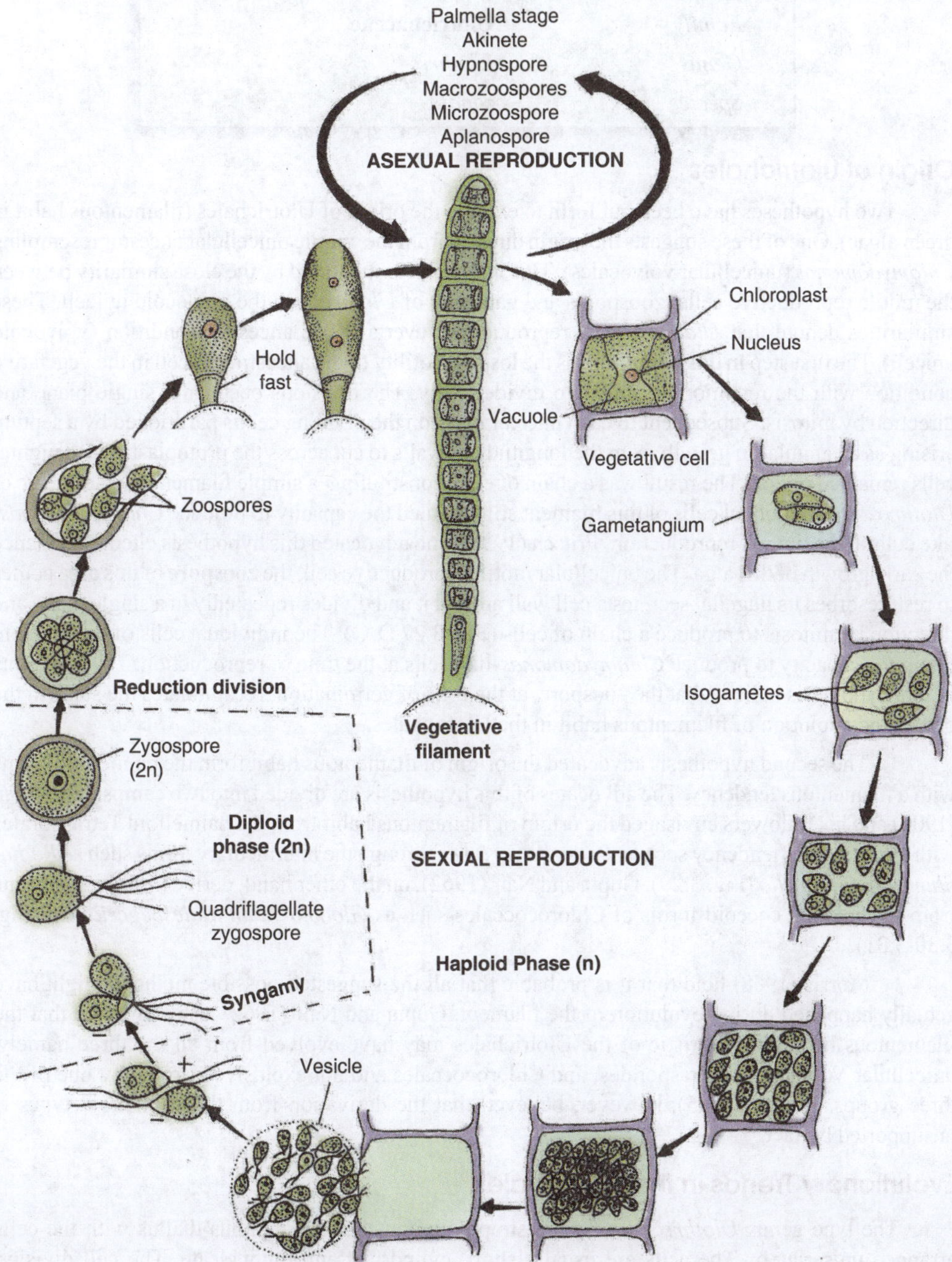

Fig. 6.10. Semi-diagrammatic representation of life-cycle in *Ulothrix.*

Taxonomic Position

Division	:	**Chlorophyta** (Chlorophycophyta)
Class	:	**Chlorophyceae**
Order	:	**Ulotrichales**
Family	:	**Ulotrichaceae**
Genus	:	*Ulothrix*
Species	:	*zonata*

Origin of Ulotrichales

Two hypotheses have been put forth to explain the origin of Ulotrichales (filamentous habit in green algae). One of these suggests the origin directly from the motile unicellular ancestor resembling *Chlamydomonas* (unicellular volvocales). This viewpoint is supported by the close similarity between the motile reproductive cells (zoospores and gametes) of *Ulothrix* and the volvocoid unicell. These similarities denote that *Ulothrix* in its reproduction reverts to the ancestral condition (volvocoid unicell). The first step in this hypothesis is the loss of motility of the ancestral unicell in the vegetative condition with the retention of ability to divide freely. The divisions occur in a single plane and direction by mitosis. Subsequent to each nuclear division the dividing cell is partitioned by a septum arising as an annular in growth from the longitudinal walls to cut across the protoplast. The daughter cells remained united. The result was a chain of cells constituting a simple filament characteristic of *Ulothrix*. The individual cells of this filament still retained the capacity to produce *Chlamydomonas*-like cells at the time of reproduction. Fritsch (1935) who advocated this hypothesis cited as evidence the early growth of this alga. The unicellular motile reproductive cell, the **zoospore** of this alga comes to rest, resorbes its flagella, secretes a cell wall around it and divides repeatedly in a single plane and direction by mitosis to produce a chain of cells (Fig. 3.27 D-G). The individual cells of the filament retain the capacity to produce *Chlamydomonas*-like cells at the time of reproduction. The adherents of this hypothesis believe that the zoospore, at the time of germination, recapitulates the steps in the origin and evolution of filamentous habit in the Ulotrichales.

The second hypothesis advocated the origin of filamentous habit form the palmelloid forms with a filamentous tendency. The advocates of this hypothesis are divided into two camps. Blackman (1900) and his followers envisaged the origin of filamentous habit from the palmelloid Tetrasporales with a filamentous tendency such as *Palmella variance* through the intermediary forms such as *Radiofilum* and *Geminella* (Fig. 3.29). Gupta and Nair (1962), on the other hand, derive filamentous forms from the colonial coccoid forms of Chlorococcales such as *Gloeotaenium loitlesbegerianum* (Fig. 3.30-3.31).

Morris (1968) held that it is probable that all the suggested possible methods might have actually happened during evolution of the filament. Gupta and Nair (1962) also suggested that the filamentous habit characteristic of the Ulotrichales may have evolved from all the three namely, unicellular Volvocales, Tetrasporales, and Chlorococcales and not exclusively from any one of the three groups. Fritsch (1935), however, believed that the derivation from the palmelloid types is unsupported by fact.

Evolutionary Trends in the Ulotrichales

The type genus *Ulothrix* possesses a simple, unbranched filamentous thallus with the cells arranged uniseriately. The cells are usually short, cylindrical and uninucleate. The cell division takes place in only one direction. The single chloroplast in the cell is parietal and takes the form of a girdle either completely encircling the circumference of the cell or more often half the surface.

Sexual reproduction is isogamous. There is much variation in the form of the chloroplast, structure of the cell wall, number of nuclei in the cell and sexual reproduction in other genera of the order. These variations in cell structure and range of sexual reproduction which show lines of development or evolutionary trends in the order are briefly stated below :

1. Modifications of Cell Structure

(*a*) *Cell Wall.* In contrast to the single piece, homogeneous cell wall of *Ulothrix* and other Ulotrichales, the cell wall in many species of *Microspora* is composed of two H-shaped pieces, the ends of which overlap each other. The protoplast of the cell is enclosed by the overlapping halves of the two successive H-pieces at the equatorial region. In the possession of this type of wall structure *Microspora* shows a close resemblance to *Tribonema* and few other Xanthophyceae. The presence of starch as a food reserve and 2 or 4 equal whiplash flagella on the swarmer, however, serve to distinguish it from the Xanthophyceae.

(*b*) *Chloroplast*. There is considerable variation in the form of the chloroplast. The chloroplast in *Hormidium* and *Gemiella* differs from that of *Ulothrix* in being much smaller often occupying less than half the length of the cell. It appears as a parietal plate. In some other genera it is still more reduced and is a small disc on one side of the cell. In contrast *Cylindrocapsa* has an axial stellate chloroplast. In *Microspora*, though parietal in position, the chloroplast appears reticulate in character with a number of moniliform discs connected together. *Sphaeroplea* is unique among the Ulotrichales in possessing annular parietal chloroplasts arranged in the form of numerous narrow transverse bands.

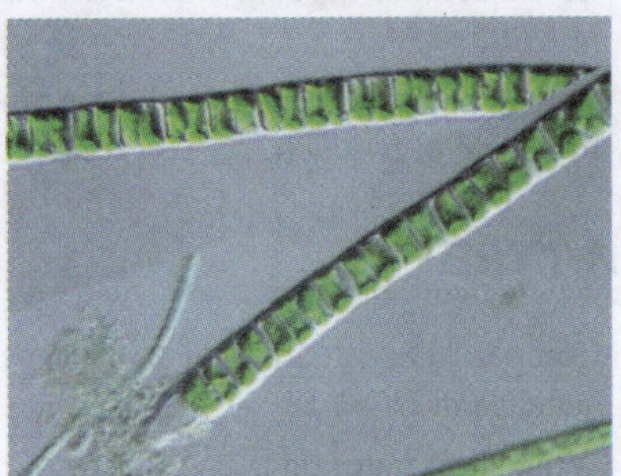

Band Shaped Parietal Chloroplasts in *Ulothrix*.

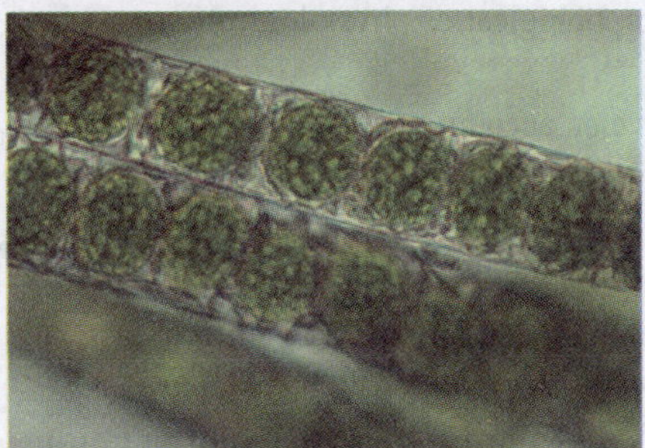

Annular band shaped Parietal Chloroplasts in *Sphaeroplea*.

(*c*) *Nucleus*. The cells in all the genera of the Ulotrichales are uninucleate. The only exception is *Sphaeroplea*. It has considerably elongated multinucleate coenocytic cells. Because of this feature (large coenocytic cells) some algologists included *Sphaeroplea* with Cladophoraceae under order Siphonocladales but assigned an isolated position because of the annular chloroplasts and characteristic oogamous sexual reproduction.

2. The cells in the filamentous thallus of Ulotrichales are arranged uniseriately. They divide in only one direction which is transverse to the longitudinal axis of the filament. The cells in the older filaments of *Cylindrocapsa* have a tendency to undergo oblique or longitudinal divisions. Consequently the cells in these regions of the filament become biseriately or multiseriately or irregularly arranged.

3. Sexual Reproduction. It is isogamous in most genera of the Ulotrichales. In *Sphaeroplea* it is oogamous though in one or two other genera of the Ulotrichales it is an advanced type of anisogamy. *Cylindrocapsa* is distinctly oogamous showing considerable resemblance to *Oedogonium.* However, this similarity does not indicate actual relationship.

Phylogency of Ulotrichales

The Ulotrichales are thought to have arisen from the volvocoid cells through the intermediary forms which were non-motile. This view point is supported by the parietal chloroplast and occurrence of volvocoid type of motile reproductive cells (gametes and zoospores) in the Ulotrichales. In methods of asexual and sexual reproduction Ulotrichales approach Cladophorales. In both, the gametes and zoospores are produced in specialized cells and are liberated in the same manner through a lateral aperture. The zoospores are quadriflagellate and gametes biflagellate. Sexual reproduction is usually isogamous rarely anisogamous. From the simple filament of *Ulothrix* are considered to have evolved later the branched types, Cladophorales and Chaetophorales and also the Ulvales with a parenchymatous thallus more than one cell thick.

QUESTIONS

ESSAY TYPE

1. Describe the habitat, mode of function and the structure of the cell in *Ulothrix*.
2. Describe the various methods of asexual reproduction in *Ulothrix*.
3. Describe the life history of *Ulothrix* commenting upon any point of morphological or physiological interest in it. *(Agra , 1996)*
4. Give an account of sexual reproduction in *Ulothrix*.
5. Describe the structure and life history of *Ulothrix* and mention how its sexual method of reproduction throws light on the origin of the sex cells in the Algae. *(Agra, 1998)*
6. Give a detailed account of the cell structure in *Ulothrix*.
7. With the help of diagrams only, describe the life cycle in *Ulothrix*. *(Bharathiar, 2000)*

SHORT ANSWER TYPE

8. Compare the structure of chloroplast in *Ulothrix* and *Chlamydomonas*.
9. Discuss the biological importance of *Ulothrix*.
10. Give a brief account of evolutionary trends in *Ulothrix*.
11. Discuss the evolutionary trends in Ulotrichales.
12. Describe the various types of zoospores in *Ulothrix*.
13. Compare the structure of zoospores and gametes in *Ulothrix*.
14. Write short notes on :
    (*a*) Cell structure in *Ulothrix*
    (*b*) Microzoospores
    (*c*) Isogametes
    (*d*) Asexual reproduction in *Ulothrix*
    (*e*) Sexual reproduction in *Ulothrix*.
15. Name an alga with girdle-shaped chloroplast.

OBJECTIVE TYPE

16. **Select the correct answer :**

(*i*) Girdle shaped chloroplast is found in
(*a*) *Chlamydomonas*
(*b*) *Volvox*
(*c*) *Oedogonium*
(*d*) *Ulothrix*

(*ii*) Multicellular filamentous alga is
(*a*) *Ulothrix*
(*b*) *Vaucheria*
(*c*) *Chlamydomonas*
(*d*) *Pandorina*

(*iii*) The life cycle of *Ulothrix* is

(*a*) Haplontic (*b*) Diplontic

(*c*) Haplobiontic (*d*) Diplobiontic

(*iv*) Which of the following alga occurs as an attached form?

(*a*) *Volvox* (*b*) *Zygnema*

(*c*) *Ulothrix* (*d*) *Vaucheria*

(*v*) The thallus of *Ulothrix* is filamentous and

(*a*) Branched (*b*) Colonial

(*c*) Unbranched (*d*) Solitary

(*vi*) *Ulothrix* produces

(*a*) Quadriflagellate microzoospores

(*b*) Biflagellate microzoospores

(*c*) Quadriflagellate macrozoospore

(*d*) All of the above

(*vii*) In *Ulothrix*, the zygote is pear shaped and

(*a*) Uniflagellate

(*b*) Biflagellate

(*c*) Triflagellate

(*d*) Quadriflagellate

(*viii*) Meiosis in *Ulothrix* takes place at the time of

(*a*) Germination of zygote

(*b*) Formation of zoospores in the filament cells

(*c*) Formation of gametes

(*d*) Fragmentation

(*ix*) The filaments of *Ulothrix* occur

(*a*) Only in marine water

(*b*) Only in fresh water

(*c*) Both in marine and fresh water

(*d*) On soil

(*x*) Which of the following is produced in maximum numbers in *Ulothrix*?

(*a*) Gametes

(*b*) Quadriflagellate macrozoospore

(*c*) Quadriflagellate microzoospore

(*d*) Biflagellate microzoospore

Ulvales: Ulva lactuca

CHAPTER 7

ORDER 4. ULVALES

- Family : Ulvaceae
- ULVA

The members of this order have uninucleate cells which in structure are similar to those of *Ulothrix* but divide in more than one plane (two or three) to produce parenchymatous thalli which may be hollow tubes, solid cylinders or flattened expanse of tissue (expanded sheets). The thalli are sessile and generally attached to the substratum. Asexual reproduction by the formation of asexual spores (mitospores) is absent. Sexual reproduction is by biflagellate isogametes *(Ulva)* or anisogametes (*Enteromorpha intestinalis*). The gametes are differentiated by successive bipartition of the contents of the vegetative cells. They are liberated through an aperture in the cell wall. According to Smith the order includes 5 or 6 genera with about 135 species. Two families Ulvaceae and a monotypic family Schizomeridaceae are reognisable in this order. Of these the former is treated here.

FAMILY : ULVACEAE

The family comprises exclusively marine forms in which cell division takes place in three dimensions to produce parenchymatous thalli which are either hollow tubes (*Enteromorpha*) or flattened expanse of tissue (*Ulva*) two cells in thickness. It includes 4 or 5 genera. The chief among these are *Ulva* and *Enteromorpha*. Both are discussed here in some detail. They exhibit distinct alternation of generation of **isomorphic** type in which the diploid asexual generation alternates with the haploid sexual generation. The zoospores are differentiated by meiosis from diploid contents of the cells of the sporophyte and thus are called the **meiozoospores** or **gonozoospores**.

ULVA

Occurrence

Ulva is a marine cosmopolitan genus. It occurs along the rocky ocean shores in the intertidal zone. A few species are found in saline or fresh water but grow in profusion in waters polluted by organic matter or sewage. According to Fritsch species of *Ulva* commonly occur in the upper part of littoral zone and also where supplies of fresh-water escape from the cliffs. The genus includes about 30 species. *Ulva lactuca* var. *rigida* is common on the Gujarat coast.

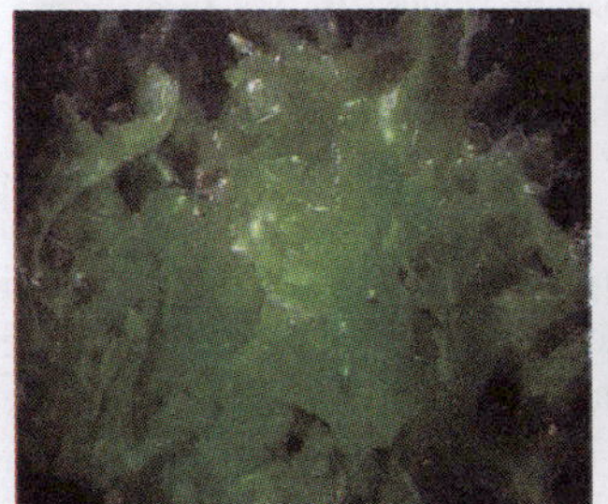

Ulva fasciata.

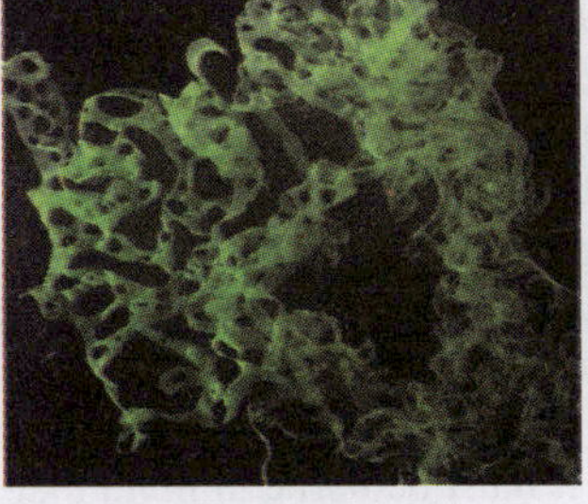

Ulva reticulata.

Ulva intestinalis.

Thallus (Fig. 7.1 A)

The parenchymatous foliaceous thallus is bright green. It is a broad and flat, expanded sheet which may be 30 cm or more in length and two cells in thickness (distromatic). It somewhat resembles a leaf of garden lettuce in form, hence the popular name 'sea lettuce' for this alga. The basal part of the bladder-like thallus is narrowed into a short **stalk**. Reaching the substratum the stalk is dilated to form an attaching disc.

In a cross section the thallus is two cell layers thick (C). The cells of the two layers are in close contact but in *U. linza* and *U. rhacodes* the two cell layers may separate from each other towards the margin or near the base. The cells are isodiametric or vertically elongated with the long axes at right angles to the surface of the thallus. The cell walls are generally distinctly stratified and the lateral walls of the adjoining cells are more or less confluent with one another. Each cell is uninucleate and has but one parietal chloroplast which is laminate to cup-shaped with deeply incised or lobed

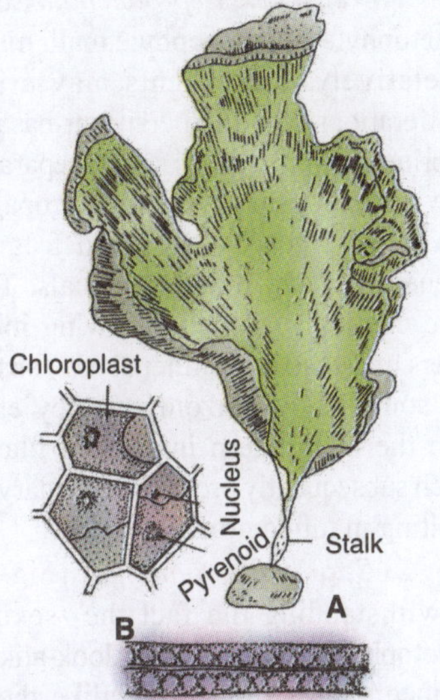

Fig. 7.1 (A-C). *Ulva lactuca.* (A), thallus attached to stone; (B), cells showing structure; (C), cross section of thallus. (A and (C) after Thuret; B, after

margins. It lies towards the outer face of the cell and has a single pyrenoid. The nucleus is located in the inner half of the cell in different positions. From some of the lower cells of the thallus (Fig. 7.2 B) grow out colourless multinucleate rhizoids one each. They intertwine with one another as they push their way down between the two thallus layers. Finally the rhizoidal strands emerge from the thallus growth down, become closely appressed to one another in the stalk region and broaden out at their tips to form the basal, pseudo-parenchymatous attachment disc. The blades perish at the end of the growing season but the attachment disc persists throughout winter and thus is perennial.

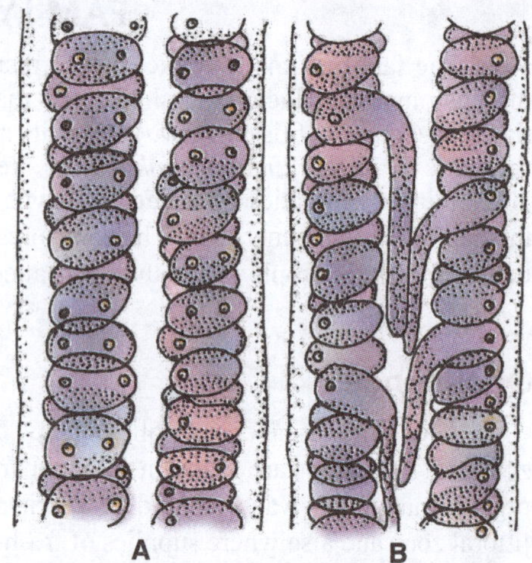

Fig. 7.2 (A-B). *Ulva.* (A), V.S. through the upper part of thallus; (B), V.S. through the lower part of the thallus (Based on Thuret).

Reproduction

The gametothallus of *Ulva* reproduces asexually as well as sexually. Asexual reproduction takes place entirely by vegetative methods. The formation of mitospores (motile or non-motile) is lacking. In this respect the life cycle of *Ulva* is interesting. It resembles that of the land plants. Genetically there are two kinds of thalli in *Ulva* namely, the haploid thalli (**gametophyte plants**) and diploid thalli (**sporophyte plants**). The former are concerned with sexual reproduction and the latter with the production of sexual spores, the **meiospores** or **gonospores**. The production of meiospores is a device to multiply the benefits of a single act of fertilization. Most of the biologists, therefore, consider it a stage under sexual reproduction. It must be borne in mind that the gametophyte and sporophyte plants are morphologically identical.

1. *Vegetative Reproduction.* Both the gametophyte and sporophyte thalli multiply vegetatively. New plants may arise as proliferationsfrom the persistent basal disc in spring. The proliferous shoots separate and grow into new plants. Vegetative propagation also takes place by detached fragments regenerating into new individuals. This is more common in plants growing in quiet waters of estuaries. Schiller (1907) observed that some of the rhizoids as they emerge from the thallus turn into green filaments which subsequently produce secondary thalli resulting in tufted growth.

2. *Sexual Reproduction* (Fig. 7.3). Notwithstanding the fact the sexual or gametophytic plants in *Ulva* look alike and produce gametes that look alike they are really of two mating types. This is shown by the behaviour of gametes at the time of fusion. Gametes from the same thallus or from sexual plants of the same kind (mating

Fig. 7.3 (A-I). *Ulva.* Sexual reproduction. (A), sectional view of a portion of thallus with cells containing gametes; (B-E) stages showing fusion of isogametes; (F-I), stages showing fusion between anisogametes (After Thuret and Borzi).

type) do not fuse with each other. Fusion takes place only between gametes coming from two different kinds of thalli. We cannot speak of such thalli as male and female. The sex difference is rudimentary. This minor difference in sex is usually denoted by the signs + and –. The species showing this rudimentary difference in sex are called **heterothallic**. All species of *Ulva* are **heterothallic**. All of them are **isogamous** except three (*U. lobata*) which are **anisogamous**. The gametes are biflagellate and are produced in the vegetative cells at the margin of the thallus (A). The gamete producing zone of the thallus is of different colour from that of the vegetative portion and is 5-15 mm broad. Every cell in the fertile zone forms gametes. Prior to gamete formation the gametangium develops a beak-like protuberance at its outer face. It extends to the surface of the thallus. Gametes are formed by successive bipartition of protoplast of the cell till 32 or 64 uninucleate daughter protoplasts are formed. The divisions are mitotic. The first division is always parallel to the surface of the thallus and the second is perpendicular to the first. Each daughter protoplast metamorphoses into a biflagellate uninucleate gamete (B). The gametes escape through an apical pore formed at the tip of the beak-like outgrowth of the gametangium. They are liberated in enormous numbers during the morning when the plants are flooded by incoming tides (Smith 1955). Smith (1947) reported them in the species of *Ulva* growing on the California coast, release of gametes and zoospores was associated with spring tidal cycles of lunar months. The gametes were released regularly with 14 days intervals early in the series of spring tides and the zoospores were liberated towards the end.

Fertilization. Fusion takes place in the surrounding water between gametes coming from plants of different mating types (C and G). The cytoplasmic fusion of two gametes is followed by the fusion of two nuclei + and – strain. The freshly formed zygote is motile at first. It has four flagella, one diploid nucleus and two chloroplasts, one contributed by each gamete. During maturation one of the chloroplasts disintegrates. Braten (1973) showed by radioactive labelling of gametes prior to copulation, that the disintegrating chloroplast is supplied through the + strain gamete. The chloroplast contributed by the – strain gamete is retained by the zygote (D and H). After a short period of activity it comes to rest, retracts its flagella and secretes a wall around it (E and I). The attaching zygote secretes a special attaching substance (Braten 1975).

Fyon (1934, A), Moewus (1938, A) and Yawada and Sato (1938) reported parthenogenetic development of gametes into haploid thalli in certain species of *Ulva*.

Germination of zygote (Fig. 7.4). The diploid zygote germinates within a day or two. It does not undergo reduction division on germination. By repeated transverse divisions which are all mitotic, in a single direction, results a simple filament resembling that of *Ulothrix* (F) in appearance. It is attached to the substratum by a basal **holdfast** and is **diploid**. During further growth the cells of the filament divide both by transverse and vertical divisions in three dimensions resulting in a blade-like, flattened expanse of tissue two cells thick. It is the **sporothallus.** The blade-like sporothalli of *Ulva* resemble the parent gametophytes (gametothalli) in all respects except being **diploid**. The haploid chromosome number for *U. fasciata* is $n = 10$ (Sarma and Chaudhari 1975).

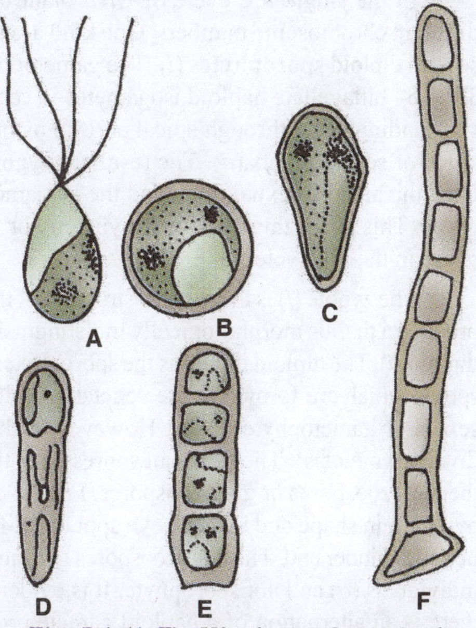

Fig. 7.4 (A-F). *Ulva lactuca.* Stages in the germination of zygote (After Thuret and Borzi).

Production of Meiospores (Fig. 7.5). Reaching maturity the diploid sporophyte thallus produces from 4-8 quadriflagellate zoospores in each cell in the same way as the gametes are produced in the gametophyte plants. However, **meiosis** occurs at the time of their differentiation so that they have a reduced (haploid) number of chromosomes and are called the **meiozoospores** or **gonozoospores**. Spore formation, at first, starts in the marginal cells and then proceeds inwards, most of the vegetative cells taking part. After their liberation through the apical pores the parent thallus remains as a bleached framework of empty cells. After a brief period of swarming in the surrounding water, each gonozoospore settles down (C), retracts its flagella and secretes a wall around it. The clothed meiospore germinates in the same way as the zygote eventually producing a sexual or haploid thallus. These thalli, in turn, will produce gametes. It is interesting to note that unlike many other green algae (*Ulothrix*), the zygote in *Ulva*, on germination, does not produce meiospores directly. On the other hand it gives rise to a multicellular sporophyte plant which, in turn, produces the meiospores. The production of meiospores is thus delayed in the life cycle.

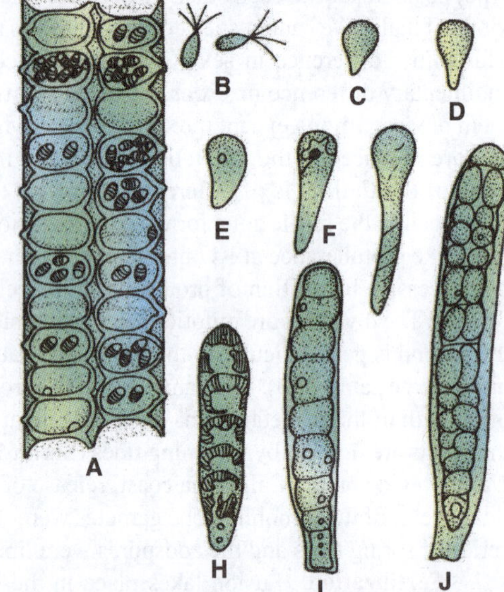

Fig. 7.5 (A-J). *Ulva lactuca.* (A), Sectional view of a fertile portion of the diploid thallus with the cells containing meiozoospores; (B), liberated meiozoospores; (C-J), stages in the germination of meiozoospores (After Thuret).

Alternation of Generations in Ulva

In the single life cycle of *Ulva* plant there occur two kinds of vegetative individuals with different chromosome numbers. One kind are the haploid **gametophytes** (*a*) or sexual plants and the second diploid **sporophytes** (*i*). The gametophyte individuals produce by repeated mitotic divisions 32 or 64 biflagellate haploid isogametes in each vegetative cell (*b*). The gametes are released in the surrounding water through apical pores. Fusion occurs between two gametes coming from different kinds of sexual plants (*d*). The resultant zygote is a diploid structure (*e* and *f*). In the life cycle of *Ulva* the haploid sexual thalli and the isogametes constitute the parent gametophyte generation or phase. This phase thus ends with syngamy or fertilization which initiates the diploid or sporophyte phase in the life cycle.

The zygote (*f*) is the pioneer structure of the sporophyte generation. It immediately germinates to produce a thallus morphologically indistinguishable from the parent sexual plants but genetically it is diploid (*i*). The diploid thallus is the sporophyte individual in the life cycle. It produces quadriflagellate spores which are formed in the vegetative cells in much the same way as are the isogametes in the sexual or gametophyte plants. However, at the time of spore differentiation (*k*) there is reduction division (meiosis). The resultant spores have thus a haploid number of chromosomes and are called the meiozoospores or gonozoospores. Four to eight gonozoospores are formed in each cell. They are pyriform in shape and have an eye spot at the forward end (*l* and *m*). The single parietal chloroplast lies at its hinder end. The gonozoospores germinate (*n-o*), in turn, to produce the sexual or gametophyte individuals (*o*) and not sporophyte. It is evident, therefore, that in the single life cycle of *Ulva* plant there is an alternation of a haploid gamete producing individual (the **gametophyte**) with a diploid spore producing individual Isporophyte). This phenomenon is called **alternation of generations**. The two critical points in the life cycle at which the life cycle switches on from one generation to the other are **fertilization (syngamy)** and **meiosis**. The two alternating individuals (gametophyte and sporophyte) in *Ulva* are morphologically identical. This type of alternation of generations is called **isomorphic or homologous**. The life cycle such as that of *Ulva* which is characterised by sporogenic meiosis and distinct alternation of two vegetative individuals (gametophyte and sporophyte) is called **diplobiontic or diplohaplontic life cycle**.

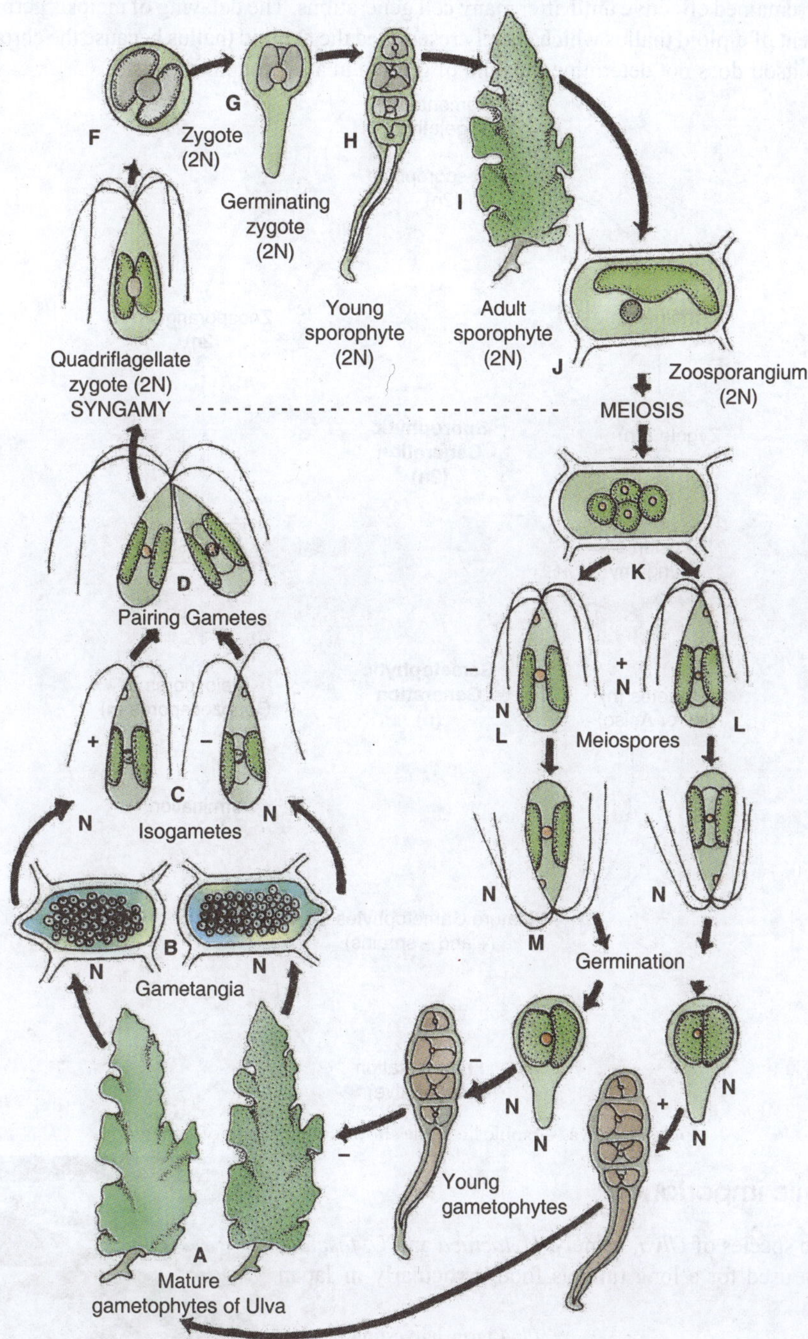

Fig. 7.6 (A-O). *Ulva.* Pictorial sketch of the sexual life cycle.

Origin of isomorphic life cycle

There are no connecting links or border line genera which could illustrate the way isomorphic life cycle might have originated in the order Ulvales. The plausible explanation one can offer is that there occurred simple mutation which inhibited zygotic meiosis at the time of zygote germination. The inhibition remained effective until after many cell generations. The delaying of meiosis permitted the development of diploid thallus which closely resembled the haploid thallus because the chromosome number in itself does not determine the kind of growth in any group of plants.

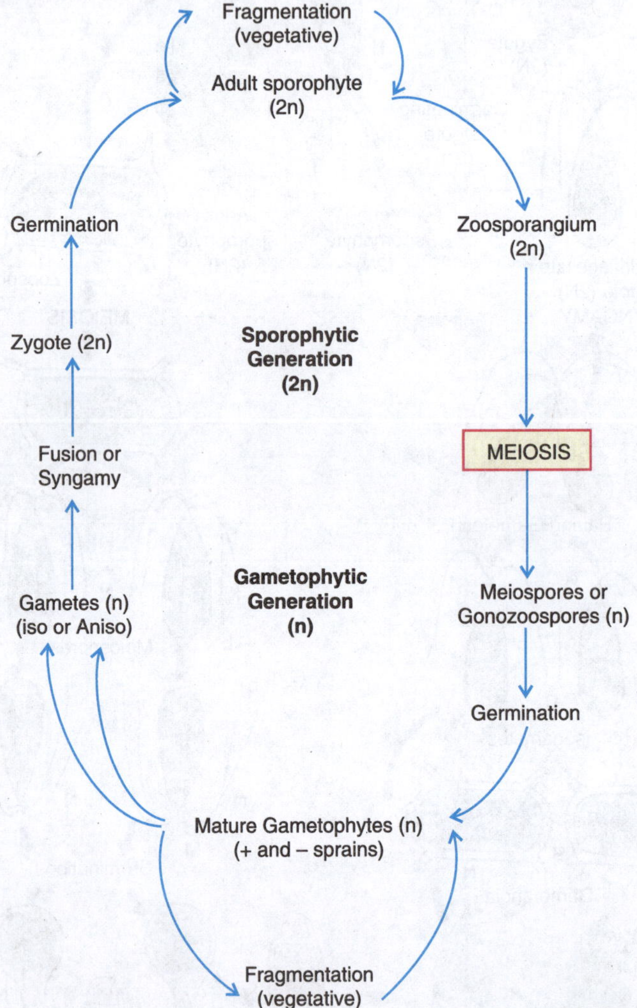

Fig. 7.7. Ulva. Graphic life cycle showing alternation of generation.

Economic importance

Two species of *Ulva*, namely, *U. lactuca* and *U. fasciata* have been used for a long time as food particularly in Japan and China.

Ulva farm harvesting.

Taxonomic Position of Ulva

The systematic position of *Ulva* is still under dispute. Some algologists place *Ulva* in the order Ulotrichales because young stages in it are uniseriate filaments closely resembling the simple filamentous thallus of *Ulothrix*. This view is further supported by the similarity of gametes and zoospores in the two genera. The general structure of the cells and reproductive processes are also very similar. The adherents of this view point consider that *Ulva* is merely a more advanced or elaborate form of *Ulothrix*. According to them the systematic position of *Ulva* is:

Division	:	**Chlorophyta** (Chlorophycophyta)
Class	:	**Chlorophyceae**
Order	:	**Ulotrichales**
Family	:	**Ulvaceae**
Genus	:	*Ulva*
Species	:	*lactuca*

Algal bloom by *Ulva*.

Smith (1955) and others placed *Ulva* in a separate order Ulvales on the basis of elaboration of vegetative structure of the thallus and the existence of a distinct sporophyte individual in the life cycle. According to this view which is held by majority of the algologists, at the present time, the systematic position of *Ulva* is :

Division	:	**Chlorophyta** (Chlorophycophyta)
Class	:	**Chlorophyceae**
Order	:	**Ulvales**
Family	:	**Ulvaceae**
Genus	:	*Ulva*
Species	:	*lactuca*

ENTEROMORPHA

Occurrence

Species of *Enteromorpha* are essentially marine and are widespread in the tropical waters. In India marine forms are of common occurrence at Madras, Okha, Dwarka, Cape Comorin and Bombay. They occur chiefly in the upper littoral, intertidal zone and also near the low tide mark usually attached to rocks or muddy substrata. A few species occur in fresh-water. Seerwani (1963) collected a fresh-water species of *Enteromorpha* from the Kundi river in Khargone, a place in the west Nimar district (M.P.). It inhabits shallow water. Later Sarma and Suryanaryana (1967) recorded *E. intermedia* attached to grasses and stones from a fresh-water stream Varuna near Varanasi.

Enteromorpha prolifera.

Enteromorpha intestinalis.

Enteromorpha compressa.

Distribution in Indian Waters

Many species of this genus occur on the shores of the mainland of India. Boergeson (1934, 1935) reported *E. flexuosa* from Dwarka and Bombay and *E. tubulosa* and *E. prolifera* from Bombay. Ramanathan (1939) recorded *E. compressa* var. *lingulata* from Madras. Dixit (1940) found *E. flexuosa* at the Malwan harbour. Krishnamurthy (154) and Umamaheswara Rao and Sreeramulu (1954) recorded *E. prolifera* and *E. prolifera* var. *tubulosa* from Madras and Visakhapatnam respectively. *E. polyclados* and *E. complanata* were reported from Andaman and Nicobar islands by Srinivasan (1960). Kole (1966) reported the occurrence of *E. clathrata* and *E. plumosa* in Porbander on the Gujarat coast. Kale (1967) described a new species *E. gujratensis* which she collected from the channels of Saurashtra saltworks, Porbander. Joshi and Krishnamurthi (1972) have listed as many as 13 taxa in this genus occurring on the shores of mainland of India. This account includes one new species *E. ovata* and one new variety of *E. linza* var. *bicornuta*. They collected *E. ovata* from Gopnath and Gujarat coast in the middle and lower intertidal zone in rock pools and margins of streamlets. *E. linza* var. *bicarnuta* was collected by them from Mahuva, Gujarat coast, attached to rocks in the intertidal zone.

E. intestinalis (L) Link is consmopolitan in its distribution and is taken as a type. Joshi and Krishnamurti (1972) collected it in the upper littoral zone from Gopnath, Rameswaran and Krusadai islands attached to rocks or free-floating.

Thallus (Fig. 7.8 A). The mature thallus is a dark green hollow tube with a wall composed of a single layer of cells. The fronds may be 15 cm high and 5-6 cm in width. The thallus is anchored to the substratum by a basal rhizoidal portion. The older thalli are free, mature specimens to give the thallus an intestine like appearance. The fronds may be simple or branched.

In surface view the cells are rounded, polygonal (Fig. 7.9 A) and irregularly disposed and in cross section are radially elongated (Fig. 7.8 B). The uninucleate cells have a single parietal laminate chloroplast which fills the cell.

The chloroplast has 3-4 pyrenoids. Henckel (1926) reported that the chloroplast in *E. intestinalis* lies against the lower side of the cell. The thallus branches arise as lateral proliferations. The growth is partly apical and partly by intercalary division.

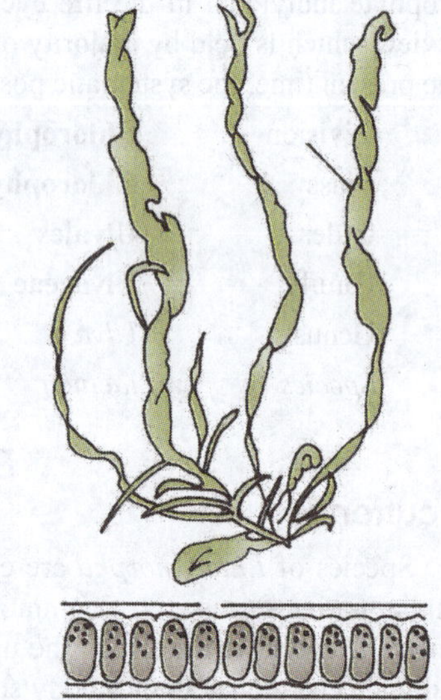

Fig. 7.8 (A-B). *Enteromorpha intestinalis.* (A) Plant showing habit; (B) T.S. thallus. (After Joshi and Krishnamurthy).

Reproduction. *Enteromorpha* reproduces through vegetative and sexual methods.

Vegetative Reproduction. *Enteromorpha* multiplies vegetatively by detached fragments and separation of proliferous shoots arising from the attached rhizoidal portion. According to Eaton, *et al* (1966) the detached segments show polarity. Upon regeneration rhizoids arise from the basal detached end and green tubular portion from the upper or distal end.

Sexual Reproduction (Fig. 7.9). It takes place by biflagellate gametes produced by the haploid thalli (A). The fusing gametes may be alike or unlike in size (C-D). In *E. intestinalis* the fusing heterogametes come from different plants (A-B). This shows a tendency to develop differentiation of sex. According to Kylin (1930) the male plants in *E. intestinalis* can be distinguished by the orange yellow colour of the fertile portion which in the female plants is yellowish green. The latter, however, is not so distinctive. The gametes are pear-shaped in form. The microgamete is small, narrow and has an eye spot, a pale, yellow green chloroplast with a rudimentary pyrenoid (D). The larger macrogamete has a green chloroplast with a distinct pyrenoid (C).

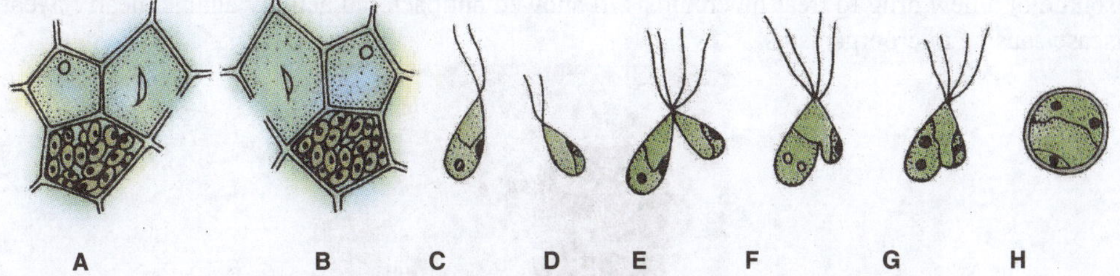

Fig. 7.9 (A-H). *Enteromorpha intestinalis.* Sexual reproduction (after Kylin).

The zygote, at first, is quadriflagellate (G). It has two chloroplasts but soon that contributed by the microgamete disorganises. After a short period of swarming the zygote comes to rest, retracts its flagella and secretes a wall around it (H). According to Hartman (1929) and Kylin (1930) it then germinates very soon. There are others who speak of a resting period. Parthenogenesis has been reported in some species such as *E. clathrata* (Bliding, 1933).

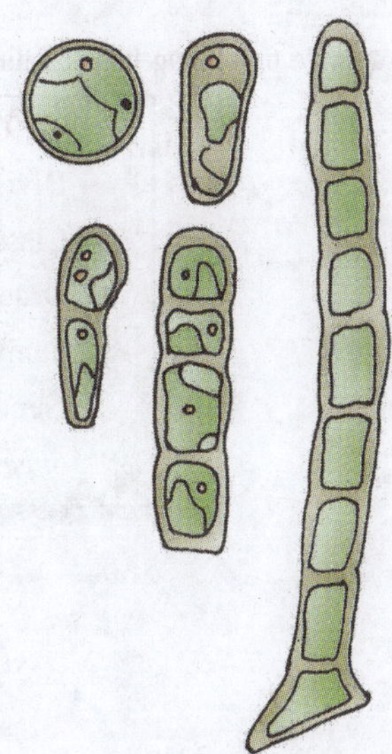

Fig. 7.10 (A-E). *Enteromorpha intestinalis.* Stages in germination of zygote (After Kylin).

Germination of zygote (Fig. 7.10 A-E). By repeated divisions, which are all mitotic and in one direction (C-D), a simple filament consisting of several **diploid** cells is formed (E). It is attached to the substratum by a basal **holdfast**. During further development the cells in the filament divide both by transverse and vertical divisions to produce a solid cylinder two cells thick. The two layers of cells separate from each other from top to bottom to produce a hollow tube one cell thick. It is the diploid **Sporophyte**.

Fig. 7.11 (A-F). *Enteromorpha intest-inalis.* (A) Meiozoospore; (B-F) stages in the germination of meiozoospore to form the thallus.

Sporophyte. Morphologically the diploid sporophyte plant is similar to the haploid parent gametophyte plants. Reaching a certain stage of maturity the diploid protoplasmic contents of some of the cells divide to produce 4-8 quadriflagellate spores. The division is **meiotic**. The spores are thus haploid and are called the **meiozoospores** or **gonozoospores** (Fig. 7.11 A).

The gonozoospores, on germination, produce the haploid gametophyte plant which, in turn, produce the gametes. There is thus a regular alternation between a haploid sexual (gametophyte) generation and a diploid sporophyte generation. The vegetative plants of the two generations (gametophyte and sporophyte) resemble one another morphologically. *Enteromorpha* thus exhibits isomorphic alternation of generations as *Ulva* does (Fig. 7.12).

Economic Importance. It has recently been reported that scientists have extracted a potential anti T.B. drug from *Enteromorpha*. It is still in semi-purified fractions. When purified it may become a source of a new drug to treat tuberculosis. It showed antibacterial activity against nearly twenty disease causing microorganisms.

Wrinkle preventing homeostatin contains natural botanical ingredient by enteromorpha compressa.

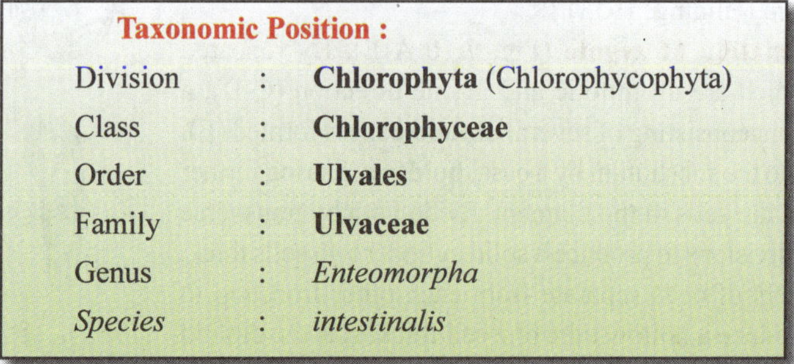

Taxonomic Position :

Division	:	**Chlorophyta** (Chlorophycophyta)
Class	:	**Chlorophyceae**
Order	:	**Ulvales**
Family	:	**Ulvaceae**
Genus	:	*Enteomorpha*
Species	:	*intestinalis*

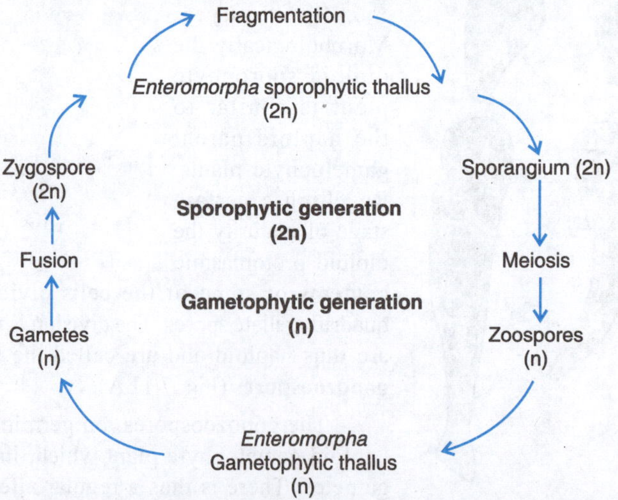

Fig. 7.12. *Enteromorpha.* Graphic life cycle.

QUESTIONS

ESSAY TYPE

1. Give an account of structure and reproduction in *Ulva*.
2. With the help of labelled diagrams, describe the alternation of generation in the members of Ulvales.
3. What do you understand by the term Isomorphic alternation of generation? Explain it with the help of the life cycle of an alga studied by you. (*Bharathiar*, *2000*)
4. Describe the occurrence, thallus structure, reproduction and life cycle in *Enteromorpha*.
5. Describe the characters of the order Ulvales. Give an account of reproduction in a member of the order.

SHORT ANSWER TYPE

6. **Write short notes on:**
   - (*a*) Sea lettuce
   - (*b*) Reproduction in *Ulva*
   - (*c*) Isomorphic alternation of generation
   - (*d*) *Enteromorpha*
   - (*e*) Systematic position of *Ulva*. (*Lucknow, 2001*)
   - (*f*) Thallus of *Ulva* (*Lucknow, 2003*)
   - (*g*) *Ulva*. (*Lucknow, 2007*)
7. What is the botanical name of sea lettuce?
8. Distinguish between *Ulva* and *Enteromorpha*.
9. Write the characteristic features of *Ulva* or *Enteromorpha*.
10. Where does *Ulva* occur in India?
11. Give an account of the modes of reproduction in *Enteromorpha*.
12. Write an account of the thallus structure in *Enteromorpha*.
13. Compare the sexual reproduction in *Ulva* and *Enteromorpha*.
14. Give a comparative account of the cell structure in *Ulothrix* and *Ulva*.
15. Describe the thallus structure in *Ulva*. Write ten sentences on *Ulva*? (*Lucknow, 2005*)
16. With the help of diagrams only, describe the life cycle in *Ulva*.

OBJECTIVE TYPE

17. **Select the correct answer:**

(*i*) While *Ulothrix* is a fresh water alga, *Ulva* is
- (*a*) Marine
- (*b*) Fresh water
- (*c*) Terrestrial
- (*d*) Sometimes fresh water, sometimes marine

(*ii*) The life cycle in *Ulva* shows
- (*a*) Isomorphic alternation of generation
- (*b*) Heteromorphic alternation of generation
- (*c*) Dimorphic alternation of generation
- (*d*) Absence of alternation of generation

(*iii*) Vegetative reproduction in *Enteromorpha* is by
- (*a*) Budding
- (*b*) Fragmentation
- (*c*) Aplanospores
- (*d*) Chlamydospores

(*iv*) The thallus in *Ulva* is
- (*a*) Unicellular

(*b*) Filamentous

(*c*) Coenobium

(*d*) Parenchymatous foliage like

(*v*) Which of the following is commonly known as *sea lettuce*?

(*a*) *Volvox*

(*b*) *Ulva*

(*c*) *Enteromorpha*

(*d*) *Sargassum.*

(*vi*) Adult plant of *Ulva* is

(*a*) Gametophyte

(*b*) Sporophyte

(*c*) Both of the above

(*d*) None of the above

(*vii*) The gametes in *Ulva* are pear-shaped and

(*a*) Uniflagellate

(*b*) Biflagellate

(*c*) Tetraflagellate

(*d*) Aflagellate

(*viii*) Isomorphic alternation of generation means

(*a*) alternation between a large sporophyte and a small gametophyte

(*b*) alternation between a small sporophyte and a large gametophyte

(*c*) alternation between a sporophyte and a gametophyte similar in appearance

(*d*) alternation between plants morphologically similar but physiologically different

(*ix*) Which of the following is of economic importance?

(*a*) *Ulothrix*

(*b*) *Chlamydomonas*

(*c*) *Ulva*

(*d*) *Volvox*

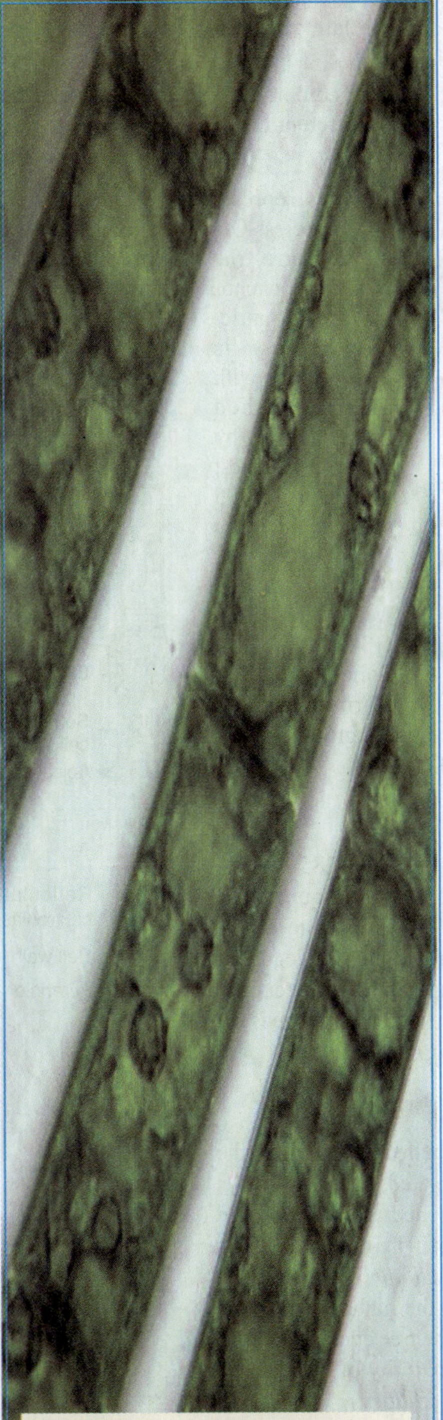
Oedogoniales: *Oedogonium cardacum*

CHAPTER 8

ORDER 5. OEDOGONIALES

- Oedogonium
- Bulbochaete

Oedogoniales are essentially a group of fresh-water green algae which love to grow in quieter situations and usually shun flowing waters. They often occur attached by a special holdfast cell. The **distinctive** features of the order are:

(*i*) The thallus is a branched (*Oedocladium* and *Bulbochaete*) or unbranched filament (*Oedogonium*) made up of uninucleate, cylindrical cells. The nucleus is parietal.

(*ii*) The vegetative cells usually are broader near the anterior end and thus exhibit apical basal polarity.

(*iii*) The chloroplast is a parietal network (close or loose) with the padded portions enclosing pyrenoids.

(*iv*) The method of cell division is unique and by annular splitting of the lateral cell wall, thus forming apical caps and the cells bearing caps are called cap cells.

(*v*) There is a subapical ring of numerous flagella around the anterior end of the reproductive cells (zoospores, androspores and male gametes).

(*vi*) Asexual reproduction is by means of multiflagellate zoospores formed singly per zoosporangium which is an unspecialized cell.

(*vii*) Sexual reproduction is by advanced type of oogamy comprising specialized sex organs (oogonium and antheridium) formed on the same or different filaments.

(*viii*) Some species produce dwarf male plants.

The order includes a single family *Oedogoniaceae*. It is represented by three genera, *Oedocladium, Bulbochaete* and *Oedogonium* with about 400 species. Of these *Oedocladium* is mainly terrestrial in which the thallus is a heterotrichous filament. The other two are aquatic. *Oedogonium* is the only one with an unbranched filament. It is by far the commonest with about 285 species. In India abut 114 species of *Oedogonium* were recorded by Gonzalves and Sonad (1961) in Karnataka State alone. The genus is best known and thus taken as a type of the group. A short account of genus *Bulbochaete* has also been presented.

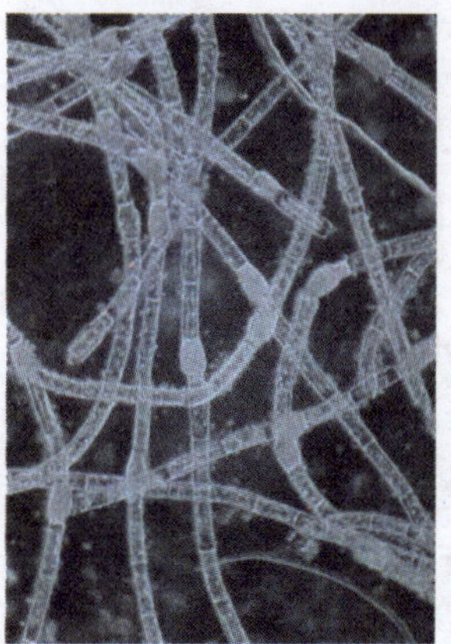

Oedogonium sp.

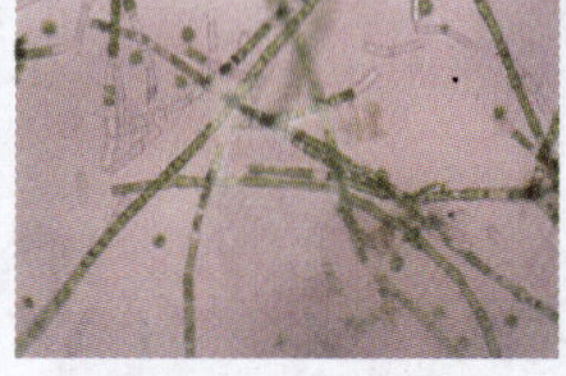

Oedogonium foveolatum.

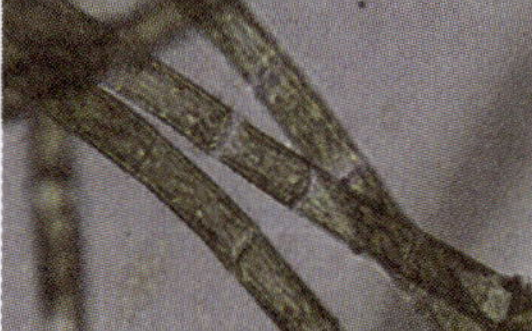

Oedogonium capillare.

OEDOGONIUM

It is a common, submerged aquatic alga comprising about 285 species. It may occur attached to the solid objects like stone or wood in quiet fresh water. Often it grows epiphytically on the larger green algae or upon the leaves, petioles and stems of aquatic angiosperms in fresh-water ponds, tanks, lakes and quiet streams. The mature filaments are free floating but the younger ones are attached. It is less common in the running water. The attaching organ is the basal cell differentiated especially for this purpose. (Fig. 8.1).

Organisation of thallus (Fig. 8.1)

The plant body is a thallus which is a long, unbranched thread called the **filament**. The filament consists of a single row of elongated, cylindrical cells arranged end to end. The filament usually occurs attached at the lower end by means of a basal cell, the **rhizoidal cell** or **holdfast**. The rhizoidal cell is modified to form a more complex holdfast than that of *Ulothrix*. The holdfast is expanded into a flattened disc with outgrowths. In the mature condition, however, the filaments float in yellowish green mats. The free end of the distal cell of the filament is broadly rounded in the case of majority of the species. In a few, it ends in a fine, slender, hairlike process (*O. ciliata*).

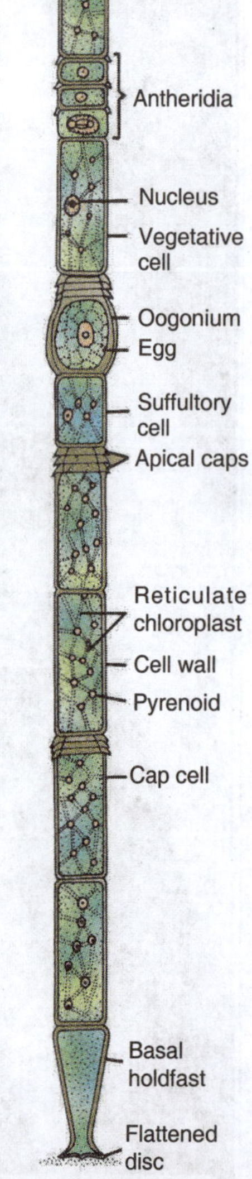

Fig. 8.1. *Oedogonium*. A monoecium filament.

Cell Structure (Fig. 8.2)

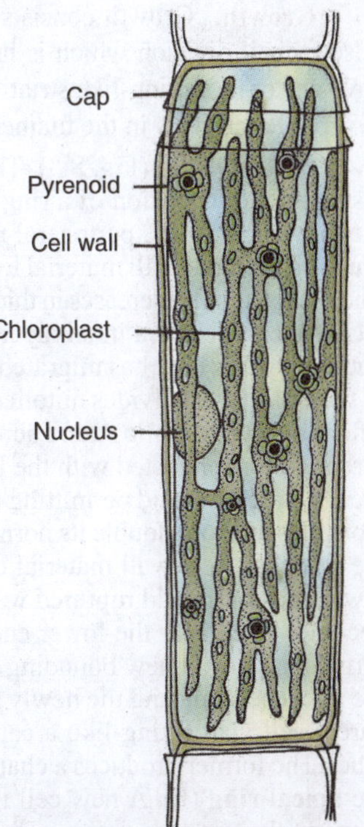

Fig. 8.2. *Oedogonium.* Cell structure (Based on Prings-heim).

The cells are elongated and cylindrical with a more or less dilated upper end in some species. The green vegetative cell consists of a thick or not conspicuously thickened, rigid cell wall enclosing the protoplast. The **cell wall** is differentiated into two concentric layers. The inner layer which is next to the protoplast is **cellulose** in nature. The other layer, according to Tiffany consists of **pectic substance** (pectose). External to the pectic layer in *Oedogonium* is a surface investment of **chitin.** It is often referred to as the third layer. It prevents the dissolving away of the pectic layer so that the filaments feel like wet threads. The use of electron microscopy has, however, revealed the absence of cellulose in the cell wall.

The **protoplast** is differentiated into a thin plasma membrane, cytoplast, a single large reticulate chloroplast with numerous pyrenoids and a parietal nucleus. The cytoplast covered by the plasma membrane is closely adherent to the cellulose layer of the cell wall and forms a lining layer. It encloses a large **central vacuole** containing the **cell sap**. There is a single, large, reticulate parietal **choroplast** with a number of pyrenoids (reminiscent of the Cladophorales) embedded in the cytoplasm. The chloroplast extends from one end of the cell to the other. It is parietal in position and lies in the primordial utricle (lining layer of cytoplasm) which it completely encircles. The chloroplast has the form of a hollow-cylindrical network with narrow or broad or broad sub-parallel meshes. The numerous pyrenoids lie at the intersections of the reticulum. By the use of electron microscope Hoffman (1967) reported the occurrence of microtubules in the chloroplasts of *Oedogonium cardiacum*. Each microtubule consists of two spirally wound subunits. The microtubules furnish support to the large chloroplast. There is a single large parietal **nucleus**. It lies near the middle of the cell embedded in the cytoplasm just within the chloroplast. There is generally no slime around the filament. Certain cells in every filament possess one or more ring-like markings of hemicellulose, the so-called **apical caps**, at their distal ends. Such cells are called the **cap cells**. The presence of cap cells at intervals in the filament of *Oedogonium* is a safe criterion for its recognition from other unbranched green algae. The mode of nutrition is the same as in the other green algae.

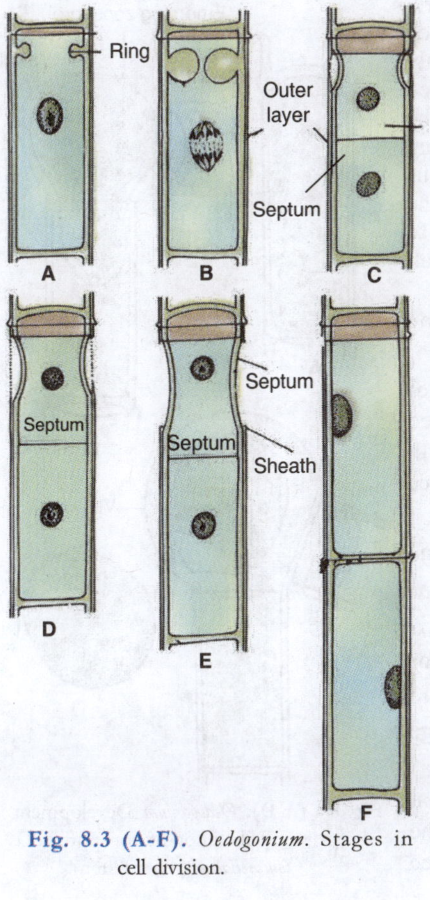

Fig. 8.3 (A-F). *Oedogonium.* Stages in cell division.

Growth. Growth consists in increase in the number of cells in the filament. The new cells arise by cell division which is largely intercalary. Only certain cells in the filament divide. These have one or more ring-like striations at their apical ends and are called the **cap cells**. The latter arise at variable intervals in the filament.

Cell Division (Fig. 8.3). The mode of cell division in *Oedogonium* is peculiar and unique. It results in the formation of a ring-like scar near the anterior end of the cell. The process starts with the movement of the peripheral nucleus to the centre of the cell and the appearance of a thickened transverse ring of wall material towards the upper end of the cell on the inner face of lateral wall (A). The ring gradually increases in thickness and becomes grooved. In an optical section the groove appears a U-shaped thickened inner layer of the wall with the open part covered by the outer layer (B). The nucleus by this time has migrated upwards. It lies at about one-third the distance from the upper end of the cell. Here it divides mitotically (B). The division of the nucleus is followed by the formation of a floating Cytoplasmic strand across the vacuole between the two daughter nuclei(C). For sometime it remains unconnected with the lateral wall. The outer and the middle layers external to the groove then rupture all round permitting the thickened portion to be stretched out (C). Consequently the cell elongates to about double its normal length with the distal half having a new membrane formed from the stretched new wall material of the thickened ring. It is intercalated between the upper and the lower ends of the old ruptured wall (D). At the same time the septum is pushed upwards and finally becomes fixed near the lower end of the intercalated membrane (E). The upper daughter cell thus formed has now a new bounding wall consisting mainly of the intercalated membrane formed from the thickened ring and the newly synthesised piece (F). There is, however, a portion of the ruptured parent cell wall fitting like a cap at its upper end and a portion forming a bottom sheath at the other. The former produces a characteristic ring-like mark, the **apical ring** (F). A new cell is thus always interposed between the two old portions of the parent cell wall. Only the cell possessing the apical ring divides again. It is called the **cap cell**. As successive divisions always occur at the same place, a number of apical rings develop there giving a characteristic striated appearance to the cap cells. Thus the number of apical rings the cap cell contains denotes the number of divisions the cell has undergone. According to Freund (1928). the deficiency of nutritive salts stops cell division.

Asexual Reproduction. It takes place by the vegetative method of fragmentation and sporulation.

(*a*) *Fragmentation*. It consists of the breaking of the filament into small segments of living cells called **fragments**. Each fragment by cell division and growth develops into a new filament. Fragmentation may take place by any of the following methods :

(*i*) Dying out of some cells here and there in the filament.

(*ii*) Through accidental breaking.

(*iii*) Through the formation of zoospores or gametes here and there in the filament.

(*b*) *Sporulation*. It takes place by means of large zoospores and sometimes by **akinetes**.

(*i*) Zoospore formation. (Fig. 8.4). It is the normal and most effective method of asexual reproduction. The formation of zoospores is said to depend on the presence

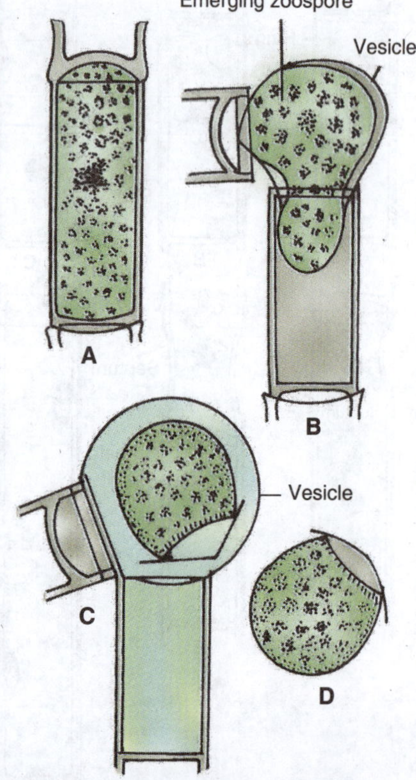

Fig. 8.4 (A-D). *Oedogonium*. Development and liberation of zoospore in *O. concatenatum* (After Hiren).

of a certain amount of free carbon dioxide in the surrounding water (Gussewa, 1931). Any cap cell of the filament usually the recently divided one may become a **zoosporangium** (A). Usually the cells at the terminus of the filament function as zoosporangia. Generally only a few cap cells in the filament produce zoospores.The zoospores are produced singly and in the cells containing abundant food reserves. In the formation of the zoospore the entire protoplast of the zoosporangium withdraws somewhat from the wall. The nucleus moves towards one side of the protoplast. The protoplast then rounds up. At the same time a colourless semicircular area appears on one side adjacent to the nucleus (A). This is caused by the receding of the chloroplast from this end of the protoplast (A). A single or double row of blepharoplast granules then appears at the base of the hyaline area. The basal granules are connected by fibrous strands to form a complete circular ring. From each granule arises a single flagellum. In this way a ring of flagella is formed around the base of the colourless beak-like area of protoplast. It is the anterior pole. With the formation of zoospore the cell wall ruptures transversely in the region of the apical cap (B). The two halves gape apart. The mature zoospore, surrounded by a delicate mucilaginous vesicle, slips out through the aperture (C). The vesicle soon dissolves allowing the typical *Oedogonium* swarmer to escape and swim about (D). The liberated zoospore (D) is a deep green spherical or pear-shaped structure. It has a ring of short flagella at the base of a colourless, beak-like forward end. This kind of flagellation is called **stephanokont**. The zoospore possesses an eye spot, a chloroplast and numerous contractile vacuoles near the periphery.

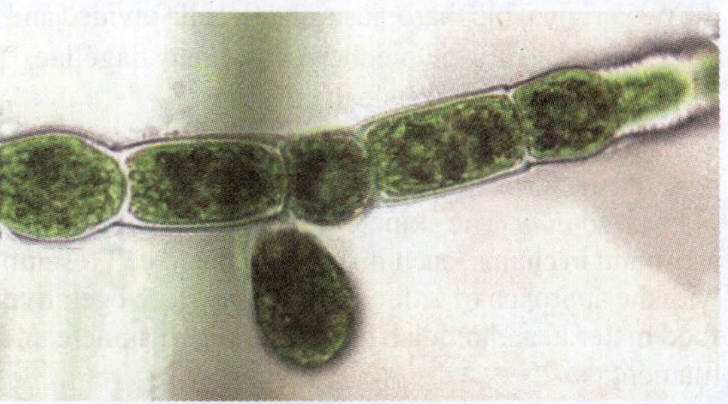

Zoospore Settled on an *Oedogonium Filament.*

Germination of zoospore (Fig. 8.5). The liberated zoospore remains motile for about an hour. Finally it settles down on some solid object with the colourless, flagellar (anterior) end downwards (A). In this state it withdraws its flagella and secretes a cell wall which lacks the superficial chitinous material (B). The colourless anterior end of the quiescent zoospore develops into a simple or branched *holdfast.* The smooth surface of the substratum induces a simple holdfast and the rough one induces a branched holdfast (C). The development of the one-celled, sessile germling, depending on the species, takes place in the following two ways :

(*i*) In most of the species the one-celled germling divides transversely by an apical ring (normal method of cell division). The basal cell cut off by the first division remains colourless and does not divide again. It persists as the attaching organ (D). The upper cell by the normal method of division and redivision of its daughter cells forms the new filament (E).

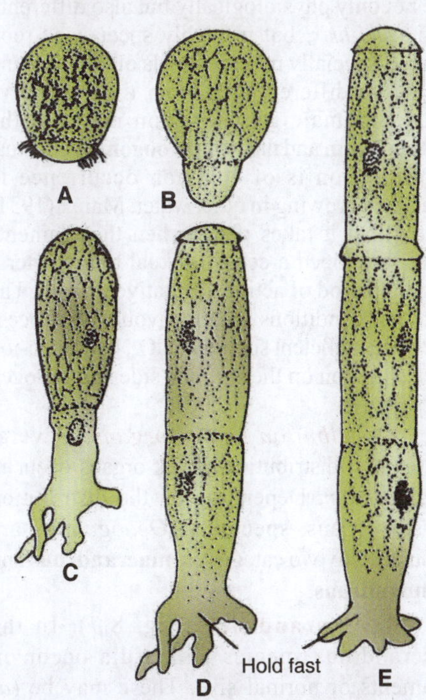

Fig. 8.5 (A-E). *Oedogonium.* Stages in the germination of a zoospore to form a filament.

(*ii*) According to Scherffel (1901) in some species, the quiescent zoospore flattens to become hemispherical (Fig. 8.6). From the free convex surface of this one-celled hemispherical germling arises a cylindrical outgrowth. At the point of origin, the cell wall of the hemispherical basal cell ruptures and becomes turned to one side. A septum appears at the juncture of the cylindrical outgrowth and the hemisphere when the former has reached a certain length. The cylindrical outgrowth subsequently grows into a new filament by the normal method of cell division described above.

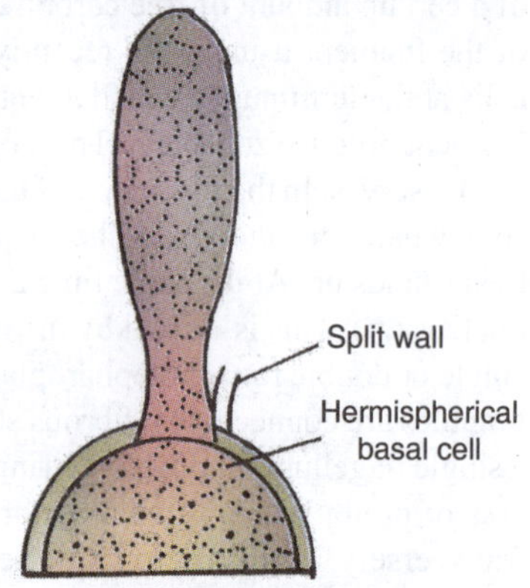

Fig. 8.6. *Oedogonium rufescens.* Germling with a hemispherical basal cell (after Scherffel).

Two views have been put forth to explain the origin of the stephanokont zoospore of *Oedogonium*. According to one, the group arose directly from a flagellate ancestor with a ring of flagellae. If this view is held valid, the Oedogoniales have no relationship with other green algae. The second view holds that the two original blepharoplasts and flagella divided and redivided resulting in a ring of blepharoplasts and flagellae. This view emphasizes the Ulotrichalean affinities.

(*ii*) Akinete formation. Wille (1883) and Handa (1928) reported the formation of resting cells (akinetes) in certain species of *Oedogonium*. The akinetes are thick-walled, reddish brown, more or less rounded structures. They are found in chains, each inside an inflated cell resembling an oogonium. The akinetes are developed with the approach of unfaourable period for vegetative growth and contain abundant starch as reserve food material and reddish orange oil. Each akinete under suitable conditions directly produces a new filament.

Sexual Reproduction. *Oedogonium* shows marked advance over *Ulothrix* in the method of sexual reproduction. It is distinctly an advanced type of oogamy. The sexual cells or the gametes are not only physiologically but also different structurally. They are produced not in vegetative cells as in *Ulothrix,* but in highly specialised reproductive organs, the **gametangia**. The latter are, of course, specially modified cells of the filament. They are differentiated from the vegetative cells. The male gametangium is called the **antheridium** and the female **oogonium**. Sexual reproduction is of common occurrence in filaments growing in quiet water. Mainx (1931) stated that it takes place when the filaments have developed a certain sexual tonus, after a certain period of active vegetative growth. The external conditions which favour the process are light, sufficient supply of CO_2, hydrogen-ion concentration on the alkaline side and nitrogen deficiency.

Distribution of sex organs. Several patterns of distribution of sex organs occur in *Oedogonium*. Depending on the distribution of sex organs, species of *Oedogonium* are grouped into two categories, **macrandrous** and **nannandrous**.

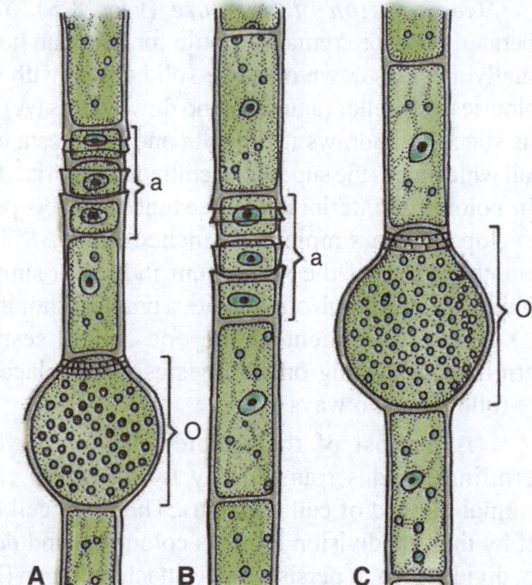

Fig. 8.7 (A-C). *Oedogonium.* Distribution of sex organs in macrandrous sp. A, Macrandrous monoecious with the antheridia (*a*) and oogonia (*o*) on the same filament ; B-C, Macrandrous dioecious with antheridia (*a*) on filament B and oogonia on filament *C.*

1. Macrandrous (Fig. 8.7). In the macrandrous species antheridia occur on filaments of normal size. These may be (*a*) *monoecius* or (*b*) *dioecious*.

(*a*)*Macrandrous monoecius* (Fig. 8.7, A.). The antheridia and oogonia in monoecious macrandrous species occur on the same filament which is thus, **bisexual** (*O. nodulosum, O. fragile* and *O. hirnii*).

(*b*)*Macrandrous dioecious* (Fig. 8.7, B-C). The species in which the two kinds of sex organs occur on separate filaments are called **macrandrous dioecious** (*O. crassum*). The filaments are thus unisexual. There are distinct male and female filaments of normal size. They are similar. In the vegetative state it is not possible to recognize which is which. Physiologically of course they are different. One bears antheridia (B) and the other oogonia(C). The common examples of this category are *O. aquaticum, O. cardiacum* and *O. gracilius*.

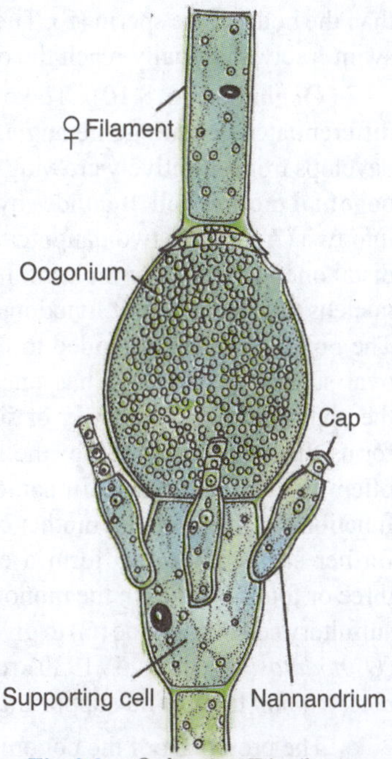

2. Nannandrous (Fig. 8.8). Some dioecious species of *Oedogonium* exhibit **dimorphism**. The male and the female filaments show distinct morphological differences. In them the antheridia (*a*) are produced by special, much reduced male filaments called the **dwarf male** plants or **nannandria** (*O. concatenatum*). The latter grow epiphytically attached to the female filaments. The latter are of normal type or usual type.

Fig. 8.8. *Oedogonium.* Distribution of sex organs in a nannandrous sp.

Antheridia (Fig. 8.9). The antheridia are flat, short, cylindrical, disc-like cells or segments of the filament. They lie in a row or series consisting of a variable number from 2 to 40 (D). The contents of each antheridium commonly develop into two sperms rarely into one. The antheridia are either terminal or intercalary in position. The antheridia in the macrandrous species are developed by the rapid and repeated transverse divisions of a vegetative cell. The latter is called the **antheridial mother cell**. It is one of the cap cells (A). It divides into two unequal cells (B), the upper much smaller **antheridium** (*a*) and the lower larger **sister cell** (*b*) divides again (C). The process is repeated a number of times os that a row of short, compartment-like cells, which are antheridia, are formed (D). The protoplast within each antheridium commonly divides by a transverse or vertical wall to form two sperms. The sperms, in the former case lie superimposed and in the latter case side by side (D) within the antheridium. Rarely they are produced singly. They escape in the same manner as the zoospores by the transverse rupture of the cell wall and are freed into a thin vesicle (E) which soon dissolves. The liberated sperms are pale green, yellowish green spherical bodies each with an apical ring of short flagella at the base of the colourless, beak-like anterior end. They resemble the zoospores in the type of flagellation, morphology and method of liberation but are smaller in size and have fewer flagella. They contain less chlorophyll. The flagella in some species (*O. crassum* and *O. kurzii*) are longer

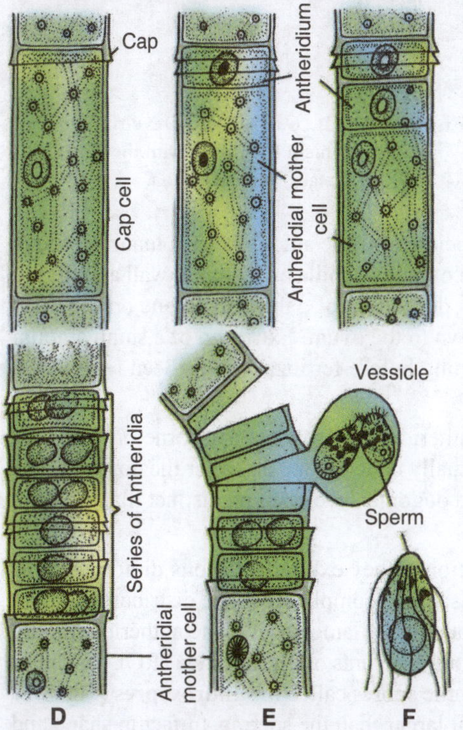

Fig. 8.9 (A-F). *Oedogonium.* Stages in the development of antheridia and liberation of antherozooids.

than the body of the sperm (F). The liberated sperms swim freely and finally reach the oogonia.

Oogonia (Fig. 8.10). The oogonia are highly differentiated female gametangia. Each oogonium develops from an actively growing cap cell called the **oogonial mother cell**. It divides by a transverse wall into two (A). Of the two daughter cells the upper or distal one is richer in cytoplasm. It contains a larger nucleus than the lower. It functions as an **oogonium**. The oogonium gets distended to form a rounded or oval structure. It always has one or more caps at the upper end (A). The lower or sister daughter cell forms the **supporting cell** or the **suffultory cell**. It often remains undivided. In some species it again functions as an oogonium mother cell and undergoes further segmentation to form a chain of two (B), three or four oogonia. In the monoecious species the suffultory cell may divide (C) to give rise to antheridia (*O. nodulosum*). Ohashi (1930) reported that in *O. americanum* there is no supporting cell.

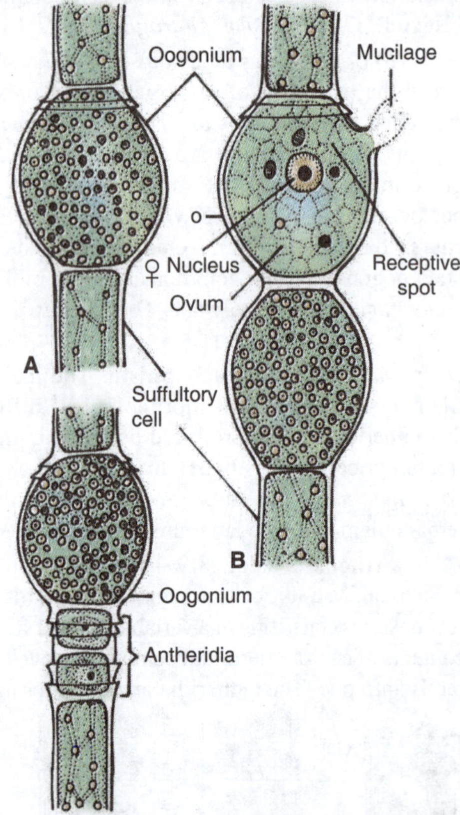

Fig. 8.10 (A-C). *Oedogonium*. Development of oogonium (A and B) with the upper one in B ready for fertilization.

The protoplast of the oogonium stores reserve food materials and forms a single egg. The egg or oosphere has a centrally located nucleus. Owing to the presence of chlorophyll the ovum is green. It is non-motile and is retained within the oogonium. In a mature ovum (B,*o*) the nucleus migrates to the periphery. Prior to fertilisation the egg protoplast slightly recedes from the oogonial wall to form a small clear patch, the **receptive spot,** external to the nucleus. The oogonial wall develops a small pore (poriferous species) or a transverse slit (operculate species) near the anterior end at maturity above the receptive spot. The pore is formed by gelatinisation of a tiny papilla on oogonial wall and the slit by a transverse split in the wall. In either case there is deposition of a thin membrane on the inner side of the exit. It forms a sort of a conduit leading down to the ovum. Extrusion of a small amount of a gelatinous substance (mucilage) through the opening before fertilization has been reported in some species (B, *o*).

In the macrandrous monoecious species, the antheridia and oogonia are borne on the same filament. To ensure cross fertilization the antheridia usually develop one day after the oogonia. The macrandrous dioecious species have the antheridia and oogonia developed on distinct filaments but the filaments are of the normal type.

The **nannandrous** species (Fig. 8.11) are dioecious. They exhibit a curious dimorphism of sexual plants. The sexual processes are more specialised and complicated. The distribution of sex organs is peculiar. The oogonia are produced on normal, large filaments (A). The antheridia (a) are produced by special, very small filaments, the **"dwarf male"** plants or **nannandria** (RC). The latter are produced by the germination of peculiar type of motile spores called the **androsopres** (Fig. 8.11 B). The androspores are smaller than the zoospores but larger than the sperms. In fact in shape and structure they are the small editions of the zoospores. Sometimes they are yellowish. The androspores are produced within cells called **androsporangia**. The latter are formed by the repeated transverse division (Fig. 8.11, B) of any vegetative cell of a large filament. If the androsporangia are borne in the oogonial filament the species is known as **gynandrosporous** (Fig. 8.11 C). *O. concatenatum* is an example. The oogonial filament bearing the nannadria is called **gynandrosporous nannandrous filament** (C). The species in which androsporangia are borne in a separate filament is named as **idioandrosporous**. *O. iyengarii, O. confertum* and *O. setigerum* are examples of this type. In this case the filament bearing the androsporangia is called **androsporangiate** (B) and the female filament bearing the nannandria is called the **idioadrosporus nannandrous filament** (A).

The androsporangia (B) are flat discoid cells. They are formed in the same manner but are rather larger. The androspores are formed singly within the androsporangia. The androspores are motile and are provided with a subpolar crown of flagella (Fig. 8.11, B). When liberated the androspore is enveloped in a vesicle which soon vansihes. The androspore swims in all directions till it reaches a female filament and becomes affixed either on the wall of the oogonium as in *O. ciliatum* (*D*) or on the supporting cell (*O. concatenatum* A). There it surrounds itself, with a cell wall. The attached androspore (L) then germinates to form a minutes dwarf male plant or filament called the **nannandrium** (RC). The nannandrium consists of a rhizoid-like elongated or attaching or stalk cell. This one-celled germling (rhizoidal holdfast or attaching cell) cuts off one or more flat cells at its tip. These are **antheridia** (*a*). The protoplast of the antheridium often divides to form two sperms. In a few species the attaching cell functions directly as an antheridium and produces two sperms from its contents. Iyengar (1951) however reported that the antheridium of the dwarf male produces a single sperm. The sperms are liberated either by the disorganisation of the antheridial cell or by the separation of a lid at the top (D). The development of oogonia in the nannandrous species is the same as in the macrandrous dioecious species.

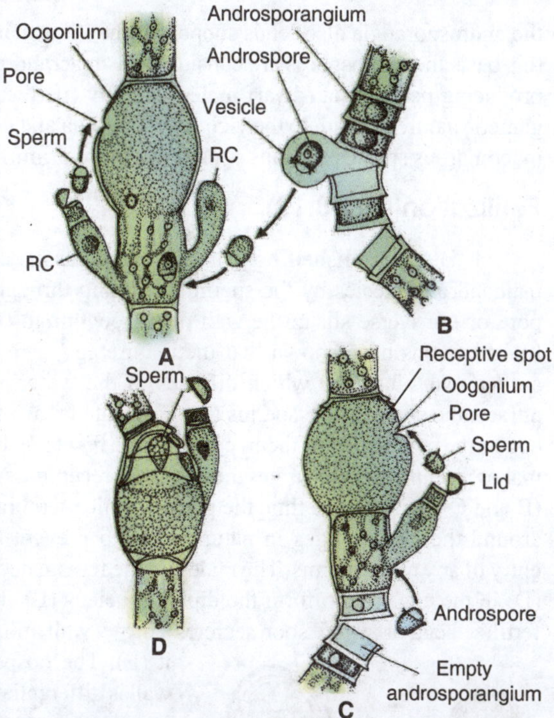

Fig. 8.11 (A-D). *Oedogonium*. Sexual reproduction in nannandrous species. A, Idioandrosporous nannandrous filament; B, Androsporangiate filament; C, Gynanndrosporous nannandrous filament; D, Oogonial filament of *O. ciliatum* with the oogonium during fertilization and nannandrium attached to it (D, after Pringsheim).

Origin of the Nannandrous species. The production of dwarf males is a striking feature of *Oedogonium*. The consensus of opinion supports the view that the nannandrous species have originated by specialization from the macrandrous species. The following facts support this view :

(*i*) Occurrence of smaller and narrower male filaments than the female in some of the isomorphic dioecious (macrandrous dioecious) species.

(*ii*) Precocious development of antheridia on the young filaments of the dioecious species produced by the germination of zoospores.

(*iii*) Similar oogonial development both in the nannandrous and macrandrous species.

The cases enumerated above indicate that the nannandrous habit in the genus has arisen by reduction in size of the male filaments of the isomorphic dioecious (macrandrous dioecious) species of *Oedogonium*. The resemblance both in structure and development between the antheridia and the androsporangia also lends support to the view that the latter have evolved from the former. On this basis the androspores are considered as macrandrous sperms that have acquired the capacity of producing nannandria (dwarf males) directly (parthenogenetically). They, however, retained their gametic nature to swim to the vicinity of oogonia and germinate there. On the basis of this hypothesis macrandrous species are thus considered more primitive.

Ferilization (Fig. 8.12)

It is accomplished both in the macrandrous and nannandrous species by the sperm swimming through a pore or transverse slit in the wall of the oogonium (B). Investigations have shown that the mature egg secretes a chemical substance which diffuses in the water and attracts the sperms. The sperms thus fascinated reach the egg or the ovum. One of them, probably the first to arrive, makes its entry and penetrates the egg at the receptive spot (B and C). At the same time the protoplasmic membrane around the egg changes in nature so as to prevent the entry of any more sperms. The male and the female nuclei (D) in the egg fuse to form the diploid nucleus (E). The fertilized egg or zygote soon secretes a heavy wall around it (E). The oospore wall is differentiated into three layers but in some species it consists of two layers. Frequently the layer external to the innermost is ornamented. It has pits, reticulations or costae, sometimes it is smooth. The oospore passes into the resting period during which its contents undergo a change. The contents change colour from green to brown or red. A large quantity of reddish oil accumulates in the cytoplasm. Eventually the oospore is liberated by the decay of the oogonial wall and rests on the mud at the bottom of the pond where it enters upon a further period of rest (F). It has a diploid nucleus formed by the fusion of the two haploid nuclei. Chilling is reported to hasten germination. In some species according to Fritsch the zygote germinates without undergoing any resting period. Parthenogenesis has been reported by some investigators. The unfertilized egg develops into a parthenopsore with a zygote-like wall.

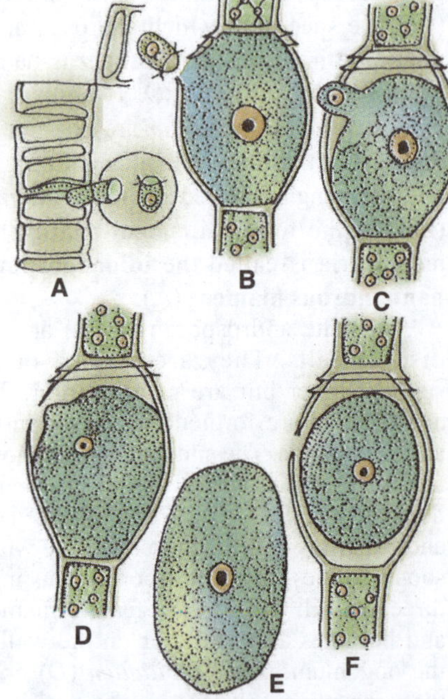

Fig. 8.12 (A-F). *Oedogonium*. Stages in fertilisation; (A-E); (F) liberated oospore (B-E) (based on Jaranyi).

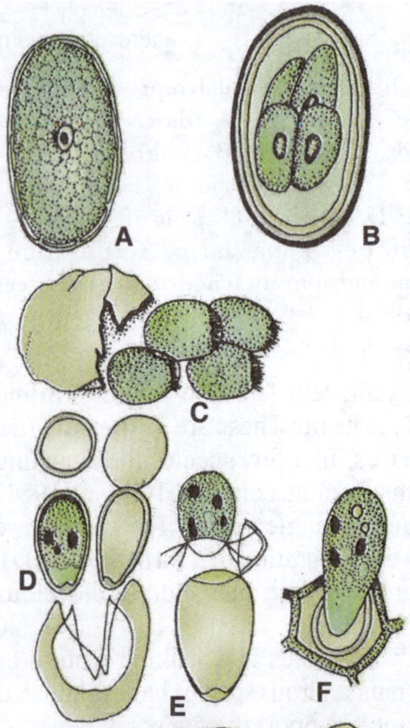

Fig. 8.13 (A-I). *Oedogonium*. Germination of oospore (D and E after Mainx; F-I based on Juranyi).

Germination of Oospore (Fig. 8.13)

According to Mainx, oospores (A) subjected to frost conditions exhibit optimum germination. Prior to germination the diploid oospore nucleus undergoes **zygotic meiosis** to form four haploid nuclei (Hoffman, 1965). The protoplast loses its red colour and turns green. The haploid nuclei are organised into four uninucleate daughter protoplasts by cleavage of the oospore protoplast (B). Soon after each of the haploid daughter protoplasts furnishes itself with a crown of flagella to become a motile spore resembling the zoospore of the asexual stage. It may be called a **meiozoospore**. According to Smith the oospore wall ruptures to liberate the mature, motile meiospores or meiozoospores (C). They are at first surrounded by a delicate vesicle. The latter soon disappears. Fritsch however, reported that in some species, the oospore wall ruptures and the naked protoplast emerges (F). It is liberated into a vesicle (G) where it divides (H). The daughter protoplasts soon develop flagella to become meiozoospores (I). The meiospores escape through an aperture in the vesicle. Under certain conditions the daughter protoplasts do not develop flagella. Each secretes a thin wall to become a meioaplanospore. These meioaplanospores are set free by the gelatinisation of the vesicle (D). Each meioaplanospore after some days liberates a swarmer (E). These swarmers give rise to new plants.

The liberated meiozoospore swims about for a while and then settles down on some solid object to germinate. It gives rise to the haploid filament of *Oedogonium* in precisely the same way as does the zoospore produced by the zoosporangium. Mainx (1931) stated that in *O. plagiosomum* var. *gracilius* two meiozoospores give rise to male filaments and the remaining two to the female filaments. Fritsch is of the opinion that this kind of segregation may be general for the dioecious species.

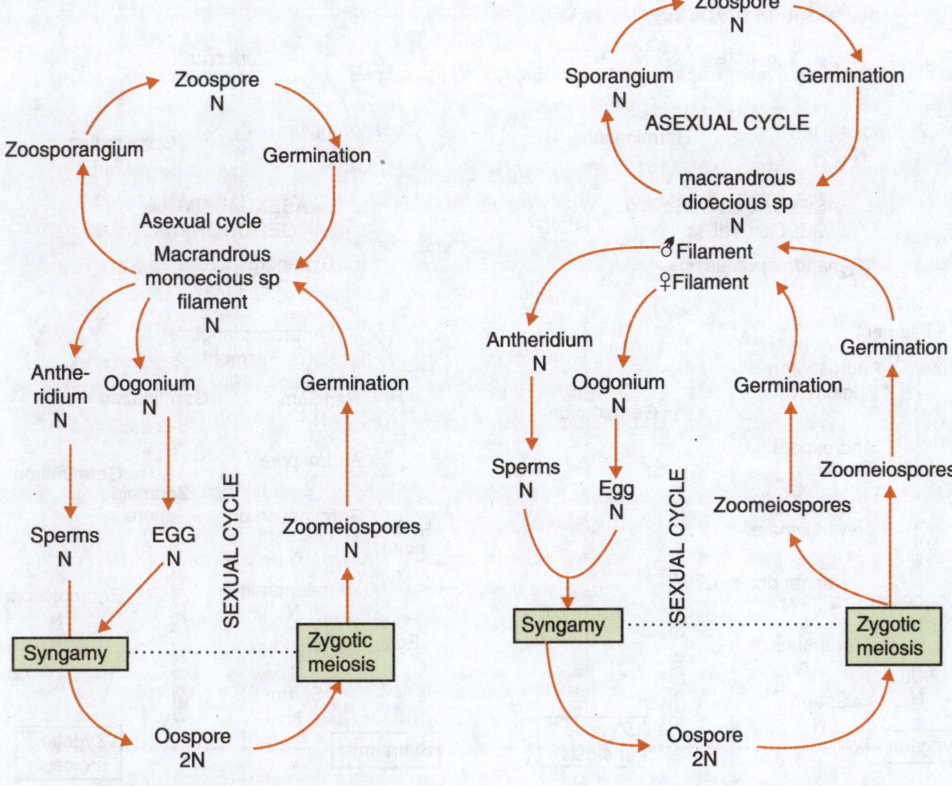

Fig. 8.14. *Oedogonium.* Word diagram of the life cycle of Macrandrous monoecious species.

Fig. 8.15. *Oedogonium.* Word diagram of the life cycle of Macrandrous dioecious species.

Salient Features.

1. The thallus, no doubt, is an unbranched filament as in *Ulothrix* but it shows more marked specialisation of cells accompanied by greater interdependence upon each other.
2. In addition to the green **vegetative cells** which form the majority and are concerned with nutrition there is the basal **rhizoidal** cell which is non-green and more specialised to function as an attaching organ.
3. There are **specialized cap cells** at variable intervals in the filament which carry on cell division and in addition may function as zoosporangia.
4. The **gametangia** are also highly specialised cells. They are not ordinary vegetative cells but arise by the specialisation of the vegetative cells.
5. The haploid filament reproduces by **fragmentation**, sporulation and **sexual reproduction.**
6. Asexual reproduction by sporulation takes place by means of **multiflagellate zoospores** and **akinetes**.
7. The zoospores are produced singly per zoosporangium and are spherical or pear-shaped objects furnished with a crown of flagella at the anterior end at the base of a colourless beak.
8. Sexual reproduction is of an advanced type of oogamy and the sex organs arise by the specialisation of the vegetative cells.

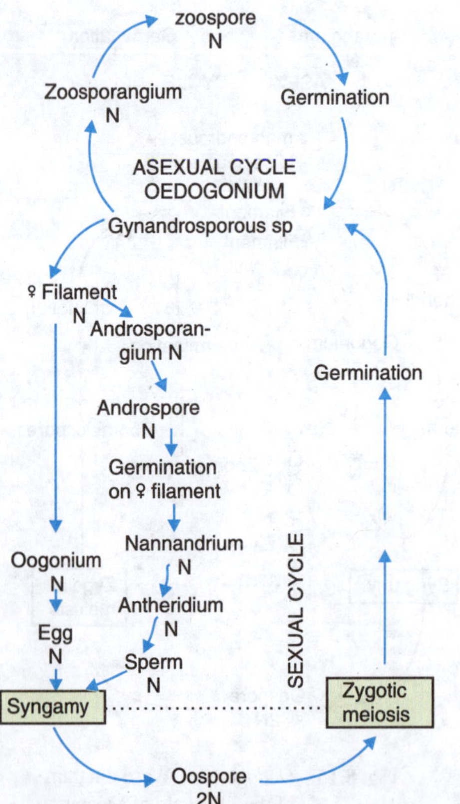

Fig. 8.16. *Oedogonium.* Word diagram of the life cycle of gynandrosporus nannandrous species

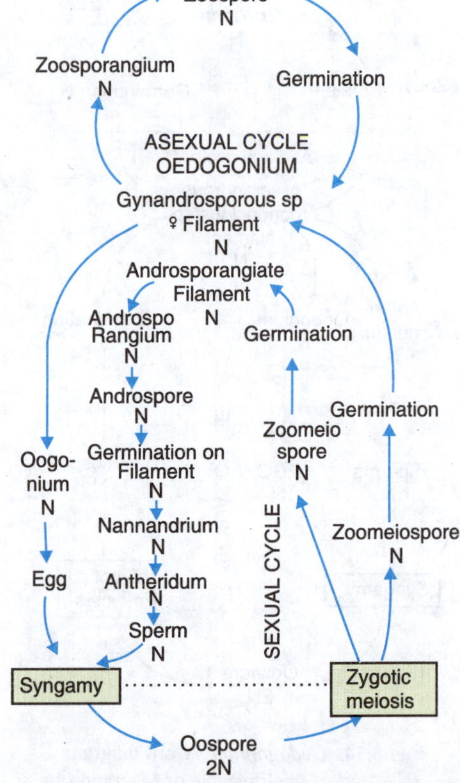

Fig. 8.17. *Oedogonium.* Word diagram of the life cycle of idioandrosporous species.

9. The male sex organs which are called the **antheridia** are short, cylindrical, disc-shaped cells formed in a row or series varying in number from 2-40.
10. The female sex organs called the **oogonia** are large, rounded or oval bodies arranged usually singly sometimes in chains of two or three in the filament.
11. The female gametes or the oospheres are produced singly within the oogonia. They are retained and fertilization takes place *in situ*. The oosphere has a definite **receptive spot** which marks the point of entry of the sperm.
12. The male gamete or the sperm is motile and multiflagellate. In shape and structure it resembles the zoospore but is smaller in size.
13. The zygote or the oospore is a heavy-walled resting structure. It enables the plant to tide over the period unfavourable for growth.
14. Upon germination its diploid protoplast undergoes **zygotic meiosis** to form four haploid, motile **meiospores** each of which on liberation produces the haploid filament.
15. On the basis of the distribution of antheridia and the size of the male filament the different species of *Oedogonium* are divided into the following three types:
    (*i*) **Macrandrous monoecious**. The antheridia and oogonia are produced on the same filament which is of normal size.
    (*ii*) **Macrandrous dioecious**. The antheridia and oogonia appear on different filaments which are morphologically alike and of normal size.
    (*iii*) **Nannandrous**. The antheridia are produced on the male filament which is much reduced and called the **nannandrium**. It is attached to the female filament. The female filaments bearing oogonia are normal.
16. A curious dimorphism and distribution of sexual plants is thus exhibited by the nannandrous species. All other species of *Oedogonium* which do not produce nannandria (dwarf male plants) are called **macrandrous**.
17. The nannandria or dwarf male plants are produced by the germination of another kind of peculiar motile spores which in shape and structure are miniature editions of the zoospores. They are called the **androspores.**
18. The androspores are produced singly in flat, cylindrical discoid cells resembling antheridia but are larger in size. These are the **androsporangia**.
19. On the basis of the distribution of androsporangia the nannandrous species of *Oedogonium* are of the following two types :-
    (*i*) **Gynandrosporous** (Fig. 8.11 C). The androsporangia are developed and consequently the androspores are produced in the same filament that bears oogonia.
    (*ii*) **Idioandrosporous** (Fig. 8.11 A-B). The androsporangia and so the androspores are produced on filaments which do not bear oogonia. Thus, the androsporangia bearing filaments and the oogonial filaments are distinct.
20. The androspore in either case swims about and finally settles on the wall of the oogonium or supporting cell and germinates to produce the male dwarf plant or the nannandrium.
21. The nannandrium is a one-celled germling. It consists of a basal attaching rhizoidal cell bearing terminally one or two discoid antheridia. In some species the basal cell directly functions as an antheridium.
22. Each antheridium usually produces two sperms, rarely one.

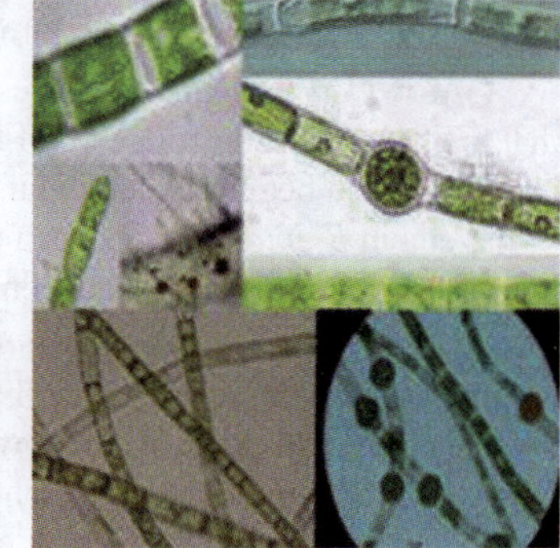

Oedogonium.

BULBOCHAETE

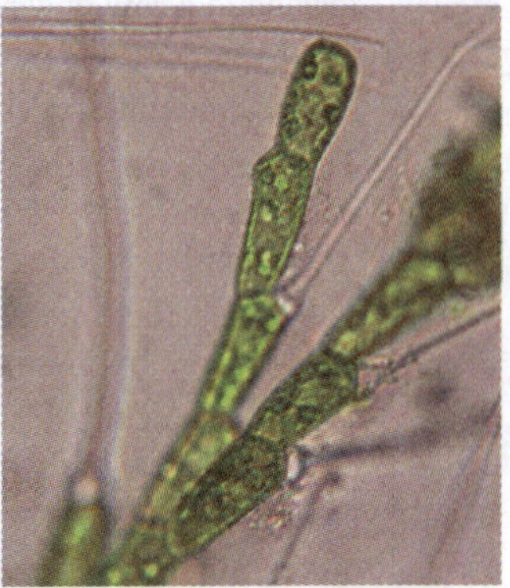
Bulbochaete sp.

Occurrence. *Bulbochaete* is usually epiphytic on the stems and leaves of aquatic plants and on other fresh water algae. The blades of grasses and sedges have often been found with the adult or young alga located on them. Some species of *Bulbochaete* have been recorded as epiphytes on species of *Spirogyra.*

Bulbochaete does not occur as frequently as *Oedogonium*, due to its small size and epiphytic habit. Over 120 species of the alga has been recorded, the most common ones are *B. intermedia* and *B. gigantea.*

Organisation of thallus

The thallus of the alga is made up of unilaterally branched filaments, the cells are generally somewhat cylindric with a plane base. The cells display a tendency to be broader at their upper end which is somewhat dome-shaped, being composed of two sloping surfaces forming an angle with each other. One of the surfaces bears a seta above while the other subtends the next upper cell of the axis. Setae are mostly colourless and may be upto 9 mm. in length. Setae are hollow and bulbous at the base, it is this characteristic that give the genus its name.

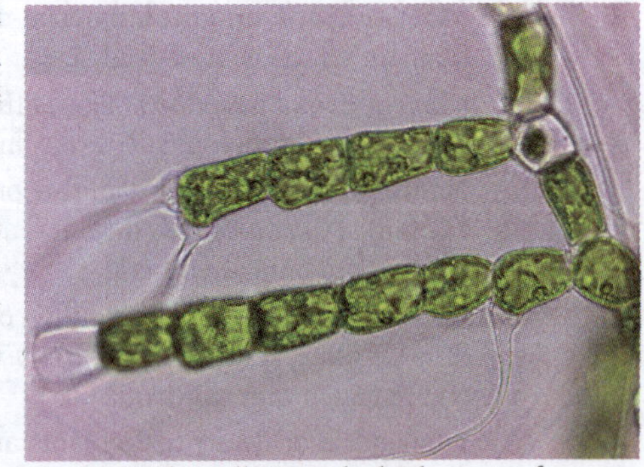
Unilaterally Branched Filaments of *Bulbochaete*.

The basal cell of the filament is usually oblanceolate, tapering towards its lower end to form the holdfast.

Cell Structure. The cell has a rigid cell wall-smooth or dotted, with punctae or granules. Inside the cell wall is the protoplasm. The chloroplast is in the form of a peripheral, reticulate cylinder extending from one end of the cell to the other. Pyrenoids are usually scattered in the chloroplast usually at the intersections of the reticulum. As many as 30 pyrenoids may be seen in the chloroplast. There is a single haploid nucleus.

The structure of the cell is generally similar to that of *Oedogonium.*

Reproduction. *Bulbochaete* reproduces by asexual and sexual methods.

Asexual Reproduction. Asexual reproduction takes place through zoospores, produced in the same way as in *Oedogonium*. Generally the cells producing zoospores are older cells and provide with caps. Cook (1962) observed that in *Bulbochaete hiloensis* a series of cells in a branch may show progressive stages in zoospore formation.

As in *Oedogonium,* zoospores in *Bulbochaete* occur singly. It has been reported by Cook (1962) that the flagella of the zoospore of *Bulbochaete hiloensis* are formed sometime before the zoospore is released and begin to beat about while the zoospore is still enclosed within the cell. The number of flagella in *B. hiloensis* has been estimated to the about 25 (Cook, 1962) while in *Oedogonium* in number of flagella ranges between 100 and 120.

On germination, the zoospore divides, the first transverse division after a secretion of a wall around itself is simple and cuts off a small colourless cell with rounded apex, without the formation of a ring. Next a rupture occurs towards the upper end of this cell and its contents enclosed within a new membrane grow out into a long hollow bristle-the seta with bulbous base. As it grows, the seta pushes aside the ruptured portion of the original wall which remains attached laterally to the basal cell for some time, like a small lid like appendage, further divisions occur in only the basal cell as usual. As a matter fact the whole primary filament is formed by the activity of this cell. The result of such exclusively basal growth is that the older cells are pushed upwards, the oldest cell being the terminal cell, and the youngest cell being the one next to the basal cell. Therefore, each cell of the filament will rarely have more than one apical cap. Generally the terminal cell has two setae, one of which is sheathed at the base which the other is without a sheath. Every other cell of the plant except the basal cell has a single seta. The setae are either fragile or they may persist.

Branches are also produced in a regular sequence.

Sexual Reproduction. Sexual reproduction in *Bulbochaete* is oogamous.

The *female reproductive organ* is oogonium. Any cell except the basal cell may act as oogonial mother cell which divides into two cells with ring formation, and the septum between the cells is fixed in the middle of the lower cylindric part. Thus, the membrane of the upper cell consists of the cap at the top, the stretched thickening ring and a small portion of the short cylindrical sheath. The stretched portion then bulges and gets converted into an *oogonium*. This oogonium also called primary oogonium by Pringsheim (1858) contains a good amount of the contents of the mother cell chiefly starch. A second thickening ring now forms near the centre of the primary oogonium. Nuclear division occurs followed by a transverse rent in the wall and a stretching of the ring. The transverse wall developed during this second division takes a place on a level with the top of the original cylindric sheath within which to saffultory cells are now present. In this way the wall of oogonium consists of usually three pieces.

The two suffultory cells may be equal or unequal in length. The oogonia are generally globose, depressed globose or ellipsoid and are borne singly.

The *male reproductive organ* is antheridium and is formed by the division of autheridial mother cell. The antheridium may be erect or patent. The antherdial protoplast divides to form two spermatozoids.

The spermatozoids resemble the zoospores, but the former are light green or yellow in colour with little chlorophyll and fewer flagella. According to Cook (1962), the spermatozoids may have eight or nine flagella.

Fertilization. When mature the oogonium has a single oosphere or egg cell with a centrally placed nucleus. According to Cook (1962) a refractive exudate comes out through a weakened portion of the upper part of the oogonial wall and an open round pore is visible after the exudate dissolves that part of oogonial wall. A gelatinous substance which is believed to attract spermatozoids is secreted by the oogonium through this part. A large number of spermatozoids are attracted, out of which only one spermatozoid enters the oogonium and fertilization takes place. Even after fertilization, one, two or three spermatozoids enter the oogonium but soon turn away from the fertilized egg, increase in size and burst.

The oospore. Soon the fertilized oosphere matures into oospore, the green contents decrease as also the starch contents. A thick wall envelops the oospore. The wall may be colourless, or pale yellow brown or black. The oospore is round in shape filled with oil and an orange red pigment.

The oospores, after release sink to the bottom of water in which the plant is growing or lodges on stones, aquatic plants or any other convenient site. The oospores generally undergo a period of rest after which they germinate.

Germination of the oospore. Prior to germination the diploid nucleus divides meiotically, after which the oospore looses its red colour and increases in size alongwith increase in chlorophyll content. The contents also divide and four zoospores are liberated into a vesicle formed by an opening in the outer wall of the oospore. The vesicle soon widens and the zoospores now develop flagella. The vesicle burst, and the zoospores escape. The zoospores are like the asexual zoospores and germinate in the same way as the asexual zoospores.

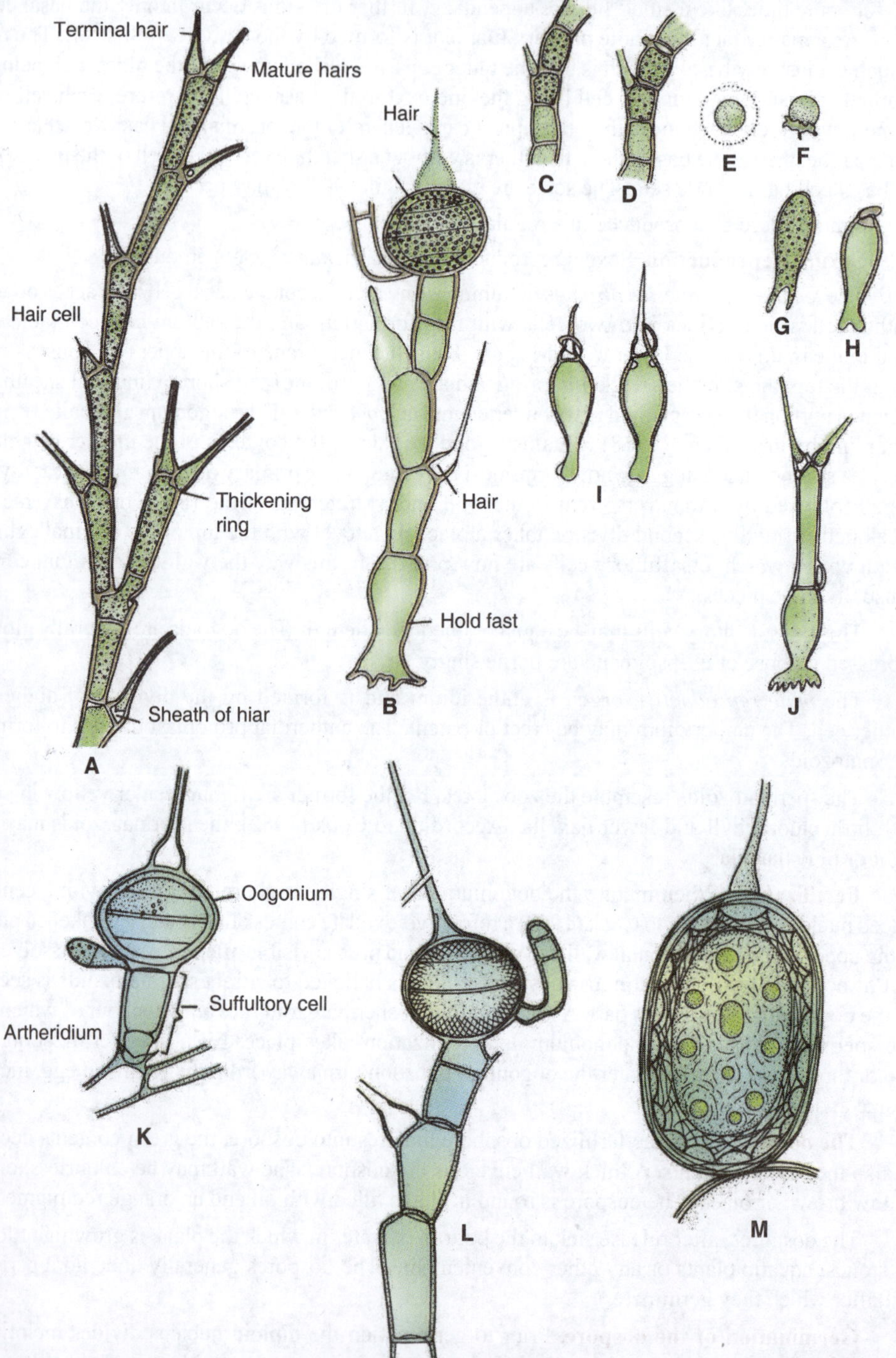

Fig. 8.18. *Bulbochaete.* (A) young plant; (B), portion of a plant showing glower portion.; (C-D), Formation of zoospore; (E), Zoospore in a vesicle; (F-J), Germination of Zoospore; (K-M), Sexual reproduction; (K), A young oogonium with two suffultory cells and developing antheridium; (L), fertilized oogonium; (M), Mature oospore.

Nannandrous species also occur in *Bulbochaete* and the sexual reproduction in these species is similar to those of *Oedogonium.*

Phylogeny and Systematic Position of Oedogoniales. *Oedogonium* with its other two related genera is now placed in a distinct order Oedogoniales. The members of this order possess a characteristic green colour, have pyrenoids with a starch sheath, store reserve food in the form of starch and have non-jacketed unicellular sex organs. These features, the Oedogoniales share with the green algae. On the basis of this, the modern algologists place this special, small but advanced order in the division Chlorophyta (Green algae). In the division Chlorophyta, the Oedogoniales hold an isolate deposition. They are considered to have neither known ancestors nor descendants.

The distinction of the simple unbranched filament into base and apex, the presence of holdfast, uninucleate condition of cell with its single, peripheral chloroplast are features reminiscent of the Ulotrichales. These resemblances prompted some algologists to assume that the order was derived from a ultotrichaceous ancestor which disappeared in the course of evolution. Smith (1938) who supported this theory explained the origin of multiflagellate swarmer of the Oedogoniales by the division of two or four original blepharoplasts and flagellae of the bi-or quadriflagellate swarmers found in the Ulotrichales. He thus considered the Oedogoniales to be a remote offshoot from the Ulotrichaceous species.

However, in their cytology, peculiar manner of cell division, and growth of cell wall, multiflagellate reproductive cells and dimorphism in the sexual filaments, the Oedogoniales are unique. The differences are so distinctive that some algologists consider Oedogoniales as representing a different phyletic line of evolution from the Ulotrichales but which has paralleled the latter in certain features. Many algologists thus hold that the Oedogoniales evolved directly from some motile green unicellular flagellate with a ring of flagellae and represent a line of evolutionary development (parallel to the Ulotrichales) which terminated with this group. They have no real connection with the other members of the green algae.

Some algologists strike a different note. To them the ring-shaped chloroplast, presence of hairs on the thallus of *Bulbochaete*, dwarf males and heterotrichous habit of *Oedocladium* are features showing distant affinity with the Chaetophorales. Fritsch (1945) on the basis of these resemblances suggested origin of Oedogoniales from a common ancestry with Chaetophorales. The consensus of opinion, at present, supports the following classification of *Oedogonium* and *Bulbochaete*:

OEDOGONIUM

Division	:	**Chlorophyta** (Chlorophycophyta)
Class	:	**Chlorophyceae**
Order	:	**Oedogoniales**
Family	:	**Oedogoniaceae**
Genus	:	*Oedogonium*
Species	:	*nodulosum*

BULBOCHAETE

Division	:	**Chlorophyta** (Chlorophycophyta)
Class	:	**Chlorophyceae**
Order	:	**Oedogoniales**
Family	:	**Oedogoniaceae**
Genus	:	*Bulbochaete*
Species	:	*setigera*

QUESTIONS

ESSAY TYPE

1. Describe the habit, structure and differentiation of cells in the filament of *Oedogonium*.
2. Give an account of the vegetative (thallus) structure and reproduction in *Oedogonium*. *(Agra)*
3. Describe the thallus structure and reproduction in *Bulbochaete*.
4. Describe the habit, structure and methods of reproduction in *Oedogonium* and indicate the features of special interest in its life history. *(Agra)*
5. Give an illustrated account of life-cycle of *Bulbochaete*.
6. Write in detail the life history/cycle of *Oedogonium*. *(Punjab 1945, 1947; Agra, 1993; Gorakhpur, 1998; Trichy, 1995; Bihar, 1991, 1992)*
7. Describe the sexual reproduction in *Bulbochaete* with labelled diagrams.
8. Give an illustrated account of structure and reproduction in *Oedogonium/ Polysiphonia*. *(Kumaon, 2000; Gorakhpur, 1992, 1996; Awadh, 1992, 1999)*
9. Describe the features of special interest in the structure and reproduction of *Oedogonium*. Explain the mode of formation of cap cells. *(Allahabad, 1992)*
10. Describe mode of sexual reproduction in *Oedogonium*. *(Allahabad, 1999; Punjab, 1991; Rohilkhand, 1992)*
11. With the help of labelled diagrams describe the method of sexual reproduction in *Oedogonium*. *(Gorakhpur, 1990)*
12. Describe the life cycle/sexual reproduction of nannandrous species of *Oedogonium*. *(Gorakhpur, 1994; Bundelkhand, 1996; Rohilkhand, 1994, 1997)*
13. Give an account of the special features of structure and reproduction in *Oedogonium*. *(Kanpur, 1998)*
14. Give diagrammatic representation of the life-cycle illustrating the relative strength of haploid and diploid phase in *Oedogonium*. *(Himachal Pradesh, 1996)*
15. Give an account of structure and reproduction in Macrandrous species of *Oedogonium*. *(Bhagalpur, 1996)*
16. How does the Macrandrous species of *Oedogonium* differ from nannandrous species in the method of sexual reproduction? *(Bhagalpur, 1990)*

SHORT ANSWER TYPE

17. **Write short notes on :**

(*i*) Dwarf males or Nannandria or Nannadrous species. *(Kumaon, 1995; Allahabad, 1991, 1993, 1996, 2001; Gorakhpur, 1991, 1996, 1999; Awadh, 1998; Lucknow, 1992, 1994, 1997, 1999, 2001, 2003, 2006; Bundelkhand, 1995)*

(*ii*) Cell division in *Oedogonium*. *(Agra, 1954; Kumaon, 1998, 1999; Allahabad, 1994, 2001, 2005; Lucknow, 1995, 2000 2002, 2005, 2007)*

(*iii*) Asexual reproduction in *Bulbochaete*.

(*iv*) Zoospores of *Oedogonium* and *Vaucheria*. *(Allahabad, 1999; Lucknow, 1992)*

(*v*) Cell structure in *Oedogonium*. *(Gorakhpur, 1994; Poorvanchal, 2006)*

(*vi*) Reproduction in *Oedogonium*. *(Awadh, 1997)*

(*vii*) Cap cells in *Oedogonium*. *(Kanpur, 1997; Meerut, 1999; Bundelkhand, 1997; Himachal Pradesh, 1997)* (*viii*) Cell division in *Oedogonium*.

(*Bundelkhand, 1993; Himachal Pradesh, 1993; Rohilkhand, 1993*)

(*ix*) What is a Nannandrium?

(*Kerala, 2001*)

18. Describe the following :

(*a*) Asexual reproduction in *Oedogonium* with diagrams

(*Allahabad, 1991, 2004; Poorvanchal, 2004*)

(*b*) Sexual reproduction in *Oedogonium with* diagrams.

(*Allahabad, 1993; Awadh, 1992, 1995; Poorvanchal, 2005*)

(*c*) Nannandrium.

(*Allahabad, 1995; Meerut, 1998; Bhagalpur, 1992; Rohilkhand, 1996*)

(*d*) Nannandrous and Macrandrous species. (*Agra, 1991*)

(*e*) Dwarf male in *Oedogonium* with diagrams. (*Trichy, 1995*)

(*f*) Macrandrous species.

(*Kanpur M.Sc., 2002*)

19. Identify the following giving one important characteristic

(*a*) *Oedogonium* (*b*) *Bulbochaete.*

20. What is cap cell in *Oedogonium*?

(*Rajasthan, 1998*)

21. Differentiate between :

(*i*) *Oedogonium* and *Bulbochaete*

(*Bharathiar, 2000*)

(*ii*) Antheridia of Macrandrous and Nannandrous species of *Oedogonium*

(*Rohilkhand, 1993*)

(*iii*) Macrandrous and Nannandrous species

(*Allahabad, 2003; Bundelkhand, 1994; Rohilkhand, 1991*)

(*iv*) Thallus structure of *Oscillatoria* and *Oedogonium*

(*v*) Sexual reproduction in *Bulbochaete* and *Oedogonium*

(*vi*) Zoospores of *Oedogonium* and *Ulothrix*. (*Allahabad, 2004*)

22. Give one important character of the following :

(*a*) *Oedogonium*. (*Bundelkhand, 1998*)

(*b*) *Bulbochaete.*

23. Elucidate the mode of cap formation in *Oedogonium*. (*Himachal Pradesh, 1990*)

24. Explain the nannandrous forms in *Oedogonium*. (*Himachal Pradesh, 1991*)

25. (*a*) Draw graphic life cycle of macrandrous species of *Oedogonium.*

(*b*) Describe cell structure and asexual method of reproduction in *Oedogonium.*

(*Himachal Pradesh, 1992*)

26. Draw neat and labelled diagrams of the following :

(*a*) Reproductive organs of *Oedogonium*. (*Rohilkhand, 1995*)

(*b*) Reproductive organs of *Bulbochaete.*

(*c*) Cell division in *Oedogonium.*

(*d*) Cell structure in *Bulbochaete.*

(*e*) Cell structure in *Bulbochaete.*

27. Write a brief account of asexual reproduction in *Bulbochaete*. (*Bharathiar, 2000*)

28. Draw graphic life-cycle of *Bulbochaete.*

29. Describe briefly the sexual reproduction in *Bulbochaete.*

30. By means of labelled diagrams only describe the life cycle of *Oedogonium* or *Bulbochaete.*

31. Draw neat and labelled diagrams of the following and mention their characters briefly.

(a) *Oedogonium* (*Lucknow, 2004*)

(*b*) *Bulbochaete.*

32. How caps are formed during cell division of *Oedogonium*? (*Allahabad, 2004*)

33. How many kinds of motile cells are produced in the life cycle of *Oedogonium*?

(*Poorvanchal, 2003*)

OBJECTIVE TYPE

34. Select the correct answer :

(*i*) Chloroplast in *Oedogonium* is
(*a*) Cup shaped (*b*) Reticulate
(*c*) Spiral (*d*) Stellate.

(*ii*) *Oedogonium* is a member of class
(*a*) Chlorophyceae
(*b*) Phaeophyceae
(*c*) Myxophyceae
(*d*) Rhodophyceae

(*iii*) *Oedogonium*, *Bulbochaete* and *Oedocladium* belong to the order
(*a*) Volvocales
(*b*) Conjugales
(*c*) Oedogoinales
(*d*) Siphonales

(*iv*) Cap cells are characteristic of the genus
(*a*) *Spirogyra* (*b*) *Oedogonium*
(*c*) *Vaucheria* (*d*) *Chara*

(*v*) *Bulbochaete* belongs to the order
(*a*) Oedogoinales
(*b*) Conjugales
(*c*) Siphonales
(*d*) Volvocales

(*vi*) The major difference between *Oedogonium* and *Spirogyra* is that in *Oedogonium*
(*a*) The cells are longer than broad
(*b*) The chloroplast is spiral
(*c*) Sexual reproduction is through conjugation
(*d*) Cap cells are absent

(*vii*) Stephanokentian type of zoospores are present in
(*a*) *Oedogonium* (*b*) *Spirogyra*
(*c*) *Vaucheria* (*d*) *Chara*

(*viii*) While *Oedogonium* and *Bulbochaete* are exclusively fresh water forms, *Oedocladium* is
(*a*) Also fresh water form
(*b*) Marine
(*c*) Terrestrial
(*d*) Halophytic

(*ix*) The attachment of *Oedogonium* to substratum is effected by
(*a*) The lowest cell directly
(*b*) Holdfast to the attachment disc
(*c*) Both of the above
(*d*) None of the above

(*x*) The nannandrium develops from
(*a*) Aplanospore (*b*) Zoospore
(*c*) Androspore (*d*) Hypnospore.

(*xi*) The zoospores of *Oedogonium* are
(*a*) Nonciliate (*b*) Uniciliate
(*c*) Biciliate (*d*) Mullticiliate

(*xii*) Idiandrosporous forms of *Oedogonium* are species bearing
(*a*) Antheridia
(*b*) Oogonia
(*c*) Androsporangia and oogonia on the same filaments
(*d*) Androsporangia and oogonia on separate filaments.

(*xiii*) The nannandrium is so called because
(*a*) It is a small plant
(*b*) It is a male plant
(*c*) It produces small sperms
(*d*) It is a small male plant

(*xiv*) During cell division in *Oedogonium* the cell that receives the cap is
(*a*) The lower cell
(*b*) The upper cell
(*c*) Both the lower and upper cells
(*d*) Sometimes the lower cell and sometimes the upper cell

(*xv*) In *Oedogonium* androspores always germinate.
(*a*) On oogonia
(*b*) On the suffultory cell
(*c*) Anywhere close to *oogonium*
(*d*) Free in water

CHAPTER

ORDER 6. CLADOPHORALES

- Family : Cladophoraceae
- Cladophora
- Pithophora

Cladophorales: Cladophora sp.

The order Cladophorales includes filamentous green algae occurring both as marine and fresh water forms. The thallus predominately consists of a branched (*Cladophora*) or rarely of an unbranched (*Urospora*) filament composed of cylindrical, coenocytic cells. The cell generally contains a single, reticulate chloroplast with many pyrenoids. It encircles the protoplasts. Asexual reproduction takes place by zoospore, aplanospores or akinetes. The meiozoospores are quadriflagellate. Sexual reproduction is isogamous or anisogamous. The order includes a single family Cladophoraceae with 12 genera and some 350 species. The important genera included in this order are *Cladophora, Pithophora, Chaetomorpha, Urospora, Spongomorpha, Rhizoclonium* and *Basicladia*.

Cladophora *Pithophora* *Chaetomorpha*

Urospora *Spongomorpha* *Rhizoclonium* *Basicladia*

REPRESENTATIVE GENERA OF ORDER CLADOPHORACEAES

FAMILY : CLADOPHORACEAE

The plants are coarse, wiry and attached by a holdfast or rhizoid-like outgrowths. Some genera occur exclusively in fresh water whereas others are strictly marine. Some forms like *Urospora* and *Spongomorpha* are met within the sea or brackish water. S.H. Koorders in 1902 reported a certain marine species of *Cladophora* from Java which occurs in symbiotic relations with a certain species of Sponges (*Ephydatia fluvitalis*). The family is characterised by the presence of elongate, coenocytic cells. They are mostly arranged in the form of branched filaments. The threads are usually attached to the substratum by means of rhizoidal cells. The chloroplast is parietal and reticulate. There are many pyrenoids. In certain species of *Pithophora* peculiar tendril-like attaching organs have been reported. The tips of some of the ultimate branches become drawn out into narrow and branched rhizoid-like structures that are green in colour and coil around foreign objects thus aiding in attachment. These branches are not separated by septa.

CLADOPHORA

Occurrence. It is a widely distributed genus comprising about 160 or so species. Some species of *Cladophora* are fresh water whereas others are marine. The former occur in the shallow water of lakes and some in the running water of streams. These are often attached to stones, logs and other

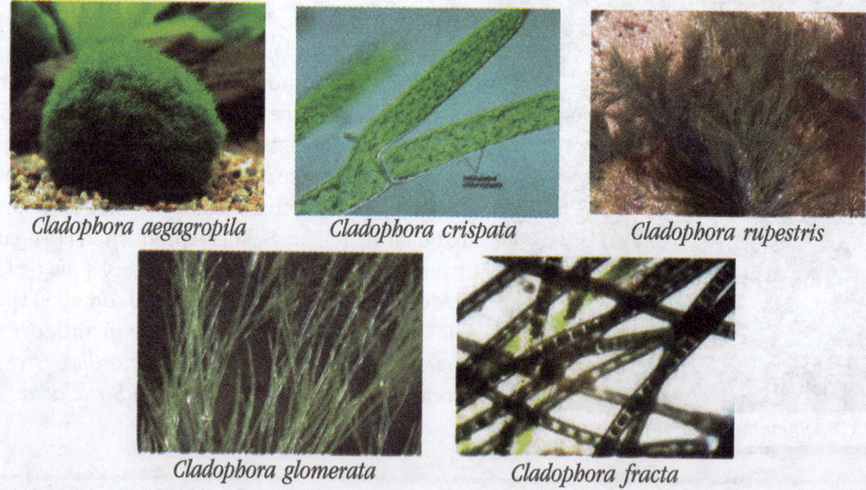

Cladophora aegagropila *Cladophora crispata* *Cladophora rupestris*

Cladophora glomerata *Cladophora fracta*

objects under water. Some of them such as *C. fracta* may become free-living later. There is a group of wholly free living species (*C. holosatica, C. profunda*) called *Aegagropila*. These frequently live as ball-like growths. Still a few species are found growing epiphytically on certain submerged water plants and shells of molluscs (epizoic). *C. crispata* is an example of the epizoic species. It occurs on the shells of snails. The common Indian species are: *C. glomerata, C. crispata* and *C. fracta*.

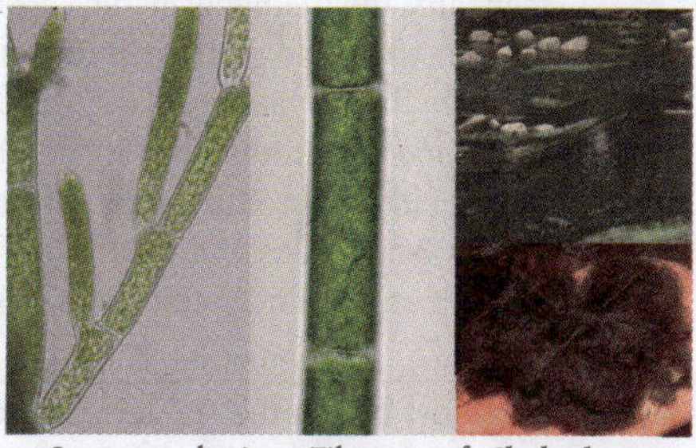

Coarse and stingy Filament of *Cladophora*

Plant body (Fig. 9.1). It is a thallus which occurs in the form of bush-like tufts of long waving, branched threads attached to rocks at the water line of lakes and streams (A). The branching filaments are usually coarse and stringy to touch. The individual cells are large, multinucleate, cylindrical and placed end to end (B). The thallus is anchored to the substratum either by means of rhizoidal outgrowths arising from the base of the plant or by a number of septate rhizoidal branches growing down from the basal cells of the filament (C). In *C. glomerata* the rhizoidal branches grow further and function as stolon-like outgrowths. From these, new upright threads arise at intervals. The stolon thus help in vegetative propagation by breaking down into several pieces. Each piece develops upright threads and function as a separate plant. In certain marine species of *Cladophora* supplementary rhizoids also arise from the lower ends of cells lying above the basal cell. They serve as an accessory means of support for the comparatively larger plants.

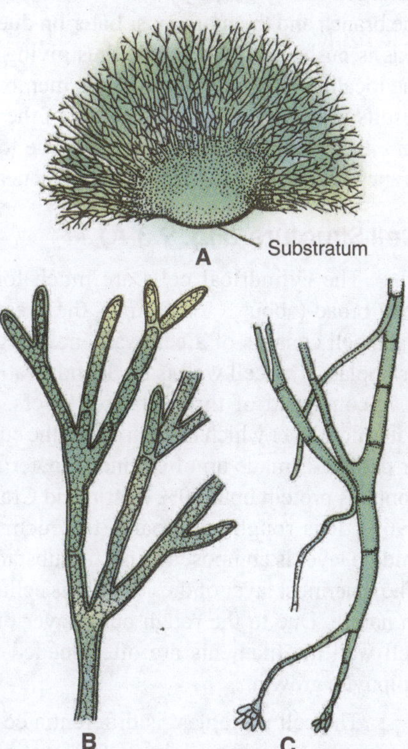

Fig. 9.1 (A-C). *Cladophora*. (A), showing habit; (B), portion of the thallus; (C), formation of rhizoids in *C. glomerata* (after Brand).

In the *Aegagropila* group (Fig. 9.2), the species growing in quiet water are found floating on the surface forming free-floating tangled masses (A). When the thallus is subjected to a continual gentle rolling motion in water the young branched filaments aggregate to grow into cushions or balls under the influence of wave action (B). These *Cladophora* balls often float and are washed up on the shore. These balls forming species are called **aegagropilous**. *C. holsatica* is an example.

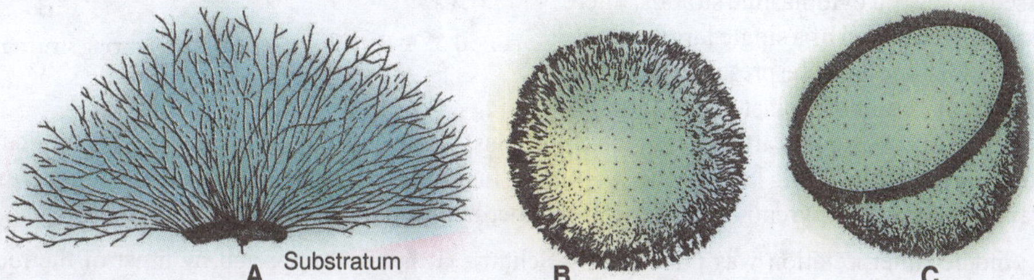

Fig. 9.2 (A-C). *Aegagropilous sp.* (A) young filament; (B), a ball of *C. holsatica*; (C), the same cut through (after Acton).

Genetically, the filaments in most species of *Cladophora* are of two kinds, **haploid** and **diploid**. The former constitute the **gametothallus** and latter **sporothallus**. Morphologically these two kinds of filaments (thalli) are alike. In the vegetative state it is not possible to differentiate between a gametothallus and a sporothallus. Physiologically, however, they are different. The haploid filaments are concerned with the production of gametes and the diploid filaments produce (*i*) quadriflagellate meiozoospores by meiosis and (*ii*) diploid zoospores. *C. glomerata* however, has only one kind of filaments which are diploid. Meiosis in this species takes place at the time of gamete formation.

Branching. The branching is lateral. A branch arises as a later outgrowth of the parent cell near its upper end just below the septum. The outgrowth is separated from its parent cell by septum at right angles to that in the main axis. It is close to the point of origin of the outgrowth. The outgrowth enlarges and grows by cell division to form a branch. At first there is a wide angle between the branch and the main axis. Later on due to evection the branch is pushed further up and the main axis is pushed on to one side thus giving the appearance of a dichotomy. Evection takes place by the localised surface growth of the membrane of the parent cell immediately beneath the branch. It results in the upward displacement of the latter. Cases are not rare when two lateral branches arise on either side from the same cell. Due to evection these present a trichotomous appearance. The branches generally arise from the cells near the apex.

Cell Structure (Fig. 9.3 A)

The cylindrical cells are much longer than broad (about 3 to 20 times the breadth). Each cell consists of a cell wall enclosing the protoplast. The cell wall is thick and stratified. It is composed of three distinct layers. The outermost layer which is often called the cuticle or pellicle is made up of chitinous material. It contains protein upto 70% (Haric and Craigie, 1969). It is rough and coarse to touch. The middle layer is composed of pectic substances. The innermost layer of the cell wall is cellulose in nature. Due to the rough outer layer of the cell wall the filaments are often loaded with epiphytic growth.

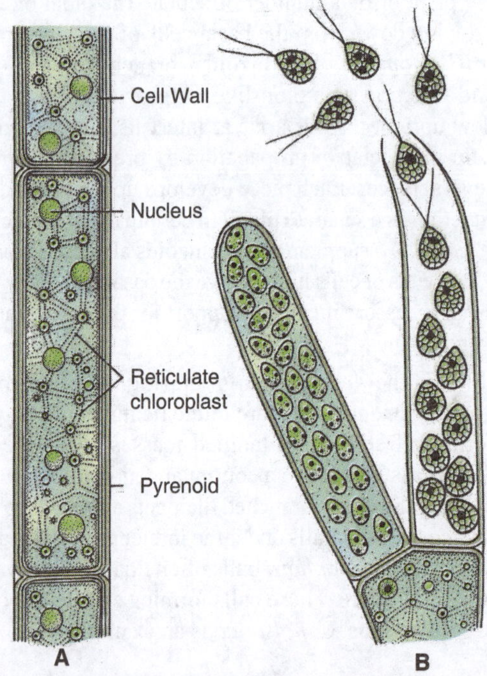

Fig. 9.3 (A-B). *Cladophora.* (A), cell showing structure; (B), *C. glomerata* forming zoospores.

The cell protoplast is differentiated into cytoplast, numerous nuclei, a large reticulate chloroplast with a number of pyrenoids. The colourless cytoplast lies next to the cell wall forming a fairly thick lining layer. It surrounds a big central vacuole. The vacuole may be traversed by several cytoplasmic strands. There are cases when instead of a single large vacuole several smaller ones may be present. Regarding the nature of the chloroplast there is much controversy. According to Carter (1919) the chloroplast is parietal and is a single elongate, reticulate structure lodged in the lining layer of the cytoplasm. In the reticulations of the chloroplast are present several pyrenoids. The pyrenoids increase in number by ordinary division.

Another interpretation was put forth by Schmitz. It is now supported by most of the recent workers. According to Schmitz there are numerous discoid chloroplasts forming the reticulum. Most of these lie next to the cell wall in the cytoplasmic lining and a few are distributed in the cytoplasmic strands crossing the central vacuole. These discoid chloroplasts are usually without pyrenoids but Schussing reported that in *C. suhriana* each discoid chloroplast contains a pyrenoid. Smith holds that the chloroplast is reticulate but in some cases the connecting strands between the intersections are very thin. Consequently the cell appears to contain numerous discoid chloroplasts. These are several nuclei present in each cell. They lie internal to the chloroplasts or the single chloroplast, as the case may be. There is more or less a regular arrangement of the nuclei. In certain species of *Cladophora* the number of nuclei is fewer and may be even two or three. Cells with a single nucleus have also been recorded. Macdonald and Pickett Heaps (1976) who studied the ultrastructure of the vegetative cell of *Cladophora* reported that the thick cell was composed of parallel microfibrils is covered by a well-defined cuticle. Within the cell wall is a very thin layer of cytoplasm containing microtubules. The latter run parallel to the long axis of the cell. Internal to this thin cytoplasmic layer is a layer of vacuoles of variable thickness. Between the latter and the large central vacuole is cell cytoplasm which contains the usual cell organelles. The chloroplasts with their bilenticular pyrenoids lie external to the nuclei. Bilenticular pyrenoids are characteristic of the Cladophorales.

Growth. Growth consists of an increase in the number of cells in the filament. The new cells are formed by the division of the pre-existing cells. In *Cladophora* all the cells are capable of division. It is partly carried out by the apical cell. The apical cell of a branch or that of the main axis elongates considerably. It then divides by a transverse septum into an upper and a lower cell. The lower cell may or may not take further part in growth. In *C. sauteri* the growth is mainly affected by the apical cell. The cells cut off by it take no further part. The formation of septa is in no way related to the nuclear division. The protoplast is withdrawn from the wall in the middle of the cell where the new septum will arise. The space becomes filled with mucilage. At the same time an annular thickening or ring of thickening material is visible within this space all round the cell in a transverse plane. It functions as the primordium of the septum. The thickening gradually extends towards the centre and fuses with the inner layer of the longitudinal wall. The core of the original annular thickening becomes continuous with the inner layer of the adjacent longitudinal walls.

Asexual Reproduction. The haploid and diploid filaments in *Cladophora* reproduce vegetatively by the following methods:

(*i*) *Storage cells.* These occur in most of the perennial species of *Cladophora* and provide a method of tiding over the unfavourable period of growth. In *C. glomerata* and also in *C. ophiophila* the rhizoids grow towards the substratum. On reaching the substratum they undergo repeated divisions to form a cellular expanse consisting of short branched filaments. The cells of these filaments become filled with food material. The walls of the food laden cells become encrusted with carbonate of lime. These storage cells survive after the other parts of the plant have died. They grow into new filaments in the next season.

(*ii*) *Stolons.* Vegetative propagation by the formation of stolon-like branches has already been described. These stoloniferous branches break up into several pieces. Each fragment or piece is capable of growing into a full-fledged plant.

(*iii*) *Akinetes.* The formation of akinetes has been reported in certain fresh-water species of *Cladophora*. They are usually formed in the erect threads of the mature plants particularly on the larger branches. The cells get gorged with food reserves. This is followed by the thickening of the parent cell walls. The food laden cells with thickened and highly resistant walls often become more or less swollen. These specially modified vegetative cells are called the akinetes. Cholonky (1930) reported the formation of akinetes in *C. glomerata*. The formation of akinetes in this species is preceded by repeated division of cells. The shorter segments thus formed becomes gorged with food materials. The walls undergo extreme thickening. The thick-walled cells become swollen and more or less pear-shaped. These cells do not get detached from the main filament. At the time of germination the contents of the akinete divide transversely to form new cells which develop into new branches. According to Ernst akinete formation takes place when there is deficiency of nutritive salts in the surrounding medium.

(*iv*) *Fragmentation.* Propagation by fragmentation is the sole means of multiplication in the aegagropilous species.

(*v*) *Spore formation* (Fig. 9.3 B). Most of the species of *Cladophora* do not produce any kind of asexual spores. The only exception is *C. glomerata*. It reproduces asexually by the formation of biflagellate zoospores produced in ordinary cells called sporangia. Usually the sporangia occur near the tips of the diploid filaments. Genetically the zoospores are diploid and are differentiated by mitosis. On germination each zoospore produces the diploid filament of *Cladophora*. The zoospores in this species thus serve to reproduce the diploid generation. They play no role in the phenomenon of alternation of generations.

Sexual Reproduction (Fig. 9.4). It is oogamous and entirely the concern of the haploid filaments (gametothallus). Most of the species, however, are reported to be heterothallic. Fusion takes place between gametes of opposite mating types designated as + and – strain. The gametes are produced in the ordinary vegetative cells called **gametangia** near the tips of the branches of the **haploid filaments**. Prior to gametic formation, the nuclei in the gametangium undergo active nuclear division which is mitotic. The only exception is *C. glomerata* in which there is gametangial meiosis. Each daughter nucleus is fashioned into a biflagellate gamete either by progressive cleavage of the cytoplast (Smith) or by accumulation of small amount of cytoplast around each daughter nucleus (Czempyrek and Schussing). The mature gametes escape through a circular pore at or near the upper end of the cell wall (Fig. 9.4). The liberated, pear-shaped biflagellate gametes are beaked in a certain species of *Cladophora* (Fig. 9.5). Shyam (1980) reported that the biflagellate gametes in *C. callicoma* are somewhat elliptical with a long narrow apex and an anterior streak-like eyespot. The gametes are smaller than the zoospores. After release they swim about actively in water. Fusion occurs between gametes coming from filaments of two different mating types termed plus and minus strains. There is thus marked dioecism in *Cladophora*. The fusing gametes, however, resemble each other morphologically and thus are called the **isogametes**. The resultant fusion cell is called a **zygote**.

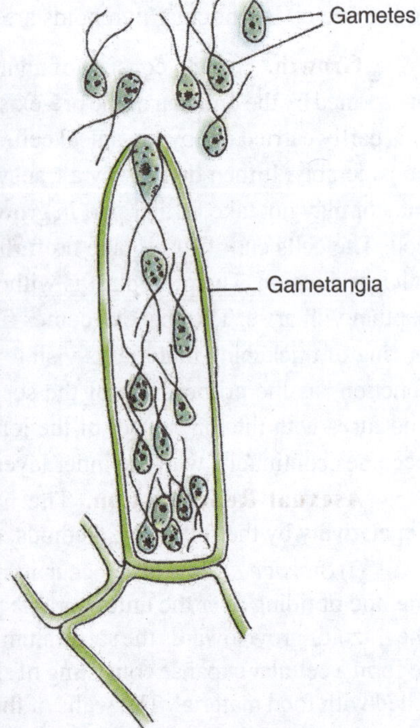

Fig. 9.4. *Cladophora.* Formation and liberation of gametes.

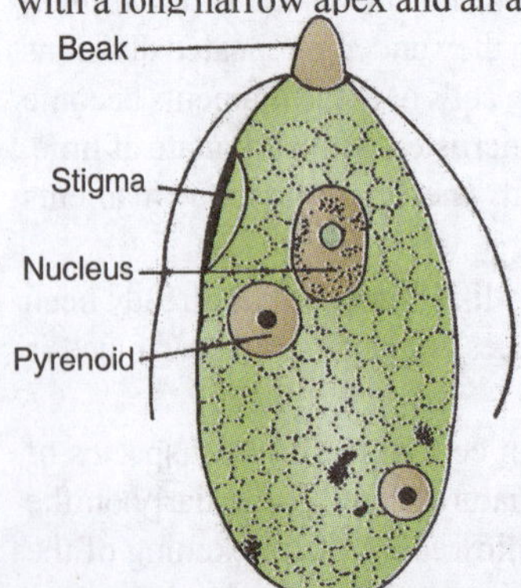

Fig. 9.5. *Cladophora.* Beaked biflagellate gamete (After Strasburger).

Parthenogenesis has also been recorded in certain species of *Chadophora*. In certain other cases the gametes that fail to fuse with their partners from the other filaments disintegrate and die.

Germination of Zygote. The zygote does not undergo any resting period. It germinates directly within a day or two. By repeated mitosis it produces a sporothallic filament which is diploid. The diploid filament (sporothallus) normally reproduces by the formation of meiospores or gonospores which are motile.

Sporothallus (Fig. 9.6). It is the diploid filament in the sexual life cycle of *Cladophora*. As mentioned before it resembles the haploid filament (gametothallus) morphologically as well structurally but differs genetically and in function. It produces the haploid swarmers called **meiozoospores**. They are differentiated by meiosis from the diploid protoplast of the sporangium which is an ordinary vigorously growing vegetative cell. The sporangia occur near the tips of the branches of the diploid filaments. Prior to spore differentiation the protoplast of the zoosporangium undergoes active nuclear division (A). The first two nuclear divisions of each nucleus constitute **meiosis**. Sporogenic meiosis is thus the rule in the life cycle of *Cladophora*. The only exception is *C. glomerata* in which there is gametangial meiosis. Consequently several **haploid daughter nuclei** are produced (A). At this stage there is a complete disappearance of the central vacuole due to gradual encroachment of the cytoplast (B). Regarding further changes there are two views. According to Czempyrek and Schussing the cytoplast aggregates around each haploid daughter nucleus (B). In this way several uninucleate bits of protoplasts are formed (C). Other investigators (Smith) reported progressive cleavage of the cytoplast into uninucleate daughter protoplasts. Each uninucleate part gets metamorphosed into a quadriflagellate meiozoospore (D). It is haploid. During meiospore formation there appears a colourless lens-shaped area at or near the upper end of the parent cell wall. A small circular pore then develops by the gelatinisation or bursting of this hyaline area (C). The mature meiozoospores are liberated singly through this pore. All of them escape into the surrounding water. According to Fritsch they are liberated through a lateral aperture in the wall.

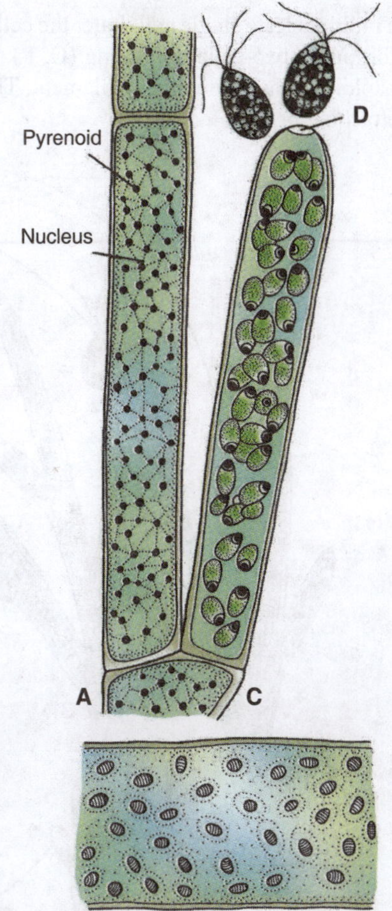

Fig. 9.6 (A-D). *Cladophora.* (A), vegetative cell in which the diploid nuclei have undergone meiosis; (B), differentiation of meiozoospores;(C), zoosporangium containing meiozoospores; (D), liberated meizoospores or gonozoospores.

The liberated meiozoospore (D) is a small pyriform structure. It has a definite anterior beak-like end with a distinct eye spot and a more or less rounded posterior portion. It is quadriflagellate. Nishimura and Kanno recorded the presence of only two flagella in the zoomeiospores of *C. sauteri.* Their results have not been confirmed. There is a single chloroplast that occupies the posterior part of the meiospore. The liberated meiozoopores swim about actively for some time.

Germination of meiozoospore (Fig. 9.3 K-P). Gradually the meiozoospores become passive and come to rest with their arterial ends towards the substratum (K, L). The flagella are resorbed. The quiescent meiozoospore secretes a wall and elongates vertrically (M, N). The single haploid nucleus undergoes several mitotic divisions forming several nuclei. This results in the formation of a multinucleate single cell. Later the cell undergoes transverse septation into two multinucleate cells forming a two-celled germling (O, P). The germling develops in the usual manner into a mature haploid or the gametophyte filament. The basal cell becomes the rhizoidal cell and functions as an attaching organ.

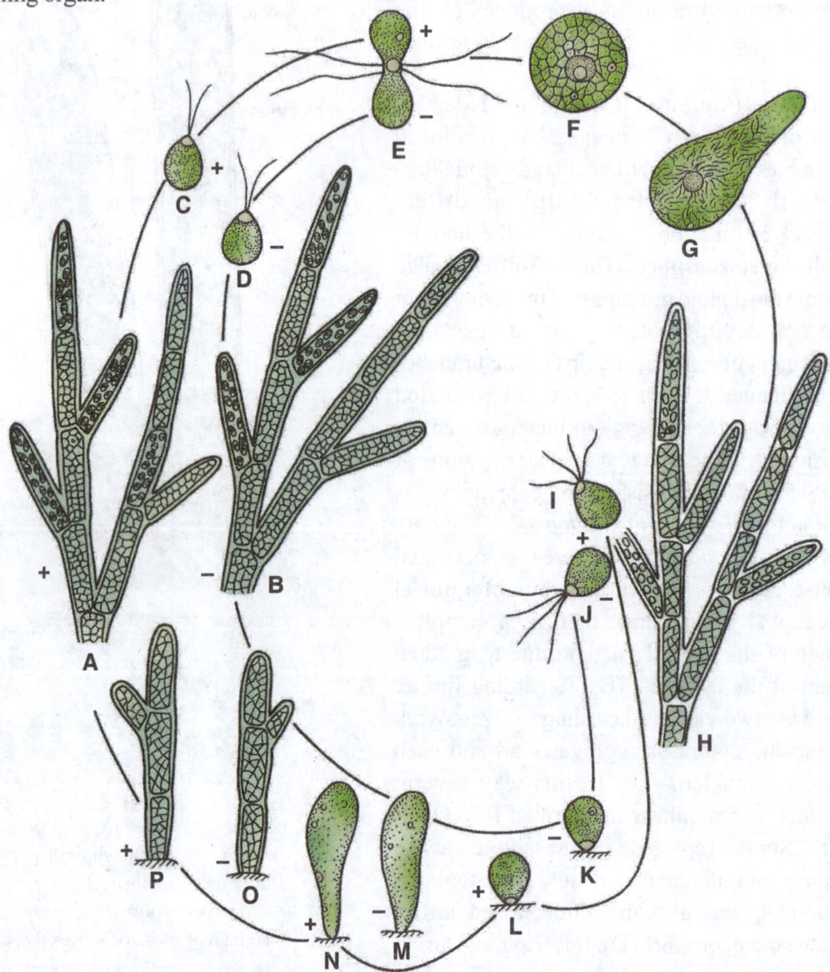

Fig. 9.7. Pictorial sexual cycle of *Cladophora.*

Life Cycle and Alternation of Generations (Fig. 9.8)

With the exception of *C. glomerata* the life cycle of all species of *Cladophora* exhibits alternation of generation which is of *isomorphic type*. Shyam (1980) collected *C. callicoma* from the river Ganges near Varanasi (U.P.) and described its life cycle and cytology. There occurs two types of vegetative filaments (thalli) in the single sexual cycle of this species. One of these is the diploid filament (sporophyte) with a mitotic chromosome count of $2n = 24$ and the other is the haploid filament (gametophyte) with a chromosome count of $n = 12$. The diploid filament (sporophyte) with 24 chromosomes may be called the **sporothallus** and the haploid filament (gametophyte) with 12 chromosomes represents the **gametothallus**.

***(a)* Sporothallus.** The terminal and subterminal cells of the branches of the diploid sporothallus with 24 chromosomes function as *zoosporangia.* The diploid nuclei of the zoosporangium undergo a number of nuclear divisions. The first two nuclear divisions of each diploid nucleus constitute **meiosis**. The resultant multinucleate haploid protoplast of the zoosporangium becomes differentiated into a number of haploid daughter protoplasts, each with 12 chromosomes. The uninucleate haploid daughter protoplasts become metamorphosed into pear-shaped quadriflagellate meiospores or gonospores known as the **meiozoospores.** The diploid filaments or sporothallus thus produces the haploid zoospores or meiozoospores. After liberation from the zoosporangium the meiospores germinate to produce the haploid filaments which are morphologically similar to the diploid filaments but genetically different being haploid.

(*b*) Gametothallus. The terminal and subterminal cells of the branches of the haploid filament or gametothallus with 12 chromosomes function as *gametangia*. The haploid nuclei of the gametangium undergo a number of nuclear divisions which are all *mitotic*. The multinucleate protoplast of the gametangium becomes differentiated into a large number of biflagellate, somewhat elliptical gametes which are smaller than the meiozospores. The gametes are released through a lateral aperture in the wall at the apical end. The gametes released from the gametangium swim about actively. Fusion occurs between gametes coming from filaments of opposite mating types (+ and –). Morphologically the fusing gametes are alike (isogametes) but physiologically anisogametes because gametes from the same filament do not fuse. After fusion the diploid zygote germinates directly to give rise to the diploid sporophyte filament or sporothallus which is morphologically similar to the gametothallus formed by the germination of the quadriflagellate meiospore. All the haploid structures namely meiozoospores, + and – filaments constituting the gametothallus, gametangia and + and – gametes constitute the sexual or gametophyte generation of the life cycle of *Cladophora*. It starts with meiosis and ends with fusion of gametes (syngamy). All the diploid structures namely, zygote, diploid filament (sporothallus) and zoosporangia constitute the asexual or sporophyte generation in the life cycle. It starts with syngamy and ends with meiosis. In the single life cycle of *Cladophora* there is regular alternation of the spore producing diploid filaments (asexual generation) with the gamete producing haplontic filaments (sexual generation). This phenomenon is called

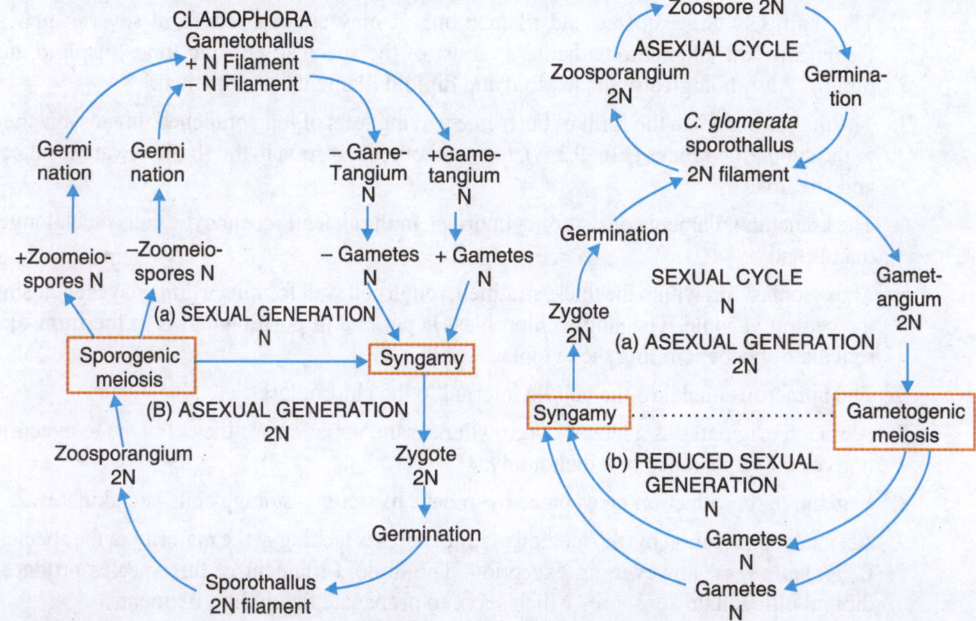

Fig. 9.8. *Cladophora.* Word diagram of the life cycle

Fig. 9.9. *Cladophora glomerata.* Word diagram of the life cycle.

alternation of generations. Since the two alternating individuals (sporothallus and gametothallus) in the life cycle are morphologically similar it is known as *isomorphic* alternation. The life cycle of *Cladophora* which is characterized by distinct alternation of generations and sporogenic meiosis is called diplobiontic or *diplohaplontic* life cycle.

Investigations on *C. glomerata* tell a different tale (Fig. 9.9). In the life cycle of this species no haploid vegetative individual is organised and thus no alternation of generation takes place. List (1930) reported that both gametes and zoospores are produced at different times of the year on the diploid filaments. Meiosis takes place at the time of gamete differentiation in spring. The diploid filaments constituting the *sporothallus* produce diploid biflagellate zoospores for a greater part of the year in the zoosporangia. Sporogenic meiosis is thus lacking in this species. On germination the diploid zoospores reduplicate the diploid generation. This goes on over a long period. In spring the zoosporangia on diploid filaments function as gametangia. Meiosis takes place at the time of differentiation of the diploid gametangial protoplast into gametes. It is termed *gametangial meiosis*. The gametes are produced in spring only. The sexual generation is thus extremely reduced. It is represented by the haploid gametes. There is no vegetative haploid plant or gametothallus in the life cycle. The gametes fuse to form the diploid zygote which germinates to produce by repeated mitosis the diploid filament or **sporothallus**. Thus *C. glomerata* shows no regular alternation of generations. Such a life cycle which is characterised by a diploid thallus (sporothallus) only and gametanic meiosis is called *diplontic.* In this respect this species (*C. glomerata*) resembles the Siphonales. Sinha and Noor (1967) observed regular alternation of haploid generation with chromosome count $n = 12$ and diploid generation with $2n = 24$ in a fresh water form of *C. glomerata* identified as *C. glomerata* var. *fasciculata* f. *fasciculata*.

Salient Features

1. *Cladophora* is unusual in its habit. Some species are exclusively freshwater and others marine.
2. The thallus is large, coarse and filamentous. It may attain a length of several inches. Genetically the filamentous thallus in most of the species is of two kinds, haploid and diploid. Morphologically the haploid and diploid filaments are identical.
3. The thallus occurs in the form of bush-like waving tufts of long branched threads attached to the submerged rocks (Fig. 9.1 A), logs and other substrata in the shallow water of lakes and streams.
4. The branching filament consists of cylindrical, multinucleate, coenocytic cells much longer than broad.
5. The cytoplast lies within the thick, stratified, rough cell wall forming a lining layer enclosing the central vacuole. The single chloroplast is parietal in position and is in the form of a reticulate sheet encircling the cytoplast.
6. The numerous nuclei of the cell lie internal to the chloroplast.
7. A branch originates as a lateral outgrowth near the upper end of the cell. Due to evection it gives the appearance of a dichotomy.
8. Vegetative reproduction takes place by means of stolons, storage cells and akinetes.
9. Asexual reproduction by the method of mitospores is lacking in the majority of the species. *C. glomerata* is, however, an exception. The diploid filament of this species produces diploid biflagellate zoospores which serve to propagate the diploid filaments.
10. The haploid filaments reproduce sexually by the formation of biflagellate gametes.
11. The fusing gametes are isogametes. In their sexual behaviour they exhibit functional disparity.

12. Only isogametes from different haploid filaments fuse.
13. The resultant zygospore undergoes no resting period. It germinates directly into a new *Cladophora* filament which is diploid.
14. In majority of the species the diploid filaments produce meiospores or gonospores by meiosis.
15. The meospores are motile, pear-shaped, quadriflagellate structures which may thus be called meiozoospores.
16. Each meiozoospore, on germination, produces an alternate haploid filament in the life cycle.
17. *Cladophora* thus exhibits isomorphic alternation of generations.
18. *Cladophora glomerata*, however, has a wholly different type of life cycle. The filament is diploid. It reproduces asexually by the formation of diploid, biflagellate zoospores. Meiosis takes place at the time of gamete formation so that thee is no haploid generation. The latter is represented only by the gametes.

Inter-relationship and systematic position of Cladophorales

The members of this order are unicellular, coenocytic algae mostly with a branched and sometimes with an unbranched filamentous thallus. Thus in the morphological nature of the thallus they show affinities with the Ulotrichales and in the coenocytic nature of the cells with the Siphonales. Many phycologists place all the coenocytic genera in the order Siphonales. Some advocated the inclusion of siphonaceous coenocytic genera in the order Siphonales and of the separate coenocytic genera in the Siphonocladales. Fritsch (1927) however, placed the septate coenocytic genera in the order Ulotrichales on the basis of the following common characteristics which they share between themselves:

1. Multicellular thallus consisting of mostly branched filaments.
2. The reported occurrence of a single nucleus in the cell in very rare cases in *Cladophora*.
3. Differentiation of gametes and zoospores in unmodelled vegetative cells.
4. Occurrence of motile, quadriflagellate zoospores and biflagellate gametes in both.
5. Liberation of reproductive cells through a lateral aperture.
6. Copulation usually isogamous rarely anisogamous.
7. Parietal chloroplast.

Later in 1935 Fritsch came to the conclusion that all septate, coenocytic genera have sufficient distinctive characters of their own to be included in a separate family Cladophoraceae under order Chadophorales of the green algae. Papenfuss (1951), Prescott (1951) and Smith (1938) supported this viewpoint. The inclusion of these genera in a separate family Cladophoraceae has been universally accepted. There is difference of opinion as to whether or not the family Cladophoraceae should be placed in a separate order Cladophorales. Boergesson (1939) and Feldman (1938-39) still opposed the removal of Cladophoraceae from the Siphonocladales. The majority of the pohycologists, however, now recognise Cladophoraceae. The taxonomic position of *Cladophora*, thus at present, is :

Division	:	**Chlorophyta** (Chlorophycophyta)
Class	:	**Chlorophyceae**
Order	:	**Cladophorales**
Family	:	**Cladophoraceae**
Genus	:	*Cladophora*
Species	:	*fracta*

PITHOPHORA (WITTROCK, 1877)

Occurrence. It is exclusively a fresh water form found in tropical and subtropical regions. It occurs submerged or in floating masses in the fresh water of ponds, pools, lakes, tanks and streams. The commonly known species of this genus are *P. oedogonia, P. kewensis* and *P. ceavana*. It resembles *Cladophora* in general appearance.

Pithophora sp.

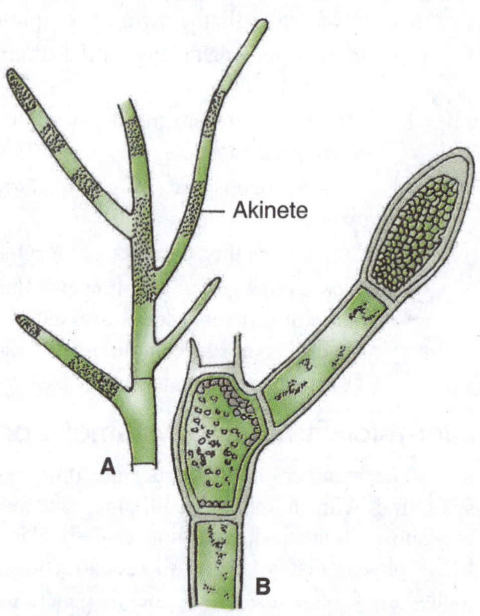

Fig. 9.10 (A-B). *Pithophora.* (A) Portion of thallus of *P. oedogonia* with terminal and intercalary akinetes; (B) akinete formation in *P. kewensis* showing structure. (A, after Smith; B, after Fritsch).

Thallus. Generally it is rough to touch and is a branched filament (9.10 A). Branching is lateral. Usually the branches arise singly either on one side or on both sides. Rarely do they arise in opposite pairs. Each branch arises as a lateral outgrowth at a short distance below the upper septum of the cell in the apical portion of the filament and mostly stands at first, at right angles, to the main axis. Later due to evection it is pushed upwards to make an acute angle with the main axis. The filament consists of long, cylindrical cells which are longer at the ends of branches than others in the thallus. In certain species of *Pithophora* peculiar tendril-like attaching organs develop when tips of some ultimate branches come in contact with a solid substrate such as a stem, a piece of wood or a dead leaf. These specialised tendril-like attaching organs, also called the rhizoidal branches, are narrow, green in colour and are not separated by a septum from the main axis. They coil around the foreign objects thus aiding in attachment.

Cell Structure. Each cell consists of a cell wall enclosing the multinucleate protoplast. The cell was is thick but lacks stratification. The cytoplasm forms a lining layer enclosing a large central vacuole which is traversed by cytoplasmic strands. Embedded in the cytoplasm is a single parietal chloroplast which has the form of a reticulate sheet (Fig. 9.3A). There are a number of pyrenoids, each situated at the intersections of the reticulum. The numerous nuclei (50-60 per cell) lie internal to the chloroplast.

Reproduction (Fig. 9.10 B). Apart from vegetative propagation by accidental breaking of the filament into fragments, akinete formation is the sole method of reproduction. Neither asexual reproduction by spores nor sexual reproduction by gametes has been recorded in this genus.

Akinete Formation. At the time of akinete formation there is contraction of the protoplast followed by migration of most of the cell contents into the upper or distal end of the cell which shows a slight swelling. A cross wall then appears separating the swollen portion forming an akinete from the almost empty longer lower part of the mother cell. The latter contains very little cell contents and is termed the *subsporal* cell. The developing akinete becomes gorged with food reserves. Thereafter additional wall layers develop on the inner side of the cell wall of the developing akinete. The akinete thus has a thick resistant cell wall. It is in fact a resting cell. The akinetes in the filaments may be terminal or intercalary in position. In the latter case they may be solitary or in short series of two or three. Mothes (1930) reported that under cultural conditions the nearly empty half of the mother cell may again become filled with protoplasm and form another akinete. Earnst (1908) found that the akinetes are formed when there is deficiency of nutritive salts. With the drying of the substratum the vegetative cells of the filament perish. The akinetes remain behind and survive.

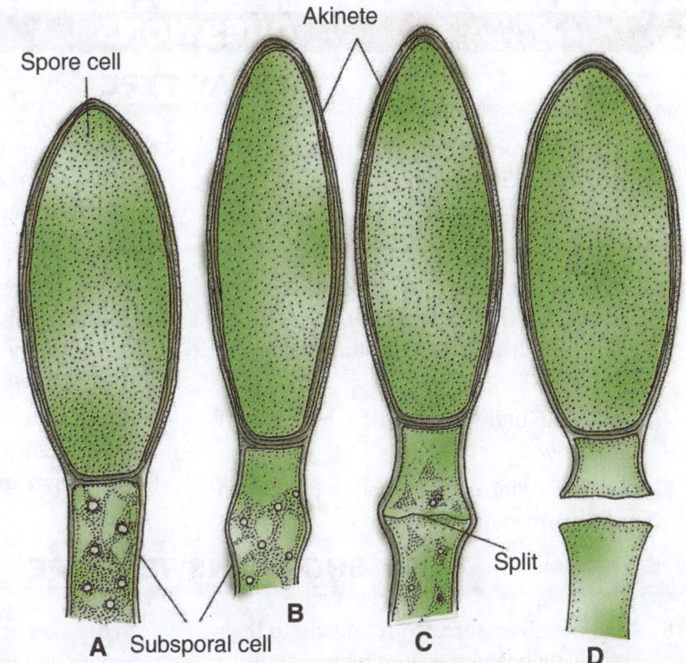

Fig. 9.11 (A-D). *Pithopora.* Showing stages in the formation with separation of akinete. (After Randhawa and Venkataraman).

Randhawa and Venkataraman hold that the akinetes of *Pithophora* are not released by the decay or disintegration of the thallus (Fig. 9.11). According to them when the akinete is nearly mature the scanty contents of the subsporal cell move upward and accumulate in the form of a ring like band a little below the akinete (C). Later the cell wall in the band region becomes swollen and a circumcissal split appears in it (D). With the widening of the split the mature akinete gets separated from the subsporal cell.

Germination of akinetes. With the return of conditions favourable for growth, the akinete germinates. The protoplast divides by a transverse wall into two daughter cells. One of these elongates to form the rhizoid and the other grows into a filament (Earnst, 1908).

Taxonomic Position

Division	:	**Chlorophyta** (Chlorophycophyta)
Class	:	**Cladophyceae**
Order	:	**Cladophorales**
Family	:	**Cladophoraceae**
Genus	:	*Pithophora*

QUESTIONS

ESSAY TYPE

1. Give an account of the structure and reproduction in *Cladophora.*
2. Write what you know about cell structure and thallus in *Cladophora.*
3. Give an account of alternation of generations in *Cladophora*. How is *C. glomerata* different from other species in this respect?
4. Describe briefly the life history of *Cladophora.*
5. Give an account of structure and reproduction in *Pithophora.*
6. Give a diagrammatic sketch of the life cycle of *Cladophora* and indicate the relative length of each generation.
7. Describe the life cycle of *Cladophora.* (*Rohilkhand, 1995*)
8. Describe the life cycle of *Cladophora.* Mention the differences in the life cycles of *Cladophora* and *Oedogonium.*
9. Compare and contrast the sexual reproduction in *Cladophora*, *Ulothrix, Spirogyra* and *Oedogonium.*

SHORT ANSWER TYPE

10. Assign with reasons the following to their systematic position according to Smith :
    (*a*) *Cladophora* (*Rohilkhand, 1991, 1996*)
    (*b*) *Pithophora.*
11. Write short notes on :
    (*a*) Branching in *Cladophora* (*Rohilkhand, 1994, 1994, 1997*)
    (*b*) Cell structure of *Cladophora.* (*Punjab, 1998*)
    (*c*) Sexual reproduction in *Cladophora.*
    (*d*) Sexual reproduction in *Pithophora.*
    (*e*) Difference in sexual reproduction of *Cladophora* and *Oedogonium.*
    (*f*) Formation and evection of Akinetes in *Pithophora.*
    (*g*) Asexual reproduction in *Cladophora* and *Pithophora.*
    (*h*) Vegetative reproduction in *Cladophora.*
    (*i*) Cell division in *Cladophora.*
12. How would you justify the position of Cladophorales as a separate order?
13. (*a*) Describe perennation in *Cladophora.*
    (*b*) What are akinetes? (*Lucknow, 2001*)
14. Comment on the alternation of generation in *Cladophora.* (*Kanpur, 2004*)
15. By mistake the material of the following got mixed up.
    (*a*) *Scytonema* (*b*) *Gracillaria*
    (c) *Cladophora* (*d*) *Ectocarpus*
    What are the criteria which you will use to separate the components? Give suitable diagrams also. (*Kanpur, M.Sc., 2005*)
16. Differentiate between cell structure of *Cladophora* and *Ulothrix.*
17. Compare *Cladophora* and *Pithophora.*

OBJECTIVE TYPE

18. **Select the correct answer :**
    (*i*) Most of the species of *Cladophora* are
    (*a*) Fresh water (*b*) Marine
    (*c*) Terrestrial (*d*) Saline
    (*ii*) The vegetative thallus of *Cladophora* may be
    (*a*) Haploid as a rule
    (*b*) Diploid as a rule
    (*c*) Haploid or diploid

(*d*) Neither haploid nor diploid

(*iii*) Besides *Oedogonium*, which of the following has reticulate chloroplast?

(*a*) *Spirogyra*

(*b*) *Cladophora*

(*c*) *Chlamydomonas*

(*d*) *Ulothrix*

(*iv*) Meiozoospores are produced on

(*a*) Gametothallus

(*b*) Sporothallus

(*c*) Both gametothallus and sporothallus

(*d*) Neither gametothallus nor sporothallus

(*v*) Gametes in *Cladophora* are pear-shaped

(*a*) And biflagellate

(*b*) And tetraflagellate

(*c*) Beaked and biflagellate

(*d*) Without beak and tetraflagellate

(*vi*) Most common method of reproduction in *Pithophora* is through

(*a*) Akinetes

(*b*) Zoospores

(*c*) Fragmentation

(*d*) Chlamydospores

(*vii*) Sexual reproduction in *Cladophora* is

(*a*) Isogamous

(*b*) Anisogamous

(*c*) Oogamous

(*d*) Neither isogamous nor anisogamous

(*viii*) Which of the following has an isomorphic life cycle?

(*a*) *Cladophora*

(*b*) *Oedogonium*

(*c*) *Zygnema*

(*d*) *Caulerpa*

(*ix*) Which of the following has diploid vegetative thallus?

(*a*) *Oedogonium*

(*b*) *Cladophora*

(*c*) *Zygnema*

(*d*) *Caulerpa*

(*x*) A reticulate chloroplast with several pyrenoids occur in

(*a*) *Chlorella*

(*b*) *Chlamydomonas*

(*c*) Coelochaete

(*d*) *Cladophora*

Chaetophorales: sp.

CHAPTER 10

ORDER 7. CHAETOPHORALES

- Family : Chaetophoraceae
- Stigeoclonium
- Draparnaldia
- Draparnaldiopsis
- Fritschiella
- Family : Trentepohliaceae
- Trentepholid

GENERAL CHARACTERISTICS

This order is characterised by a more or less complex organisation of the thallus which is **heterotrichous filament.** The heterotrichous filament consists of **prostrate** and **erect** components. The prostrate component or system is typically a flat structure attached to the substratum. From it arises the upright or erect system consisting of branched threads. These two parts of the thallus are readily recognisable in many species of *Stigeoclonium, Fritschiella* and *Iwanoffia.* The reproductive organs are borne on the erect or projecting system. In some cases the prostrate system disappears and only the erect threads are visible, *e.g., Draparnaldia* and *Draparnaldiopsis.* In other cases the erect system disappears and the prostrate system becomes more or less discoid as in some species of *Coleochaete* and species of *Aphanochaete.* The greater development of one component over the other has been shown to be dependent upon environmental factors. Investigations have shown that the excess of nitrogen in the culture medium inhibits the growth of the prostrate system whereas that of magnesium lowers the growth of the prostrate system in *Stigeoclonium*. This morphological plasticity of the thallus of Chaetophorales is considered to be of great biological significance.

Majority of the plants in this order are enveloped with mucilage and bear hairs or setae on the thallus. They are of two kinds. In some cases the hair are unicellular. They are narrow elongated structures that are colourless and are borne on the cell wall. In still other cases there may be rows of narrow elongated cells terminating the branches. As an example of the latter may be cited the members of the family Chaetophoraceae. In *Coleochaete* the hair are outgrowths of the wall. In some forms the hair are lacking altogether.

Heterotrichous habit and presence of setae or hair are the two distinctive features of this order. The cells are of ulotrichoid type with a parietal chloroplast. The number of pyrenoids varies from one to many.

Classification

Fritsch recognises five families in this order. These are :

1. Chaetophoraceae. *Draparnaldia, Draparnaldiopsis, Stigeoclonium, Chaetophora, Fristchiella, Iwanoffia* and *Aphanochaete* are the important genera included in this family.
2. Trentepohliaceae. It includes *Trentepohlia, Gongrosira* and *Cephaleuros*.
3. Coleochaetaceae is represented by *Coleochaete.*
4. Chaetosphaeridiaceae comprises *Chaetosphaeridium* and others.
5. Pleurococcaceae includes *Pleurococcus.*

There is difference of opinion among the algologists with regard to the classification of these highly differentiated forms of green algae. Many algologists including Morris and Fritsch who consider that heterotrichous type of construction warrants a distinct order for these algae put them in the order Chaetophorales. They consider the latter to be more advanced than the Ulotrichales.

Smith (1950) on the other hand does not recognise this order. He includes all the green algae with heterotrichous habit in a single family Chaetophoraceae under order Ulotrichales. The cytological evidence supports this view (Abbas and Godward, 1963). To them the heterotrichous habit of these algae denotes an advanced condition of the simple unbranched filament. In their view to establish a separate order for branched and unbranched filamentous genera of green algae with a ulotrichoid type of cell structure is unwarranted.

Prescott splits this order into two, **Chaetophorales** comprising two families, namely, Chaetophoraceae and Protodermataceae and **Coleochaetales** including families Coleochaetaceae and Aphanochaetaceae.

As stated above Fritsch (1935) has devided the order into five families : Chaetophoraceae, Trentepohliaceae, Coleochaetaceae, Chaetosphare diaceae and Pleurococcaceae. Of these the families namely Chaetophoraceae, Trentepohliacae, Coleochaetaceae and Pleurococcaceae are considered here.

FAMILY : CHAETOPHORACEAE

The family includes about 50 genera and 225 species. Usually they are aquatic but *Fritshciella* and *Iwanoffia* are terrestrial. They are distinguished from the simple filament of Ulotrichales by the thallus which is branched and filamentous. The filament shows basal, distal differentiation. Plant body in some forms is a heterotrichous filament. It consists of a branched basal and erect or projecting

Fritschiella *Draparnaldiopsis* *Chaetophora*

Stigeoclonium *Draparnaldia*

system but in some genera one or other of these is suppressed. The projecting threads are also branched. The ultimate branches are usually drawn into long sharp, hyaline points or in setae which may be several cells in length. The cells are uninucleate. The single chloroplast is laminate or may be in the form of a girdle encircling the protoplast partially or wholly. Fragmentation is rare. Zoospore formation is quite common. The zoospores may be bi-or quadriflagellate. But for the terminal and the rhizoidal cells, the zoospores may be formed in any cell. The smaller cells produce only one zoospore. The number varies from 2, 4, 8, and 16 in others. Aplanospores and akinetes are also formed under unfavourable conditions. Palmella condition is commonly found in *Stigeoclonium* but is of rare occurrence in others. Sexual reproduction is isogamous. The gametes are biflagellate. Oogamy is present in *Chaetonema.* In *Aphanochaete* the female gamete is motile but larger in size than the comparatively smaller male gamete. This shows heterogamy. The important genera included in this family are *Chaetophora, Stigeoclonium, Draparnaldia, Draparnaldiopsis,* and *Fritschiella.* Of these *Stigeoclonium, Draparnaldia, Draparnaldiopsis* and *Fritschiella* are discussed here is some detail.

STIGEOCLONIUM (KUETZING)

Occurrence. It is a common green alga found in well-aerated, fresh water. The genus comprises about 35 species. Of these 6 have been reported from India. These are *S. tenue, S. lubricum, S. lubricum formasalina, S. flagelliformis, S. nannum* and *S. attenuatum.* They occur in standing water of ponds, springs or flowing streams attached to submerged aquatic plants or stones.

Thallus (Fig. 10.1A). It is usually a heterotrichous filament frequently enclosed in a delicate gelatinous sheath which is of watery consistency and thus is not easily demonstrable. The filamentous thallus is differentiated into a **prostrate portion** and an **erect portion** (A). From the basal prostrate portion arise a number of upright filaments. The latter constitute the erect system. The upright filaments are sparingly branched in an alternate (*S. attenuatum*) or opposite (*S. flagelliformis*) manner. The branches end in long, hyaline, multicellular hairs (C) or terminate in a point. The branch arises from the upper part of the parent cell. The cells of the main filament and of the laterals are of the same size. The projecting system shows **diffuse growth** which means growth by division of any of the cells in the filament. The prostrate system anchors the thallus to the substratum. The filaments forming the prostrate system exhibit apical growth and are made up of shorter cells.

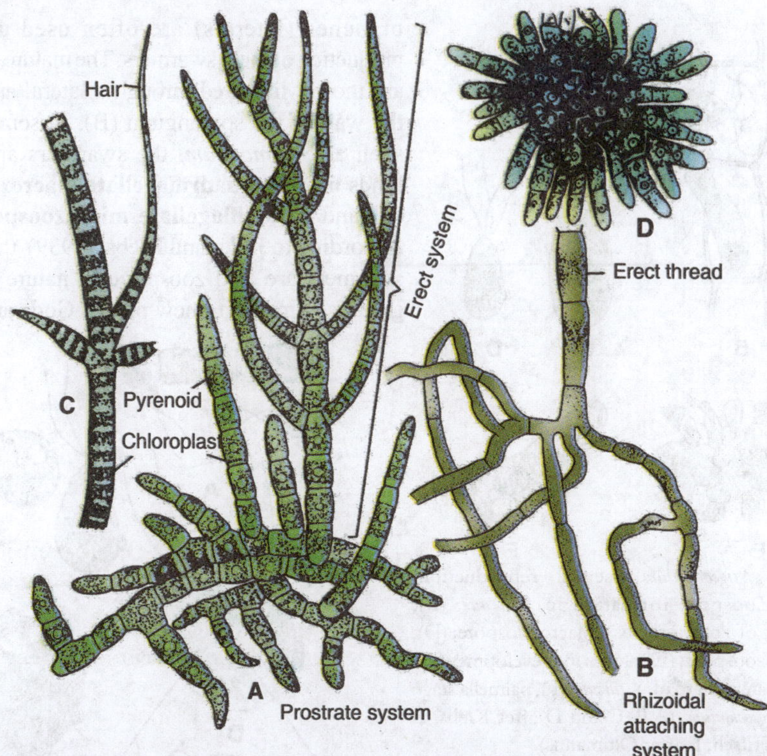

Fig. 10.1 (A-D). *Stigeoclonium.* (A), Showing habit; (B), Rhizoidal attaching system of S. *amoenum* grown in culture; (C). part of an erect threads of *S. protensum;* (D). compact disc-like system of *S. farctum;* (B, after Godward; C, after Thuret, D. after Berthold).

There is great variability in the ratio of development of prostrate to erect system in the different species. In some the former and in others the latter is very much reduced. The more developed the prostrate system, the less developed is the projecting system and *vice versa.* In case the prostrate system is poorly developed, the basal cells of the projecting filaments put out rhizoids which serve as organelles of attachment. In form the prostrate system may be a cushion or a compact disc consisting of numerous branches packed together and coalescent to form a single layered stratum (*S. farctum,* D). In some species the prostrate system is a loosely and irregularly branched thread consisting of a few cells. In still others it is a richly branched compact expanse (A).

Cell Structure. The vegetative cells are uninucleate, each with a single chloroplast which is parietal in position and girdle-shaped in form. It has one or more pyrenoids. Frequently the chloroplast occupies a part of the length of the cell.

Asexual Reproduction. (Fig. 10.2). Besides fragmentation, *Stigeoclonium* reproduces asexually by aplanospores and akinetes. The akinetes are produced in the prostrate system. Palmella stages have been reported by Livingstone (1900) in many species. In this state *Stigeoclonium* multiplies vegetatively. Exposed to desiccation, the component cells of the thallus separate, round off, become thick-walled and divide in any plane. This is the **Palmella stage** (F). In this condition the cells may separate or remain in groups. Under favourable conditions they resume activity and again form a filamentous thallus. The normal method of asexual reproduction, however, is by zoospores.

Zoospore formation (Fig. 10.2 C). It is of frequent occurrence. The zoospores are quadriflagellate and are produced usually singly in the vegetative cells called zoosporangia. The zoosporangia, in the main, are confined to the projecting system (A). All the cells in the smaller

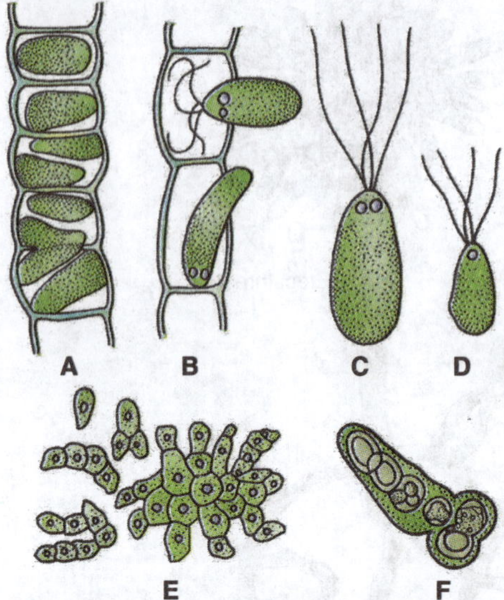

Fig. 10.2 (A-F). *Stigeoclonium*. Asexual reproduction. (A), Zoospore formation in *S. tenue;* (B), escape of zoospores; (C), Macrozoospore; (D), Microzoospore; (E), stages in Development of prostrate system of *S. farcum;* (F), palmella stage of *Stigeoclonium.* (A.B, C and D after Klebs, E after Fritsch; F after Oltimanns).

branches (laterals) are often used up in the production of these swarmers. The mature swarmers are mostly liberated through a lateral aperture in the wall of the sporangium (B). In some species such as *S. amoenum,* the swarmers are of two kinds namely, **quadriflagellate macrozoospores** (C) and **quadriflagellate microzoospores** (D). According to Juller and Klebs, (1939) the smaller swarmers are also zoosporic in nature and may give rise directly to new plants. Godward (1942) reported that in *S. amoenum,* the quadriflagellate microzoospores behave as gametes. Besides the difference in size, the two kinds of zoospores differ in other respects as well. The microzoospores are narrower, have a lateral projecting stigma towards the back end of the body, are slightly paler than the macrozoospores, remain motile for a longer period and exhibit greater sensitiveness to light. The macrozoospores have a flat lateral stigma situated in the middle of the body.

Germination of macrozoospore. After the usual swarming period, the macrozoospore comes to rest and attaches itself at its anterior end to the substratum. It withdraws its flagella and secretes a delicate wall around it. The body of the macrozoospore, in some species, undergoes a transverse division to grow directly into an erect thread. The cells of the new filament next to the substratum later develop into a prostrate system. In other species the one-celled germling first forms the prostrate system from which arise the upright branches constituting the projecting system.

Sexual Reproduction (Fig. 10.3). It is isogamous and takes place by the fusion of

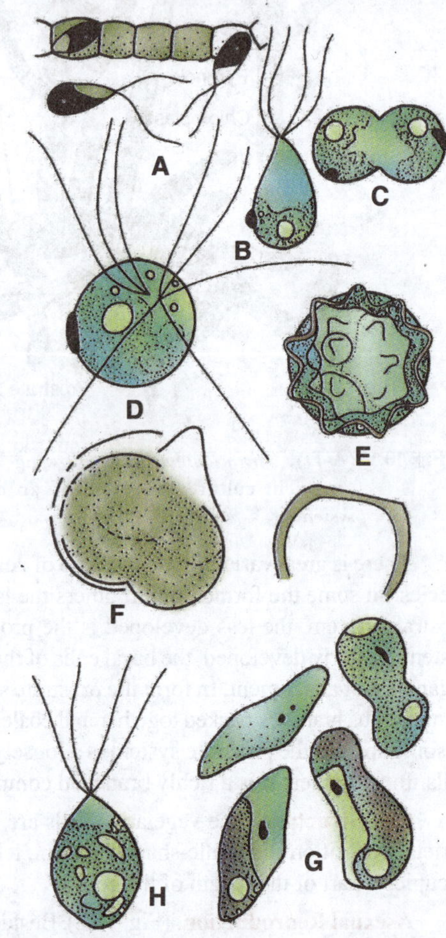

Fig. 10.3 (A-H). *Stigeoclonium*. Sexual reproduction. (A), emerging biflagellate gamete of *S. tenue;* (B), Quadriflagellate gamete of *S. amoenum;* (C). Fusion of quadriflagellate gametes; (D), Octo-flagellate zygospore; (E). Resting zygospore; (F), germinating zygospore with the outer wall layer ruptured and the contents divided into four parts still enclosed by the inner layer, (G). Empty zygospore wall with the four meiozoo-spores having escaped from it; (H), liberated meiozoospore or gono-zoospore. (A, after Smith; rest after Godward).

biflagellate gametes produced singly in the vegetative cells called gametangia (A). The zygote formed is quadriflagellate. When fusion takes place between quadriflagellate swarmers functioning as gametes (mentioned above), the resultant zygote is octoflagellate (D). Copulation between gametes occurs during the swarming period. First the two gametes get entangled by their flagella. This is followed by lateral fusion beginning at the front end. The resultant motile zygote has two eyespots and 4 or 8 flagella. After a period of motility the zygote resorbs its flagella, rounds up and secretes a wall around it (E). The eye--spots have disappeared. The green colour fades away. The zygote now becomes bright orange in colour.

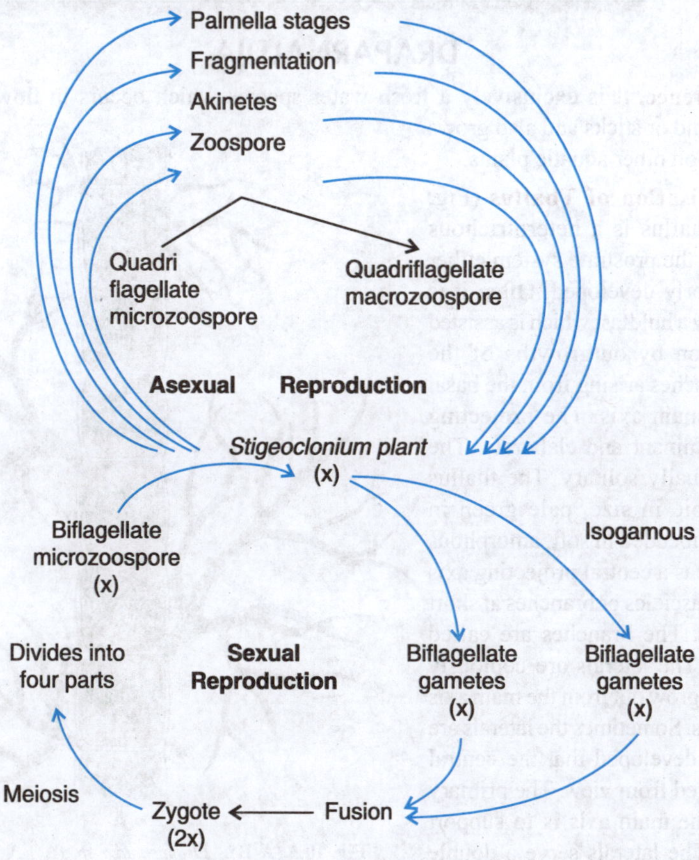

Fig. 10.4. *Stigeoclonium*. Word diagram of the life cycle.

Germination of Zygote. After the resting period the zygote germinates. The contents divide into four parts (F), each with an eye-spot. The division of the nucleus is presumed to be meiotic. Each uninucleate part develops 4 flagella to become a quadriflagellate meiozoospore (H). On liberation each meiozoospore germinates in the same way as the quadriflagellate macrozoospore to form a plant resembling the parent which produced the gametes. Ocassionally the zygote undergoes no resting period. It germinates immediately in day or so as produces directly (without the intervention of spores) a small unbranched filament of diploid cells which represents the diploid stage in the life cycle. This stage later produces meiozoospores by meiosis. This particular species of *Stigeoclonium* (*S. subspinosum*) shows a tendency towards heteromorphic condition and thus exhibits alternation of generations which is heteromorphic.

Taxonomic Position:

Division	:	**Chlorophyta** (Chlorophycophyta)
Class	:	**Chlorophyceae**
Order	:	**Chaetophorales**
Family	:	***Chaetophoraceae***
Genus	:	*Stigeoclonium*
Species	:	*tenue*

DRAPARNALDIA

Occurrence. It is exclusively a fresh water species which occurs in flowing cold water attached to sand or sticks and also grows epiphytically on other aquatic plants.

Organisation of Thallus (Fig. 10.5). The thallus is a heterotrichous filament with the prostrate system either absent or poorly developed. Often it is represented by a holdfast which is assisted in its function by outgrowths of the rhizoidal branches arising from the basal cells of the main axis. The projecting system is dominant and elaborate. The plants are usually solitary. The thallus is macroscopic in size, pale green in colour and embedded in soft, amorphous mucilage. It has a central projecting axis which bears fascicles or branches at short intervals (A). The branches are called the **laterals.** The laterals are copiously branched and grow out from the main axis usually in tufts. Sometimes the laterals are so profusely developed that the central axis is obscured from view. The primary function of the main axis is to support the laterals. The laterals serve a double purpose. They function in photosynthesis and reproduction. The projecting system of *Draparnaldia* thus shows division of labour. The apices of the branches usually terminate in pointed cells or long hair-like protrusions called the estate. From the lower region of the main axis arise numerous, multicellular, rhizoidal branches which anchor the plant to the substratum (B). In some species (*D. glomerata* and *D. villosa),* the rhizoidal branches coil round the main axis to form a covering which is often called the cortical investment.

Fig. 10.5 (A-B). *Draparnaldia sp.* (A), A segment of the thallus; (B), Rhizoid formation in *D. plumosa* (After Berthold).

The main axis consists of large, barrel-shaped cells which are all alike in form, size and structure. The cytoplasm is vacuolate and contains a single small parietal chloroplast. The chloroplast has incised edges and forms a girdle in the equatorial region. It is entire or reticulate and has numerous pyrenoids. There is a single nucleus.

The laterals are composed of much shorter cells with dense cytoplasmic contents. Each cell has a single nucleus. The parietal plate-like chloroplast wholly fills the cell and contains a single pyrenoid. The component cells of the laterals diminish in size, the more distal they are from the proximal end.

The thallus of *Draparnaldia* can easily be identified by, (*i*) the presence of a central axis consisting of large barrel-shaped cells, (*ii*) highly branched whorls of laterals consisting of much shorter cells which diminish in size, the farther they are from the proximal end, (*iii*) branches terminating in pointed cells or long, hair-like setae and (*iv*) the soft copious mucilage that bathes the entire thallus.

Reproduction. *Draparnaldia* reproduces both by asexual and sexual method of reproduction.

Asexual reproduction (Fig. 10.6). It takes place by means of quadriflagellate zoospores (A). They are produced in the branch cells usually singly. Not uncommonly all the lateral branches may be used up in zoospore formation (*D. plumosa* and *D. glomerata*). In some species the number of swarmers produced in each cell may vary from one to four. The zoospores are pyriform, quadriflagellate structures with definite anterior and posterior regions. There is an eyespot present in its anterior region. Two contractile vacuoles have also been reported. There is a single chloroplast occupying the posterior part of the zoospore. It contains one or two pyrenoids. The zoospores escape through an aperture in the lateral wall. The liberated zoospores swim about for a short while. After a short period of activity they come to rest with the anterior flagellated end downwards. The flagella are resorbed. The quiescent zoospore secretes a wall around it. The posterior part elongates. The nucleus undergoes division followed by wall formation. In this way the germling directly develops into the upright axis. The terminal cell is produced into a long seta. The germling develops a basal rhizoidal cell when it is 4 or 5 cells in height. Lateral branches arise from the main axis later when the germling becomes sufficiently older.

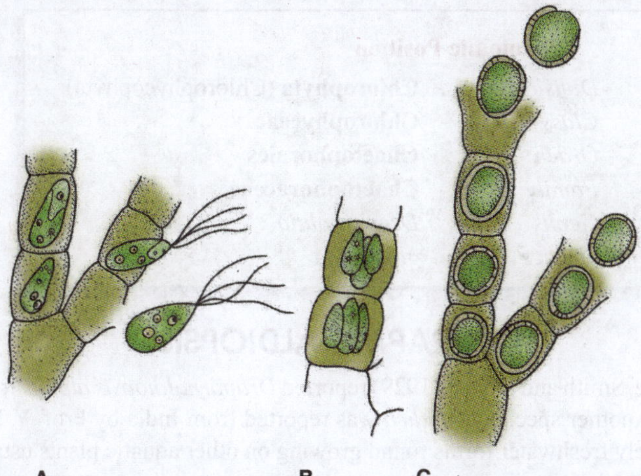

Fig. 10.6. (A-C). Asexual reproduction in *Draparnaldia.* (A), Formation and liberation of zoospores; (B), Each cell with four zoospores; (C), Formation of hypnospores.

Formation of **hypnospores** has also been reported under unfavourable conditions (C). They are formed singly. The protoplast of the hypnospore is coloured deep orange.

Several authors have reported two kinds of zoospores in *Draparnaldia,* quadriflagellated macrozoospores and quadriflagellated microzoospores. Uspenskaja reported that the formation of two kinds of swarmers can be induced by changing the hydrogen ion concentration in *Draparnaldia glomerata.* Microzoospores are formed when the waters are markedly alkaline. Certain authors like Pascher are of the view that these quadriflagellate microzoospores have taken over the sexual role by completely eliminating the biflagellate swarmers as found in *Ulothrix.*

Sexual reproduction. (Fig. 10.7). It is isogamous and takes place by the formation of quadriflagellate gametes smaller in size than the zoospores (A). They are produced in the cells of the short lateral branches. They are likewise liberated through a lateral aperture in the wall. Before union, the gametes withdraw their flagella, and become amoeboid (B). The amoeboid gametes fuse in pairs (C) to form a zygote. The zygote secretes a wall around it (D) and germinates directly into a new plant. The nuclear behavior is not known. According to Pascher (1907) it gives rise to four or five germilings. Anyhow the details are lacking. The gametes which fail to fuse may round off and develop parthenogenetically into new plants. In a certain species of *Draparnaldia* gametes are produced under alkaline conditions and zoospores under neutral and acid situations.

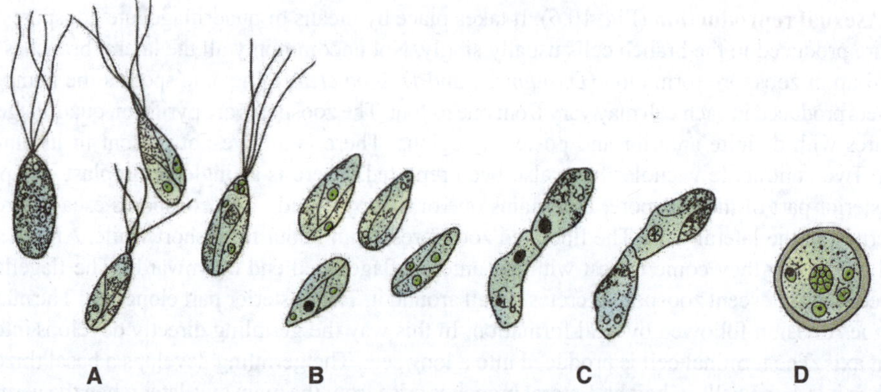

Fig. 10.7 (A-D). *Draparnaldia.* Sexual reproduction. (A), isogametes; (B) gametes after shedding flagella: (C), fusion of gametes; (D), zygospore.

Taxonomic Position

Division	:	**Chlorophyta** (Chlorophycophyta)
Class	:	**Chlorophyceae**
Order	:	**Chaetophorales**
Family	:	**Chaetophoraceae**
Genus	:	*Draparnaldia*
Species	:	*glomerata*

DRAPARNALDIOPSIS

Occurrence. Smith and Klyver (1929) reported *Draparnaldiopsis alpinis* from the Huntington lake, California. Another species *D. indica* was reported from India by Prof. Y. Bhardwaja (1933). Both are exclusively freshwater forms found growing on other aquatic plants usually in the shallow water of ponds and lakes.

Organisation of the thallus. (Fig. 10.9, A). The macroscopic filamentous thallus shows further advance in vegetative organisation than *Draparnaldia.* It is covered by a broad gelatinous matrix of thin consistency and consists of a main axis bearing laterals of limited (B) and unlimited growth. The upright main axis is attached to the substratum by means of rhizoidal branches which are multicellular and arise from the basal cell (Fig. 10.10, C). It is more or less of uniform breadth except the base and the apex where it is attenuated. The main axis is made up of two kinds of cells (A). They are the short **nodal cells** and the along **internodal cells.** The internodal cells are elongate and cylindrical but sometimes barrel-shaped. They are usually 2 or 3 times as long as the nodal cells. The latter are short, discoidal and barrel-shaped. These two types of cells are joined end to end regularly alternating with each other throughout the entire length of the filament. Sometimes there may be more than one internodal cell between the two successive nodal cells. The nodal cells along bear the laterals which are of two types :

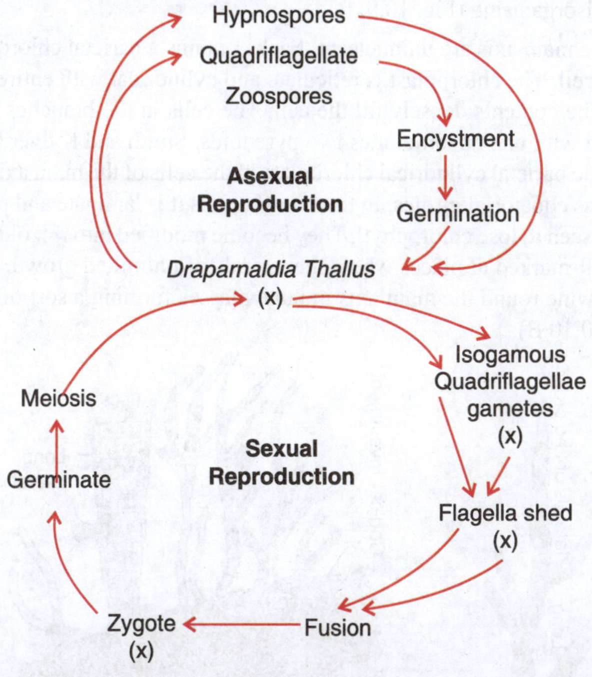

Fig. 10.8. Word diagram of the life cycle of *Draparnaldia*.

(*i*) *Long branches or branches of unlimited growth* (Fig. 10.10, A). The long branches arise at intervals from the nodal cells. Their number varies from one to four per cell. They repeat the structure of the main axis.

(*ii*) *Short branches or branches of limited growth* (Fig. 10. 9, B). The short branches normally arise in whole of four or in opposite pairs in *D. indica* from the median or any part of the nodal cell and are set at right angles to the main axis. They are profusely branched and show no evidence of an axis. The same nodal cell may bear branches of both kinds or all of limited growth or all of unlimited growth. Smith reported that in *D. alpinis* the basal cells of the short branches are cuneate and di-to trichotomously forked at the apex. The cells borne by these basal cells may also be forked dichotomously at their apices and this may continue for three or four cells formed in succession. The

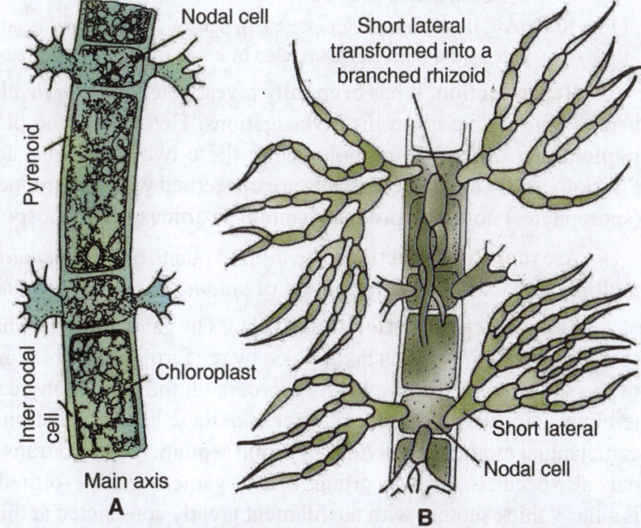

Fig. 10.9. (A-B). *Draparnaldiopsis indica.* (A), portion of main axis with alternating nodal and internodal cells; (B), portion of main axis showing origin of short laterals from the nodal cells and complete transformation of a short lateral (*b*) into a branched rhizoid (After Bhardwaja).

ultimate branches of the short laterals taper towards the apex ending in long, colourless, hair-like processes consisting of one or more greatly elongated cells in which the chloroplasts have totally disappeared or are disorganising (Fig. 10.9, B).

The cells of the main axis are uninucleate. Each contains a parietal chloroplast occupying the whole length of the cell. The chloroplast is reticulate and cylindrical with entire margins. There are several pyrenoids. The contents densely fill the cell. The cells in the branches have also a parietal reticulate chloroplast with one or sometimes two pyrenoids. Smith and Kylver (1929) reported that in *D. alpinis* the single parietal cylindrical chloroplast in the cells of the main axis and long branches is reticulate and forms equatorial girdles. In the short laterals it is laminate and parietal. In *D. indica* the short laterals are seen to lose chlorophyll. They become modified into rhizoid-like branches. This tendency is very well-marked at places where the laterals of unlimited growth arise. The rhizoidal threads sometimes twine round the main axis in huge masses forming a sort of cortical investment concealing it (Fig. 10.10 B).

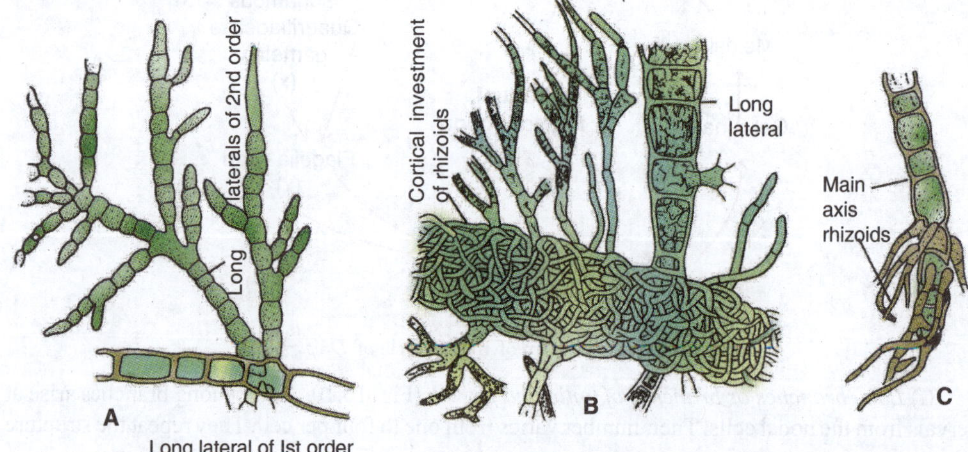

Fig. 10.10 (A-C). *Draparnaldiopsis indica.* (A), part of a long lateral bearing long laterals of the other orders; (B). part of the main axis concealed by cortical covering of rhizoids (After Bhardwaja).

Reproduction. It has been fully investigated in *D. indica* by Singh (1942). The account given here is entirely based on his investigations. Genetically the plants in *D. indica* are of two types, **haploid** and **diploid**. Morphologically these two kinds of plants are similar as in *Chladophora.* Physiologically the haploid plants are concerned with the production of gametes. The diploid plants (sporophytes) normally produce haploid swarmers (meiozoospores) by meiosis.

Asexual Reproduction. The diploid plants of *Draparnaldiopsis indica* have been reported to multiply asexually by the formation of *aplanospores* and *akinetes* which are diploid.

Sexual Reproduction (Fig. 10.11). The gametophyte or haploid filaments are concerned with sexual reproduction which takes place by the formation of biflagellate gametes produced in the cells of the short laterals of the last two orders on the gametophyte or sexual plants (A). Excepting the terminal hair cells most of the other cells of these laterals function as **gametangial mother cells.** Each gametangial mother cell divides by a thin septum. A second transverse division and rarely a third one may also occur. In this way groups of 2 - 4 gametangia are formed. At either end of such a group there is a fairly thick septum with the filament greatly constricted at this place. The septa between the cells of each group of gametangia are thin and the walls only slightly constricted. Later the gametangia in each become barrel-shaped. The stretched septa between the individual gametangia of the group become thin and almost indistinct. Their protoplasts round off and recede from the gametangial wall. The rounded protoplast of each gametangium develops two flagella and takes on the shape of a gamete. The barrel-shaped structure, at this stage, contains gametes equal in number to the gametangia in the group. The mature gametes are liberated by the gelatinisation of the longitudinal cell walls of the barrel-shaped structure. The liberated gametes are rounded or ovoid nucleated bodies with the two equal flagella of whiplash type arising anterioly (B–F). The nucleus lies more or less centrally in the colourless cytoplasm of the anterior portion. The single crescent-shaped chloroplast lies at the posterior end. It has a single pyrenoid. The linear eye-spot is located laterally at the posterior end. The fusion takes place between the isogametes (G) from two different plants. *D. indica* thus is **heterothallic.** Sometimes the fusing gametes are slightly unequal in size. They face each other by their anterior ends, become entangled by the flagella and fuse (G-I). Fusion starts at their anterior ends (C). The zygote is quadri-flagellate and motile for some time (I). At this stage it contains two nuclei, two pyrenoids and two eye-spots. Soon it comes to rest. The flagella are withdrawn. The quiescent zygote then gradually rounds off and secretes a thin wall around it. At this stage the two pyrenoids fuse and so do the two nuclei but the eye spots disappear. The zygote undergoes no rest. It germinates directly to produce the new diploid thallus by repeated mitosis as in *Cladophora*.

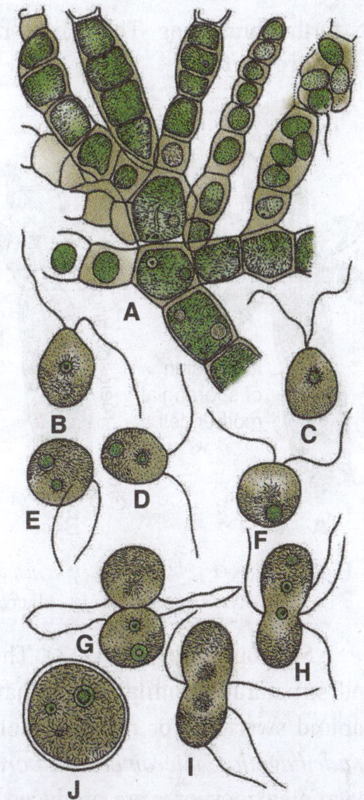

Fig. 10.11 (A-J). *Draparnaldiopsis indica.* Sexual reproduction. (A), apical part of a long lateral showing conversion of cells of short lateral into gametes; (B–F), different shapes of gametes; (G–I), fusion of gametes; (J), Zygote (After R. N. Singh).

Germination of Zygote (Fig. 10.12). At the time of germination the spherical zygote (A) elongates to become an ovoid structure. A hyaline protuberance is formed at one end (B). A septum then appears dividing the ovoid zygote into a 2-celled germling (D). The upper cell develops a chloroplast with one or two pyrenoids and thus becomes green. It divides and redivides to form the main filament consisting of numerous diploid cells. Later the component cells of the main filament become differentiated into internodal and nodal cells. From the latter begin to arise the laterals of both kinds. The lower cell of the 2-celled germling is colourless and has meagre contents (E). It becomes slender and undergoes branching to form a tuft of attaching rhizoidal branches. The rhizoidal system extends

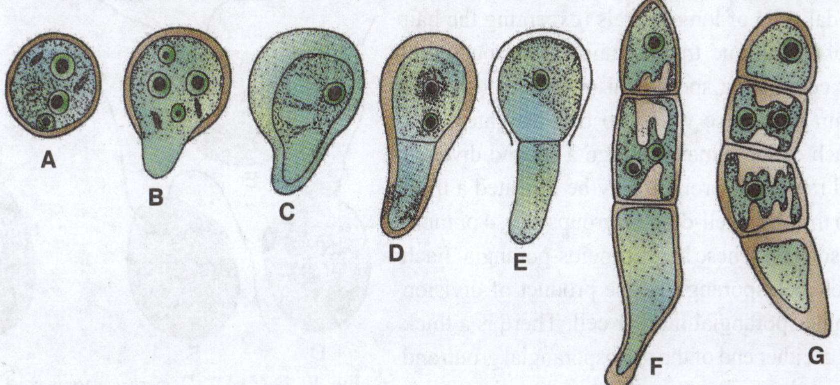

Fig. 10. 12 (A-G). *Draparnaldiopsis indica.* Germination of zygote. Explanation in the text. (After R. N. Singh).

by further branching. The plant formed by the germination of the zygote is diploid and thus called the **sporophyte.**

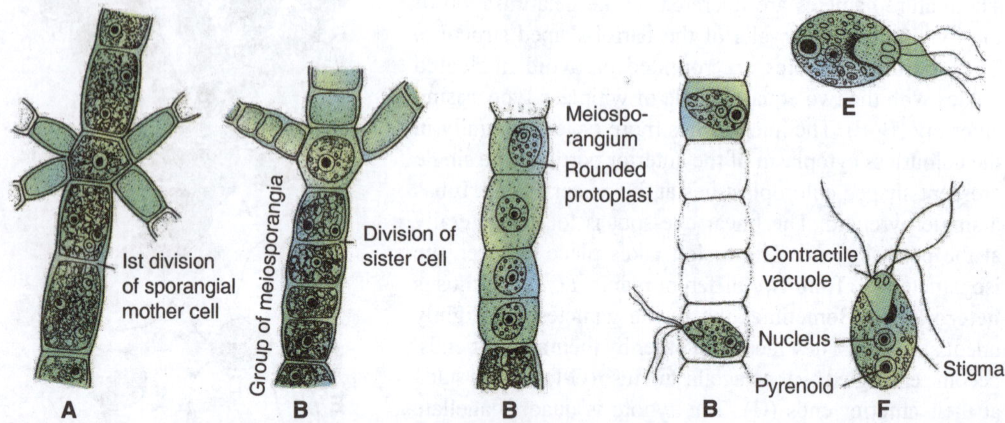

Fig. 10. 13 (A–F). *Draparnaldiopsis indica.* Stages in the development of macromeiozoospores (A–C); (D), liberation of zoospores; (E–F). Macromeiozoospores (After R.N. Singh).

Sporophyte (Fig. 10.13). The sporophyte or diploid filamentous thallus is morphologically and structurally similar to the haploid plants which bear the gametes. It, however, produces, haploid swarmers or meiozoospores of 3 kinds namely *quadriflagellate macromeiozoospores, quadriflagellate micromeiozoospores* and the *biflagellate micromeiozoospores.* The quadriflagellate macromeiozoospores are produced in the internodal cells of the long laterals of the first order (A). The quadriflagellate micromeiozoospores are formed in the internodal cells of the long laterals of the subsequent order. Excepting the hair cells, the cells of the short laterals of all orders produce only the biflagellate micromeiozoospores. The meiozoospores of all kinds are differentiated from the diploid protoplast of their respective sporangia by **meiosis**. Excepting the nodal cells of long branches and terminal hair cells of short laterals, most of the cells take part in the formation of meiozoospores.

(*a*) Development of Sporangia. Most of the internodal cells of long laterals (excepting the hair cells) of the diploid thallus function as sporangial mother cells. Each sporangial mother cell divides by a thin transverse wall into two daughter cells (A). Each of these may undergo a second division (B) and rarely the process may be repeated a third time. In this way well-defined groups of 2, 4 or more cells result (B). These are the meios-porangia. Each group of meiosporangia is the product of division of a single sporangial mother cell. There is a thick septum at either end of the meiosporangial group and the lateral shows an appreciable constriction at this point. Subsequently each group of meiosporangia

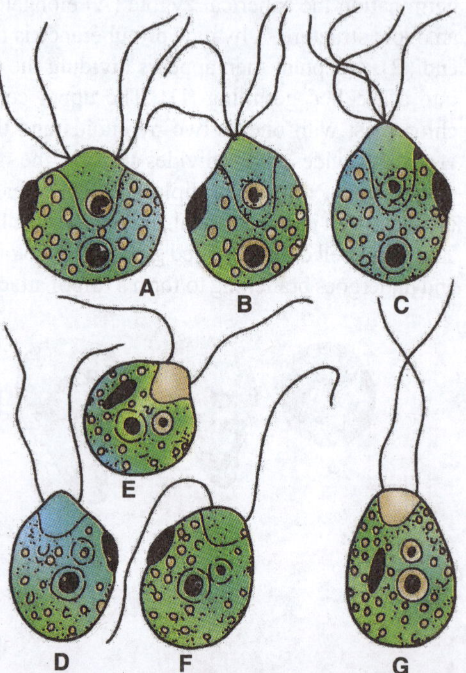

Fig. 10. 14 (A–D). *Draparnaldiopsis indica.* Different kinds of micro meiozoospores (After R.N. Singh).

becomes barrel-shaped due to the gradual elongation and turgidity of their protoplasts. The walls between the individual meiosporangia become very thin and indistinct.

(*b*) Differentiation of meiozoospores. At this stage the protoplast of each meiosporangium recedes from the cell wall and rounds off (C). The rounded protoplast develops flagella at its anterior end and takes on the shape characteristic of the swarmer (D). The mature meiozoospores are finally liberated by the gelatinisation of longitudinal walls of the meiosporangia (D). The quadriflagellate macromeiozoospores are ovoid or elliptical in shape (E and F). The nucleus is located in the centre. The anteriorly inserted four whiplash flagella are of equal length and are nearly sub-terminal in position. The flattened red stigma is situated somewhat medianly but more towards the anterior end. The massive cup-shaped chloroplast fills the major part of meiospore and has a single pyrenoid at its posterior end. The two contractile vacuoles lie at the anterior end immediately beneath the points of insertion of flagellate.

The quadriflagellate micromeiozoospores are nearly spherical or pyriform (Fig. 10. 14 A–C). The former are flattened and slightly bulged in the median portion (A). The four flagella are inserted around the anterior protoplasmic beak-like papilla. In the pear-shaped form, the anterior end is pointed (B–C). The four flagella arise from the pointed end. The nucleus is located more or less centrally in the colourless cytoplasm of the cup of the basin-shaped massive chloroplast with a pyrenoid at its posterior end. The projecting eye-spot is situated laterally nearer its anterior end. The biflagellate micromeiozoospores vary in shape and may be ovoid, obovoid, spherical or subspherical in form (Fig. 10.12 D–G). Each has a single massive slightly notched chloroplast with a single pyrenoid situated at the centre and the nucleus slightly above. There are two anteriorly inserted whiplash flagella one on either side of the protoplasmic beak. The projecting eye-spot lies in the middle.

Germination of meiozoospore (Fig. 10.15). The liberated meiozoospores swim about in water (A). Finally they come to rest (B), the macromeiozoospore after 8 hours and micromeiozoospores after 24 hours. Each of them attaches itself to some convenient object by its anterior end. It withdraws its flagella and secretes a delicate wall around it. The quiescent, clothed meiozoospore forms a colourless protuberance at its attached end (C–D). The protuberance elongates and becomes cut off near its base by a septum (E). A 2-celled germling is thus formed. The lower colourless cell functions as the rhizoidal cell. It elongates into a slender rhizoid and has meagre contents. It bears the branched rhizoids (H). Meanwhile the upper cell of the germling becomes green. It divides a number of times to form a short filament of a few cells (F). Each cell of the filament develops a band-shaped chloroplast with 2 or 3 pyrenoids. In due course of timely repeated cell division of green cells a main filament consisting of numerous cells is formed (G). All the component cells are, at first, alike. Later on differentiation into nodal and internodal cells takes place. Laterals of both kinds begin to arise from the nodal cells. The rhizoid by this

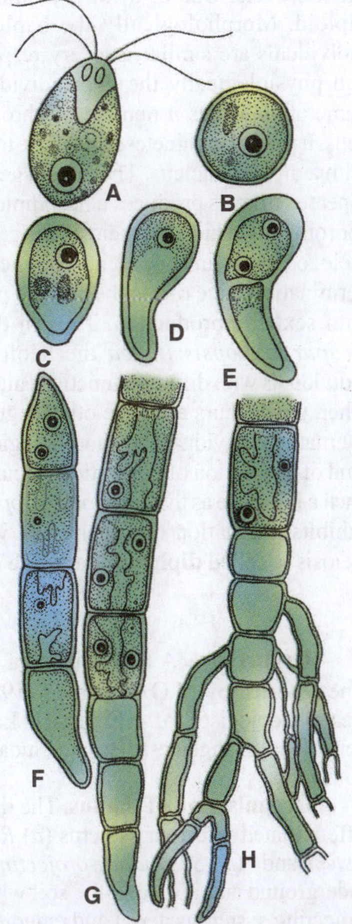

Fig. 10.15 (A-H). *Draparnaldiopsis indica.* Germination of macromeiozoo spores (After R. N. Singh).

time has developed into a rhizoidal system by branching (H). The biflagellate micromeiozoospores germinate into plants with narrower filaments. The haploid thallus formed by the germination of meiozoospores reproduces by gametes.

Taxonomic Position :

Division	:	**Chlorophyta** (Chlorophycophyta)
Class	:	**Chlorophyceae**
Order	:	**Chaetophorales**
Family	:	**Chaetophoraceae**
Genus	:	*Draparnaldiopsis*
Species	:	*indica*

Alternation of Generations (Fig. 10.16). As in *Cladophora,* the life cycle of *Draparnaldiopsis indica* exhibits alternation of generations which is isomorphic. There occurs two distinct vegetative individuals in the life cycle. One of them is haploid and the other diploid. Morphologically the haploid and diploid individuals are similar in every respect. Genetically and physiologically the two individuals differ. The gametothallus has *n* number of chromosomes in the cells. It produces gametes which fuse to form the zygote with a diploid nucleus. The zygote germinates and by repeated mitosis produces a new diploid individual or sporothallus which by meiosis differentiates haploid meiozoospores in sporangia. The meiozoospores, on germination, give rise to the haploid plants concerned with sexual reproduction. Thus in the life cycle of *Draparnaldiopsis indica* the haploid and diploid individuals with different functions alternate with each other. One occurs after the other regularly. Since the alternating individuals are morphologically similar this kind of alternation of generations is called isomorphic. Such a life cycle as that *Draparnaldiopsis indica* which exhibits alternation of generations with sporogenic meiosis is called **diplohaplontic life cycle**.

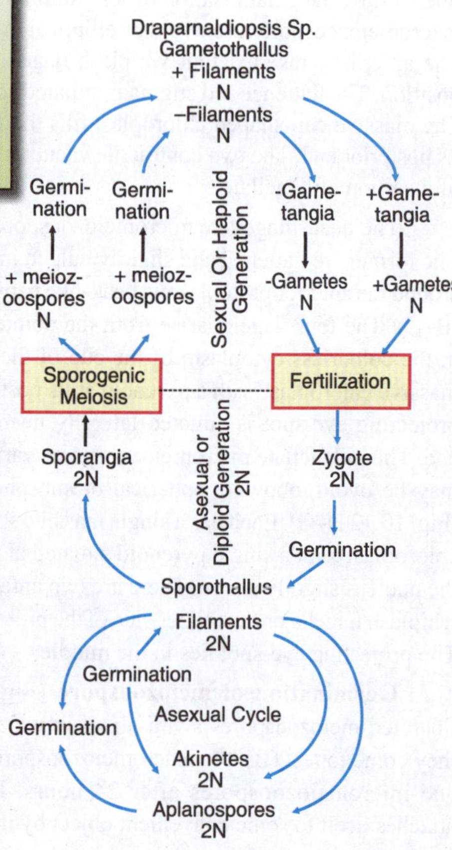

Fig. 10.16. *Draparnaldiopsis indica.* Word diagram of life cycle.

FRITSCHIELLA

Occurrence. A terrestrial alga which grows usually on moist alkaline soils was for the first time reported by M.O.P. Iyenger (1932) from South India and was named after his teacher and a great phycologist Prof. F.E. Fritsch. Later on it was also collected from other parts of India as well as from Nepal, Bangladesh, Burma, Sudan and Japan. *Fritschiella tuberosa* is the only reported species from India.

Organisation of thallus. The thallus of *Fritschiella* shows heterotrichous organisation and is differentiated into four systems (*a*) *Rhizoidal system,* (*b*) *Prostrate system,* (*c*) *Primary projecting system* and (*d*) *Secondary projecting system.* Out of these, rhizoidal and prostrate systems are underground and buried in the soil whereas primary projecting system is subaerial. Only secondary projecting system is aerial and emerges out of the soil surface.

The *rhizoidal system* consists of one or more septate rhizoid like elongated structures. It grows downwardly and does have colourless cells due to the absence of chloroplasts. However, it is absent in *Fritschiella simplex,* a species reported from Bangladesh.

The *prostrate system* is made up of rounded or irregularly swoller cluster of cells. This is branched filamentous, tuberous or parenchymatous and in mature thallus comprises of short congested branches which are differentiated into nodes and internodes. Certain nodal cells of prostrate system give rise to rhizoids towards lower side (which penetrate into the soil) and primary projecting system towards upper side. The *prostrate system* in young plants is not well-developed but gradually it increases in size with the growth of the thallus and at maturity the thalli have very well-developed prostrate system.

The primary projecting system develops from prostrate system and is erect. It consists of uniseriate or biseriate filaments which may be simple or branched, sub-aerial and green. The cells are small and round in shape and resembles the cells of prostrate system.

The secondary projecting system is aerial and consists of freely branched uniseriate filaments. It is given out from the primary projecting system and has elongated cells. It is less developed in the thalli growing in exposed areas and is absent in *F. simplex*.

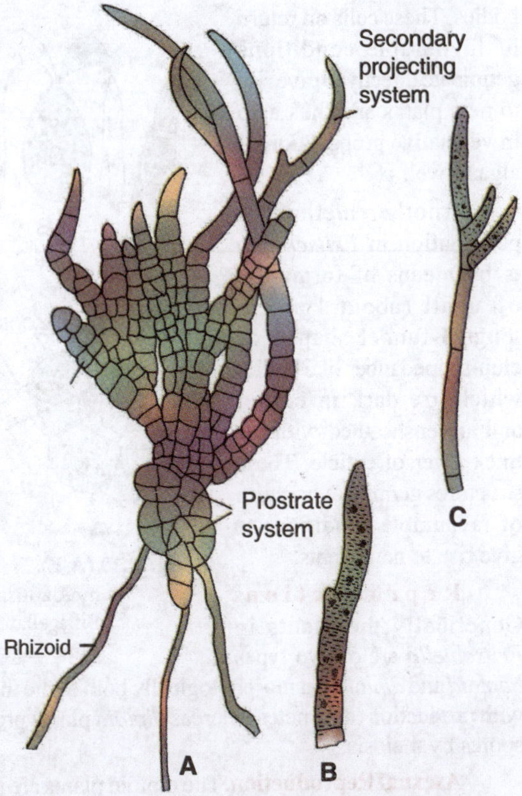

Fig. 10.17 (A-C). *Fritschiella.* (A), showing habit of plant; (B and C), segment of thallus showing secondary projecting system.

Cells structure. The cells are uninucleate and thin walled. Except the cells of rhizoidal system all the cells are green due to the presence of chloroplasts. The cells of secondary projecting system posses *curved plate-like* chloroplast which have 2-8 pyrenoids. The cells of primary projecting system and prostrate system have usually less developed chloroplasts with 2-4 pyrenoids and dense cell contents. However, the cells of rhizoidal system do not have chloroplasts and are colourless.

Methods of Perennation. During unfavourable conditions sub-aerial primary projecting system and aerial secondary projecting system alongwith the rhizoidals system usually degenerates and remaining prostrate system serves as preventing structure. The cells of the prostrate system perennate during dry conditions without undergoing any remarkable change in structure and on return of favourable conditions germinate directly to give rise to new plants. However, the nodal cells of prostrate system sometimes get detached from the parent thallus due to death and decay of internodal cells and serve as perennating

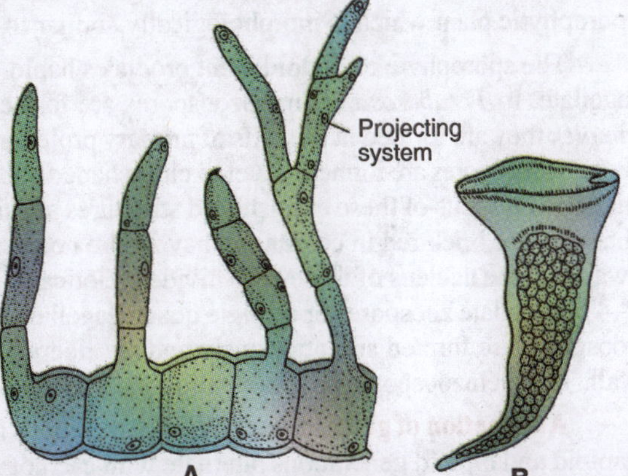

Fig. 10.18 (A-B). *Fritschiella.* Methods of Perennations. (A), Germination of pesennating prostrate system to produce primary projecting systems; (B), Funnel-shaped tuberous body.

bodies. These cells on return of favourable conditions germinate directly to give rise to new plants and thus help in vegetative propagation of alga as well.

Another method of perennation in *Fritschiella* is by means of formation of small (about 1mm in length) funnel shaped or club-shaped tuber like bodies which are dark in colour and are ensheathed with the thick layer of cuticle. These structures germinate or return of favourable conditions to give rise to new plants.

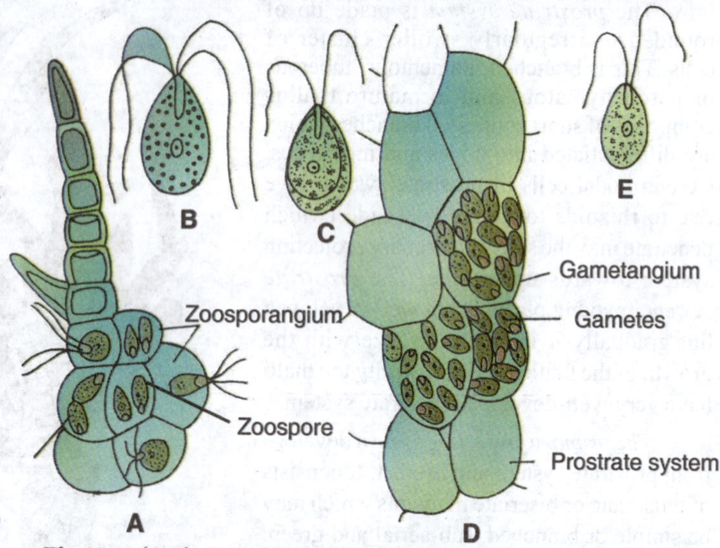

Fig.10.19 (A-E). *Fritschiella.* (A), Segment of thallus showing meiozoospores; (B). Quadriflagellate meiozoospores; (C), Biflagellate meiozoospore; (D), Segment of thallus showing gametangia; (E), Biflagellate gamete.

Reproduction. Genetically the plants in *Fritschiella* are of two types *haploid* and *diploid* but morphologically both of the them are similar. The *haploid* plants are concerned with production of gametes whereas *diploid* plants produce biflagellate and quadriflagellate meiozoospores by meiosis.

Asexual Reproduction. The diploid plants are reported to multiply asexually by akinetes which are usually produced by the cells of ultimate branchlets. However, sometimes these structures are also found in any cell of the thallus.

Sexual Reproduction. The sexual reproduction is isogamous and biflagellate gametes develop in the cells of prostrate system of haploid plants. During the gametes formation the protoplast of gametangium divides repeatedly and large number of biflagellate gametes are produced. The gametes are morphologically similar to meiozoospores except that they are smaller in size. The gametic union results in the formation of diploid zygote which germinates immediately to give rise to diploid sporophytic plant which is morphologically similar to the haploid gametophyte.

The sporophyte or diploid plant produces haploid swarmers which are biflagellate or quadriflagellate. In *F. tuberosa* the meiozoospores are formed in the cells of prostrate system while in *F. simplex* they are formed in the cells of primary projecting system. The filaments, in the cells of which the meiozoospores are formed, develop club-shaped endings and get covered with a thick dark brown pellicle. The cells of these club-shaped structures are similar to the other cells of the filament except that they are brick red in colour and have more homogenous cell contents. During the formation of swarmers the nucleus of these cells divide meiotically to form four haploid meiozoospores. Usually 2–4 biflagellate zoospores or a single quadriflagellate zoospore per cell is formed. Both the types of zoospores are formed simultaneously and are liberated through a lateral aperture in the sporangial wall. The meiozoospores on germination produce haploid plants (Fig. 10.20).

Alternation of generation. *Fritschiella* exhibits isomorphic *alternation of generation* in which haploid and diploid generations alternate with each other and have morphologically identical thalli. The sporophytic thallus produces through meiosis haploid meiozoospores which on germination give rise to the gametophyte. The gametophyte reproduces sexually and produces gametes of similar shape and size (isogametes) which after fusion form diploid zygote. The zygote after germination produces diploid sporophyte which is morphologically identical with the haploid gametophyte.

Projecting system
Rhizoid
A
B
C
D
E
F
G

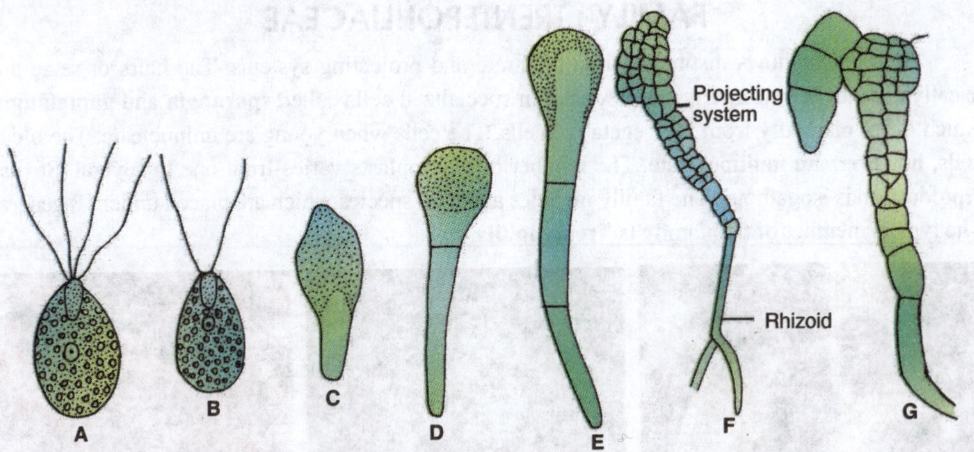

Fig. 10.20 (A-G). *Fritschiella.* Different stages of germination of meiozoospores and formation of thallus.

Taxonomic Position :

Division	:	**Chlorophyta** (Chlorophycophyta)
Class	:	**Chlorophyceae**
Order	:	**Chaetophorales**
Family	:	**Chaetophoraceae**
Genus	:	*Fritschiella*
Species	:	*tuberosa*

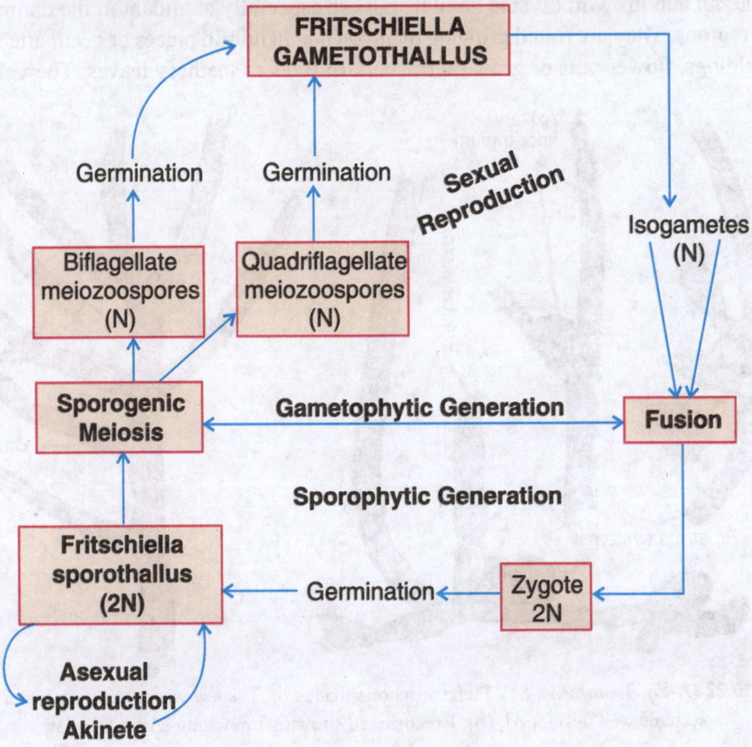

FRITSCHIELLA GAMETOTHALLUS
Germination
Germination
Sexual Reproduction
Isogametes (N)
Biflagellate meiozoospores (N)
Quadriflagellate meiozoospores (N)
Sporogenic Meiosis
Gametophytic Generation
Fusion
Sporophytic Generation
Fritschiella sporothallus (2N)
Germination
Zygote 2N
Asexual reproduction Akinete

Fig. 10.21. *Fritschiella tuberosa.* Word diagram of life cycle.

FAMILY : TRENTEPOHLIACEAE

The thallus shows distinction into prostrate and projecting systems. The hairs or setae are usually absent. The swarmers are produced in specialized cells called **sporangia** and **gametangia** which differ markedly from the vegetative cells. The cells when young are uninucleate. The older cells, however, are multinucleate. The number of chloroplasts varies from one to several. Sexual reproduction is isogamous. The family includes about 80 species which are placed under 18 genera. The typical member of this family is **Trentepohlia**.

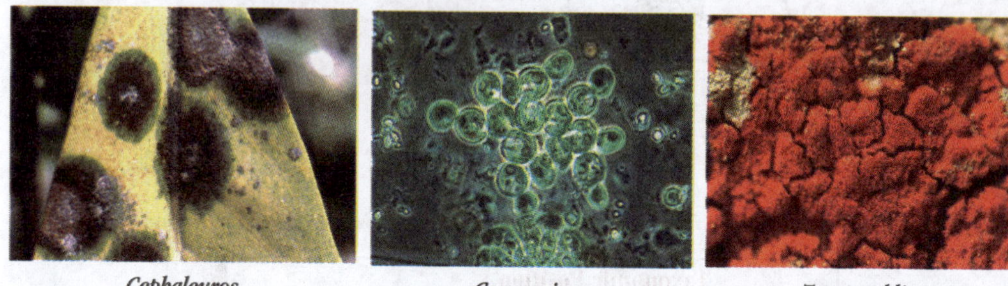

Cephaleuros *Gongrosira* *Trentepohlia*

MEMBERS OF FAMILY TRENTEPOHLIACEAE

TRENTEPOHLIA

The genus comprises about 50 species. Eleven species of this genus have been reported mostly from the North-Eastern region of India (Bruhl and Biswas, 1923). Jose and Chowdary (1980) also added 7 more to the list raising the number of Indian species of this genus to eighteen. All the species are strictly aerial and grow in diverse habitats but are especially abundant in the damp tropical and subtropical regions. They are found growing on moist soil in humid places or occur attached to rocks, walls of buildings, flower pots or grow on the bark of trees or leathery leaves. The yellow to range

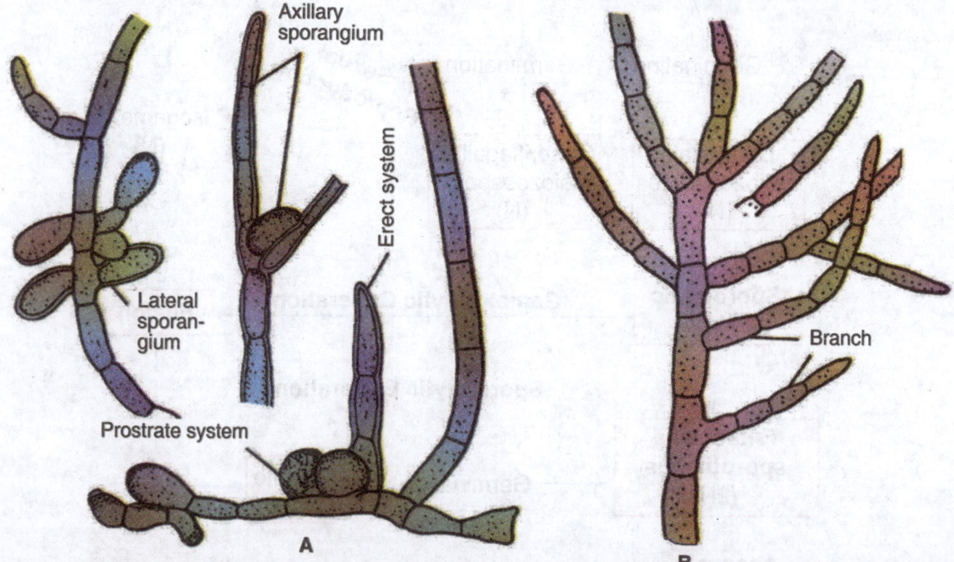

Fig. 10.22 (A-B). *Trentepohlia.* (A), Heterotrichous thallus of *T. aurea* with both the prostrate and erect systems well developed; (B), Erect thread showing branching (After Chodat).

red colouring of the thallus, due to the presence in the cells of abundant haematochrome, makes it quite conspicuous. Some species occur as an algal component of many lichen thalli.

Organisation of Thallus (Fig. 10. 22). It is filamentous and branched. Usually it shows distinction into prostrate and erect systems (A). In many species both the systems are well-developed and branched (*T. aurea*). The branching in the erect threads (B) may be alternate, opposite or unilateral. The laterals arise from the top, middle or subterminally from the parent cell. In some species (*T. umbrina*), the prostrate portion is well-developed but the erect threads are reduced. Some other species have the erect system more extensive than the prostrate system. The growth is apical.

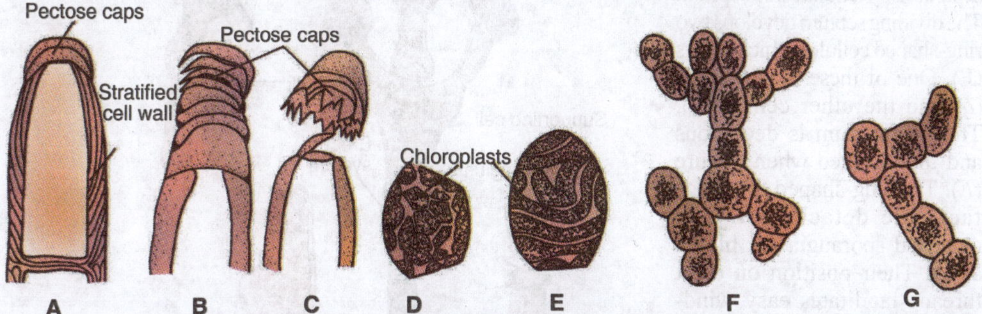

Fig. 10.23 (A-E). *Trentepohlia.* (A), cell with a stratified cell wall and terminal pectose caps; (B) and (C), mode of formation of caps; (D), cell containing discoid chloroplasts; (E), cell containing band shaped chloroplasts; (A–C), after West and Hood; (D-E), after Geitler; F-G after Gobi).

Cell structure (Fig. 10. 23 A-E). The cells may be cylindrical or barrel-shaped rarely moniliform (*T. monilia*). They usually have thick plainly stratified cellulose walls. The successive wall layers may be parallel or divergent. In the latter case the free end of the terminal cell of each branch bears a pectase cap (A) or a series of caps (B). The caps are periodically shed or pushed aside and replaced by new ones (C). West and Hood (1911) held that the caps are due to a secretion. Their function may be to reduce transpiration from the exposed tip or to afford protection. It may be a simple device to remove the waste materials. The septa between the cells may have each a single large pit penetrated by a proto-plasmic strand. The protoplast in older cells is multinucleate but in young cells it is uninucleate. There are several chloro-plasts in each cell. They are parietal in position and discoid (D) or band-shaped in form (E). Often the discs are arranged in bands (Geitler, 1923). The chloroplasts lack pyrenoid.

Asexual Reproduction. The thallus reproduces asexually both by *vegetative methods* and *spore formation*. Vegetative propagation takes places by the following methods :

(*i*) *Fragmentation.* Vegetative multiplication by wind borne segments or fragments of the thallus is common. The rounded or ellipsoidal cells of the prostrate system of *T. umbrina* separate from one another (Fig. 10. 23 F-G) to form a fine dust. The latter is blown by wind. On falling on a suitable substratum the one-celled fragments germinate to form new thalli.

Irgang (1927-28) reported that in *T. iolithus* most of the cells of the thallus perish in the dry season leaving behind only a few which are rich in haematochrome. The surviving cells serve as a means of perennation and propagation.

(*ii*) *Akinetes.* The vegetative cells of the prostrate system may form akinetes. The akinetes have thick walls and may occur in several successive cells. Each akinete which is a resting cell germinates on the return of conditions suitable for growth. It directly forms a new thallus.

Spore formation (Fig. 10.24). The asexual spores are flagellated swarmers (zoospores). They are differentiated in specialized ovoid or ellipsoidal cells called sporangia. The latter are of three types, pedicellate, funnel-shaped and sessile sporangia.

(*a*) *Pedicellate sporangia* (B). These are terminal or lateral in position on the branches of the erect system. The pedicellate sporangium consists of a lower stalk cell terminating in a spherical body cell or the sporangium proper (B). The stalk cell frequently becomes bent in a typical knee-shaped manner. It is differentiated into a broad subspherical proximal portion and a narrow, cylindrical stalk-like distal portion. The latter ends in a sporangium. The pedicellate sporangium arises as a tubular outgrowth from a somewhat enlarged vegetative cell (C). The subtending vegetative cell is called the supporting cell. It may give rise to more than one sporangia (A). The tubular outgrowth swells at tis tip to form the sporangium proper (D). The sporangium and the stalk cell are separated by a transverse septum. The dividing septum develops two ring-shaped cellulose thickenings (E). One of these is peripheral (*o*) and the other central (*i*). The sporangium is deciduous and thus is shed when mature (*F*). The ring-shaped thickings facilitate detachment. The detached sporangia are blown away. Their position on erect threads facilitates easy wind-dispersal. On wetting they germinate to produce swarmers. Each sporangium produces numerous swarmers which in several species of *Trentepohlia* are quadriflagellate. In a few others they are reported to be biflagellate. The zoospores thus produced are liberated through a protruded terminal or lateral aperture. The liberated zoospores are somewhat flattened. Each germinates straightaway to produce a new individual.

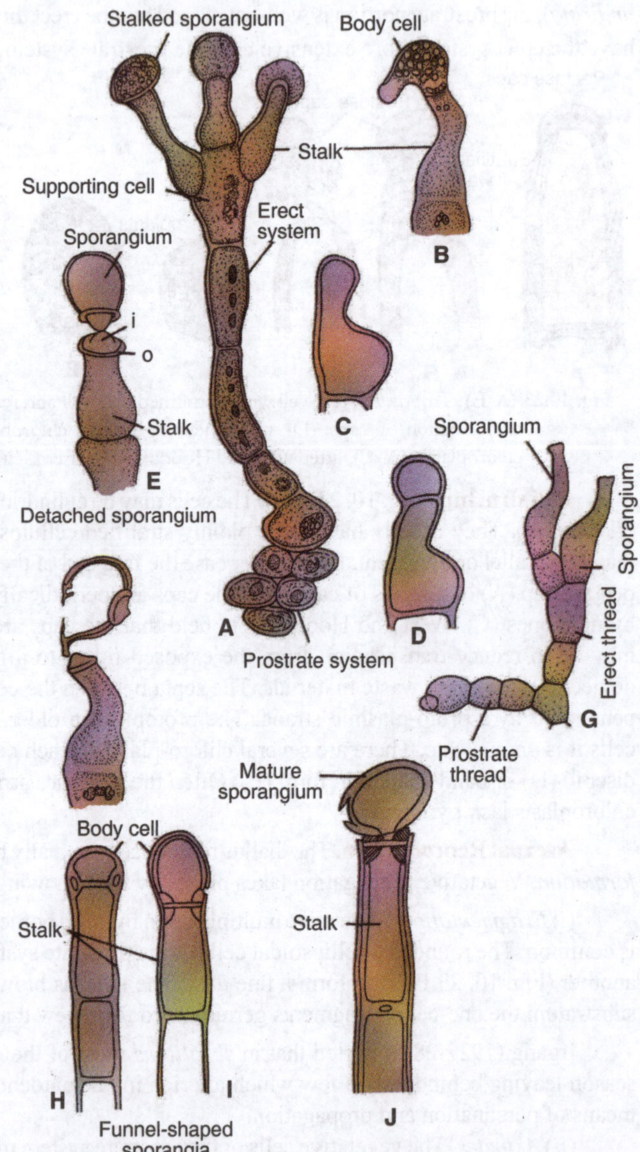

Fig. 10. 24 (A-J). *Trentepohlia.* (A), Erect thread of *T. umbrina* showing the supporting cell bearing two lateral and one terminal stalked sporangia. (B), Pedicellate sporangium; (C-D), development of stalked sporangium; (E-F), Detachment of sporangium; (G), Repeated formation of stalked sporangia in *T. monilia;* (H-J), Development of funnel-shaped sporangium (A,B,E,F, after Brand).

Aplanospores. In one species the contents of the sporangium before detachment are reported to divide to form aplanospores.

Jeejli Bai (1962) observed that in *T. monilia,* the subtending cell resumes activity after the detachment of the mature sporangium. It grows and gives rise to another terminal pedicellate sporangium. The persistent stalk of the old sporangium is pushed to one side. This event may be repeated once, twice or more times (G).

(*b*) *Funnel-shaped sporangia* (H-J). Brand (1910)

reported the occurrence of funnel-shaped sporangia in *Trentepohlia.* The funnel-shaped sporangium is cut off from the distal portion of a cylindrical cell by a transverse septum which lies between two superposed thickening rings. Towards maturity the parent cell walls splits at the level of the septum to detach the sporangium (J). What kind of swarmers the detached sporangium differentiates is not known at present. The pedicellate and funnel-shaped sporangia may occur on the same or separate plants.

(*c*) *Sessile sporangia* (Fig. 10.22, A). These are special reproductive cells that never become detached. They are formed by the increase in size of a cell that may be terminal, lateral, intercalary or rarely axillary in position. It is customary to call these reproductive cells as *sessile sporangia.* They produce only biflagellate swarmers without being detached from the thallus. The liberated swarmers have been reported to function as isogametes. Fritsch (1935), however, reported that they can germinate without fusion and thus function as zoospores. This double behaviour of the biflagellate swarmers to behave as zoospores or conjugate in pairs is a primitive characteristic.

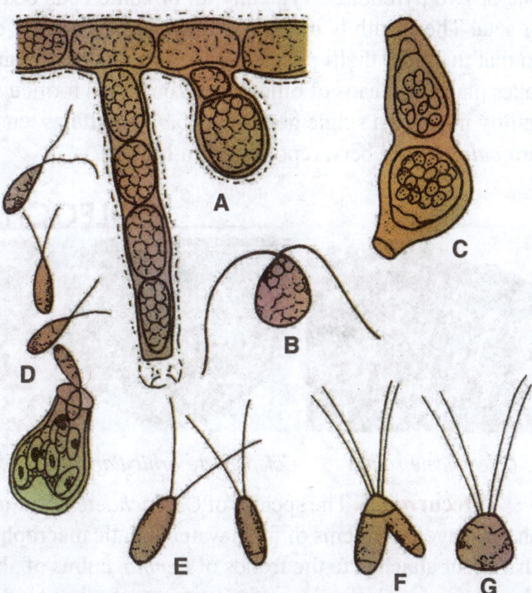

Fig. 10.25 (A-G). *Trentepohlia.* Sexual reproduction. (A), *T. aurea* with a lateral branch bearing a terminal gametangium; (B), liberated gamete; (C), gametangia of *T. umbrina*; (E-F), stages in sexual fusion in *T. bleischii*; (G), Quadriflagellate zygote (A-B, after Smith; C-D after Karsen; E-G after Wille).

Sexual Reproduction (Fig. 10.25). It is isogamous. The isogametes are biflagellate (D) and produced in gametangia which are intercalary or terminal in position (A). They are formed like sessile sporangia by mere enlargement of the cell. Often they are produced on the basal part of the plant (Fig. 10.22, A). The gametangia often differ from the sporangia of the same species. The protoplast of the gametangium produces a number of biflagellate gametes which are liberated through a somewhat protruded terminal aperture (D). The position of gametangia on the basal parts of the thallus favours inundation which admits of an easy sexual fusion between the gametes. Parthenogenetic development of gametes into filaments has been observed in some species (Meyer, 1936).

Taxonomic Position

Division	:	**Chlorophyta** (chlorophycophyta)
Class	:	**Chlorophyceae**
Order	:	**Chaetophorales**
Family	:	**Trentepohliaceae**
Genus	:	*Trentepohlia*
Species	:	*umbrina*

FAMILY : COLEOCHAETACEAE

A few representatives of this family possess a typical heterotrichous habit. In some forms only the projecting system is present. Generally, however, the prostrate system is dominant and the thalli are disc-like. The vegetative cells are uninucleate and have a single laminate parietal chloroplast with one or two pyrenoids. Typically all or some cells bear a single long, cytoplasmic, sheathed bristle or seta. The sheath is in the form of a basal cylinder of firm mucilage. The setae readily break off so that the older thalli possess only the sheaths. The growth is always apical. Asexual reproduction takes place by means of biflagellate zoospores formed singly. Sexual reproduction is oogamous. The family includes a single genus *Coleochaete*. It has ten species. Of these *C. soluta, C. scutata* and *C. orbicularis* have been reported from India.

COLEOCHAETE

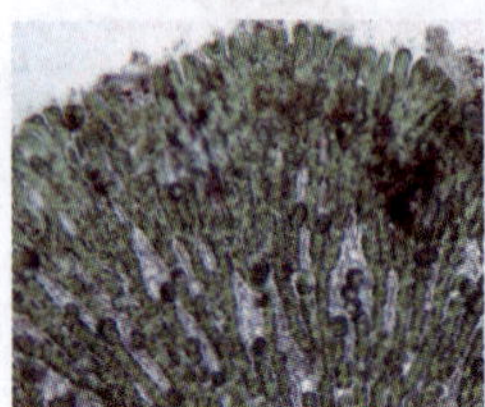

Coleochaete soluta

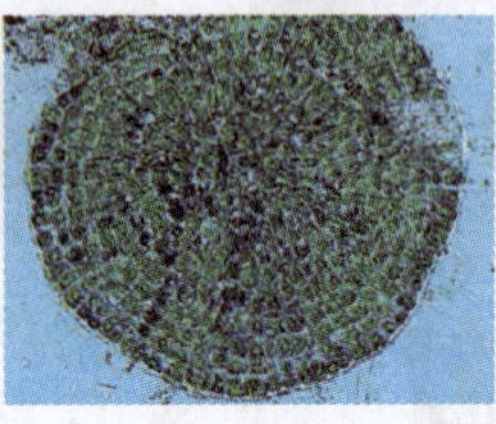

Coleochaete orbicularis

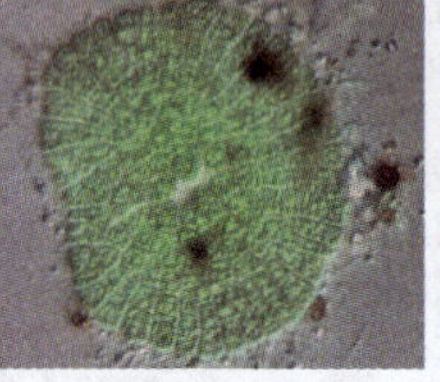

Coleochaete scutata

Coleochaete Irregularis

Occurrence. The species of *Coleochaete* occur in similar habitats. They all grow epiphytically on the leaves and stems of fresh water aquatic macrophytes such as *Hydrilla* and *Potamogeton.* They also occur attached to the fronds of *Lemna,* culms of alpha and undersurface of the floating leaves of *Nymphoea* and *Nelumbo* or the surfaces of other algae. However species with disc-like thalli have been reported to occur also on shell, sides of laboratory aquaria and other organic substrates in the shallow water of the fresh water littoral zone. A species of *Coleochaete* (*C. nitellarum)* has been reported to grow endophytically beneath the cuticle of thallus of *Nitella* with the seta extending through the host wall.

Organisation of the thallus. *Coleochae*[*te*] as the species of this genus present a range of plant body types. Some species of *Coleochaete* have a very much branched filamentous thallus which is distinctly *heterotrichous* in habit. *C. pulvinata* (Fig. 10.26) is the best example. The thallus in this species consists of a definite prostrate system giving rise to the projecting system. The projecting system consisting of erect branches is well-developed and forms a hemispherical cushion usually enveloped in mucilage. The prostrate system is represented by closely set branches that form a sort of a disc. In other species such as *C. divergence* and *C. nitellarum* only the prostrate system is present. It is composed of loosely arranged branching threads. However, occasionally erect branches may arise from the prostrate system at short intervals. In certain other species the projecting system has altogether disappeared, only the prostrate system is present and is dominant. This is found in the thalli of *C. irregularis, C. soluta, C. scutata* and *C. orbicularis.*

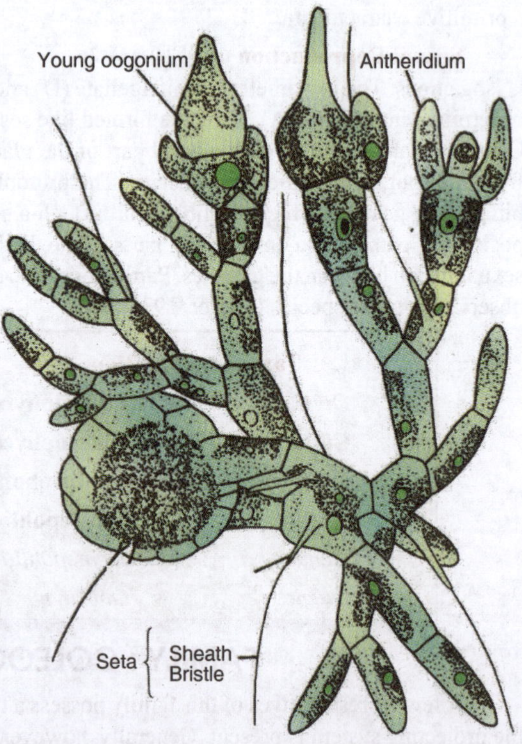

Fig. 10. 26. *Coleochaete pulvinata.* Erect thread bearing a spermocarp and young sex organs.

In *C. irregularis* the thallus is distinctly filamentous. The filaments grow more or less radially from a central point and branch irregularly to form a monostromatic plant representing the prostrate system. The latter consists of loosely adherent branched filaments. The branches like other green algae are produced by protrusion from the main filament. This type of branching together with radial growth results in empty spaces between the filaments. The matrix maternal between the filaments holds them in position.

The thalli of *C. soluta, C. scutata* and *C. orbicularis* are more regular and circular. In them the thallus is closely appressed to the substratum. The cell divisions in these species are restricted to the peripheral cells only.

In *C. soluta* (Fig. 10. 27), the prostrate system is in the form of monostromatic disc. It is composed of loosely spreading filaments which remain distinct. They are readily recognisable from one another and radiate from the centre. The thallus is cemented together with matrix which extends around the thallus periphery as well.

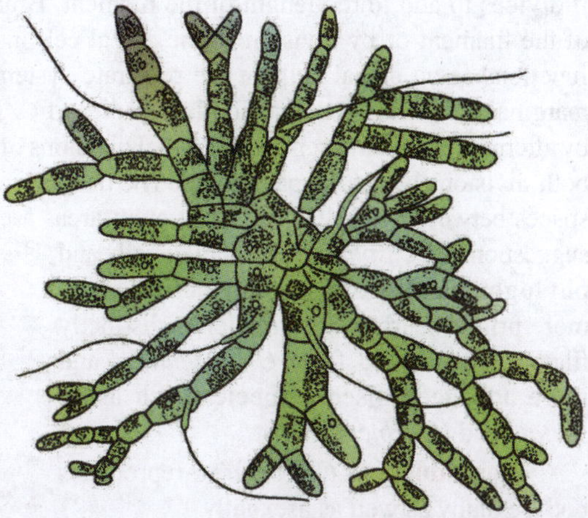

Fig. 10.27. *Coleochaete soluta.* Thallus represented by the prostrate system consisting of loosely arranged branching filaments.

In *C. scutata* and *C. orbicularis* (Fig. 10. 28), the filaments composing the discoid prostrate portion have more or less lost their identity. They are laterally united to form a parenchymatous disc which is circular but in *C. scutata.* It may be uniform in outline (Fig. 10.31, A). The filaments radiate from a common centre and are fused from the very cell beginning to form a compact stratum one cell layer in thickness.

Cell Shape. The shape of the cell varies in the prostrate and erect branches. In the prostrate portion of the species like *C. scutata* and *C. pulvinata,* the cells are usually hexagonal or polygonal. Near the margin they are usually four-sided with more or less convex outer walls. The cells in the erect threads of *C. pulvinata* are longer than broad and are more or less cylindrical in shape. The apical cells are elongated and have blunt ends.

Cell Structure. Each cell has a single nucleus and a single large curved chloroplast with one or two distinct pyrenoids. The chloroplast is parietal and laminate. It is irregular in shape and partly or wholly encircles the protoplast. A characteristic feature of the vegetative cells in *Coleochaete* is the presence of ensheathed bristles (Fig. 10.28). Certain cells in all the thalli bear a single bristle. These cells are called the seta cells. In the older thalli the bristles may be broken off. Wesley in 1928 studied the development of the **bristles.** According to him the development of the bristle starts with the formation of a pore in the outer wall of the cell. Formation of the **pore** is followed by the secretion of a new membrane. The latter extends out through the pore in the form of a protrusion. Opposite the pore there appears a deeply staining granule **(blepharoplast)**. This granule issues a dense granular mass, which passes into the newly

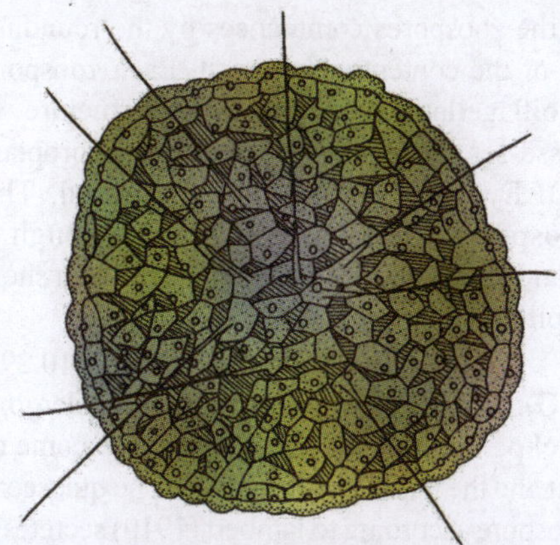

Fig. 10. 28. *Coleochaete orbicularis.* Thallus represented by the disc-like prostrate system consisting of filaments united laterally to form a compact disc circular in outline.

formed protrusion. The protrusion now ruptures at its distal end. The granular cytoplasmic contents elongate through the pore in the form of a thread which ultimately forms the bristle. The ruptured protrusion forms a sheath or collar around the base of bristle or the hair. The sheath gets slightly swollen at its base. The swollen knob-like base extends into the cell. The bristles project at right angles to the thallus. They are very delicate and readily fall off as the cells grow older leaving behind the basal sheath. The specialized seta cells characteristic of *Coleochaete,* probably, are an anti-herbivore adaptation. Marchant (1937) suggested that the snails are repelled by seta cells.

Growth. The growth is always apical. In the erect branches there is a definite **apical cell.** It divides to add to the length of the filament. Branches arise as lateral outgrowths from the cells of the filament or by division of the apical cell into two equal daughter cells. Each daughter cell functions as an apical cell. In the prostrate system the growth takes place by the activity of the marginal meristem. The discoid plant bodies of *C. soluta, C. scutata* and *C. orbicularis* are formed by alternating circumferential and radial divisions of peripheral cells. In *C. scutata* and *C. orbicularis* both divisions bisect the parent cell. The daughter cells remain intimately connected. There are no spaces between the cells. The substratum areas are thus completely covered with cells. There is a suggestion that *C. soluta* is morphologically and phylogenetically transitional between the more primitive irregularly branched distinctly filamentous discoid form *C. irregularis* and more advanced discoid species such as *C. scutata* and *C. orbicularis.*

Reproduction. *Coleochaete* reproduces both sexually as well as asexually.

Asexual reproduction (Fig. 29, A-C). It takes place by means of motile asexual spores called **Zoospores.** Zoospore formation occurs during spring or in early summer. They are produced singly in vegetative cells of the thallus (A). In fact any cell of the thallus is capable of producing them but usually the terminal or the sub-terminal cells are involved. The formation of the zoospores commences by the rounding up of the contents of the cell. Each zoospore is biflagellate ovoid or rounded structure. It possesses a laterally placed single chloroplast (B). Eye spot has not been reported. The zoospores when mature escape through a rounded opening formed at the apex of a short papilla-like outgrowth.

Germination of zoospore C (Fig. 10.29), D-G). On liberation the zoospores undergo a brief period of flagellation. Later they come to rest and the flagella are resorbed. The quiescent zoospore according to lambert (1910) secretes a wall around it. This one-celled germling starts dividing. The first wall is either horizontal (D) or vertical. In the former case the zoospore divides into two cells one above the other. The upper cell develops into a hair. The lower cell

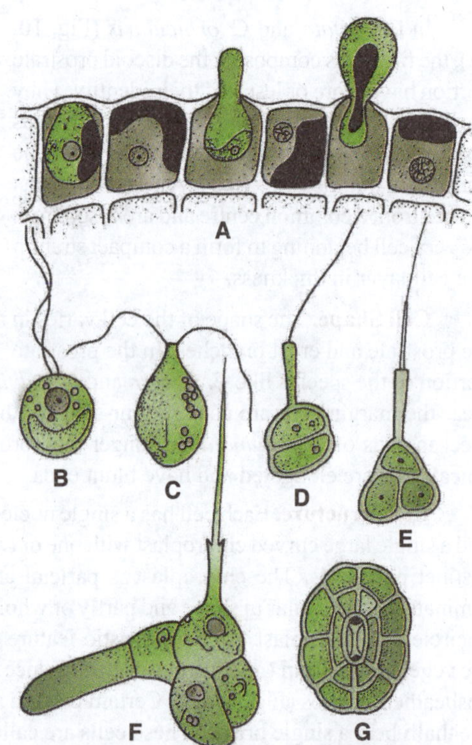

Fig. 10.29 (A-G). *Coleochaete.* Asexual reproduction. (A), Few marginal cells of the discoid thallus showing the formation and emergence of zoospore; (B), Liberated zoospores; (C), zoospore of *C. nitellarum* with an apical tubular prolongation; (D-E), side view of early stages of development of thallus by the germination of zoospores of *C. scutata;* (F), young thallus of *C. pulvinata;* (G), Later stage of the young thallus of *C. scutata* (A-B, after Westley; C, after Jost; F, after Chodat, rest after Pringsheim).

undergoes segmentation in two directions at right angles to the substratum to form the disc (E). When the division is vertical the two resultant cells lie side by side. Growth and division of these cells and of their derivatives result in the formation of a prostrate system of *Coleochaete* thallus (G). In the heterotrichous species (*C. pulvinata*) the threads of the erect system arise from the prostrate system. The zoospore in *C. nitellarum* (C) at the time of germination, forms a tubular outgrowth. The latter punctures the cuticle of *Nitella* and gains an entry. Inside, the tube forms the thread.

Formation of thick-walled **aplanospores** has also been reported by Wesley (1928). They are formed under unfavourable conditions. The aplanospores are produced singly within each cell.

Sexual Reproduction (Fig. 10.30). Unlike other Chaetophorales sexual reproduction in *Coleochaete* is **oogamous** and of a highly specialized type. The plants may be monoecious (*C. pulvinata*) or dioecious (*C. scutata*). The male sex organs are called **antheridia** and the female **oogonia** or **carpogonia.**

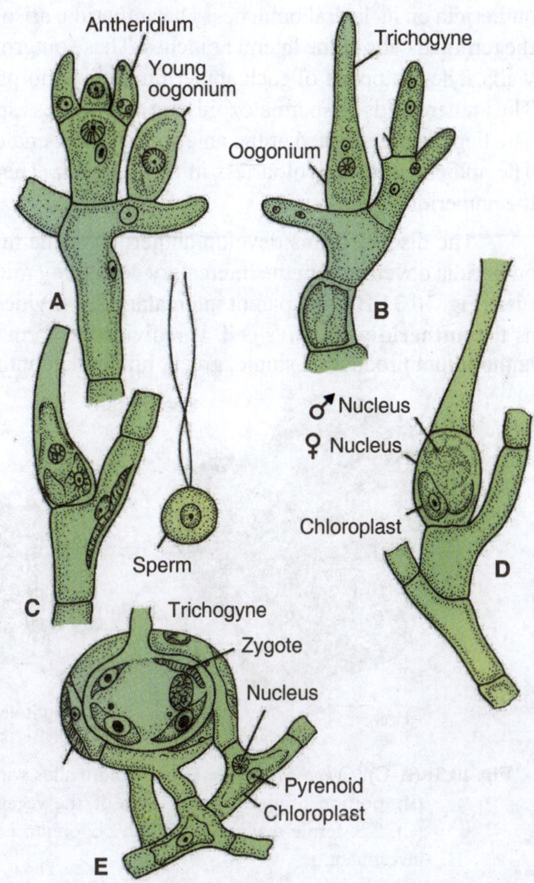

Fig. 10.30 (A-E). *Coleochaete pulvinata.* Sexual reproduction. (A), portion of a projecting thread with a cluster of antheridia and a young oogonium or carpogonium; (B), Oogonium with trichogyne, (C). Oogonium with trichogyne opened ready for fertilization; (D), Oogonium after fertilization but before nucelar fusion; E, Zygote with the enclosing sheath (Young spermocarp).

Carpogonia or Oogonia (Fig. 10.30). In the heterotrichous species such as *C. pulvinata,* the oogonia are borne terminally on short, lateral branches of the erect filaments (A). The terminal oogonium is later pushed to one side by a branch arising from the cell lying below it (C). Thus ultimately it appears lateral in position. The carpogonium consists of an enlarged basal portion and a long neck, the **trichogyne** (as in the red algae). The basal enlarged portion contains a prominent nucleus and one or more chloroplasts with distinct pyrenoids. The trichogyne is filled with colourless cytoplasm. Towards maturity the protoplast of the basal portion rounds off to form an **oosphere** or **egg.** At this stage the tip of the trichogyne breaks open (C). A little amount of cytoplasm is exuded in the form of a drop. The ovum or the oosphere is now ready for fertilization.

In *C. scutata* (Fig. 10.31) and other discoid species, oogonia originate from certain peripheral meristematic cells of the discoid thallus. These cells grow in size and each puts out a short protuberance on one side. The protuberance represents the neck or trichogyne. In mature oogonia the trichogynes open to make an opening for the spermatozoids to enter and fertilize the egg.

Antheridia. (Fig. 10.30A). The antheridia in *C. pulvinata* are usually borne terminally in groups on the branches of the erect system. The main filament bearing oogonia may also develop antheridia on its lateral branches. The antheridia arise as colourless bluntly conical outgrowths from the terminal cells of the lateral branches. These outgrowths are cut off from the parent cell by distinct walls. The protoplast of each antheridium metamorphoses into a single colourless antherozoid (S). The antherozoids or spermatozoids are more or less spherical or oval in shape. They are biflagellate. The flagella are inserted at the anterior or apical end of the antherozoid. There is a distinct nucleus. The antherozoids are colourless in *C. pulvinata.* They escape by the breaking down of the wall of the antheridium at the apex.

The discoid forms develop antheridia at the margin of the disc. In *C. scutata,* however, the antheridia develop from the intercalary cells lying midway between the margin and the centre of the disc (Fig. 10.31,B). The parent intercalary cell divides into two. One of the daughter cells functions as the **antheridial mother cell.** It redivides to form a series of antheridia. The protoplast of each antheridium produces a single, green, biflagellate antherozoid (B).

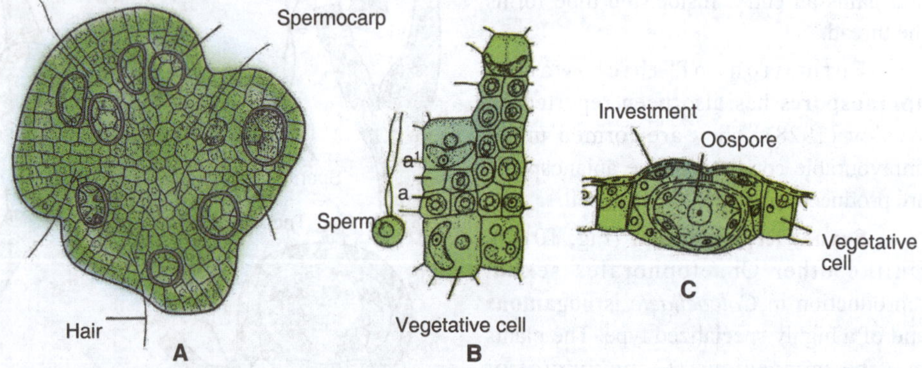

Fig. 10.31 (A-C). *Coleochaete scutata.* (A), discoid thallus with spermocarps overgrown by the surrounding cells; (B), portion of thallus with some of the vegetative cells having divided to form small antheridia; (C), C.S. fertile thallus through an oogonium containing a zygote and the whole surrounded by an investment.

Fertilization (Fig. 10.30, D). The liberated antherozoids (S) swim towards the opened necks of the oogonia. Presumably they are attached chemotactically toward the opened tip of the trichogyne. Only one antherozoid enters the neck and brings about fertilization. The male nucleus for sometime lies by the side of the female nucleus (D). The male pronucleus is smaller in size. It enlarges and fusion with the female pronucleus takes place when the two nuclei are of the same size (Lewis, 1970).

Post-fertilization Changes (Fig. 10.30, D-E). The three obvious changes that take place after fertilization are : (*i*) development of sheath, (*ii*) formation of spermocarp, and (*iii*) germination of zygote (oospore).

Development of sheath. Following fertilization the basal portion of the oogonium enlarges and is then cut off from the trichogyne by the formation of a septum (D). The fertilised egg or zygote increases in size. It rounds off and secretes a thick wall around it to become an **oospore.** The oospore is retained within the enlarged basal portion of the oogonium which remains attached to the haploid thallus. Meanwhile the vegetative cells adjacent to the oogonium and those underlying it are stimulated to growth by the act of fertilization. They develop short branches. In *C. pulvinata* these branches grow up and completely envelop the oogonium forming a sort of a continuous parenchymatous investment or sheath (E) around it. The discoid forms have the sheath developed only on the side away from the substratum. Graham and Wilcox (1983) observed that in *C. orbiculari* following fertilization marginal cells, on both sides of the oogonia containing young zygotes, continue to divide. Consequently the oogonia are submerged in the plane of the thallus and thus become intramarginal. They appear as prominent hemispherical objects in *C. scutata* (10.31C). This covering layer of vegetative cells around the oogonia containing the mature zygotes reminds one of the protected zygotes and young sporophytes of the lower embryophytes (Bryophyta). Continued marginal growth formation of oogonia and their subsequent fertilization result in the formation of concentric rings of oogonia containing zygotes. They are completely covered on the exposed sides by a layer of vegetative cells. During maturation, the zygote increases in size considerably, accumulates starch and lipid and the zygote wall thickens. The nucleus is located in the centre. The plastids are peripheral in position. With high voltage microscopy Graham and Wilcox (1983) observed localized wall ingrowths in the vegetative cells adjacent to the zygotes. These ingrowths were present only on the walls immediately adjacent to the zygotes and were conspicuous in the vegetative cells surrounding the mature zygotes. These investigators attach great significance to this feature and compare these cells with the gametophytic placental transfer cells present around the foot region of the sporophytes of the Bryophytes such as *Funaria* to which they resemble in their morphology, location and the time of development. They presume that these furtative placental cells in *C. orbicularis* facilitate transfer of photosynthates from the haploid thallus to the developing zygotes.

Formation of Spermocarp. The wall of the investing cells in the mature oogonia take on red or reddish brown colour. The whole structure or fructification consisting of the oogonium with its investment or sheath is called the *spermocarp.* It is red or reddish brown in colour. Oltman reported that later in the course of development the sheath disintegrates but the inner membranes of the investing cells and the oogonial wall jointly form a thick brown covering around the resting oospore.

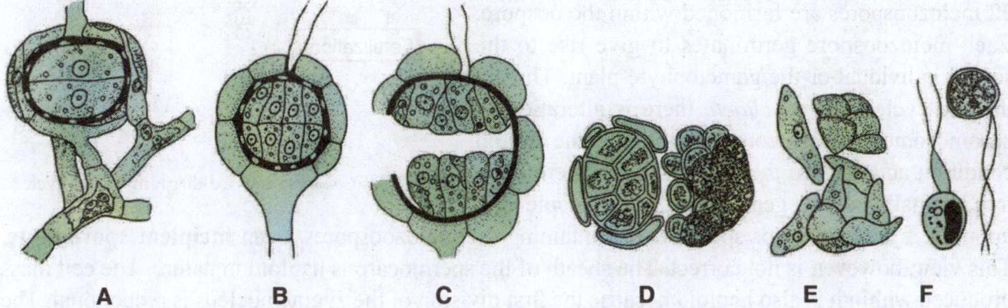

Fig. 10.32 (A-F). *Coleochaete.* Germination of Oospore. (A-B), early divisions of oospore; (C), sectional view of spermocarp at a later stage consisting of a sheath enclosing a cellular mass formed by the germinating oospore; (D), two oospores at a later stage: (E), Early stage in the formation of meiozoospores from the cell mass; (F), liberated meiozoo-spores. (A, B, after Engler, C, after Oltmans; D and F after Prigsheim).

Germination of oospore (Fig. 10.32). The oospore perennates over the winter season enclosed by the double membrane consisting of its own wall and the oogonial wall with the inner membrane of perished investing cells. When the favourable conditions set in, the protoplast of the hibernating oospore rejuvenates. It turns green and divides. The first two divisions of the diploid nucleus are reported by Allen (1905) to constitute **meiosis.** Each series of nuclear divisions is followed by cytokinesis. It takes place by the formation of a cell plate across the microfibrils parallel to the spindle axis at telophase (phragmoplast). The first wall is laid down in a plane perpendicular to the long axis of the original oogonium (A). Two more walls appear at right angles to each other dividing the oospore into 8 cells. This is the **octant stage**. By further division of the octant 16 (B) or 32 haploid, wedge-shaped cells are formed within the oospore. The protoplast of each cell metamorphoses into a biflagellate meiozoospore or gonozoospore. Graham and Tylor (1986) report that each meiozoospore develops within a chamber branded by a chamber wall comparable to the 'special walls' which typically surround the sporocysts (spore mother cells, of members of the spore tetrad in land plants). The envelope of the oospore, which consists of the spermocarpic and oospore walls, ruptures (C) circumferential. Each hemisphere contains half the number of meiozoospores. The meiozoospores are thus set free through the split (F). Each meiozoospore resembles a zoospore in all respects (Pringsheim). The liberated meiozoospores swim about for some time. Finally they become quiescent. The quiescent meiozoospore withdraws its flagella and secretes a wall around it. It then germinates in the same way as the zoospore (Fig. 10.29, D-G) to give rise directly to a new thallus which is a sexual or gametophyte plant in the life cycle.

Life Cycle (Fig. 10.33)

There is no distinct alternation of generations in the life cycle of *Coleochaete*. The adult morphologically organized plant is haploid and represents the gametophyte or the sexual generation. It bears the most highly developed sex organs among the Chlorophyceae. They are called the antheridia and the oogonia. The diploid condition is established in the oospore. In fact, single-celled oospore is the only diploid structure in the life cycle. There is no diploid plant. On germination the diploid nucleus of the oospore undergoes meiosis. The resultant haploid nuclei undergo mitotic divisions. Ultimately 16 or 32 meiozoospores are fashioned within the oospore. Each meiozoospore germinates to give rise to the sexual individual or the gametophyte plant. Thus in the life cycle of *Coleochaete* there is alteration of chromosome numbers from the haploid to the diploid condition and back to the haploid state. There is no true alternation of two generations. Some people look upon the fructification or spermocarp containing the meiozoospores as an **incipient sporophyte.** This view, however, is not correct. The sheath of the spermocarp is haploid in nature. The cell mass produced within it is also haploid because the first division of the zygote nucleus is reductional. The life cycle of *Coleochaete* which is characterised by a haploid adult and zygotic meiosis is described as *haplontic* or *haplobiontic, haploid.*

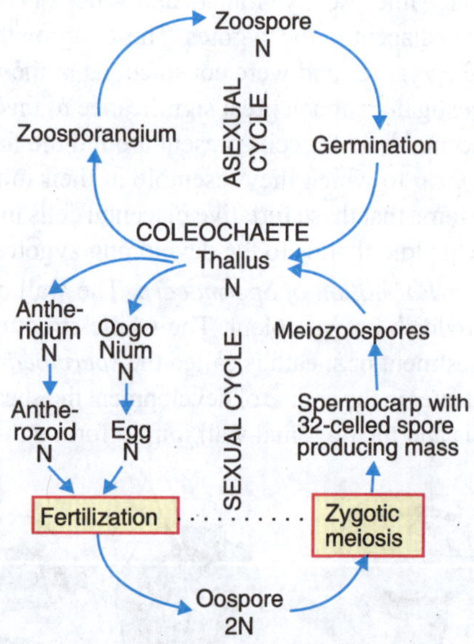

Fig. 10.33. *Coleochaete.* Word diagram of life cycle.

Affinities of Coleochaete

Coleochaete possesses green colour and pigments similar to the other green algae. Like the latter it stores reserve food as starch, has non-jacketed unicelullar sex organs, haplontic (haplobiontic and haploid) life cycle, haploid genetic constitution of gamete bearing thallus, zygotic meiosis and lack of two multicellular generations. Among the green algae it resembles the Chaetophorales in its cell structure and heterotrichous thallus. However, it is more advanced than all other green algae including the chaetophorales because (*i*) of its oogamous sexual reproduction, (*ii*) retention of the zygote on the haploid plant (feature unusual), (*iii*) presence of a protective layer of vegetative cells around the mature zygotes - a feature which reminds one of the protected zygotes and sporophytes of Bryophytes like *Riccia,* (*iv*) zygote being green at the time of meiosis, and (*v*) following meiosis it consists of a multicellular structure, the *spermocarp* from which emerge 16 or 32 biflagellate microspores instead of 4 typical of the green algae. Bower (1908) compared spermocarp to a simple or primitive sporophyte.

Investigations of some eminent algologists during the last quarter of the 20th century revealed that *Coleochaete* possesses certain features which are common among land plants but are lacking or rare in the green algae. These are presence of a phragmoplast during cell division (Marchant and Picket-He APS, (1973), a multilayered structure and cytoskeleton spline in zoospores and sperms

(Pickett Heaps and Marchant, 1972, Graham and McBride, 1979), Glycolate oxidase (frederick, Grauleber and Tollert, 1973), peroxisomes (Marchant, 1973), parenchyma in some species (Graham, 1982), and presence of parenchyma cells with growths around the zygotes of *C. orbicularis* resembling placental transfer cells which are common in the embryophytes (Graham and Wilcox, 1983).

A close study of the process of sexual reproduction and life cycle of *Coleochaete* and certain Nemalionales (*Batrachospermum*) has revealed a series of analogies between the two. In the face of these in relation of this genus to the other green algae is by no means clear. The development of a tubular outgrowth, the **trichogyne** from the apex of the oogonium of *Coleochaete* gives it the appearance of a carpogonium of the red algae. The retention of the oospore within the oogonium and development of sheath around it forms a sort of a fructification called **spermocarp** which is analogous to the cystocarp of *Batrachospermum* (a red alga). Furthermore Pringsheim and Lambart reported that the swarmers (meiozoospres) of *Coleochaete* do not give rise directly to the sexual plant. Instead they produce a number of generations of dwarf asexual plants (Fig. 10.33, D) which reproduce by zoospores only before a new sexual plant is produced. The same kind of life cycle (**haplobiontic**) is exhibited by a red alga *Batrachospermum*. These similarities do not throw any light on the phylogenetic relationships of the two. It may be a case of parallel development or homoplasy.

The occurrence of a discoid thallus in some species of *Coleochaete* is of special significance. It reminds one of the characteristic flat thalloid gametophyte of a liverwort such as *Riccia* and the spermocarp apparently looks very much like the sporogonium. This apparent resemblance tempted early investigators Bower (1908) to suggest that a form similar to *Coleochaete* could well have been progenitor of liveworts. He thus suggested that bryophyte alternation of generations and sporophyte originated in green algae with a life cycle similar to that of *Coleochaete.* The discovery of the presence of localized ingrowths by Graham and Wilcox (1983) lend support to this view. It shows that nutritional relationship between two different phases of the life cycle (gametophyte and sporophyte) similar to that which occurs on the bryophytes may have evolved in the green algae *Coleochaete.*

Salient Features

1. All the species are exclusively fresh water forms.
2. With the exception of *C. nitellarum,* which is endophytic on *Nitella,* all others grow epiphytically on the submerged aquatics or other algae. They are attached to the host plant by small outgrowths from the basal walls of the cells.
3. In some species the thallus is a true heterotrichous filament (*C. pulvinata*) whilst in others only the prostrate system is present. The prostrate system is in the form of a circular disc which in *C. soluta* is composed of loosely branched threads and in *C. scutata* it is a pseudo-parenchymatous disc.
4. The upright filaments grow by means of an apical cell and the prostrate cushion by means of the marginal meristem.
5. Each cell is a uninucleate structure and contains a single parietal chloroplast with one or two pyrenoids.
6. Certain cells in all the thalli bear a single sheathed bristle. In the older thalli the bristles may break off. The sheaths are only left behind.
7. *Coleochaete* reproduces asexually by means of zoospores.
8. The zoospores are **biflagellate** structures which are produced singly. They lack an eye-spot and are ovoid in form.
9. Sexual reproduction is an advanced type of **oogamy.** Some species are monoecious (*C. pulvinata*) and others dioecious (*C. scutata*).

10. The male sex organs are called **antheridia** and female **carpogonia** or **oogonia**.
11. The carpogonium has a basal enlarged portion and a short neck or trichogyne.
12. Each antheridium produces a single biflagellate colourless (*C. pulvinata*) or green (*C. scutata*) *antherozoid*.
13. The protaplast of the basal enlarged portion of the carpogonium rounds off to form a single **oosphere**.
14. The antherozoid enters the open neck of the trichogyne and travels down to the **oosphere** to fuse with it.
15. After fertilization the trichogyne is cut off by a septum and the basal part containing the fusion cell or zygote enlarges.
16. The zygote secretes a wall around it to become an oospore. Meanwhile branches arise from the adjacent and underlying vegetative cells and grow round the oogonium forming a continuous parenchymatous investment around It. The whole structure thus formed is called the **spermocarp**. It hibernates during winter.
17. The cells of the investment of the spermocarp die and oospore develops a thick brown wall.
18. Following spring, the protoplast of the oospore turns green and undergoes segmentation. The first two divisions are meiotic. By subsequent divisions thirty-two wedge-shaped cells are finally formed. Each cells is converted into a **meiozoospore** which is a biflagellate swarmer.
19. Each motile swarmer germinates to form the haploid sexual plant.
20. *Coleochaete* thus does not exhibit distinct alternation of generations.

Taxonomic Position :

Division	:	**Chlorophyta** (Chlorophycophyta)
Class	:	**Chlorophyceae**
Order	:	**Chaetophorales**
Famly	:	**Coleochaetaceae**
Genus	:	*Coleochaete*
Species	:	*scutata*

FAMILY : PLEUROCOCCACEAE (PROTOCOCCACEAE)

This is a monotypic family which includes non-zoosporic, unicellular, aerial green algae. The solitary cells have the ability to divide. Owing to the slow separation of daughter cells, the alga is usually found in small temporary associations or flattened packets of 2,3,4 or more cells (occasionally up to 50 cells). Cell division is the sole method of reproduction. It is very rapid under favourable environments. There are no motile cells in the life cycle. *Pleurococcus naegelii* Chod. (=*Protococcus viridis*) is one of the commonest green alga with a world wide distribution.

PLEUROCOCCUS (= PROTOCOCCUS)

Occurrence. It is a widely distributed terrestrial green alga which grows in aerial habitats that are moist or usually houses, damp soil and in other similar situations where humidity is high enough for it to survive. The alga occurs forming a green coating on the surface of the substratum. A small bit of this coat examined under the microscope is seen to consist of innumerable small rounded to ellipsoidal green bodies crowded together.

Thallus (Fig. 10. 34, A-B). The *Pleurococcus* plant is a unicell which, at maturity, is more to less globose to broadly ellipsoidal in shape (A). It consists of a spherical protoplast enclosed by a definite cell wall which is cellulose in nature. The cell wall is firm and fairly thick. It is differentiated into two layers, the inner cellulose layer and outer of cellulose and pectose. It has no gelatinous evelope external to it. The protoplast lacks vacuoles and thus the cell sap is highly concentrated. The small nucleus is centrally located (B). The single laminate chloroplast is parietal in position and is more or less lobed at the margins. The lobes curve within the periphery of the protoplast. Frequently it is so greatly

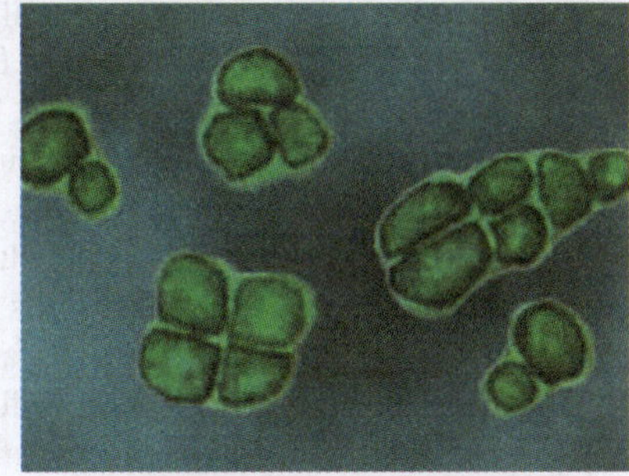
Pleurococcus vulgaris.

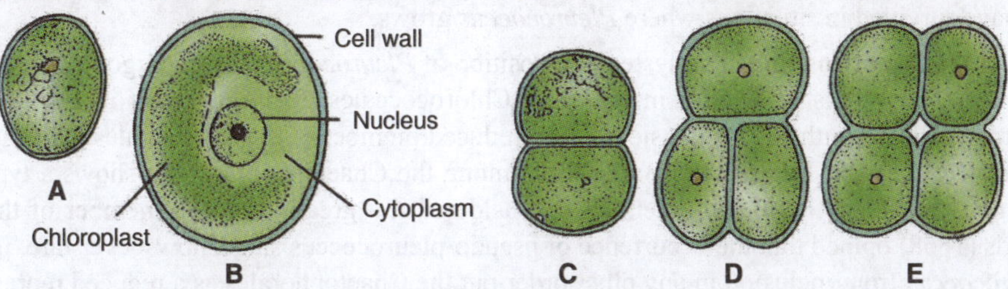

Fig. 10. 34 (A-E). *Pleurococcus.* (A) single cell; (B), the same highly magnified to show structure; (C-E), two, three and four-celled temporary colonies.

lobed that there appear to be two or more of them. The chloroplast usually lacks pyrenoids. The *Pleurococcus* unicell is so constructed as to grow in situations where it is exposed to the dangers of prolonged desiccation. The resistance of the unicell to desiccation depends on the resistant cell wall with its capacity to absorb moisture directly from the air. It is assisted by a highly concentrated cell sap and special nature of protoplast.

Reproduction (Fig., 10. 34 C-E). No asexual spores or gametes are formed. The unicell merely reproduces vegetatively by cell division which may occur in three planes (C-E). The first division is transverse (C). The resultant daughter cells may separate from each other immediately and become spherical. Frequently they remain united and divide again one or more times to form small, temporary colonies. The second division is at right angles to the preceding one (D-E). The succeeding divisions may be in the third plane. In its method of cell division a cross wall is formed across the unicell as in multicellular (filamentous) algae. It is in sharp contrast to the primitive unicellular forms in which the parent cell wall disintegrates and the daughter protoplasts form completely new cell walls.

Is Pleurococcus (Protococcus) a primitive or reduced form? Under moist conditions or when submerged but more often in cultures, the cells in the temporary colony may divide and become arranged in the form of small simple or branched filaments (Fig. 10.35, B-D). This is the **pseudopleurococcus** stage. This tendency of *Pleurococcus* cell to grow into filaments under certain conditions is of no evolutionary

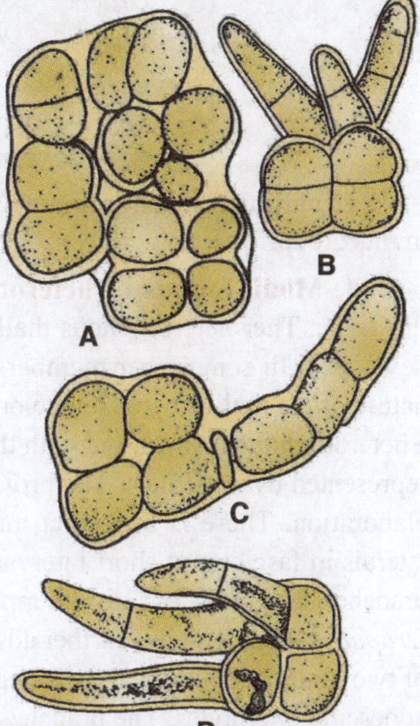

Fig. 10.35 (A-D). *Pleurococcus.* (A), Normal colony; (B-D), Reversion to filament formation. (After Chodat).

importance because it is now generally considered not as a primitive but a reduced or simplified form derived from a highly developed filamentous ancestor by retrogression or reduction of both the vegetative and reproductive functions. This simplification probably was in response to its change to a terrestrial mode of life. This viewpoint is supported by the more elaborate method of cell division of *Pleurococcus* unicell by the formation of a cross wall which is recognised as a distinct advance in plant phylogeny. The primitive unicellular algae divide by disintegration of the parent cell wall and formation of completely new cell walls around the daughter protoplasts. Thus its advanced method of cell division in more than one plane, failure to develop a multicellular thallus and lack of zoospores and gametes which even the primitive forms like *Chlamydomonas* possess indicate that *Pleurococcus* has descended from a fairly advanced multicellular ancestor by retrogression. If pseudopleurococcus state is considered to represent a temporary reversion to the ancestral conditions the ancestor must have been a member of the order Chaetophorales. The latter with its delicate branched thallus could not have survived in situations where *Pleurococcus* grows.

Taxonomic Position. The systematic position of *Pleurococcus* has undergone many changes. The earlier algologists placed it in the order Chlorococcales. Some placed it in a special group Pleurococoales. Smith (1955) considered it a reduced member of the Ulotrichales. Chodat (1894), Oltmans (1922) and Fritsch (1955) placed it among the Chaetophorales as it shows a typical cell structure characteristic of this order. They considered it a greatly reduced member of this order. Morris (1868) opined that the occurrence of pseudo-pleurococcus stage, howsoever rare, precludes *Pleurococcus* from inclusion in any other order but the Chaetophorales as a reduced representative of this order. The consensus of opinion, at present, is in favour of this view. Thus taxonomic position of *Pleurococcus* is as under :

Division	:	**Chlorophyta** (Chlorophyophyta)
Class	:	**Chlorophyceae**
Order	:	**Chaetophorales**
Family	:	**Pleurococcaceae**
Genus	:	*Pleurococcus*
Species	:	*naegelii*

Evolutionary trends in the Chaetophorales. The following evolutionary trends are noticeable in this group.

1. Modifications of heterotrichous plant body. *Stigeoclonium* is the simplest genus of this order. Ther heterotrichous thallus of this genus has both, the prostrate and erect systems, well developed. In some other members of Chaetophorales it is difficult to identify the heterotrichous nature of the thallus because one or other of the systems is reduced. The thallus in *Draparnaldia* is a heterotrichous filament but with the prostrate system either absent or poorly developed. Often it is represented by a holdfast. The projecting system is distinct and shows an important morphological elaboration. There is a distinct main axis (absent in *Stigeoclonium*). It bears the short or long laterals in fascicles at short intervals. The cells of the main axis are much larger than those of the branches. The main axis gives support and the laterals function in photosynthesis and reproduction. *Draparnaldiopsis* shows further advance in the organisation of erect system. The main axis consists of two kinds of cells, short and long. They alternate with each thus dividing the axis into so-called nodes and internodes. The branches arise from the short nodal cells only.

On the contrary in *Aphanochaete,* the erect system is extremely reduced. It is represented only as a few hairs whereas in *Endodermal* it is completely absent. Some forms merely have the prostrate system. For example, in *Coleochaete scutata, C. soluta* and *C. orbicularis* only the prostrate system is present. In *C. scutata* it is compact and discoid. The presence of a distinct, flat, plate-like prostrate thallus in species of *Coleochaete* is of special significance. It tempted the earlier algologist to suggest that a form similar to *Coleochaete* could well have been the progenitor of the liverworts.

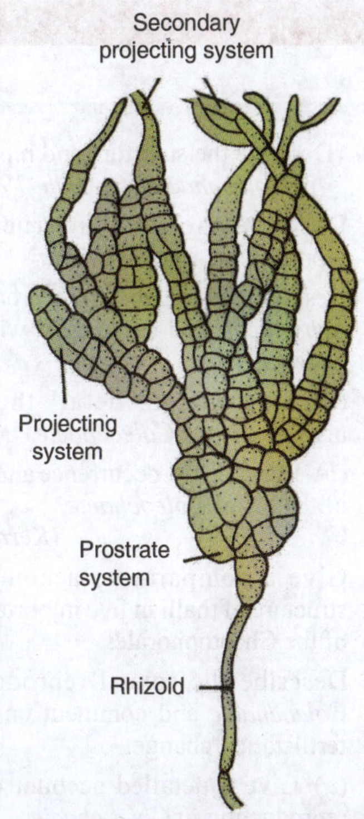

Fig. 10.36. *Fritschiella tuberosa.* A chaetphoraceous alga with a 3-dimensional fleshy body and a branching habit (After Iyengar).

To one school of phylogeneticists the presence of similar pigments and starch as the end product of photosynthesis in both the green algae and land plants suggest the possibility of primitive land plants (bryophytes) having originated from the green algal ancestors. The Chaetophorales provide the clue. The plant body in some of them (*Stigeoclonium*) is a heterotrichous filament. It consists of a prostrate system of green branching threads from which arise the erect threads constituting the erect aerial system. The erect system is considered ancestral to the leafy stem of a moss plant. The prostrate system has been retained as a green, thread-like branching protonema. Fritsch (1935-45) was the ardent supporter of this hypothesis. He held that the Chaetophorales represent the surviving descendants of forms which migrated to the land in the remote past. He visualized that the prostrate gametophyte and the erect aerial sporophytes of land plants evolved from the heterotrichous terrestrial green alga, the former by the suppression of erect system and latter by the suppression of the prostrate system. The discovery of a terrestrial chaetophoraceous green alga *Fritschiella tuberosa* in 1932 by Iyengar furnished a great impetus to this hypothesis. This alga tends to build up a three dimensional fleshy thallus with a branching habit and thus has proper combinations of prerequisites for a land plant ancestor. In fact Fritsch and Smith found in *Fritschiella tuberosa* the closest approach to the early progenitor of land plants.

2. Development of specialized sporangia. Most of the genera of this order produce spores in the vegetative cells. In *Endoderma* the spores are produced in enlarged cells which function as sporangia. The sporangia in *Trentepohlia* are highly elaborate structures. They are at least of two kinds, **pedicellate** and **sessile** sporangia. The sessile sporangia are simply enlarged vegetative cells. They produce biflagellate swarmers which may also fuse. The pedicellate sporantia are specialised structures. Each is formed by the swelling of the tip of a special outgrowth. The pedicellate sporangia thus consists of a stalk ending in a capsule-like structure, the sporangium proper. The latter when mature gets detached from the stalk and liberates quadriflagellate or biflagellate zoospores.

3. Transition from isogamy through anisogamy to an advanced type of oogamy. The more primitive members of Chaetophorales (*Stigeoclonium*) show isogamy. *Aphanochaete,* however, exhibited a pronounced type of anisogamy. The latter eventually leads to a complete loss of motility in the large female gamete and its retention within the oogonium. This type of sexual reproduction which is called oogamy is found in *Coleochaete.* It has most highly developed sex organs in the green algae.

QUESTIONS

ESSAY TYPE

1. Describe the structure and reproduction of *Coleochaete*. *(Awadh, 1993, 1996)*
2. Describe the thallus structure and reproduction in *Fritschiella*.
3. Describe the methods of reproduction in *Fritschiella* and compare it with that of *Coleochaete*.
4. (*a*) Describe in detail the thallus organisation in *Coleochaete*.

   (*b*) Describe the occurrence and structure of thallus of *Coleochaete*. *(Kerala, 2001)*
5. Give a comparative account of the structure of thalli in five important genera of the Chaetophorales.
6. Describe the sexual reproduction in *Coleochaete* and comment on the post-fertilization change.
7. (*a*) Give a detailed account of sexual reproduction in *Coleochaete*. *(Lucknow, 2000)*

   (*b*) Describe the vegetative structure in *Coleochaete*. *(Kerala, 2001)*
8. Describe the various grades of heterotrichy met within the order Chaetophorales.
9. Describe the thallus structure and reproduction in *Stigeoclonium*.
10. Write a short account of organisation of thallus in *Draparnaldia* and *Draparnaldiopsis*. Compare cell structure in both the genera.
11. Give a brief account of asexual and sexual reproduction in *Draparnaldia* and *Draparnaldiopsis*.
12. Write about the organisation of thallus in various species of *Coleochaete*. *(Bharathiar, 2000)*
13. Discuss the plenomenon of alternation of generation in *Coleochaete* and compare it with that of *Batrachospsermum*.
14. Describe briefly the life history of *Coleochaete*.
15. Discuss the affinities and systematic position of Chaetophorales.
16. Give an account of structure and reproduction in *Trentepohlia*.
17. Describe the structure of thallus and reproduction in *Pleurococcus*.
18. Describe the asexual reproduction or methods of perennation in different person of Chaetophorales.
19. Give the systematic position and description of the following.

    (*a*) *Fritschiella* *(Lucknow M. Sc., 2006, 2004)*

    (*b*) *Draparnaldiopsis*
20. Write a concise account of Chaetophorales. Discuss its affinities and its role in the evolution of land plants. *(Lucknow M.Sc, 2005)*

SHORT ANSWER TYPE

21. Draw labelled diagrams of reproductive organs of *Coleochaete*. *(Awadh, 1991)*
22. Give the most important characteristic of the following :

    (*i*) *Coleochaete* (*ii*) *Stigeoclonium* *(Awadh, 1992)*

    (*iii*) *Draparnaldia*

    (*iv*) *Draparnaldiopsis*

    (*v*) *Fritschiella*

    (*vi*) *Trentepohlia* (*vii*) *Pleurococcus*.
23. Write short notes on :

    (*i*) Thallus structure of *Coleochaete* *(Awadh, 1997)*

    (*ii*) Heterotrichous habit *(Awadh, 1999; Agra, 1994)*

    (*iii*) Sexual reproduction in *Coleochaete*

    (*iv*) Asexual reproduction ins *Stigeoclonium*

    (*v*) Methods of perennation *Fritschiella*

    (*vi*) Cell structure in *Draparnaldia*

(*vii*) Sexual reproduction in *Draparnaldiopsis*

(*viii*) Thallus structure of *Fritschiella*

(*ix*) Characteristic features of *Trentepohlia* and *Pleurococcus*

(*x*) Spermocarp.

(*xi*) *Trentepohlia*.

(*Lucknow M. Sc., 2006*)

24. Draw well-labelled diagrams of the thallus and mention systematic position of the following.

(*i*) *Coleochaete* (*Awadh, 1998*)

(*ii*) *Fritschiella*

(*iii*) *Stigeoclonium*

(*iv*) *Draparnaldiopsis*

(*v*) Thallus of Coleochaete.

(*Lucknow, 2005*)

25. Draw labelled diagrams of the thallus of *Coleochaete* with spemocarp.

(*Lucknow, 1992, 1994, 2001*)

26. Write down systematic position salient features and draw labelled diagrams of:

(*a*) *Coleochaete* (*Lucknow, 2007*)

(*b*) *Fritschiella* (*d*) *Draparnaldia*

27. Write arconcise account of reproduction of *Coleochaete*. (*Madras, 1997*)

28. Discuss in brief the importance of alternation of generation in *Fritschiella*.

29. Describe the salient features of Chaetophorales.

30. Write an explanation note on the sexual reproduction in *Coleochaete*.

31. Name the fruit body of *Coleochaete*.

32. Explain the most fertilisation changes in *Coleochaete*.

33. Who is credited with the discovery of *Fritschiella* ?

34. (*a*) Name the families of Chaetophorales.

(*b*) Name the alga where the thallus consists of erect and prostrate systems.

OBJECTIVE TYPE

35. Select the correct answer :

(*i*) The genus *Fritschiella* was discovered by

(*a*) V.J. Chapman

(*b*) TV. Desikachary

(*c*) BN Prasad

(*d*) M.O.P. Iyengar

(*ii*) The heterotrichous habit means

(*a*) Prostrate and erect branches

(*b*) Long and short branches

(*c*) Branches modified into leaves and air bladders

(*d*) Rhizoidal and photosynthetic branches

(*iii*) Fruiting body in *Coleochaete* is called

(*a*) Ascocarp

(*b*) Basidiocarp

(*c*) Spermocarp

(*d*) Cremocarp

(*iv*) The spermocarp of *Coleochaete* produces

(*a*) Spermatozoids

(*b*) Meiozoospores

(*c*) Oospores

(*d*) Mitospores

(*v*) Which of the following is produced due to post-fertilization changes in *Coleochaete* ?

(*a*) Cystocarp

(*b*) Mericarp

(*c*) Pericarp

(*d*) Spermocarp

(*vi*) Which of the following is terrestrial?

(*a*) *Fritschiella*

(*b*) *Draparnaldiopsis*

(*c*) *Draparnaldia*

(*d*) *Stigeoclonium*

(*vii*) The chloroplast in *Stigeoclonium* is
(*a*) Cup shaped
(*b*) Girdle shaped
(*c*) Reticulate
(*d*) Ribbon shaped

(*viii*) In *Stigeoclonium* the asexual reproduction takes place by
(*a*) Quadriflagellate macrozoospores
(*b*) Quadriflagellate microzoospores
(*c*) Both of the above
(*d*) None of the above

(*ix*) The thallus in Chaetophorales is
(*a*) Unicellular
(*b*) Filamentous
(*c*) Discoid
(*d*) Heterotrichous

(*x*) In *Draparnaldia* the gametes are
(*a*) Isogametes and quadriflagellates summers
(*b*) Heterogamous and quadri-flagellate swarmers
(*c*) Isogametes and biflagellate swarmers
(*d*) Heterogametes and quadriflagellate swarmers

(*xi*) In which of the following diploid thallus reproduce asexually by aplanospores and akinetes?
(*a*) *Volvox*
(*b*) *Draparnaldia*
(*c*) *Draparnaldiopsis*
(*d*) *Ectocarpus*

(*xii*) Macrozoospores and micromeiozoo-spores are produced in
(*a*) *Stigeoclonium*
(*b*) *Draparnaldia*
(*c*) *Draparnaldiopsis*
(*d*) *Volvox*

(*xiii*) Curved plate like chloroplasts with 2-8 pyrenoids are present in the cells of
(*a*) *Stigeoclonium*
(*b*) *Draparnaldia*
(*c*) *Draparnaldiopsis*
(*d*) *Fritschiella*

(*xiv*) The common method of perennation in *Fritschiella* is through
(*a*) Daughter colonies
(*b*) Protonema
(*c*) Funnel shaped tuberous bodies
(*d*) All of the above

(*xv*) Quadriflagellate and biflagellate meiozoospores one produced in
(*a*) *Stigeoclonium*
(*b*) *Draparnaldia*
(*c*) *Fritschiella*
(*d*) *Draparnaldiopsis*

(*xvi*) Which of the following cells has pectose caps and stratified cell walls in the cells ?
(*a*) *Stigeoclonium*
(*b*) *Fritschiella*
(*c*) *Trentepohlia*
(*d*) *Draparnaldia*

(*xvii*) Pedicellate, funnel shaped and sessile sporangia are produced in
(*a*) *Trentepohlia*
(*b*) *Draparnaldia*
(*c*) *Stigeoclonium*
(*d*) *Fritschiella*

(*xviii*) In *Coleochaete* the chloroplasts are
(*a*) Cupshaped
(*b*) Collar shaped
(*c*) Parietal and laminate
(*d*) Reniform

(*xix*) The sexual reproduction in *Coleochaete* is
(*a*) Isogamous
(*b*) Anisogamous
(*c*) Oogamous
(*d*) All of the above

(*xx*) All the species of *Coleochaete* are
(*a*) Marine
(*b*) Fresh water
(*c*) Terrestrial
(*d*) *Epiphytic*

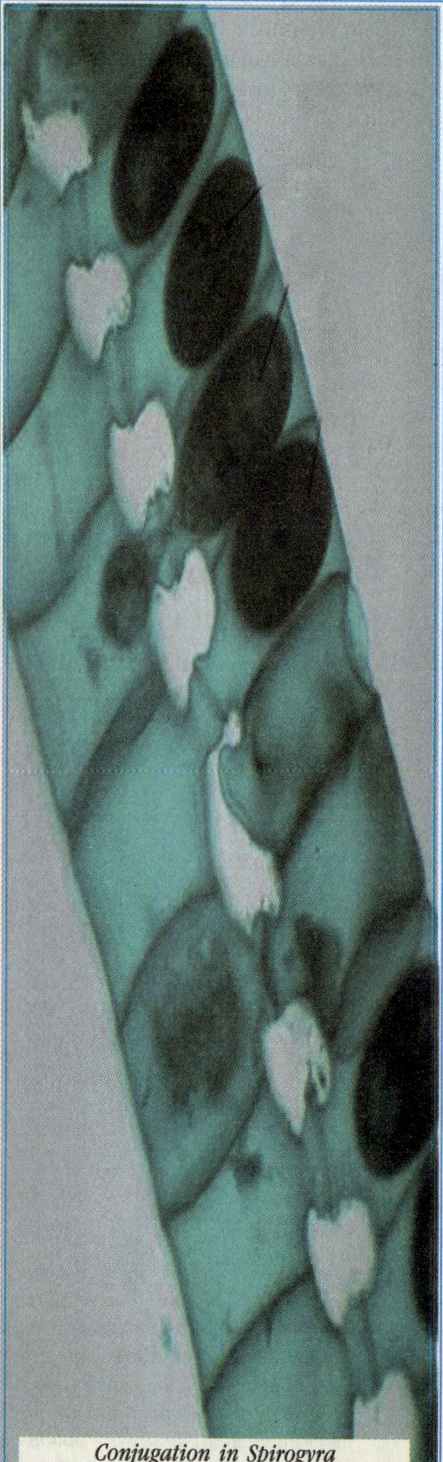
Conjugation in Spirogyra

CHAPTER 11

ORDER 8. ZYGNEMALES (Conjugales)

- Family : Zygenemaceae
- Sub-family : Zygnemoideae
- Sub-family Spirogyroideae
- Family : Desmidiaceae
- Unconstricted Desmid – Closterium
- Constricted Desmid - Cosmarium

Smith calls this order **zygnematales**. It includes strictly fresh water, unicellular (Desmids) and multicellular filamentous forms (*Spirogyra*). The filament is unbranched. In the number of species it forms the largest group of the Chlorophyta but is very well-defined in its methods of reproduction and chloroplast morphology. The Conjugales are quite distinct from the other green algae in the total absence of motile cells in the life cycle. Even the reproductive cells are without flagella (non-motile). The lack of flagella gives the group the name **Akontae**. The chloroplasts are large, few in number per cell and elaborate with a definite shape. Each has numerous, large pyrenoids. In shape the chloroplast may be a flat plate (*Mougeotia*), a helicoid band (*Spirogyra*) or consists of two axial stellate portions (*Zygnema*) or a ridged column. The cell wall has a mucilage covering.

Asexual reproduction by zoospores or any other kind of asexual spores is generally lacking. Sexual reproduction is accomplished by **conjugation** - a feature which furnished the basis for the descriptive name of the order conjugales. The gametes are non-flagellate and move toward one another by amoeboid action. In fact Round (1963), on the basis of marked elaboration of the chloroplast, complete absence of flagellate stages and sexual reproduction process, suggested to separate these members of the Conjugales from the other green algae and placed them in a separate class **Conjugatophyceae**. The order includes about 3,000 species which are placed under 40 genera. They are fresh water forms.

Classification

Chapman recognises four major families in the order Conjugales. These are Zygnemaceae (includes filamentous forms), Gonatozygaceae, Mesotaeniaceae (saccoderm desmids) and Desmidiaceae (placoderm desmids). Both the desmid families probably have evolved from the filamentous ancestors by over specialization of segments. Smith (1933) divided Zygnematales into three families — Zygnemataceae which included all the filamentous forms, Mesotaeniaceae and Desmidiaceae. Transeau (1951) followed Smith. Fritsch (1935) and Pascher group saccoderm desmids (Mesotaeniaceae and Zygnemaceae) together in one suborder Exconjugatae and the placoderm Desmids in a separate suborder Desmidioideae. Fritsch (1935) thus suggested the following synopsis of classification of the order Conjugales :

SUBORDER I. **Exconjugatae**. It includes 4 families:

1.Mesotaeniaceae. *Mesotaenium Roya* and *Netrium*

2.Zygnemaceae. Zygnema, *Spirogyra* and *Debarya*

3.Mougeotiaceae. *Mougeotia, Sirogonium* and *Zygonium*

4.Gonatozygaceae. *Gonatozygon*

SUBORDER II. **Desmidioideae**. It includes one family.

5.Desmidiaceae. *Closterium* and *Cosmarium*

According to Prescott (1969) this group of Chlorophyta merits the status of a subphylum. Following Oxada (1935) he recognised two orders as follows :

Subphylum Conjugatae. It includes :

ORDER I. **Desmidiales**. It comprises 5 families, namely, Gonatozygaceae, Mesotaeniaceae, Cosmariaceae, Hyalothecaceae, Closteriaceae.

ORDER II. **Zygnematales**. It includes two families viz. Mougeotiaceae and Zygnemataceae.

Randhawa (1959) suggested that the classification of Zygnematales could be based with greater certainty on the character of the chloroplast. On this basis he divided

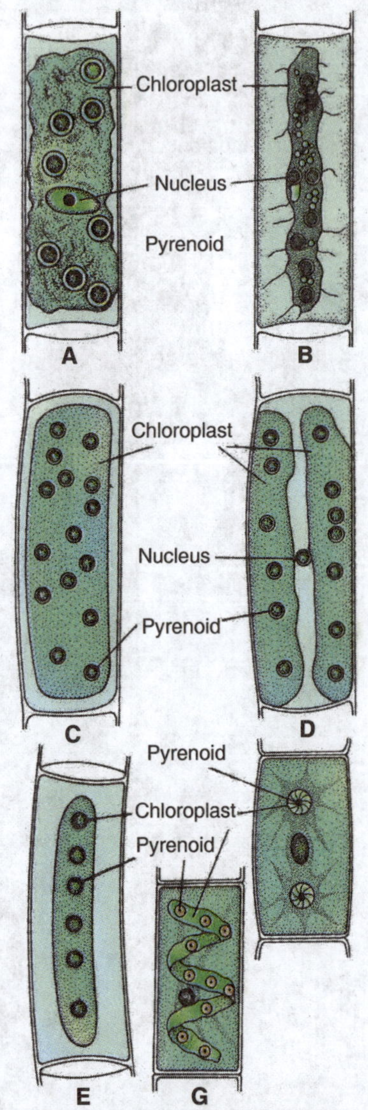

Fig. 11.1 (A-G). *Zygnemaceae.* Four types of chloroplast. (A), Axial or axile, plate type as in *Mougeotia*; (B) the same in profile position; (C), two parietal plate type as in *Sirocladium*; (D), the same in side view; (E), axile ribbon-type as in *Debarya*; (F), stellate as in *Zygnema*; (G), ribbon-type spirally twisted (After Randhawa).

the order into three families, namely Zygnemaceae, Mesotaeniaceae (Saccoderm desmids), and Desmidiaceae (Placoderm desmids). Family Zygnemaceae was further divided into four sub-families as follows :

Sub-family 1. **Mougeotioideae** including *Mougeotiopsis, Debrarya, Mougeotia* and *Tamnogametum.*

Sub-family 2. **Zygnemoideae**. It includes *Hallasia, Zynemopsis, Zygnema, Pleurodiscus, Zygogonium* and *Entransia.*

Sub-family 3. **Spirogyroideae**. It includes *Spirogyra* and *Sirogonium*.

Sub-family 4. **Sirocladioideae** comprising a single genus *Sirocladium*.

The classification of Zygnemales as proposed by Randhawa is followed in this edition of the text.

FAMILY : ZYGNEMACEAE

The representatives of this family normally have an unbranched, filamentous thallus consisting of cylindrical cells placed end to end. There is no basal-distal differentiation. The adult filament is free floating. The cell walls have no pores nor any external markings. Sexual reproduction takes place by the method of **conjugation** and the union of amoeboid gametes generally through the **conjugation tubes** and rarely through a pore in the intervening walls between the cells. The zygospore is either round, oval or quadrangular and has a 3-layered wall. The Zygospores are formed either in the conjugation tube or within one of the gametangia. The chloroplasts are of four types, axial plate-like (A-B) or ribbon-like (E), stellate (F), parietal plate-like (C-D) and ribbon-like spirally twisted (G). Randhawa (1959) reported 580 species and 13 genera. Most of the species grow unattached in quiet water. A few, however, have a holdfast which attaches the thallus to stones or other objects in water. *Mougeotia, Zygnema* and *Spirogyra* are the three best known genera of this family.

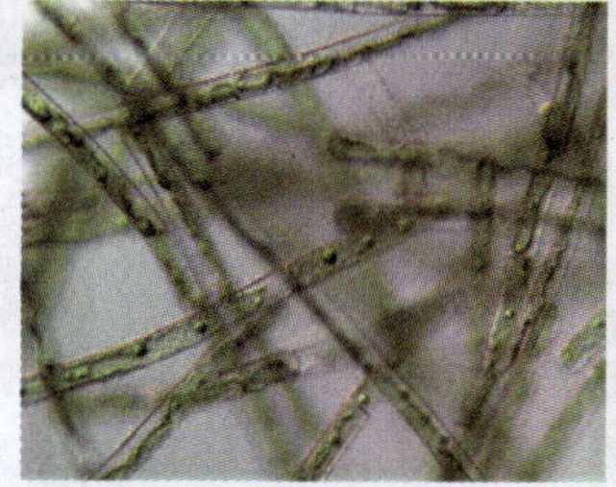

Mougeotia

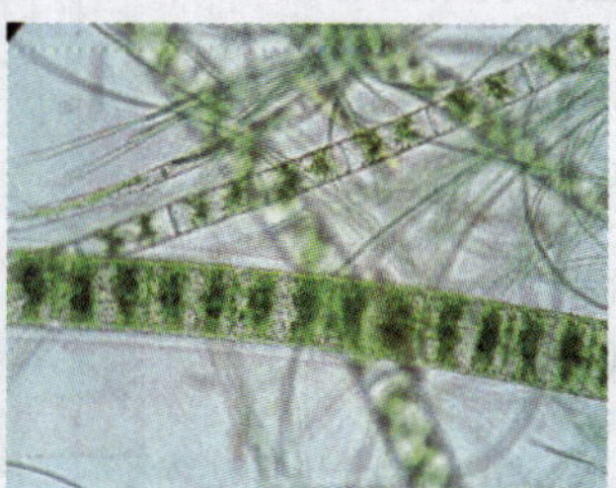

Zygnema

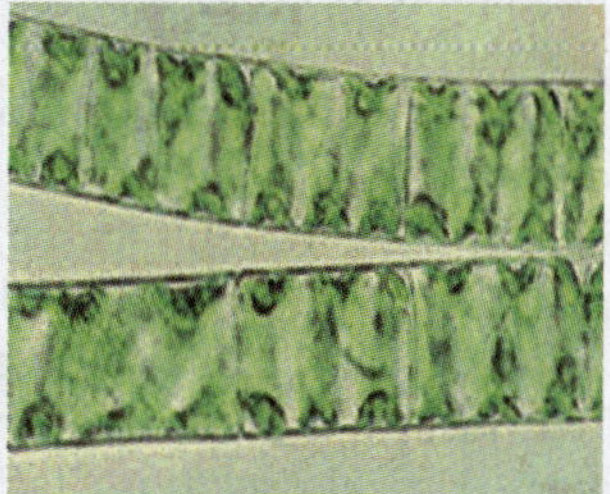

Spirogyra

MEMBERS OF FAMILY ZYGNEMACEAE

The characteristic feature of this sub-family is the presence of a single, axile, plate-like chloroplast with or without pyrenoids. It comprises four genera. Of these *Mougeotia* is discussed here.

MOUGEOTIA (AGARDH 1824)

Habitat. It is a fresh water, filamentous green alga found in ponds, semi-permanent pools, lakes and slow-flowing streams as bright green, free-floating masses. In all 108 species have been described so far in this genus. Of these 18 have been recorded from India.

Thallus (Fig. 11.2,A). It is a simple filament consisting of long cylindrical cells placed end to end. Rarely the filament develops unicellular (D-E) or multicellular outgrowths which attach the filament to the substratum. The cells are much longer (at least 4 times) than broad and have plane end wall (A-B).

Cell Structure. The cell consists of a cell wall enclosing the protoplast (B). The cell wall is a single piece. It is thin and consists of two layers, the inner cellulose and outer of pectic material. The cellulose layer forms a complete envelope around the protoplast. Ultrastructurally the cellulose layer consists of fine, thread-like structures, the **microfibrils** which form a network. Marchant (1978) detected microfibrils attached to the cytoplasmic face of the detected microfibrils attached to the cytoplasmic face of the plasma membrane. The orientation of the microfibrils reflects that of the microfibrils in the cell wall. He thus suggested that it is probable that these cortical (peripheral) microtubules influence orientation of microfibrils in the cell wall.

The outermost portion of the pectose layer changes into pectin which dissolves in water to form a thin slimy sheath around the filament. The cell wall is thus differentiated into three layers. The outer is mucous and amorphous. The two inner layers of pectose and cellulose are fibrillar. The cross wall consists of a middle septum of pectic material with a layer of cellulose on each side of it. The middle septum superficially resembles the middle lamella of higher plants. The cross wall in *Mougeotia* splits into two circular discs at an early stage. They are separated by a gelatinous substance. The protoplast is differentiated into the cell membrane, cytoplasm, nucleus and a single chloroplast with more than one pyrenoids. The central vacuole is modified by the presence of axile chloroplast.

The cell membrane is a fine, delicate, elastic protoplasmic covering forming the outermost limit of the protoplast. It is capable of growing as the cell enlarges and is differentially permeable. The cytoplasm along with the cell membrane forms a lining layer within the cell wall (C). In the centre of the cell is a single flat, axile plate-like chloroplast. It is nearly as broad as the cell and as long (C). The remainder of the vacuole is filled with a vacuolar sap rich in tannins. The chloroplast is attached to the peripheral cytoplasm by delicate strands of cytoplasm. Ordinarily the chloroplast presents its surface to the light but exposure to strong bright light causes profile position in which it appears a linear thread occupying a small part of the cell (C). The narrow-celled species have 2,3 or more pyrenoids arranged in a single row. In the species with broad cells the chloroplast has numerous irregularly scattered pyrenoids. The single nucleus is found in the centre of the cell and pressed to one side of chloroplast apposed to the cell wall.

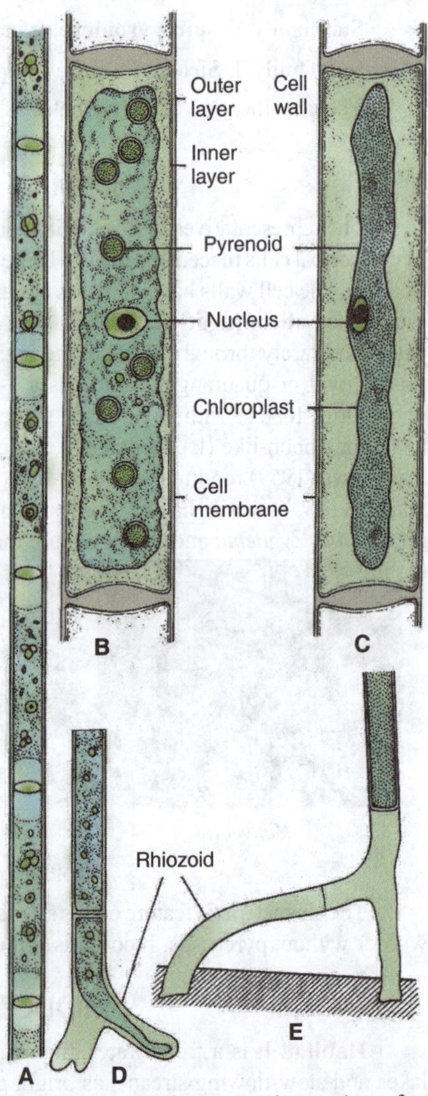

Fig. 11.2 (A-E). *Mougeotia.* (A), portion of a filament; (B), Face view of the cell showing structure, and the chloroplast in surface view; (C), chloroplast in profile position; (D-E), basal cell showing rhizoid formation (D-E, after Salisbury, B-C, based on Morris).

Mechanism of chloroplast orientation. Haupt and Schonobohm (1970) suggested that the chloroplast of *Mougeotia* orients within the cell in response to light. The light induced alterations of chloroplast orientation and shape in response to the intensity of incident light serves a double purpose. Firstly, it enables the alga to carry on its maximum photosynthetic activity under less than optimum conditions of illumination. Secondly, it serves to shield the chloroplast from destructive high light intensities.

The early hypothesis advocated by Seven (1908) suggested that the chloroplast itself is both light sensitive and capable of movement or orientation through the cytoplasm. The evidence that has accumulated during the last two decades goes against this hypothesis. It has been demonstrated experimentally that the primary photon receptor is not located in the chloroplast but resides within the cortical region of the cytoplasm of the cell (Fischer-Arnold, 1963, Haupt, 1970). In *Mougeotia* the photoresponse involves directed movement of the chloroplast in response to unilateral illumination. It is mediated by phytochrome. Unidirectional red light establishes a gradient of the physiologically active phytochrome form Pfr in the cell. It triggers chloroplast movement (Haupt, 1959). The chloroplast edge moves away from the region of the highest Pfr concentration. Kramel *et al* (1984) showed that dichroic change of phytochrome orientation in the algal cell from Pr orientation (parallel to the cell surface) to pfr orientation (normal to the cell surface) occurs during the time period 5 ms to 30 ms after inducing red flash. Foos (1971) suggested involvement of microfilaments in chloroplast movement. Marchant (1967) identified contractile protein known as **actin** in the cytoplasm of *Mougeotia*. Its intracellular location still remains uncertain. Actin consists of thin microfibrillae. He proposed and the indirect evidence indicates that the microfilament of F-actin stretching between the chloroplast and cortical cytoplasm are involved in light induced reorientation and movement of chloroplast in the cytoplasm of *Mougeotia*. The active shearing force involved may result from the interaction between the F-actin microfilaments and the organelle linked myosin.

Kramel *et al* (1984) reported that the orientation movements of *Mougeotia* chloroplast from profile to face position is mediated by phytochrome.

Reproduction. The genus reproduces vegetatively, asexually and sexually.

1. *Vegetative Reproduction* (Fig. 11.3, A). It takes place by the method of **fragmentation.** It is a process in which the individual filaments either break up or dissociate into several short, one to few-celled segments or fragments. The breakage or dissociation normally takes place between two vegetative cells. The breakage is accidental and due to mechanical damage by the bites from the aquatic animals or injury caused by water currents. Dissociation is brought about by the mechanism of cell disjunction (A). There is splitting of the end walls or septa (*d*) and subsequent dissolution of the middle lamellae into mucilage. This is due to conversion of pectose middle lamellae into pectin that dissolves in water to form mucilage. According to Llyod (1926), the cross wall between the two cells splits into two circular discs at an early stage. A mucilaginous jelly develops in between by the dissolution of middle lamella (a^2). The two parts bulge apart by the pressure exerted by the intervening mucilage and appear biconvex (a^2). This condition is maintained

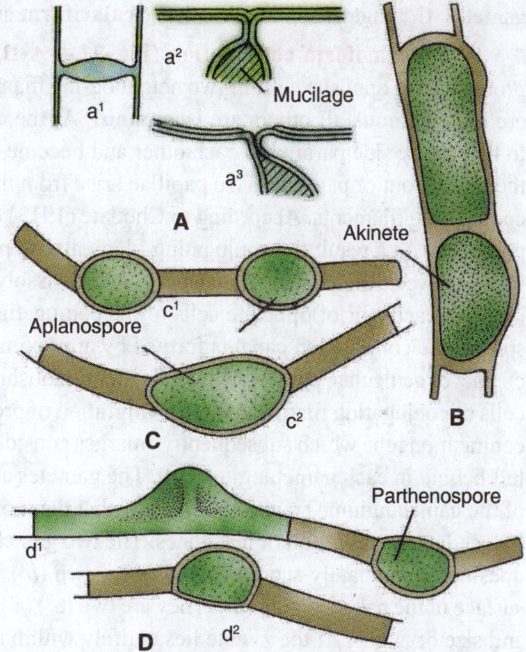

Fig. 11.3. (A-D). *Mougeotia.* Asexual reproduction. (A) Mechanism of cell disjunction; (B), Akinetes of (M). *capucina*; (C), Aplanospore formation in M. *boodlai;* (D), Parthenospore formation in *Gonatonema ventricosum* (A, after Fritsch; B, after Borge; C and D, after West).

as long as the adjoining cells remain turgid. With decrease in the turgidity of one of the adjacent cells, the septum becomes bulged into it by higher turgid pressure in the other cell. The simultaneous dilation of the intervening mucilage increases the curvature (a^3). The shearing strain exerted by these happenings causes the rupture of the common longitudinal wall and forces apart the adjacent cells.

2. *Asexual Reproduction* (Fig. 11.3, B-D). Fritsch (1948) reported the formation of **akinetes** (B) in species inhabiting mountain tarns and lakes with relatively low temperature. Starch and fat begin to accumulate in the protoplast of cells destined to form akinetes. The walls become markedly thickened. In form the akinetes differ but little from the vegetative cells. Filaments producing akinetes often have an envelope of mucilage.

Aplanospore formation is not common in *Mougeotia* (C). It has been reported only in seven species. The protoplast of the vegetative cell contracts to become rounded or ovoid and then secretes a wall around it to become an **aplanospore**. Residual cytoplasm is left behind in the aplanosporangium. Formation of parthenospores is of frequent occurrence in this genus (D). In fact in some species parthenospores are the only reproductive structures known. The protoplast contracts into the middle portion of the cell (d_1). The contracted protoplast then secretes a wall around it to become a **parthenospore** (d_2). The cells concerned with parthenospore formation generally exhibit geniculation and the cell containing the parthenospore is somewhat enlarged. The species which exclusively reproduce by parthenospores are placed in a separate genus *Gonatonema* for which there is not much justification because otherwise they resemble *Mougeotia* in vegetative structure. Furthermore some of the parthenosporic species have been reported to form zygotes.

3. *Sexual Reproduction.* It takes place by conjugation which involves the fusion of amoeboid gametes. Conjugation is of two types, **scalariform** and **lateral**.

(*a*) Scalariform conjugation (Fig. 11.4, A-D). In this case fusion takes place between two protoplasts of opposite cells of two neighbouring filaments. With the exception of three species which are **anisogamous** all others are **isogamous**. At the commencement of the process filaments come to lie side by side parallel to each other and become intimately glued by mucilage. The next step is the putting out of papillae. Two papillae arise from the opposite cells at the corresponding points of conjugating filaments. According to Chodate (1913) these papillae elongate to form the conjugating processes. As a result the conjugating filaments are pushed apart. Reaching their full length the end walls between the conjugation processes are dissolved and a narrow **conjugation** tube is formed between each pair of opposite cells of the conjugating filaments (a^7). According to Fritsch in some species the conjugation canal is formed by approximation and subsequent fusion along the point of contact of geniculate pairs of cells (b^1). After establishment of conjugation canals between the opposite cells of conjugating filaments there is migration of protoplasts from the opposite gametangia into the conjugation tube which subsequently enlarges considerably in diameter. A portion of the cytoplasm is left behind in each gametangium (a^2). The gametes are thus formed from only a part of the contents of the gametangium. Transcau suggested that the movement of the gametes into conjugation tube is induced by the secretion of hormones. The two gametes fuse to form a zygote. The zygote, however, does not immediately secrete a wall of its own (b^1). Special walls are, at first, laid next to the free surface of the naked zygote (b^1). They are two (a^2) or four (b^2) in number depending upon the position and size of zygote. If the zygote lies entirely within the conjugation canal there is formation of two special walls and the zygote is said to be "adjoined by two cells" (a^2). In case it protrudes into the gametangia four special walls are laid (b^2) and the zygote is said to be "adjoined by four cells". After completion of this outer envelope consisting of the conjugation canal and special walls, the zygote secretes a wall of its own (zygote wall b^2) around it. Gametes which fail to fuse often function as azygospores and develop parthenogenetically into new plants.

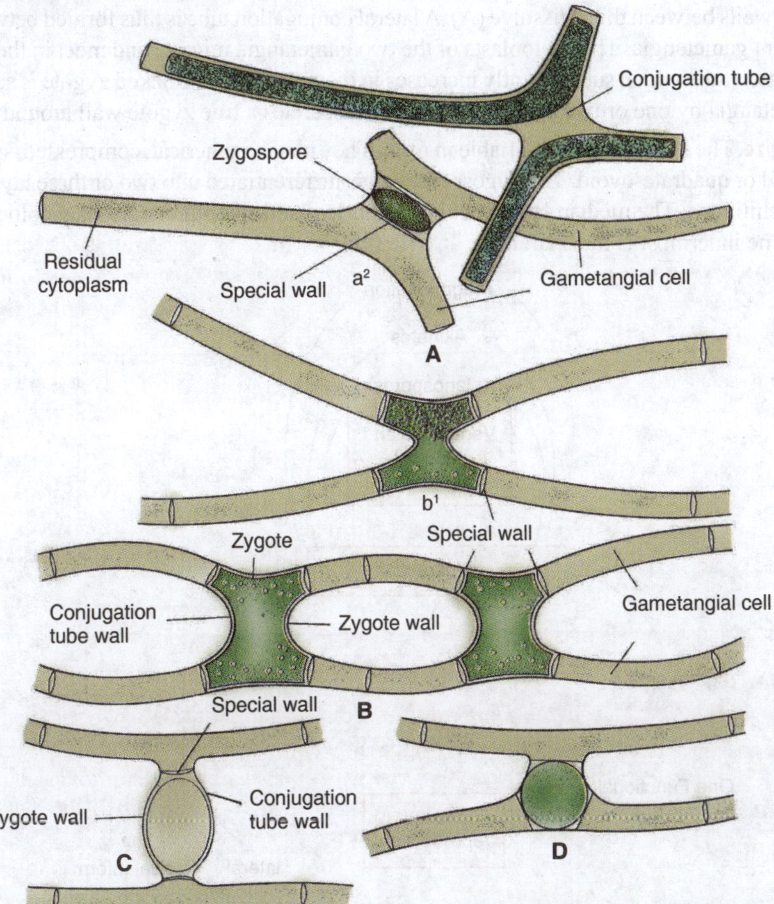

Fig. 11.4 (A-D). *Mougeotia.* Scalariform conjugation. (A), Formation of conjugation tube at a^1 and formation of a zygospore in the conjugation tube at a^2; (B), Scalariform conjugation in *M.*viridis (b^1) with the zygospore projecting into the gametangia (b^2); (C), *M.* scalaris with the zygospore formed in the conjugation tube; (D), *M.* tenuis with zygospore mainly in one of the conjugating cells (C,D, after Transcau).

(b) **Lateral Conjugation.** In this case fusion takes place between two protoplasts of the adjacent cells of the same filament. Lateral conjugation has been reported in *M. jogensis* (Fig. 11.5A-B), *M. genuflexa* (Fig. 11.5C) and *M. oedogonioides*. It is of **isogamous** type. In the first two species two tube-like protuberances which arise from the adjacent cell one on either side of separating septum,

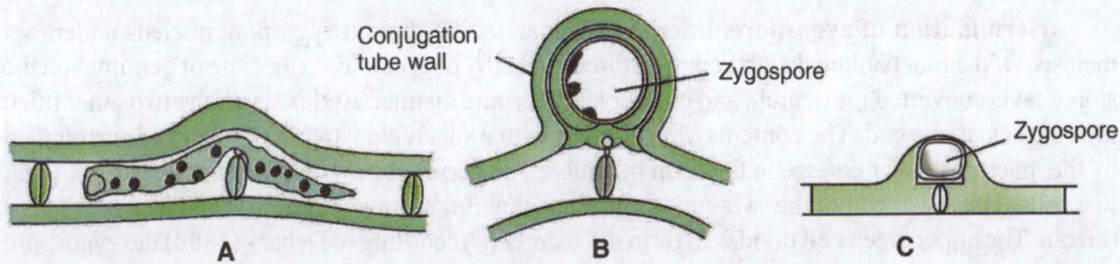

Fig. 11.5 (A-C). *Mougeotia.* Lateral conjugation. Isogamous lateral conjugation in *M. jogensis* (A-B) and *M. genuflexa* (C). (A-B, after Iyengar; C, after Wittrock).

meet. The end walls between them dissolve (A). A lateral conjugation tube is thus formed between the ends of adjacent gametangia. The protoplasts of the two gametangia migrate and meet in the middle of the conjugation tube which subsequently increases in diameter (B). The naked zygote is separated from the gametangia by one or two special walls. It then secretes a true zygote wall around it.

Zygospore. The zygospores are variable in form. They may be spherical, compressed, spheroid, ovoid, ellipsoid or quadrate-ovoid. The zygospore wall is differentiated into two or three layers. The outer layer is chitinous. The median layer is either colourless and smooth or variously coloured and ornamented. The innermost is transparent.

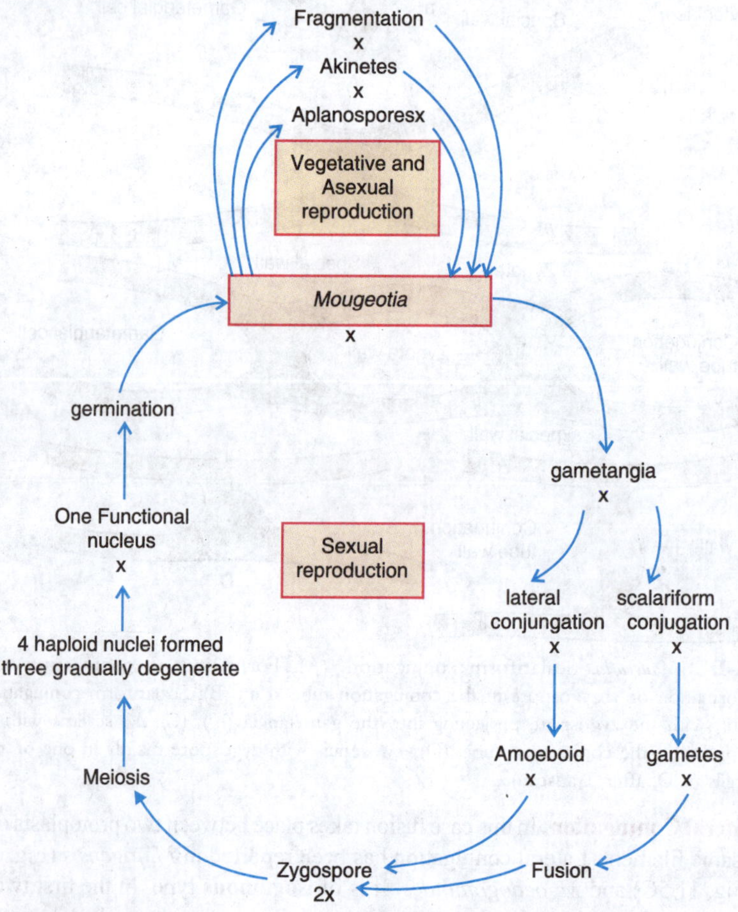

Fig. 11.6 Word diagram of the life-cycle of *Mougeotia*.

Germination of zygospore. Prior to germination, the diploid zygospore nucleus undergoes meiosis. Of the four haploid daughter nuclei, three gradually degenerate. At the time of germination the stored fat is converted into starch, and chloroplasts become distinct. At this stage, the two outer thick layers crack at one end. The contents of zygospore with a surviving haploid nucleus and surrounded by the innermost layer emerge in the form of a tube. The germ tube divides by a transverse wall into two cells, The lower cell of the two-celled germling contains scanty chlorophyll and develops into a rhizoid. The upper green cell divides to form the filament. According to Debary (1858) the zygospore sometimes germinates by a lid-like opening of the outer two layers of the zygospore wall.

Taxonomic Position

The genus *Mougeotia* was first established by Agardh in 1824. West (1916) divided the family into three sub-families, Mesocarpeae, Zygnemaeae and Spriogyreae on the basis of the structure of the chloroplast and position of zygospores. He placed *Mougeotia* in the sub-family Mesocarpeae on the basis of the presence of zygospores in the conjugation tube. Smith (1933) placed it in the family Zygnemataceae of order Zygnematales. Transeau (1951) followed Smith. Fritsch (1935) placed it in the family Mougeotiaceae of suborder Euconjugatae, order Conjugales. On the basis of the character of the chloroplast Randhawa (1959) divided Zygnemaceae into four sub-families namely, Mougeotiodeae, Zygnemoideae, Spriogyroideae and Sirocladiodeae. He placed *Mougeotia* in the sub-family Mougeotioideae. Thus the taxonomic position of *Mougeotia* according to Randhawa is as follows :

Division	:	**Chlorophyta** (Chlorophycophyta)
Class	:	**Chlorophyceae**
Order	:	**Zygnemales**
Family	:	**Zygnemaceae**
Sub-family	:	**Mougeotioideae**
Genus	:	*Mougeotia*
Species	:	*indica*

SUB-FAMILY : ZYGNEMOIDEAE

There are two axile chloroplasts either with irregular processes or stellate in each cell. The nucleus lies in the cytoplasmic bridge between the chloroplasts. The sub-family includes six genera of which *Zygnema* is described here.

ZYGNEMA (AGARDH, 1824)

Occurrence. *Zygnema* is a common green alga with about 100 species. Of these Randhawa (1959) reported 23 from India. All are aquatic except *Z. terrestre* Randh., they are found in considerable quantities in quiet freshwater of ponds, pools, lakes and also in slow flowing streams as yellowish green free-floating masses on the surface of water.

Thallus (Fig. 11.7). It is a simple, fine, unbranched thread or a filament consisting of cylindrical cells arranged one above the other in a row. The cells are longer than broad usually not more than twice the breadth. There are a few exceptions in which the cells are 2 to 5 times as long as broad. Occasionally short, few-celled laterals are developed from the filament. All the cells in the filament are alike.

Cell Structure (Fig. 11.7). Each cell consists of a tiny mass of protoplast surrounded by the cell wall. The cell wall as in *Mougeotia* and *Spirogyra* consists of a single piece. It is differentiated into two concentric layers. The inner gives the *cellulose* reaction. External to it is the thick pectose layer covered by the mucilage sheath. The cross walls between the cells in the filament are plane. The protoplast is differentiated into cell or plasma membrane, cytoplasm, nucleus and two stellate or star-shaped chloroplasts. In *Z. terrestre* the chloroplasts

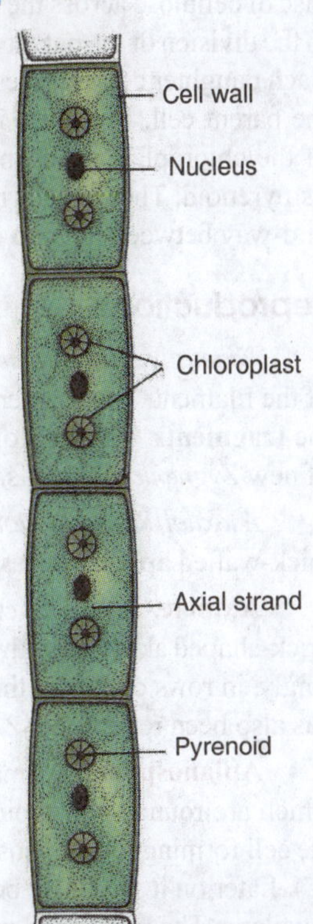

Fig. 11.7. *Zygnema.* Part of the filament showing cell structure.

are spherical bodies. Randhawa reported that in some cases (*Z. himalayense*) the individual cells contain 4 chloroplasts. The chloroplasts are embedded in the cytoplasm and lie axial to each other in the longitudinal axis of the cell. Each chloroplast consists of a central body with a single pyrenoid at its centre. A number of delicate strands radiate out from the central body of the chloroplast and extend to the plasma membrane. At times the chloroplasts may draw in their processes. The single centrally situated nucleus lies embedded in the middle of the broad strand of cytoplasm connecting the two chloroplasts.

Cell Division (Fig. 11.8). It increases the number of cells in a filament. The freshly formed cells grow in size. As a result the filament increases in length. The division of the cell is transverse and is preceded by the division of the nucleus which is mitotic (A). The nuclear division starts early in the evening. It is completed by midnight. With the formation of two nuclei in the cell there appears an annular projection of cytoplasm in the peripheral layers. The wall separating the two daughter cells arises as an annular ingrowth of the plasma membrane midway between the two ends of the cell. A delicate membrane then suddenly appears in the annular projection (B). It gradually grows inwards covered by the cytoplasm. Finally it forms a complete disc of cellulose across the cell cavity (C). This results in the division of the parent cell into two daughter cells. Each daughter cell receives one of the chloroplasts of the parent cell. It divides into two (D). The division of the chloroplast is accompanied by the division of its pyrenoid. The nucleus now migrates to a position mid-way between the two chloroplasts (E).

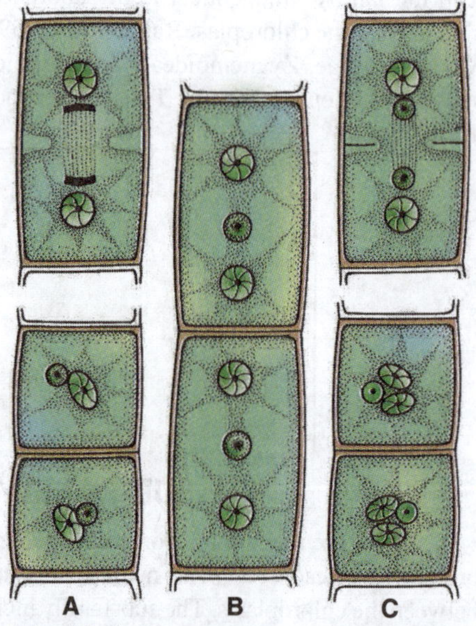

Fig. 11.8. (A-E). *Zygnema.* Stages in cell division.

Reproduction

Vegetative Reproduction. The vegetative method of propagation consists in the **fragmentation** of the filament. The filament breaks into short segments of living cells or into individual cells called the **fragments**. The cells of the fragments by cell division and growth make possible the formation of new *Zygnema* filaments.

Asexual Reproduction (Fig. 11.9 C-E). It is accomplished by the formation of **akinetes** and thick-walled **aplanospores**.

Akinete. It is a thick walled, resting vegetative cell. Randhawa reported the formation of brick-shaped akinetes with orange coloured thick wall in *Z. giganteum*. The akinetes may be formed singly, in rows of two or three and in later stages in long chains of many. The formation of akinetes has also been reported in *Z. peliosporium. Z. sterile* reproduces by akinetes only.

Aplanospores. As many as twelve species of *Zygnema* reproduce by means of **aplanospores**, which are rounded or ovoid in form and are produced singly in the vegetative cells. In *Z. terrestre* the cell forming the aplanospore shows a slight swelling on one side and the cell protoplast contracts (C). Later on it swells on both sides (D). The swollen part of the aplanosporangium is covered with mucilage. The contracted protoplast secretes a wall around it. The wall is thick, variously coloured and ornamented (F-H). The mature aplanospores, according to Randhawa 91938) are of dark greenish blue colour. They are of diverse shapes and may be oval, pyramidal, or barrel-shaped (H).

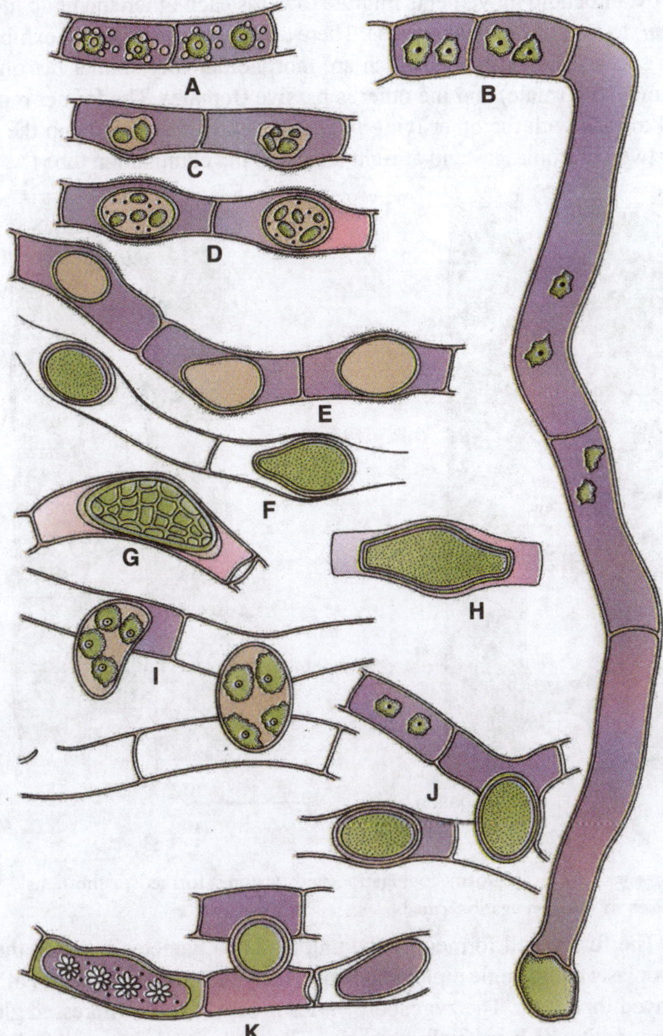

Fig, 11.9 (A-K). *Zygnema terrestre.* (A), cells showing structure, (B), lower portion of the filament consisting of 3 rhizoidal cells and two sub-aerial cells; (C-E), stages in the formation of aplanospores; (F-J), different shapes of aplanospbres, stages in scalariform conjugation; (K), fusion of contents of two adjacent cells (By courtesy of Late Dr. M.S. Randhawa).

Sexual Reproduction. It takes place by the method of conjugation both **scalariform** and **lateral**.

(*a*) Scalariform conjugation (Fig. 11.10). The process is similar to that of *Spirogyra*. The onset of the process is indicated by the filaments lying side by side in pairs (A). They secrete abundant mucilage. Small dome-shaped protuberances grow towards each other from the opposite pair of cells (B,a). The protuberance of the male cell is reported to be longer than that of the female cell. The protuberances elongate and finally meet (A, a). The conjugating filaments are gradually pushed apart as the protuberances grow. At the point of contact the intervening walls between the two protuberances or outgrowths dissolve (A, b). With this the two outgrowths form a **conjugation tube.**

An open passage between the two opposite cells is thus formed (A, c). It is known as the **conjugation canal**. Most species of *Zygnema* are usually strictly **isogamous**. In them (*Z. indicum*) both the gametes by amoeboid movements migrate towards each other and unite in the conjugation canal (*Z. areolatum*) to form the **zygospore** (B). There are other species which exhibit **physiological anisogamy**. Such species have gametes which are morphologically similar but one of the uniting pairs is actively amoeboid (male) and the other is passive (female). The former moves through the conjugation canal to fuse with the other lying *in situ* (A, c). Before migration the male protoplast rotates so that the two chloroplasts stand at right angles to the conjugation tube (A, b).

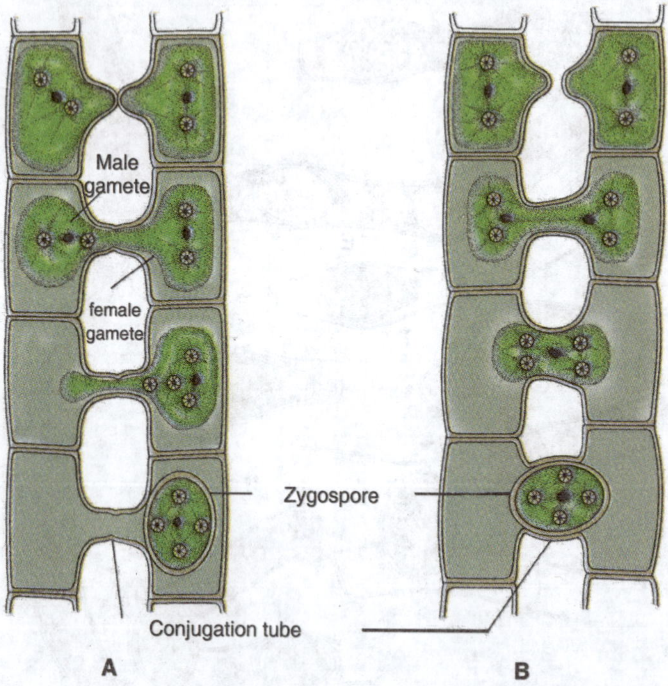

Fig. 11.10 (A-B). *Zygnema.* (A), Scalariform conjugation with zygotes formed in the female cells; (B), Fusion of gametes in the conjugation canals.

Zygospore. The fusion cell formed by the union of two gametes is called the **zygospore.** It contains four chloroplasts and a single diploid nucleus (A,d). Sometimes the fusion of the two gamete nuclei may be delayed for a time. The zygospore varies in form from compressed globose to ovoid. It secretes a wall around it which gradually thickens. The thick zygospore wall is differentiated into three layers (A, d). The thin, hyaline outer layer is cellulose or pectose in nature. It is named the **exospore.** The thick middle layer or **mesospore** is made up of cellulose. According to Tiffany (1924) it may be more or less chitinized. It is either smooth or ornamented depending on the species. The thin delicate inner-most layer of the zygospore wall is called the **endospore.** It is entirely cellulose in nature. The colour which is developed in the mesospore (median layer) varies from pale yellow to chestnut brown. In some cases it is bright blue to blue black. The zygospores sink to the bottom of the pond after disintegration of the walls of the female cells or conjugation tubes. The released zygospores enter upon a period of rest. Two out of the four chloroplasts which are located on the short axis of the zygospore disintegrate.

Czurda (1930) reported that in *Z. circumcarinatum* the papillae break down at their tips before they meet. The protoplasts escape from the open ends and fuse. Zygospores are thus formed between the opposite papillae.

(*b*) Lateral Conjugation (Fig. 11.11). It is commonly found in *Z. heydrichii.* Rao (1937) reported lateral conjugation in *Z. gangeticum.* The adjoining cells of the same filament give out tubelike protuberances one on either side of the septum. The protuberances finally meet. The separating septum eventually breaks in the region of the protuberance (A). A communication or side link is thus established between the two cells. The contents of one cell may migrate into the other where the fusion of two protoplasts takes place to form a zygospore. Often the contents of two gametangia fuse in the conjugation canal above the cross-wall septum (B). The walls of two gametangia bulge in the fused region to form a dome-like structure. Randhawa reported that lateral conjugation is the commonest mode of reproduction in *Z. czurdae* and *Z. himalayense.* Fritsch reported scalariform and lateral conjugation both taking place simultaneously in the two conjugating filaments of *Z. pectinatum* at (*a*) and (*b*) respectively in Fig. 11.12. *Zygnema stellinum* is reported to exhibit sexual dimorphism. Steinecke has observed that the female cells in this species are longer and broader. They have longer chloroplasts and pyrenoids than the male. Sometimes conjugation does not occur. The rounded gametes secrete thick walls and behave as **azygospores**.

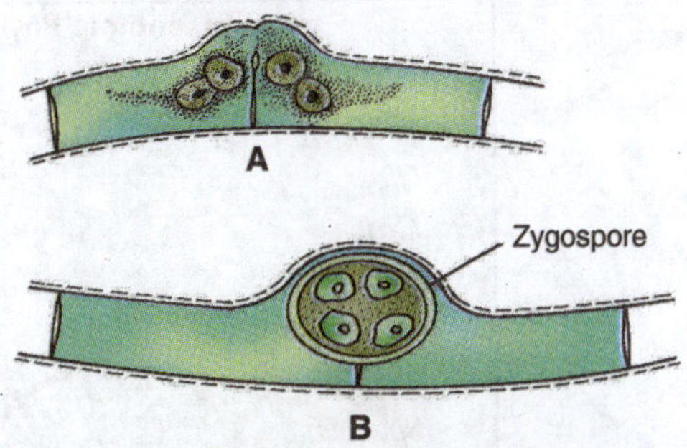

Fig. 11.11 (A-B). *Zygnema gangeticum*. Lateral conjugation. (A), Early stage showing migration of gametes; (B), Formation of Zygospore in the side link (After Rao).

Germination of the Zygospore (Fig. 11.13). At the beginning of the next growing season the zygospore lying at the bottom of the pond germinates to form a new plant. Prior to germination diploid zygospore nucleus undergoes meiosis to form four haploid nuclei (A-C). Three of these nuclei disintegrate and only one remains (D). This is the functional haploid nucleus. On germination the two outer layers of the zygospore wall rupture (E). The contents of the zygospore surrounded by the innermost layer emerge either completely or partially and divide transversely (F). The daughter cells divide and redivide. In this way a new filament is formed (H). The nucleus in these divisions undergoes equational mitosis.

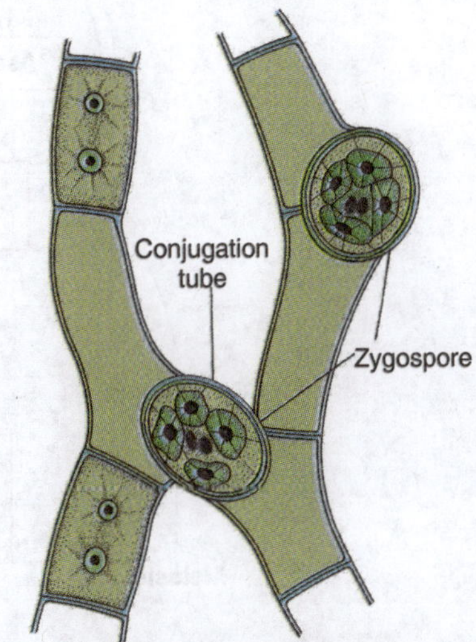

Fig. 11.12. *Zygnema pectinatum* showing scalariform and lateral conjugation simul-taneously (After Fritsch).

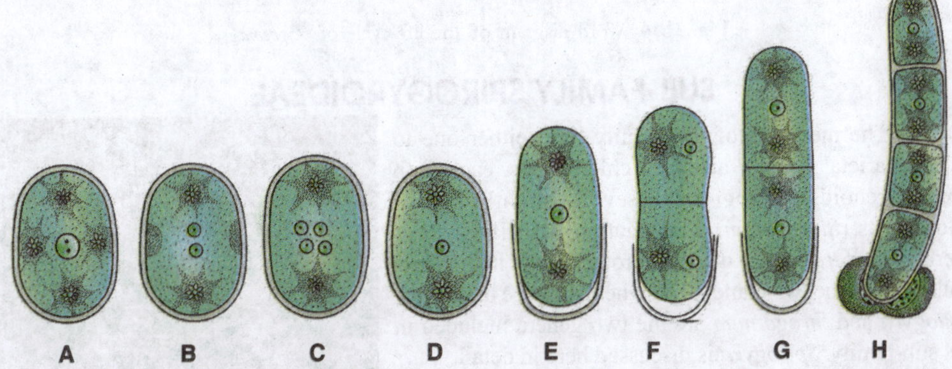

Fig. 11.13 (A-H). *Zygnema.* (A-D), Stages showing zygotic meiosis; (E-H), Germination of zygospore (A-G after Smith and H after Hallas).

Taxonomic Position

Division (Chlorophycophyta)	:	**Chlorophyta**
Class	:	**Chlorophyceae**
Order	:	**Zygnemales**
Family	:	**Zygnemaceae**
Sub-family	:	**Zygnemoideae**
Genus	:	*Zygnema*
Species	:	*czurdae*

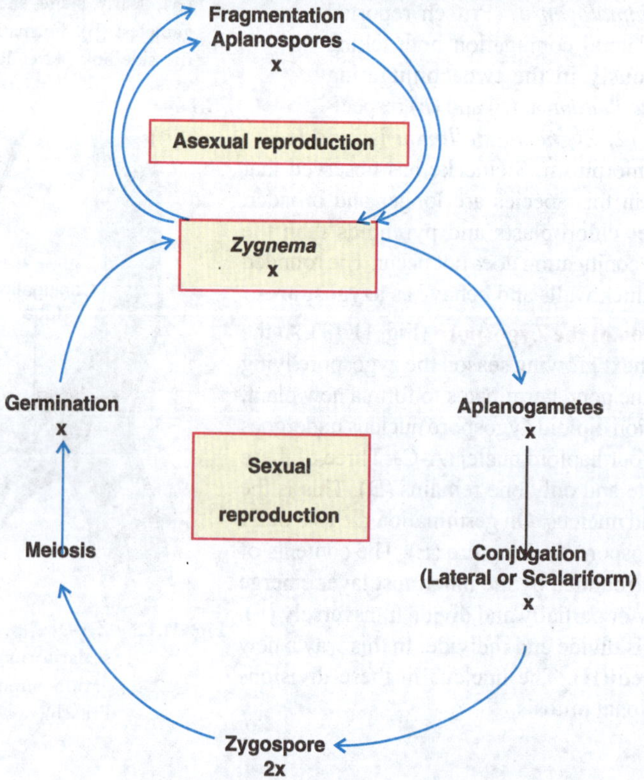

Fragmentation
Aplanospores
x

Asexual reproduction

Zygnema
x

Aplanogametes
x

Sexual reproduction

Conjugation
(Lateral or Scalariform)
x

Zygospore
2x

Meiosis

Germination
x

Fig. 11.14. Word diagram of the life-cycle of *Zygnema.*

SUB-FAMILY SPIROGYROIDEAE

The members of this family have either one to several parietal spirally arranged chloroplasts, each with many pyrenoids (*Spirogyra*) or several slightly curved chloroplasts (*Sirogonium*). Conjugation is scalariform or lateral but in *Sirogonium* directly through a pore in the cross wall between the two gametangia where they are in contact. *Spriogyra* and *Sirogonium* are the two genera included in this sub-family. *Spirogyra* is discussed here in detail.

Floating mass of Spirogyra and Mougeotia ►

SPIROGYRA (LINK 1820)

Occurrence. It is one of the commonest of green algae abundant in spring. It is found in bright green, free-floating masses in the still water of fresh water ponds, pools, lakes and ditches and also in slow flowing streams called *Nadees*. Because of the slippery feel of the threads *Spirogyra* is often called the pond-scum or water silk. According to Randhawa (1959) genus includes about 289 species and of these 94 have been reported from India. They are of universal distribution but are relatively rare in the tropics. Of the various species *S. condensata* is commonly met with.

Plant body. It is thallus which consists of a long green cylindrical thread about 1/10 mm across and several centimetres long. It is silky, hair-like, unbranched and unattached and is often called a **filament**. Normally the filaments are free living, free floating, slimy and each consists of a few hundred similar cells placed end to end in a single row. There is thus no basal distal differentiation. *S. adnata* and *S. jogensis* (Fig. 11.16) however, are exceptions. The basal cell in these species functions as an attaching organ. It develops rhizoidal outgrowths which assist in attaching the filament to the substratum. Jao (1936) reported the occurrence of rhizoids in *S. rhizopus* and *S. rhizobrachialis, S. dubia* and *S. affinis* also have rhizoids.

Spirogyra varians

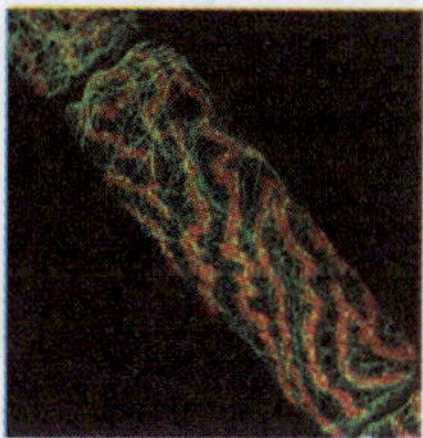

Spirogyra porticalis

Spirogyra crassa

Fig. 11.15. *Spirogyra.* A portion of free floating filaments.

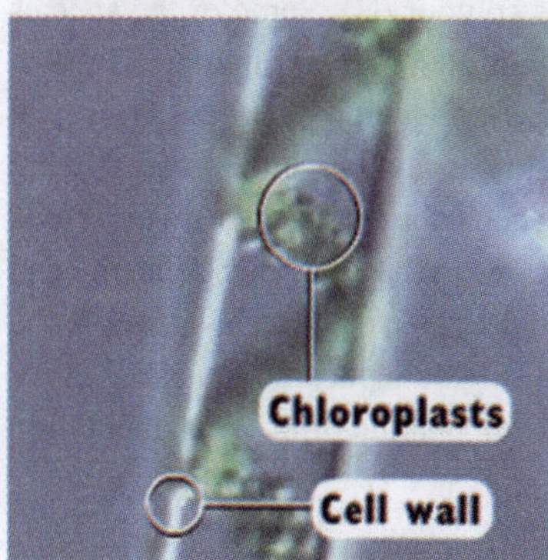

Cell structure of Spirogyra

Cell structure (Fig. 11.16). The cells comprising the filament are cylindrical in form with transverse end walls. They are usually much longer than broad (A). The cell has the structure typical of a parenchymatous cell. It consists of two parts, the cell wall and the protoplast.

(*a*) Cell Wall (Fig. 11.17 C-F). It surrounds the protoplast and thus forms its boundary. It is protective in function. The **Cell Wall** in *Spirogyra* is a single piece. There are neither any pores in the wall nor external markings on it. It consists of two concentric layers. The inner is *cellulose* in nature. Dawes (1965) reported that it consists of cellulose-1 microfibrils. It forms a complete envelope around the protoplast. External to it is the *pectose* layer. The outermost portion of pectose changes into pectin. The latter dissolves in water to form a *slimy sheath*. This mucilage sheath is sometimes referred to as the third layer of the cell wall. It makes the filament feel slippery

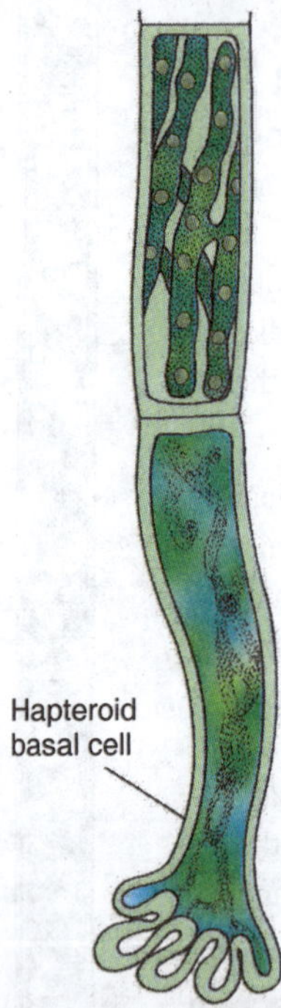

Fig.11.16. *Spirogyra jogensis.* A lower portion of the filament showing a single veget-ative cell and a hapteroid basal cell.

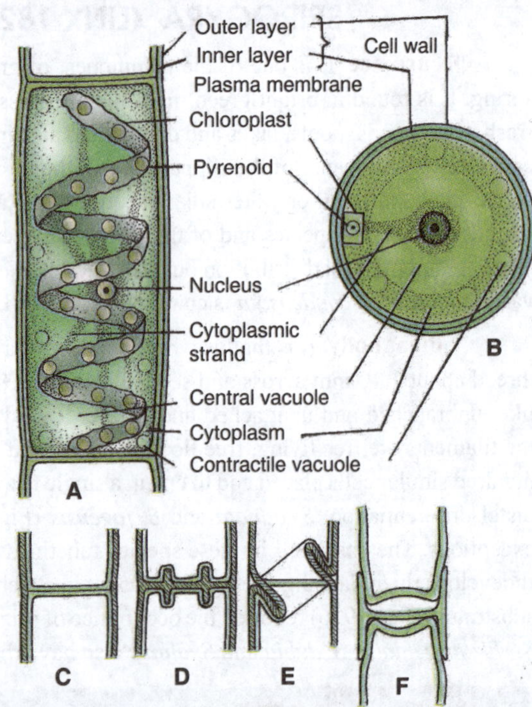

Fig. 11.17 (A-F). *Spirogyra.* (A), cell showing structure; (B) the same as seen in crosssection; (C), plane septum; D, replicate septum; (E), semireplicate septum; (F), Colligate septum.

to touch. In contrast the filaments of *Ulothrix* or *Oedogonium*, in which there is no slimy layer, feel like a group of wet threads. The end or cross wall is three layered. There is the middle septum or lamella of pectose with a primary wall or cellulose layer on either side. Many species of *Spirogyra* have plane and single cross walls. The latter have homogeneous contour (C). Some of the species with relatively narrow filaments have the cross wall split in the middle into two circular discs and thus appears shaped like a biconvex lens (Fig. 11.8 A). Other species develop peculiar cylindrical ring-like ingrowths on the lamella over which is deposited a cellulose layer on either side. If there are two ring-like ingrowths in opposite directions in the adjoining cells, the cross walls are described as **replicate** (D). In case there is a single infold in alternate positions in the adjoining cells, the end walls are known as **semireplicate** (E). Lloyd (1926) reported the presence of short **H-pieces** in the cell walls near the septa. Such septa are known as **colligate** (F). Colligate septa are seen clearly in *S. colligata.* Whether **H**-pieces occur normally in the cell wall or develop at the time of fragmentation only is not clear.

(*b*) The **protoplast** is differentiated into *plasma membrane, cytoplast, nucleus*, one or more *chloroplasts* with *pyrenoids* and a large *central vacuole* (A). The plasma membrane is differentially permeable and surrounds the cytoplast. In fact, it is an integral part of cytoplast which it invests and thus is living. The cytoplast forms a thin lining or a peripheral layer in contact with the cell wall closely invested by the plasma membrane. The cytoplast encloses a large **central vacuole** filled with cell sap which is often rich in tannins and other organic and inorganic compounds in solution. The sap vacuole is separated from the surrounding cytoplast by a thin, semipermeable membrane, the **tonoplast**. Embedded in the cytoplast and parietal in position are the large conspicuous **chloroplasts**. They vary in number in different species from one to as many as fifteen in each cell. A single chloroplast was recorded in *S. venkataramanii sp.* nov. by R.S. Rattan. Each chloroplast is an elongated, green helical band lying against the wall (parietal). It runs in a spiral manner and anticlockwise direction from one end of the cell to the other. Spiral chloroplast or chloroplasts is a feature from which the genus takes its name. All the chloroplasts in the cell may be loosely or tightly coiled and run spirally in parallel. The band-shaped chloroplast is either narrow or broad. The former has a smooth margin and the latter serrated. Situated in the spiral chloroplasts are a series of small, spherical, rounded bodies, the **pyrenoids** (A). They lie equidistant from one another and are the centres of starch formation. The cell has a single large **nucleus**. It lies in the middle of the cell. In the narrow forms it is angular. The corners touch the adjacent pyrenoids. The broader forms have alenticular (double convex) or flatly cylindrical nucleus. The nucleus is embedded in a central mass of little cytoplasm and hangs in the centre of the sap cavity. It is held in position by delicate, radiating cytoplasmic strands. They run across the vacuole to the region of pyrenoids in the chloroplasts. Dawes (1965) who studied the fine structure of *Spirogyra* cell reported that the nucleus has a complex nucleolus. In addition, the cell has numerous dictyosomes, endo-plasmic reticulum and mitochondria. The chloroplast consists of photosynthetic bands each consisting of 4-12 lamellae (thylakoids).

Growth. It is intercalary and sharply contrasts with the apical growth of the higher plants. It takes place by the division of any of the cells of the filament except the holdfast in the attached species. The daughter cells elongate. There is thus an increase in the length of the filament.

The above account shows that the multicellular filament of *Spirogyra* shows no differentiation of cells. Even the distinction into base and apex met with in *Ulothrix* is absent. All the cells in the filament are alike there being neither any specialisation of function nor differentiation of structure. Each cell is independent of the other. It is capable of carrying on all the vital functions such as nutrition, growth, cell division and reproduction. Each cell may, therefore, be looked upon as an individual, unicellular plant, and the whole filament as an assemblage or a group of unicellular organisms held together as filaments by the superficial delicate sheath containing cutin which swells to become mucilaginous and thus holds the cells together. Such groups of independent individuals are called **colonies**. *Spirogyra*, therefore, has little justification to be called a true multicellular plant. It may be a colony in which the cells are arranged end to end. It represents a filamentous colony in contra-distinction to the spherical, disc-like or irregular colonies of single-celled algae. The formation of a filamentous colony is the result of successive cell divisions occurring in parallel planes so that a chain of cells is formed.

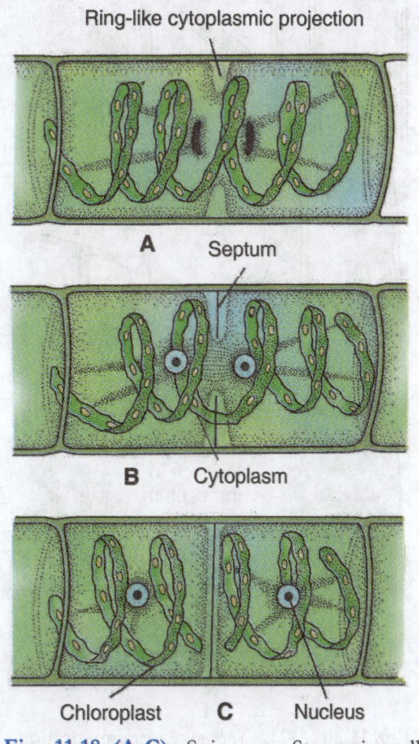

Fig. 11.18 (A-C). Spirogyra. Stages in cell division.

Cell Division (Fig. 11.18). The nucleus divides mitotically into two. With the formation of two daughter nuclei, the spindle instead of disappearing persists into the anaphase. It bulges outwards at the equator towards the side wall of the parent cell. Simultaneously with this an invagination of the plasma membrane appears in the lining layer of the cytoplasm midway between the two ends of the cell. It advances inwards to fuse with the bulging spindle (B). Suddenly a delicate septum appears in the sulcus of the plasma membrane. The septum arises from the inner surface of the longitudinal wall of the cell. Both invagination of plasma membrane and the septum grow inwards toward the centre like the closing of an iris diaphragm. It bridges across the cell cavity dividing the parent cell into two daughter cells (C). The chloroplasts are simply cut into two by the septum on either side of which is secreted the cellulose wall material. Thus the differentiation of dividing wall in *Spirogyra* and many other filamentous algae is **centripetal** in contradistinction to centrifugal differentiation typical of more highly organized plants. Cell division in *Spirogyra* and other filamentous forms increases the number of cells in the filament. It does not increase the number of filaments.

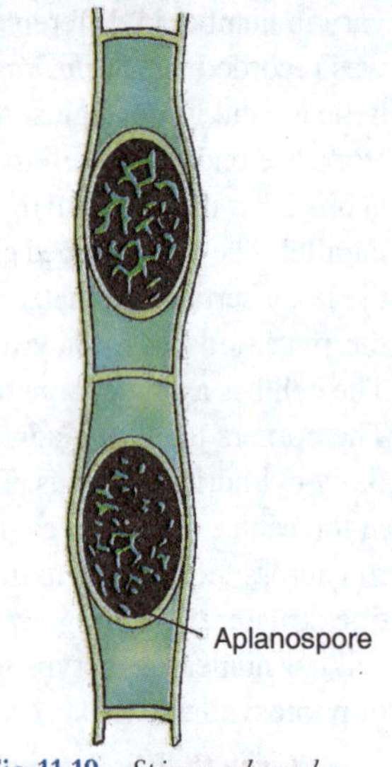

Fig. 11.19. *Spirogyra karnalae.* Aplanospore formation (After Rand-hawa).

Reproduction

Asexual reproduction. Asexually *Spirogyra* reproduces commonly by the vegetative method of fragmentation. Asexual reproduction byspores was normally considered to be absent. Till recently eight species of *Spirogyra* out of 300 known have been recorded to form aplanospores. The total absence of zoospsore formation in this aquatic alga is surprising. Randhawa (1938) reported that *S. alpanospora* exclusively reproduces by means of aplanospores formed singly in the vegetative cells. Formation of akinetes with thick lamellated walls was reported in *S. farlowii* by Transeau in 1915.

Fragmentation (Fig. 11.20). Vegetatively *Spirogyra* multiplies by the method of **fragmentation** when the conditions are exceptionally favourable. Fragmentation takes place by the dissociation of the filament at the cross walls into one or more short lengths or segments called the **fragments**. Each fragment may consist of a few living cells or even a single cell. It grows into a new filament by repeated cell division and growth. Fragmentation my be brought about in the following ways :

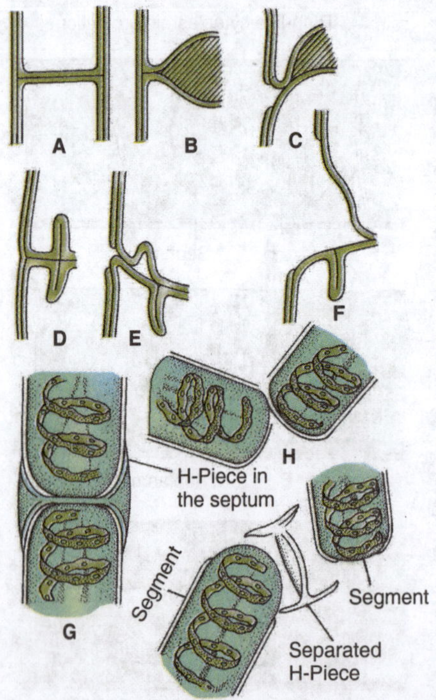

Fig. 11.20 (A-I). *Spirogyra.* Fragmentation by different modes of cell disjunction (After Llyod).

(*a*) By mechanical injury caused by the bite of fish or other aquatic animals or action of water currents.

(*b*) Softening and subsequent dissolution of the middle lamellae of the end walls. By sudden changes in the habitat factors such as temperature or acidity of water, the middle lamellae of the end walls become gelatinous. Consequently as turgor difference arises between the adjacent cells, causing one cell to bulge into the other. The strain on the junctions between the cells finally reaches the breaking point (A-C).

(*c*) In some species with replicate end walls (Fig. 11.20 D) the rings of one cell get evaginated forcing apart the cells. The rings of the other cell evaginate after separation to give it a distinct rounded extremity (E-F).

(*d*) Some species develop H-shaped pieces (*S.colligata*) in the end walls of adjacent cells (G).

These make fragmentation relatively simple. As the wall inverts owing to increased turgor of the cell, the H-piece slips off (H-I). Consequently the two cells come apart.

Sexual reproduction. It is accomplished by **conjugation** which is a very primitive method of isogamous type of sexual reproduction. It involves the fusion of the entire contents of the two similar unspecialized vegetative cells when the latter become yoked. According to Tiffany (1924) the change of pectose into water soluble pectin ceases in the conjugating threads. Pessony (1968) reported that low concentration of nitrogen and high light intensity enhance conjugation. Sexual conjugation takes place towards the end of the growing season. It is of two types:

(*a*) Scalariform (ladder type) **conjugation**. It is of common occurrence and takes place between the opposite cells of the two neighbouring filaments.

(*b*) Lateral conjugation. It is rarely found and takes place between two adjacent cells of the same filament.

(*a*) Scalariform conjugation (Fig. 11.21). It takes place at night between the recently divided cells. Chapman divides this process into three distinct phases: *maturation phase, gametic union phase* and *zygospore contraction phase.*

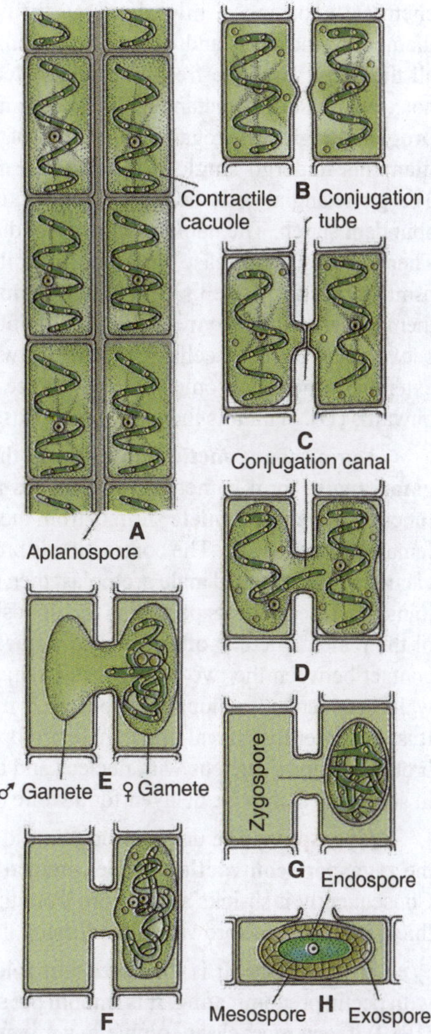

Fig. 11.21 (A-H). *Spirogyra.* Stages in scalariform conjugation (A-G). (H), Female cell containing a mature zygospore.

During the **maturation phase** the participating filaments of the same species of *Spirogyra* come together. Morphologically these filaments are alike. There are no visible differences between them. Functionally, however, they differ. These differences become apparent only at the time of fusion of gametes. During the night filaments of the two mating types come close together, line up side by side and cell by cell (A). The single filaments do not move at all. The presence of others perhaps acts as a stimulus. The adjusting movements between the conjugating filaments may either be brought about by localized secretion of mucilage or by changes in the surface tension. The close contact induces the secretion of more mucilage which holds the filaments together. The sticking together here and there of pairs of filaments is the putting out of papillae (B). They are formed on the recently divided cells which are naturally shorter. There is difference of opinion with regard to the development of papillae. According to Saunders, a short, bud-like papilla arises from one of each pair of opposite cells towards the other. Usually the papillae appear one of the conjugating filaments. On touching, these papillae stimulate the development of similar outgrowths from the opposite and corresponding points on the other filament. The papillae are thus in contact from the very beginning. The other view is that the papillae (p) simultaneously grow from the young and recently formed cells of both the conjugating filaments (B). These papillae elongate to form conjugation processes. The conjugation processes of the opposite cells meet immediately to form the **conjugation tubes** (C). The latter elongate and push apart the two conjugating filaments, which are connected all along their length in a ladder-like arrangement. Lloyd (1928) stated that the conjugation papilla of the female cell is usually shorter and thicker than the male. In some species the male papilla forms the entire conjugation canal. The development of papillae according to some is the result of contact stimulus (thigmotropism) and thus compared to the rhizoids. Others hold that they are the result of a chemical stimulus. Grote (1977) reported that in *S. majuscula*, the growth of the conjugating papillae (copulation tubes) is controlled not by thigmotropic (touch) stimuli but probably by interaction between hormone-like chemical substances, migrating by diffusion from one filament to the other and depends on a single division of all the cells. Thus the freshly agglutinated filaments do not yet consist of gametangia. These cells must be termed **progametangia**. All progametangial cells of the conjugating filaments undergo single division to form gametangia. The gametangial cells united by conjugation tubes store abundant starch. The nuclei and nucleoli diminish in size. There is also a decrease in the permeability and in the osmotic pressure of cell sap. With the completion of conjugation tubes the double wall (*w*) between them dissolves by enzyme action. A continuous passage way, the **conjugation canal** (D) is formed between the opposite cells. The arms of two protoplasts (of the opposite gametangial cells) which extend into the canal come in direct contact. The two protoplasts remain in contact from this moment onwards (D). With this the maturation phase ends.

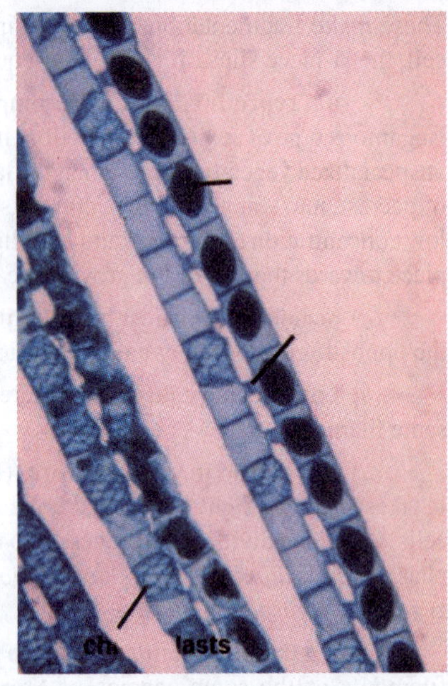
Conjugation in *Spirogyra*

During the **gametic union phase** the protoplasts of the conjugating cells, now called the **gametangia**, form gametes. The gametes are formed singly. The protoplast which is destined to function as a **male gamete** shrinks from the parent cell wall without losing contact with the other or female protoplast (D). The contraction is brought about by the removal of water from the contractile vacuoles. The rounded male protoplast then migrates through its respective conjugation canal slowly, joins and fuses with its partner in the opposite cell which can now be called **female**. The movement of the male gamete is often labelled **amoeboid**. Its mechanism, however, is obscure. Possibly the contact between the two protoplasts having been established by the dissolution of the intervening walls between the conjugation tubes, their fusion and rounding may be chiefly due to surface tension. It is only after the arrival of the male protoplast that the female contracts from its cell wall. The two protoplasts fuse, nucleus with nucleus and cytoplasm with cytoplasm. The fusion of the two nuclei, in some cases, may be delayed for a while (F).

Zygospore. The united protoplasmic mass of the two gametes is called a **zygospore**. During the **zygospore contraction phase** contractile vacuoles of the zygospore protoplast lose more water. Consequently it shrinks away more from the cell wall and assumes a rounded or oval form. A new thick, multistratose brown wall is formed. It is usually differentiated into the following 3 layers :

(*i*) Exospore. It is the outer thin colourless layer of the zygospore wall, cellulose in nature with pectin predominating. It is smooth but sometimes sculptured. According to Ashraf and Godward (1980) it is in most cases soluble in a mixture of cellulose and pectinase.

(*ii*) Mesospore. It is the thickest middle layer of the zygospore wall, acetolysis, and enzyme resistant presumably due to its sporopollenin content (Ashraf and Godward, 1980). It is pale yellow to chestnut brown in colour usually with a dehiscence line. It may be smooth but sometimes has a sculptured outer surface.

***(iii)* Endospore.** It is the thin colourless inner layer of the zygospore wall, cellulose in nature. The green zygospore which is usually ellipsoidal in form rarely ovoid or lenticular, later on becomes darker as the individual chloroplasts disappear. The male chloroplasts are usually the first to disappear. The zygospore is now capable of resisting unfavourable conditions. It is brownish or red in colour and is densely filled with starch and oil.

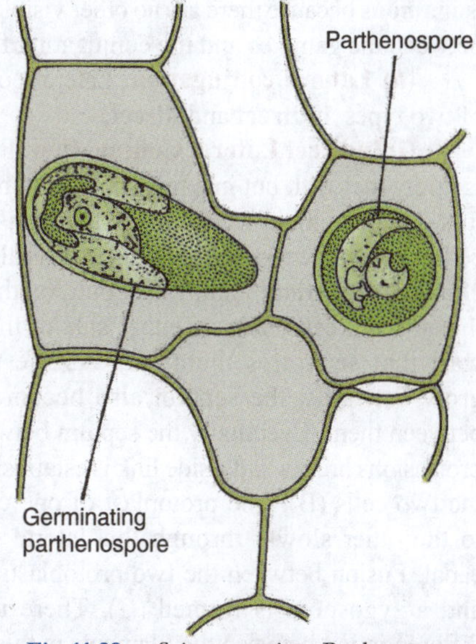

Fig. 11.22. *Spirogyra varians.* Parthenospore formation (After Klebs).

Parthenogenesis (Fig. 11.22). Sometimes the cells begin to form gametes but for some reason conjugation does not occur. It appears to be interrupted or to fail at some stage. For example in *S. groenlandica* the conjugation papillae are linked but the protoplasts in the conjugating cells round off and secrete cell walls around them before the intervening double wall between the conjugation tubes dissolves. These rounded structures behave as spores. They grow into new filaments. Such spores may be called **parthenospores** or **azygospores**. Parthenospores are also formed in *S. mirabilis* with the preliminaries of conjugation. This phenomenon, when a gamete grows into a new individual without fertilization, is called **parthenogenesis.**

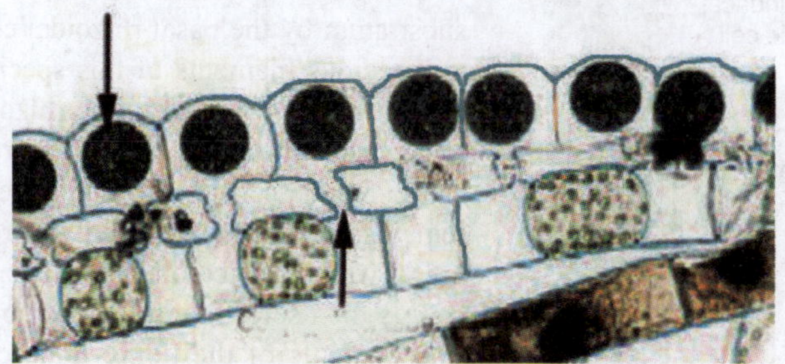

Figure should be change

Zygospores in *Spirogyra*

Differentiation of Sex

The gametes in *Spirogyra* do not show any morphological differences. They are similar in size and form. They are as a matter of fact **isogametes** which differ only in their behaviour. One of them is active and moves from one filament to another where the second gamete awaits it. This is a case of primitive anisogamy or physiological heterogamy. The active gamete is known as the '**male**' and the passive as '**female**'. The slight distinction of sex between the gametes is accompanied by sexual differentiation between the filaments. The protoplasts of all the cells of one conjugating filament act as male gametes and those of the other as female gametes. The filaments are thus **unisexual** and the species **dioecious.**

The filament that gives up its own protoplast is called the **male** and the one which receives it is called the **female.** Despite this slight distinction of sex the reproduction in *Spirogyra* is considered

isogamous because there are no other visible differences between the gametes and the conjugating filaments.

(*b*) Lateral conjugation. Lateral conjugation is of two types, **indirect** and **direct**.

(*i*) Indirect Lateral Conjugation (Fig. 11.23). It is rarely met with but in a few species such as *S. affinis* it is quite common. In lateral conjugation the adjacent cells of the same filament conjugate. It is also known as **chain conjugation**. Papilla-like outgrowths arise from the adjacent cells, one on either side of the transverse wall that separates them (A). As the protrusions grow outwards, the septum also becomes stretched between them. Eventually the septum between the two protrusions breaks and a side link is established between the two cells (B). The protoplast of one cell migrates to the other slowly through this lateral conjugation canal. Fusion between the two protoplasts takes place and a zygospore is formed (C). There is no sexual differentiation between the filaments in this case. They are **bisexual**. The species exhibiting this method of conjugation and **monoecious**. The filament in this case has a cell containing a zygospore alternating with an empty cell along its entire length.

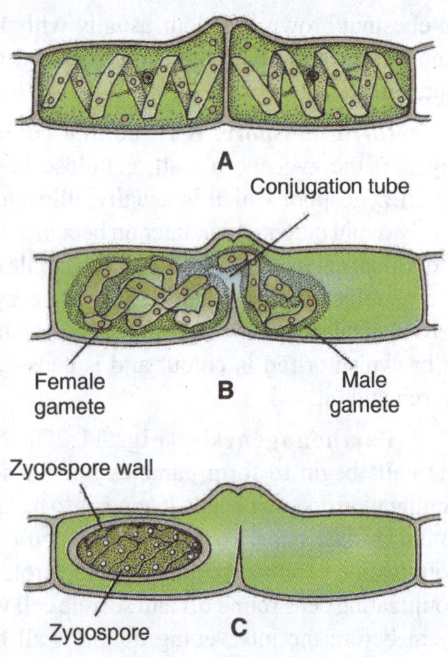

Fig. 11.23 (A-C). *Spirogyra sp.* Stages in lateral conjugation of indirect type. Klebs).

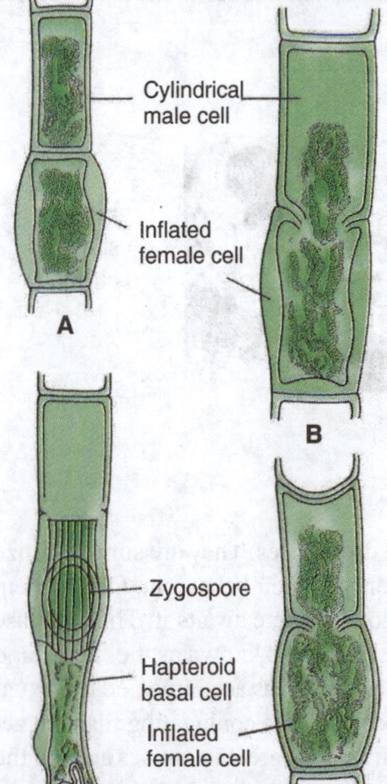

Fig. 11.24 (A-D). *Spirogyra jogensis.* Stages in direct lateral con-jugation. (After iyen-gar).

(*ii*) Direct Lateral Conjugation (Fig. 11.24 A-D). Iyengar (1958) reported direct lateral conjugation in *S. jogensis* in which the filaments are attached to the substratum by the basal rhizoidal cell. Conjugation in the monoecious filaments of this species usually takes place between two cells next to the rhizoidal cell. Of these, the lower which becomes swollen all round always functions as the female cell (A, b). The upper cylindrical male cell remains unchanged. A thick highly refractive mucilaginous layer, which is ribbed longitudinally in the female cell (D), is then secreted by the protoplasts of both the cells. The contents of the upper male cell develop a conical protuberance on the side next to the female cell (A, a). This is accompanied by a corresponding visible contraction of the protoplast of the male cell at the opposite end. The conical protuberance comes in contact with the separating septum and presses against it. It gradually elongates and exerts more and more pressure on the septum. Finally it bores its way right in the middle of the septum (B). Through this pore the male gamete migrates into the female cell (C). The male and the female gametes fuse in the female cell to form a zygospore (D). Direct lateral conjugation was also recorded by R.S. Rattan (1967) in *S. mirabilis* which he collected from Kapurthala.

Transeau (1938) reported that in *S. gratiana* both scalariform and lateral conjugation take place simultaneously (Fig. 11.25). Certain cells of the conjugating filaments resort to scalariform conjugation (*a*) and certain others to lateral conjugation of indirect type (*b*).

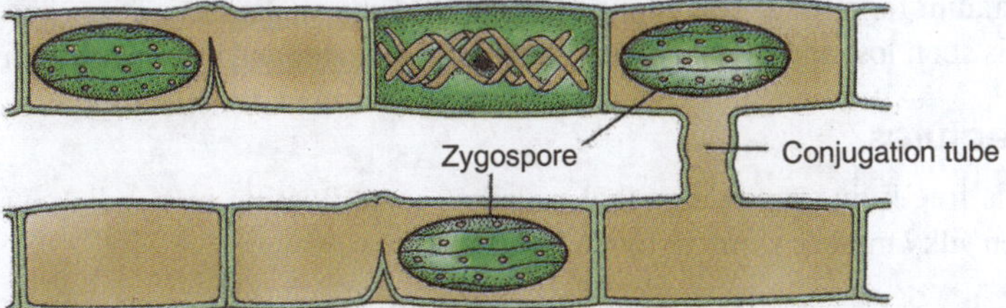

Fig. 11.25. *Spirogyra gratiana.* Scalariform and lateral conjugation taking place simultaneously in the conjugating filaments.

Conditions favouring conjugation. According to Banecke (1898) nitrogen deficiency is an important conditioning factor. Czurda (1933) considered that conjugation takes place when the active vegetative period is passed and there is increase in the *p*H value. Altitude has also been found to influence conjugation. Transeau (1916) suggested that fruiting in *Spirogyra* is correlated with the ratio between the surface and volume of the cell.

Maturation of Zygospore (Fig. 11.26 A-E). The zygospores, in either case, are liberated by the destruction or decay of the cell walls of the female cells. They sink to the bottom of the pool.

Their thick resistant walls enable them to withstand desiccation. They may be blown by the wind from the bottom of the dried up pond to another. Prior to germination the diploid zygospore (A) nucleus undergoes **meiosis.** It is known as **zygotic meiosis**. Meiosis as usual comprises two successive divisions. The first is **reduction division** (B) and the second **mitotic** (C). Both constitute **meiosis**. At the end of meiosis the zygospore protoplast contains four haploid nuclei. (C) Three of these abort (D). There is no **cytokinesis**. The surviving haploid nucleus is functional. The zygospore with the surviving haploid nucleus (E) germinates to give rise to a single filament.

Germination of zygospore (Fig. 11.26 E-G). The zygospore with the surviving functional haploid nucleus germinates only when in water (E). The two outer layers of the thick zygospore wall burst. The contents with a haploid nucleus still enclosed by the inner layer grow out as a long cylindrical germ tube in which the bright green colour reappears (F). You remember that the large spiral chloroplasts disappear in the formation of gametes and their subsequent union. However the zygospores contain minute granules named **proplastids**. The proplastids are considered instrumental in the formation of new chloroplasts. The green germ tube divides transversely to form a two-celled **germling** (G). The lower cell of

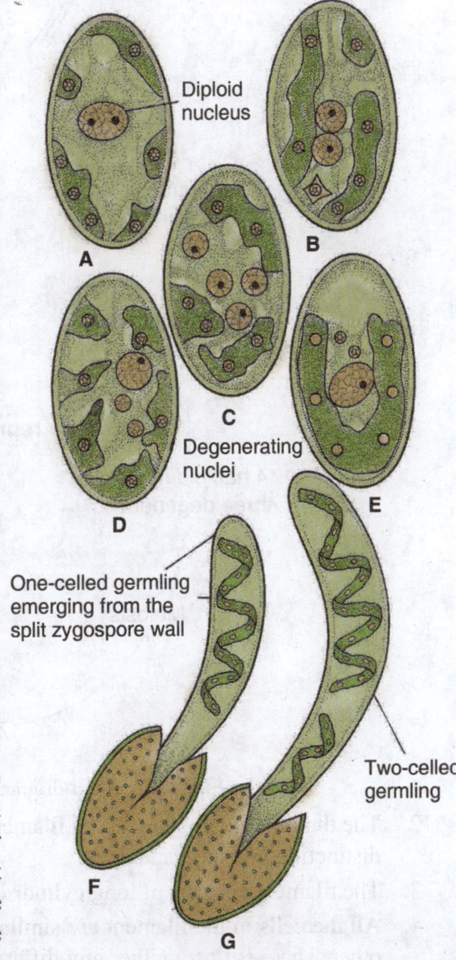

Fig. 11.26 (A-G). *Spirogyra.* Stages in zygotic meiosis (A-D) and germination of zygospore (E-G) Based on Troudle.

the germling is colourless or little green and rhizoidal in character. The upper cell is green. The later divides repeatedly to form a new filament. The colourless cell attaches the young *Spirogyra* filament to the substratum for a time. The young filament thus shows distinction into base and apex but this distinction is soon lost and the filament floats freely in water and becomes free living.

Salient Features

1. It is free living green alga that usually occurs floating in standing water in the form of bright, green silky masses slimy to touch.

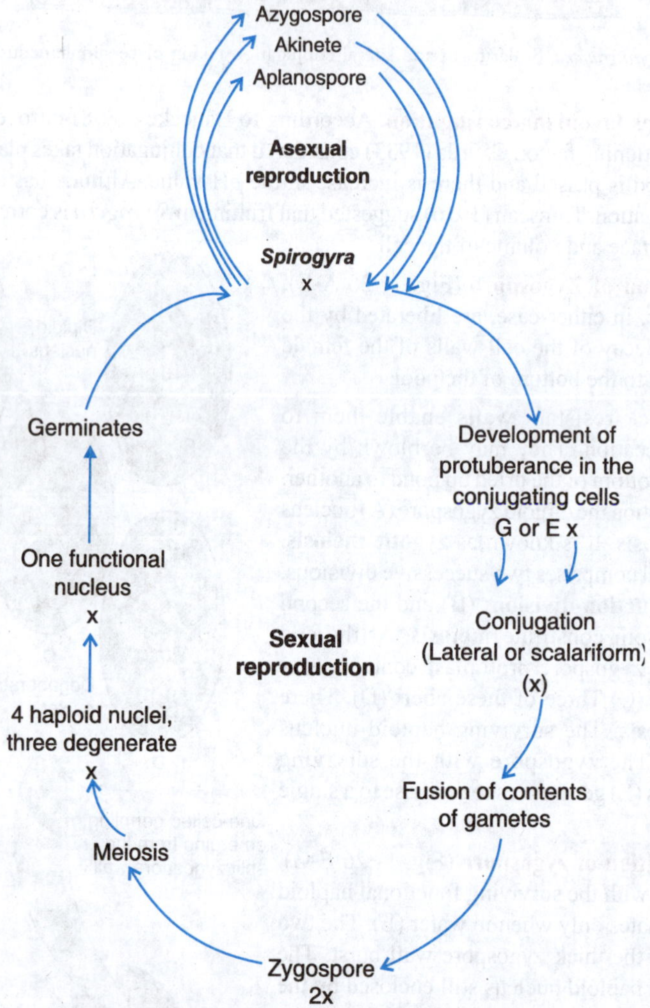

Fig. 11.27. Word diagram of the life cycle of *Spirogyra.*

2. The thallus is filamentous. The filament is a long hair-like unbranched thread. It shows no distinction into base and apex.
3. The filament consists of long cylindrical cells placed end to end in row.
4. All the cells in the filament are similar in structure, function and are independent of each other. There is thus neither any differentiation of cells in the filament nor any division of labour.
5. Each cell consists of a cell wall enclosing the protoplast.

6. The cell wall consists of two concentric layers. The inner is cellulose in nature. The outer is a pectose layer covered with a mucilage sheath. The latter gives a slimy feel.
7. The protoplast is differentiated into a plasma membrane, cytoplasm, nucleus, one or more chloroplasts with pyrenoids and a central vacuole.
8. The cytoplasm forms a utricle around the large, central vacuole containing the cells sap.
9. The single large nucleus surrounded by a tiny mass of cytoplasm remains suspended in the centre of the vacuole by a number of cytoplasmic strands.
10. The number of chloroplasts varies from one to several. Each chloroplast is a green spiral band and is parietal in position. It is embedded in the cytoplasm and coils round from one end of the cell to other in an anticlockwise direction.
11. Each chloroplast bears a number of pyrenoids equidistant of each other.
12. The filament during the growing season multiplies by the method of fragmentation.
13. Asexual reproduction is isogamous and accomplished by the method of conjugation which may be scalariform or lateral.
15. In the former, conjugation tubes connect the young and recently divided opposite cells of the two conjugating filaments.
16. The gametes are non-motile, similar in size and structure but differ in their behaviour. The male gamete migrates through the conjugation tube by amoeboid movements and joins the passive female gamete in the opposite cell to form a zygospore.
17. Lateral conjugation is of two types, indirect and direct.
18. In indirect lateral conjugation male gamete migrates through the conjugation canal formed between the adjoining cells of the same filament.
19. In direct lateral conjugation the male gamete migrates through a pore in the intervening cell wall between the two adjacent cells.
20. The zygospore is a drought resisting structure. It germinates after a period of rest and develops directly into a new filament.

Taxonomic Position

Division	:	**Chlorophyta** (Chlorophycophyta)
Class	:	**Chlorophyceae**
Order	:	**Zygnemales**
Family	:	**Zygnemaceae**
Sub-family	:	**Spirogyroideae**
Genus	:	*Spirogyra*
Species	:	*condensata*

FAMILY : DESMIDIACEAE

The representatives of this family are **placcoderm desmids**. They are single-celled microscopic organisms of great variety and beauty of form. Often they are exquisitely symmetrical and practically all are fresh-water forms. They are by far the commonest of all the members of the Conjugales and number about 2500 species which are placed under 23 genera. These are unicellular forms. The cell wall consists of two pieces and generally has pores. The unicell consists of two symmetrical halves known as the semi-cells. The two semi-cells are joined together by a connecting zone, the **isthmus**. In the isthmus region lies the nucleus. The isthmus region is often constricted but in a few it is unconstricted. The placcoderm desmids with unconstricted isthmus are called the

unconstricted desmids, while others are called the **constricted desmids**. The best example of the former is *Closterium* and of the latter *Cosmarium*.

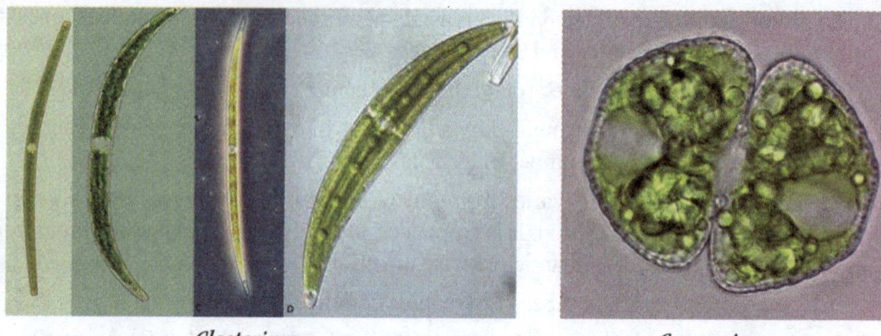

Closterium *Cosmarium*

MEMBERS OF FAMILY DESMIDACEAE

UNCONSTRICTED DESMID – CLOSTERIUM

Occurrence. *Closterium* is an example of the **placoderm**, **unconstricted desmids**. It is a widely distributed solitary desmid found at the bottom of ponds and drains. It has nearly 200 species. They usually occur mixed with slime. Some species occur intermingled with other free floating fresh-water algae.

Plant body (Fig. 11.28). The **thallus** is an elongated cell tapering from the middle towards the ends (A). It is generally circular in cross-section (B). Most species are straight and spindle shaped. A few others have a curved cell which is crescentric in lateral view and may be distinctly lunate or arcuate (A). The *Closterium* cell consists of two symmertrical halves called the **semi-cells**. The **isthmus** region which connects the two semi-cells is not constricted so that the plant body is distinctly a single cell.

Cell structure. The *Closterium* cell is composed of the cell wall and the protoplast. The **cell-wall** is made up of two halves. One half is older than the other. It belongs to the previous generation. The two halves fit closely over one another. In the mature cell, the cell wall is highly perforated and differentiated into two layers, the **inner** and the **outer**. The inner layer is thin, structureless and is mainly cellulose in nature. The outer layer has a substratum of cellulose impregnated with iron and pectic compounds. It is firm and somewhat thicker. The surface may be smooth or have delicate striations. In many species the cell wall has longitudinal manner. There are narrow grooves between the fine ridges. The cell wall is perforated by vertical pores through which mucilage is secreted. The pores are arranged in longitudinal rows in the narrow grooves. This mucilage layer is often referred to as the third layer of the cell wall. The secretion of mucilage causes slow, gliding somewhat jerky movements of the cell in the direction of light.

Vacuole
Pyrenoid
Cell wall
Nucleus
B
Chloroplast
A

Fig. 11.28 (A-B). *Closterium sp.* (A), showing cell structure; (B), T.S. cell (After Fritsch).

The *Closterium* cell is not divided into two semi-cells by a median constriction but the protoplasmic contents are divided. There are two beautifully lobed **chloroplasts**, one in each semicell. The chloroplast is axial in position. It is in the form of a curved cone with longitudinal ridges of green, radiating from a comparatively slender axis. Thus in cross-section the chloroplast looks like a hub with radiating spokes. The **pyrenoids** are arranged in a row or an axial series on the chloroplast. In some species the pyrenoids are numerous and scattered in an irregular manner. The single **nucleus** lies in the isthmus region embedded in a bridge of cytoplasm connecting the two chloroplasts (Fig. 11.28, A). Near each tip of *Closterium* cell is a hyaline cytoplasmic region. The latter contains a small spherical **polar vacuole**. Each vacuole contains one or more particles of gypsum (Calcium sulphate) which are by-products of metabolism. The gypsum particles exhibit brownian movements which cease with the death of the individual. The number of particles is always the same in the two polar vacuoles of the same plant. However it is not constant for the species. *Closterium* differs from other desmids in the following respects :

1. Elongated spindle-shaped thallus tapering towards both ends from the middle. It may be straight or crescent-shaped.
2. Absence of constriction in the isthmus region which connects the semi-cells so that the plant body is more distinctly a single cell.
3. Presence at each end of a small vacuole containing one or more particles of gypsum which exhibit brownian movements.

Movements. *Closterium* cell often performs slow oscillatory movements about one fixed end. Sometimes the free end may swing through 180° and become attached whereas the other end is released and repeats the event. In this way the cell moves over the substratum by a series of somersaults.

Reproduction. *Closterium* reproduces vegetatively and sexually. There is no asexual reproduction by spores.

Vegetative reproduction (Fig. 11.29). It takes place by the method of cell division which is transverse. A transverse constriction (c) in the chloroplast of each semicell is an indication of the initiation of cell division (A). It takes place much earlier than the division of the nucleus. Mitosis is however, completed before the chloroplasts are fully divided. At the commencement of the process the inner layer of the cell wall develops an annular thickening which projects slightly into the cell as a septum (A). A slight constriction may appear at the surface of dividing cell marking the position of this structure. When the nuclear division is nearly complete the septum grows inwards as cylindrical strip demarcating the two new daughter halves (B). Each daughter nucleus now moves to the point where each chloroplast is dividing (B). These points become the mid-regions of the new daughter cells. According to Van Wisselingh a cellulose layer is deposited over the dividing septum before it finally splits to form the two daughter individuals (C). The daughter individuals are cone-shaped (D). This cellulose layer in each daughter cell bulges out to form the new semi-cell (E). By this way the new desmid comes to consist of one old semi-cell and the other a newly regenerated semi-cell (E). By this time the original chloroplast of the old semi-cells has divided into two daughter chloroplasts. The

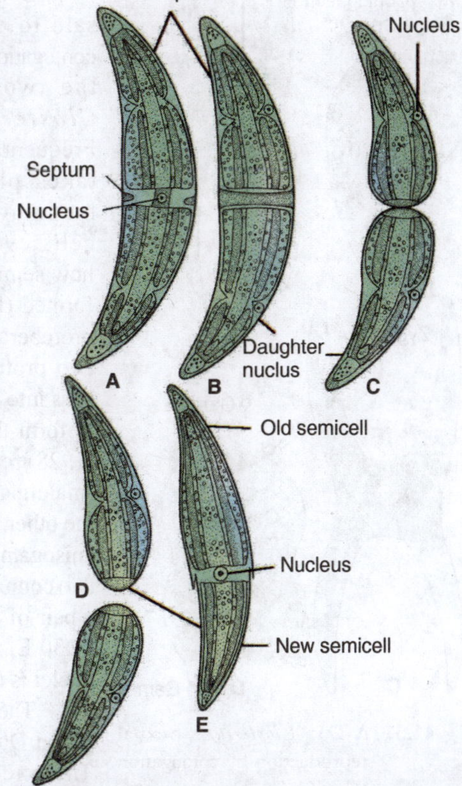

Fig. 11.29 (A-E). *Closterium.* Asexual reproduction by the vegetative method of cell division (Based on Fritsch).

daughter chloroplasts enlarge. One of these remains in the old semi-cell and the other moves to the new or regenerated semi-cell. The nucleus by this time has migrated to the isthmus region of the newly formed desmid.

Sexual reproduction (Fig. 11.30). It is accomplished by the method of **conjugation**. At the onset of conjugation the desmids come together and become enveloped by a common gelatinous sheath (A). Within the sheath they arrange themselves parallel to each other in pairs. In a majority of the species splitting of the cell wall at the isthmus of the two conjugating desmid takes place. The semi-cells than gape apart to allow the protoplasts to escape (B). The protoplasts are amoeboid in nature. They move towards each other by their slow amoeboid movements and meet mid-way between the two parent cells. Gametic union takes place between the released protoplasts outside the parent cell walls which finally disintegrate (C). The resulting fusion cell or the zygote secretes a heavy wall around it to become a **zygospore**.

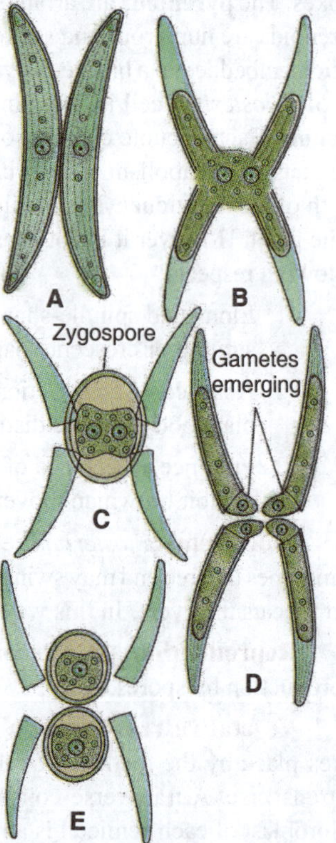

Fig. 11.30 (A-E). *Closterium.* Diagrams illustrating sexual reproduction, (A-C and D, after Oltmanns).

Some species are said to develop small conjugation tubes between the two conjugating *Closterium* cells. Frequently conjugation takes place between immature *Closterium* cells even before the new semi-cells are fully formed (Fig. 11.31 A). The conjugation tubes arise as small protuberances from the isthmus of each conjugating cell. The two protuberances join to form the tube. The two gametes pass into the conjugation canal (Fig. 11.31 B). There they fuse to form the thick-walled zygospore (Fig. 11.31 C). Scherffel (1928) reported that in *C. parvulum* the protoplast of one cell (male gamete) passes into the conjugation canal earlier than the other. This suggests sexual differentiation or physiological anisogamy (Fig. 11.31 D). In some species each semi-cell of the two conjugants produces a gamete (Fig. 11.30 D). Consequently a pair of zygospores if formed between the two parents (Fig. 11.30 E). In some species the fusion between the two gamete nuclei is delayed for some time (C).

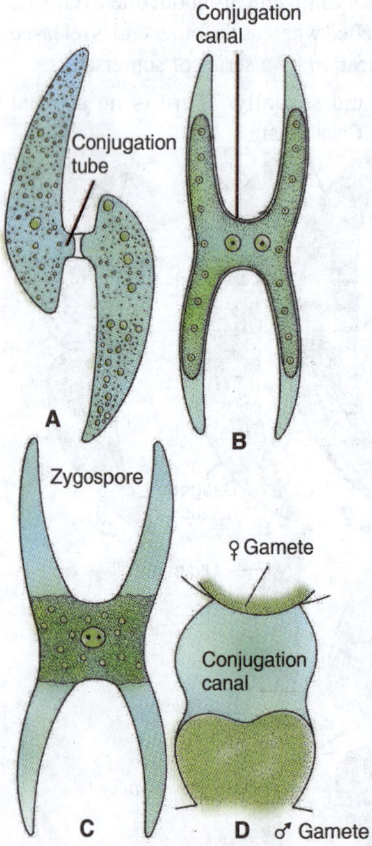

Fig. 11.31 (A-D). *Closterium.* Sexual reproduction by conjugation tubes (A-C); (D), *C. parvulum* show-ing physiological anisogamy. (A after Fritsch and D after Scherffel).

The zygospore is a resting spore. It enters upon a resting period. During the resting period it changes colour from green to red. The spore coat is differentiated into three layers predominantly made up of cellulose and pectin. Of these the inner is thin and colourless. The outer layer, as a rule, is smooth but in a few species develops warty or spiny processes distinctive of the species.

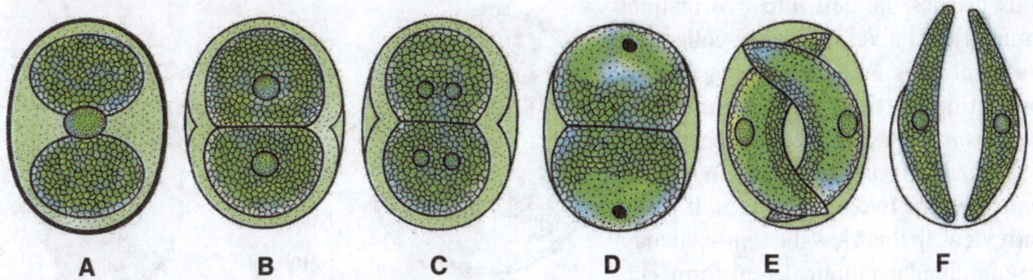

Fig. 11.32 (A-F). *Closterium* (A-C) Zygotic meiosis; (D-F), Stages in germination (After Klebahn).

Maturation and Germination of Zygospore (Fig. 11.32). Two of the four chloroplasts contributed by the gametes degenerate. The diploid zygospore undergoes two nuclear divisions forming four daughter nuclei (A-C). These two divisions constitute **meiosis**. The zygospore protoplast which by now has escaped from the outer layers of the zygospore wall divides into two parts. The latter are still enclosed by a thin membrane. Each of the two daughter protoplasts has one chloroplast and two haploid nuclei, one of which is functional and the other non-functional (D). The latter gradually disappears and the former enlarges. Each daughter protoplast fashions into a young *Closterium* cell (E). The investing membrane finally ruptures and the two *Closterium* individuals are released (F). They gradually assume the shape characteristic of the species.

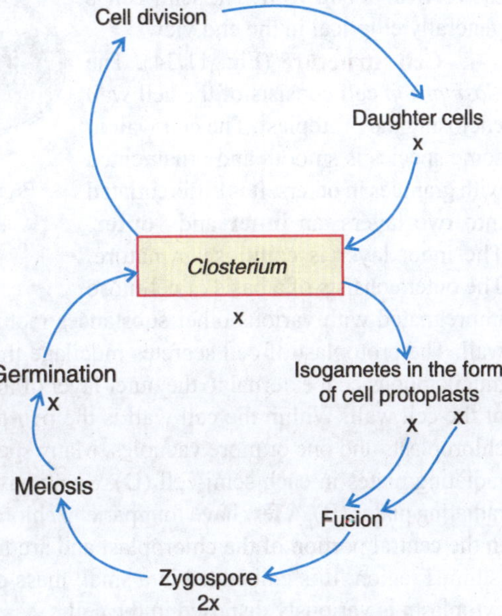

Fig. 11.33. Word diagram of the life-cycle of *Closterium*.

In some species three of the daughter nuclei formed in meiosis degenerate. The zygospore protoplast with the surviving nucleus develops into a vegetative cell. In still a few others all the four nuclei may be functional.

Taxonomic Position

Division	:	**Chlorophyta** (Chlorophycophyta)
Class	:	**Chlorophyceae**
Order	:	**Zygnemales** (Conjugales)
Family	:	**Desmidiaceae**
Genus	:	*Closterium*
Species	:	*cynthia*

CONSTRICTED DESMID—COSMARIUM

Occurrence. It is free floating, solitary desmid. It occurs in fresh water ponds intermingled with other algae. Usually it occurs in abundance in mucilaginous masses along the walls of tanks and reservoirs in winter. The genus includes over 800 species.

Plant body (Fig. 11.34). It consists of a small, flat cell (A). The length of the cell is one and a half times greater than the breadth. The unicell has a deep median constriction, the **sinus**. The sinus divides the cell into two distinct symmetrical halves, the **semi-cells**. The two semi-cells are joined by a narrow connecting portion, the **isthmus**. The *Cosmarium* cell has three principal planes of symmetry. It naturally rests in a plane containing the two longest axes. It is the **front view**. In this view the semi-cell may be semi-circular, elliptical, reni-form, etc. The others are the **lateral** or **side view** and the **vertical** or **end view**. The semi-cell is generally elliptical in the end view.

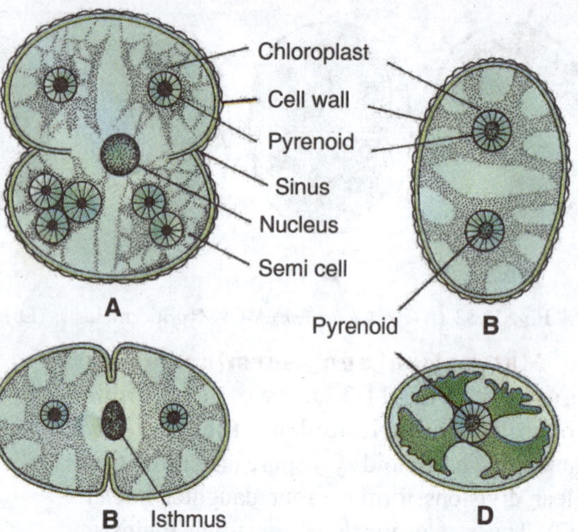

Fig. 11.34 (A-D). *Cosmarium.* (A), Two chloroplasts in each semi-cell; (B), T.S. of (A); (C), One chloroplast in each semi-cell; (D), T.S. of (C).

Cell Structure (Fig. 11.34). The *Cosmarium* cell consists of the cell wall enclosing the protoplast. The cell wall in some species is smooth and ornamented with granules in others. It is differentiated into two layers, an **inner** and **outer**. The inner layer is cellulose in nature. The outer consists of a basis of cellulose impregnated with various other substances including iron compounds. There are pores in the cell wall. The protoplast of cell secretes mucilage through these pores. The secreted mucilage forms a mucilaginous coat external to the outer layer of the cell wall. It is sometimes called as the third layer of the cell wall. Within the cell wall is the **protoplast**. It is differentiated into cytoplasm, nucleus, chloroplasts and one or more vacuoles. Many species have a single large axial chloroplast with four radiating plates in each semi-cell (D). Some have two chloroplasts in each semi-cell with several radiating plates (B). A few have four parietal chloroplasts in each semi-cell. The pyrenoids are situated in the central portion of the chloroplast and are frequently numerous. The single nucleus lies in the isthmus region. It is embedded in a small mass of cytoplasm and lies in the centre. The remaining cytoplasm is variously disposed in the cell.

Reproduction. It take places by the following two methods:

Cell division (fig. 11.35). Vegetatively *Cosmarium* multiplies by the method of **cell division**. At the commencement of cell division the nucleus divides mitotically into two daughter nuclei (A). The division of the nucleus is followed by elongation of the isthmus region (B). A constriction then appears on the wall midway between the two daughter nuclei (C). An annular in growth arises at the constriction. It grows inwards across the isthmus region dividing the parent cell into two daughter cells (C). Each daughter cell possesses a portion of the isthmus (D). The latter grows into a new semi-cell. As the younger semi-cell grows in size the chloroplast or the chloroplasts in the older semi-cell divide along with their pyrenoids. Half the number of daughter chloroplasts then migrate into the new semi-cells. Owing to their peculiar cell division one of the semi-cells in the desmid's plant body is always younger than the other. When multiplication is rapid numerous desmids occur embedded in a mass of jelly.

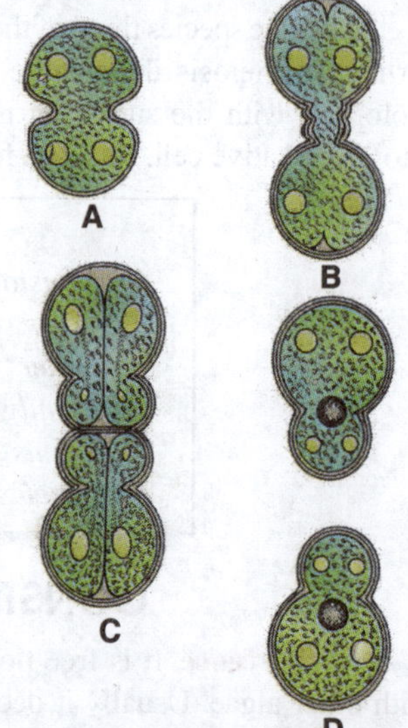

Fig. 11.35 (A-D). *Cosmarium* sp. Stages in division (After De Bary).

Sexual reproduction (Fig. 11.36). It is **isogamous**. It takes place by the method of conjugation in which two entire cell contents behave as gametes. The first indication of conjugation is the coming together of the desmids. They get embedded in a mucilaginous mass. Here they lie side by side in pairs (A). The isthmus region of each of the two conjugating desmids in the envelope splits. The semi-cells gape apart to allow an amoeboid escape of the protoplasts (B). The entire protoplast of the conjugating desmid functions as a single gamete. The gametes of the two conjugating desmids come out of their cell walls (C). They move towards each other by slow amoeboid movements. The walls of the four empty semi-cells lie nearby disintegrating in mucilage. Finally the gametes meet midway between the two conjugants (D).

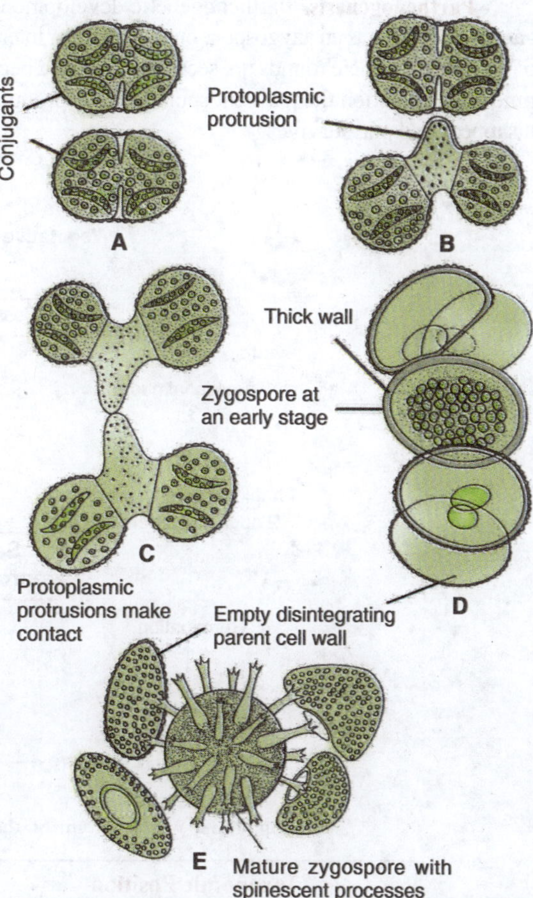

Fig. 11.36 (A-E). *Cosmarium.* Diagram illustrating sexual reproduction (C after Engler and the rest after De Bary).

In some species a **conjugating tube** is developed between the two conjugating *Cosmarium* cells. It is formed in the isthmus regions of the conjugating cells. The two amoeboid gametes creep towards each other and fuse in the conjugation canal. The union of the cytoplasm of the two gametes is followed by the union of their nuclei which in a few species is delayed. Eventually it takes place. The resultant zygospore secretes a heavy wall around it (D). It is composed of three layers. The outer is called the **exospore**, middle **mesospore** and the inner **endospore**. The exospore is either smooth or spiny (E). The zygospore then sinks to the bottom of the pond and enters upon a dormant state.

Germination of zygospore (Fig. 11.37). At the beginning of the next growing season the zygospore germinates (A). The diploid zygospore nucleus divides by meiosis to form four haploid daughter nuclei (B–C). Of these two are functional and the other two gradually disintegrate (C). The zygospore protoplast now divides into two daughter protoplasts. Each daughter protoplast contains one functional nucleus, one disintegrating nucleus and a chloroplast (D). The zygospore wall ruptures and the two daughter protoplasts are liberated (E). Each daughter protoplast, after liberation, assumes the shape characteristic of the genus.

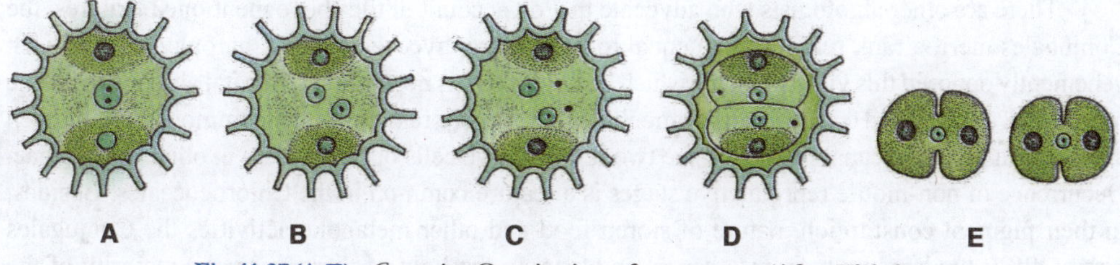

Fig. 11.37 (A-E). *Cosmarium.* Germination of zygospores (After Klebahn)

Parthenogenesis. Parthenogenetic development of the gametes has also been reported. Each gamete functions as an azygospore on the failure or interruption of conjugation. In this case the protoplast of the desmid gamete rounds up, secretes a wall and becomes a spore called the **parthenospore**. At the time of germination the parthenospore nucleus produces 4 daughter nuclei. Three of these gradually disappear and one survives.

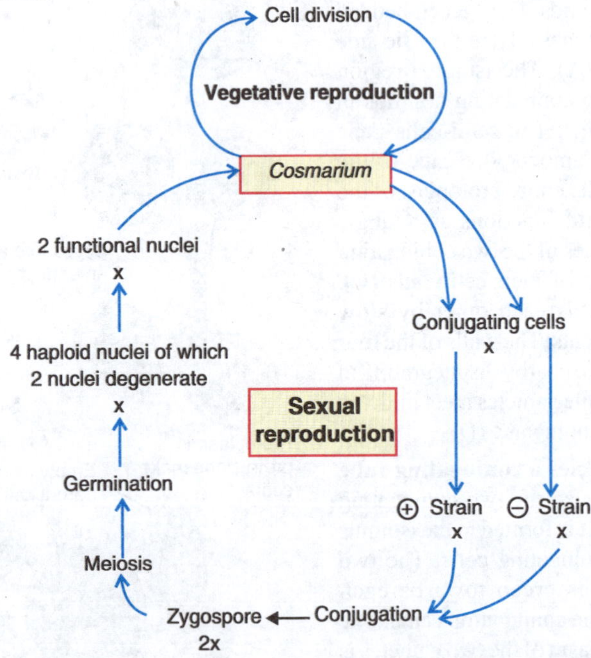

Fig. 11.38. Word diagram of the life-cycle of *cosmarium*

Taxonomic Position

Division	:	**Chlorophyta** (Chlorophycophyta)
Class	:	**Chlorophyceae**
Order	:	**Zygnemales** (Conjugales)
Family	:	**Desmidiaceae**
Genus	:	*cosmarium*

Origin, evolution and affinities of Conjugales (Zygnemales). This is a distinct and well-defined group of green algae characterises by its own chloroplast morphology, conjugation type of sexual reproduction and total absence of motile reproductive stages. The earlier algologists, on the basis of certain superficial resemblances, related the conjugales to the Bacillariophyceae (Diatoms).

This view was long discarded and is now of historical interest only.

There are other algologists who advocate that on account of the above mentioned features, the Conjugales merit a rank (class status) equal to the Chlorophyceae and the Charophyceae. Fritisch vehemently opposed this view-point for which he holds there is no warrant. The total absence of motile reproductive stages and conjugation of amoeboid gametes are features not uncommon in other green algae. Formation of a canal linking up the two reproductive cells occurs widely in other green algae. Occurrence of non-motile reproductive stages is a feature common in the Chlorococcales. Besides, in their pigment constitution, nature of stored food and other metabolic activities the Conjugales do not differ fundamentally from other green algae. In the light of these facts, the majority of the algologists, at present, are in favour of giving Conjugales an order status under the Chlorophyceae. They consider this group to have diverged very early from primitive green algae with the total loss of flagellate stages. Sehussing (1925) on the basis of the absence of motile reproductive stages was prompted to suggest the evolution of Conjugales from the Chlorococcales as the latter have also a tendency to suppress motility. Others, however, think that there is no adequate evidence to support this view.

Smith (1938-54) advocated the evolution of this order from the motile, unicellular progenitors of the order Volvocales. He advanced the following arguments in favour of his view :

(*i*) Reported occasional occurrence of amoeboid gametes in some species of *Chlamydomonas*.

(*ii*) Sexual reproduction in certain species such as *C. eugametos* is similar to conjugation (Fig. 11.39 A–B). In the interval between loss of flagella and fusion between gametes, the latter lie opposed and adjacent to each other resembling those of some conjugales (Mesotaeniaceae).

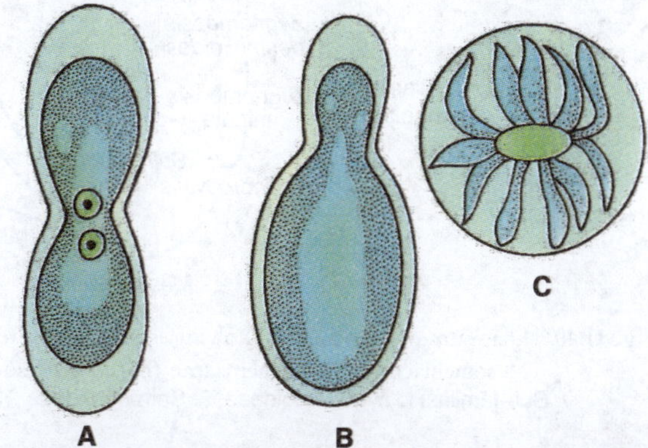

Fig. 11.39 (A-C). *Chlamydomonas.* Isogamous and anisogamous sexual fusion in *C. eugametos*; C, *C. arachne* with an axile and stellate chloroplast (A and B after Moewus and C after Eul).

Fritsch (1938) supported Smith's hypothesis advocating the origin of the Conjugales from Chlamydomonas-like ancestors. He argued that in species of *Chlamydomonas* such as *C. braunii* in which the gametes are enclosed in a membrane and the latter is not discarded prior to sexual fusion, a sort of a communication is established between the two cells. This canal-like structure linking up the two cells resembles the conjugation tube and depicts the process of conjugation in Mesotaeniaceae. The only objection against this view-point is how to explain the presence of an axile plate-shaped chloroplast of the Conjugales which is not seen in the Volvocales. However, Pascher (1930) reported that in *Chlamydomonas arachne* (Fig. 11.39 C) and *C. eradians* the chloroplast is axile and almost stellate. Randhawa (1959) suggested that the evolution of axile chloroplast of Conjugales from the more complicated parietal chloroplast of *Chlamydomonas* may be the result of loss of motility in the former. The consensus of opinion among the algologists, thus supports the view that the Conjugales have evolved from the plexus of primitive algae among the Volvocales which are no longer extant. Among the Zygnemales the members of the family Mesotaeniaceae (Saccoderm desmids) with their relatively simple chloroplasts and wall structure and the normal production of four daughter individuals from the zygospore are considered primitive. The similarity of structure of cell wall, chloroplasts and mode of conjugation indicate close relationship between Mesotaeniaceae and Zygnemaceae of order Zygnemales. According to Randhawa (1959) forms like *Zygnemopsis minutua* and *Z. desmidiodes* bridge the gap between Saccoderm desmids (Mesotaeniaceae) like *Cylindrocystis* and filamentous Zygnemaceae. In fact they may be taken as colonial forms of *Cylindrocystis* which have developed the filamentous habit. Fig. 11.40 shows the probable relationship between the genera of these two families. There is a possibility that the Placcoderm Desmids (Desmidiaceae) have a separate origin from the hypothetical *chlamydomonas*-like ancestral stock at an early stage. This view is represented in Fig. 11.41 which also represents the interrelationships between the families and sub-families of order Zygnemales.

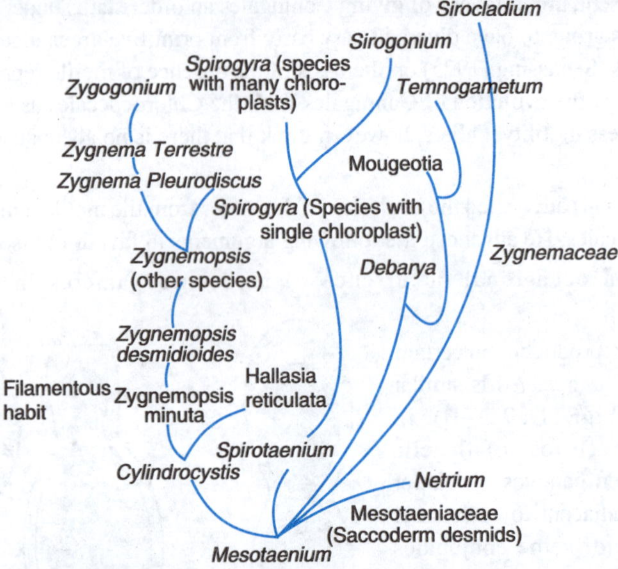

Fig. 11.40. Diagram of a scheme showing suggested interrelationships among the genera of families Mesotaeniaceae and Zygnemaceae (Based on Randhawa).

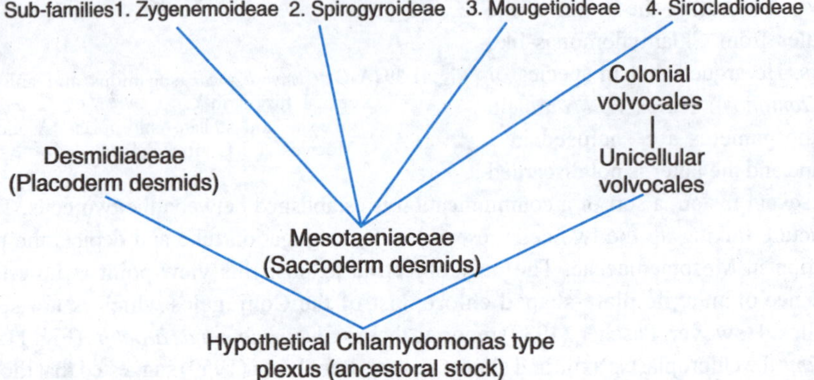

Fig. 11.41. Diagram of a scheme showing the suggested inter-relationships between the sub-families of order Zygnemales (After Randhawa).

QUESTIONS

ESSAY TYPE

1. Give an illustrated account of the life history of *Zygnema*. *(Kanpur, 1997)*
2. Describe the vegetative structure of the thallus and mode of scalariform conjugation in *Spirogyra*.
3. Give a brief account of lateral conjugation in *Spirogyra*. How will you distinguish a filament containing zygospores formed as a result of lateral conjugation from the one containing zygospores formed as a result of scalariform conjugation?
4. Enumerate the salient features in the life history of *Zygnema, Mougeotia* and *Spirogyra*.
5. Compare the thallus structure and life history of *Zygnema*, *Mougeotia* and *Spirogyra*. Point out the common characteristics in them.

6. Give an account of thallus structure, mode of reproduction and life history of *Closterium*.
7. Describe the thallus structure and reproduction in *Cosmarium*.
8. Compare the thallus structures of *Cosmarium* and *Spirogyra*.
9. Name any constricted desmid and describe its life history in detail.
10. Compare the life histories of constricted desmid with that of an unconstricted desmid.
11. Describe in detail the structure and life cycle of *Zygnema*.
12. Describe the cell structure of *Cosmarium* and explain the method of cell division with the help of suitable diagrams.
13. How does *Zygnema* reproduce sexually? Describe various methods.
14. Compare the cell structure in *Cosmarium*, *Spirogyra, Ulothrix* and *Oedogonium*.

SHORT ANSWER TYPE

15. Identify the following by giving one important character. Refer the following to their respective systematic positions giving reasons.
    (*a*) *Spirogyra* (*Kanpur, 1996, 1994*)
    (*b*) *Zygnema* (*Kanpur, 1999*)
    (*c*) *Mougeotia*
    (*d*) *Closterium*
    (*e*) *Cosmarium* (*Lucknow M. Sc., 2004*)
16. Draw neat and labelled diagrams of any three of the following and mention some of their characters briefly :
    (*a*) *Spirogyra* (*Kanpur, 1998*)
    (*b*) *Mougeotia* – cell structure
    (*c*) *Zygnema* thallus
    (*d*) Cell structure of *Cosmarium*
    (*e*) Cell structure of *Closterium*
17. (*a*) Describe the conjugation in *Zygnema*. (*Kerala, 2001*)
    (*b*) List the diagnostic features of the order Zygnematales or Conjugales.
18. The filaments of *Spirogyra* loose their colour after conjugation. Why?
19. Write short notes on :
    (*a*) Lateral conjugation (*Allahabad, 1991, 1997, 2004*)
    (*b*) Differences between zygospores and oospores (*Allahabad, 1995*)
    (*c*) Chloroplast in *Zygnema* (*Lucknow, 1992*)
    (*d*) Differentiation of sex in *Spirogyra*
    (*e*) Conjugation (*Bharathiar, 2000*)
    (*f*) Cell structure in *Zygnema* and *Spirogyra*
    (g) Sexual reproduction in Desmids (*Lucknow M. Sc.*, 2005)
20. Name an alga which shows lateral conjugation. (*Allahabad, 1992,2001*)
21. Name an alga which shows scalariform conjugation. (*Allahabad, 1993*)
22. Name a constricted desmid.
23. Name an unconstricted desmid.
24. Compare *Mougeotia* and *Sirogonium*.

OBJECTIVE TYPE

25. **Fill in the blanks :**
    (*i*) The botanical name of Pond Silk is ....
    (*ii*) The characteristic feature of the order Conjugales is ....
    (*iii*) In .... the chloroplast is stellate.
    (*iv*) In .... the chloroplast is axile plate type.
    (*v*) Two parietal plate type chloroplasts are present in ....
    (*vi*) The common method of vegetative reproduction is through ....

(*vii*) The desmids are .... organisms having two .... chloroplasts.

(*viii*) Spiral chloroplast is the characteristic feature of ....

(*ix*) Conjugation between two morphologically similar and physiologically dissimilar gametes results in a structure called ....

(*x*) .... is the example of unconstricted desmid.

(*xi*) The example of a constricted desmid is ....

(*xii*) Azygospores are produced as a result of ....

(*xiii*) In *Cosmarium* the chloroplast is single, large and................ .

(*xiv*) Generally numerous pyrenoids are present in the chloroplast of the members of ....

(*xv*) The median construction in *Cosmarium* is known as ....

26. Select the correct answer :

(*i*) *Cosmarium* belongs to the order

(*a*) Volvocales
(*b*) Ulotrichales
(*c*) Conjugales
(*d*) Siphonales

(*ii*) Differentiate between

(*a*) Zygospore and oospore (*Allahabad, 2006*)

(*iii*) The shape of the chloroplast in *Zygnema* is

(*a*) Ribbon shaped
(*b*) Girdle shaped
(*c*) Reticulate
(*d*) Cup shaped

(*iv*) Scalariform conjugation takes place in

(*a*) *Oedogonium*
(*b*) *Zygnema*
(*c*) *Ulothrix*
(*d*) *Volvox*

(*v*) The shape of the chloroplast in *Cosmarium* is

(*a*) Axial with radiating plate
(*b*) Spiral
(*c*) Reticulate
(*d*) Cup shaped

(*vi*) The zygospores of *Cosmarium* after germination gives rise to

(*a*) One daughter cell
(*b*) Two daughter cells
(*c*) Three daughter cells
(*d*) Four daughter cells

(*vii*) What is the diploid stage in *Spirogyra* called?

(*a*) *Zygote*
(*b*) Oospores
(*c*) Aplanospores
(*d*) Zygospores

(*viii*) Vegetative reproduction in *Zygnema* takes place by

(*a*) Zygospores
(*b*) Aplanospores
(*c*) Fragmentation
(*d*) Oospores

(*ix*) Pond silk or 'water silk' is the common name of

(*a*) *Spirogyra*
(*b*) *Ulothrix*
(*c*) *Mucor*
(*d*) *Oscillatoria*

(*x*) Sometimes a ladder like structure is exhibited in filaments of *Spirogyra.* This is due to

(*a*) Direct conjugation
(*b*) Scalariform conjugation
(*c*) Lateral conjugation
(*d*) Asexual reproduction

(*xi*) The members of conjugales are

(*a*) exclusively fresh water
(*b*) exclusively marine
(*c*) 50 percent fresh water
(*d*) 90% fresh water

(*xii*) The name Pond Silk is given to *Spirogyra* filaments because

(*a*) The filaments secrete a silky substance

(*b*) The cellulose layer secretes a mucilage like substance

(*c*) The pectose layer of filaments become mucilaginous

(*d*) The filaments are not rough

(*xiii*) Important feature of *Spirogyra* that distinguishes itself from *Oedogonium* is

(*a*) The nucleus

(*b*) Its habitat

(*c*) The size of vacuole

(*d*) The spiral chloroplast

(*xiv*) One of the important features of *Maugeotia* is

(*a*) The spiral chloroplast

(*b*) The plate-like axial chloroplast

(*c*) The reticulate chloroplast

(*d*) The cup-shaped chloroplast

(*xv*) When the zygospore in *Spirogyra* germinates

(*a*) All the nuclei produced are functional

(*b*) Only one haploid nucleus is functional

(*c*) Only one nucleus is inactive and non functional

(*d*) The diploid nucleus does not divide

(*xvi*) On germination the zygospore nucleus in *Zygnema* divides meiotically to produce

(*a*) 4 haploid nuclei and all are functional

(*b*) 4 haploid nuclei and only one nucleus is functional while others abort

(*c*) 4 haploid nuclei, out of which two nuclei abort and two nuclei are functional

(*d*) 4 haploid nuclei out of which only one nucleus aborts and three nuclei are functional.

(*xvii*) Nucleus in *Spirogyra* is suspended in the cell with the help of

(*a*) Vacuole

(*b*) Cytoplasmic strands

(*c*) Chloroplast

(*d*) All of the above.

(*xviii*) Which of the following statements is correct about *Spirogyra*?

(*a*) Asexual reproduction takes place by zoospores

(*b*) Filaments showing scalariform conjugation are homothallic

(*c*) Filaments showing lateral conjugation are homothallic

(*d*) Filaments showing lateral conjugation may be homothallic.

(*xix*) Meiosis in *Spirogyra* takes place in

(*a*) Zygospores at the time of their germination

(*b*) Zygospores at the time of their formation

(*c*) Zoospores when they are formed

(*d*) *Zoospores when they germinate.*

Caulerpa racemosa

CHAPTER 12

ORDER 9. CAULERPALES (Old Siphonales)

- Family : Caulerpaceae
- Caulerpa

The order derives its name from the Greek word meaning tube. It denotes the acellular condition of the thallus. The outstanding feature is the diploid plant body. It is filamentous, vesicular or elaborately differentiated, generally multinucleate and thus termed coenocytic. The coenocyte generally consists of a moderately thick layer of cytoplasm lining the wall. It encloses a central vacuole. Numerous disc-shaped chloroplasts lie embedded in the cytoplasm. Pyrenoids are commonly present. The numerous nuclei lie internal to the chloroplasts. The arrangement is reversed in the growing apices. The cross walls remain absent until the formation of reproductive structures. Sexual reproduction ranges from isogamy to anisogamy. According to Strain, the chloroplasts contain two distinctive xanthophylls, in addition, to the pigments characteristic of the green algae. These are **siphonein** and **siphonoxanthin**. The other differences are that the thallus is diploid, α-*carotene* replaces β-*carotene* of the green algae, contains numerous discoid chloroplasts and all are marine especially abundant in the warm seas, tropical or subtropical.

The order is divided into a number of families. Fritsch recognises nine families including Protosiphonaceae and Vaucheriaceae. Smith includes Protosiphonaceae and Vaucheriaceae. Majority of the modern phycologists question the practice of placing family Vaucheriaceae with its single genus *Vaucheria* in the class Chlorophyceae, order Siphonales and place Protosiphonaceae with its single genus *Protosiphon* in the order Chlorococcales. The remaining seven families included in this order are Caulerpaceae, Derbesiaceae, Dasycladaceae, Codiaceae, Valoniaceae, Chaetosiphonaceae, and Phyllosiphonaceae. We shall consider *Caulerpa* of family Caulerpaceae.

FAMILY : CAULERPACEAE

General features. The coenocytic thallus exhibits high degree of morphological elaboration. It consists of rhizome-like creeping portion. The rhizome bears root-like appendages, the rhizoids on its under surface and upright shoot-like appendages on its upper surface. The two important features of the family are (*i*) the occurrence of cylindrical, branched ingrowths (trabeculae) of wall material traversing the large central vacuole in all parts of the coenocytic thallus and, (*ii*) occurrence of two types of plastids namely the chloroplasts and the amyloplasts. Vegetatively the members of this family reproduce by the method of **fragmentation**. Asexual reproduction by spores is lacking. Sexual reproduction is either **isogamous** or **anisogamous**. According to Smith the family is represented by a single genus *Caulerpa*. Fritsch adds two more *Bryopsis* and *Pseudobryopsis*. Here *Caulerpa* is considered in detail.

Caulerpa · *Bryopsis* · *Pseudobryopsis* · *Caulerpa cactoides* · *Caulerpa prolifera*

Caulerpa Filiformis · *Caulerpa Mexicana* · *Caulerpa Cupressoides*

Caulerpa Verticillata · *Caulerpa Serrulata* · *Caulerpa brachypus* · *Caulerpa Racemosa*

CAULERPA

Distribution. The genus includes about 73 species. All of these are exclusively marine. They are chiefly confined to the warm (tropical) seas. *C. prolifera* and *C. olliveri* are the exceptions. They occur in the Mediterranean. Many species are found along the Indian coast. They have been reported by Boergesen (1932) from most varied localities. They occur in sheltered as well as in exposed places, in shallow waters and in deep sea, on sandy-muddy bottoms and on stones and coral reefs as lithophytes. Caulerpas show marked adaptations in each case to the habitats in which they grow. A few grow epiphytically on the roots of mangrooves. On the ecological basis Boergesen (1907) and Svedelius classified the various species of *Caulerpa* into the following three categories:

(*i*) Mud collecting species. Some of them are lithophytes and others grow epiphytically upon the roots of mangrooves. *C. verticillata* belongs to this category. It is a lithophyte. It collects a lot of mud and fine organic debris amidst its felt of thread-like rhizoids.

(*ii*) Sand and mud bottom species. The species constituting this group grow rooted in the mud or sand at the bottom. The rhizome usually has a pointed tip with which it bores its way through the soft substratum. The simple rhizoids reaching a depth of a few centimetres branch repeatedly. *C. cupressoides* is the best example of this group.

(*iii*) Rock and coral reef species. The rhizoids of these species branch immediately after their origin. *C. racemosa* is the common representative of this group.

Organisation of the thallus

The multinucleate coenocytic thallus of the marine algae *Caulerpa*, in the majority of the species, is very elaborate. It attains a high degree of morphological differentiation. The elaboration is in form and complexity of structure. In size and external form it simulates an angiospermic plant with a root-stock. In some species it attains a size which may be as much as 30 cms. in height and a metre in length. The thallus has a definite external form. It is differentiated into three organs during its vegetative growth. There is the cylindrical, **rhizome-like** creeping portion. From the upper surface of the horizontally growing rhizome arise a number of erect branches which resemble foliage shoots and thus are often called the **assimilatory shoots**. From the lower side of the rhizome arise the numerous, branched, thread-like, colourless **rhizoids**. The rhizome-like creeping portion and the attaching rhizoids exhibit no variation in form. They are much the same in all the species. The upright assimilatory shoots, however, exhibit a great diversity of form. In *C. fastigiata* (Fig. 12.1. C) the vertically growing shoot is made up of irregularly branched threads. In many species the erect branches have a central, cylindrical axis. The axis bears lateral outgrowths which are known as the **assimilators**. The assimilators are frequently flattened but vary considerably in their form and

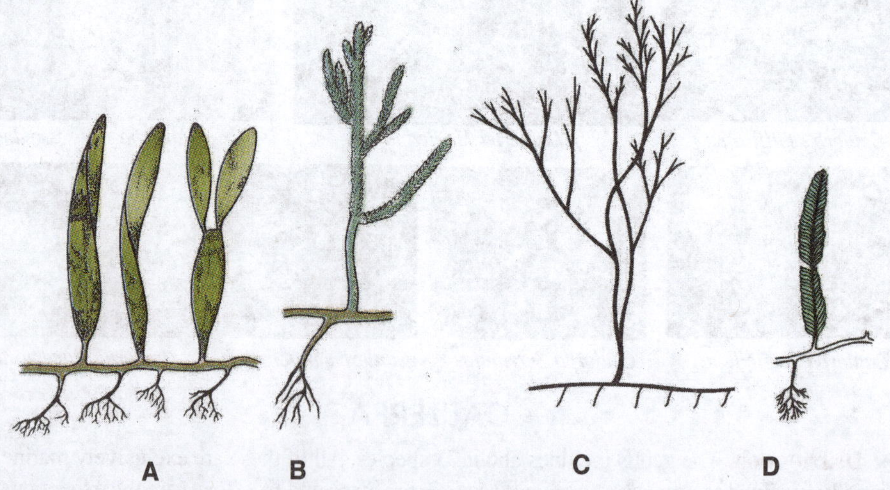

Fig. 12.1 (A-D). Species of *Caulerpa*. (A), *Caulerpa prolifera*; (B), *C. cupressoides*; (C), *C. fastigiata*; (D), *C. tasifolia*.

arrangement. Consequently the upright branches resemble in appearance to a considerable extent the shoots of many higher plants such as cacti, yews, cupresses, mosses, lycopods, and the like. The different species of *Caulerpa* have been named accordingly. For example in *C. verticillata* the assimilators are cylindrical and arranged in a *verticillate* manner on the erect branches as is the case in *Chara*. The assimilators of *C. selago* are long, subulate and imbricate. They are compactly arranged on the upright axes. All these species have radially organised upright axes. In other species the upright axes are bilaterally symmetrical. In *C. taxifolia* (Fig. 12.1. D) the assimilators are arranged in two rows as in the shoots of Yew (*Taxus*). In some species such as *C. prolifera* (Fig. 12.1. A) the upright branches are flattened, shortly stalked, leaf-like structures which directly arise from the upper surface of the rhizome in two rows. *C. crassifolia* and *C. scalpellformis* posses flattened, upright branches that are pinnately divided. Genetically the thallus is a *diploid* structure. The turgor and thickened cell walls afford sufficient mechanical support to keep the assimilatory shoots upright.

Boergesen (1927) and Svedelius (1906) stated that the radically symmetrical forms are primitive and the bilateral ones more advanced. They cite the following proofs in favour of their view :

(*i*) The bilaterally organised axes are often radial at the base.

(*ii*) Occurrence of both types of shoots on the same plant.

(*iii*) In several species (*C. cupressoides*) both the radial and bilateral forms occur in different habitats.

Internal Structure (Fig. 12.2, A-B). The coenocytic thallus is unseptate. The form of the thallus is maintained by turgor and thickness of the cell wall. *C. hypnoides* is an exception. A septum appears at the tip of each assimilator in this species. It cuts off a small apical cell. The wall is made up of callose, pectin, pectic acids and a polymer of pentose sugar. There is no cellulose. The wall gradually increases in thickness behind the apex by successive deposition of these materials in successive strata. Within the thick wall is a thick, lining layer of **cytoplasm** which surrounds a large, **central vacuole**. Part (1924) stated that the vacuole contains colloids. The latter coagulate in fixatives. The peripheral cytoplasm contains numerous **nuclei** internal to the wall. Embedded in the multinucleate layer of cytoplasm are many discoid **chloroplasts**. The chloroplasts are devoid of **pyrenoids**. They, in addition to the usual pigments characteristic of the green algae, contain two distinctive xanthophyll pigments. They are the **siphonein** and **siphonoxanthin**. β-carotene typical of the green algae is replaced by α-carotene. All these structures are continuous throughout the plant. Numerous transverse and longitudinal, cylindrical interconnecting strands are found running across the cavity in all parts of the thallus. They from a network of cross supports which protects the branches from undue swelling and bursting. These skeletal rods are made up of callose and pectose materials and are called the **trabeculae**. The trabeculae are the branching cylindrical ingrowths of the wall

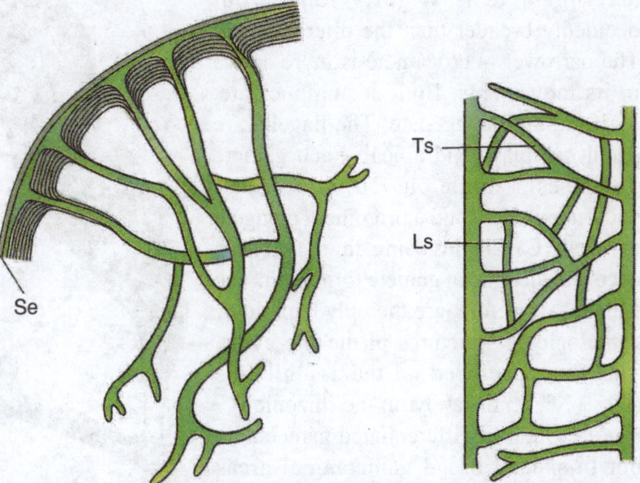

Fig. 12.2 (A-B). Internal structure of rhizome and aerial branch of *Caulerpa sp.* (A), Portion of a transverse section through the rhizome. *Se,* Surface layer; (B), Longitudinal section through the aerial part of thallus; *Ts.* Transverse skeletal strand; *Ls.* Longitudinal skeletal strand.

materials. They are arranged perpendicularly to the surface. In addition there arise small, peg-like ingrowths from the wall. The trabeculae perhaps serve as mechanical devices that increase the rigidity of the plant. Noll (1888) reported that the diffusion of the solution of mineral salts is quicker through the trabeculae than through cytoplasm. The other suggestion is that the trabeculae increase the surface of the cytoplasm as the latter is everywhere in contact with them. The trabeculae are strongly developed in the rhizome portion (Fig. 12.2, A) and are absent or poorly developed in the rhizome. In the flat assimilators (Fig. 12.2, B) they run from surface to surface in an irregular fashion.

Growth. It is apical. The protoplasm is densely aggregated at the growing points.

Reproduction. *Caulerpa* reproduces by two methods, vegetative and sexual. Asexual reproduction by spores is lacking.

1. Vegetative Reproduction. (*i*) *Fragmentation*. Vegetatively *Caulerpa* reproduces by the method of **fragmentation**. There is sporadic separation of proliferous shoots as well. The progressive death and decay of the older parts of rhizome sets free the branches. Each branch or fragment becomes an independent plant. This leads to rapid, local propagation. Sometimes dispersal of detatched fragments may take place. Reaching a suitable place each grows into a new plant.

(*ii*) *Regeneration*. Small (about 3 cm) pieces of the blade or rhizome of this green coenocytic algae are capable of regenerating completely new plants. According to Jacobs (1964, 1970) and Matilsky and Jacobs (1983) the regenerating segments exhibit definite polarity relating to the orientation of the newly formed organs. In the blade segment, randomly positioned upright or inverted the rhizoids and rhizome will grow at the end and the new blade premordia will form at the apical end. The rhizome stigments, however, do not exhibit strict polarity of regeneration with respect to growth.

2. Sexual Reproduction (Fig. 12.3, A-C). Majority of the species are **dioecious**. Iyengar (1940) reported one monoecious species from India. Sexual reproduction is generally **anisogamous**. It consists in the fusion of **gametes** that are not alike. In some species they are distinguishable into **micro**-and **macrogametes**. The macrogamete is slightly longer but decidedly broader than the microgamete. The narrower microgamete is more active in its movements. Both are uninucleate, pyriform and biflagellate. The flagella are equal, whiplash and apical. Each gamete possesses a single curved **chloroplast** without pyrenoid and a prominent elongate **eye-spot** (C). Thallus being diploid **meiosis** occurs at the time of gamete formation. The gametes, therefore, are the only haploid or gametophytic structures in the life cycle. They are developed on the assimilators (Fig. 12.3, A) or rarely on the rhizome (*C. prolifera*) not in differentiated gametangia but in special broad gametangial areas separated from the rest of the protoplast by a membrane. The gametes are liberated shortly after day-break in a mass of mucilage through the apices of narrow elongated discharge papillae, the **extrusion papillae**

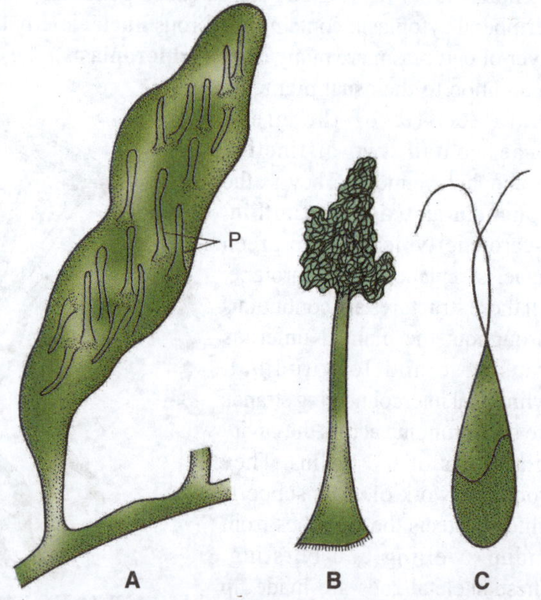

Fig. 12.3. Sexual reproduction in *Caulerpa* sp. (A), Aerial branch (assimilator) with extrusion papillae; (B), Extrusion papillae with the swarmers coming out through the exit; (C), swarmers or gamete.

(Fig. 12.3, B). The papillae about 2 mm long are developed on the assimilators simultaneously with the formation of gametes. Soon after the liberation of gametes, the fertile parts of the thallus die and decay. The gametes fuse in pairs in the open sea to produce a zygote which retains all the four flagella for a time. Soon it retracts its flagella, rounds off and secretes a wall around it. According to Miyake and Kunieda (1973) the germinating zygote increases in size. The number of chloroplasts goes up to 30 or more.

Salient Features

1. The coenocytic thallus simulates to a remarkable degree, the external appearance and size of a vascular plant with a creeping root stock.
2. Owing to the absence of septa the multinucleate lining layer of cytoplasm containing numerous, minute, round nuclei and disciform chloroplasts and the single, large central vacuole is continuous throughout the plant.
3. Numerous cylindrical skeletal strands traverse the central vacuole. They serve as mechanical supports.
4. Vegetative propagation is by means of fragmentation.
5. Asexual reproduction by spores is unknown.
6. Sexual reproduction is generally anisogamous and most of the species are dioecious.
7. The thallus being diploid, meiosis takes place at the time of gamete formation.
8. The gametes are motile uninucleate and biflagellate structures. They are pyriform in shape and each has a single chloroplast and a conspicuous eye-spot. They are distinguishable into micro- and macrogametes.
9. They escape through the ends of extrusion papilla in a mass of mucilage.
10. The fertile portion of the thallus dies soon after the liberation of gametes.
11. The zygote retains all the four flagella for a time. Soon it resorbs the flagella, rounds off, and secretes a wall around it.
12. Because of gametangial meiosis the life cycle of *Caulerpa* is described as haplobiontic diploid or diplontic.

Relationships of Siphonales. The affinities of the Siphonales are still obscure. Divergent views have been expressed as regards their origin. However, the consensus of opinion among the algologists, at present, favours their origin from the Chlorococcales. The obliteration of the zoosporic Chlorococcales such as *Characium* have led to the evolution of a form like *Protosiphon.* The latter is considered a border line genus between the Chlorococcales and the Siphonales. Some algologists place it in the Chlorococcales and others favour its inclusion in the Siphonales. The more complex Siphonales are considered to have evolved from an ancestor similar to *Protosiphon.*

Taxonomic Position

Division	:	**Chlorophyta** (Chlorophycophyta)
Class	:	**Chlorophyceae**
Order	:	**Caulerpales** (Siphonales)
Family	:	**Caulerpaceae**
Genus	:	*Caulerpa*
Species	:	*prolifera*

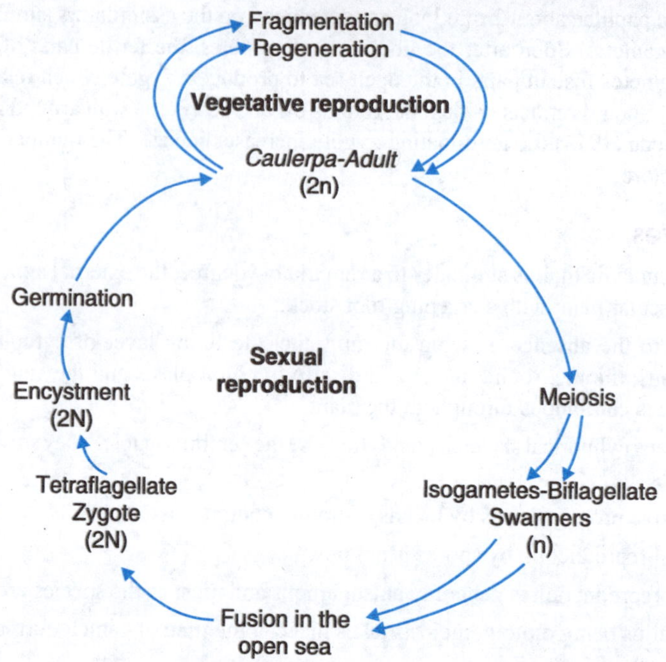

Fig. 12.4. Word diagram of the life-cycle of *Caulerpa*.

QUESTIONS

ESSAY TYPE

1. Describe the structure and life history of *Caulerpa* with special reference to the Indian species.
2. Describe the organisation of the thallus in *Caulerpa*.
3. Describe the internal structure of the thallus and modes of reproduction in *Caulerpa*.
4. Give a detailed account of sexual reproduction in *Caulerpa*.
5. Write an account of structure, reproduction and life-cycle of *Caulerpa*.
6. Describe the life-cycle of *Caulerpa*. (*Madras, 1995*)
7. Give a detailed account of vegetative reproduction in *Caulerpa*.

SHORT ANSWER TYPE

8. Describe the thallus of *Caulerpa*. (*Bharathiar, 1995*)
9. Write short notes on :
   (*a*) Cell structure in *Caulerpa*
   (*b*) Vegetative reproduction in *Caulerpa*
   (*c*) Sexual reproduction in *Caulerpa*
   (*d*) Organisation of thallus in *Caulerpa*
10. *Caulerpa* has been placed in the order Siphonales. Why?

OBJECTIVE TYPE

11. **Select the correct answer :**
   (*i*) What is peculiar about the chloroplasts in *Caulerpa?*
   (*a*) Two distinctive xanthophylls *i.e.*, siphonein and sephonoxanthin are present.

(*b*) It does not have chloroplasts
(*c*) β-carotene is present
(*d*) No pigments are present

(*ii*) The thallus in *Caulerpa* is
(*a*) Haploid
(*b*) Diploid
(*c*) Triploid
(*d*) Tetraploid

(*iii*) In which of the following the thallus is coenocytic, branched and diploid?
(*a*) *Spirogyra*
(*b*) *Oedogonium*
(*c*) *Caulerpa*
(*d*) *Vaucheria*

(*iv*) In which of the following alga, asexual reproduction is lacking?
(*a*) *Oedogonium*
(*b*) *Chlamydomonas*
(*c*) *Vaucheria*
(*d*) *Caulerpa*

(*v*) The sexual reproduction in *Caulerpa* is
(*a*) Isogamous
(*b*) Anisogamous
(*c*) Heterogamous
(*d*) Oogamous

(*vi*) The trabeculae of *Caulerpa* arise from
(*a*) Plasma membrane
(*b*) Cell wall
(*c*) Cytoplasm
(*d*) Nucleus

(*vii*) *Caulerpa* is
(*a*) Exclusively marine
(*b*) Exclusively fresh water
(*c*) Sometimes marine
(*d*) Sometimes fresh water

(*viii*) Which one of the following has diploid coenocytic thallus?
(*a*) *Hydrodictyon*
(*b*) *Caulerpa*
(*c*) *Vaucheria*
(*d*) *Oedogonium*

(*ix*) Cellulose is absent in
(*a*) *Zygnema*
(*b*) *Cladophora*
(*c*) *Caulerpa*
(*d*) *Cosmarium*

(*x*) *Caulerpa* belongs to Siphonales because
(*a*) It has tube like construction.
(*b*) There are so many chloroplasts
(*c*) It is a marine alga
(*d*) There are many sporangia in it

Chara sex organs

CHAPTER 13

CHAROPHYTA (Charophycophyta) Class : Charophyceae

- Chara
- Nitella Agardh

GENERAL FEATURES

The green algae included in this division are best known as the Characean algae. Commonly they are called the *stoneworts*. The stoneworts have a worldwide distribution. They occur more or less in still, clean waters, fresh-water or brackish. A few species are found in both the habitats. Typically they form subaquatic meadows in shallow waters growing in soft mud. Light favours growth of the Charophytes which flourish from the months of August to March disappearing in the hot summer months. They are abundant during the cold season in northern India. Drying up conditions hasten the formation of sex organs. This division includes both living as well as fossil forms. So far about 294 living species have been recorded. They have been placed under 7 genera. About

69 species have been recorded in India, Burma, Ceylon and Pakistan. They belong to five genera. These are *Chara* (27 species), *Tolypella* (3 species), *Nitella* (37 species), *Nitellopsis* (1 species) and *Lychnothamnus* (1 species). The other two living genera are *Lamprothamnium* and *Protochara*.

The plant body presents a great elaboration of vegetative structures. It is practically always erect and consists of a long, slender, jointed, green or grey main axis with a regular succession of nodes and internodes. The central axis is branched. At each node arises a whorl of lateral branchlets. Sexual reproduction is oogamous and very complex. Antheridia and oogonia differ considerably from the corresponding organs in the other green algae in structural complexities and elaboration. They are large and can be seen even with the naked eye. There is no asexual reproduction by the formation of asexual spores. The zygote, on germination, forms a protonema from which a *Chara* plant is developed.

Chara *Tolypella* *Nitellopsis*

Nitella sp *Lychnothamnus* *Lamprothamnium*

Chara corallina

Chara baltica

Chara zeylanica

Classification

The division includes a single class, **Charophyceae.** The diagnostic features of this class are equisetoid habit, attachment to the substratum by multicellular rhizoids, occurrence of many discoid chloroplasts without pyrenoids in each cell, lack of asexual spores of any kind, presence of a jacket of sterile cells around the sex organs, biflagellate, spirally shaped sperms and occurrence of protonemal stage in the life cycle. The Charophyceae comprises a single order, the **Charales.** The order Charales comprises four families. All the 7 living genera are included in a single family, the **Characeae**. The other three families include only fossil forms. The family Characeae is divided into the following two sub-families :

Sub-family 1. Nitelloideae. The oogonium has *ten coronal* cells *Nitella* and *Tolypella* are the two genera included in this sub-family.

Sub-family 2. Charoideae. The oogonium has *five coronal* cells. This sub-family comprises five genera, *Chara, Protochara, Nitellopsis, Lychnothamnus* and *Lamprothamnium.*

The best known among the living-genera are *Chara* and *Nitella.* We shall study both.

CHARA

Occurrence. It is a submerged aquatic alga which grows attached to the soft mud at the bottom, along the margins of fresh water pools, lakes and slow-flowing streams forming thick masses. Generally it prefers clear, fresh, hard and still shallow water. A few species such as *C. baltica* grow in salt water containing a very small percentage of salt (less than 1%). The species growing in water charged with calcareous materials become encrusted with calcium carbonate. *Chara* is represented by 27 species in India. Of these *C. wallichii, C. zeylanica* and *C. corallina* are fairly common.

A. Organisation of the Thallus (Fig. 13.1). The plant body is a thallus attached in the mud by multicellular rhizoids. The individual plants generally attain a length of 20 to 30 cms. The maximum height achieved however, is 90 cm. The thallus has a long, slender, flexuous upright branched main axis which is differentiated into a well-marked series of alternating short nodes and long internodes. The internode consists of a single, elongated, multinucleate and cylindrical cell, several times longer than broad. The internodal cell is in some species, surrounded by a jacket of narrow, elongated cells constituting the **cortex**. Half of the cortical cells investing the internode are derived from the node below. They grow in opposite directions to form an investment around the internodal cell. The node remains short and is made up of a cluster of several, small isodiametric cells. There are two central cells surrounded by 6-20 peripheral cells in the cluster. From each node arise the following four types of appendages :

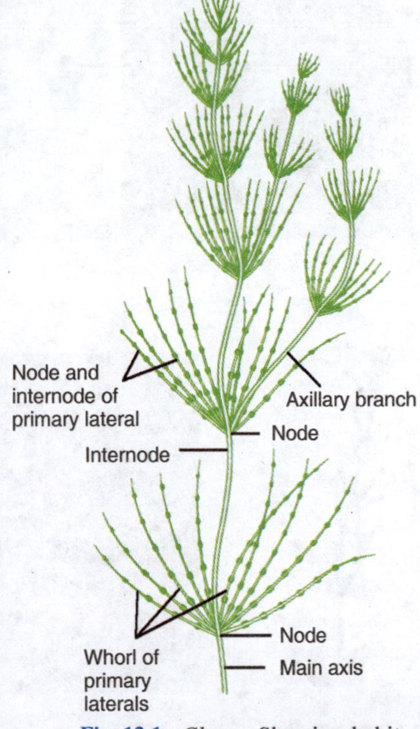

Fig. 13.1. *Chara*. Showing habit and organisation of thallus (After Fritsch).

1. Branchlets (Fig. 13.2). A whorl of short branches of limited growth arise from each node of the central axis (Fig. 13.1) These are called the **branchlets**. Some authors call them the **primary laterals** or **leaves**. They alternate

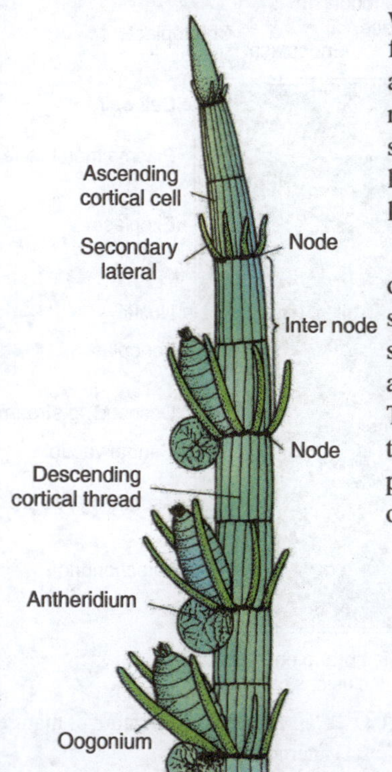

Fig. 13.2. *Chara.* A fertile primary lateral.

with one another at the successive nodes. Each branchlet arises from the peripheral cell of the node which functions as its apical cell. The short branches of branchlets consist of a limited number of nodes and internodes characteristic of a particular species. Usually the number varies from 5 to 15. The primary laterals in turn often develop much shorter, one called, spine-like branches called **secondary laterals** at their nodes.

2. Long branches. In addition, the stem node may bear one or more branches of unlimited growth. They arise usually singly at some of the older nodes of the main axis on the inner side of the oldest primary lateral in the whorl. Being apparently axillary in position they are often called the **axillary branches.** The axillary branch continues the growth of the thallus and has the same structure as the main axis. It takes its origin from the primary internodal cell which lies below the basal node of the oldest branchlet in the whorl.

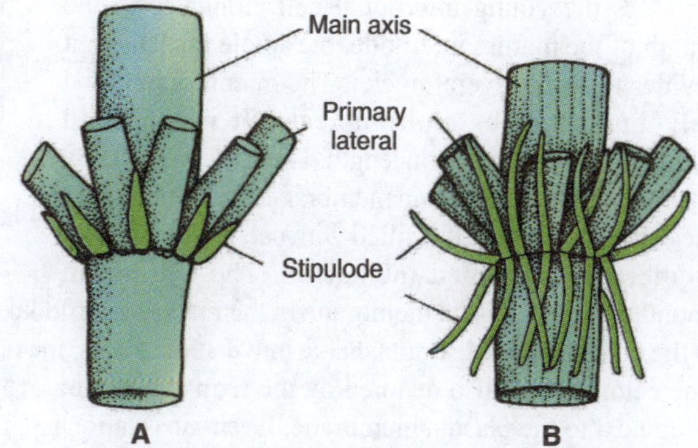

Fig. 13.3 (A-B). *Chara.* Stipulodes. (A), Haplostephanous stipulodes of *C. braunii*; (B), Diplostephanous stipulodes of *C. baltica* var. *affinis*(After Groves).

3. Stipulodes. (Fig. 13.3.). These are unicellular outgrowths that arise from the basal node of each branchlet. In the majority of species of *Chara* there are two stipulodes at the base of each branchlet, one on either side. They are arranged in a single row or circle at the stem node. Such species are named **Bistipulatae**. In some species (*C. braunii,* and *C. nuda*) one of the two stipulodes remains absent so that there are as many of them as the number of branchlets (**Unistipulatae**). They alternate with the latter. The species of *Chara* with the stipulodes in a single whorl are called **haplostephanous** (A). Some species have the stipulodes arranged in two whorls (B) and are called **diplostephanous** (*C. baltica* and *C. tomentosa*). There are species in which the stipulodes remain rudimentary (*C. nuda* and *C. wallichii*). In *C. pashanii* the stipulodes are absent.

4. Cortex. In many species the internodal cell, as mentioned above, is covered by a sheath of vertically elongated narrow cells constituting the **cortex.** Such species are described as **corticated** (*C. zeylanica*), others **ecorticated** (*C. wallichii, C. corallina, C. braunii,* and *C. nuda*). In the ecorticated species the internodal cell is naked. Half of the cortical cells ensheathing the internode are developed from the node above and half from the node below. The nodes of the central axis in *Chara* may thus bear four kinds of appendages *viz.,* branchlets or primary laterals, axillary branch, stipulodes and cortex.

Rhizoids. The *Chara* thallus is fixed to the substratum by multicellular, branched **rhizoids.** The septa between the cells are oblique. The rhizoids arise from the lower nodes of the main axis and may not show any differentiation into nodes and internodes. They mainly function as organs of attachment but also take a prominent part in the absorption of salt solutes. Production of bulbils and secondary protonema are their subsidiary functions.

Cell Structure. The nodal and unelongated internodal young cells are small in size and have similar structure. They are filled with dense granular cytoplasm which lacks conspicuous vacuoles. The single nucleus is centrally located. The numerous, small discoid chloroplasts lacking pyrenoids are distributed evenly throughout the cytoplasm. The cell protoplast at its periphery is differentiated into a thin, delicate plasma membrane. It lines the inner surface of the cell wall. The cell wall is cellulose in nature and possesses a superficial gelatinous layer of doubtful composition. Ultrastructurally the cellulose consists of fine fibrils (microfibrils) embedded in a homogeneous matrix.

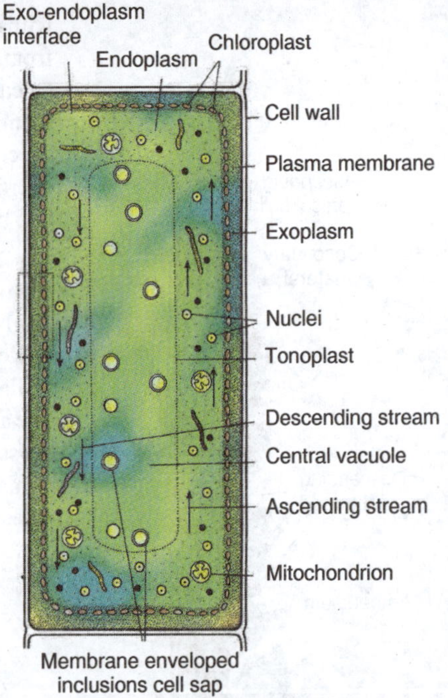

Fig. 13.4. *Chara.* Showing structure of mature internodal cell.

As the young internodal cell elongates to the length of the mature internode, the single nucleus in it divides to form several nuclei. The mature internodal cell thus becomes multinucleate. It is long and cylindrical about 2-5 cm in length. There is a single large vacuole delimited by a thin membrane, the **tonoplast** in the centre of the cell. It is filled with cell sap containing membrane enveloped inclusions. The cytoplasm bounded by the plasma membrane in the mature internodal cell is restricted to a lining layer internal to the cell wall. It is distinguishable into distinct zones, the outer **ectoplasm** and the inner **endoplasm**. The ectoplasm is also denoted by the terms **cortex or exoplasm**. It is a dense and firm plasmagel lying next to the plasma membrane. It remains stationary. The ectoplasm contains *microtubules* and numerous *chloroplasts.* The microtubules occur just internal to the plasma membrane near the cell wall. The discoid or ovoid chloroplasts devoid of pyrenoids are normally located in the stationary ectoplasm in longitudinally spiral and parallel series towards the interface with the endosplasm. Ultrastructurally each chloroplast is bounded by a double membraned *chloroplast envelope.* Embedded in the matrix within the chloroplast envelope are 40-100 *discs* or *thylakoids.* (Nitelle). All of them may form single *stack* or *band* occupying most of the volume of the chloroplast.

Directly within and surrounded by the ectoplasm is the more fluid and less denser flowing endoplasm. It forms a thin sleeve around the large central vacuole. Several nuclei with distinct nucleoli are uniformly dispersed in the endoplasm which, in addition, contains ribosomes, mitochondria, endoplasmic reticulum, dictyosomes and many other particles. The number of cisternae in the dictyosome ranges from 5-8.

Cyclosis. Normally endoplasm is in a constant state of rotation. It moves as a giant twisted belt which follows a spiral pathway along with length of the cell parallel to the files of the stationary cortical chloroplasts. Many particles including nuclei, mitochondria, sphaerosomes and the like are transported in this cytoplasmic stream called *cyclosis.* It flows up on one side and down on the other. The two opposing cytoplasmic streams (ascending and descending) adjoin a colourless strip of stationary cytoplasm (indifferent zone) located on either flank of the cell. The phenomenon of cyclosis was first discovered about a little more than two hundred years ago by Corti (1774).

Mechanism of streaking. Investigators have come to the conclusion that cytoplasmic streaming in the Characean cell is based on an actin-myosin system. The classical investigations of Kamiya and Kurdo (1956, 1959) revealed that the site of motive force generation for endoplasmic streaming in the characean internodal cells is located at the narrow border between the stationary ectoplasm and moving endoplasm. Later thin, linear, stationary fibrillar structures were discovered by Kamitsubo (1966, 1972) and Nagai and Rebhum, (1966) at this site. The location of fibrils at this site fitted well with Kamiya and Kurodi's conclusion. These fibrils were found on the inner face of the spirals of chloroplasts (Fig. 13.5) located in the ectoplasm. They run parallel to the direction of endoplasm streaming. Each of these subcortical fibrils is bundle of 50-100 microfilaments shown to be similar to muscle F-actin (Palevitz *et al*, 1974 and Kersey *et al* 1976). It is now generally held that these subcortical structures are responsible for generating active shearing force at the exoplasm-endoplasm interface for endoplasmic streaming in *Chara* internodal cells. It has been proposed that the active shearing force results from the interaction between an endoplasmic factor and the subcortical firbils composed of microfilaments of F-actin. Several investigators have studied this factor. Bradley (1973) suggested that the motive force results from the interaction between actin microfilaments and myosin which may be anchored at suitable sites on the endoplasmic reticulum. Cheu and Kamiya (1975) suggested that the mysoin factor is localized in the endoplasm. Williamson (1975) came to the conclusion that the motive force for streaming is produced by the interaction between the microfilaments of actin and myosin-like protein linked with the endoplasmic organelles. The consensus of opinion, therefore, is that the myosin factor which interacts with microfilaments in the presence of ATP and Mg^{2+} resides in the endoplasmic organelles. Nagai and Hayama (1979) studied the ultrastructure of the endoplasmic myosin factor arranged on the endoplasmic organelles. They found the organelles to be equipped with fine filaments with globular bodies which according to them interact with subcortical F-actin and microfilaments to produce the motive force.

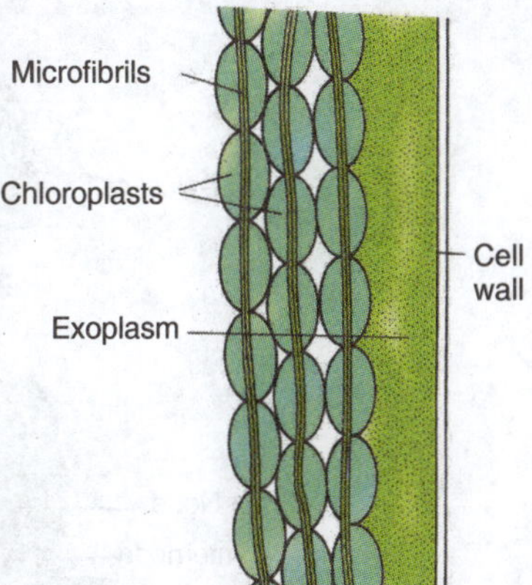

Fig. 13.5. Showing subcortical fibrils affixed to each row of chloropllast on the side facing the endoplasm (Diagrammatic).

Growth (Fig. 13.6). The growth of the main axis and the branches is brought about by means of a single, large, dome-shaped **apical cell** (Fig. 13.6) located at the tip of the axis. It cuts off segments parallel to its flat bottom or posterior face (B). The segment thus cut off divides by a transverse wall into a bi-convex, lower **internodal initial** and a bi-concave, upper **nodal initial (C).** The former remains undivided and elongates considerably to form the internode. The nodal initial

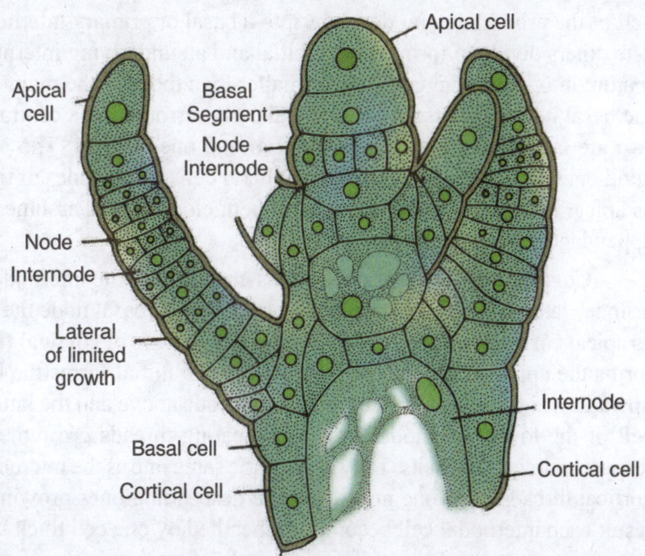

Fig. 13.6. *Chara.* Vertical section through the apical portion of the thallus (After Sachs).

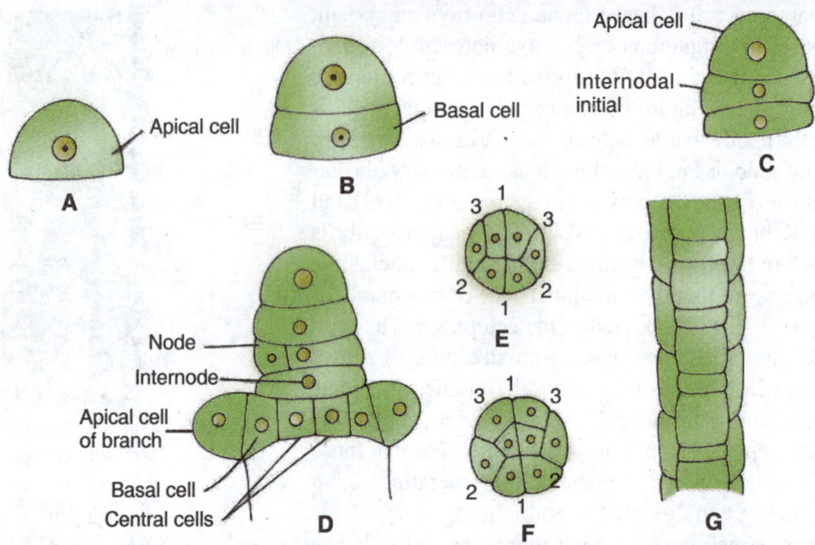

Fig. 13.7 (A-G). Segmentation of apical cell of *Chara* and subsequent changes leading to growth.

divides a number of times to form the nodal elements. It divides by a vertical halving wall into two semicircular cells (E1-1). This is followed by a vertical division of each of the daughter cells by a curved septum in a plane intersecting the first plane of division (E2-2). The succeeding divisions are also vertical. Each intersects the preceding plane of division. In this way a definite number of **peripheral cells** varying from 6-20 but usually six are cut off. They surround a pair of **central cells** (F). The peripheral cells at the node protrude and function as **apical cells** of the branchlets or primary laterals (D). The number of primary laterals at a node thus is determined by the number of peripheral cells at the node. The primary laterals (Fig. 13.2) exhibit the same type of apical growth but with limited activity. The internodes are relatively shorter and the apical cells after cutting off 5 to 15 segments ceases to grow and ends in an elongated process. The first segment cut off by the apical cell of the primary lateral develops into a basal or primary internodal cell (D) which remains small. The others divide to form a nodal initial and an underlying internodal initial. The internodal initials mature into internodal cells much smaller than those of the main axis. According to Sundaralingam the basal node consists of 4 peripheral cells surrounding a central cell. Of the four peripheral cells two are **lateral** in position, one **adaxial** and one **abaxial**. The primary laterals alternate with one another at successive nodes. The peripheral cells at the nodes of the primary laterals do not function as apical cells. Instead, all or some of them elongate and assume the form of one-celled, spine-like appendages called the secondary laterals.

Cortication. The corticating cells ensheathing the internode arise from the basal nodes of the primary laterals. Of the four peripheral cells at the basal node the adaxial and abaxial ones function as apical **cortical initials**. They give rise to two rows of **cortical threads**. The adaxial cortical initial forms the upper thread and the abaxial cortical initial forms the **lower threads.** The former grows upward over the internodal cell of the internode above and the latter grows down over the internodal cell of the lower internode. As the corticating threads grow they remain closely applied to their respective internodal cells. They grow at the same rate as the internodal cells. Eventually the ascending cortical threads from one node meet the descending ones growing from the next node above. As a result each internodal cell becomes ensheathed by one cell thick layer of cortex developed from the two nodes, upper and lower. The cortex of the main axis shows nodes and internodes, whereas that of the primary laterals does not show any differentiation.

Reproduction. *Charales* reproduce by **vegetative** and **sexual** methods only. **Asexual** reproduction by spores is entirely lacking.

1. Vegetative Reproduction. It involves the separation from the parent plant of various kinds of regularly produced vegetative outgrowths. Each of these develops into a new plant. The common vegetative methods are :

(*i*) Amylum stars (Fig. 13.8, (A-C). These are star-shaped aggregations of cells developed on the lower nodes of the main axis. The cells of the amylum star are laden with amylum starch. The detached amylum star grows into a new plant.

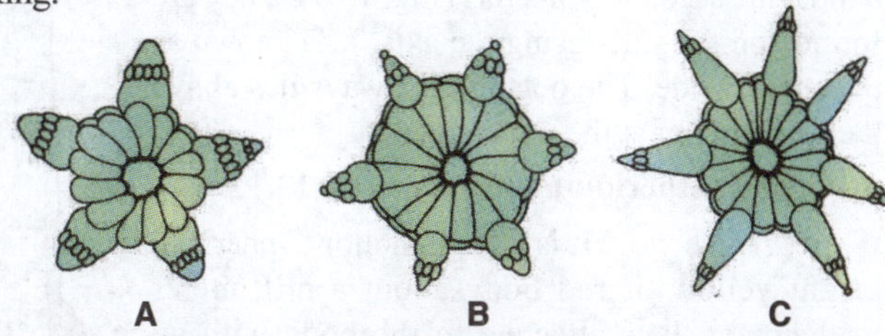

Fig. 13.8 (A-C). Vegetative reproduction in Charales. Amylum stars (Star-shaped stem bulbils in *Nitellopsis obtusa* (A, aftger Groves: B and Cafter Megula).

(*ii*) Bulbils (Fig. 13.9, A-B). These are small, rounded tuber-like structures (A) developed upon the rhizoids (*C. aspera*). Unilateral stembulbils are developed on the stem-nodes of *C. baltica* (B). The bulbils are perennating bodies which carry on vegetative propagation. The detached bulbil forms a new plant.

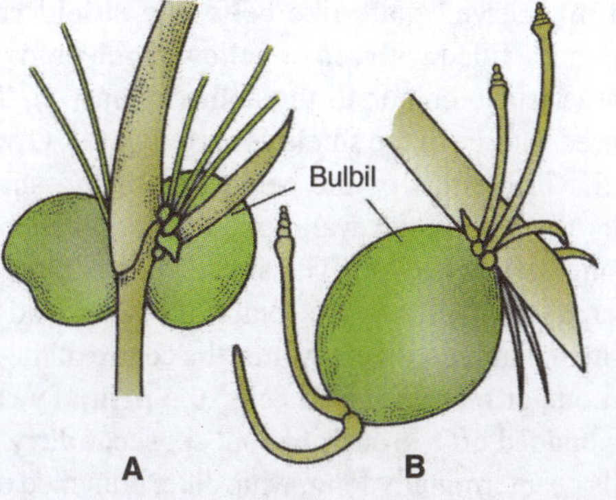

Fig. 13.9 (A-B). *Charales.* Vegetative reproductive structures (Contd.), (A) Spherical rhizoidal bulbil of *C. aspera*; (B) Unilateral stem bulbil of *C. baltica* (After Griesenhagen).

(*iii*) Amorphous bulbils (Fig. 13.10, A-B). These are clumps of several small cells laden with food materials. They are produced on the lowest stem nodes (*C. delicatula,* A) or nodes of rhizoids (*C. fragifera* and *C. baltica* B) as lateral outgrowths.

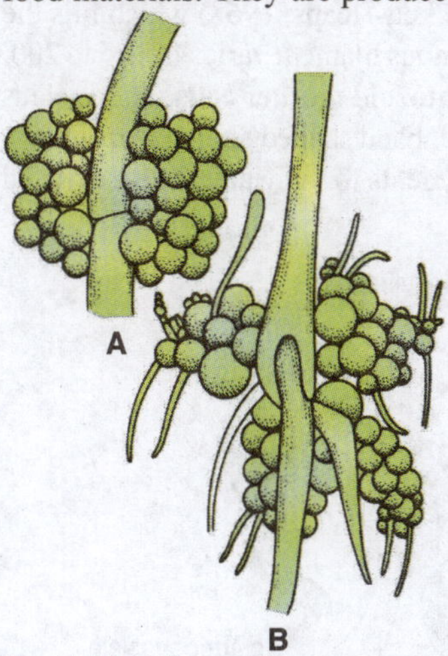

Fig. 13.10 (A-B). *Charales.* Vegetative reproductive structures. Amorphous bulbils. (A) root bulbils (*C. fragifera*); (B) multicellular root bulbil (*C baltica;* After Giesenhagen).

(*iv*) Secondary protonemata. Sometimes naked adventitious protonema-like branches arise from the surviving nodes of the older plants after hibernation of some of the nodes of the parent plants. They may also develop from the primary rhizoid ring or dormant apices. The secondary protonemata give rise to new plants like the primary protonema.

2. Sexual Reproduction (Fig. 13.11). It is **oogamous**. The sex organs display a high degree of specialisation and are far more complicated than among any other thallophytes. In fact they bear a superficial resemblance to the multicellular sex organs of archegoniates. They are visible to the naked eye when mature. The male sex organ is a large, round, bright yellow or red structure. It is commonly called the **antheridium**. Some call it a **Globule**. The female sex organ or the oogonium is a large, oval body covered with a multicellular envelope. Smith calls it a **nucule**. Most of the species are **monoecious** or **homothallic** but a few are **dioecious** or **heterothallic** (*C. wallichii*). The former are **protandrous.** In the homothallic species both the sex organs develop in pairs at the nodes of the primary laterals amidst the secondary laterals (Fig. 13.9). They are formed on the side facing the axis. It is known as the adaxial side. The oogonium always lies above the antheridium at the same node.

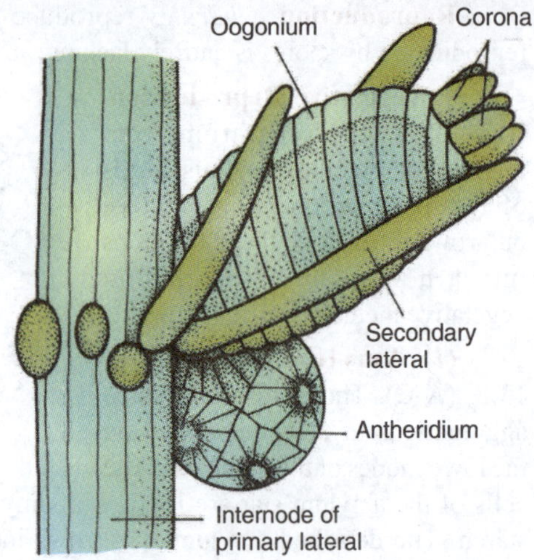

Fig. 13.11 (A-F). *Chara.* A node of fertile primary lateral showing sex organs.

The antheridium. (Globule Fig. 13.12.)

(*a*) *Structure.* It is a large, hollow, spherical, bright yellow or red body about a millimeter in diameter. It is attached to the node with an inconspicuous pedicel and regularly has the oogonium above it (Fig. 13.9). The antheridium has a wall composed of eight closely fitting large, hollow, curved plate-like cells, the **shield cells** which are filled with red or yellow pigment giving characteristic colour to the antheridium (A). The curved surface of the shield faces outwards. Owing to the infoldings of the cell walls of the shield cells the antheridial wall appears many-celled in a longitudinal section. The shield cells enclose an internal cavity. From the centre of each shield cell arises a rod-shaped cell, the **manubrium.** The eight manubria project towards the centre of the antheridial cavity. Each manubrium bears at its inner end one or more rounded cells, the **primary capitula** or **head cells**. Each primary capitulum cell has budded off a group of about six **secondary capitula** on its inner side. The secondary capitulum cells bear terminally long, whip-like, branched or unbranched, many-celled sperm producing threads called the **spermatogenous filaments** or the **antheridial filaments.** Sometimes these filaments are borne on tertiary or quaternary capitula. The antheridial filaments are intertwined (B). They form a dense tangled mass immersed in a fine mucilaginous material (Pickett-Heaps, 1968) which fills the antheridial cavity. The number of cells in a fully grown spermatogenous filament varies from 5 to 200. Each of these small thin walled discoid cells functions as a **spermatozoid mother cells**. The nucleus of the spermatozoid mother cell elongates to form a spirally coiled, band shaped **spermatozoid** (C). At maturity the shield cells fall apart exposing the antheridial filaments to the water. The expanded

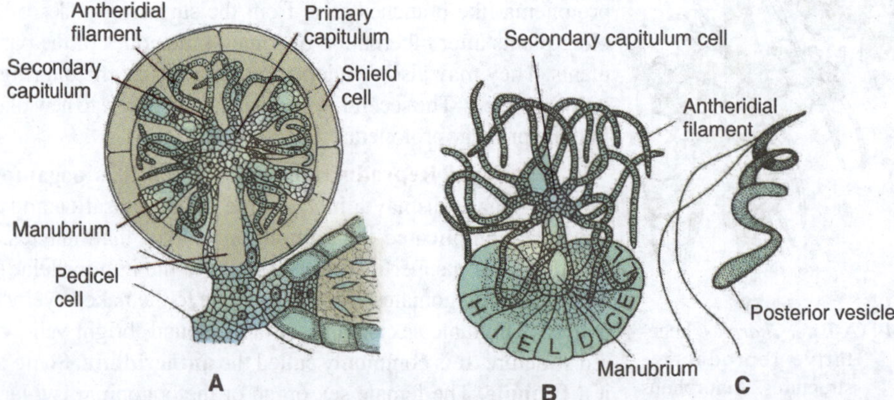

Fig. 13.12 (A-C). *Chara.* (A), L.S. mature antheridium; (B), shield cell with manubrium projecting from it and bearing at its tip primary capitulum which has cut off secondary capitula cells, each bearing a pair of antheridial filaments; (C), a sperm (A, After Smith and C, after Strasburger).

tangled spermatogenous filaments attached to the capitula upon the manubrium look like a many thronged whip (B). According to Sundaralingam (1946) the antheridium of *C. zeylanica* consists only of 4 shield cells and corresponding number of manubria and head cells. It is termed **quadriscutate**.

The mechanism of escape of spermatozoids is disputed. Fritsch (1935) reported that the spermatozoids escape by falling apart of the shield cells followed by gelatinisation of the walls of spermatozoid mother cells. Rose (1987) stated that the spermatozoid escapes through an aperture in the wall of the spermatozoid mother cell. Groves and Bullock-Webster (1920) believed that the spermatozoids burst from the mother cells and swim about. According to Stewart (1937) and Sundaralingam (1946) the antherozoid dashes against the side wall which thus ruptures at the spot and the antherozoid emerges through the aperture with the posterior end first and flagellated anterior end afterwards. Coccuci and Caceres (1976) reported that the posterior end of the sperm before taking on the final shape released flattened pellet-like structures into the fine mucilaginous matrix of the spermatogenous cell. These pellets, proteinaceous in nature reach in the vicinity of the cell wall and release a kind of enzyme which brings about localized lysis of the cell wall to form a pore which permits the sperm to escape.

Development of antheridium (Fig. 13.13). Both the sex organs in *Chara* develop from an adaxial peripheral cell at the node of a primary lateral. It divides periclinally into two cells, *outer* and *inner*. The outer cell functions as the **antheridium initial**. The inner undergoes another periclinal division. Of these the one next to the antheridium forms its basal node of five peripheral cells. The upper peripheral cell of the basal node of the antheridium functions as the **oogonium initial** whereas others form secondary laterals.

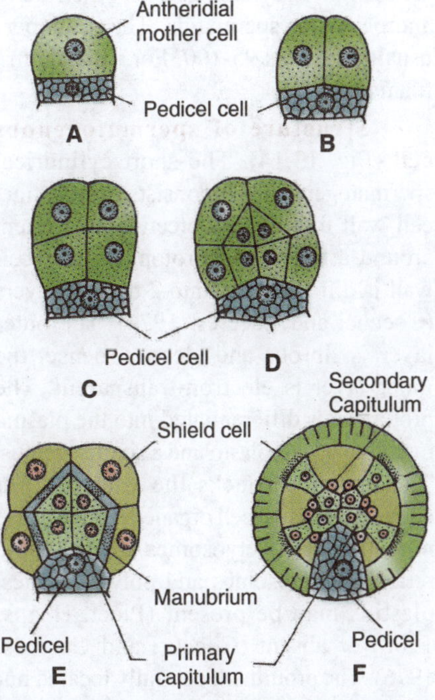

Fig. 13.13 (A-F). *Chara.* Successive steps in the development of antheridium (Based on Smith).

The antheridium initial divides transversely into a basal **pedicel cell** and a terminal **antheridial mother cell** (A). The pedicel cell undergoes no further division. The antheridial mother cell assumes a spherical form and divides by two vertical walls at right angles to one another to form four quadrately arranged cells. Only two are seen in section (B). It is the **quadrant stage** of the antheridium. Each of these cells divides transversely to give rise to the **octant stage** (C). In section only four cells are seen. Each octant cell undergoes a periclinal division to form an **outer** and **inner cell** (D). According to the old view of Sachs (1882) and Migula it is the inner cell which divides again periclinally but Walter (1929), Sundaralingam (1954, 1960) and Sundaralingam and Frances (1958) hold that it is the outer cell which divides periclinally. Anyhow two periclinal divisions appear in each octant cell to form a row of three cells lying one above the other on the same radius (E). At this stage the antheridium consists of 24 cells arranged in either diagonal series of these cells each. The antheridium consisting of 8 shield cells with a corresponding number of manubrium and head cells is called **Octusculate**. Only the innermost cell of each series undergoes further division. The outermost cell is known as the **primary capitulum** or **head cell**. Considerable enlargement of the young antheridium takes place in the course of further development. Consequently the inner segments gradually become separated from one another. The eight shield cells expand laterally and develop into curved plates with convex surface outwards. They develop the characteristic infoldings on their inner wall. Towards maturity the shields acquire red-coloured contents due to change in the colour of the chloroplasts. With the expansion of shield cells in a lateral direction, the inner cells (manubria) separate from each other and cavities result (F). The antheridial wall is completed at the base by the **pedicel cell**. The latter usually protrudes markedly into the cavity of the antheridium. The middle segment of each primary diagonal series undergoes considerable radial elongation and forms the rod-shaped **manubrium**. The manubria arise from the centre of the shield of eight-shield cells and bear at their inner end the **primary capitula cells**. At this stage the antheridium consists of eight-shield cells, eight manubria and eight primary capitular cells. Each primary capitulum cell divides to form about six **Secondary capitula** on its inner surface. The secondary capitular cells may function as the initials of the **antheridial filaments**. They undergo repeated mitosis to form the latter. The antheridial or the spermatogenous filaments may be branched or unbranched. As the antheridium enlarges the antheridial filaments immersed in the fine mucilaginous material gradually fill up the antheridial cavity with a dense tangle (Fig. 13. 12 A). In some cases the secondary capitular cells may divide to form tertiary capitular cells which in turn bear the antheridial filaments.

Spermatogenous filament. Each antheridial or spermatogenous filament consists of a single series of short, cylindrical cells which are variously termed as the spermatogenous cells, spermatocytes, androcytes or spermatids. Their number in each filament ranges from five to two hundred. The usual number is 25-100. For instance in *C. zeylanica* there are 60-80 spermatogenous cells in the filament.

Structure of spermatogenous cell (Fig. 13.14). The short cylindrical spermatogenous cell consists of a distinct cell wall forming a protective investment around a tiny bit of protoplast. The cell wall is differentiated into 2 distinct layers (Coccuci and Caceres, 1976). The outer layer is fibrous and electron-dense, the inner layer is electron-transparent. The protoplast is differentiated into the plasma membrane, cytoplasm and a single nucleus. The cytoplasm lacks the vacuoles but contains the usual cell organelles such as the mitochondria, dictyosomes, endoplasmic reticulum, ribosomes and polyribosomes, plastids may be present (Picett-Heaps, 1968) or absent (Coccuri and Caceres, 1976). The prominent centrally located nucleus has a usual double-membraned envelope with pores. Pickett-Heaps (1968) reported the presence of an indistinct centrosome consisting of a pair of co-axial rod-like structures, the **centrioles** at an early stage. The centrioles become conspicuous subsequently. The cross walls separating the adjacent spermatogenous cells have minute pores transversed by plasmodesmata. The protoplast of each spermatogenous cell metamorphoses into a single biflagellate sperm. The events leading to the transformation of the protoplast of spermatogenous cell into a sperm is called *spermatogenesis*.

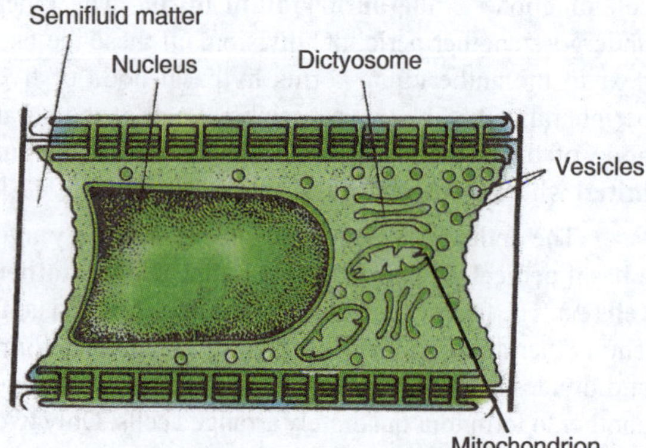

Fig. 13.14. *Chara.* Diagrammatic representation of spermatogenous cell with equaterially tightened protoplast (Based on Coccuci and Caceres).

Spermatogenesis. The process starts with the dispersion of the cytoplasm in the spermatogenous cell. It recedes from the sides of the cell except the cross walls. Consequently the cell protoplast becomes equatorially tightened. (Fig. 13.14). The centrioles become conspicuous (Pickett-Heaps, 1968). A zone of denser homogeneous cytoplasm termed the *manchette adjunct* becomes recognizable near the proximal end of the centrioles. The centrioles and the manchette adjunct move as an integral unit to the periphery and come in contact with the plasma membrane. After certain modifications both

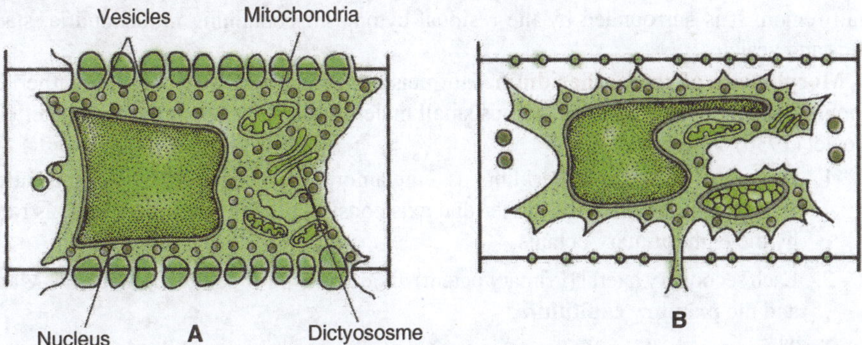

Fig. 13.15. *Chara.* Stages in the elongation of the nucleus (Based on Coccuci and Caceres).

structural and functional, the centrioles become the basal bodies which start extending the flagella. A single broad belt of closely adjacent microtubules with which flagellar bodies are associated organizes itself beneath but close to the plasma membrane at the equatorial region of the cell. The microtubules appear to be inserted at one end into the manchette adjunct. Meanwhile nucleus has migrated to one (dorsal) side of the cell and the other cell organelles to the other (ventral) side of the side (*b*). The nuclear membrane lies quite close to the plasma membrane but does not touch it because of the presence of broad band of machette microtubules. In fact the nuclear membrane presses firmly against the microtubeles of the manchette which serve as a cytoskeleton or supporting framework for the elongation of the nucleus. The manchette tubules grow in length assuming a spiral course in the cell. The spiral growth of the microtubule, gives the elongating nucleus a spiral course in the cell.

Morphology of the sperm. The mature sperm (Fig. 13.16) is an elongated, spirally coiled (helical) structure with the two flagella, which are slightly different in length and directed to the rear emerging from a point some distance behind the anterior tip. The anterior region of the sperm is called the head or rostellum. It mainly consists of the flagella, basal bodies and a layer of microtubules. The flagella are covered with tiny diamond-shaped scales arranged in close packed spiral lines.

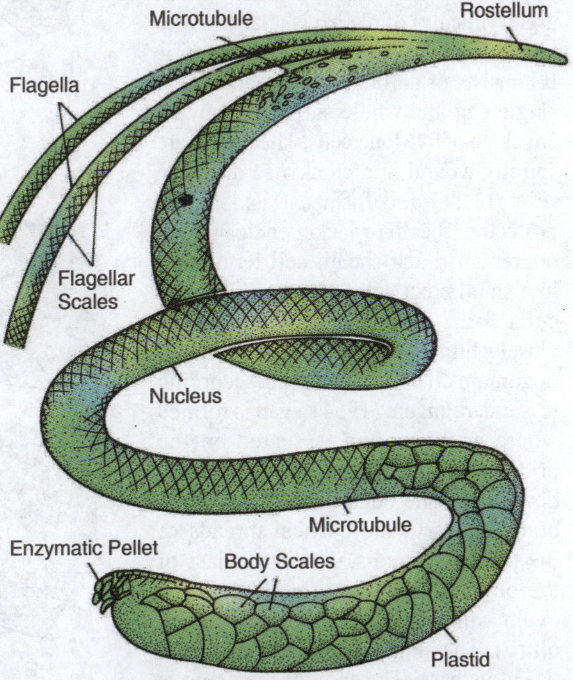

Fig. 13.16. *Chara.* Diagrammatic representation of the fine structure of a mature sperm. (Based on Coccuci and Cacere).

The anterior head region is followed by the major middle region constituting the body region of the sperm. It is mainly nuclear in origin and coiled in a compact helix of about 2- 1/2 to 3 turns. It is covered by a thin layer of cytoplasm containing microtubules on one side. It separates the nuclear body from the plasma membrane. The helical body region is also covered by small diamond- shaped scales arranged in close packed diagonal lines attached to the plasma membrane. Moestrup (1970) reported that the elongated, spirally-coiled middle region of sperm ends posteriorly into an intensely opaque tip , constituting the tail region. It is surrounded by the residual cytoplasm containing mitochondria, starch filled plastids and vesicles.

Morphology of the antheridium. Hofmeister (1852) suggested that the antheridium is a compound structure made up of numerous small male organs. This hypothesis was later elaborated by Goelel (1930) as follows :

1. He looked upon the antheridium as a metamorphosed vegetative tissue and called it is as an **antheridial axis**. The antheridial axis consists of eight secondary laterals represented by the eight primary octants.
2. Each secondary lateral (Primary octant) divides into three cells, the **shield,** the **Manubrium** and the **primary capitulum**.
3. The outer shield cell corresponds to the apical cell, the middle manubrium cell to the internode and the innermost primary capitulum to the basal node of the secondary lateral.
4. The antheridial or spermatogenous filaments arising from the primary capitulum cells are homologous to the cortical threads arising from the basal nodes of the vegetative parts.

According to another view the antheridium is considered to be a metamorphosed lateral. The capitulum cells represent the upper node, manubrium, the internode and the shield cell, lower node. The antheridial filaments represent the undifferentiated branchlets arising from the upper node. On the basis of this view each cell of the antheridial filament corresponds to an antheridium which is thus a single-celled structure as in the other algae but is surrounded by a sterile tissue which is vegetative in origin.

The oogonium (Nucule, Fig. 13.17F). (*a*) *Structure*. The female sex organ (nucule) of *Chara* is a large ovoid structure about 7 mm long at maturity. It consists of an oogonium containing a single egg, the whole surrounded by an envelope of five long cells. These cells are spirally wound in a clockwise direction (Fig. 13.11) and form a flask-shaped protective sheath or jacket enclosing the oogonium. Each sheath cell terminates in a small erect cell. Together the small cells form a tier or a crown of five closely fitting cells capping the mature oogonium. It is the **corona**. According to Sundaralingam (1954) a gap is formed where an oogonium is seated on the primary lateral. Here the cortex is not developed. The oogonium, therefore, is homologous to the cortical threads of the primary laterals. The protoplast of the oogonium produces a single large ovum which is packed with starch and oil (F). The single nucleus in the ovum lies at its base. The apex of the ovum is occupied by a colourless finely granular cytoplasm constituting the **receptive spot**. The presence of sheath around the

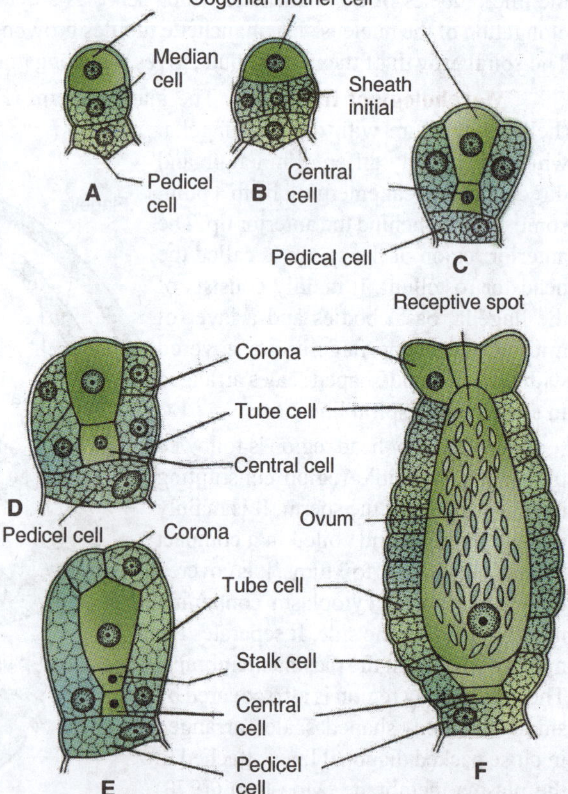

Fig. 13.17 (A-F). *Chara*. Development of oogonium. (A-E), Successive steps in the development of oogonium; (F), oogonium showing structure (Based on Smith).

oogonium in *Chara* distinguishes it from all green algae. Because of the sheath some compare it with the archegonium of bryophytes.

Development of oogonium (Fig. 13.17, A-E). The upper peripheral cell of the basal node of the antheridium facing the mother axis (adaxial) becomes an **oogonium initial**. The oogonium initial undergoes two divisions in a transverse plane to form a short filament of three cells (A). Of these, the lowermost undergoes no further division. It simply enlarges to form the one-celled pedicel subtending the oogonium. The middle or median cell divides vertically a number of times to give rise to five **sheath initials** surrounding a **central cell** (B). The terminal cell functions as the **oogonial mother cell**. It elongates vertically and divides by a transverse wall to cut off a small lower **stalk cell** and a vertically elongated terminal **oogonium** (E). The latter enlarges considerably and forms within itself a single uninucleate egg or ovum. The ovum before fertilization is packed with starch grains (F). The single central cell constitutes the node of the oogonium. The sheath initials grow upwards into five threads. They form a protective sheath around the oogonium. Each sheath cell divides by a transverse wall forming two tiers of five cells each (D). The cells of the upper tier remain small and erect constituting the **corona** or **coronula**. The cells of the lower tier are called the **tube cells**. The tube cells elongate considerably and thicken their walls (E). They gradually become spirally twisted in a clockwise direction about the oogonium as the latter elongates (F). The ovum in the oogonium becomes laden with a large amount of starch and oil. The cytoplasm at the top of the ovum is hyaline and constitutes the *receptive spot.* The nucleus, by this time, has migrated to the base (F).

Morphology of Oogonium. Of the three cells of the short filament formed during initial stages of the development of the oogonium, the basal cell represents the internode, the median cell represents the node and the upper is considered equivalent to the internode. According to this view the oogonium of *Chara* is a unicellular structure with the enclosing sterile sheath representing the modified vegetative tissue.

Fertilization. Shortly before fertilization the tube cells separate slightly from one another below the corona to form five narrow slits. Through these slits the sperms gain entrance into the sheath. One of the sperms makes its way through the gelatinised apex of the oogonial wall. It penetrates the ovum at the receptive spot and fuses with the egg nucleus.

Oospore. After fertilization the zygote secretes a cellulose membrane and becomes an oospore. The nucleus within the oospore travels to the upper pole. The oospore is retained for a time within the oogonium external to which is the sheath. The sheath hardens to form a nut-like dark or white coloured case around the oogonium. The white colour is due to the deposition of carbonate of lime. Eventually the sheath decays except the periclinal walls of the tube cells which persist as an investing layer around the oospore. The oospore falls to the bottom of the pond. There it germinates after a dormant period which may extend over weeks or months.

Germination of oospore (Fig. 13.18). Prior to germination the investing remains of the sheath around the oospore undergo decay. The diploid oospore nucleus migrates to the apical pole. There it divides by two successive divisions as in meiosis (A). Thus the only diploid structure in the life

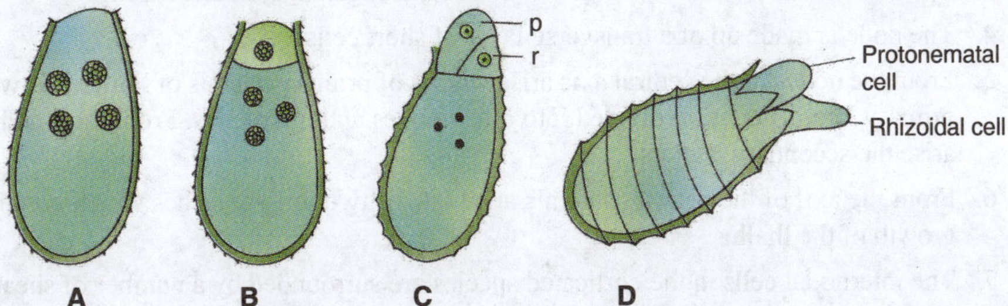

Fig. 13.18 (A-D). *Chara.* Germination of oospore. (A), Zygote at 4-nucleate stage; (B), Zygote with uninucleate upper cell and 3-nucleate basal cell; (C), Formation of protonematal cell (p) and rhizoidal cell (r); (D), Enlargement of protonematal and rhizoidal cells.

cycle is the oospore. The *Chara* plant is therefore **haploid** (Oehlkess, 1916). Sundaralingam (1946) has also supported this view. Turtle (1926), however, considered that *Chara* plant is **diploid** and the haploid stage is represented by the gametes only. A cell wall appears at the quadrinucleate stage and divides the oospore into two unequal haploid cells. The upper or the distal cell is small and uninucleate. The lower or basal cell is large and contains three nuclei (B). It is packed with reserve food. The three nuclei in the lower cell generally degenerate. The oospore wall now bursts open at the apex to expose the upper cell. The latter divides longitudinally into a **rhizoidal initial** (C) and a **protonematal initial** (C). The two grow into knob-like structures in opposite directions (D). The rhizoidal initial elongates and develops into a colourless, first or primary **rhizoid**. The protonematal initial forms an erect green filament, the **primary protonema** which is composed of nodes and internodes (Fig. 13.19, A). The peripheral cells of the basal or lower node develop into secondary rhizoids (Fig. 13.19, C). The peripheral cells of the second node except the modest one grow into a whorl of short appendages (Fig. 13.19, D). The oldest peripheral segment of this node develops into the future plant. The development of *Chara* plant as a lateral outgrowth from the protonema is an interesting feature. In this respect it approaches the mosses.

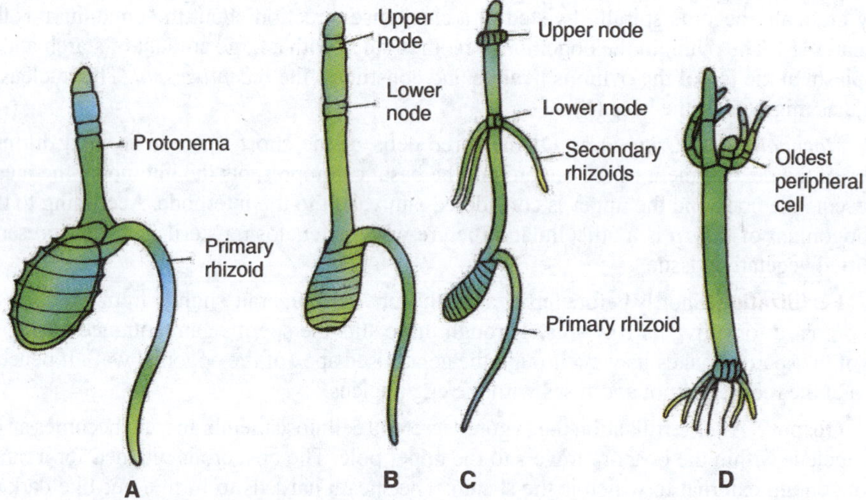

Fig. 13.19 (A-D). *Chara.* Later stages in the germination of oospore.

Salient features of Chara

1. The thallus shows an elaborate organisation and may attain a height of 20 to 30 cms. It has the appearance of miniature horse-tails (*Equisetum*).
2. The main axis consists of a series of alternating nodes and internodes.
3. The internode consists of a single, undivided, elongated, cylindrical cell.
4. The node is made up of a transverse layer of short cells.
5. From the nodes of the central axis arise whorls of primary laterals of limited growth. The primary laterals are also divided into a few nodes and internodes. From their nodes may arise the secondary laterals.
6. From the axil of the primary laterals arise one or two long branches which continue the growth of the thallus.
7. The internodal cells in the corticated species are surrounded by a number of sheath cells arising from the nodes above and below.
8. The growth in length both of the main axis and the branches takes place by means of a single, dome-shaped apical cell.

9. Asexual reproduction by spores is entirely absent.
10. Vegetative reproduction takes place by the formation of amylum stars, bulbils and secondary protonemata.
11. Sexual reproduction is oogamous. The sex organs are large, highly specialised and complicated in structure.
12. The orange or red antheridia and the oval oogonia lie in pairs. The oogonium always lies above the antheridium at the fertile nodes of the short, primary laterals in the monoecious species.
13. The oval oogonium is surrounded by a protective sheath. It consists of five long spirally wound cells ending either in a five celled corona (*Chara*) or ten-celled (*Nitella*).
14. The oogonium produces a large uninucleate egg.
15. The antheridium consists of a wall composed of eight long, curved, plate-like cells, the **shield cells**. The shield cells enclose an internal cavity. From the middle of each shield cell arises a rod-like **manubrium**. It bears at its inner end **capitulla cells** ending in long, whip-like, branched or unbranched **spermatogenous** filaments. The latter fill the antheridial cavity with a dense tangle.
16. Each spermatogenous filament consists of 5 to 200 short, discoid sperm mother cells.
17. The egg is retained within the oogonium and fertilization is internal.
18. Each spermatozoid mother cell gives rise to a single biflagellate sperm.
19. The zygote or the oospore germinates after a resting period.
20. Prior to germination the zygote nucleus divides to form four daughter nuclei. It is suggestive of zygotic meiosis. Three of the haploid nuclei gradually degenerate.
21. The *Chara* plant is thus haploid. The only diploid structure in the life cycle is the dormant zygote.
22. The zygote, on germination, produces the haploid protonema. From it the *Chara* plant arises as a lateral bud-a feature approaching some of the bryophytes (mosses).

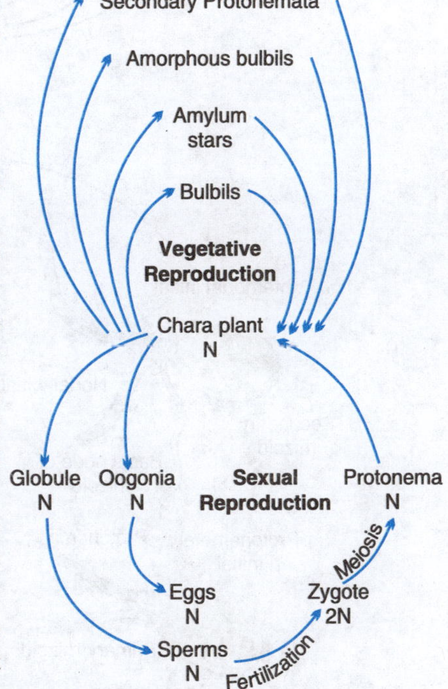

Fig. 13.20. Graphic representation of life cycle of *Chara sp.*

Taxonomic Position

Division	:	**Charophyta** (Charophycophyta)
Class	:	**Charophyceae**
Order	:	**Charales**
Family	:	**Characeae**
Sub-family	:	**Charoideae**
Genus	:	*Chara*
Species	:	*zeylanica*

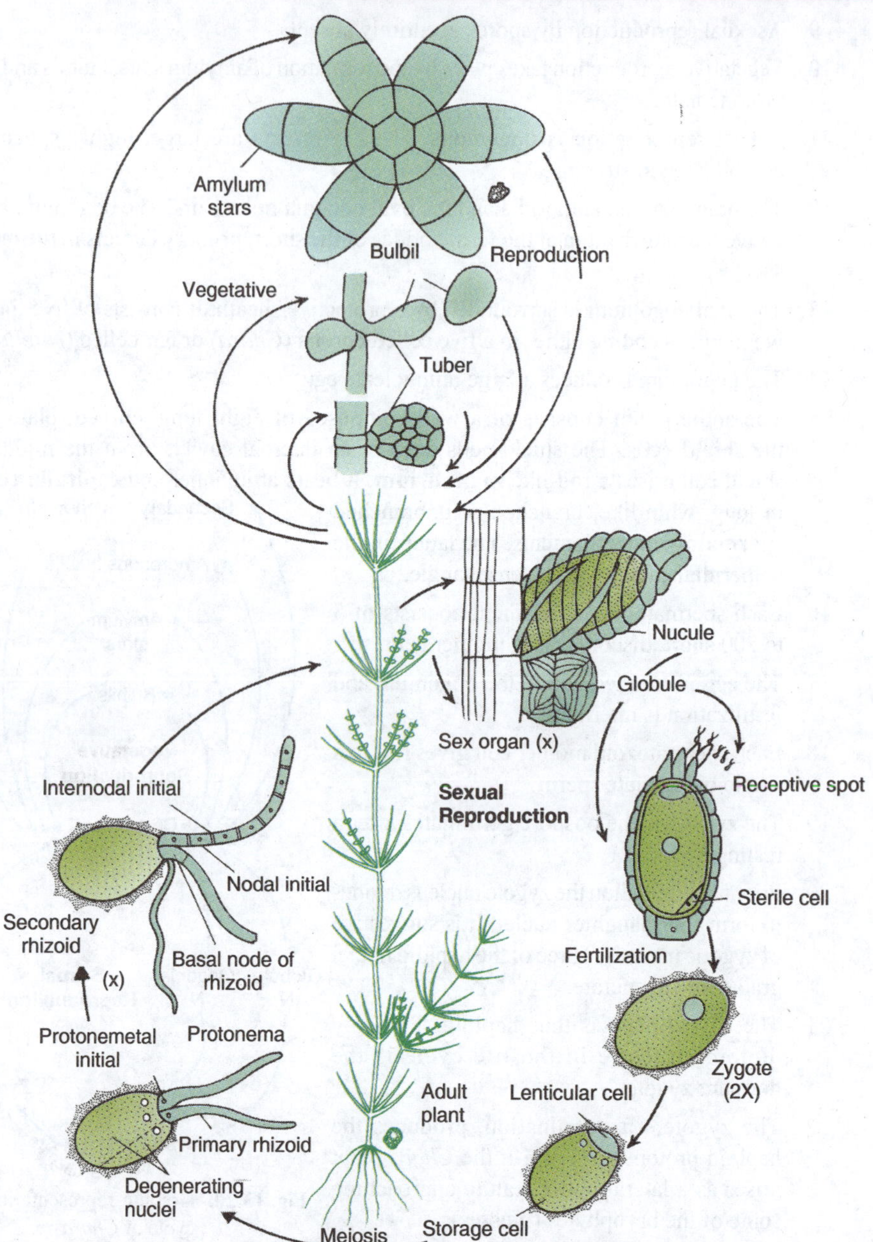

Fig. 13.21. Diagrammatic life cycle of *Chara.*

NITELLA (AGARDH)

Occurrence. *Nitella* is represented by 37 species. All of them with the exception of *N. terrestris* which occurs on damp ground, are aquatic or subaquatic. They occur in shallow water along the edges in pools, lakes and slow flowing streams attached to the sandy or muddy substratum. The account of *Nitella* in this text is mainly based on the investigations of Dr. Sundaralingam (1962 and 1967) on the Indian species of the genus.

Nitella mucronata

Nitella translucens

Nitella partita

Organisation of Thallus (Fig. 13.22). The *Nitella* plant like *Chara* has a jointed, central main axis with a whorl of branches arising from each joint (node). It is anchored to the soft substratum by multicellular, branched, colourless rhizoids. The plant thus has the appearance of miniature horsetails (*Equisetum*). In their equisetoid habit *Nitella* and *Chara* differ from all green algae. The upright, tall main axis often called the stem varies in height in different species. *N. terrestris* is very small in size as compared to *N. cernua* in which the main axis may attain a height of a metre or more. The main axis (so-called stem) is differentiated into a series of well-marked alternating **nodes** and **internodes**. The internodes is a single, elongated, undivided cylindrical cell several times longer than broad. In *C. cernua* it may reach a length of 25 cm. The internodal cells are naked as there is no ensheathing layer of narrow, elongated cells constituting the **cortex** around them. All species of *Nitella* are thus **ecorticate**. (Fig. 13.23)

The node consists of a ring of usually 6, rarely 7 or 8 **peripheral cells** enclosing two **central cells**. A whorl of branches of limited growth arise from each node. Their number in the whorls usually varies from 4 to 7. These branches are called by various names such as **branchlets, laterals** or **leaves**. In addition, the stem node bears a long branch of a branchlet in the whorl. In case there are two long branches as in *N. acuminata*, the second branch arises on the inner side of the next oldest branchlet in the whorl. The origin of a long branch or branches on the inner side of branchlets produces a fictitious appearance of axillary branching. For the reason the long branches came to be known as **axillary branches**. The axillary branch has the same structure as the main axis. There are no stipulodes at the nodes in *Nitella*.

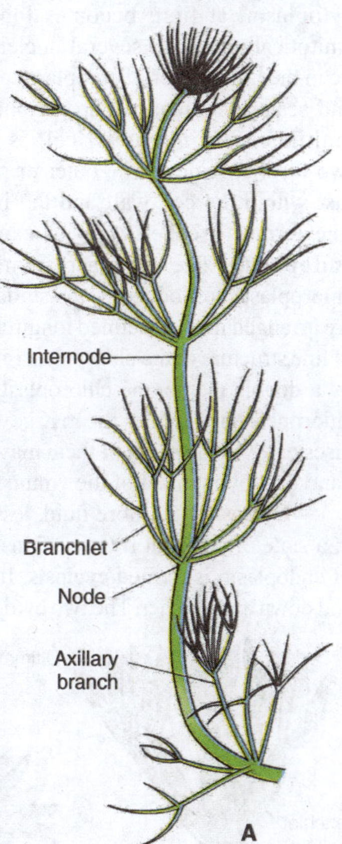

Fig. 13.22. *Nitella batrachosperma.* Showing habit (After Fritsch).

The branchlets in *Nitella* have a characteristic furcate appearance. In *N. accuminata* (Fig. 13.24 A) they are once furcate, twice furcate in *N. terrestris* (Fig. 13.24 B). They are 2 to 3 times furcate in *N. flagelliformis* and 3 times, rarely 4 times in *N. furcata.* The furcations are called the **rays** or **branchlet rays** of first, second, third or higher order depending on the species. The ultimate branchlet rays are called the **dactyls**. The dactyls may be one-called as *N. flagelliforms* and *N. terrestris* (D). They may be elongate as in *N. acuminata* but are abbreviated in some other species. In *N. furcata* the abbreviated dactyls consist of two small cells each (E).

Cell Structures. The nodal cells and young internodal cells have dense protoplasmic contents. The single nucleus is usually located in the centre of the cell. There are a number of small discoid chloroplasts, oval in shape and devoid of pyrenoids. They are uniformly distributed in the cytoplasm. Also dispersed in the cytoplasm are a number of small vacuoles. The reserve food is stored as starch. The cell wall is cellulose in nature. It has more or less a gelatinous superficial layer of unknown composition external to it.

The elongated internodal cell has a large central vacuole with the cytoplasm forming a lining layer within the cell wall. The nucleus which lies in the peripheral cytoplasm, at first, becomes lobed and then divides amitotically to form several nuclei. The resultant nuclei lie in the lining layer of cytoplasm and have large nucleoli and scanty chromatin. The peripheral cytoplasm of the multinucleate internodal cell is distinguishable into two distinct regions, the outer or parental denser region just within the cell wall and the inner less denser part. The former is called *ecto-* or **exoplasm** and the latter **endoplasm**. The exoplasm remains stationary. The chloroplasts are lodged in the stationary exoplasm. They are arranged in well-defined longitudinal series. The study of fine structure of the chloroplast revealed that it is limited by a double membrane chloroplast envelope. Within the chloroplast envelope is the matrix which contains 40-100 discs or thylokoids. All of them may form a single stack or band occupying most of the volume of the chloroplast.

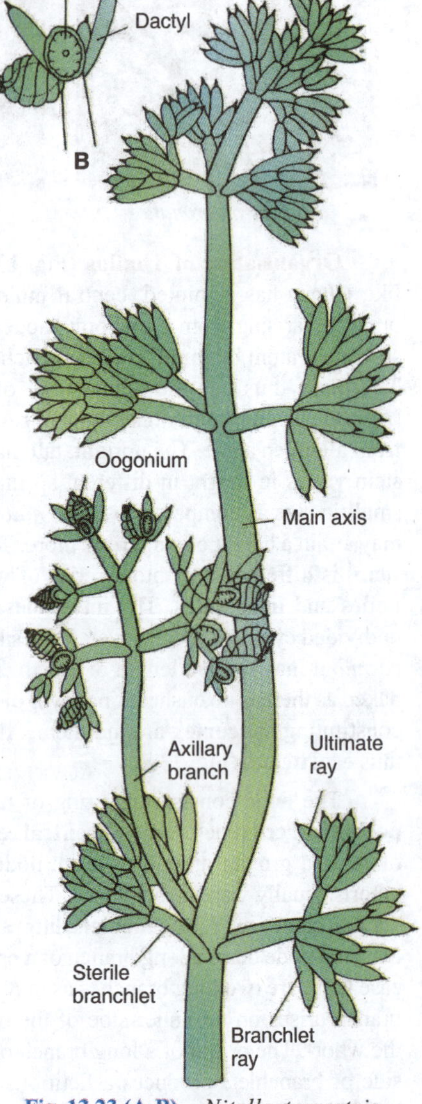

Fig. 13.23 (A-B). *Nitella terrestris*. (A) Showing habit; (B), A fertile branchlet (After Sundaralingam).

Cyclosis. The more fluid, less denser endoplasm is in a state of constant rotation. This streaming movement of endoplasm is termed **cyclosis**. It flows up on one side and down on the other. The two cytoplasmic streams adjoin a colourless strip or streak of stationary cytoplasm on either flank of the cell.

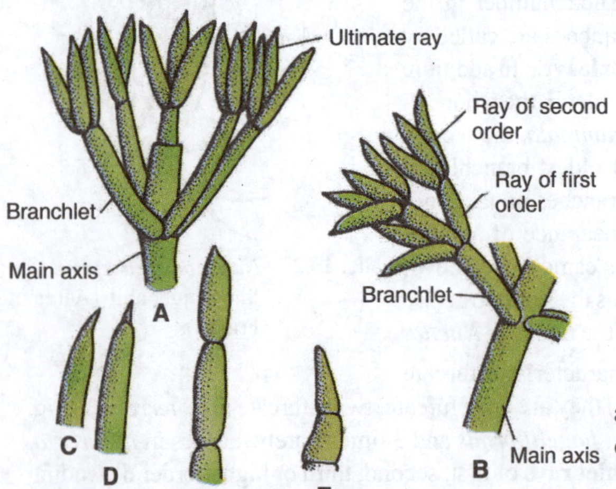

Fig. 13.24 (A-F). *Nitella*. (A), whorl of sterile branchlets of *N. acuminata* (once furcate); (B), whorl of sterile branchlets of *N. terrestris* (twice furcate); (C-F), different kinds of dactyls (After Sundaralingam).

Mechanism of Cyclosis. Kamiya and Kurto (1956) found that endoplasm has no velocity gradient in itself. It slides as a whole on the cortex-endoplasm interface by a force generated by an unknown mechanism. This force is generated in a narrow zone at the ectoplasm-endoplasm interface. It is transmitted radially inwards and is held responsible for streaming of endoplasm. Vorob Eva and Poglazov (1963) reported having isolated a contractile protein in *N. flexilis*. Palevitz *et al* (1974) reported the presence in *Nitella* extracts of microfilaments and provided evidence that the microfilaments consist of a contractile protein called *actin* which is similar to the major protein of the thin filaments or muscles in animals. Palevtiz and Helper (1975) demonstrated that the microfilaments containing actin are organized as bundles. These microfilament bundles which are often referred to as subcortical actin bundles, are located at the ectoplasm-endoplasm interface and are associated with the chloroplasts. The chloroplasts in a file oscillate or vibrate as if they are linked. In fact Kamitsubo (1980) reported, that in *Nitella axellaris* 3-4 protoplasmic fibrils, ca. 0.1 μm, are affixed in parallel, at intervals of ca. 0.5 μm, to each file of chloroplasts, on the surface facing the endoplasm. The fibrils are oriented parallel to the direction of streaming. Kamitsubo postulated that these subcortical fibrils of actin located at the solgel interface are the structures responsible for cytoplasmic streaming in characean cells. The assumption that generation of motive force for the endoplasmic stream resides exclusively in the subcortical fibrils of actin (microfilament bundles) which lies parallel to, and on the inner surface of spiral rows of chloroplasts in the cortex is supported by a considerable body of information discovered in recent years. The molecular mechanism of this movement, however, is still far from being clear at the present time.

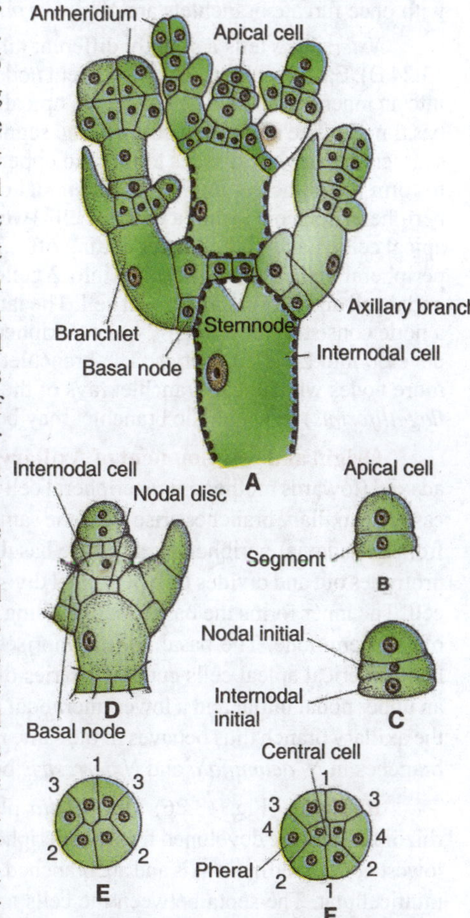

Fig. 13.25 (A-F). *Nitella.* (A), L.S. apical region showing structure of stem node and basal node of branchlet with an initial of axillary branch-developing sex organs and branchlets; (B-C), successive stages of segmentation of apical cell and differentiation of nodal and internodal initials; (D) L. section through the apical region of main axis; (E-F), C.S. young nodes showing the sequence of walls (A based on Sundaralingam but slightly modified and diagrammatic, E-F after Fritsch).

Growth (Fig. 13.25). The main axis and the axillary branches show indefinite growth. It takes place by the activity of a dome-shaped (hemispherical) **apical cell** located at the tip (A). It cuts off a series of segments parallel to its posterior face (B). Each segment divides by a curved transverse wall into two halves (C). The lower half which is known as the **internodal initial** is biconvex. It undergoes no division but elongates gradually during further growth to the length of a mature internode. The upper half, which is biconcave, is known as the **nodal initial**. It divides in a characteristic manner to form the nodal elements (D). The nodal initial first divides by a halving wall (1-1) into two semicircular cells (E). The subsequent divisions (2-2, 3-3, 4-4) take place by curved septa occurring in planes intersecting the previous one. In this way a ring of usually 6 rarely 7 or 8 peripheral cells enclosing the two central ones is formed (F). This mass of cells constitutes the node.

Origin and Development of Sterile Branchlets. The peripheral cells at the stem nodes protrude to the outside. Each then divides by a periclinal wall into an inner cell and an outer cell. The inner cell which is called the *basal nodal initial* divides to develop directly into a basal or basilar node of the branchlet. It consists of an incomplete ring of two peripheral cells

(*N. acuminata*). The ring is open on the abaxial side. The outer cell functions as an **apical cell** and forms the branchlet. In *N. acuminata* it cuts off only one segment, then ceases to divide and elongates to form an elongated dactyl with an acuminate or pointed tip. Meanwhile the segment cut off by the apical cell divides transversely into an upper nodal cell and lower internodal cell. The latter elongates to form the single internode of the branchlet. The nodal cell divides to form the upper node consisting of an incomplete ring of 3 or 4 peripheral cells. It is open on the abaxial side. The peripheral cells of the upper node of the branchlet protrude to the outside and grow into one-celled ultimate rays (dactyls). Each branchlet in *N. acuminata* thus consists of a basal node, and one more node (upper) bearing a whorl of ultimate rays or dactyls. It is thus once **furcate** (Fig. 13.24 A). The other species with once furcate branchlets are *N. syncarpa, N. cernua, N. opeca* and *N. flexilis.*

N. terrestris tells a slightly different tale. The branchlets in this species are twice furcate (Fig. 13.24 B). Each peripheral cell at the stem node protrudes to the outside and divides by a periclinal wall into an inner basal nodal cell and outer apical cell. The inner cell as a rule in *Nitella* directly forms the basal node. The apical cell cuts off one segment and becomes converted into a 2-celled dactyl. The segment meanwhile divides to form an upper nodal cell and a lower internodal cell. The latter grows to form the branchlet internode. The nodal cell divides to form the upper node consisting of 5 or 6 peripheral cells enclosing a central cell. Two or three peripheral cells at this node again function as apical cells. Each of these after cutting off a segment changes into a 2-celled dacytyl. The remaining peripheral cells directly develop into 2-celled dactyls. The segment in turn divides into an upper nodal cell and a lower internodal cell. The latter forms the internode and the nodal cell develops into a node consisting of a ring of 4 or 5 peripheral cells enclosing a central cell. These peripheral cells develop into 2-celled dactyls. The branchlet in *N. terrestris* thus consists of a basal node, and two more nodes which bear branchlet rays of the first and second order. There are species of *Nitella* (*N. flagelliformis*) in which the branchlet may be two to three times furcate as in *E. furcata.*

Origin and Development of Axillary Branches. The axillary branch takes its origin from the adaxial (towards mother axis) peripheral cell of the basal node of the oldest branchlet in the whorl. In case two axillary branches arise from the same node as in *N. acuminata,* the second branch originates from the adaxial peripheral cell of the basal node of the next oldest branchlet. The peripheral cell protrudes out and divides by a periclinal division into an inner cell and the outer hemispherical apical cell. The inner forms the basal node dividing by halving wall followed by curved septa as in the case of the stem node. The basal node comprises of 2-4 peripheral cells enclosing 2 central cells. The hemispherical apical cells cuts off a series of successive segments each of which differentiates into an upper nodal initial and a lower internodal initial as in the case of the main axis. The apical cells of the axillary branch thus behaves in the same manner as the apical cell of the central axis. The axillary branches in *N. acuminata* and *N. terrestris* bear sex organs on special fertile whorls.

Rhizoids. (Fig. 13.26) The *Nitella* plant like *Chara* is anchored to the soft substratum by rhizoids. They are developed from the peripheral cells of the lowest node of the main axis and are branched, colourless and multicellular. The septa between the cells are oblique (A). The rhizoids grow by the activity of an apical cell but show no differentiation into nodes and internodes. At the septa the ends of the adjoining cells protrude in opposite directions to form a double foot joint (B). The protruded dilated tip (C) of the upper cell is cut off by an oblique septum to form a segment which divides into a quadrant (D and E). From this plate of cells arises a cluster of branch rhizoids (F). The latter may branch further in the same manner.

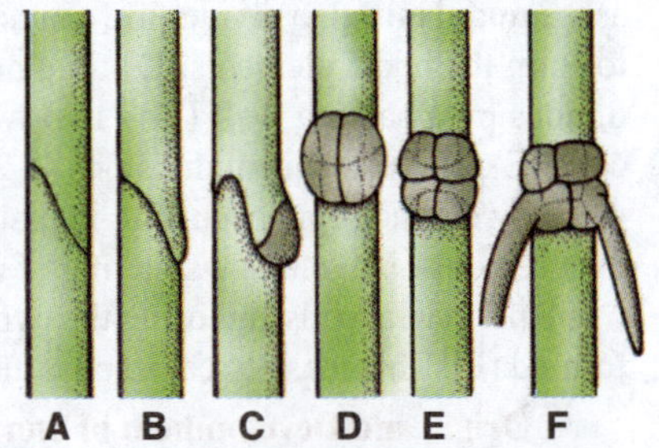

Fig. 13.26 (A-F). *Nitella asper*. Successive stages in the development of rhizoids and mode of branching. Explanation in the text (After Giesenhagen).

Sexual Reproduction. It is an advanced type of oogamy. Light intensity plays an important role in the development of sex organs. Some species (*N. acuminata, N. furcata, N. terrestris* and *N. hyalina*) are *monoecious* and some (*N. flagelliformis, N. opaca, N. superba*) are *dioecious*.

(*a*) *Position of sex organs* (Fig. 13.27). In some species there is no distinction into fertile and sterile branchlets. They are similar. In *N. acuminata* and *N. terrestris,* however, the sex organs are formed on special fertile branchlets which are borne on axillary branches and are different from the sterile branchlets (Fig. 13.24 A). In the monoecious species the antheridium is terminal at the centre of furcation and oogonium lateral in position below the antheridium (A). The antheridium is developed from the apical cell of the branchlet at the furcation. The oogonia are developed from the adaxial peripheral cells of the upper node of the branchlet. It is also called the basal node of the antheridium. The remaining peripheral cells of the node form the branchlet rays. Thus mor-phologically the oogonium is a modified branchlet ray of the higher order than that of the antheridium. In monoecious forms with more than one furcation (*N. terristris*) all the furcations many bear both the sex organs (B). *N. furcata* which is also monoecious, however, has only oogonia at the first furcation and both antheridia and oogonia on other furcations above the first (C). Final furcation is sterile. The dioecious species (*N. flagelliformis*) bears either an antheridium or an oogonium at each furcation.

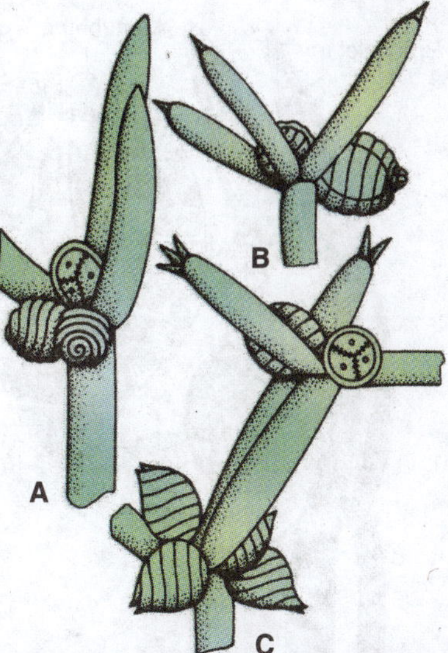

Fig. 13.27 (A-C). *Nitella.* Position of sex organs. (A), *N. acuminata;* (B), *N terrestris;* (C), *N. furcata.* (A, after Paul; B, after Iyengar; C, after Agharkar and Kundus).

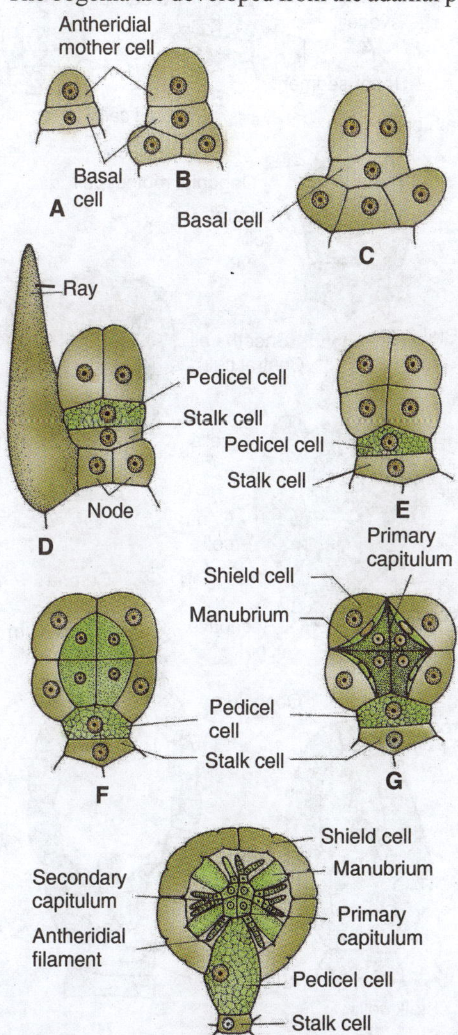

Fig. 13.28 (A-H). *Nitella.* Successive stages in the development of antheridium. (After Sundaralingam).

Antheridium (Fig. 13.28). In monoecious species of *Nitella* the antheridium is above the oogonium at the node. Except a slight difference in its early development, the structure and development of antheridium of *Nitella* is similar to that of *Chara.* It consists of a jacket layer of 8 closely fitting **shield cells**. From the middle of each shield projects inwards towards the centre of the antheridium and elongated cell, the **manubrium**. The latter at its distal end bears a terminal cell known as the **primary capitulum**. The primary capitulum buds about six **secondary capitula**. Each secondary capitulum terminally develops a pair of long filaments called the **antheridial filaments**. Each cell of the antheridial filament produces a single **antherozoid**. The mature antheridium appears stalked

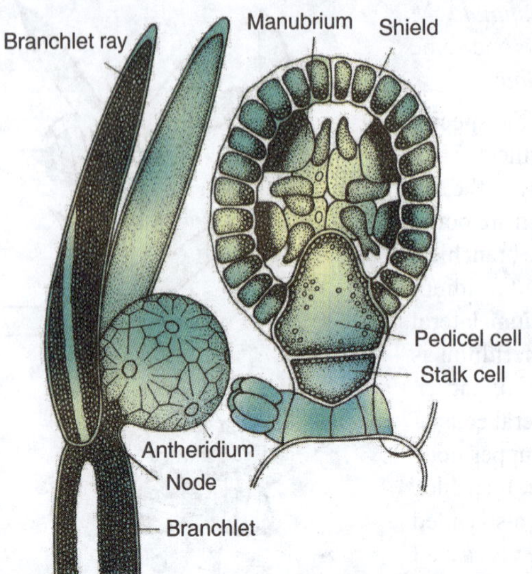

Fig. 13.29 (A-B). *Nitella flexilis*. Fertile branchlet. (A), Terminal antheridium; (B), L.S. antheridium showing structure.

Development of antheridium (Fig. 13.27). The antheridium primordium as in *Chara* cuts off a cell at its base (A) and then assumes a spherical form to function as an **antheridial mother cell** (B). The latter undergoes two longitudinal divisions at right angles to one another. The resultant four cells constitute the **quadrant stage** (C). In section only two cells are seen. Each quadrant divides transversely to form the **octant stage** (E). In section only four cells are seen. Each octant undergoes a periclinal division (F). The outer segment divides again by periclinal wall. In this way a row of three cells is formed in each octant (G). The outer cell of each row grows laterally to form a **shield cell.** The middle one elongates radially and projects inwards to form the **Manubrium**. The innermost cell of the row constitutes the **primary capitulum** or **head cell**. The latter cuts off **secondary capitular cells** which form **antheridial filaments** (H). The basal cell in *Chara* undergoes no further division. It grows pushing into the antheridial cavity to form the pedicel cell. In *Nitella*, as development of the antheridium proceeds, the basal cell divides transversely into two segments upper and lower (D). The lower segment remains short and discoid to form the **stalk cell**. The upper sister segment grows and elongates to form the **pedicel cell**. A part of the latter remains outside the antheridium. (Fig. 13.29 H). The stalk consists of a **pedicel cell** which partly projects into the cavity of the antheridium and the **stalk cell**. There is no stalk cell in the antheridium of *Chara. The* antheridium with a jacket of 8 shields and corresponding number of manubria and head cells is called **octoscutatte**. In *N. terrestris* there are only 4 shields with a corresponding number of manubria and head cells. Such an antheridium is called **quadriscutate**.

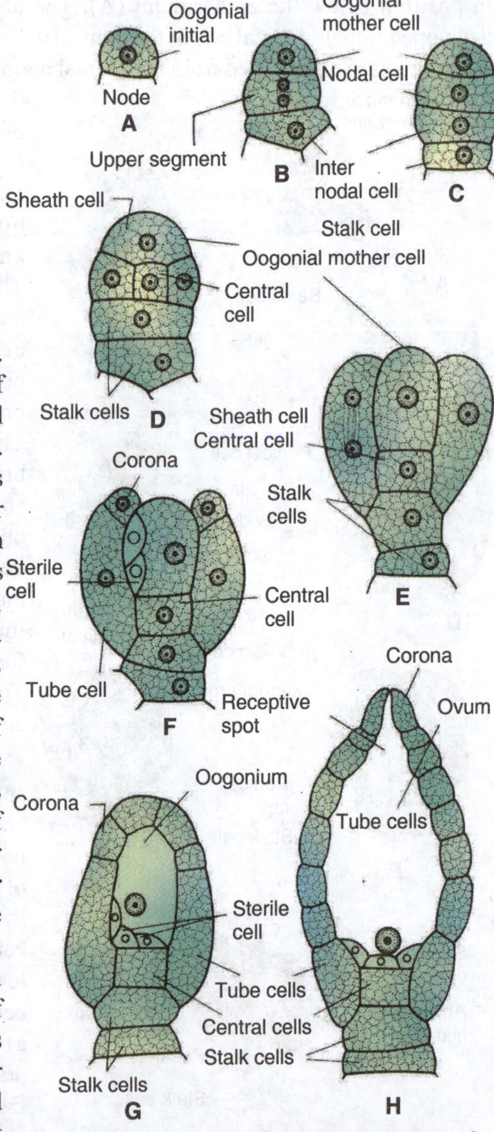

Fig. 13.30 (A-H). *Nitella*. Development of Oogonium. Explanation in the text (After Sundaralingam).

Oogonium (Fig. 13.30 H). As in *Chara* the oogonium is oval in form and is enclosed by a sheath of five long cells spirally wound around it in a clockwise direction. There are ten small closely fitting cells capping the mature oogonium. They are arranged in two tiers of five cells each and constitute the **corona**. The corona in *Nitella* is colourless, small, persistent or deciduous and two-tiered. In *Chara* it consists of five cells only. They are larger and the corona is persistent, erect and convergent, or divergent and one-tiered. Within the sheath is the oogonium. The protoplast of the oogonium forms a single ovum packed with starch and oil. The nucleus lies at the base. At the apex of the ovum is colourless, granular cytoplasm constituting the **receptive spot**.

Development of oogonium (Fig. 13.28). It develops from the adaxial peripheral cell of the node below the antheridium. The cell in question protrudes to the outside and functions as the **oogonium initial** (A). After cutting off two segments at its base it becomes the **oogonial mother cell** (B). The upper segment undergoes a transverse division to form upper **nodal cell** and a lower **internodal cell** (C). The latter along with the lower segment form the **stalk** of the future oogonium. It consists of two stalk cells (D). The upper stalk cell is also called the **pedicel cell**. The nodal cell as in *Chara* divides vertically a number of times to give a **central cell** surrounded by 5 sheath or enveloping cells (D). Even before sheath cells grow over the oogonial mother cell, the latter cuts off two lens-shaped sterile cells (F) on the side next to the antheridium (adaxial) and one flat sterile cell at its base (G). As the egg or ovum advances towards maturity, the three sterile cells come to lie near its base. Meanwhile each sheath cell grows vertically upwards and cuts off a small conical cell, the **coronal cell** at the top. These five coronal cells constitute the upper tier. Sometimes after each sheath cell cuts off another coronal cell below the upper one. These five small discoid coronal cells constitute the lower tier. The oogonium in *Nitella* thus has a two-tiered **corona**. The cells of the lower tier are smaller and discoid and those of the upper tier are comparatively larger and conical. After cutting off the coronal cells the sheath of tube cells, as they grow, twist round the egg or ovum. The spirals run from right to the left. Except for a small hyaline area, the **receptive spot** at the top, the mature egg is packed with reserve food. The nucleus lies at the base (H).

Fertilization. As the sperms mature, the shield cells separate from each other exposing the antheridial filaments. The sperms usually escape in the morning into the surrounding water in which they may continue to swim until the evening. Meanwhile the enveloping tube cells slightly separate from one another at the apex just below the corona of the mature oogonium forming five small slits (De Bary, 1871). Through these slits the sperms penetrate the sheath and the gelatinised wall of the oogonium and thus gain entrance into the ovum at the receptive spot. Within the ovum the sperm nucleus moves down to fuse with the female nucleus at its base.

How do the sperms escape from the sperm mother cells is a matter of dispute. Fritsch (1935) believed that they escape by the gelatinisation of the walls of sperm mother cells. According to Groves and Bullock-Webster (1920) the sperms burst from their walls. Steewart, (1937) and Sundaralingam, (1946) observed that in *Chara zeylanica* the mature sperm dashes against the side wall of the sperm mother cell. It ruptures and the sperm shoots out through the pore thus formed with the posterior end coming out first and flagellated end next.

Oospore. The fertilized egg or zygote enlarges and secretes a thick wall around it to become an **oospore**. The sheath decays except the thickened inner periclinal walls of tube cells which persist as spiral markings on the oospore wall. The oospore with the investing remains of the sheath gets detached and sinks to the bottom of water. It is a hard nut-like body which enters upon a period of rest at the bottom of the pond. It varies in colour from light yellow to black and is rounded to ellipsoid in form.

Germination of Oospore. Germination has been thoroughly studied in *Chara* and is thought to be the same in *Nitella*. After a period of rest which may extend over weeks or months the zygote, as in *Chara*, germinates. Prior to germination, the diploid oospore nucleus migrates to the apical pole. There it divides into four daughter nuclei. It is presumed that the division is meiotic. On this presumption the thallus plant is considered to be haploid (gametophyte). The division of the nucleus is followed by asymmetrical division of oospore into a small, uninucleate distal cell and a large, basal trinucleate cell packed with food reserve materials. The basal cell later degenerates. The oospore bursts at the apex exposing the uninucleate cell which now divides vertically to form a **rhizoidal initial** and a **protonematal initial**. The two grow in opposite directions, the former developing into a colourless, filamentous first **rhizoid** and the latter into an erect, green, filamentous **primary protonema**. Both are differentiated into nodes an internodes. The adult plant arises as a lateral branch from the second node of the protonema.

Taxonomic Position

Division	:	**Charophyta** (Charophycophyta)
Class	:	**Charophyceae**
Order	:	**Charales**
Family	:	**Characeae**
Sub-Family	:	**Nitelloideae**
Genus	:	*Nitella*
Species	:	*acuminata*

Affinities and Inter-relationships of Charales

The stoneworts (Charophytes) are a group of green-plants that have uncertain affinities with other plant groups. In fact their phylogeny remains very much a mystery even today. All the views expressed are highly speculative. The fossil evidence throws no light on their origin and affinities. It simply shows that the Charales are an extremely ancient type of algal organisation. All evidence is thus based on the living forms. The following three hypotheses have been put forth :-

1. This specialised group of plants should be placed in a separate division "Charophyta" intermediate between the Thallophyta and the Bryophyta. This view is based on the following characteristic features exhibited by the members of the group :-

(*a*) Equisetoid or anisation of the thallus with apparent root-like, stem-like and leaf-like portions.

(*b*) Nodal arrangement of the branches in whorls.

(*c*) Growth apical rather than diffuse as in the Chlorophyta.

(*d*) Sex organs complex in structure and surrounded by jackets of sterile cells.

(*e*) The spirally coiled, biflagellate sperms resembling those of Muscineae.

(*f*) Development of a young plant as a lateral branch from the protonema just like the bryophytes.

These features have led some botanists to consider the stoneworts distinct from the green algae. They elevate the order Charales to the status of a division Charophyta co-ordinate with the other great division. It is placed intermediate between the Thallophyta and the Bryophyta. Some have even suggested transference of this group to the Bryophyta on the basis of the complexity of sex organs. However, the stoneworts lack many important traits which the bryophytes possess. The chief among them are :

(*i*) Absence of alternation of generations.

(*ii*) Marked difference in the structure of the antheridium and female sex organ from those of the bryophytes. These are, in fact, unicellular structures with the surrounding jackets as modified vegetative tissue. The stoneworts therefore cannot readily be included among the bryophytes.

2. Many algologists still hold that the difference between the stoneworts and the main body of the Chlorophyceae are not so great as to place them, as a separate division. The sex organs, though apparently complex, are in actual practice simple and single celled, as suggested by Goebel and Hofmeister.

Smith holds that inspite of the differences, the Charales (stoneworts) in many of their characteristic biological and biochemical traits are similar to the green algae. For this reason he gives stoneworts the status of a class 'Charophyceae' and places it along with Chlorophycophyta) on the basis of two distinctive features. These are the equisetoid body organisation and the presence of a sterile sheath around the sex organs.

3. Fritsch treated stoneworts as an order Charales of the cells Chlorophyceae. Iyengar supported this view. They based their argument on the similarities between the stoneworts and the green algae. Adherents of this hypothesis argue that in their algal simplicity, cellulose cell wall, starch as the customary food reserve, presence of both chlorophyll *a* and *b* as the photosynthetic pigments, haploid genetic constitution of the plant, flagellar morphology of the motile cells and haplontic life cycle, the stoneworts resemble the typical green algae. In view of these common features they hold that if any affinity is to be sought it must be with the green algae. Of course they represent a highly specialized group, which diverged very early in the course of evolution from the rest of the green algae. Church (1919) considered Charales as a remnant of one of the many evolutionary lines which attempted to migrate to the land. The intermediate forms were subsequently lost. In his opinion the equisetoid habit, cortex-like covering derived from the basal cells of the laterals, limited growth of the laterals, morphology of the sex organs and restriction of sex organs to the laterals indicate that the Charales might have originated from the Chaetophorales especially, from forms like *Draparnaldiopsis* or had a mutual origin with them. *Draparnaldiopsis* has all the above-mentioned tendencies suggestive of the Charophytes. Having branched off from these green algal ancestors the Charales by specialisation and differentiation developed features not shared by rest of the algae. So most of the algologists, still place them in the Chlorophyta at the end of the line. The opponents of this view point to the apical growth and complexity of the sex organs. These features go against the possible origin of stoneworts from the Chaetophorales.

Desikachary and Sundaralingam (1962) opined "the understanding of the development of the Charophyte antheridium in terms of homologies with vegetative structures (Goebel) tends to remove a barrier to the inclusion of stoneworts in the Chlorophyta but the compound structure of the fructifications among other features, favours the recognition of a class (Charophyceae) within the Chlorophyta or, in our opinion, a separate phylum for the Charophytes." The latest trend, however, is to give the stoneworts a phylum rank because of the distinctive features of the stoneworts which make them distinct from the Chlorophyta. These distinctive features are the equisetoid habit, attachment to the substratum by multicellular rhizoids, occurrence of many discoid chloroplasts without pyrenoids in each cell, complex sex organs each with a jacket of sterile cells around it, absence of asexual reproduction by mitospores of any kinds, an elongated, biflagellate spirally coiled male gamete or antherozoid and presence of a protonematal stage from which the young *Chara* plant arises later as a lateral branch.

In the absence of asexual reproduction by mitospores, presence of numerous small discoid chloroplasts without pyrenoids in each cell, fine structure of chloroplasts, and apical growth, the stoneworts are more like the higher plants than the green algae. However, they did not develop the vascular tissue and sporangial organ essential for the progenitor of terrestrial plants. There is thus no evidence that vascular plants evolved from *Chara*-like ancestors. In view of the conflicting opinions expressed above, Bell and Woodcock (1968) considered that the stoneworts are a highly specialized group of aquatic plants long removed from the main trends of algal evolution.

As to the evolutionary trends among the Charophytes themselves there are two views. The older view is that the two subfamilies Charoideae and Nitelloideae diverged early along two parallel lines in the evolution of this group. The ecorticate forms are considered primitive with the corticate forms having evolved from them. Thus the evolutionary sequence in the Charoideae starts with the ecorticate species such as *Protochar,* passes through *Nitellopsis,Lamprothmnion, Lychnothamnus* and culminates in the corticated species of *Chara. The* other view is that *Chara* is more primitive than *Nitella* and the ecorticate forms have evolved from the corticate species. This view has been suggested by Desikachary and Sundaralingam (1962). They base their hypothesis on two factors, namely, sympodial branching of laterals and greater amount of sterile tissue in the oogonium of *Nitella.* This hypothesis paves the way for the origin of the Charophytes from forms like *Draparnaldiopsis* as suggested by Church (1919).

Economic Importance of Charales

The Charales are not of great economic importance. Zaneveld (1940), however listed nine uses for these plants, as food for aquatic animals, as fertilizer, insect control and sugar clarification, fisheries, water purification, polishes, mud baths, and therapeutic value. Caballero (1919) considered them to be useful in the destruction of mosquito larvae. It is attributed to certain larvicidal properties of *Chara*. There are others who deny this. They hold that the Charales have no larvicidal properties. The growth of *Nitella* and *Chara* in turbid water makes it very clear. The Charales serve as food for aquatic animals especially water fowl. For this reason the plants are recommended for indoor aquaria. They support abundant growth of epiphytes and small animal herbivores. The latter serve as food for the aquarium fish. In certain parts of the world the dried plants are spread over the agricultural lands to keep off insects from the soil. The pungent odour of these dried plants serves as an insect repellant. These plants are also used as manure. Their high lime encrustation serves to ameliorate dense clayey soils and prevent sourness of the soil. They are a cheap source of calcium in acidic conditions and serve as a good experimental material for physiological investigations.

Taxonomic Position of Charales

The taxonomic position of stoneworts is still debatable. Even their inclusion among the algae has been questioned. In fact the earlier botanists placed them under 'Equisetum'. Adansen (1963) and others placed them under monocotyledonous Phanerogams Naiadeae). Some pointed to their affinity to Dicotyledonous Phanerogams (Ceratophyllaceae). Berkely regarded them as a separate class, Charales intermediate between Muscineae and Thallophytes. Sachs (1875) considered them distinct from the Thallophytes and thus raised them to the rank of a division by the side of Thallophytes and Muscineae. The algal nature of the stoneworts is evident from the fact that they grow in water, have little internal differentiation and are haploid almost throughout the whole life cycle. In the nature of photosynthetic pigments, food reserves, chemical composition of cell wall, flagellar morphology, haploid genetic constitution of the plant and haplontic life cycle they resemble the green algae. Fritsch (1935) thus treated them as an order, Charales of class Chlorophyceae. Iyengar (1958) followed Fritsch. Smith (1938, 1951 and 1958) placed them in a separate class Charophyceae under division Chlorophyta (=Chlorophycophyta) on the basis of distinctive features, the chief among which are equisetoid habit, and jacket of sterile cells around the sex organs. The latest trend, however, is to elevate stoneworts (Charales) to the rank of a distinct division Charophyta (Charophycophyta) because of the distinctive features in which the stoneworts differ from the Chlorophyta. This is the view taken in the text.

QUESTIONS

ESSAY TYPE

1. Write an illustrated account of the sexual organs of *Chara* or *Nitella* and the germination of its oospores. Mention how it differs from other green algae.
2. Give an illustrated account of the method of sexual reproduction in *Chara* or *Nitella.* (*Bhagalpur, 1996*)
3. Describe the structure and development of sex organs of *Chara* or *Nitella.*

4. Describe the structure and reproduction (life history) of *Chara* and discuss its position in algae. *(Punjab, 1999)*
5. Describe the structure and reproduction in *Chara* or *Chlamydomonas* or *Ectocarpus*.
(Kanpur, 1995; Kumaon, 1999; Awadh, 1993, 1995; Ravi Shankar, 1992)
6. Give an illustrated account of the sexual reproduction in *Chara*.
(Kumaon, 1996; Guru Nanak Deo, 1991)
7. Describe the life cycle of *Chara* or *Ectocarpus* with the help of labelled diagrams only. *(Awadh, 1997)*
8. Give an account of the life history of *Chara*. *(Bihar, 1992, Magadh, 1992)*
9. Describe reproduction in *Chara* with the help of diagrams. *(Bhagalpur, 1995)*
10. With the help of labelled diagrams, describe the sex organs of *Chara*. *(Madras, 1995)*
11. Describe the sexual reproduction in *Nitella* with the help of labelled diagrams.
12. Describe the structure and reproduction in *Nitella*.
13. Describe the life cycle of *Nitella*. How does *Nitella* differ from *Chara?*
14. Describe the sex organs of *Nitella*.
15. Discuss the evolutionary trends in *Chara* and compare it with that of *Chaetophorales*. *(Lucknow M.Sc, 2006)*
16. Give an illustrated account of reproduction in *Chara* and give its systematic position. *(Poorvanchal, 2006)*

SHORT ANSWER TYPE

17. **Write short notes on :**

(*a*) Structure of sex organs in *Chara* and its systematic position
(Agra, 1947; Gorakhpur, 1997; Rohilkhand, 1992, 1994; Nagpur, 1996)

(*b*) Sex organs of *Chara* (Globulc and Nucule)
(Allahabad, 2006; Agra, 1977; U.P. College, 1997; Agra, 1991, 1996; Lucknow, 1996, 2001, 2006; Awadh, 1998; Nagpur, 1995; Meerut, 1998, 1999; Kanpur, 2005)

(*c*) Antheridium or globule of *Chara*
(Agra, 1973; Purvanchal, 1999; Kumaon, 1995; Agra, 1993; Bihar, 1990; Magadh, 1990; Ravi Shankar, 1995; Meerut, 1997; Punjab, 1998)

(*d*) Characteristic features of *Chara*
(Kanpur, 1996; Kumaon, 1995; Allahabad, 19992)

(*e*) Structure and morphology of antheridium in *Chara*
(Guru Nanak Deo, 1992)

(*f*) Thallus structure of *Chara*
(Gorakhpur, 1999)

(*g*) Charales *(Lucknow M. Sc., 2005)*

(*h*) Nucule of *Chara*
(Poorvanchal, 2007; Ravi Shankar, 1996, 1994; Punjab, 1996, 1997)

(*i*) Differences between Globule and Nucule. *(Purvanchal, 1997)*

(*j*) Sexual reproduction in *Chara*
(Bundelkhand, 1998)

(*k*) Structure of male and female sex organs of *Chara*
(Purvanchal, 1998, Bundelkhand, 1999)

(*l*) Vegetative reproduction in *Chara*.
(U.P. College, 1996; Rohilkhand, 1996)

(*m*) Systematic position of *Chara*
(Agra, 1995; Lucknow, 1995)

(*n*) Female reproductive structure of *Chara* *(Awadh, 1991)*

(*o*) Cortication of *Chara*.
(Bhagalpur, 1994)

(*p*) Thallus structure in *Nitella*

(*q*) Sexual reproduction in *Nitella*

(*r*) Post-fertilisation changes in *Nitella*

(*s*) Sex organs in *Nitella*
(Kanpur M. Sc., 2005)

(*t*) Nucule *(Bangalore, 2001)*

(*u*) Affinities of *Chara* (*Kanpur M. Sc., 2004*)

18. (*a*) Draw neat and labelled diagrams of *Chara*. (*Kanpur, 1998*)

(*b*) Draw the diagram of nucule of *Chara* and label the parts. (*Kerala, 2001*)

19. Draw neat and labelled diagrams of the following :

(*a*) Structure of globule in *Chara*. (*Kanpur, 2006; Gurakhpur, 1993*)

(*b*) Sex organs of *Chara*. (*Kanpur M. Sc., 2006; Gurakhpur, 1995, 1998; U.P. College, 1995; Awadh, 1992*)

(*c*) L.S. Globule of *Chara*. (*Purvanchal, 2000, 2003*)

(*d*) Thallus structure of *Chara*. (*Awadh, 1999*)

(*e*) Branch of *Chara* with reproductive organs. (*Rohilkhand, 1992*)

(*f*) Globule and Nucule in *Chara*. (*Rohilkhand, 1995; Bundelkhand, 1996*)

(*g*) L.S. sex organs of *Chara*. (*Kanpur, 2003*)

20. Why *Chara* is commonly called stonewort? (*U.P. College, 1998*)

21. Describe post fertilisation changes in *Chara*. (*M.D.S. Univ., 1998*)

22. Distinguish between nodal and internodal cells of *Chara*. (*Rohilkhand, 1993*)

23. What are the functions of crown cells in *Chara?* (*Bundelkhand, 1999*)

24. Mention the characteristic features of *Nitella*.

25. Compare the thallus structure of *Nitella* with that of *Chara*.

26. How does *Chara* differ from *Nitella?*

27. Comment upon the features of special interest in *Chara* and *Nitella*.

28. Describe the structure or either male of female reproductive structure of *Chara*. (*Kanpur, 2007*)

OBJECTIVE TYPE

29. Fill in the blanks :

(*i*) Stonewort is the common name for ....

(*ii*) .... is the common name for *Chara*.

(*iii*) In globule of *Chara* the centre of each shield cell extends towards node-like cylindrical cell known as ....

(*iv*) There is no .... formation in *Chara*.

(*v*) Amylum stars are densely filled with .... in *Chara*.

(*vi*) Sex organs in *Chara* are produced at ....

(*vii*) *Chara* is a .... aquatic algae.

(*viii*) The chloroplast in *Chara* lack ....

(*ix*) The common mode of vegetative reproduction in *Chara* is by ....

(*x*) In *Charales* the antheridium is known as .... and the oogonium is known as ....

30. Select the correct answer :

(*i*) The oogonium in *Chara* has five coronal cells while the number of coronal cells in *Nitella* is

(*a*) 5 (*b*) 6

(*c*) 8 (*d*) 10

(*ii*) The disc shaped chloroplast lack pyrenoids in

(*a*) *Chara*

(*b*) *Chlamydomonas*

(*c*) *Spirogyra*

(*d*) *Oedogonium*.

(*iii*) The male sex organ in *Chara* is commonly known as

(*a*) Antheridium
(*b*) Globule
(*c*) Nucule
(*d*) Zygote

(*iv*) The female sex organs in *Chara* is known as

(*a*) Antheridium
(*b*) Globule
(*c*) Nucule
(*d*) Zygote

(*v*) Reproduction by Amylum stars, and secondary protonemata occurs in

(*a*) *Chara*
(*b*) *Nitella*
(*c*) *Oedogonium*
(*d*) *Spirogyra*

(*vi*) Manubrium is found in antheridium of

(*a*) *Chara*
(*b*) *Chlamydomonas*
(*c*) *Oedogonium*
(*d*) *Spirogyra.*

(*vii*) The members of Charales are popularly called as stoneworts because

(*a*) They grow on stones.
(*b*) They have been recorded in Devonian and Silurian period.
(*c*) The thallus bears an incrustation of lime
(*d*) There are cortical species.

(*viii*) Discoid chloroplast occurs in

(*a*) *Ulothrix*
(*b*) *Chara*
(*c*) *Chlamydomonas*
(*d*) *Zygnema*

(*ix*) Asexual reproduction by zoospores is completely absent in

(*a*) *Oedogonium*
(*b*) *Cladophora*
(*c*) *Coleochaete*
(*d*) *Chara*

Light Micrograph of *Vaucheria* with Zygotes

CHAPTER 14

DIVISION : XANTHOPHYTA (Xanthophycophyta) Class : Xanthophyceae

- Class I. Xanthophyceae (Yellow-Green Algae)
- Order : Heterosiphonales (Vaucheriales)
- Family I. Botrydiaceae
- Family 2. Vaucheriaceae

The members of this division are recognised by their yellow green colour instead of grass green characteristic of Chlorophyta. Oil and fat are the usual reserve food products. Starch is not formed. The division includes 375 species and 75 genera. With a few exceptions all the members are fresh water. They are placed in a single class **Xanthophyceae**.

CLASS 1. XANTHOPHYCEAE (YELLOW-GREEN ALGAE)

General Characteristics

The plastids are yellow green. They contain chlorophyll *a* but not chlorophyll *b*, very little chlorophyll *e* and *B*-carotene. There is but one unknown xanthophyll. There is no lutein or fucoxanthein. The carotenoids are usually in excess over the chlorophylls. The chromatophores are discoid and many in each cell. The pyrenoids are absent. Starch is not formed. Oil, fat (lipid) and a glucose polymer known as **leucosin** or **chrysolaminarin** are the normal food reserves. **Leucosin** is an insoluble white substance of proteinaceous chemical composition resembling albumin (Smith, 1955). The cell wall is often absent but when present it has higher content of pectic compounds than the cell walls of green algae. Some cellulose may occasionally be present. The cell wall is silicified in a few species and appears to consist of two equal overlapping halves. The motile bodies contain more than one chromatophore and are provided with one, usually two flagella. The two flagella are of unequal length and are inserted at the anterior end. The longer flagellum is of **tinsel** or **pantonematic type**. It bears numerous fine flimmer hairs in two rows. The shorter **whiplash** or **acronematic** type has a smooth surface. Sexual reproduction is rare, if present it is isogamous. In *Vaucheria* it is oogamous. The plant body is unicellular or multicellular. In the latter case it consists of a simple filament. The class includes about 375 species. Mostly they are fresh water forms. Some are subaerial and a few terrestrial.

Classification. The primary classification of Xanthophyceae is chiefly based on thallus structure. It exhibits morphological diversity which parallels that found in the green algae but has not reached the level of morphological elaboration attained by the advanced members of Chlorophyceae. Thus the Xanthophyceae includes the motile and coccoid forms, as well as palmelloid, filamentous and siphoneous forms but parenchymatous and heterotrichous forms are conspicuous by their absence. Fritsch recognised four orders in this class. These are:

1. Heterochloridales. It includes motile forms and is analogous to the Volvocales.
2. Heterococcales (Mischococcales). It includes coccoid forms and is analogous to the Chlorococcales.
3. Heterotrichales (Tribonematales). It includes the filamentous forms and parallels the Ultorichales.
4. Heterosiphonales (Vaucheriales). It includes the siphoneous forms and is analogous to the Caulerpales (old Siphonales).

Smith added two orders more to the list raising the number to six. These are:

5. Heterocapsales (Heteroglocales). It includes the palmelloid forms and is considered analogous to the Tetrasporales.
6. Rhizochloridales. It includes the amoeboid forms and has no parallel in the Chlorophyceae.

Of the above-mentioned 6 orders we shall describe Heterosiphonales.

ORDER : HETEROSIPHONALES (VAUCHERIALES)

This order includes the coenocytic siphoneous forms and thus is analogous to the order Caulerpales (Siphonales) of green algae. The multinucleate thallus is not partitioned into cells. Such forms are more appropriately called **acellular** rather than unicellular. The tubular multinucleate thallus is of diverse forms. The order comprises two families, **Botrydiaceae** and **Vaucheriaceae**. Both are described here.

Botrydium

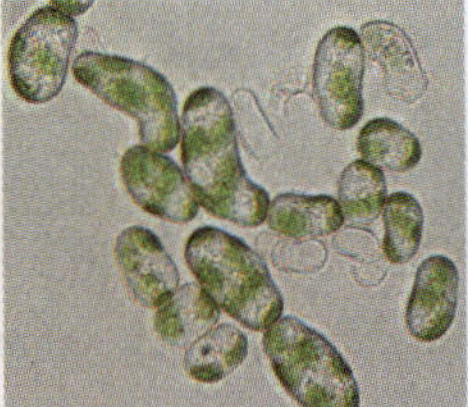
Botrydium cystosum

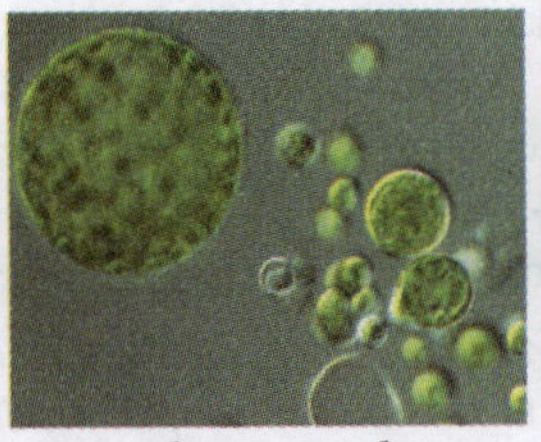
Botrydium granulatum

Botrydium botrys

FAMILY 1. BOTRYDIACEAE

The family includes the most primitive siphoneous bladder-like coenocytic forms anchored to the substratum by a rhizoidal system. Asexual reproduction takes place by biflagellate zoospores and sexual reproduction is either isogamous or anisogamous. The family includes a monotypic genus *Botrydium*.

BOTRYDIUM

Occurrence. It is taken as an example of the family Botrydiaceae. *Botrydium* is a common terrestrial alga found growing on the mud or damp soil on the bank of pools, ponds and streams. (Fig. 14.1 A). It has six species. Of these *B. granulatum* is of wide occurrence. It is found on bare damp soil or exposed mud of drying ponds and pools. Under favourable conditions it appears in countless numbers. *Botrydium* frequently grows intermingled with a green alga *Protosiphon* to which it bears a superficial resemblance. However, it differs from *Protosiphon* in its branched rhizoid, inability to divide vegetatively and many discoid chromatophores besides pigmentation, food reserves and flagellar morphology.

Thallus (Fig. 14.1). The bladder-like thallus of *Botrydium* consists of a yellow-green pear-shaped or spherical, aerial portion which is anchored to the substratum by a colourless, branched rhizoidal portion (B). The subterranean rhizoid is dichotomously branched. The tiny, balloon-like overground portion, which is 1-2 mm. in diameter, is called the **vesicle**. It has a relatively tough wall chiefly cellulose in nature. The wall is, sometimes, encrusted with carbonate of lime. Internal to the wall is a lining layer of **cytoplasm** surrounding a large, central **vacuole** (C). The cytoplasm contains numerous small scattered **nuclei** and several discoid or fusiform **chromatophores** which are distributed evenly in one or more layers within the cell membrane wall external to the nuclei. They are often connected with one another by strands of dense cytoplasm. Each chromatophore has a naked pyrenoid located in the centre (*D*). There is never any starch in the protoplast. The photosynthetic reserves accumulate as oil and fat (Lipid). Smith (1955) reported the occurrence of **leucosin**. Within the lining layer of cytoplasm is the **central vacuole**. The central sap-vacuole and the lining layer of cytoplasm are continuous throughout the plant. There are no chloroplasts in the rhizoidal portion but the latter contains numerous nuclei scattered throughout the cytoplasm. The coenocytic siphonaceous thallus-like plant body of *Botrydium* has all the essentials of a multicellular organism but it is not divided into cells. It is more appropriate to call it **acellular** rather than **unicellular**.

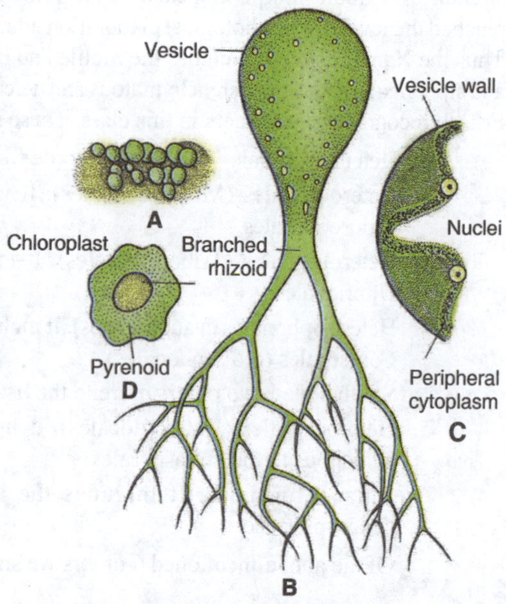

Fig. 14.1 (A-D). *Botrydium granulatum.* (A), several indivudals showing habit; (B), thallus showing differentiation into balloon shaped overground part and colourless dichotomously branched underground part; (C), portion of young vesicle; (D), chromatophore of young thallus with a pyrenoid (C-D after Korchikoff).

Besides *B. granulatum* three other important species have been reported from different parts of the world. These are *B. tuberosum*, *B. wallrothii* and *B. divisum*. *B. wallrothii* (Fig. 14.4 A) differes from *B. granulatum* (Fig. 14.1 B) in (*i*) small size of the vesicle and its rough stratified wall, (*ii*) deposition of carbonate of lime uniformly over the vesicle and (*iii*) monopodially branched underground rhizoidal system.

B. tuberosum (Fig. 14.2 A). This species was reported from India by Iyengar (1925) and from Russia by Miller (1927). The thallus consists of a small aerial vesicle attached to the substratum by a dichotomously branched rhizoidal portion. There is no deposition of lime on the vesicle. The rhizoidal branches bear small cyst-like structures at their tips. Iyengar (1925) described a new species of *Botrydium* from India. It is *B. divisum* (Fig. 14.2 B). It differs from all other species of *Botrydium* in having a branched aerial vesicle which is not covered with lime.

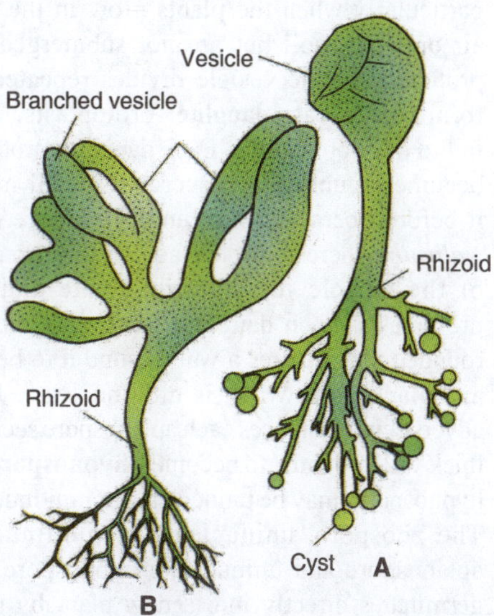

Fig. 14.2 (A-B). *Botrydium.* (A) *B. tuberosum* with rhizoidal branches bearing terminal tuber like cysts; (B), *B. divisum* with a branched vesicle (After Iyengar).

Reproduction. *Botrydium* reproduces both asexually as well as sexually. Vegetative reproduction is not known.

(*i*) Asexual Reproduction. It takes place by the formation of diverse types of spores. They may be zoospores, aplanospores or resting spores depending upon the conditions in which the alga grows.

Zoospores (Fig. 14.3). The zoospores are formed when the plants are submerged under water. The protoplast of the vesicle divides into innumerable small uninucleate parts (A). Each uninucleate daughter protoplast mctamorphoses into an ovoid or pyriform zoospore furnished with two unequal flagella inserted a little to one side of its anterior end (C). The longer flagellum is of **tinsel** (or **pantonematic**) type. It bears a double row of numerous fine lashes along its entire length giving the flagellum a feathery appearance. The shorter flagellum is of **whiplash** (or **acronematic**) type. It lacks the lashes and has a smooth surface. The zoospore has two, sometimes more chromatophores which are lateral in position. There is no eye-spot. They are liberated through an apical aperture formed by the gelatinisation of the wall of the vesicle (B). Each zoospore after the usual swarming period comes to rest, resorbs its flagella and secretes a wall around it. It then germinates by giving out a tubular colourless rhizoid at its attached end.

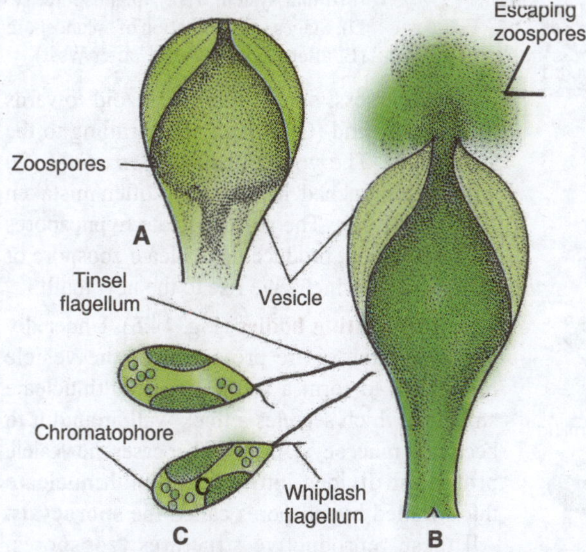

Fig. 14.3 (A-C). *Botrydium*. Zoospore formation. (A), vesicle proto-plast forming numerous zoos-pores; (B), vesicle discharging zoospores; (C), two liberated zoospores (A-B after Rostaf-inskiand, Woronin and C, after Kolkwitz).

(*ii*) Aplanospores (Fig. 14.4.). The aplanospores are formed under certain conditions particularly when the plants grow in the damp air on a wet soil but are not submerged. The protoplast of the vesicle divides repeatedly to form uninucleate daughter protoplasts. These fail to develop flagella. Each daughter protoplast becomes rounded and secretes a wall around it before liberation. It is an aplanospore. In *B. wallrothii* there is a cleavage of the protoplast of the vesicle into multinucleate daughter protoplasts. Each daughter protoplast becomes rounded and secretes a wall around it to become an aplanospore which is multinucleate. Under adverse circumstances each aplanospore secretes a thick wall around it to become a **hypnospore**. The hypnospores may be uninucleate or multinucleate. The zoospore, uninucleate or multinucleate aplanospore and uninucleate hypnospore, each germinates directly into a new plant by giving out a colourless, simple tubular rhizoid towards the attached end (C). It fixes the germling to the substratum. The young plant of *Botrydium* (D) with an unbranched hizoid is thus often mistaken for *Protosiphon*. The multinucleate hypnospores on germination, produce uninucleate zoospore or aplanospores which give rise to the new thalli.

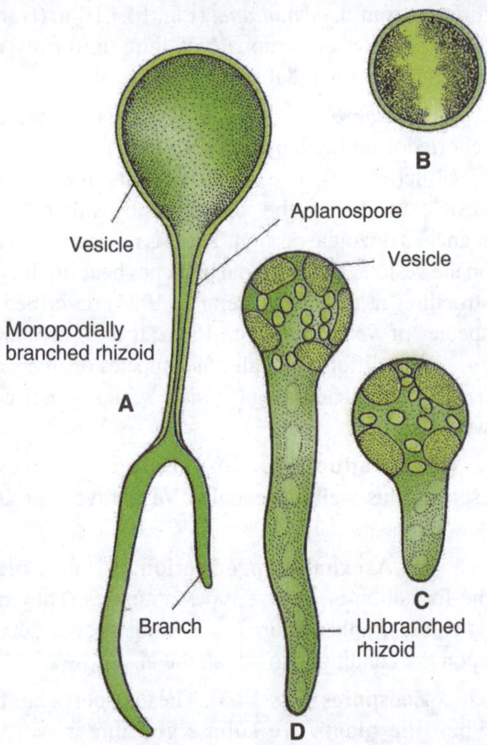

Fig. 14.4 (A-D). *Botrydium*. (A), A thallus of *B. wallrothi* with a mono-podially branched rhizoidal system; (B), Aplanospore; (C-D), Stages in germination of aplanospore; (B, after Korschoikoff, C after West)

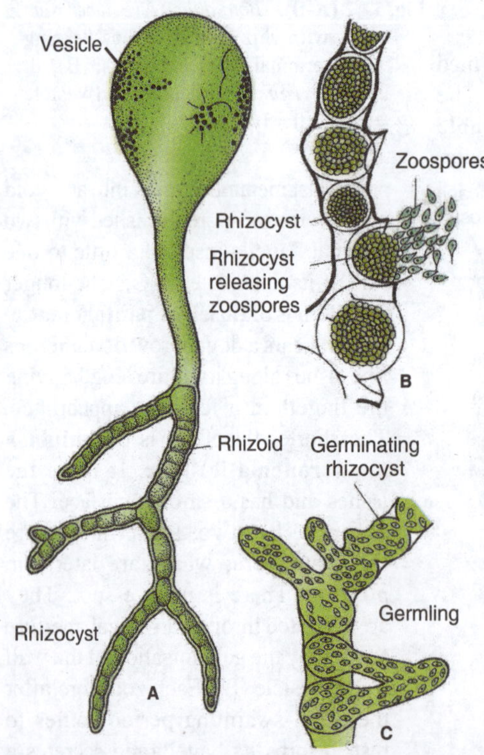

Fig. 14.5 (A-C). *Botrydium*. (A), formation of rhizocysts in *B. granulatum;* (B), rhizocyst releasing zoos-pores on germination; (C), rhizocysts directly germinating into new plants (A-C after Rostafiniski and Woronin).

(*iii*) Resting bodies (Fig. 14.5). Under dry conditions the entire protoplast of the vesicle rounds off to form a single large multinucleate structure which secretes a thick wall around it to become a **macrocyst**. In still other cases the vesicle protoplast divides into several, multinucleate thick-walled hypnospores called the **sporocysts**. All these reproductive structures (zoospores, aplanospores, hypnospores, macrocysts and sporocysts) are formed in the overground vesicle of the plant. Sometimes particularly during periods of scarcity of water the protoplast of the vesicle in *B. granulatum* passes down into the rhizoidal portion of the plant. There it divides to form several, thick-walled multinucleate resting cysts called the **rhizocysts** (A). They are globose or ellipsoidal in form and are serially arranged in the rhizoids. With the return of condition favourable for growth the protoplast of the rhizocyst divides to form uninucleate zoospores (B) or aplanospores each of which gives rise to a new plant. Sometimes the rhizocyst directly germinates into a new plant (C).

The protoplast of the aerial vesicle in *B. tuberosum* migrates into the rhizoidal portion. There it collects at the tips of the rhizoids which get inflated and swollen into round, thick-walled multinucleate hypnospores or cysts often erroneously called the **tubers** (Fig. 14.6).

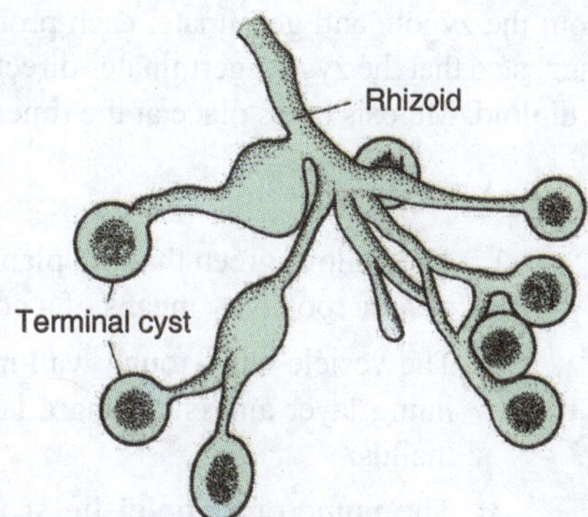

Fig. 14.6. *Botrydium tuberosum.* Rhizoidal branches bearing terminal cysts singly (after Iyengar).

Sexual Reproduction (Fig. 14.7). *B. granulatum* is monoecious (homothallic) and sexual reproduction is isogamous. Some species are reported to be dioecious (heterothallic) and may be anisogamous (Moewus, 1940). The gametes are formed during the rainy season. The protoplast of the vesicle divides into uninucleate parts each of which metamorphoses into a biflagellate gamete. The gametes are ob-pyriform to broadly ellipsoidal in form (A). Each gamete has 1 to 3 or 4 chromatophores. The eye-spot may be present or lacking. The two unequal flagella are attached to the broader anterior end. The longer flagellum is of tinsel-type and shorter is of whiplash type. The gametes may escape through an apical aperture formed by gelatinisation of the vesicle apex. Usually isogametes conjugate before they are set free. They meet and fuse by their non-flagellate posterior ends (B, Rosenberg, 1930). The gametes which fail to fuse may develop parthenogenetically. The spherical zygote secretes a wall around it (E) and germinates. It undergoes no resting period. The division of the zygote nucleus is meiotic. According to Moewus (1940) the heterothallic species produce the gametes which are of unequal size (anisogametes) and are liberated through an apical pore formed by the gelatinisation of the vesicle apex. The liberated gametes become opposed in pairs at their anterior ends and then fuse laterally.

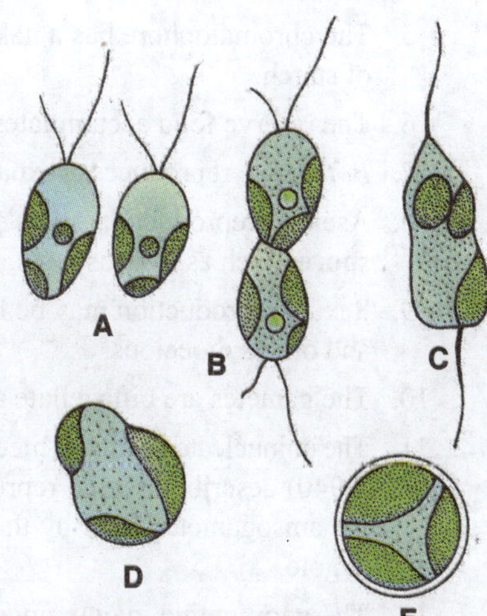

Fig. 14.7 (A-E). *Botrydium.* Sexual reproduction. (A), isogametes; (B-D), stages in gametic union; (E), zygote (C-E after Rosenberg.)

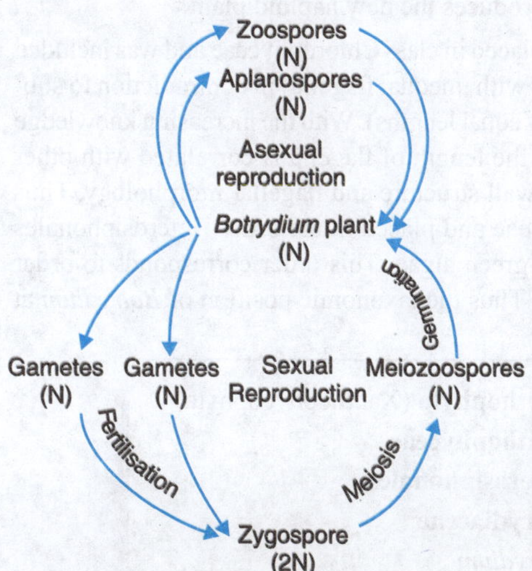

Fig. 14.8. Graphic representation of the life cycle of *Botrydium.*

Formation of Meiospores. The diploid nucleus of the spherical zygote undergoes meiosis. The protoplast divides to form 4 or 8 motile meiospores which are biflagellate. The meiozoo-spores escape from the zygote and germinate, each producing a new plant. On the other hand Rosenberg (1930) suggested that the zygote germinates directly and by mitosis produces a new *Botrydium* thallus which is diploid. Meiosis takes place at the time of differentiation of gametes.

Salient Features

1. The yellow, green thalloid plant body consists of a small, green, pear-shaped overground **vesicle** rooted by means of a colourless dichotomously branched **rhizoidal** portion.
2. The vesicle has a tough wall mainly of **cellulose**. Within it is the **cytoplasm** forming a lining layer and enclosing a large central **vacuole**. Both are continuous throughout the thallus.
3. The numerous **nuclei** lie scattered throughout the cytoplasm. The several discoid **chloroplasts** are confined only to the cytoplasm of the vesicular portion.
4. In addition to chlorophyll *a* and little of chlorophyll *c,* the chromatophores contain excess of b-carotene and one xanthophyll.
5. The chromatophore has a naked pyrenoid located in the centre but there is no formation of starch.
6. The reserve food accumulates in the form of oil, fats (lipid) and **leucosin**.
7. *Botrydium* reproduces asexually. There is no vegetative reproduction.
8. Asexual reproduction takes place by means of **zoospores, aplanospores**, and **resting spores** such as hypnospores, macrocyst, and rhizocysts.
9. Sexual reproduction may be **isogamous** or **anisogamous.** Some species are **monoecious** and others dioecious.
10. The gametes are biflagellate and the flagella are of **unequal** lengths.
11. The uninucleated gametes meet and fuse by their posterior ends (Rosenberg, 1930). Moewus (1940) described sexual reproduction in an anisogamous species of *Botrydium* in which the anisogametes meet by their anterior ends and then fuse laterally to form a spherical zygote.
12. The germinating zygote undergoes meiosis to form 4-8 motile, biflagellate **meiospores** called the **meiozoospores**, each of which produces the new haploid plant.

Taxonomic Position. *Botrydium* was formerly placed in class Chlorophyceae and was included in the sub-group 'Heterokontae' (meaning motile cells with unequal flagella) in contradiction to sub-group "Isokontae" (meaning motile cells with flagella of equal lengths). With the increasing knowledge about its structure it was found that the difference in the length of flagella is correlated with other differences such as pigmentation, food reserves, cell wall structure and flagellar morphology. Thus the genus was shifted to a separate class Xanthophyceae and placed in the order Heterosiphonales which includes the multinucleate siphoneous yellow-green algae. This order corresponds to order Caulerpales (old Siphonales) of class Chlorophyceae. Thus the taxonomic position of *Botrydium* at present is :

Division	:	**Xanthophyta** (Xanthophycophyta)
Class	:	**Xanthophyceae**
Order	:	**Heterosiphonales**
Family	:	**Botrydiaceae**
Genus	:	*Botrydium*
Species	:	*granulatum*

FAMILY 2. VAUCHERIACEAE

General Features. The thallus consists of a branched, aseptate, coenocytic filament with apical growth. Usually it is attached to the substratum by means of branched rhizoids. The numerous chloroplasts are oval or elliptical without pyrenoids. Typically the reserve food is oil or fat (lipid). Asexual reproduction takes place by the formation of multiflagellate zoospores, aplanospores or akinetes. Sexual reproduction is oogamous. The family includes a single genus *Vaucheria*, which is described here in detail.

Vaucheria

VAUCHERIA (DECANDOLE)

Occurrence. The genus comprises more than 40 species most of which occur in abundance in the temperate regions. Of these nine have been reported from India. The majority are terrestrial (adapted to a life on land) or fresh water aquatics. A few species are marine (*V. piloboloides*). The terrestrial forms occur during winter months frequently on damp soil and mud-flats exposed on drying up of ponds and puddles. The aquatic forms are found in very shallow water of ponds and ditches or near the bank of slow flowing streams. The terrestrial species form extensive yellowish or deep green, dense felt-like covering or mat on wet soil and in flower pots in green houses. A portion of the covering, under the microscope, is seen to consist of tangled or inter-woven threads which are rather coarse, sparingly branched, tubular filaments. The two widely distributed species are *V. sessilis* and *V. geminata*. They are common during the winter months from December to February in N. India. *Vaucheria* aquatic is unique. *V. sessilis* occurs both on land and in water. *V. amphibia* is amphibious.

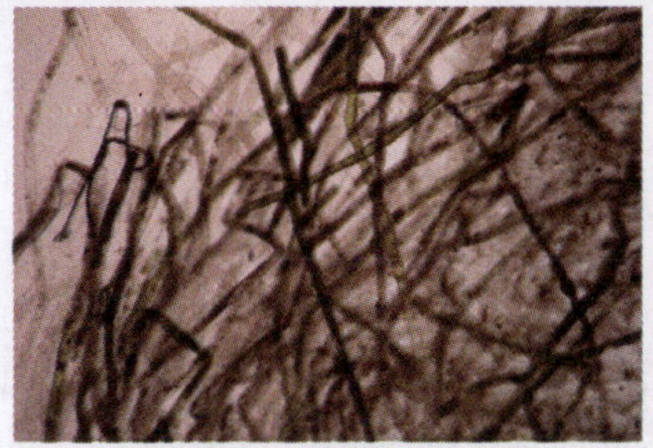

Vaucheria sessilis

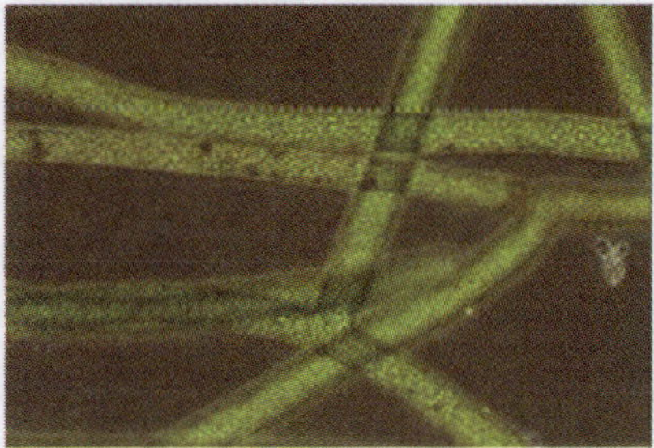

Vaucheria litorea

Thallus (Fig. 14.9). The thallus is composed of yellowish green, cylindrical or tubular rather coarse filaments branched at irregular intervals. In the terrestrial species the thallus is often attached to the substratum by small tufts of colourless rhizoids or a lobed hapteron. The aerial, erect green filaments exhibit monopodial branching and apical growth. The filaments in some species branch sparingly and in others profusely. The branching is irregular. Cross walls are not present to separate the cells of the filament and the branches. Thus the protoplasm is continuous along its entire length and extends without a break into the branches. Septa, however, appear in connection with the formation of reproductive structure or sealing of an injury.

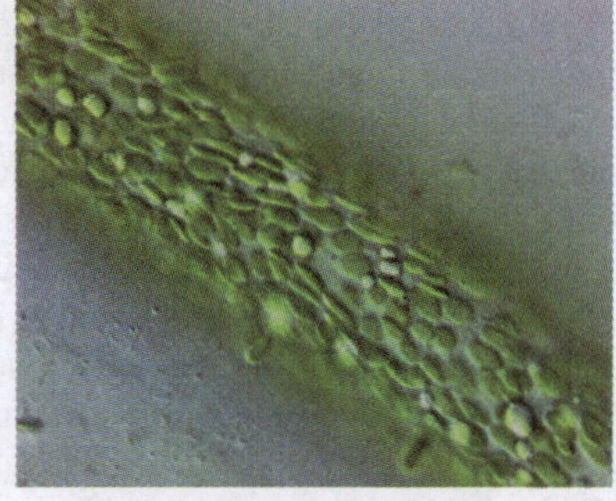

Siphonaceous Coenocytic Filaments of *Vaucheria*

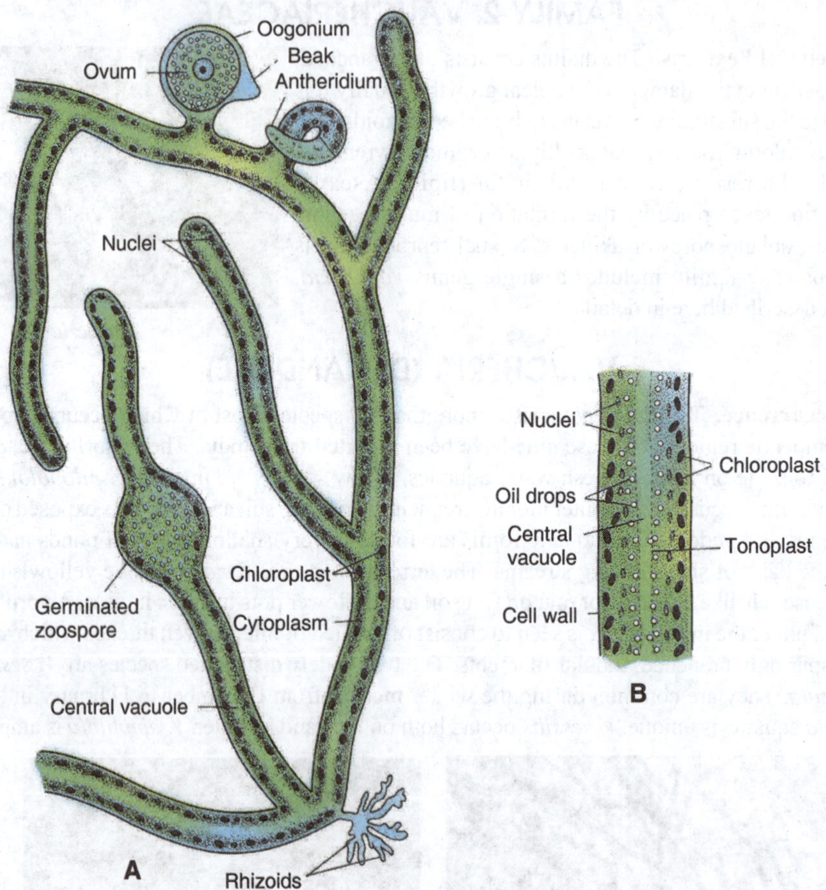

Fig. 14.9 (A-B). *Vaucheria sp.* (A), Filament with rhizoids; (B). Part of vegetative filament showing arrangement of nuclei and chloroplasts.

Structure (Fig. 14.9). The filament wall is relatively thin and weak. It lacks elasticity. It consists of an inner layer of cellulose and an outer of pectic substances. The native or mercerized cellulose which is characteristic of the cell wall of green algae is absent in *Vaucheria.* Just within (close to) the wall is a thin or thick continuous layer of **cytoplasm** which contains the usual membrane bound organelles such as the mitochondria, small vesicles and others. It contains numerous, very small oval, circular or elliptical chromatophores, arranged in an outer layer. Internal to the chromatophores are innumerable, minute **nuclei** forming an inner layer. Internal chromatophores contain all the pigments characteristic of the Xanthophyceae. These are chlorophyll *a,* chlorophyll *e* and more than the normal amount of the carotenoid pigments and contain one unknown xanthophyll. The chlorophyll *b* typical of the green algae is lacking. The two xanthophylls, **siphonein** and **siphonoxanthin** characteristic of the Caulerpales (Old Siphonales) are absent. There are no **pyrenoids**. Christensen (1952) however reported the presence of a pyrenoid in each fusiform chromatophore in *V. medusa.* Oil rather than starch is the principal reserve food. It accumulates in the form of countless drops in the cytoplasm. Tiffany (1924) opined that the photosynthetic product in *Vaucheria* can be changed to starch under continuous illumination. A large **central vacuole** filled with sap occupies the centre of the filament. We thus notice that *Vaucheria* possesses all the essentials of cells structure. The cytoplasm and nuclei, however, are not partitioned into distinct cells. Such a condition is known as **Coenocytic** and such filaments whose protoplasm is continuous, multinuclear and not partitioned into separate protoplasts are called **coenocytes**. The absence of cross walls makes the thallus vulnerable. From the point of infection a parasitic fungus may spread readily throughout the plant body. Moreover, a slight injury may cause the emptying of the entire thallus. To guard against such an emergency a septum is formed to localise the injured part and to save the rest of the filament. In addition a siphoneous thallus suffers from mechanical disadvantages as compared with a thallus divided into cells.

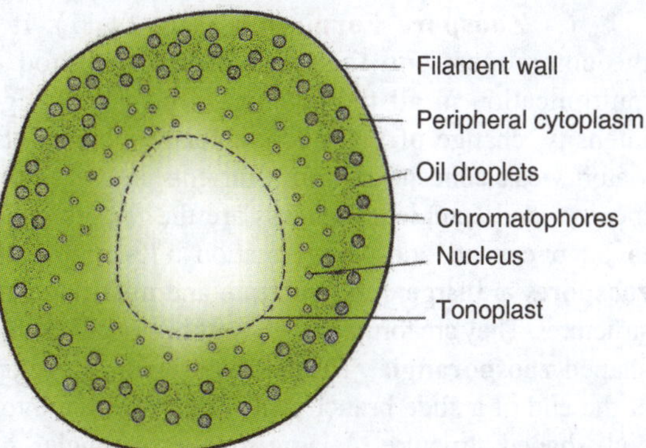

Fig. 14.10. *Vaucheria*. C.S. Thallus showing relative position of the contents (diagrammatic).

Cytoplasmic Streaming. Fischer Arnold (1963) found that the chloroplasts of *Vaucheria* move accompanied by small vesicles and mitochondria with the streaming of cytoplasm along the longitudinal axis of the filament in the dark. Otto and Brown (1974) reported the occurrence of cytoplasmic fibrils in the filaments of *V. litorea* but they failed to notice any directional chloroplast movement associated with the fibrils. They on the contrary described the chloroplasts as being in random motion. Thus the mechanism of streaming in *Vaucheria* remained a mystery. Blatt and Briggs (1978) observed longitudinally oriented cytoplasmic fibrils on the cortical region of the cytoplasm in *V. sessilis*. They found that the fibrils are associated with the chloroplasts, mitochondria and the vesicles. They suggested that these fibrils appear to guide the streaming of these organelles. In 1980 Blatt *et al* (1980) with the help of electron transmission microscope discovered that these fibrils appear to be composed of bundles of F-actin microfilaments and cytoplasmic myosin linked with streaming organelles. The active shearing force responsible for cytoplasmic streaming in *Vaucheria* results from the interaction between organelle linked myosin and F-actin microfilaments.

Blatt and Briggs (1980) further discovered that the streaming organelles aggregate in the region of the filament of *V. sessilis* illuminated with low intensity blue light. Organelle aggregation is accompanied by the formation of cortical fibre reticulation in light. In fact the blue light induces cortical fibre reticulation in which the organelles become trapped as they stream into the blue light. The photoreceptor is located at the plasma membrane (Blatt *et al* 1981).

Is *Vaucheria* a multicellular or a unicellular form? It is incorrect to describe *Vaucheria* as a unicellular alga. It is a coenocyte. It possesses all the essentials of a multicellular organism but the cytoplasm and the numerous nuclei are not partitioned into distinct cells. The septa remain suppressed in the vegetative condition. Their appearance in connection with the formation of reproductive organs or when the filament is injured confirms the view. In *Vaucheria pseudohamata* the filaments are often septate. Moreover in unicellular forms growth consists in increase in size of the entire cell. In *Vaucheria* growth is apical where the protoplasm is colourless and transparent. It is thus more appropriate to call *Vaucheria*, in general as an **acellular**, coenocytic organism rather than a unicellular or multicellular form.

Reproduction. *Vaucheria* reproduces by all the methods; *vegetative, asexual* and *sexual.* Vegetative reproduction is secured through **fragmentation** in which the thallus accidentally breaks up into short segments, each of which becomes thick-walled. It is, however, not common.

1. Asexual Reproduction. It takes place in a variety of ways depending upon the habitat in which the alga lives in. In the aquatic species it occurs by the formation of **zoospores**. The terrestrial species form zoospores only when flooded.

(*i*) **Zoospore Formation** (Fig. 14.11). It is the commonest and the most prolific method of multiplication in all the aquatic species. Low light intensity, change of the medium for running to still water, weak concentration of nutrients in the culture medium and complete darkness are the factors which favour or enhance zoospore formation in *Vaucheria*. The **zoospores** are large, multinucleate and multiflagellate structures. They are formed singly within elongated club-shaped **zoosporangia**. The zoosporangium develops at the end of a slide branch which gets swollen into a club-shaped structure. A large number of nuclei and chloroplasts along with the cytoplasm stream into and accumulate in the swollen tip before it is separated from the rest of the filament by a transverse septum (A). Consequently the vacuole in the region destined to form the zoosporangium diminishes in size. As a result the zoosporangium appears deep green. At this stage the nuclei in the young zoosporangium come to lie external to the chromatophores in the colourless cytoplasm at the periphery within the plasma membrane. The entire protoplast of the zoosporangium then contracts to form an oval multinucleate mass, the incipient zoospore. The mature zoospore escapes through a narrow aperture which is formed by the gelatinisation of the wall at the distal end of the zoosporangium (B). The zoospores are generally liberated early in the morning. Pairs of centrioles appear *de novo* in the colourless peripheral cytoplasm just within the plasma membranes. Each pair becomes associated with a nucleus. A pair of flagella located in the vesicle then emerges from each pair of centrioles opposite each nucleus. Eventually the flagellate vesicles move to the periphery of the incipient zoospore and fuse with the plasma membrane. In this way numerous pairs of flagella becomes situated on the surface of zoospore cytoplasm opposite the nuclei (Ott and Brown 1974 B).

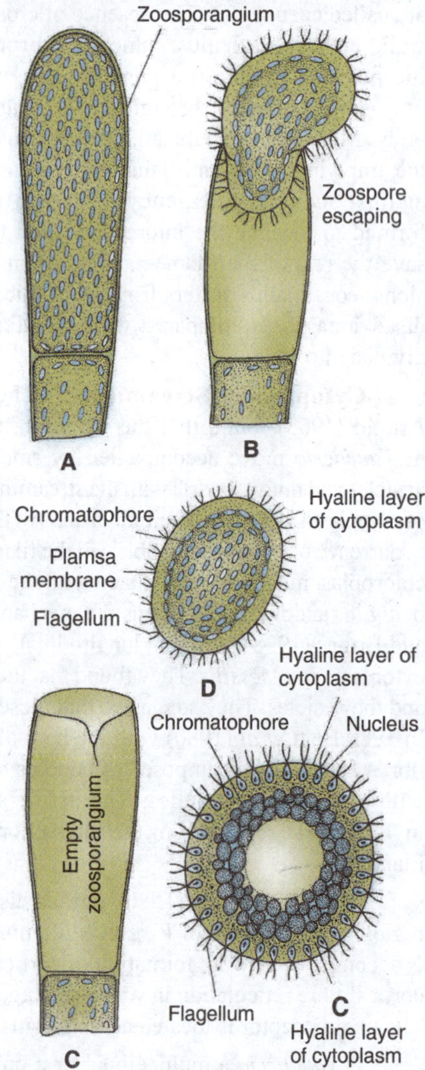

Fig. 14.11 (A-E). *Vaucheria*. zoospores formation. (A) zoosporangium; (B) zoospore escaping; (C) empty zoosporangium; (D) liberated zoospore; (E) cross section of zoos-pore showing structure.

Structure of Zoospore. The liberated zoospore is a large, yellow-green, ovoid structure (D). It has a superficial hyaline layer of cytoplasm containing numerous nuclei. Internal to this are the numerous chromatophores embedded in the cytoplasm which also contains the contractile vacuoles. In the centre is the sap vacuole which may be traversed by strands of cytoplasm (E). The surface of the zoospore is covered with numerous flagella, a pair opposite to the pointed end of each pear-shaped nucleus. At the base of each pair of flagella are two-blepharoplasts (centrioles). According to Dangeard (1946) the flagella in each pair are of equal lengths. Koch (1941) found slight difference in the length of the flagella in the pair but stated that they are of whiplash type. Greenwood, Manton and Clarke (1957) reported that the flagella of a pair are typically heterokont. The shorter flagellum of each pair is directed towards the front end of the zoospore.

Morphology of the zoospore. The multinucleate and multiflagellate zoospores are peculiar to *Vaucheria*. All other genera related to it produce small, biflagellate uninucleate zoospores developed in large number in each zoosporangium. It is possible that the zoospore of *Vaucheria* is a compound structure formed as a result of the failure of the protoplast within the zoosporangium to divide into uninucleate, biflagellate zoospores. According to this view it may be more appropriate to term it a **synzoospore**. This interpretation is supported by two facts, (*i*) presence of the central vacuole of the parent zoosporangium and (*ii*) paired dispositions of the flagella opposite the nuclei.

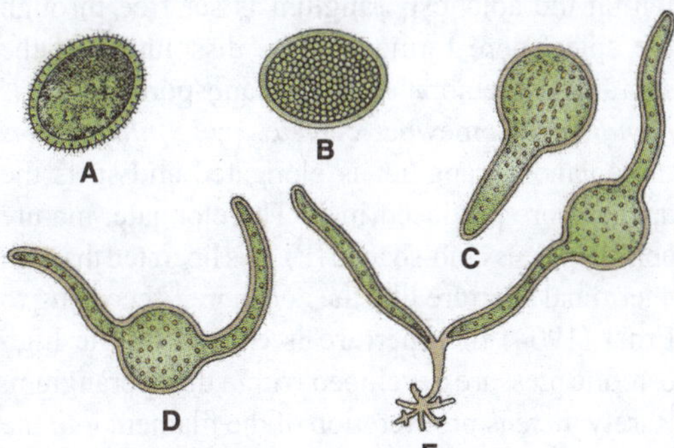

Fig. 14.12 (A-B). *Vaucheria sp.* Stages in the germination of a zoospore.

Germination of the zoospore (Fig. 14.12). After a short period (5-15 minutes) of sluggish motility, the zoospore comes to rest. Thuret (1842) observed that when the zoospores finally comes to rest, the flagella first become motionless and then vanish completely. They are resorbed. The quiescent zoospore rounds off and becomes invested with a thin, cellulose wall (B). At this stage the chromatophores move outwards and the nuclei inwards. The quiescent zoospore then elongates in one (C) or two opposite directions, in the form of tubular outgrowths (D). One of these undergoes branching to form the colourless, lobed holdfast (E) while the other continues to grow indefinitely to produce the yellowish green, tubular filament.

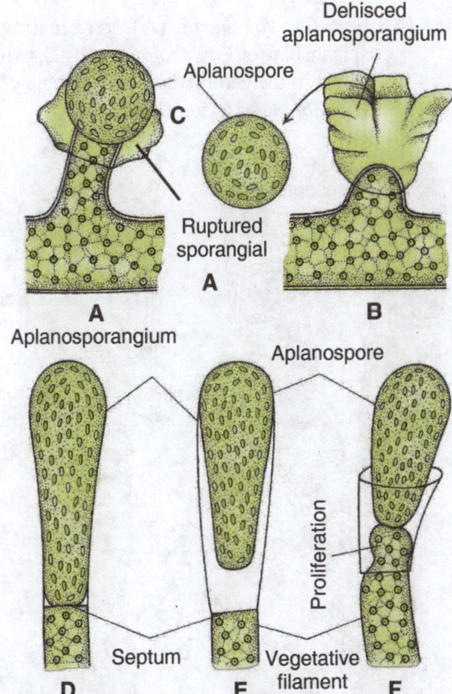

Fig. 14.13 (A-F) *Vaucheria.* Aplanospore formation. (A), aplanospore of *V. uncinata* surrounded by the irregularly ruptured sporangial wall; (B), empty aplanosporangium; (C), liberated aplanospore; (D), club-shaped aplanosporangium of *V. piloboloides*; (E), liberation of aplanospore through the apical pore; (F), proliferation of the filament into the dehisced sporangium pushing the aplanospore in front (D-F after Ernst).

(*ii*) Aplanospores (Fig. 14.13). These are non-motile asexual spores produced normally by the terrestrial species. The aquatic species produce them when the plants are exposed to drought or transferred from light to darkness or from running to still water. The species of *Vaucheria* which commonly reproduce by this method are *V. uncinata, V. hamata* and *V. piloboloides*. The last is marine. In some species (*V. uncinata* and *V. geminata*) the aplanospores are more or less rounded (C). In others (*V. piloboloides*) they are elongate (F). The aplanospores are developed at the ends of short laterals (A) or terminal branches (D). The cells in which they are produced are called the **aplanosporangia**.

The terminal aplanosporangium is cut off by a septum from the branch. It is rounded in shape in *V. uncinata* (A). The protoplast of the aplanosporangium is converted into a single rounded, thin walled **aplanospore** which is liberated by the irregular rupture of the sporangial wall. The aplanospores simply drops out of it (C). The aplanosporangia in *V. geminata* are ovoid. The single aplanospore produced within the aplanosporangium is set free through the apical pore formed by the dissolution of the sporangial membrane. The aplanosporangia in *V. hamata* are somewhat clavate. In *V. piloboloides* the aplanosporangium is elongated and so is the aplanospore produced in it. The elongate, mature aplanospore is club-shaped (F). It is liberated through a terminal aperture like the zoospore. According to Ernst (1904) the aperture is caused by the high osmotic pressure developed within the sporangium. Rarely there is proliferation of the filament into the dehisced sporangium pushing the aplanospore in front of it (F). Smith reported that the protoplast of the aplanosporangium in some aquatic species may sometimes divide into numerous small, thin-walled aplanospores called the **micro-aplanospores**. The aplanospore or micro-aplanospore, after liberation, germinates to give rise to a new thallus. Sometimes it germinates *in situ* (*V. piloboloides*). Very rarely the aplanosporangium directly grows to form a new thallus (Fig. 14.14 B-C).

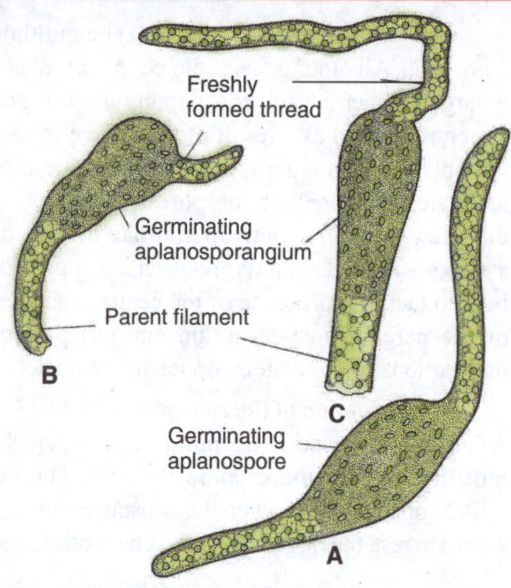

Fig. 14.14 (A-C). *Vaucheria.* (A), germinating aplanospore of *V. piloboloides*; (B-C), two stages in the germination of sporangia of *V. piloboloides* (After Ernst).

(*iii*) Akinetes (Fig. 14.15). In some aquatic and terrestrial species when exported to greater desiccation or low temperature, the branched filament divides into rows of short segments by thick, gelatinous cross walls. Their protoplasts become laden with oil. These resting multinucleate thick-walled segments are called the **cysts, hypnospores** or **akinetes**. For a time the cysts in the chain may remain connected by the parent membrane of the filament, giving it the appearance of another alga *Gongrosira*. This state of *Vaucheria* has thus come to be known as **gongrosira stage** (A). With the return of conditions favourable for growth each cyst may directly grow into a new filament (B). Sometimes the protoplast of the cyst divides into small protoplasmic bits which are liberated as small amoebae through a pore in the cyst wall (C). Each liberated amoeba-like structure (D) moves about for some time in an amoeboid manner (D). Eventually it comes to rest and assumes a spherical form. It then secretes a cell wall around it and develops into a new filament.

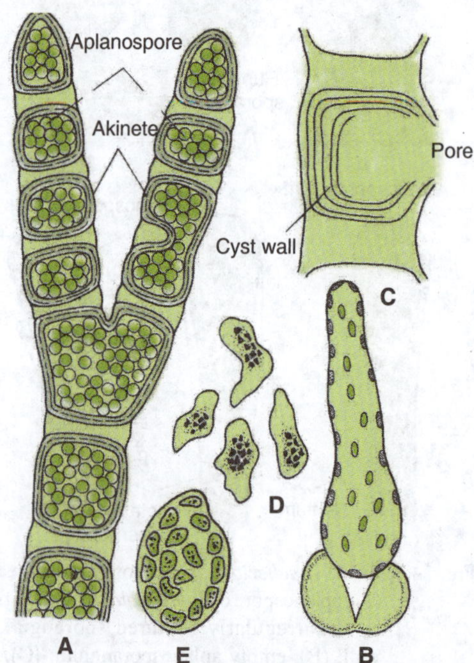

Fig. 14.15 (A-E). *Vaucheria.* (A), Gongrosira stage; (B), direct germination of cyst; (C), thick-walled cyst germinates to liberate amoeba-like bodies (D) through a pore in the cyst wall; (E) thin-walled cyst (After Stahl).

In *V. geminata* the cell walls of the filaments develop thickenings of mucilaginous material at particular places prior to akinete formation. These thickenings grow centripetally to form the dividing septa. In this way chains of akinetes are formed.

2. Sexual Reproduction. All species of *Vaucheria* reproduce sexually. Sexual reproduction is **oogamous**. It is of rare occurrence in plants growing in flowing water. Most of the species are **monoecious** (homothallic A). They are exclusively fresh water or terrestrial forms. *V. dichotoma, V. mayganadensis* and *V. litotrea* which are marine are dioecious (heterothallic). The male and female gametes differ greatly in size, form and structure. They are produced in distinct rather specialized sex organs. The sex organs are not merely modified vegetative cells as in the algae previously studied, but are specialised for gamete production. The male sex organ is called the antheridium and the female oogonium.

(*a*) *Position of sex organs* (Fig. 14.16). In the monoecious (homothallic) species the *antheridia* and *oogonia* usually occur close to one another at intervals along the same filament arising at lateral outgrowths. *V. sessilis* (A) and *V. aversa* are the common examples. The sex organs may be sessile or stalked. *V. geminata* and *V. terrestris* have the sex organs developed on special side branches. The side branch consists of a pedicel either with a terminal antheridium (C) and a number of lateral

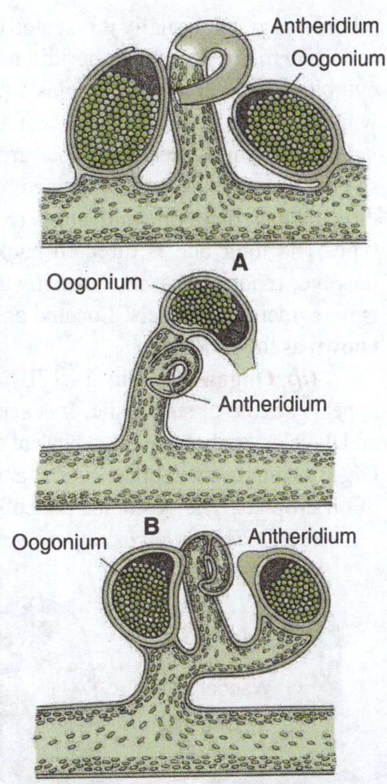

Fig. 14.16 (A-C). *Vaucheria.* Position of sex organs.

oogonia (*V. hamata*) or a terminal oogonium with an antheridium seated laterally below it (B). In a few species such as *V. sessilis*, the antheridium is situated high up on a stalk or a pedicel. It is separated by a cross wall from the stalk (A). The oogonium is sessile or very shortly stalked and rests on the filament. There is thus a great variation in the arrangement of sex organs in the different species and even in different individuals of the same species.

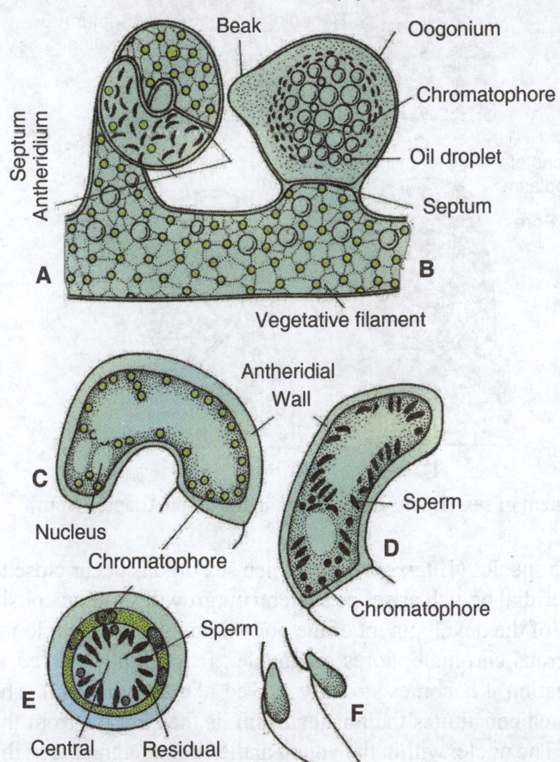

Fig. 14.17 (A-F). *Vaucheria.* (A), Structure of sex organs and differentiation of sperms (C-E); (F), liberated sperms.

(*b*) *Structure of sex organs* :

(*i*) Antheridia (Fig. 14.17 (A). Typically the mature antheridium is a cylindrical, slender, tubular structure. It may be strongly curved like a horn or hooked. In *V. aversa* it is straight and not hooked. The mature antheridium in *V. Thureti, V. piloboloides* and others is reported to be colourless but in some others it is green. Usually it is seated high up on a branch and is separated from the latter by a septum. In *V. aversa* the straight antheridium is sessile. It rests on the main filament. When young it contains cytoplasm, nuclei and chloroplasts (A). At maturity it contains numerous male gametes or sperms which in some species (*V. aversa*) are liberated through a single apical aperture and in others (*V. debaryana*) through several apertures. The liberated sperm is an extremely minute, oval, spindle-shaped (*V. sessilis*) or pear-shaped, colourless structure with two laterally inserted and oppositely directed flagella of unequal length (F). The short one points forwards and is of tinsel (pantonematic) type. The long one is directed backwards and is of whiplash (acronematic) type. Koch (1951) however reported that in *V. sessilis* the short anterior flagellum is thick with irregular edges. There is no evidence of tinsels. Located at the anterior end of the sperm cell is an emergent process. It is known as the *proboscis*.

***(ii)* Oogonium** (Fig. 14.17B). The oogonium is a spherical or ovoid structure which usually appears sessile or subsessile. It is separated from the supporting filament by a cross wall at its base and develops a short rounded beak at the tip towards maturity. The mature oogonium contains a single large nucleus located in the centre, and numerous chromatophores. Reserve food is stored in the form of oil droplets. The protoplasmic contents round off to form a single large egg cell or ovum.

(*c*) *Development of sex organs* (Fig. 14.18)

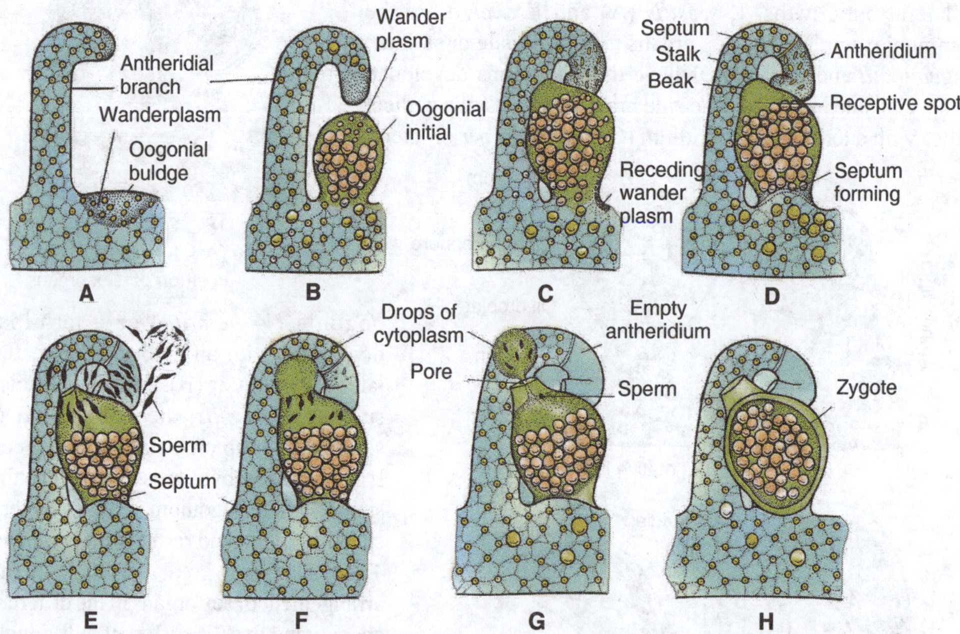

Fig. 14.18 (A-H). *Vaucheria sessilis.* Development of sex organs. Explanation in the text. (Adapted from Couch).

***(i)* Antheridium.** In the monoecious species (*V. terrestris*) in which sex organs occur close to one another on the same filament, the antheridial branch arises as a lateral outgrowth simultaneously with or slightly before the commencement of the development of the oogonium (A). At first it looks like a vegetative branch containing numerous chromatophores and nuclei. It is slightly curved at its tip (A). With further growth and elongation it becomes strongly curved like a horn (B-C). The terminal curved portion of the branch, which constitutes the **antheridium**, is then cut off from the rest of the branch by a transverse wall (D). The nuclei within the young antheridium aggregate in the central region and divide. Finally each of these becomes surrounded by a small mass of cytoplasm. Each uninucleated part becomes metamorphosed into a minute yellowish **sperm**. The residual cytoplasm containing the chromatophores migrates to the periphery in contact with the wall. The spindle-shaped sperms in *V. sessilis* become arranged radially between the central vacuole and the residual peripheral cytoplasm. The mature sperms are liberated shortly before day break through a single terminal aperture in the wall of the antheridium (E) and in some species through more than one aperture (*V. debaryana*). In *V. glomerata* the antheridium is reported to bear two to eight lateral projections each of which opens through a terminal pore.

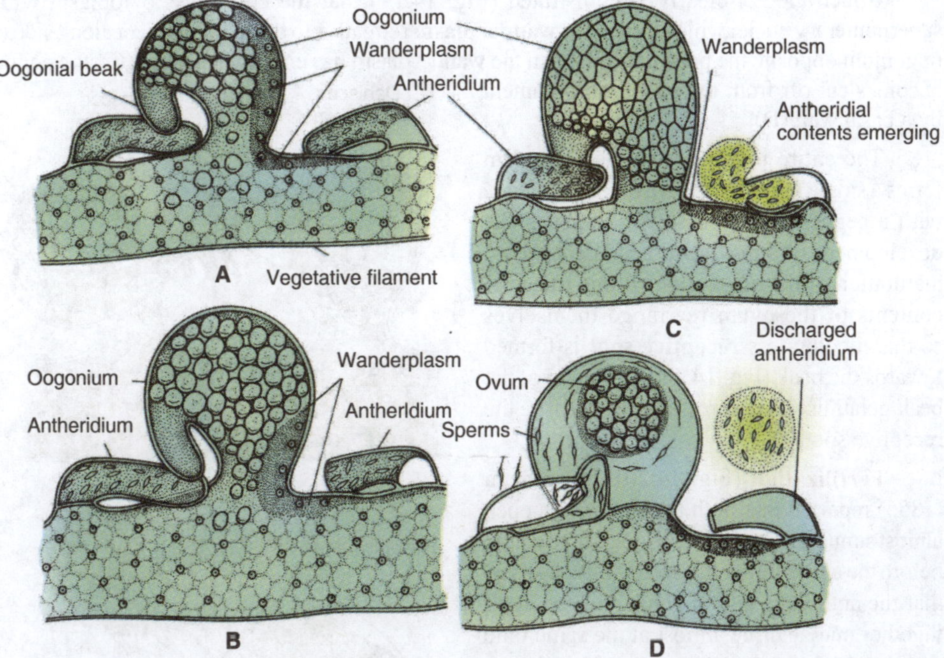

Fig. 14.19 (A-D). *Vaucheria aversa.* Diagrams showing the fate of wanderplasm; (A), wanderplasm started receding from the oogonium and sperms forming in the antheridium; (B), much of the wanderplasm out of the oogonium; (C), wanderplasm completely out of the oogonium, septum at the base of the oogonium forming and contents of antheridium passing out; (D), oogonium containing an ovum ready for fertilization and surrounded by sperms (Adapted from Couch).

***(ii)* Oogonium** (Fig. 14.18). Just near the base of an antheridial branch accumulates a colourless mass of cytoplasm. It has many nuclei but no chromatophores (A). Couch (1932) termed this colourless, heaped multinucleate mass of cytoplasm as "wanderplasm". It is the place from which the oogonium arises as a lateral outgrowth. At the wanderplasm the filament bulges out to form the **oogonium** initial. Several nuclei and chromatophores migrate into the latter. The oogonium initial increases in size and becomes rounded or ovoid in shape with the wanderplasm occupying its distal growing region (B). It is the young multinucleate **oogonium**. Towards maturity it becomes cut off from the main branch by a septum near the base (E). It is densely filled with oil and chromatophores but has a single nucleus. How it happens is not definitely known. Three hypotheses have been put forth to account for the uninucleate condition of the mature oogonium. These are :

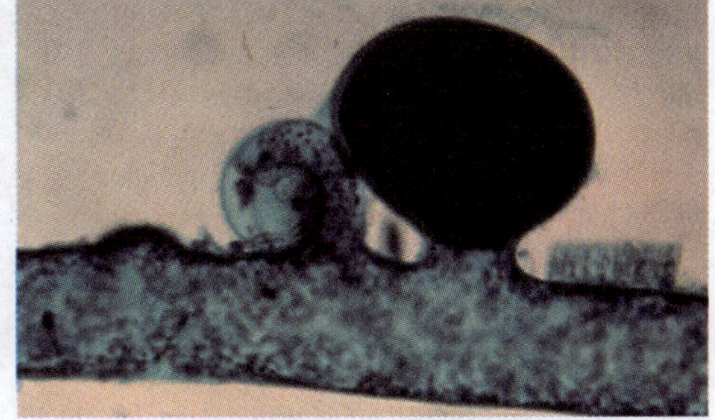

Vaucheria Antheridia and Oogonia

(*i*) The numerous nuclei in the oogonium fuse to form a single nucleus (Brehens, 1890).

(*ii*) All but one of the nuclei migrate back to the vegetative filament before the formation of the transverse septum (Oltmanns, 1895).

(*iii*) Degeneration of all the nuclei (supernumerary nuclei) but one in the oogonium (Davis 1904).

Couch (1932) clearly demonstrated (Fig. 14.19) that the colourless cytoplasm with the supernumerary nuclei which he termed **wanderplasm** retreats into the main filament along side of the oogonium opposite the beak (A-B). When the wanderplasm has entirely emerged (C), the oogonium becomes cut off from the supporting filament by a basal wall (D).

The entire protoplast of the oogonium forms a single large spherical *ovum* or *oosphere* with a centrally located single nucleus. The development of a one-sided beak denotes the maturation of the oogonium. At this stage the contents of the ovum rearrange themselves so that a colourless **receptive spot** is formed towards the beak (Fig. 14.18 C). The tip of the beak gelatinises and forms a pore opposite the receptive spot (D).

Fertilization (Fig. 14.20). Pringsheim (1855) reported that both the sex organs open almost simultaneously (A), the oogonium a little before the antheridium. Couch (1932) observed that the antheridia and oogonia situated close together may dehisce almost at the same time or either of them may open from a few minutes to 2 hours before the other dehisces. The latter device tends to check self-fertilisation. As the tip of the beak of the oogonium gelatinises to form a pore, a small drop of colourless cytoplasm oozes out through the apical aperture (B). At this stage the chromatophores and oil drops move to the centre of the ovum which is consequently surrounded by a colourless lining layer of cytoplasm. Several sperms emitted through the apical pore in the wall of antheridium gather round it (B). Many of them enter the oogonium through the pore in the oogonial wall. Only one of them is able to pierce the oosphere at the receptive spot (C). The small male nucleus lies near the female nucleus and increases in size until it has swollen to nearly the same volume prior to fusion with the latter. After fertilization a membranes is formed across the oogonial aperture (D). The zygote secretes a several layers thick wall around itself to become an **oospore** (E). The

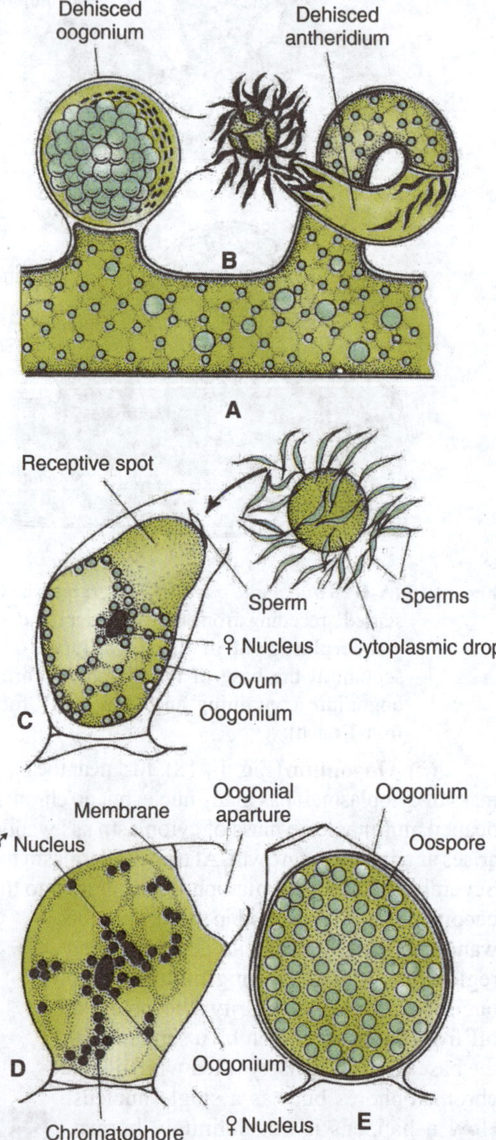

Fig. 14.20 (A-E). *Vaucheria.* Stages in fertilization; (A-B), dehisced sex organs; (C-D), two stages in fertilization; (E), oogonium containing an oospore (B-D after Oltmans).

Fertilisation in *Vaucheria* Sex Organs Dehisce to Effect Fertilization

oospore is liberated by the decay of the oogonial wall. It then enters upon a period of rest. The chloroplasts disappear. The resting zygote retains a number of reddish or brownish bodies.

Germination of oospore (Fig. 14.21). After the dormant period, the oospore germinates directly into a new filament without the formation of meiospores. It is held that reduction occurs in the first nuclear division in the germinating oospore (A). In that case the thallus plant is **haploid**. Mundie (1929) held that *Vaucheria* thallus is diploid. According to him meiosis occurs at the time of gamete formation. The thick oospore wall ruptures. Through the split emerges the colourless germ tube. The term tube turns green and forms the aerial portion of the green thallus (B). The colourless rhizoidal portion develop either directly from the oospore as an independent colourless germ tube or as a lateral outgrowth of the germ tube which forms the aerial system.

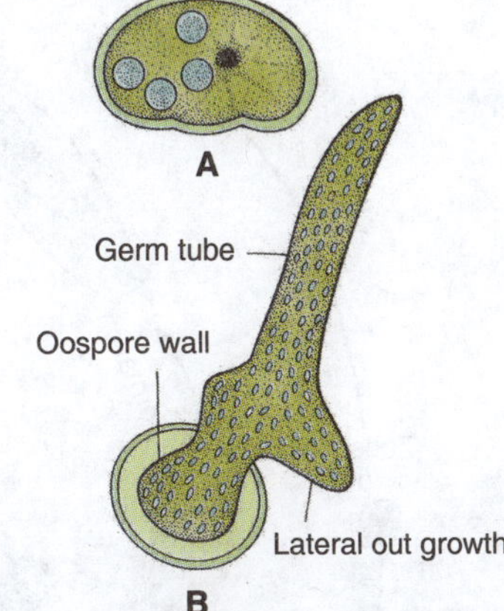

Fig. 14.21 (A-B). *Vaucheria.* (A), oospore containing 4 haploid daughter nuclei (after Hanatschek); (B), germinating oospore (after Oltmans).

Salient Features of Vaucheria

1. The plant body consists of an unseptate tubular, irregularly branched coenocytic filament. It is usually attached to the substratum by means of a branched, colourless holdfast.
2. The protoplast containing numerous nuclei and the discoid chloroplasts is continuous throughout the entire length. It extends without a break in the branches, so does all central sap vacuole.
3. Asexual reproduction is entirely by the method of sporulation. In the aquatic species the spores are motile and produced singly within zoosporangia developed at the tips of branches.
4. The motile zoospores are multiflagellate and coenocytic. There is a pair of flagella opposite each nucleus. They are looked upon as compound structures. Each synzoospore germinates to give rise to a new filament.
5. The terrestrial species multiply by the formation of aplanospores. Thick-walled akinetes have also been reported.
6. Sexual reproduction is oogamous. The sperms and eggs develop in specialised sex organs which are not merely modified vegetative cells.

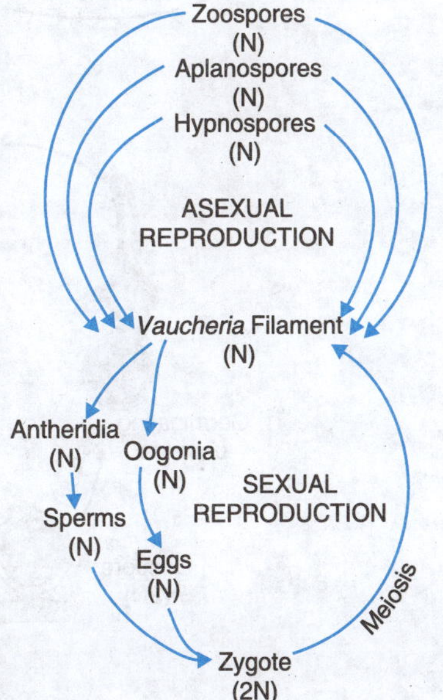

Fig. 14.22. Graphic representation of the life cycle of *Vaucheria sp.*

7. The oogonium develops a single large, uninucleated egg or oosphere within it, while the antheridium gives rise to numerous small, biflagellate sperms.
8. The zygote formed by the fusion of the egg and sperm is, as usual, a resting spore. It enters upon a period of rest. After the dormant period it undergoes zygotic meiosis and germinates to give rise directly to the new haploid *Vaucheria* filament without the intervention of meiospores.

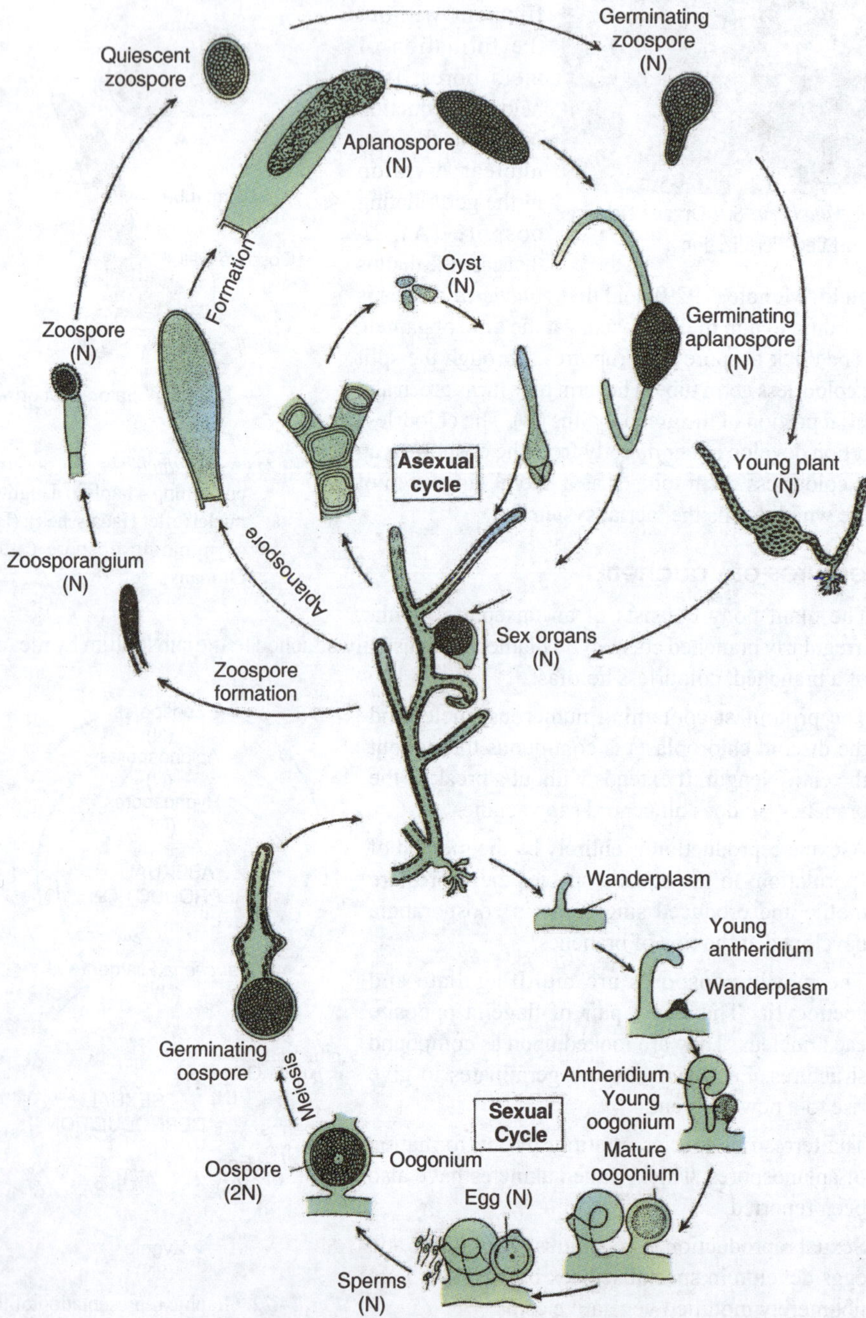

Fig. 14.23. *Vaucheria sp.* Pictorial representation of the life cycle.

Taxonomic position of Vaucheria

This has been a matter of controversy for many years. The plant was reported for the first time by Vaucher in 1803. He named it *Ectosperma*. Two years later in 1805, Decandole changed its name to *Vaucheria* after the name of its discoverer, *Vaucheria* along with three other genera *V. aucheriopsis, Dichotomosiphon* and *Pseudodichotomosiphon* was placed in the family Vaucheriaceae, order Siphonales of class Chlorophyceae. These four genera of Vaucheriaceae resemble in the siphonaceous, coenocytic, acellular organisation of the thallus and oogamous sexual reproduction. The resemblance is so great that Fritsch considered them to have arisen from a common ancestry. The ardent supporters of this classification which continued to hold sway till the middle of the present (20th) century were Fritsch, Iyengar and many others.

The investigation carried on during the last many years, however, tell a different tale. The exact relationship of *Vaucheria* with the other 3 genera of Vaucheriaceae came under suspicion because of some of its features point to a more nearer approach to the yellow-green algae (Xanthophyceae) than to the Chlorophyceae. Strain (1948) made a critical study of the pigments of a few species of *Vaucheria* and came to the conclusion that they possess pigments characteristic of Xanthophyceae. On the basis of Strain's conclusion on the pigment constitution of *Vaucheria,* Smith (1950) transferred this genus to the class Xanthophyceae. Many of the algologists, at present favour inclusion of family Vaucheriaceae with its single genus *Vaucheria* in the class Xanthophyceae. The arguments advanced in favour of the view are :

(*i*) The absence of native or mercerized cellulose (typical of the green algae) in the filament wall.

(*ii*) Absence of chlorophyll *b* and presence in the zoospores of *Vaucheria* of chlorophyll *e,* a pigment characteristic of class Xanthophyceae.

(*iii*) Presence of carotenoid pigments in excess of the chlorophylls.

(*iv*) Lack of two xanthophylls, *siphonein* and *siphonòxanthin* which are of universal occurrence in the Siphonales.

(*v*) Absence of pyrenoids in the chromatophores.

(*vi*) Oil and fat rather than starch as the principal food reserves.

(*vii*) Presence of paired flagella of unequal lengths on the zoospore.

(*viii*) Presence of two oppositely directed, unequal flagella, one of whiplash and the other of tinsel type on the sperm, a feature typical of the motile cells of all Xanthophyceae.

On the basis of the features mentioned above the inclusion of *Vaucheria* in the class Xanthophyceae has practically become a settled fact. Pascher (1914) suggested the inclusion of 3 classes Xanthophyceae, Chrysophyceae and Bacillophyceae in a single division Chrysophyta. The algologists in general agreed to Pascher's suggestion. This classification is still followed in many text books. The latest trend, however, is to segregate the three classes and elevate each of them to the rank of a division. Thus according to the latest view the systemic position of *Vaucheria* is as follows:

Division	:	**Xanthophyta** (Xanthophycophyta)
Class	:	**Xanthophyceae**
Order	:	**Heterosiphonales**
Family	:	**Vaucheriaceae**
Genus	:	*Vaucheria*
Species	:	*geminata*

There are some phycologists who still feel that the evidence put forth is not quite sufficient to warrant the segregation of *Vaucheria* from the order Siphonales, class Chlorophyceae. The chief objection is the oogamous sexual reproduction in *Vaucheria* which is unlike any process in the other Xanthophyceae.

Affinities of Vaucheria. In the siphonaceous, coenocytic acellular organisation of the thallus, photosynthetic pigments, discoid chloroplasts, cell wall composition, flagellar morphology of motile cells, and principal food reserves, *Vaucheria* indicates relationships with other members of the Xanthophyceae. The multinucleate, aseptate thallus and oogamous sexual reproduction are the features in which *Vaucheria* approaches the green algae. For a long time it was classified as a green alga and placed in the order Siphonales of the division Chlorophyta.

In the coenocytic nature of the thallus, chemical composition of cell wall and oogamy in the sexual phase, Vaucheriaceae exhibit a striking resemblance with **Oomycetes**. The resemblance is so great that the Saprolegniaceae are believed to have close affinities with the Vaucheriaceae (*Vaucheria*). The discovery of tinsel and whiplash type of flagella in the motile stages of *Vaucheria* is indicative of the Phycomycetes having been derived from *Vaucheria*-like ancestors. Whether these resemblances are suggestive of real affinities or the result of convergent development is at present difficult to decide.

Origin and relationships of the Xanthophyceae

Most of the species of the Xanthophyceae occur in fresh water and have a flagellate phase in the life cycle. The existence of the flagellate phase is indicative of the fact that Xanthophyceae have a motile unicell ancestry. The evolution from the motile ancestral form took place along lines parallel to those found in the Chlorophyceae. Many of them remain at the flagellate level of organisation, others, exist in colonial, palmelloid and a few even in filamentous states. Despite these diverse evolutionary developments showing a close parallel with the Chlorophyceae, the rate of evolution has been much slower in the Xanthophyceae. As a result the advanced members of this class could not reach the level of elaboration and differentiation attained by the advanced members of Chlorophyceae. Consequently the parenchymatous and heterotrichous forms of thalli typical of the green algae are conspicuous by their absence in the Xanthophyceae. The student might ask why the rate of evolution has been slow in this case. It is not possible to give a straight answer to this question. At the best one can attribute it to different pigmentation coupled with biochemical features restraining variation and evolutionary success. Besides, sexual reproduction is rare in the Xanthophyceae. The few cases reported so far do not pass beyond the isogamous rarely the anisogamous stage. *Vaucheria* is the only exception which exhibits oogamy. The poor development of sexual reproduction indicates that the Xanthophyceae are still in course of evolution because of the lower range of sexual reproduction, limited genetic interchange and the range of variation open to natural selection.

Steinecke (1932) suggested derivation of the Xanthophyceae by retrogression from the green algae like *Microspora*. The reduction series commenced with the filamentous Xanthophyceae. Fritsch (1935) considered this view based on too slender a foundation to deserve serious consideration.

Pascher (1921), on the basis of certain common features, suggested a close relationship between the Xanthophyceae, Chrysophyceae and Bacillariophyceae and advocated the inclusion of these three classes in a single division Chrysophyta.

QUESTIONS

ESSAY TYPE

1. Describe the life cycle of *Vaucheria* with the help of suitable diagrams. (*Kanpur, 1996; Gorakhpur, 1990, 1995, 1999; Agra, 1994; Lucknow, 1998*)
2. Give an account of structure and life history of *Vaucheria* and mention the salient features in it. (*Kanpur, 1998; Purvanchal, 1999; Ranchi, 1991*)
3. Describe/Discuss the important features in the life cycle of *Vaucheria* with suitable diagrams. (*Gorakhpur, 1997; U.P. College, 1998; Bundelkhand, 1994*)
4. Give an illustrated account of sexual reproduction in *Vaucheria* and add a note on its systematic position. (*Purvanchal, 2004; Agra, 1991; Madras, 1955; Trichy, 1995*)
5. Discuss/describe about the systematic position of *Vaucheria*. (*Agra, 1995; Allahabad, 1992*)
6. Describe the structure and mode of reproduction in *Vaucheria*. (*Allahabad, 1991, 1997; U.P. College, 1996; Bundelkhand, 1997*)
7. With the help of suitable sketches, describe the life history of *Vaucheria*. Assign it to its systematic position giving reasons. (*Lucknow, 1992, 1994, 1999; Ravi Shankar, 1995*)
8. Give an account of reproduction and systematic position of *Vaucheria*. (*Meerut, 1997*)
9. With the help of labelled diagrams only, illustrate the life cycle of *Vaucheria*. (*Meerut, 1999*)
10. Give a diagrammatic sketch of the life cycle of *Vaucheria*. (*Punjab, 1996*)
11. Describe the various modes of reproduction in *Vaucheria*. (*Punjab, 1996*)
12. Write an account of resemblances and differences in the life history of *Albugo* and *Vaucheria*. (*Punjab, 1991; Himachal Pradesh, 1993*)
13. Give an account of thallus structure and reproduction in *Vaucheria*. (*Punjab, 1992*)
14. Enumerate the structural and nutritional peculiarities of *Vaucheria* which distinguish it from green algae. (*Himachal Pradesh, 1992*)
15. Describe structure and life history of *Vaucheria*. How does it differ from green algae in its nutritional peculiarities? (*Himachal Pradesh, 1996*)
16. Give the distinguishing features of the class Xanthophyceae. Also give its classification.
17. Give an account of habitat and structure of thallus of *Botrydium*.
18. Describe the modes of asexual reproduction in *Botrydium*.
19. Give a complete account of structure and life history of *Botrydium*.
20. (*a*) Give an illustrated account of thallus structure and modes of reproduction in *Botrydium*.

    (*b*) Describe the life history of *Vaucheria*. Assign it to its systematic position giving reasons. (*Lucknow, 1999*)

SHORT ANSWER TYPE

21. Identify the following and mention their important characters with the help of diagrams. Give their classification as well:

    (*i*) *Vaucheria* (*Kanpur, 1999; Kumaon, 1996; Rohilkhand, 1996; Bundelkhand, 1998; Lucknow, 2007*)

    (*ii*) *Botrydium*
22. Draw neat and labelled diagrams of the following :

    (*i*) Sex organs of *Vaucheria* (*Gorakhpur, 1991, 1993, 1994; Purvanchal, 2000; Lucknow, 2000*)

(ii) Antheridium and oogonium of *Vaucheria* (Gorakhpur, 1992)

(iii) *Vaucheria* filaments with sex organs (Rohilkhand, 1991)

(iv) Thallus structure of *Botrydium*

(v) Sex organs of *Botrydium*

23. Differentiate between *Vaucheria* and *Botrydium*.

24. Describe the salient features of *Botrydium*.

25. Describe the thallus structure of *Botrydium*.

26. Write short notes on :

(a) Synzoospore

(Gorakhpur, 1996, 1998; Kumaon, 1995, 1997, 2000; Agra, 1993, 1996; Allahabad, 1992, 2003; Lucknow, 1996, 2000, 2007; Purvanchal, 1996; Ravi Shankar, 1994; Bihar, 1993; Meerut, 1998)

(b) Sexual reproduction in *Vaucheria*.

(Kanpur, 2004; Gorakhpur, 1999; Allahabad, 1999; Bundelkhand, 1993)

(c) Fertilisation of *oogonium* in *Vaucheria*. (Agra, 1995)

(d) Mode of reproduction in *Vaucheria*. (Allahabad, 1993)

(e) Asexual reproduction in *Vaucheria*. (Allahabad, 1994, 2005; U.P. College, 1999; Rohilkhand, 1992)

(f) Salient distinguishing features of Xanthophyceae. (Purvanchal, 2006; Allahabad, 1996)

(g) Taxonomic position of *Vaucheria*. (Allahabad, 2002; Lucknow, 1995; Rohilkhand, 1994)

(h) Sex organs of *Vaucheria* (Lucknow, 2002; Purvanchal, 1996, 97; Magadh, 1990)

(i) *Vaucheria* (Bihar, 1991)

(j) Zoospore formation in *Vaucheria* (Bhagalpur, 1990)

(k) Filament of *Vaucheria* (Bhagalpur, 1991)

(l) Asexual reproduction in *Botrydium*.

(m) Sexual reproduction in *Botrydium*.

(n) Resting bodies.

(o) Formation of spermocarp (Lucknow, 2003)

27. Describe a siphonaceous thallus. (Gorakhpur, 1992)

28. (a) Write two characters to differentiate the thalli of *Vaucheria* and *Ectocarpus*. (Gorakhpur, 1994)

(b) Distinguish between Vaucheria and Valonia (Lucknow, M. Sc., 2004)

29. Name an alga which produces synzoospores. (Allahabad, 1993)

30. Compare the distinguishing feature of Cyanophyceae, Xanthophyceae and Phaeophyceae. (Lucknow, 1995)

31. (a) Name any syphonaceous alga and give its systematic position. (U.P. College, 1995)

(b) Write the name and systematic postion of coenocytic alga studied by you. (Lucknow, 2004)

32. Give the structure of chloroplast/ chromatophore of *Vaucheria*. (U.P. College, 2000)

33. Give the botanical name of the alga in which main thallus is branched coenocytic and siphonaceous. (Purvanchal, 1997)

34. Differentiate between

(a) Aplanospore and synzoospore (Purvanchal, 2000)

(b) *Vaucheria and Spirogyra* (Allahabad, 2003)

35. (a) Name the synzoospore producing alga. (Rohilkhand, 1991)

(b) Name two algae which produce multiflagellate zoospores. (Kanpur, 2006)

36. Give the systematic position of *Vaucheria*. (Guru Nanak Deo, 1991; M.D.S. Jaipur, 1998)

37. Describe synzoospore of *Vaucheria*. (*M.D.S. Univ., Jaipur, 1998*)

38. Describe sexual reproduction in *Vaucheria*. (*Kanpur, 2006; Trichy, 1955*)

39. Why is *Vaucheria* called 'Golden yellow Algae'? Discuss its systematic position. (*Punjab, 1999*)

40. Explain the structure of plant body in *Vaucheria*. (*Himachal Pradesh, 1991*)

41. Define synzoospore. (*Himachal Pradesh, 1993*)

42. Give an account of haploid and vegetative structure of *Vaucheria*. (*Himachal Pradesh, 1997*)

43. Is the thallus of *Vaucheria* a gametophyte or sporophyte? Justify your statement.

44. Why are *Caulerpa* and *Vaucheria* put under different orders even though thallus of both are regarded as unicellular or siphonaceous?

45. *Vaucheria* is like a phycomycetous fungus but for chromatophores. Why?

46. Why is *Vaucheria* not placed under Chlorophyceae although it contains chlorophyll?

47. Classify *Botrydium* and describe its salient features.

48. What is the difference between aplanospore and synzoospore? (*Kanpur, 2006*)

49. Comment on alternation of generation in *Vaucheria*. (*Kanpur, 2005*)

50. With the help of labelled diagrams only compare the cell structure of *Vaucheria* and Oedogonium? (*Kanpur, 2003*)

51. How many flagella are found in male gametes of *Vaucheria*? (*Allahabad, 2004*)

52. Write the name and systematic position of a coenocytic alga studied by you. (*Allahabad, 2006*)

53. Write explanatory notes on Gongrosira stage. (*Purvanchal, 2003*)

OBJECTIVE TYPE

54. **Fill in the blanks :**

(*i*) *Vaucheria* belongs to class ....

(*ii*) Synzoospores are produced in ....

(*iii*) *Botrydium* belongs to family ....

(*iv*) The rhizoid branches bear cyst like tuberous structures in ....

(*v*) The zoospores in *Botrydium* is ....

(*vi*) In *Vaucheria* the most common method of asexual reproduction is through multiflagellate multinucleate structures known as ....

(*vii*) In *Botrydium* sexual reproduction is ....

(*viii*) In *Vaucheria* chlorophyll .... is lacking.

(*ix*) The food reserve in class Xanthophyceae is ....

(*x*) The sexual reproduction in *Vaucheria* is ....

(*xi*) Flaments are siphonaceous in ....

55. **Select the correct answer.**

(*i*) Which of the following is coenocytic and multinucleate?

(*a*) *Vaucheria*

(*b*) *Polysiphonia*

(*c*) *Spirogyra*

(*d*) *Sargassum*

(*ii*) Reserve food material in *Vaucheria* is

(*a*) Starch

(*b*) Protein

(*c*) Oil

(*d*) All of the above

(*iii*) In *Vaucheria* cross walls are

(*a*) never present

(*b*) sometimes present
(*c*) always present
(*d*) present only at the time of sexual reproduction

(*iv*) Oil droplets are found in
(*a*) *Sargassum* (*b*) *Cladophora*
(*c*) *Polysiphonia* (*d*) *Vaucheria*

(*v*) Asexual reproduction in *Vaucheria* takes place by zoospores which are
(*a*) Uniflagellate
(*b*) Biflagellate
(*c*) Quadriflagellate
(*d*) Multiflagellate

(*vi*) Which of the following has coenocytic thallus?
(*a*) *Hydrodictyon* (*b*) *Vaucheria*
(*c*) *Ectocarpus* (*d*) *Spirogyra*

(*vii*) *Botrydium* belongs to class
(*a*) Chlorophyceae
(*b*) Charophyceace
(*c*) Xanthophyceae
(*d*) Bacillariophyceae

(*viii*) Synzoospores are present in
(*a*) *Botrydium*
(*b*) *Oedogonium*
(*c*) *Vaucheria*
(*d*) *Chlamydomonas*

(*ix*) Siphonaceous thallus is found in
(*a*) *Vaucheria* (*b*) *Zygnema*
(*c*) *Spirogyra* (*d*) *Ectocarpus*

(*x*) The alga in which thallus is branched, coenocytic and siphonaceous is
(*a*) *Vaucheria*
(*b*) *Ectocarpus*
(*c*) *Chlamydomonas*
(*d*) *Zygnema*

(*xi*) The multiflagellated zoospore in *Vaucheria* is generally interpreted as
(*a*) Synzoospore
(*b*) Sporangium
(*c*) Zoosporangium
(*d*) All of the three

(*xii*) Resting bodies are present in
(*a*) *Botrydium* (*b*) *Vaucheria*
(*c*) *Ectocarpus* (*d*) *Spirogyra*

(*xiii*) In *Botrydium* the sexual reproduction is
(*a*) Isogamous
(*b*) Anisogamous
(*c*) Heterogamous
(*d*) Oogamous

(*xiv*) In *Vaucheria* the sexual reproduction is
(*a*) Isogamous
(*b*) Anisogamous
(*c*) Heterogamous
(*d*) Oogamous

(*xv*) *Vaucheria* is usually terrestrial or
(*a*) Fresh water forms
(*b*) Marine form
(*c*) Saline forms
(*d*) Alkaline forms

Tight microscopic image of *Melosira arctica* an Arctic ice diatom

CHAPTER 15

DIVISION : BACILLARIOPHYTA (Bacillariophycophyta) Class : Bacillariophyceae (Diatoms)

- Pinnularia
- Economic Importance
- Phylogeny of Diatoms
- Navicula

General Characteristics

The division Bacillariophyta comprises a single class Bacillariophyceae, the members of which are popularly known as the **diatoms**. They number about 16,000 species grouped under about 200 genera. The number of individuals is beyond reckoning. The majority are single-celled but in some the cells form pseudofilaments and colonial aggregations.

The diatoms are both aquatic and terrestrial. In the latter case they occur in subaerial habitats where there is periodic moisture. The aquatic forms are both fresh water and marine. Many occur in brackish water. In habit some are planktonic and others benthic. The benthic or attached species are formed on mud, sand or rocks. Some are epiphytic on plants and a few occur or animals (epizoic). Epiphytic forms may be attached by the mucilage covering of the diatom frustule or by a mucilaginous stalk secreted by the diatom cell. The planktonic free floating forms are found both in fresh and marine water. *Melosira, Nitzschia, Navicula* and *Cocconeis* are the common fresh-water genera found in pools, puddles, ponds, lakes and streams. The diatoms are microscopic, variously coloured, and of diverse forms. Some look like tiny boats. (Fig. 15. 1 A). Some resemble half moon (*Melosira*). Others are seen as triangles (*Triceratium*), rectangles (Fig. 15.1 A) and circles (Fig. 15.2 B). The diversity of form is based either on a bilateral or a radial plan. The chief characteristic features of this class are : (*a*) the diploid thallus; (*b*) the silicified cell wall consisting of two overlapping halves bearing characteristic secondary structures; (*c*) storage of food reserves as oil and chrysolaminarin or a protein-like food material called **volutin**; (*d*) motile stage (sperms) with a single pantonematic (tinsel) flagellum sometimes two; (*e*) Production of unique type of spores known as auxospores and (*f*) the presence of chlorophyll *a* and *c* but not *b*, together with fucoxanthin as the photosynthetic pigments. Generally the diatoms are yellow brown, golden-yellow or olive green. The colour is due either to the presence of a very large proportion of the carotenoids (normal constituents of chlorophyll) or to the presence of an accessory brown pigment called the **diatomin**. The latter forms the principal colouring matter. It is a xanthophyll pigment variously named as fucoxanthin or **isofucoxanthin** or **diatomin**. Fucoxanthin is a characteristic pigment of the class Phaeophyceae (Brown Algae). For this reason some algologists consider the diatoms to be related to the brown algae. Others hold that diatomin is not identical with the fucoxanthin of the brown algae. The diatoms have a unique physical structure. They differ from all the other algae in the following respects :-

(1) *Nature and structure of the cell wall.* The diatom cell wall is composed of two parts, an inner continuous pectin membrane and an outer highly silicified armour of hydrated silica. The silicified cell wall is often called the *shell* or *frustule*. The silica envelope shows characteristic secondary structures. A completely enclosing wall of silica (frustule) would hinder cell division. Hence it is made up of two overlapping halves. One of these is the older of the two. The older half fits closely over the younger half like the lid of a box (Fig. 15.1 A). The overlapping side walls of the two halves are called the **girdles**. The word diatom literally means "cut into two". This refers to the characteristic double nature of the silica wall.

(2) *Production of Auxospores.* These are specialized enlarged spores, also known as the **growth spores**. They are zygotic in nature and serve to restore to the normal, the reduction in the size of the diatom cell occurring during vegetative propagation. For this reason they are also called the **rejuvenescent cells** or **renewal spores**.

Primary Classification. The diatom unicells are of various forms. They are placed under the following two taxonomic groups :-

1. Order Centrales. The unicells are radially symmetrical *e.g., Cyclotella* (Fig. 15.2 A).

2. Pennales. The unicells exhibit bilateral symmetry (*Pinnularia*, Fig. 15.1 A,B).

Occurrence

There are about 16,000 species of diatoms. These are widespread and occur in abundance, isolated or in colonies in almost every aquatic situation. Many form the major part of the bottom flora of lakes, pools, puddles and ponds. Some forms occur in aerial habitats such as the damp soil., brick work, rocky cliffs, among the mosses and on the bark of trees. Some terrestrial species survive desiccation for months. The diatoms are extremely abundant in the marine waters. The common examples of marine forms are *Triceratium* and *Hyalodiscus.* They occur as epiphytes on seaweeds. They constitute an important part of the fresh water or marine **plankton** which forms the basic food of the aquatic animals. Many occur attached to various substrata or grow intermingled with or as epiphytes on other algae and plants. A few are epizoic. Some have been found in the hot springs. In fact they are ubiquitous organisms and are found nearly everywhere where life can exist. In their mode of nutrition the diatoms are **autotrophic** but can utilize organic substances. *Nitzschia puttida* is, however, colourless. It lives saprophytically.

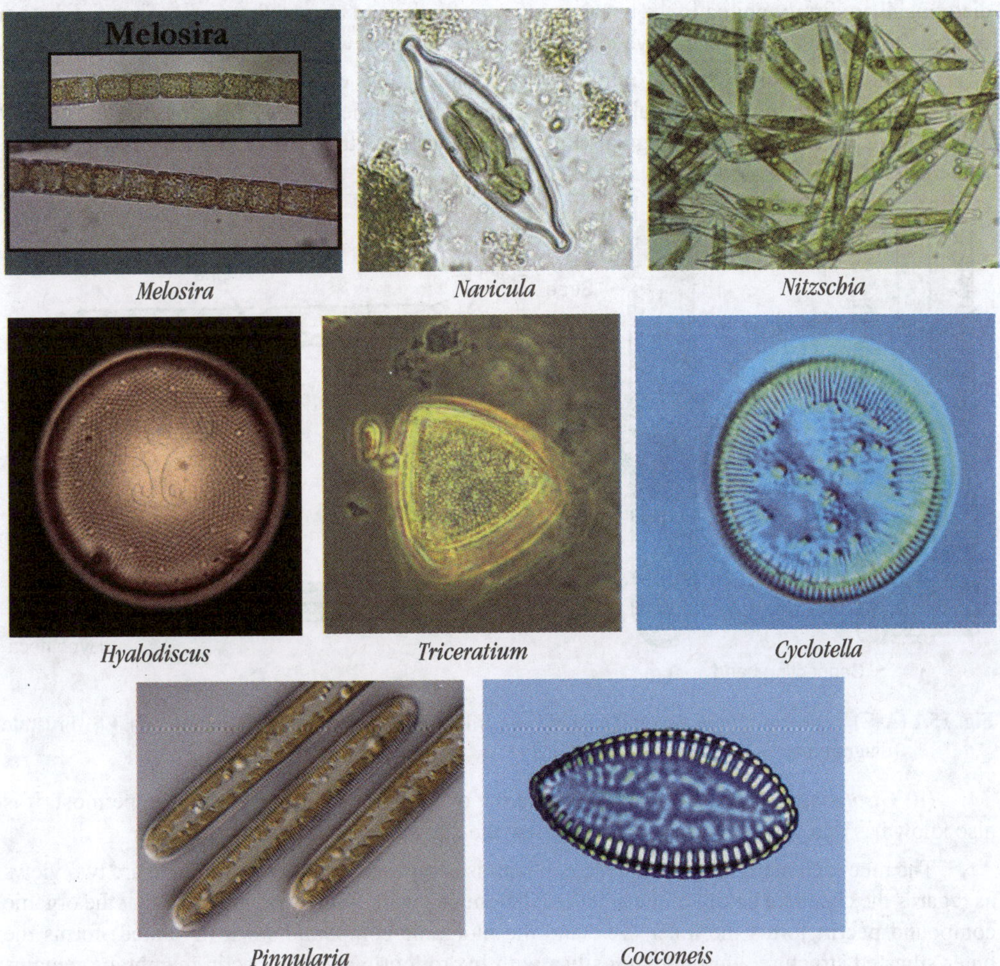

Melosira *Navicula* *Nitzschia*

Hyalodiscus *Triceratium* *Cyclotella*

Pinnularia *Cocconeis*

Thallus (Fig. 15.1). The diatom thallus is essentially a diploid **unicell**. The tiny unicells occur isolated. Often the unicellular thalli adhere by their mucilaginous sheaths to form colonies. The colonies may be filamentous or are enclosed in a common gelatinous envelope with no specialization of parts.

Cell Structure. Structurally the diatom cell consists of two parts, the **cell wall** and the **protoplast**.

1. Cell Wall (Fig. 15.1 A-C). It is differentiated into two layers. The inner layer of the cell wall is in the form of a thin continuous pectin membrane which encloses the protoplast. The outer layer of the diatom cell is highly characteristic. It forms an armour of hydrated silica which consists of two overlapping halves (A and C). They fit closely together like the two parts of a candy box or petri dish (A). So the relation of the two halves is much the same as that of a box and its lid. The lid corresponds to the outer half called *epitheca*. It is always older than the inner half. The latter corresponds to the main box and is known as the *hypotheca*. Each half or theca, as it is called, consists of two parts namely (*i*) the *valve* forming the main surface corresponding to the top or bottom of the candy box, and (*ii*) connecting bands comparable to the uncurved sides. The margin of each valve is firmly attached to its connecting band. In the diatom cell the connecting band of the outer half of the cell wall (epitheca) overlaps that of the inner half (hypotheca). The two connecting bands in the diatom cell are firmly united in the overlapping region which is termed the *girdle.* Thus the diatom cell can be seen in the following two different aspects when viewed from surface :

(*i*) *Valve view* (Fig. 15.1 B). In this view the valve side is uppermost. The diatom cell is seen from the top or bottom. It is also called the **top view**. The centric diatoms in valve view are usually circular (Fig. 15.2 A), rarely triangular (Fig. 15.3 C) whereas the pennate diatoms may be linear, lanceolate, ellipsoid, sigmoid or ovoid in form.

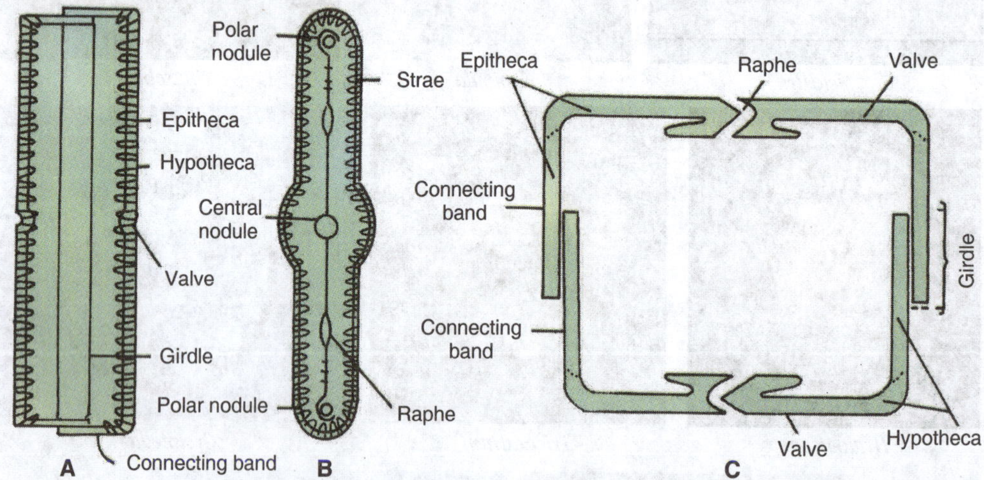

Fig. 15.1 (A-C). *Diatoms* (*Pinnularia*). (A) Side or girdle view. (B) Top or valve view. (C) T.S. Frustule (diagrammatic). (A and B based on Pfitzer).

(*ii*) *Girdle view* (Fig. 15.1 A). In this case the overlapping bands or girdle is uppermost. It is also known as the side view because in this view the diatom cell is seen from the side.

The mucilage material covering the cell wall is composed of pectic acid. There are two views as regards the chemical nature of the silicified diatom cell wall. Some workers think that the organic compound **pectin** forms the inner layer and the inorganic compound silica (hydrated) forms the outer silicated structure. On removing silica with hydroflouric acid the pectin membrane remains intact. There are others who hold that the cell wall is mainly made up of pectin which is strongly impregnated with silica. The connecting bands lack all kinds of markings. All kinds of ornamentation such as ridges, pits and fine pores are confined to valve portion of the cell wall. They are arranged in beautiful geometric patterns. In some forms they are so fine that they can only be made out with the help of excellent microscopes. Because of this the diatom shells or frustules are used to test the objectives of microscopes. The fine markings or ridges on the valves are called the **striae**. These striations appear to be composed of fine dots which represent small pits called the **punctae**. The striations on the valves are arranged in the following two patterns :-

(*i*) *Pennate type*. (Fig. 15.2, B). The striations are arranged in a pinnate manner. They appear in two regular rows or series, one on either side of the axial strip. The axial strip may either be a plain area in between and called the **pseudoraphe** or it has a longitudinal slot called the **raphe**. The raphe extends from one end of the valve to the other. It also bears three round spots called the **nodules**. There is one nodule at each end and the third near the mid-point. The former are called the **polar nodules** (Fig. 15.1, B). The median one is called the **central nodule**. The nodules are the structural modifications of the cell wall. The diatoms with pennate type of striations are mainly freshwater.

They have elongate valves which may be boat or needle-shaped. They are placed in the order **Pennales** or **Pennatae**. Commonly they are called the **pennate diatoms**. They are bilaterally symmetrical, not circular and have a raphe. A typical example of a fresh water pennate diatom is *Navicula halophila*. It is boat-shaped in valve view and rectangular in girdle view.

(*ii*) *Centric type* (Fig. 15.2, A). The striations are arranged radiately. The dots are usually large and valves circular. They have no raphe. The centric diatoms are radially symmetrical. The diatoms of this category are placed in the order **Centrales** or **Centricae.** They are commonly called the **centric diatoms.** They are generally marine forming important constituents of phytoplankton.

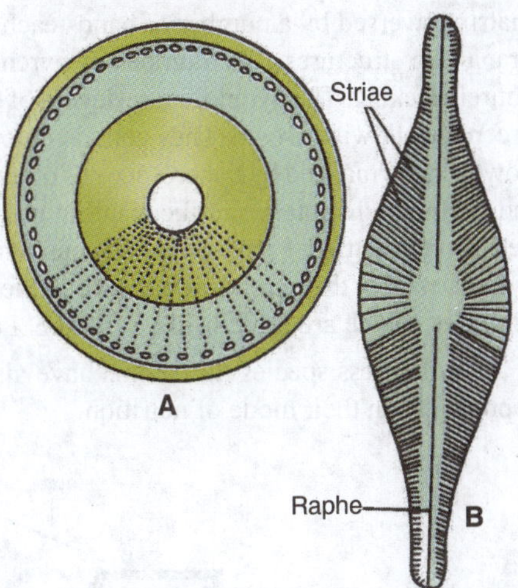

Fig. 15.2 (A-B). *Diatoms*. (A), Top view of a centric diatom with markings arranged radially, (B), Valve view of a pennate diatom *Navicula* with marking arranged in a pinnate manner.

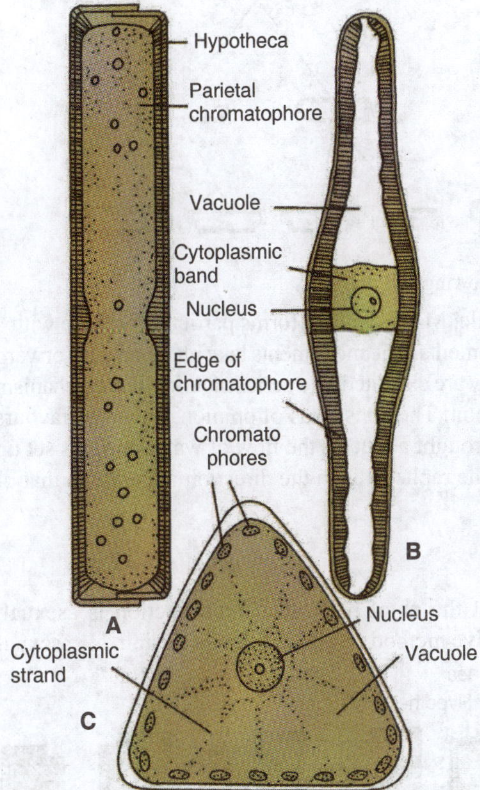

Fig. 15.3 (A-C). *Diatoms*. Cell structure. (A), girdle view of a pennate diatom *Pinnularia* showing plate-like parietal chromatophores; (B), the same in valve view showing nucleus and edges of two lateral chromatophores; (C), Top view of a centric diatom with several chromatophores.

2. Protoplast (Fig. 15.3). It is differentiated into plasma membrane cytoplasm, a single nucleus, one to several chromatophores with or without pyrenoids, the **cytoplasm** forms a thick layer just within the cell wall and encloses a large **central vacuole** (B). The single large **nucleus** in the Pennales lies embedded in the **cytoplasmic band** that runs across the middle of the central vacuole and is connected with the lining layer of cytoplasm next to the cell wall (B). The nucleus is bounded by a nuclear membrane and has one or more nucleoli. In the centric diatoms the nucleus is generally found in the cytoplasm lining the cell wall. Included in the cytoplasm are the mitochondria, golgi bodies, endoplasmic reticulum and the **chromatophores**. The number and shape of chromatophores varies according to the species. In the **centric** diatoms there are a few medium sized to several small chromatophores (C). They are discoid or irregular in shape. The **Pennales** have one or commonly two large parietal chromataphores (B). When single it is irregularly lobed and has perforations. It extends the whole length and the girdle breadth. In case there are two they are laminate and situated longitudinally along the opposite sides of the protoplast. Under the electron microscope, the double membrane limited chromatophore is seen to consist of a matrix traversed by a number of bands each consisting of 4-6 lanellae which are not organised into grana-like structures. The number of pyrenoids varies from one to several per chromatophore or entirely lacking. The pyrenoids are devoid of the starch sheath. Typically the chromatophores are olive green to yellowish brown. They contain chlorophylls *a* and *c* but no chlorophyll *b*. The chlorophylls, however, are masked by the presence of other pigments such as β-carotene and xanthophylls. Among the xanthophylls lutein and fucoxanthin predominate. The nature of pigments in the diatom cells is yet scarcely settled. Fatty oils are the customary products of photosynthesis. They are present in the cytoplasm near the pyrenoids but outside the chromatophore envelope. Besides there occur **volutin-globules**, which are protein-like in nature. Polysaccharide leucosin may also be present.

Colourless species of diatoms have also been reported. They lack chromatophores. They are saprophytic in their mode of nutrition.

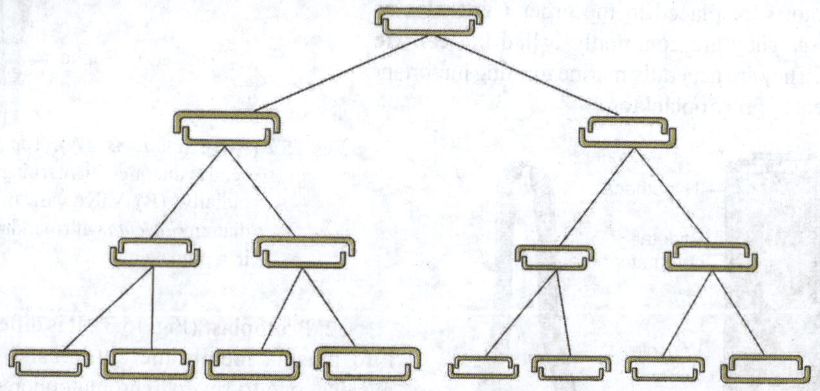

Fig. 15.4. *Diatom.* Showing cell division.

Movements. The centric forms are non-motile. Many pennate forms particularly those with a raphe exhibit slow, jerky, spontaneous gliding movements. The movements take place only in forward and backward direction in the longitudinal axis. They are executed by a series of jerks. The mechanism of locomotion in the diatoms is still a debatable point. The consensus of opinion, however, favours Muller's theory. According to him locomotion is brought about by the flow of water currents set up by the streaming movements of the cytoplasm in the raphe. It is in the direction opposite to that of the diatom movement.

Reproduction (Fig. 15.5, A-E)

Diatoms reproduce by a sexual process but the chief method of reproduction is asexual. Asexually they multiply exclusively by the vegetative method of cell division or fission. No asexual spores are formed ordinarily. Resting spores have been recorded in some Centric diatoms.

Cell Division or Fission in Diatoms (Fig. 15.4 and 15.5). It usually occurs at night and is of a peculiar type and fairly rapid. At the commencement of the process the cell protoplast increases in diameter (A). The overlapping thecae lesson their hold and slightly separate at the girdle. The nucleus divides by mitosis in a plane perpendicular to the longitudinal axis which means parallel to the valve surfaces (B). This is followed by the

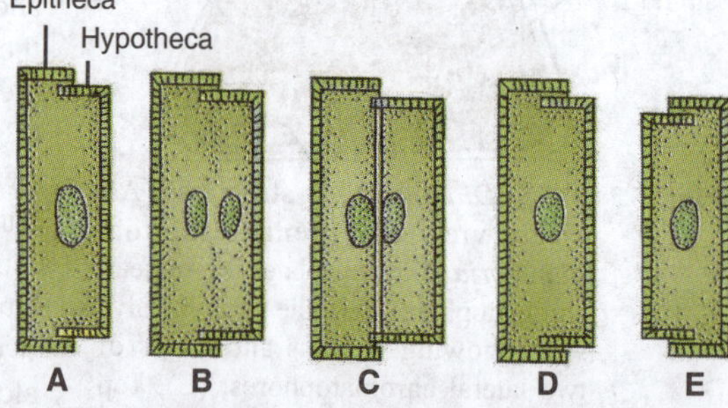

Fig. 15.5 (A-E). *Diatoms*. Asexual reproduction. Explanation in the text.

longitudinal cleavage of the cytoplasm into two parts along the median line and in a plane parallel to the valve faces (C). One daughter protoplast now lies in the epitheca and the other in the hypotheca (C). Their opposite sides are naked. Each daughter protoplast assisted by the nucleus covers its naked face by secreting a new theca fitting inside the old one (D and E). The new theca, therefore, always becomes the hypotheca of the daughter cell. The daughter cells thereafter separate. The one which carries the epitheca of the parent cell remains the same size (D). The other which takes the hypotheca with it is slightly smaller (E).

Sexual Reproduction. As a result of the unique method of cell division described above most the diatom cells progressively become reduced in size. They will be unable to survive if they shrink to a size incompatible with physiological processes. Usually well before they reach their lower size limit, they imitate processes which restore them to their original size. These processes are sexual. The sexual process in diatoms is thus linked to a reduction in diatom cell size. Sexual reproduction thus does not occur in diatoms in which there is no reduction in cell size (Deffner, 1940). The resulting zygote which plays a significant role in this restorative process is a special rejuvenescent cell known as the **auxospore**. The prefix **auxo** means "grow". The auxospore thus is a growth spore. Perhaps under the rejuvenating stimulus of sex, the zygote (auxospore) in the diatoms photosynthesizes actively and grows rapidly in size. The zygotic membrane generally ruptures and is replaced functionally by another thin supporting membrane generally secreted by the zygotic protoplast. It is known as the **perizonium**. Enveloped by the perizonium, the auxospore grows rapidly into a large cell greater in size than the vegetative cells which take part in its formation. The fresh frustule is secreted by the auxospore within the perizonium to organise a new diatom cell of a normal size. A auxospore may directly become a new individual called a firstling or may divide to form two firstling cells of the size characteristic of the species. The details of auxospore formation in the Pennales and Centrales vary considerably and thus are discussed under separate heads.

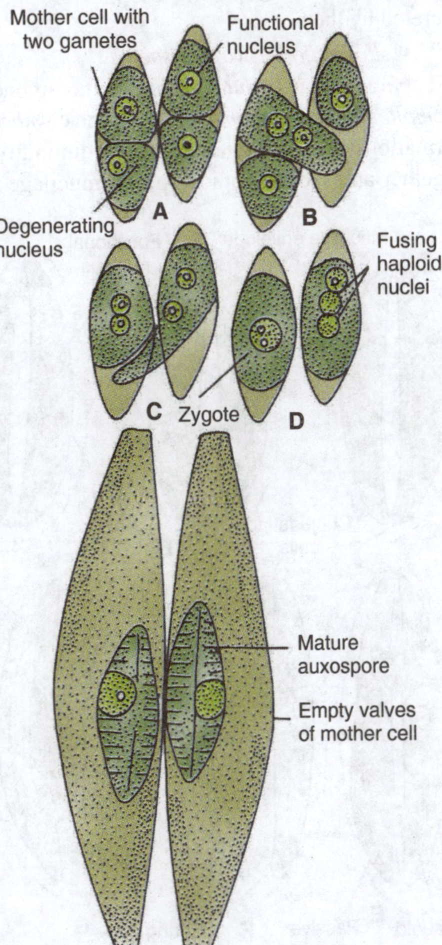

Fig. 15.6 (A–E). *Pennales.* Auxospore formation by physiological anisogamy in *Gomphonema parvulum.* (A), two conjugating mother cells, each with two gametes; (B-C), fusion of gametes; (D), formation of one zygote in each mother cell; (E), mature auxospores under the valves of mother cells (After Geitler).

Auxospore Formation in the Pennales. The auxospores in the Pennales are formed by gametic union rarely by autogamy. The gametic union may be isogamous, anisogamous or pedogamous. Most of the species of pennate diatoms are **monoecious** (homothallic). Some are **dioecious** (heterothallic), *Navicula halophila*). The fusing isogametes or anisogametes are non-flagellated. Neither ova nor sperms are formed as a rule. *Rhabdonema adriaticum,* however, is an exception. It is dioecious and oogamous. In certain diatoms auxospores may be formed parthenogenetically. The auxospores in pennate diatoms are thus formed in diverse ways, the chief among these are narrated below :-

A. Auxospore formation by the fusion of nonflagellated iso- or anisogametes. In this method of auxospore formation either two auxospores or one is formed from the two conjugating individuals. In either case, the two conjugating individuals become enclosed in a common mucilaginous envelop secreted by them.

Cymbella

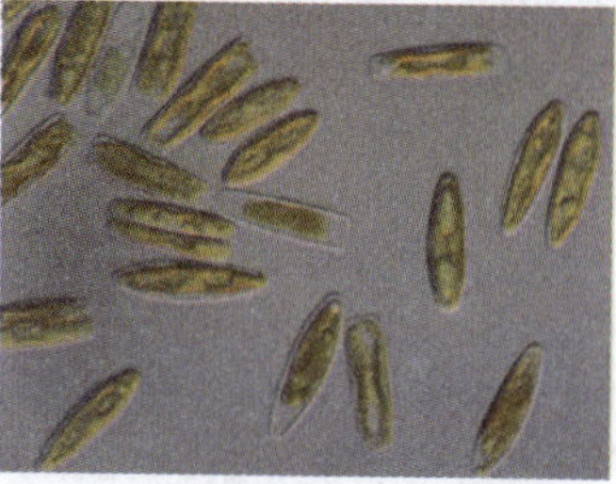

Gomphonema

1. *Formation of two auxospores by conjugating individuals.* This is the commonest method of auxospore formation and is illustrated by *Gomphonema parvulum* (Fig. 15.6) and *Cymbella lanceolata* (Fig. 15.7). In this method of auxospore formation two diatom individuals of diminutive size of a species conjugate (Fig. 15.7). They creep together and secrete copious pectic mucilage which forms a complete envelope around them (A). The diploid nucleus of each conjugate undergoes meiosis to form four haploid nuclei (B-C). Sooner or later, two of these begin to abort in each cell (C). The protoplast of each conjugate then divides either longitudinally or transversely to form only two instead of four gametes. Each gamete contains one large (functional) haploid nucleus and one degenerating haploid nucleus. Depending on the nature of gametes, fusion between them takes place in either of the following ways :

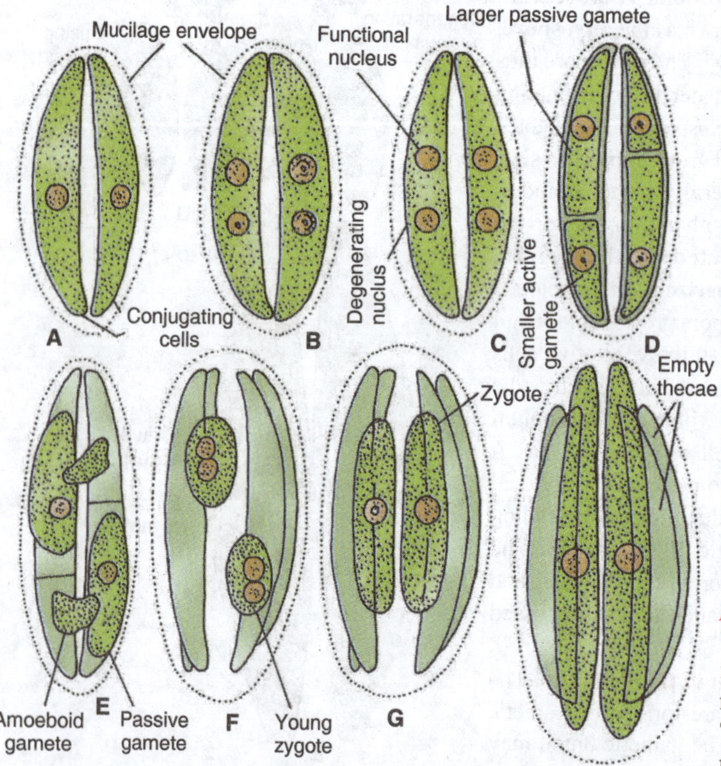

Fig. 15.7 (A-H). *Pennales.* Auxospore formation by anisogamy in *Cymbella lanceolata.* (A), two mother cells within a common gelatinous envelope; (B-C), meiosis and degeneration of two haploid nuclei in each conjugate; (D), delimitation of anisogametes; (E), fusion between anisogametes; (F), formation of two zygotes, one in each conjugate; (G-H), elongation of zygotes to form auxospores (Adapted from Gentler).

(*i*) *Gomphonema parvulum* (Fig. 15.6). The two gametes formed in each conjugate are morphologically alike (A). The division of the cell protoplast is symmetrical. The isogametes formed in each cell, however, exhibit functional disparity. One of them is amoeboid and thus active and the other passive. The amoeboid gamete of each conjugate emerges through the open valves and passes over to the other to fuse with its passive counterpart (B-C). *Gomphonema parvulum* thus provides an example of physiologically anisogamous type of gametic union. *Cymbella lanceolata* (Fig. 15.7) is example of anisogamous type of gametic union. The division of the protoplast into gametes is asymmetrical. The two resultant gametes in each mother cell are unequal in size (D). They are dissimilar in function as well. The smaller gamete in each conjugate is amoeboid and thus active whereas the larger gamete is passive. Copulation in *Cymbella lanceolata* thus takes place between anisogametes. The smaller active gamete of each conjugate may lie opposite the larger passive gamete of the other (D). The gametic union in both of these examples results in the formation of one zygote in each of the two conjugating mother cells.

(*ii*) Subrahamanyan (1948) cited a case of a dioecious *Navicula halophila* in which the two gametes formed in one conjugate are amoeboid (active) and the two formed in the other are passive or immobile. The amoeboid gametes emerge through the open valves of the parent frustule and creep into the open shell of the other to fuse with opposite passive gametes to form two zygotes in one shell. The other is empty. This is a typical example of physiological anisogamy.

The diploid fusion cells or zygotes escape from the enclosing frustules. They remain dormant for a time. Later each elongates considerably to function as an auxospore. The auxospore is enclosed in a slightly silicified pectic membrane which is known as the **perizonium**. It may be secreted by the auxospore or represents the remains of the zygotic membrane. The auxospore secretes a new frustule around itself inside the perizonium. The reconstituted diatom cell is of a normal size.

(*iii*) In some species of pennate diatoms both the gametes of the two conjugating cells are amoeboid and thus active. These isogametes come out through the open valves of their respective conjugating cells and copulate in opposite pairs midway between them. Thus two zygotes are formed outside the parent frustules.

2. *Formation of a single auxospore by two conjugating individuals* (Fig. 15.8). This method of auxospore formation is illustrated by *Cocconeis pediculus, C. placentula* var. *Klinoraphis* and

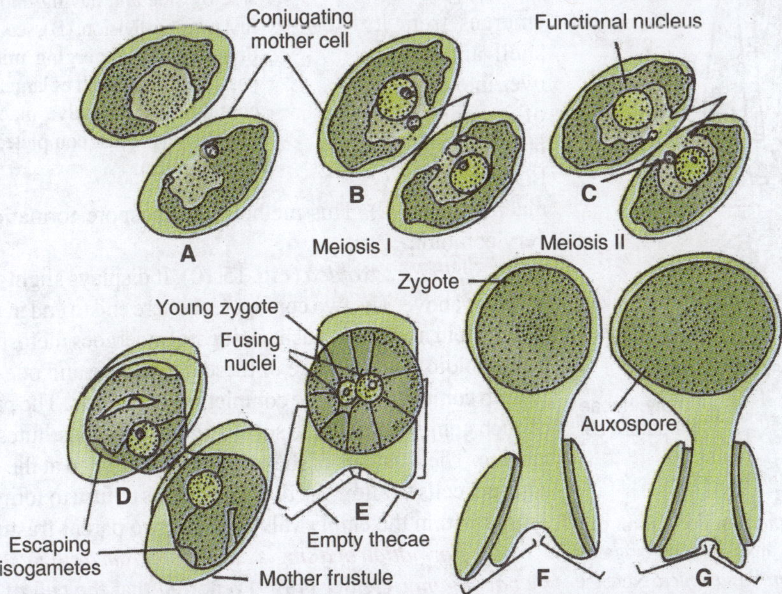

Fig. 15.8 (A-G). *Pennales.* Auxospore formation by isogamy in *Cocconies placentula* var. *Klinoraphis*. (A), two conjugating cells lying side by side; (B-C), meiosis and degeneration of one daughter nucleus after each meiotic division; (D), escape of isogametes from the mother cells and their fusion; (E), formation of a young binucleate zygote between the empty mother frustules; (F), zygote with a diploid nucleus; (G), enlarged zygote forming an auxospore. (Adapted from Geitler).

Surirella saxonica. The two conjugating cells of the first two species lie side by side in the common mucilage envelope secreted by them (A). They may or may not be sister cells. During meiosis there is immediate degeneration of one sister nucleus after the first meiotic division (B). The persisting sister nucleus undergoes the second meiotic division to form two haploid daughter nuclei in each conjugate. Of these one aborts (C). The protoplast of each conjugate containing the surviving haploid nucleus constitutes a single gamete. The two gametes one from each conjugate escape from the parent frustules (D), exhibit amoeboid movements and fuse midway between the empty parent frustules (E). The two nonflagellated fusing gametes are similar in all respects and are called the **isogametes**. The diploid zygote thus formed (F) enlarges and elongates in the plane parallel to the long axis of the parent frustules to form an auxospore (G). It secretes a fresh frustule around it within the perizonium.

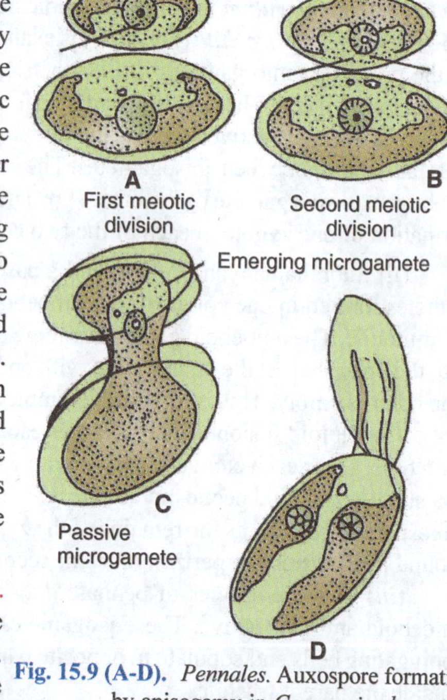

Fig. 15.9 (A-D). *Pennales*. Auxospore formation by anisogamy in *Cocconeis placentula* var. *pseudolineata*. (A), one smaller and the other larger con-jugating cells lying side by side and having undergone the first meiotic division; (B), second meiotic division; (C), emerging microgamete passing into the shell of larger conjugant containing the passive macrogamete; (D), fusion process completed (Adapted from Geitler).

Cocconeis placentula var. *pseudolineata* (Fig. 15.9). It also belongs to this category of auxospore formation but the sexual fusion is of anisogamous type. Of the two conjugating cells one is smaller than the other (A). It forms the microgametes which is amoeboid (B). The active microgamete emerges from its shell and passes over into the shell of its partner to fuse with the larger, passive macrogamete (C). This method of auxospore formation is not very common.

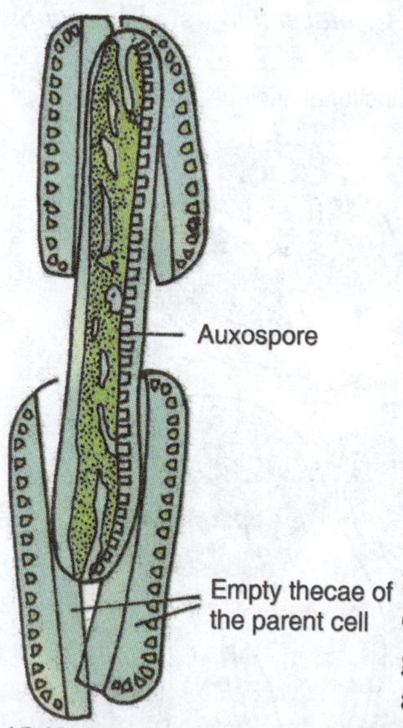

Fig. 15.10. *Pennales*. Formation of a single auxospore by isogamy in *Surirella saxonica*. Note the auxospore within the empty valves of the conjugating cells lying end to end (After Karsten).

Surirella saxonica (Fig. 15.10). It displays slight variation from the above. The two conjugates cohere end to end in this case. The diploid nucleus of each conjugate undergoes meiosis to form four haploid nuclei. Three of these degenerate and one survives in each conjugate after the completion of meiosis. The protoplast of each conjugate with the surviving nucleus constitutes a single gamete. The frustules of the two conjugates open at the opposite adjacent cells to allow the two isogametes to fuse to form a single zygote within the empty valves of the two parent frustules.

3. Formation of a single auxospore from a single individual by paedogamy. Geitler (1939) reported that the cell of a single diatom becomes enveloped in mucilage. Its diploid nucleus divides by **meiosis**. Two of the 4 resultant haploid nuclei partially degenerate. The protoplast then divides into two. Each daughter protoplasm with one functional and one non-functional haploid nuclei behaves as a gamete. The sister gametes fuse to form the fusion cell or zygote. This method of gametic union in which the two sister cells fuse to form a zygote is called **paedogamy** (pedogamy). It escapes from the parent frustule to function as an auxospore. The auxospore increases in size and becomes enveloped in the **perizonium.**

B. Auxospore formation by oogamy (Fig. 15.11). A pennate diatom *Rhabdonema adriaticum* forms auxospores by oogamy. It is **dioecious**. According to Stosch (1958), the female diatom cell (B) which functions as a oogonium is large. Its diploid nucleus undergoes meiosis. Subsequently the protoplast of the oogonium cleaves into a single large functional ovum (B). The latter occupies the upper third of the oogonium. The functional ovum has two nuclei. One of these degenerates towards maturity. The diatom cells which function as antheridia are small and numerous (A). The diploid nucleus of an antheridium undergoes meiosis followed by cleavage of the male protoplast into two microgametes. Each microgamete is a globular, naked, non-flagellate amoeboid binucleate structure. One of these haploid male nuclei is fertile. The other aborts. The antheridia containing the microgametes float passively in water and chance to be carried up by water currents to the oogonia to which they become attached by the mucilage pad (C). The two thecae of the attached antheridium gape apart to liberate the amoeboid microgametes which creep up to the region of the cleft in the oogonium and finally one of them injects only its functional nucleus into the egg cell or ovum to accomplish fertilization. The zygote thus formed functions as an auxospore. Sexual reproduction in this pennate diatom is reminiscent of the centrales.

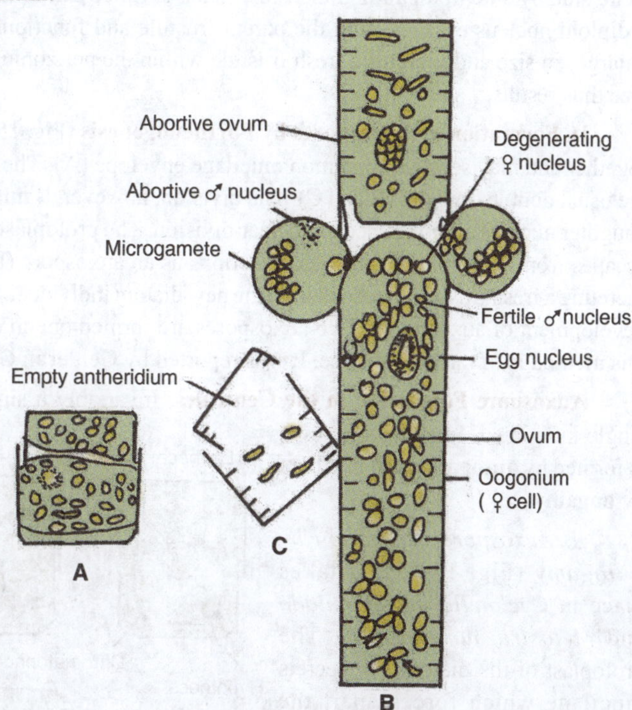

Fig. 15.11. (A-C). *Pennales*. Diatom *Rhabdonetna adriaticum* (girdle view). Sexual reproduction. (A), antheridium with two microgametes; (B), oogonium with a large functional ovum and a smaller abortive ovum; (C), attached empty antheridium (Based on Von Stosch).

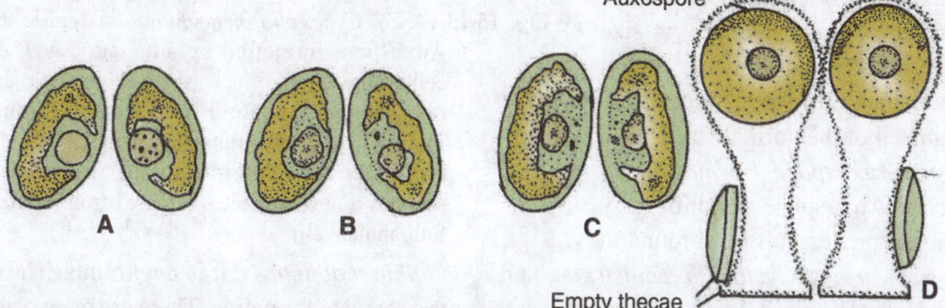

Fig. 15.12 (A-D). *Pennales*. Auxospore formation in *Cocconeis placentula*, var. *lineata* by parthenogenesis. (A), Two cells creeping together; (B-C), showing double nuclear division which is mitotic and with one daughter nucleus degenerating after each division; (D), protoplast of each cell with the surviving diploid nucleus escapes from the parent frustule and enlarges to function as auxospores (After Geitler).

C. Auxospore formation by autogamy (Fig. 15.12). This is illustrated by *Amphoranormani*. Geitler has found that the diatom cell becomes enveloped in mucilage. The diploid nucleus undergoes the first meiotic division which is reductional. The two haploid daughter nuclei in the protoplast come to lie side by side in a pair and then fuse. This is called **autogamous pairing**. The protoplast with a diploid nucleus escapes from the parent frustule and functions as an auxospore. The auxospore enlarges in size and secretes a fresh frustule within the perizonium. A new individual of the normal size thus results.

D. Formation of Auxospores by Parthenogenesis (Fig. 15.12). The diatom individuals come together and may secrete a common mucilage envelope (A). The diploid nucleus of each undergoes the usual double division (B and C). The division, however, is mitotic. There is disintegration of one daughter nucleus after the second nuclear division. The protoplast with the surviving diploid nucleus escapes from its parent frustule and functions as an auxospore (D). It enlarges considerably before secreting a fresh frustule to reconstitute the new diatom individual. In this case there is parthenogenetic development of auxospores. Two auxospores are formed one in each cell. Auxospore formation by this method is of rare occurrence. It was reported by Geitler in *Cocconeis placentula* var. *lineata*.

Auxospore Formation in the Centrales. Invariably a single auxospore is formed within a single individual. In some species it is formed by **autogamy** and in others by **oogamy**.

A. Auxospore Formation by Autogamy (Fig. 15.13). It takes place in *Cyclotella meneghiniana* and *Melosira nummuloides*. The protoplast of the diatom cell secrets mucilage which forces apart the thecae. The diploid nucleus then undergoes meiosis (B and C). Of the 4 resultant haploid nuclei two degenerate. The two surviving nuclei approach near each other and fuse (D). This is **autogamy**. The protoplast with the diploid nucleus escapes from the parent frustule (E). It functions as an auxospore or renewalspore. The latter increases considerably in size and secretes a fresh frustule within the perizonium to reconstruct a new diatom cell of the normal size.

Fig. 15.13 (A-E). *Cyclotella meneghiniana* (Centric diatom) Auxospore formation by autogamy. (A), diatom cell with a diploid nucleus; (B-D), meiotic division resulting in 4 haploid nuclei and beginning of fusion of the two functional ones; (E), diploid protoplast enlarging to form an auxospore which secretes a new frustule. (Adapted from Iyengar and Subrahmanyan).

B. Auxospore Formation by Oogamy. Auxospore formation by the sexual process has been found in *Melosira varians, Cyclotella tenuistriata,* and *Biddulphia mobiliensis*. It is **oogamous**. The diatom individuals are diploid. Meiosis takes place at the time of gamete formation. There are many variations in the development of sex organs in different species of centric diatoms.

Antheridia (Fig. 15.14). The male diatom cells in *Melosira varians* directly function as **antheridia** (A). The diploid nucleus of antheridium undergoes meiosis followed by cleavage of the

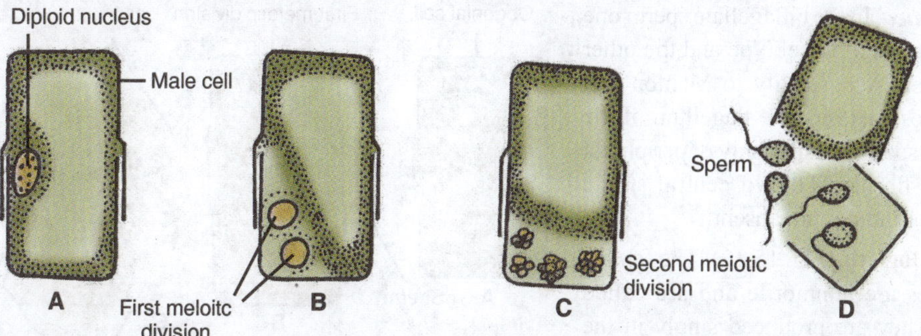

Fig. 15.14 (A-D). *Melosira varians* (Centric diatom). Stages in the development of sperms. (A), male diatom cell with a diploid nucleus; (B-C), meiotic nuclear division followed by cleavage of cytoplasm; (D), antheridium releasing 4 sperms. (After Von Stosch).

protoplast around the four haploid nuclei (B-C). Each uninucleate daughter protoplast metamorphoses into a sperm. The mature antheridium thus released four sperms each furnished with a simple flagellum of pantonemetic or tinsel type (D).

The protoplast of the male cells in certain other species of centric diatoms, on the other hand, divides by mitosis to form smaller cells (Fig. 15.15 A). These daughter cells function as simplified antheridia and are termed the **spermatogonia**. For example in *Melosira varians,* the mother cell gives rise to 4-8 spermatogonia. In *Biddulphia granulata* (A) and *Lithodesmium undulatum* (B) several spermatogonia are formed. Each spermatogonium forms 4 sperms by meiosis. The diatom mother cell in this case releases a large number of sperms. The total may go up to 128. Formerly these were considered to be asexual in nature and called microspores.

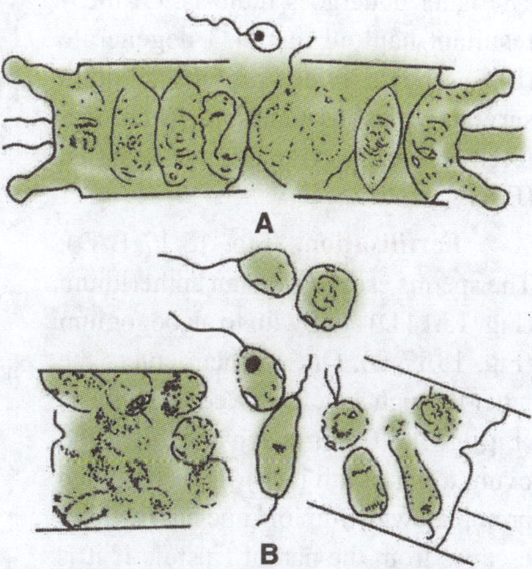

Fig. 15.15 (A-B). *Centrales.* Formation of sperms, (A), *Biddulphia granulata;* (B), *Litho-desmium undulatum* (after Von Stosch).

Schulz and Traitior (1968) reported the occurrence of both uniflagellate and biflagellate sperms in cultures of *Cyclotella meneghiniana* and

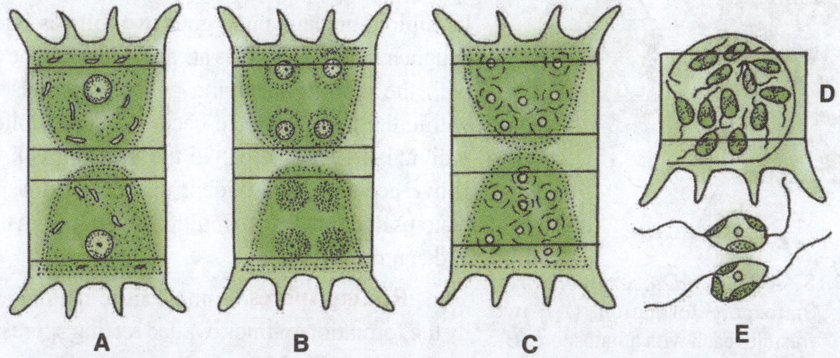

Fig. 15.16. (A-E). *Centrales.* Formation of sperms (microspores) in *Biddulphia mobiliensis*. (A) vegetative cell divides into two daughter cells; (B-D), stages showing repeated division of protoplast of each daughter cell; (E), liberated sperms (After Bergon).

C. cryptica. In the biflagellate sperm one flagellum is of tinsel type and the other whiplash. Accordingly to Manton and Von Stosch (1966) the flagellum of the diatom sperm consists only of peripheral double fibrils. The two central single fibrils or tubules are absent.

Oogonia. Fig. 15.17). The female gametes are nonmotile and are called **eggs**. They are produced singly in the **female** diatom cell or **oogonium** (A) which is generally a larger diatom cell with an elongated diploid nucleus. The latter undergoes meiosis. Of the 4 resultant haploid nuclei 3 degenerate (B-C). The undivided protoplast with a surviving nucleus functions as an egg. It remains in *situ* (B) or may be released (E-F).

Fertilisation. (Fig. 15.17 B-D). The sperms escape from an antheridium (Fig. 15.14 D) and swim to an oogonium (Fig. 15.17 B). One of them enters the latter through a cleft between the thecae (Fig. 15.17 C) or reach the liberated ovum to fuse with it (F). The combined protoplast with a diploid nucleus (zygote) escapes from the parent frustule if it is still present and germinates forth with. It is the renewal spore or auxospore. The auxospore increases in size. Its diploid nucleus undergoes two mitosis but only one daughter nucleus survives at each division. The auxospore with the diploid surviving nucleus secretes a frustule within the perizonium to become a new dipliod diatom individual. It is also called the **firstling cell**. From the above account it is evident that the sexual process in diatoms does not lead to multiplication in both the pennate and centric diatoms.

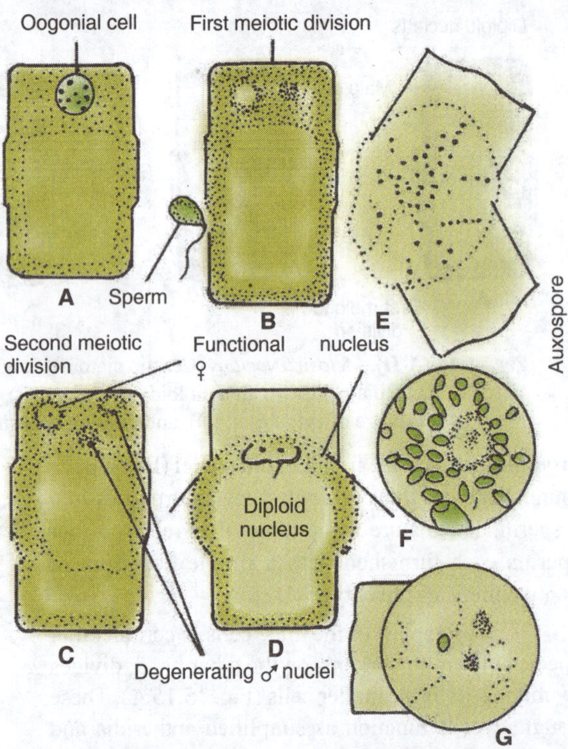

Fig. 15.17. (A-G). *Centrales.* Stages in the formation of auxospore by oogamy in *Melosira varians* (A-D). (A), oogonium with a diploid nucleus; (B), first division and entry of sperm; (C), Second meiotic division prior to union of male and surviving female nucleus, (D), fusion of male and female nuclei in the young auxospore. (A-D after V. Stosch); (E), ovum emerging from oogonium in *Cyclotella* (after Geitler); (F), free ovum in *Lithodesmium undulatum* (after V. Stosch); (G), zygote in *Cyclotella* (after Geitler).

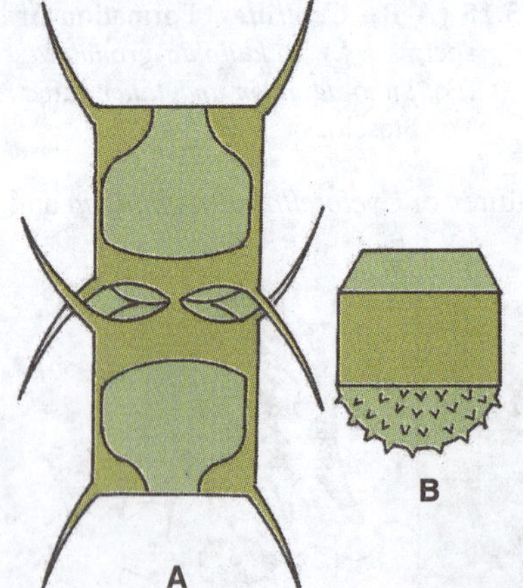

Fig. 15.18 (A-B). *Chaetoceros elmorei.* Statospore formation. (A), two frustules each with a statospore (B), liberated statospore (After Boyer).

Resting spores. Some centric diatoms reproduce by the formation of thick-walled resting spores known **as statospores** (Fig. 15-18 A-B), or **cysts**. These are formed under adverse conditions. In *Melosira* two statospores are formed in each cell endogenously.

PINNULARIA

Salient Features

1. It is a fresh water alga occurring in ponds, pools, and ditches and is usually found floating on the surface of water.
2. *Pinnularia viridis* is a unicellular species whereas *P. socialis* and *P. debesii* are colonial. The individual cells are elongate-elliptical in shape or in the form of an oblong box and measures about 140μ × 20μ.
3. The silicious cell wall or *frustule* is made up of two halves or *Valves* one fitting into other like a box (Fig. 15.1). The outer valve is older one and is called *epitheca* while inner and younger one is called *hypotheca.* Each valve has two portions-the flat top (Valve face) and the margin or *Girdle*. The top corresponds to the upper side or lower side (bottom) of the box and the girdle to the overlapping sides of the box. The girdles or mantles of two valves overlap each other and are together called girdle band. When viewed from side it is called girdle view (Fig. 15.1 A) and when looked from top it is called valve view (Fig. 15.1 B). The valve view reveals the detailed structure of the organism.
4. In the valve view very prominent markings the **costae** are seen which are arranged in the form of parallel transverse ribs. These markings arise from margin but do not reach the centre. These costae in the frustule are made up of rows of small dots having fine cavities. Along the centre of the valve there is long line known as **raphae** which is a long slit in the wall and is not straight but is sigmoid. The raphe divides the valve symmetrically.
5. There are three spots in the raphe which are thickenings in the cell wall. The thickening in the centre is called **central module** while thickenings at two ends are called **polar nodules**.
6. The raphe is supposed to be associated with the movements of the alga and *Pinnularia* shows creeping movements. The exact mechanism is not completely known but the movement appears to be related with the friction caused by the streaming of protoplasm.
7. The protoplast consists of colourless cytoplasm, the primordial utricle which surrounds a large central vacuole and two large parietal olive green or golden brown coloured plastids each extending to the length of the cell. The pyrenoids may be present in each chloroplast and photosynthetic product is fatty oil or leucosin. Pigments in the plastids comprises of chlorophylls *a* and *c,* xanthophyll, and -carotene and the presence of diatomin renders golden brown colour to the alga.
8. Spherical or oval nucleus lies in a bridge like mass of dense vacuolated cytoplasm extending across the centre.
9. Reproduction takes place by mitotic cell division and *auxospore formation*. During cell division increase in the size of cell at the right angles to the girdle take place and nucleus divides mitotically. The nuclear division is followed by the separation of two valves and cleavage of cytoplasm resulting in the formation of two equal parts. Each half of the divided cell thus possesses one valve of the parent cell which becomes *epitheca* while other side remains naked and develops soon a new valve which functions as hypotheca. The newly developed valve fits in accurately inside the old valve.
10. As a result of the utilization of old parent valve by the daughter cells during mitotic cell division and multiplication of alga, one cell maintains the size of the parent cell while the other cell becomes slightly smaller and thus during successive generations the size of one of the daughter cells is progressively reduced. However, gradual reduction in the size of cells is compensated by auxospore formation after sometimes and normal size is regained.

11. Auxospore formation is an important process in diatoms because it sets the rejuvenescence and cells attain their normal (parent) size. During the process abundant mucilage is secreted by the cell which results in pushing the valves apart and liberation of the protoplast. After being liberated, it grows to the maximum size and after attaining the size of the parent cell surrounds itself with a silicified pectin membrane called *perizonium.* There is no nuclear division. The fresh valves and connecting bands along with other structures of fully grown cell develop afterwards and thus a large new individual is formed.
12. Sexual reproduction has not been reported in *Pinnularia*.

Taxonomic Position

Division	:	**Bacillariophyta** (Bacillariophycophyta)
Class	:	**Bacillariophyceae**
Order	:	**Pennales (Pennatae)**
Family	:	**Naviculoideae**
Genus	:	*Pinnularia*
Species	:	*viridis*

Economic Importance

The diatoms constitute an important source of food for aquatic animals including fish and whale. Oil is the principal result of diatom photosynthesis. The oil contents of the whale blubber, of fish and shark liver oils with their vitamins A and D, are regarded as the condensates from oil droplets originally stored in the diatom cells. They were passed along through the food chain to be finally stored in the fish liver. Vitamin A is abundant in plankton diatoms. Diatom *Nistochia* is extremely rich in this vitamin. The diatoms, in fact, perform a distinct service in the economy of nature. They trap and conserve the life-given nutrient materials washed from the land into the sea and thus keep them into circulation.

Diatomaceous Earth

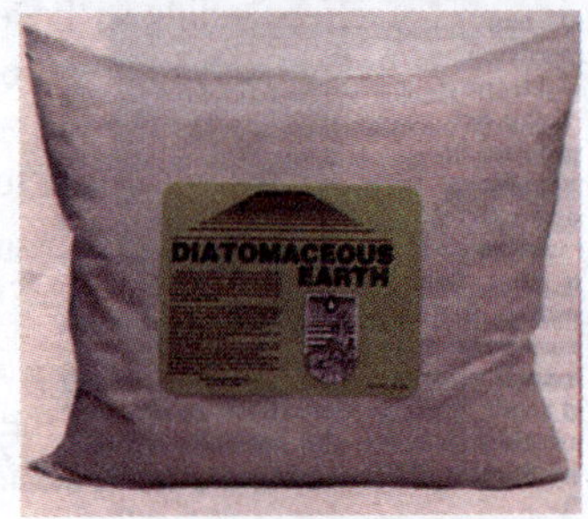

Diatomaceous earth for natural snail control

The diatoms have lived for may ages. The indestructible siliceous frustules of these past diatoms have formed oceanic sediment. The beds of ancient ocean are full of this oceanic sediment. Such thick beds form the **diatomaceous earth**. This is a rock-like deposit. It extends hundreds and thousandth of feet in depth in some localities. The diatomaceous earth is mined in several parts of the world to obtain a whitish powder (**diatomite**) which is put to the following uses:

1. The diatomite is fire proof and highly absorbent. Because of these properties it is used in filters in industries (sugar refineries and brewing industries); in packing corrosive chemical liquids; in polishes for metals, insulation for pipes and furnaces, in the manufacture of dynamite, insulation or refrigerators, houses and in making the latter sound proof.

2. Sprinkling of diatomaceous material on the walls of the mines reduces the dangers of dust explosions.

3. It is also used in tooth paste.

According to one view significant percentage of the world's oil supply (petroleum) is considered to be of diatom origin. When the diatom cell dies, the unused oil collects with billions of droplets resulting from the dead and decayed bodies of diatoms. This oil collects in certain pockets under proper geophysical conditions. Much of the petroleum used in industry today is probably derived from the oils synthesized and stored by diatoms of past ages.

Phylogeny of Diatoms

The relationships of Bacillariophyta are rather obscure. On the basis of similar pigmentation, nature of food reserves, chemical composition of cell wall and formation of an encapsulated dormant spore (**Statospore**), Pascher suggested relationship of the Bacillariophyceae (diatoms) with the Chrysophyceae and Xanthophyceae. In fact, he suggested the inclusion of the three classes in the same phylum Chrysophyta. The opponents of this view hold that the similarities, no doubt, are significant but in the absence of any clear connecting forms, a full evaluation of the evidence is difficult. Because of the presence of fucoxanthin, a typical pigment of the brown alga, some algologists consider the diatoms to be related to the brown algae.

As to the evolutionary trends among the diatoms themselves the centric diatoms are considered to be primitive and the pennate forms derived from them. The fossil evidence supports this view. Owing to the presence of highly silicified cell wall, fossil record of diatoms of past ages is available. Fossils of centric forms have been reported from the Jurrasic period whereas the pennate forms make their first appearance in the early Tertiary. This denotes that the pennate diatoms originated from the centric forms. The pennate forms with no raphe are considered the immediate descendants.

NAVICULA

Occurrence: *Navicula* is one of the most common genera of freshwater habitats of diatoms. It is cosmopolitan and ubiquitous in distribution and forms a major component of the planktonic vegetation. It is very important as primary producer in the food web of aquatic ecosystems. In India *N. halophila* is the most common species of the genus.

Thallus Structure: *Navicula* is isobilaterally symmetrical unicellular diatom and occurs in isolated form. The cell wall consists of two overlapping halves, the upper half, known as the *epitheca* and the lower, the **hypotheca**. Each theca is further divided into two parts-the main surface and its incurved margins which are termed the **valve** and the **connecting band**, respectively. The term girdle is applied to the side pieces collectively. When fitted together, the connecting band of the epitheca overlaps that of the hypotheca and the two bands remain united in the overlapping region by a cementing organic substance present between them. Accordingly, a cell can be seen from two different aspects, the **valve view** and the **girdle view** (Fig. 15.19 A-B).

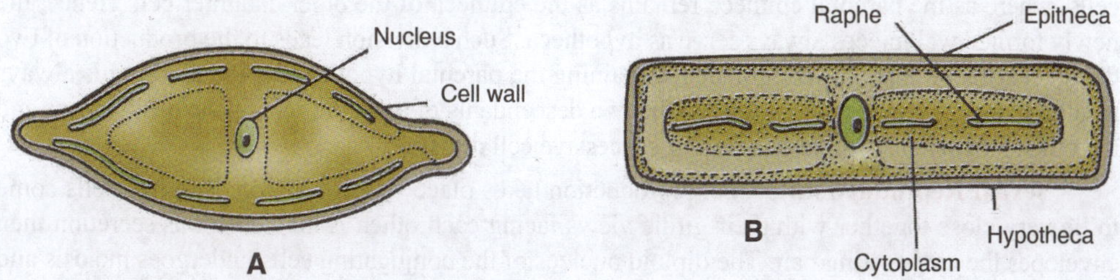

Fig. 15.19 (A-B). Cell in valve and girdle view respectively.

The cell wall is enriched with amorphous silica and has an organic component of unknown nature.

In *Navicula,* there is a longitudinal slit, termed the **raphe**, which is interrupted in the middle by a **central nodule** formed by internal thickening of the valve. Each pole of the valve also contains a **nodule (polar nodule)** which is formed in the same manner as the central nodule. The valve markings of *Navicula* consists of fine pores bridged from the inside by sieve membranes containing numerous fine pores. The raphe situated in the central rib is also called the apical rib. The rows of pores situated on the lateral sides of the apical rib and separated from the other by solid silica ribs are termed **transapical ribs**. The pores within a row are seperated by siliceous connections which link the adjacent transapical ribs.

A thin plasma membrane, which lies close to the inner face of the cell wall, surrounds the protoplast. The protoplasm surrounds a central vacuole. There are two brownish yellow large plate-like frequently lobed chromatophores, adpressed to the girdle side in the resting cell. The nucleus lies in a massive plasmatic bridge which seperates the vacuoles and connects the two chromatophores. The cytoplasm also contains mitochondria, endoplasmic reticulum and golgi bodies. The reserve food material is in the form of fatty oils volutin globules.

Reproduction: Both vegetative and sexual reproduction occur in *Navicula.*

Vegetative Reproduction: (Cell division, Fig. 15.20) Before division the two thecae of the parent cell become slightly apart due to increase in size of nucleus and volume of cell. The nucleus then divides mitotically and karyokinesis is followed by cytokinesis of the protoplast resulting in two uninucleate parts along the valvar plane. With the completion of cytokinesis each daughter cell receives one chromatophore which further divides transversely to form two chromatophores.

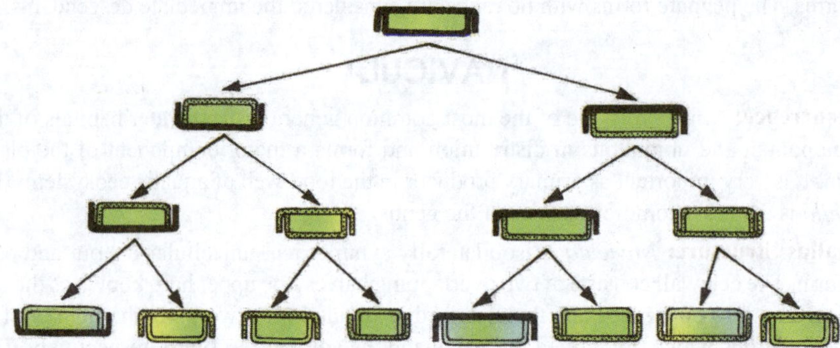

Fig. 15.20. Vegetative reproduction (diagrammatic representation).

Soon after cytokinesis, new halves and connecting bands are deposited along the freshly formed protoplasmic surfaces. The parental hypotheca serves as the epitheca of one of the two daughter cells, where as the parental epitheca remains as the epitheca of the other daughter cell. Hence, the newly formed wall pieces always serve as hypotheca. Such a division leads to the production of two daughter cells of unequal size; the cell containing the parental hypotheca as the epitheca is always smaller than its sister cell. Thus, one of the two descendants of a diatom retains the original size and the other shows reduction in size during successive cell divisions.

Sexual Reproduction: Sexual reproduction takes place by conjugation. The two cells come to lie very close together with their girdle views facing each other. A mucilagenous secretion then envelopes the conjugating pair. The diploid nucleus of the conjugating cells undergoes meiosis and produces four haploid nuclei of which two form amoeboid gametes and the remaining two degenerate. Each gamete with a single chromatophore shows some amoeboid movement inside the valves and then round up. The valves of the mother cell are pushed apart and the two gametes of each conjugating cell become free within their respective valves. The two gametes of a conjugating cell moves towards the other cell where they fuse in pair and forms two zygotes. Physiologically, the active gametes were considered as male and less active or stationary ones as female. One of the two male gametes moves first and fuses with the female gamete and then the second one follows the same process. The zygote thus formed is spherical, large and termed as auxospore.

After a resting period of few hours the zygote germinates. The protoplast of the zygote elongates and the two gametic nuclei from two pairing gametes now fuses. New valves are then secreted. The new cells which lie parallel to old mother cell valves, divide vegetatively and this marks the beginning of the new vegetative phase.

Taxonomic position

Division	:	Bacillariophyta
Class	:	Bacillariophyceae
Ords	—	Pennales
Family	—	Naviculoideae
Genus	—	*Navicula*
Species	—	*halophila*

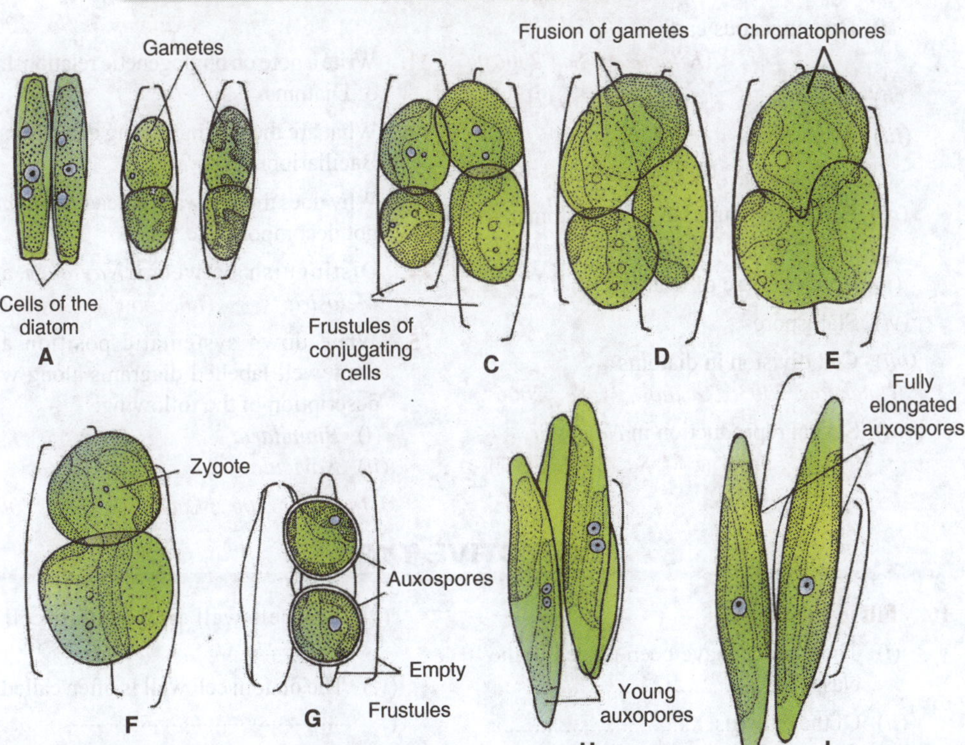

Fig. 15.21. Sexual reproduction in *Navicula*: A-G. conjugation of gametes and formation of the auxospore, II-I. A stage in the enlargement of auxospore, 1. fully elongated auxospore (After Subrahmanyam, 1946).

QUESTIONS

Essay Type

1. Describe habitat, structure and reproduction in *Pinnularia.* (*Kanpur, 1997*)
2. Give a general account of Diatoms.
3. Describe the modes of reproduction in Pennate diatoms. (*Meerut, 1993*)
4. Give an account of auxospore formation in Pennates.
5. Describe the distinguishing characters of Bacillariophyceae. (*Kakatiya, 1998; Nagarjuna, 1988; Osmania, 1991*)
6. Describe the structure of pennate diatoms. (*Meerut, 1996*)
7. Describe cell structure of cell wall and reproduction in Diatoms. (*M.S. Univ. 1995*)
8. Write a short essay on economic importance of diatoms.
9. Describe sexual reproduction in diatoms. (*Kanpur M. Sc., 2005*)

SHORT ANSWER TYPE

10. **Write short notes on the following:**
    (*i*) Diatomaceous earth (*Kanpur M. Sc., 2006*)
    (*ii*) Diatoms (*Kumaon, 1997*)
    (*iii*) Auxospores (*Kumaon, 1988; Nagarjuna, 1998*)
    (*iv*) Economic importance of Diatoms. (*Nagarjuna, 1990; Osmania, 1993*)
    (*v*) Cell structure of Diatoms
    (*vi*) Statospores
    (*vii*) Cell division in diatoms (*Lucknow, 2003; Lucknow, M. Sc., 2006*)
    (*viii*) Sexual reproduction in *Navicula* (*Lucknow, M. Sc., 2005, 2007*)
    (*ix*) Movements in diatoms. (*Kanpur M. Sc., 2004*)
11. Write a note on phylogenetic relationships of Diatoms.
12. What are the distinguishing characters of Bacillariophyceae ?
13. Why does the cell wall of dead *Pinnularia* not decompose?
14. Distinguish between: *Navicula* and *Melosira.* (*Lucknow M. Sc., 2004*)
15. Write down systematic position and draw well labelled diagrams along with description of the following:
    (*i*) *Pinnularia*
    (*ii*) *Navicula*
    (*Lucknow, 2005; Lucknow, M.Sc., 2006*)

OBJECTIVE TYPE

16. **Fill in the blanks :**
    (*i*) The diatoms have been placed in the class ....
    (*ii*) Of the diatoms, ... forms are non motile.
    (*iii*) The example of diploid unicell is ....
    (*iv*) The saprophytic diatom is ....
    (*v*) The cell wall in a diatom cell is ....
    (*vi*) The diatom cell wall is often called as ....
    (*vii*) In diatoms the cell division results in the formation of unequal ....
    (*viii*) .... are the characteristic spores of diatoms.

(*ix*) Sexual reproduction is absent in ....

(*x*) The asexual method of reproduction in diatoms is ....

(*xi*) Statospores are usually formed in the genus ....

17. Select the correct answer :

(*i*) Auxospore formation is exhibited by

(*a*) Desmids (*b*) Diatoms
(*c*) Green algae (*d*) Red algae

(*ii*) The position of chromatophores in *Pinnularia* is

(*a*) Axial
(*b*) At the two ends
(*c*) Opposed to the walls
(*d*) Opposed to the girdle

(*iii*) Raphe in diatoms is visible in the

(*a*) girdle view (*b*) side view
(*c*) valve view (*d*) end view

(*iv*) The shape of the chromatophores in *Pinnularia* is

(*a*) Discoid (*b*) Plate like
(*c*) Star like (*d*) Rod like

(*v*) The structure which is responsible for locomotion in diatoms is

(*a*) Costa (*b*) Raphe
(*c*) Valve (*d*) Girdle

(*vi*) In which of the following Statospores are produced

(*a*) *Melosira* (*b*) *Cosmarium*
(*c*) *Cladophora* (*d*) *Spirogyra.*

(*vii*) In which of the following the daughter cells produced as a result of division are unequal ?

(*a*) Chlorophyceae
(*b*) Phaeophyceae
(*c*) Xanthophyceae
(*d*) Bacillariophyceae

(*viii*) Fresh water or marine planktons are largely made up of

(*a*) Desmids (*b*) Diatoms
(*c*) Green algae (*d*) Brown algae

(*ix*) Oil chrysolaminarin and volutin are food reserves of

(*a*) Desmids (*b*) Diatoms
(*c*) Green algae (*d*) Brown algae

(*x*) Diatomite which is economically important is obtained from the algae commonly known as

(*a*) Green algae
(*b*) Blue green algae
(*c*) Yellow algae
(*d*) Yellow brown algae

CHAPTER 16

Fucales : *Cystoseira*

PHAEOPHYTA (Phaeophycophyta, Brown Algae)

- General Features
- Distribution

General Features

The brown algae are distinguished by their colour which varies from olive green through light golden to a rather deep shade of brown. This is due to the presence of a golden brown xanthophyll pigment **fucoxanthin** (C_{40} H_{54} O_6) in their chromatophores. This is in addition to chlorophyll *a*, chlorophyll *c*, B-and C-carotene and other xanthophylls (such as lutein, flavoxanthin and violaxanthin). The fucoxanthin, however, is usually sufficiently strong to partially mask the chlorophyll and carotenoid pigments in the cells. The pigments are located in the plastids which usually lack pyrenoids. Hasex and Blinks (1950) showed that fucoxanthin is active as an accessory pigment in photosynthesis.

Typically the cells are uninucleate. The large thalli are tough, leathery or rubbery in texture. Usually the thalli of brown algae secrete abundant mucilage which readily absorbs moisture and retains it tenaciously. This keeps the plants moist at low tide when they are exposed. Mannitol and laminarin are the reserve photosynthetic products, rarely some fats. The presence of fucosan vesicles in the cells is another characteristic. The plant body is always immobile and multicellular. Architecturally it is either a uniaxial branched filament or a large elaborately parenchymatous thalloid organisation usually of the defined form. In the majority the general construction includes a **holdfast**, a long or short **stipe** and an expanded **blade**. The latter function both, as a photosynthetic and a reproductive structure.

Motile reproductive cells are commonly found in the brown algae, generally these are pyriform or spindle-shaped and biflagellate. The two flagella are of unequal lengths and inserted laterally. One of these is of whiplash type and the other tinsel. Sexual reproduction ranges from isogamy to oogamy through anisogamy. Oogamy, however, is the general rule. Alternation of generations is frequently present. Fucales are the only exceptions. Generally alternation of generations is of isomorphic type but developed to a point where the gametophyte and sporophyte plants begin to diverge morphologically (*Laminaria*). The number of species composing the class is over 1500. They are placed under more than 250 genera.

Distribution

Unlike the Cyanophyceae and the Chlorophyceae which are mainly fresh water forms, the brown algae are almost exclusively marine. There are only few rare ones which are fresh water and usually grow in the streams that drain directly into the ocean. These are *Pleurocladia lacustris, Sphacelaria, Pseudobodanella lithoderma,* and *Heribaudiella*. The marine forms are found in the shallow waters along the coasts of all seas but they attain their greatest development both in number and large size in the cold waters of oceans and seas of northern latitudes. The majority grow in shallow water in the intertidal zones of rocky coastlines. They are benthonic and grow as lithophytes attached by holdfasts to the rocks, stones or timbers beneath the surface. They usually develop air bladders to buoy up the free portions of their thalli. At low tide they are exposed to air for several hours. The algin coating protects them from desiccation by retaining sufficient amount of water. Besides desiccation, the plants growing in the intertidal zone are exposed to continuous buffeting of the waves as the tides ebb and flow. It may be mentioned here that most seas afford two distinct habitats for attached flora. These are, (*a*) the part of the shore which is twice covered and uncovered by the tides and (*b*) the region below low water mark. The former is called the **littoral** or **intertidal** and the later **sub-littoral** zone. The sub-littoral zone is more extensive and the algal vegetation of this region always remains submerged. The littoral or intertidal zones or rocky shores often show distinct zonation of species forming distinct horizontal layers or bands. Each layer has an almost pure stand of a member of the Fucales. In both the littoral and sub-littoral zones the brown algae occur in very definite areas. Covering the rocks of the upper littoral zone, which are under water during the highest tides, grow *Fucus* and other allied forms. In the lower littoral zone are found the kelps (*Laminaria*). Between these two areas occur brown algae as *Leathesia*. The sub-littoral zone abound the giant kelps.

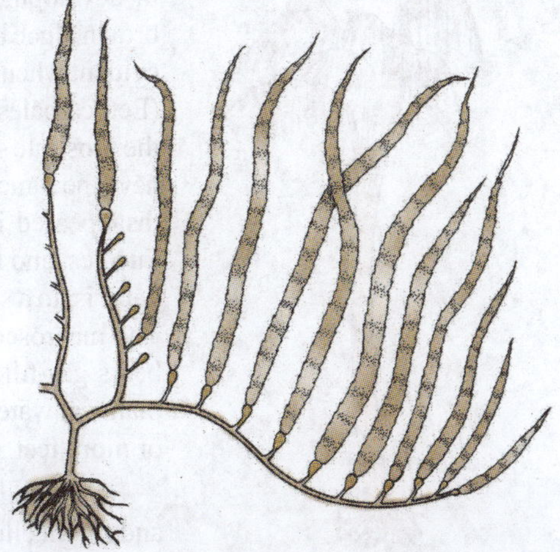

Fig. 16.1. *Phaeophyta. (Macrocystis pyrifera)*

The brown algae predominate in the littoral flora of the cold water of the temperate, arctic and

antarctic seas. Hence they are abundant along the northern stretches of Atlantic and Pacific coasts. Dictyotales and *Sargassum* are, however, unorthodox genera. They are fairly abundant in the warm tropical seas (Atlantic ocean). They are best developed in the shallow water between low and high tide marks. Of course, a few can penetrate to very considerable depths more than 300 ft. below the surface of water. Normally they grow attached to rocks as lithophytes and other inanimate substrata by means of extensively branched *holdfasts* or grow as epiphytes on other algae and some are endophytes. Sometimes they become detached and form floating masses.

Range of thallus organization. The Phaeophyta are an assemblage of most highly evolved types of Algae. In them the vegetative organisation of the thallus surpasses that of any of the algae so far considered. All are multicellular and sessile. They have a definite form and are both the largest and the most rugged of the algae. They display the highest degree of body differentiation. Unicellular and colonial forms (motile or non-motile) are unknown at present. The unbranched filament so common in the Chlorophyceae has also not been reported. The simplest type of the plant body is represented by a **heterotrichous** filament which is the highest stage of development in the green algae. Morphologically therefore the brown algae begin where the green algae finish. Heterotrichy is only evident when the plants are examined in early stages of development (Ectocarpales). Older plants may either have the erect system or only the prostrate system. In this they resemble the Chaetophorales which have the same type of thallus organisation. Heterotrichy has gradually disappeared in the higher types of brown algae *e.g.,* Dictyotales, Fucales, and Laminariales. Among the Laminariales heterotrichous habit is, to a slight extent, exhibited by the gametophyte generation. The macroscopic frond-like thallus of the advanced forms usually bears gas-filled bladders or floats which buoy up the free ends of plants in water. In a few species the plant may attain a length of 200 or more feet.

Fig. 16.2. *Phaeophyta. (Lessonia flavicans.)*

Among the higher members like *Dictyota* there is a complex and multicellular plant body (Fig. 18.1 A). It is differentiated into a lower more or less cylindrical portion and an upper dichotomously branched flattened portion which is commonly known as the **frond**. The cylindrical portion is fixed to the substratum by means of a tuft of **rhizoids**. Internally the thallus lacks any differentiation of tissues. It is composed of three layers of cells (Fig. 18.3). The upper and the lower layers form the peripheral part. They enclose a central layer whose cells are filled with **fucosan vesicles** and contain reserve food materials. The cells of the peripheral layers are packed with chromatophores.

Next in the order of complexity are the Fucales. They exhibit marked morphological and anatomical complexity. The plant body (Fig. 10.1 A) consists of a forked, flattened, band-shaped upper part known as the **blade** and a more or less cylindrical lower part (**stipe**). The latter has a distinct attaching **disc**. In *Fucus* branching is dichotomous. Some of the members of the order Fucales show a complex system of branching (*Sargassum*). The plant body in *Sargassum* consists of a central axis from which arise lateral branches which may be flattened in a vertical plane and look like "leaves" (Fig. 20.20). The so-called leaves may again subtend further branches. The thallus is furnished with air bladders. The branching thalli of *Sargassum* give it a bushy appearance.

So far as the internal structure of the Fucales is concerned, they exhibit elaborate differentiation of tissues. A section through the thallus (Fig. 20.3 A) reveals an outer epidermal or the meristoderm layer followed by an extensive cortical region. Cortex may have definite outer and inner cortical layers. In the centre there is the medulla composed of loosely arranged cells followed by the lower cortex. As compared with the Dictyotales, the Fucales show much greater differentiation of tissues.

The more advanced forms such as the Laminariales show much greater morphological and anatomical differentiation. The main plant represents the sporophytic generation. It is the largest known among the Algae. The gametophytes are comparatively minute, filamentous structures. The sporophytes are differentiated into two main regions. These are the cylindrical stipe or the stalk and the flattened blade or lamina (Fig. 19.2). The stipe gradually passes into the flattened lamina and the growing region lies in the transitional zone between the two. At the base the stipe gives rise to a richly branched holdfast. The smallest plants reach a metre in length. The largest may be 200 feet in length. *Macrocystis pyrifera* (Fig. 16.1) attains a maximum length of 60 metres.

Plants of *Lessonia flavicans* (Fig. 16.2) bear resemblance to miniature trees. The thick stipe divides several times near its upper end and the ultimate branches end in long lanceolate blades. It reaches 4 metres in length.

Postelsia palmaeformis (Fig. 16.3), a native of the Pacific shores of North America, resembles in appearance the palm trees. It is commonly known as the *sea palm.* The thallus is composed of a short, thick flexible stipe or axis bearing at its apex a crown of blades or laminae. It is anchored to the substratum by a well-developed and much branched holdfast. Why does the thallus reach such massive dimensions in the Brown algae and how do these massive thalli solve the mechanical problem ? The brown algae occur in the sea. The sea water is rich in mineral contents and thus is able to supply them at the rate which growth of the thallus demands. It is not possible in fresh water. The problem of mechanical support to the massive thallus is solved by upthrust of the sea water. It affords support. The thallus resists the sweeping action of waves by attaching itself firmly to the substratum. It withstands the tearing action and tension of pressure by the tough, hard, slimy nature, leathery and rubbery texture of the thallus. Presence of cross-connections and the trumpet hyphae in the thallus with perforated walls clearly speak of an advanced type of anatomical differentiation.

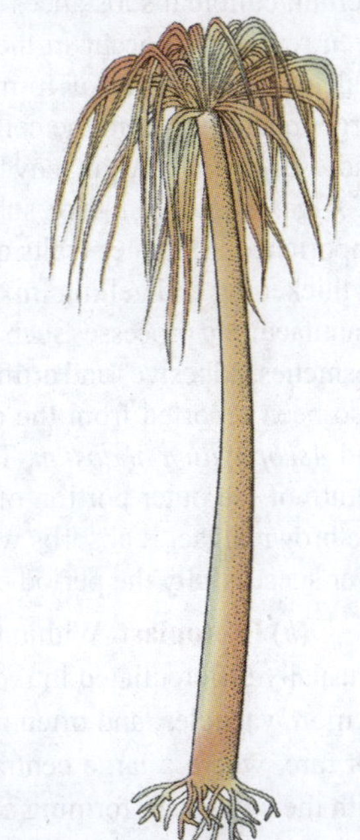

Fig. 16.3. *Phaeophyta. (Postelsia palmaeformis.)*

Macrocystis

Lessonia

Postelsia

Structure of the Cell (Fig. 16.4)

The phaeophycean cell has an organisation similar to that of higher plants. It consists of a cell wall enclosing a tiny mass of protoplast.

(*a*) Cell wall. It consists of two layers, the inner of cellulose and outer of gelatinous and pectic mateiral. The outer layer is usually thin and reticular in nature. The cellulose layer consists of fine parallel microfibrils. It contains glucose and xylose as sub-units. Due to the presence of the pectic compounds in the outer layer, the cell wall becomes gelatinous or gummy. Certain colloidal substances like **'algin'** and **'fucoidin'** have been reported to occur in the outer pectic portion of the cell wall of the majority of forms. These have been isolated in larger quantities from the cell walls of *Laminaria, Sargassum* and *Ascophyllum*. Algin may form about 24% of the dry weight of Alga. These gum-like substances are of great economic importance. Alginates (salts of alginic acid) have the property of thickening and gelling mixtures and thus finds use in many manufacturing processes such as in the preparation of ice cream, cosmetics, adhesives and artificial silk. Presence of callose, has also been reported from the cell walls of *Laminaria digitata* and *Ascophyllum nodosum*. The mucilaginous and gelatinous nature of the outer portion of the cell wall is of much use to the brown algae. It absorbs water and helps the plants to pass over successfully the period of exposure when water recedes.

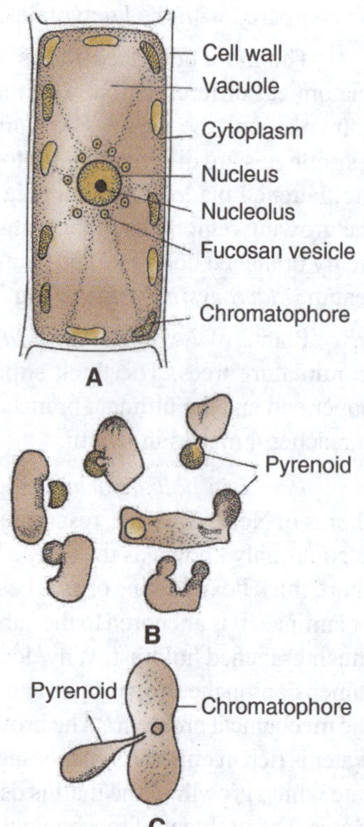

Fig. 16.4 (A-C). *Phaeophyta*. (A), brown algal cell showing structure; (B), chromatophores with protruded pyrenoids (*Ectocarpus siliculosus*); (C), single chromatophore of *Mesogloea* (B after Knight and C after Chdefaud).

(*b*) Protoplast. Within the cell wall is the **protoplast**. It is usually differentiated into cytoplasm, a single nucleus, one or more vacuoles, and often more than one plastid. Cases are not rare, where a large central vacuole is present (*Dictyota*) with the cytoplasm forming a lining layer within the cell wall (A). Presence of **centrosomes** and **mitochondria** has also been recorded. The study of fine structure of brown algal cell has, in addition revealed the presence of ribosomes, endoplasmic reticulum and dictyosomes (golgi apparatus). This cytoplasm contains non-living inclusions which are of the nature of food reserves. The important food reserve of the brown algae is a complex carbohydrate, **laminarin**. It is a unique soluble polysaccharide reserve regarded by some to arise from the simple sugar of photosynthesis. It is present in great abundance in certain species of *Laminaria*. On extraction, laminarin appears as a water soluble and a tasteless powder. *Mannitol* is another photosynthetic product. It is a kind of complex alcohol widely distributed in the cells of almost all the brown algae. The simple sugars are converted into **mannitol** type of alcohol. It is because of this that the sugars are rare in Phaeophyceae. Starch is entirely absent. Certain members of the order Fucales (*Ascophyllum*) contain fats. Proteins and iodoamino acid are also present. High concentrations of iodine, potassium, magnesium and other solutes from sea water accumulate in the cells of brown algae. As a matter of fact certain brown algae (*Laminaria*) are a source of commercial iodine.

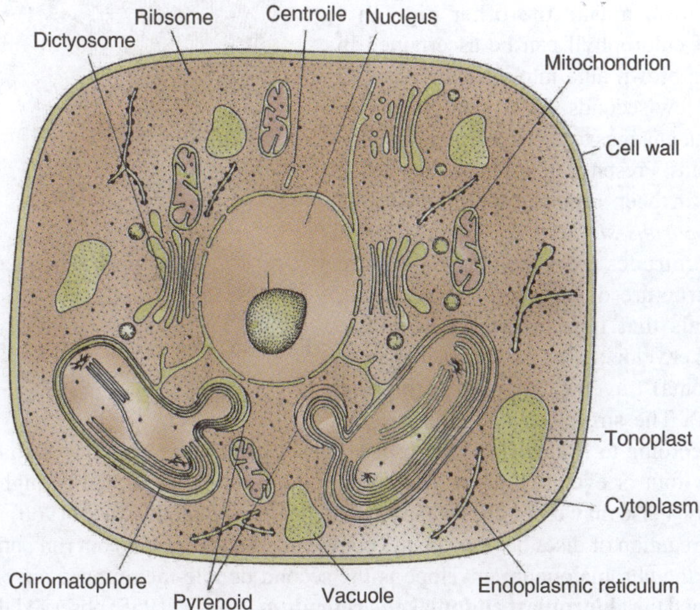

Fig. 16.5. *Phaeophyta.* Diagram of a brown algal cell showing fine structure.

Embedded in the cytoplasm are usually more than one plastids. These are known as the **chromatophores**. Some algologists prefer to call them **phaeoplasts**. They are usually parietal in position. There may be single large plate-shaped chromatophore or several small discoid chromatophores arranged in the parietal layer of the cytoplasm. The shape of the chromatophores varies with genera and species. There may be a single or few plate-like, ribbon-like (*Ectocarpus*) or band-shaped chromatophores per cell. In *E. confervoides* the ribbon is spiral and narrow. The older cells of a certain species of *Ectocarpus* may even contain several discoid chromatophores. So in this genus there is a great variation in the size; shape and number of chromatophores. In certain genera the chromatophores may even be branched. *Pilayella fulvescens* has stellate axile chromatophores. The brown colour of the chromatophores is due to the presence of a golden-brown xanthophyll pigment known as **fucoxanthin** or **phycoxanthin**. In addition, carotenes, other xanthophylls (diatoxanthin) and chlorophyll *a* and a little chlorophyll *c* are present. There is no chlorophyll *b*. Fucoxanthin masks the other pigments. Presence of chlorophyll can be ascertained by immersing a brown alga into a basin of boiling water for a few seconds. As a result of this the brown colour changes to green. Chromatophores lack pyrenoids. Presence of pyrenoid-like bodies has, however, been reported in certain brown algae (*Ectocarpus siliculosus*). When present it usually protrudes from the chromatophore. The ultrastructure of the chromatophore (Fig. 16.6) reveals that it is bound by a double-membraned envelope which encloses a granular matrix (stroma) traversed by a series of **discs** (thylakoids). The stroma in addition contains globuli. According to Bouck (1965) discs are stacked into bands each of which comprises usually three, rarely four or even two discs. The discs in the band are separated by highly uniform space. Occasionally a disc may cross the stroma to join an adjacent band. Evans (1966) reported that there is close aggregation of discs but they do not cohere into bands (stacks) but run parallel to each other. External to the chromatophore envelope is the second double-membraned envelope. The latter is usually termed the **chloroplast endoplasmic reticulum**. Bouck (1965) observed that the chloroplast endoplasmic reticulum (Fig. 16.6) is continuous with the nuclear envelope. The extension of the nuclear envelope around the plastids has been confirmed by Neushul and Dahl (1972). When there are more than one chromatophores in the cell, all are linked with one another through the chloroplast endoplasmic reticulum. This nuclear envelope-chloroplast association may involve the dictyosomes and pyrenoid and thus provide an internal channelling system in the cytoplasm. The projected pyrenoid lacks photosynthetic thylakoids (Evans, 1966) or rarely may contain chromatophore discs traversing the granular or fibrous matrix (Bouck, 1965). According to Bouck (1965) the pyrenoid is surrounded by three double-membraned envelopes. The outer envelope which is greatly dilated forms a cap-like sac over the pyrenoid body and is called the **pyrenoid sac**. Internal to this forming the middle envelope is the **chloroplast endoplasmic reticulum**. The innermost one is the **chloroplast envelope**.

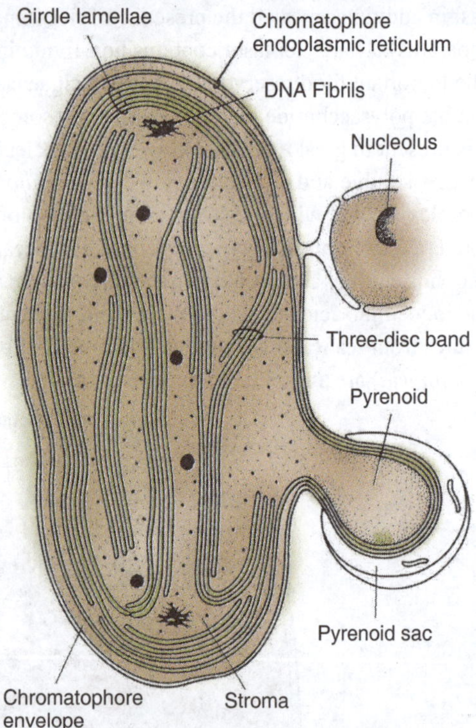

Fig. 16.6. *Phaeophyta.* Diagram showing fine structure of a chromatophore of a brown algal cell.

The mitochondria may occur scattered throughout the cytoplasm (*Fucus*), lie along the margin of the cell (*Chorda*) or may be situated immediately outside the nuclear envelope (*Zonaria*). They are double membraned sacs variable in shape. The inner membrane invaginates to form tubular or villus-like cristae. There is no apparent association of mitochondria with the other organelles (Bouck, 1965). Dictyosomes consisting of stacks of 4 to 8 cisternae may occur in the perinuclear region (*Zonaria, Giffordia* and *Chorda*) but in some brown algae they are scattered throughout the cell (*Fucus*). Neushal and Dahl (1972) observed dictyosomes backed by endoplasmic reticulum from which vesicles appeared to form and merge with the dictyosomes.

Chorda *Giffordia* *Cystoseira*

The brown algal cell has a prominent nucleus. It has the usual structure. There is the limiting double-membraned nuclear envelope enclosing the nuclear sap containing one or a few nucleoli. It divides by the usual mitotic process. The nuclear envelope has pores which occur at irregular intervals. It is continuous with the endoplasmic reticulum of the cytoplasm at many places. In the resting nuclei of many brown algae (*Cystoseira* and *Fucus*) are seen Feulgen-positive bodies of about 0.6 in diameter. They are known as the *chromocentres.* Except the brown algae chromocentres have not been reported in any other algal group. Their significance is not yet known. The location of a rod-like centriole in a depression in the nuclear surface at the apical end is another important feature of the brown algal cell.

Aggregated around the nucleus are present certain highly refractive and colourless vesicles. They contain a chemical substance known as **fucosan** which is considered a waste product by some algologists. It is present in the form of a fluid. It shows acidic reactions. These vesicles are termed as **fucosan vesicles**. They are usually abundant at the sites of high metabolic activity.

The peculiar biochemical characteristics unique to the brown algae are :-

1. Accumulation of reserve food as fats and oils rather than starch.
2. Mostly the microfibrils in the cell walls contain polysaccharides, mannan and xylan in addition to cellulose.
3. The presence in the cell walls of colloids like fucine and fucoidin unknown outside the brown algae.
4. The occurrence of alginic acid in the middle lamellae and primary walls.

Asexual Reproduction (Fig. 16.7). It takes place both by *vegetative methods* and *spore formation.*

(a) Vegetative Reproduction

***(i)* Fragmentation** is the usual and universal method of vegetative propagation in the brown algae. A portion of the thallus known as the **fragment** breaks away from the parent thallus and grows into a new individual. The best example of multiplication by detached fragments is a free-floating species of the genus *Sargassum (S. natans)*. It exclusively multiplies by this method. In certain other brown algae the entire thallus gets split lengthwise into a number of separate parts which remain anchored to the substratum. Each separate part becomes a new plant.

***(ii)* Propagula.** *Sphacelaria* produces special, adventitious branches called the **propagules** or **propagula**. The propagule (A) may be wedge-like to di- or triradiate. It has long or short stalk. Each propagule starts as a normal outgrowth. Later the apical cells divide vertically and the propagule becomes a bi- or triradiate branch system. The propagula are abscissed from the parent plant. The detached propagule grows into a new plant. Many species of larger sea-weeds are able to regenerate a new thallus from the cut stump or stipe.

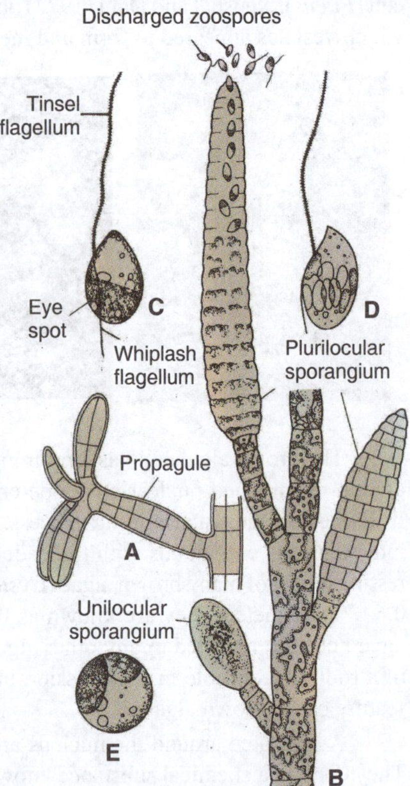

Fig. 16.7 (A-E). *Phaeophyta.* Asexual reproductive structures. (A), propagule of *Sphacelaria* (based upon Olt-mans); (B), diploid filament of *Ectocarpus siliculosus* bearing both plurilocular and unilocular sporangia; (C), zoospore of *Chorda filum* (after Reineke), (D), zoospore of *Ectocarpus globifer* (After Kuckuck), (E), zoospore prior to germination.

(*b*) Spore formation. Asexual reproduction in the Ectocarpales and Sphacelariales takes place by the formation of spores which are differentiated by **mitosis** and thus belong to the category of *mitospores*. Being motile they are called **zoospores**. The zoospores are produced within well-defined multicellular asexual reproductive structures called the **plurilocular zoosporangia** borne upon the diploid or sporophyte plants (B). Genetically the zoospores are diploid and serve to multiply the sporophyte generation. They play no role in alternation of generations and are concerned only in the asexual cycle of the plant and its reduplication. Morphologically the zoospores of brown algae are peculiar. They are pyriform (C) sometimes reinform (D) in outline and are usually with an **eyepot** which is situated near the point of origin of flagella. The zoospore bears two laterally inserted flagella each arising from a basal granule. One of these is of **Whiplash type** and the other **tinsel.** They are of unequal lengths. The longer tinsel flagellum bears a double row of mastigonemes (hairs) and is directed forward in motion and the other backwards. In some species one of the two flagella may be rudimentary. At the broader hinder end of the zoospore is one rarely more brown chromatophores devoid of pyrenoids. According to Manton (1957) who investigated the fine structure of the zoospore of *Scytosiphon lomentarius,* the chromatophore is a membrane bound structure. Its matrix is traversed by a number of lamellae forming discs. He could not estimate the number of discs per band. There were no grana-like structures.

Scytosiphon

The sporophyte plants of Ectocarpales and Sphacelariales, in addition, bear another kind of sporagium which is not partitioned. It is called *unichambered* or *unilocular* sporangium. The unilocular sporangia are borne either on the same plant (A) or on separate plants. In all other zoosporic orders of Phaeophyceae the diploid or sporophyte plants bear only the unilocular sporangia but no plurilocular sporangia. The diploid contents of the unilocular sporangium undergo meiosis. The resultant 4 haploid nuclei undergo mitosis. After the completion of nuclear division the surrounding cytoplasm separates into as many uninucleate portions. The uninucleate daughter protoplasts metamorphose into biflagellate motile haploid elements similar to the diploid zoospores in structure. These are known as the **meiozoospores**. The meiozoospores, on germination, produce the alternate haploid plants (gametophytes concerned with sexual reproduction). Many biologists consider meiospores formation as a stage in sexual reproduction. It serves to multiply the beneficial effects of a single act of fertilization. The meiospores are thus not the true asexual spores. Their formation is contingent upon sexual reproduction. In Dictyotales the meiospores are non-motile. They may be called the *meioaplanospores.* The Fucales do not reproduce by the formation of any kind of spores.

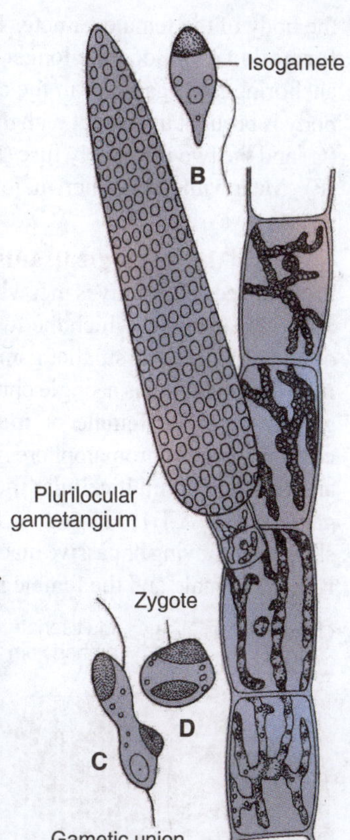

Fig. 16.8 (A-D). *Phaeophyta.* Isogamous sexual reproduction in *Ectocarpus.* (A), haploid filament bearing a nearly mature plu-rilocular gametangium in which isogametes are produced; (B), an isogamete; (C), fusion of isogametes; (D), the resultant zygote.

Sexual Reproduction

It is the concern of the haploid plant which in most of the brown algae is a multiceullar, free living individual. The only exception are the Fucales in which the sexual plant is diploid. Oogamy is the general rule in most of the Phaeophyta. Thus the remark that the brown algae being where the green algae end also applies to sexual reproduction besides vegetative organisation. There are, however, many instances where sexual reproduction ranges from simple isogamy through anisogamy to oogamy. The gametangia exhibit various degrees of specialization.

***(a)* Isogamy** (Fig. 16.8). In the majority of members of order Ectocarpales and Sphacelariales sexual reproduction is isogamous. The gametophyte plants may be homo-or heterothallic. The isogametes which look and behave alike are of equal size and resemble zoospores in morphology (B). The fusing gametes show marked morphological similarity and both the gametes are actively motile at the time of gametic union (C). The isogametes are produced in multichambered structures called plurilocular gametangia (A). The gametic union results in the formation of a zygote (D) which germinates immediately to give rise to the diploid or sporophyte plant. Gametes which fail to fuse disintegrate. According to Hygen (1934) and Papenfuss (1935) they may sometimes develop partheno-genetically into sexual filaments.

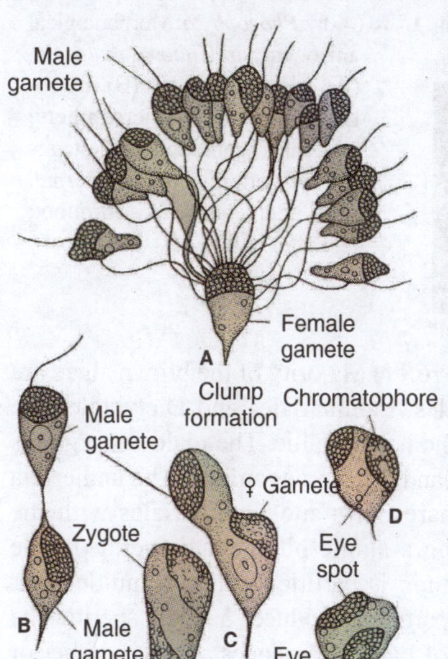

Fig. 16.9 (A-F). *Phaeophyta.* Sexual reproduction in *Ectocarpus siliculosus.* (A), clump formation; (B-E), stages in gametic fusion; (F), zygote with a wall (After Berthold).

***(b)* Physiological anisogamy** (Fig. 16.9 A-F). In certain members of orders Ectocarpales and Sphacelariales the gametes are morphologically similar but show distinction in their behaviour at the time of gametic union and thus exhibit **Physiological anisogamy**. One of these, the so-called **female gamete** soon becomes passive and motionless. The more active male gamete swims to it and fuses with it. In *Ectocarpus siliculosus* a number of active male gametes swim to and cluster around each passive female gamete. Each male gamete in the cluster attaches itself by the longer flagellum to the body of the female gamete. This is known as **clump formation** (A). Soon the longer flagellum of one of the anchoring male gametes in the cluster contracts (B). Its body is brought in contact with that of the female gamete (C) and the two ultimately fuse (D-E) to form the zygote (F). Meanwhile the other male gametes in the clump swarm away.

(*c*) Morphological anisogamy (Fig. 16.10). Pronounced anisogamy is met with the *Cutleria multifida* and a few others in which the fusing motile gametes are of unequal size. The smaller gamete is called the male or **microgamete**. It has a single chromatophore. The larger gamete is called female or **macrogamete**. It usually contains several chromatophores (D). Both are motile and are produced in multicellular morphologically different gametangia (A-B). The active microgamete seeks the sluggishly moving or passive macrogamete and fuses with it (E). The male and the female nuclei fuse within a few hours after fusion of the two gametes (F). The zoospore and motile gamete or sperm have the same structure.

Fig. 16.10 (A-F). *Phaeophyta.* Morphological anisogamy in *Culteria multifida*. (A), microgametangia; (B), macrogametangia; (C), microgamete; (D), macrogamete; (E-F), stages in fertilisation which is external. (A-F after Thuret and Bormmet; rest after Kuckuck).

Fig. 16.11. (A-D). *Phaeophyta*. Oogamy. (A) part of a female sorus of *Zonaria* in section showing two oogonia; (B), part of a male sorus of *Zonaria* in section showing two multiseriate antheridia; (C-D), two stages in fertilization in Dictyota. (A-B after Haupt; (C-D) after Williams).

Ccutleria

(*d*) Oogamy (Fig. 16.11) Majority of the brown algae are oogamous. Desmarestiales, Laminariales and Dictyotales are exclusively oogamous and heterothallic. The male sex organ is called the **antheridium** and female **oogonium**. The antheridia are unicellular in Desmarestiales and Laminariales with the entire protoplast forming a single biflagellate sperm. In the Dictyotales the antheridium is partitioned into a multiceullar structure (B). Each compartment produces a single sperm. The unicellular antheridium of Fucales produces a large number of sperms (about 64). The oogonium is a one-celled structure (A). With the exception of Fucales it produces a single ovum. The ova are invariably extruded from the oogonia prior to fertilization which is thus external. In the Dictyotales and Fucales, the ova are discharged into the surrounding water (C) where fertilization takes place (D). In the Laminariales and Desmarestiales, the single ovum, after extrusion, remains attached at its lower portion with the apex of the oogonium. It is fertilized *in situ*. It even germinates while still attached to the portion of the female gametophyte (Fig. 19.15). Occasionally the gametes develop parthenogenetically into new plants.

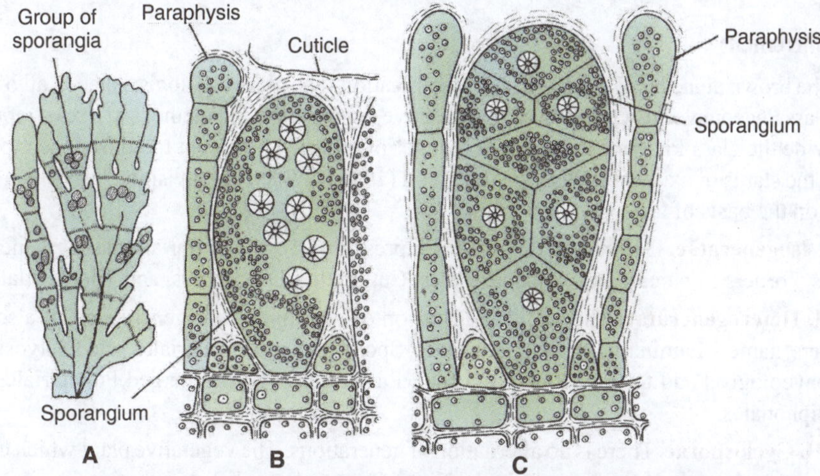

Fig. 16.12 (A-C). *Phaeophyta.* Meiospore formation in Zonaria. (A), part of a diploid thallus bearing sporangia in groups; (B), Young Sporangium containing 8 haploid nuclei as a result of mitotic division after meiosis: (C), nearly mature sporangium containing 8 meioaplanospores delimited by walls (After Hau Haupt).

Meiospore formation (Fig. 16.12). The zygote formed as a result of fertilization usually germinates immediately without undergoing a period of rest. There is no zygotic meiosis. The resulting individual is a diploid sporophyte. In all the zoosporic brown algae, except the Ectocarpales and Sphacelariales, the diploid thallus (sporophyte plant) bears only unilocular sporangia. The diploid contents of the unilocular sporangium after the completion of nuclear division, the first two of which are always meiotic, metamorphose into biflagellate meiozoospores. The diploid thallus of Dictyotales (A) bears a special kind of unilocular sporangium. It is called the tetrasporangium. The diploid contents of the tetrasporangium undergo meiosis to produce four meioaplanospores. In *Zonaria,* there is a third nuclear division resulting in the formation of 8 haploid nuclei (B). After the completion of nuclear division the cytoplasm undergoes cleavage to form 8 meioaplanospores (C). Before being released the meioaplanospores become rounded. The released meiospore secretes a wall around it and germinates to form a haploid thallus.

Alternation of Generations

There is distinct alternation of generations in the sexual life cycle of the Phaeophyta with meiosis during spore formation. In the majority of the cases it is **isomorphic** (*Dictyota* and *Zanardinia*). In certain orders such as Laminariales and *Cutleria* it is distinctly **heteromorphic** with the reduced and filamentous gametophytes alternating with the huge sporophyte plant. The sporophyte generation in the brown algae may be perennial or annual but the gametophyte is always annual. The life cycle in the Fucales exhibits no morphological alternation of generations. The sex organs are borne on the sporophyte

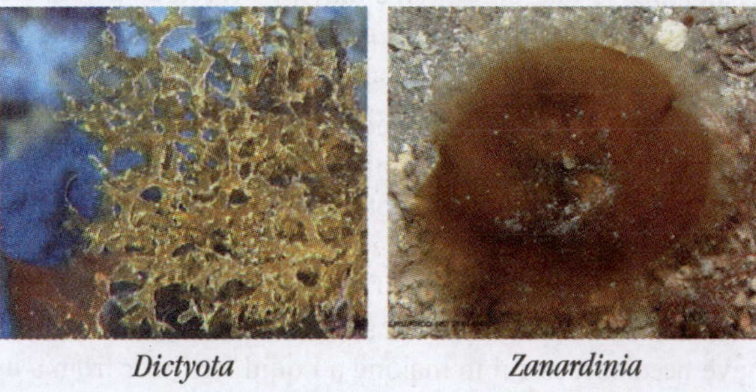

Dictyota *Zanardinia*

plant. The haploid condition in the Fucales is restricted to the gametes only making the life cycle as *Diplontic*.

Classification

The brown algae have a complicated classification. The older algologists placed all of them in a single class Phaeophyceae. On the basis of vegetative organisation and methods of sexual reproduction, they divide the class Phaeophyceae into a number of orders. The recent trend has been to raise this class to the status of a division Phaeophyta. Kylin (1933) divided the division Phaeophyta into three **classes** on the basis of the type of life cycle :-

I. Isogeneratae. It is characterised by the presence of two similar alternating generations. It includes 5 orders : Ectocarpales, Sphacelariales, Cutlariales, Dictyotales, and Tilopteridales.

II. Heterogeneratae. There is an alternation of two dissimiliar generations. It also includes five orders, namely Laminariales, Desmarestiales, Sporochnales, Chordariales, and Dictyosiphonales. Some phycologists add to the list a sixth order, Punctariales. Kylin merged Punctariales with the Dictyosiphonales.

III. Cyclosporae. There is no alternation of generations. The vegetative plant which constitutes the dominant phase in the life cycle is the diploid sporophyte. It bears the sex organs which by gametogenic meioses produce the haploid gametes. There is no vegetative haploid stage. This class includes the order Fucales.

Fritsch (1935) placed the brown algae in the class Phaeophyceae of the subdivision **Algae** and classified it into 9 orders. These are Ectocarpales, Sphacelariales, Cutleriales, Tilopteridales, Dictyotales, Sporochnales, Desmarestiales, Laminariales, and Fucales. Some algologists add Chordariales and Dictyosiphonales to the list to make 11 orders. Smith (1955) recognised three subclasses viz. Isogeneratae, Heterogeneratae and Cyclosporae which have been raised to the level of class by Kylin (1933). However, the classification proposed by Kylin (1933) has been followed in this text.

Economic Importance

The brown algae provide food and shelter for fish and other marine animals. In some parts of the world and Northern Europe they serve as a feed for farm animals. Many species of the brown algae are rich in minerals, and contain carbohydrates and vitamins. Consequently they furnish food for the people of the oceanic islands and the Orient, the Chinese and the Japanese. The Japanese alone use more than twenty species for food and many others as fertilizers. They make sweet cake from a certain kelp and serve others as vegetables. Kombu which is a standard food in Japan and a product of *Laminaria* and other kelps may be served as a vegetable or made into a confection by coating with sugar. Young stipes of *Laminaria* are eaten and the stipes and bladders of *Nereocystis* are used to make a candied citron called **seatron**. In S. America species of *Durvillea* is dried, salted and sold.

Kelps, which is the general name for large, brown sea weeds constitute an important source of iodine, mineral salts, bromine and potash. *Laminaria, Fucus* and some other sea weeds are the sources of iodine. Potash is obtained from *Nereocystis, Macrocystis, Pelagophycus* and a few others. Some are used as fertilizers because of their high content of nitrogen, potassium and other minerals. They are low in phosphorus. The sea weeds are spread in the fields and ploughed under or dried and burned. The ash is employed as a soil fertilizer. It has been reported that scientists of C.S. M. CRI have been successful in making a liquid fertilizer from a brown sea weed *Sargassum*. The fertilizer contains all the micronutrients required for plants.

Extracts of kelps are put to many industrial uses. A viscous, colloidal preparation, **algin** obtained from the brown algae (*Macrocystis* and *Laminaria*) is used in a number of industrial processes. Algin has the property of thickening and gelling mixtures. It serves for sizing cloth and imparts smooth texture to the frozen product (ice cream). It is also used in the manufacture of fertilisers. Its use in the manufacture of fertilisers, showing creams and lotions, is not less important. It is a very important component of impression materials in dental work.

Diagnostic Features of Brown Algae

1. Unicells and colonial forms (both motile and nonmotile) are completely absent nor the unbranched filament typical of the Chlorophyceae has so far been reported in the brown algae.
2. The cells are uninucleate and the cell wall besides the cellulose and pectic substances contains a gel-like polysaccharide **algin** (alginic acid) and the carbohydrate **Fucoidin** (fucinic acid).
3. The photosynthetic pigments in the chromatophore are chlorophyll *a* and little *c* but no chlorphyll *b*, b-carotene and xanthophyll such as fucoxanthin and diatoxanthin. The fucoxanthin masks the other pigments.
4. The presence of whitish granules termed the **fucosan vesicles** in the cells is another characteristic of the brown algae.
5. The reserve food is stored in the form of mannitol, laminarin and rarely as fat droplets. They are formed in the pyrenoids outside the chromatophores and diffuse into the vacuoles.
6. The zoospores and motile gametes are furnished with two unequal flagella inserted laterally on the pear or spindle-shaped body. The longer anterior tinsel flagellum which is directed forward in motion bears flimmer hairs on two sides, the posterior shorter whiplash flagellum is directed backwards.
7. Zygotic meiosis, which is common among the green algae, is completely absent in the browns.
8. All the browns with the excpetion of Fucales undergo alternation of generations with meiosis during spore formation.

QUESTIONS

Essay Type

1. Mention the general characteristics of Phaeophyceae and discuss the various types of life cycles in the group. *(Rohilkhand, 1989)*
2. Describe alternation of generations in Phaeophyceae. *(Rohilkhand, 1992)*
3. Describe various types of life cycles in Phaeophyceae. *(Rohilkhand, 1994, 1996)*
4. Discuss the phenomenon of alternation of generations in the members of Phaeophyceae studied by you. Discuss their inter-relationships and evolutionary tendencies. *(Rohilkhand, 1999)*
5. Write an account of the range of thallus oranisations in brown algae and discuss their evolutionary tendencies. *(Rohilkhand, 2000)*
6. Describe methods of sexual reproduction and structure of sex organs in Phaeophyceae.
7. Give an account of cell structure and modes of reproduction in Phaeophyta.
8. Write what you know about range of vegetative structure in brown algae.

Short Answer Type

9. Write short notes on :

(*a*) Distinguishing features of Phaeophyceae *(Gorakhpur, 1996; U.P. College, 1999; Allahabad, 1993)*

(*b*) General characters of Phaeophyceae *(Lucknow, 1997, 1999, 2006)*

(*c*) Economic importance of brown algae *(Awadh, 1997)*

(*d*) Modes of sexual reproduction in Phaeophyceae

(*e*) Unilocular sporangium *(Tamil Nadu, 1995)*

(*f*) Zoospore

(*g*) Heterotrichous condition

(*h*) Cryptoblasts

(*i*) Heterotrichous filament *(Poorvanchal, 2004; Kanpur, 2006)*

(*j*) Alternation of generation in phaeophyceae *(Kanpur M.Sc., 2006)*

(*k*) Fresh water phaeophyceae *(Lucknow M.Sc., 2006)*

(*l*) Life cycles patterns in Brown algae *(Lucknow M.Sc., 2005)*

(*m*) Thallus organisation in phaeophyceae *(Lucknow M.Sc., 2003)*

(*n*) Growth patterns in phaeophyceae *(Lucknow M.Sc., 2003)*

10. Name the reserve food materials found in Phaeophyceae. *(Gorakhpur, 1998)*

11. Compare the distinguishing features of Myxophyceae, Xanthophyceae and Phaeophyceae. *(Lucknow, 1995)*

12. Distinguish between Phaeophyceae and Xanthophyceae. *(Lucknow, 1998)*

13. Name the main pigments and reserve food material in the class Phaeophyceae. *(Purvanchal, 1996)*

14. Name two reserve food materials found in Phaeophyceae. *(Rohilkhand, 1992)*

15. Name the pigments of class Phaeophyceae. *(Himachal Pradesh, 1993)*

16. What are unilocular and plurilocular reproductive bodies? Where do they occur? What is their importance ?

17. What are the advances that the plant body in Phaeophyceae shows over that of Chlorophyceae?

18. Describe the structure of a typical cell in Phaeophyceae.

19. Are members of Phaeophyceae fresh water or marine ?

20. What is the function of plurilocular sporangia ?

21. What are the spores produced by unilocular sporangia ?

22. Write a brief account of anisogamy in Phaeophyceae.

23. Write what you know about fragmentation.

24. Name a genus of Phaeophyceae found in India and possessing male and female conceptacles. *(Allahabad, 2004)*

Objective Type

25. Fill in the blanks:

(*i*) In Phaeophyceae reserve food material is ....

(*ii*) The general term for large brown algae is ....

(*iii*) The carbohydrate food reserve of brown algae is ....

(*iv*) The principal pigment importing brown or olive brown colour to the thallus of Phaeophyceae is ....

(*v*) Alginic acid is present in the cell walls of class ....

(*vi*) In Phaeophyceae .... vesicles are present.

(*vii*) Algin is obtained from ....

(*viii*) All the brown algae except .... show distinct alternation of generation.

(*ix*) Phaeophyceae are usually ....

(*x*) Sea palm is the common name given to ....

(*xi*) Agar is derived from ....

26. Select the correct answer :

(*i*) The principal pigment importing distinctive brown or olive brown colouration to the thallus of Phaeophyceae is

(*a*) Siphonoxanthin

(*b*) Fucoxanthin

(*c*) Necoxanthin

(*d*) Flavoxanthin

(*ii*) In class Phaeophyceae the reserve food is stored in the form of

(*a*) Laminarin (*b*) Glucose

(*c*) Fructose (*d*) Glycogen

(*iii*) Which are the two pigments seen in Phaeophyceae ?

(*a*) Chlorophyll and Xanthophyll.

(*b*) Chlorophyll and Fucoxanthin

(*c*) Phycocyanin and Phycoerythrin

(*d*) Phycocyanin and Xanthophyll

(*iv*) A pigment that is absent in Phaeophyceae is

(*a*) Chlorophyll *a*

(*b*) Chlorophyll *b*

(*c*) β-Carotene

(*d*) Chlorophyll *c*

(*v*) Heterogeneratae is so called because

(*a*) It has two different types of sporangia

(*b*) It has heteromorphic alternation of generations

(*c*) It has two different types of sex organs

(*d*) It has two types of zoospores.

(*vi*) Unilocular sporangia is so called because it has only

(*a*) One zoospore

(*b*) Haploid zoospores

(*c*) One meiotic division in it

(*d*) One chamber.

(*vii*) Plurilocular sporangia is so called because they

(*a*) Produce numerous zoospores

(*b*) Have numerous compartments

(*c*) Have many divisions of their contents during zoospore formation.

(*d*) Have numerous vacuoles

(*viii*) Which of the following is commonly called sea palm ?

(*a*) *Ectocarpus*

(*b*) *Postelsiapalmaeformis*

(*c*) *Fucus*

(*d*) *Dictyota*

(*ix*) Algin is produced by the numbers of class

(*a*) Chlorophyceae

(*b*) Myxophyceae

(*c*) Phaeophyceae

(*d*) Rhodophyceae

(*x*) The reserve food is stored in the form of manitol and laminarin in

(*a*) Chlorophyceae

(*b*) Myxophyceae

(*c*) Rhodophyceae

(*d*) Phaeophyceae

(*xi*) Plurilocular sporangia are produced on

(*a*) Haploid plants

(*b*) Diploid plants

(*c*) Triploid plants

(*d*) Polyploid plants

(*xii*) The zoospores are pyriform and
- (*a*) Uniflagellate
- (*b*) Laterally biflagellate
- (*c*) Apically biflagellate
- (*d*) Aflagellate

(*xiii*) Unilocular sporangia produce
- (*a*) Mitozoospores
- (*b*) Meiozoospores
- (*c*) Both of the above
- (*d*) None of the above

(*xiv*) The members of class Phaeophyceae are
- (*a*) Freshwater forms
- (*b*) Exclusively marine
- (*c*) Both freshwater and marine
- (*d*) Terrestrial

Plurilocular Gametangia in *Ectocarpus*

CHAPTER 17

CLASS : ISOGENERATAE Order : 1. Ectocarpales

- Family : Ectocarpaceae
- Ectocarpus

This order includes the simplest and the least specialised members of the brown algae. The thallus is a heterotrichous filament or a frond composed of interlacing adjoining filaments. The growth is apical or intercalary. The zoospores (mitospores) are produced by the diploid plant and the isogametes by the haploid plants. Comparatively simple organisation of the thallus and lack of true oogamy are the chief characteristics of this order. The Ectocarpales thrive best in the temperate and the polar shores. In the tropics the representatives of the order are rare. The order includes about 50 genera which are grouped under two families, Ectocarpaceae and Ralfsiaceae.

FAMILY : ECTOCARPACEAE

The thallus is a simple heterotrichous filament consisting of uniseriate, branched threads. The cells are uninucleate with one or more chromatophores which may be disc-like or ribbon-shaped. The sporophyte plant bears two kinds of zoosporangia, the **unilocular** and **plurilocular**. The former by meiosis produce the motile meiospores or gonospores which germinate to form the haploid or gametophyte plants. The plurilocular sporangia by mitosis produce motile spores which are diploid and called **zoospores**. The latter germinate to form the sporophyte plants. The haploid plants bear the multicellular cone-shaped gametangia which produce the gametes. The zygote formed by the fusion of gametes germinates directly to form a diploid sporophyte. Sexual reproduction is isogamous and the life cycle exhibits isomorphic or homologous type of alternation of generations. Ectocarpaceae includes a number of genera. Of these *Ectocarpus* is the best known.

Ectocarpus sp.

Ectocarpus littoralis

Ectocarpus shaerophorus

Ectocarpus tomemtosus

ECTOCARPUS (LYNGBYE, 1819)

Distribution. This marine alga (Fig. 17.1 A) is the most primitive of all the brown algae and comprises many species. A few of them (about 6) have been reported to occur in freshwater (Chapman, 1964). The marine species are of worldwide distribution but are found in abundance in the colder seas of the temperate and polar regions. They occur in plenty along the Atlantic coast but are scarce along the Pacific coast. The plants occur attached to the rocks and stones along coasts both in the littoral and sublittoral zones. Some species occur in shallow water on the sides of the tidal pools. Some species like such as *E. coniferus* and *E. breviarticulatus* grow as epiphytes on other algae especially the Fucales and Laminariales. *E. fasciculatus* is found growing on the fins of a certain fish in Sweden. A few species are endophytic or endozoic (wholly or partially). About 16 species of this genus have been reported from India (Misra, 1966) and common species are *Ectocarpus coniferus, E. indicus, E. arabicus and E. dermonematus*. These occur mostly along the East and West costs in the supralithoral zone.

Habit. The plant grows attached to the substratum and consists of tufts of brownish, delicate highly branched filaments which wave freely in water.

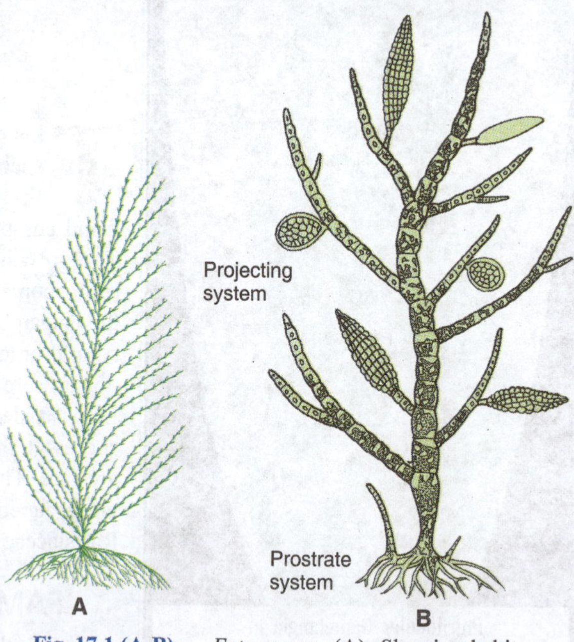

Fig. 17.1 (A-B). *Ectocarpus.* (A), Showing habit; (B), Heterotrichous filament.

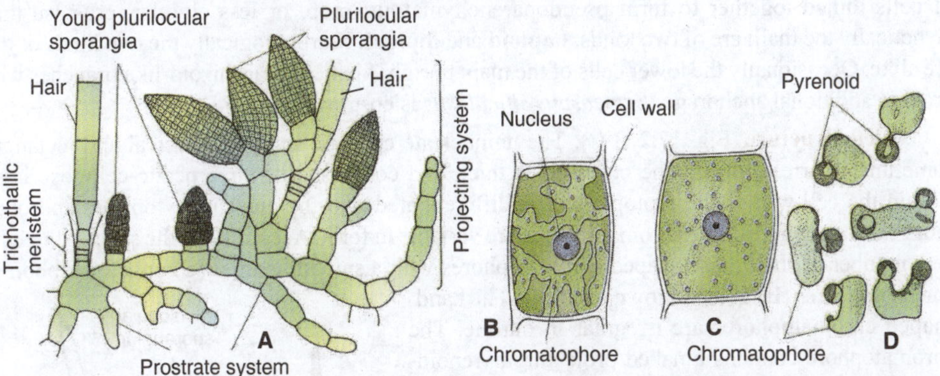

Fig. 17.2 (A-D). *Ectocarpus.* (A), *E. reptans* with the thallus showing differentiation into prostrate system and erect or projecting system, the latter bearing plurilocular sporangia and showing trichothallic meristem; (B), cell showing structure and band-shaped chromatophores; (C), cell with discoid chromatophores, (D), chromatophores with projecting pyrenoids. (A-B after Taylor and D after Knight)

Thallus (Fig. 17.2 A). It is a small heterotrichous *filament* (A) differentiated into two parts (*i*) the prostrate or creeping portion and (*ii*) the erect or projecting portion.

The prostrate portion is sparsely or irregularly and frequently branched. It either sticks to the substratum as in the epiphytic species or is firmly attached to the substratum by rhizoids. The erect portion which waves freely in water consists of a number of upright copiously branched uniseriate or monosiphonous filaments arising from the prostrate part. In some cases (e.g. *E. granulosus*) the older portions of the main axis of the thallus may be surrounded by the descending rhizoid-like branches arising from the lower cells of lower branches and thus become corticated. A few species are sparingly branched. The branching is always lateral. The branches arise just beneath the septa. Occasionally due to evection which pushes the branch upwards and the main axis to one side, apparent dichotomy may result. The branches end either in a point (Fig. 17.4 B) or taper into a series of narrow, elongated tapering hyaline, vacuolated cells forming a colourless hair (Fig. 17.4 A). The branch grows by means of a basal meristem. The cells in the main filament and the branches are usually arranged one above the other in a single strand to form uniseriate, branched filaments. Occasionally (*Pylaiella*) the cells divide longitudinally. In that case the filament consists of several rows

Ectocarpus granulosus

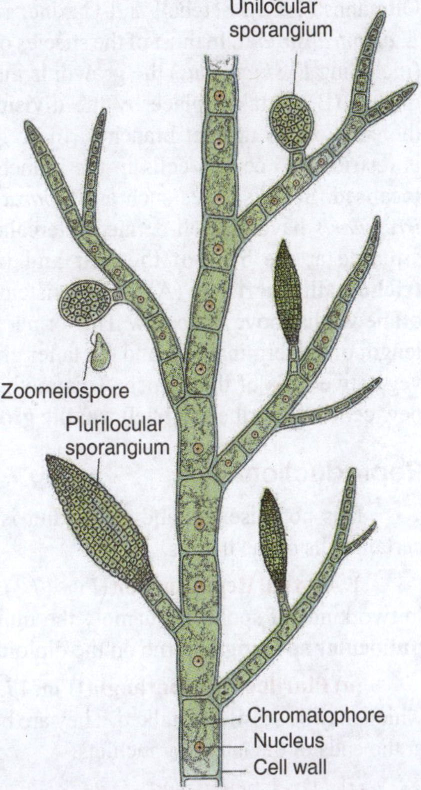

Fig. 17.3. *Ectocarpus.* A portion of the upright filament showing structure.

of cells joined together to form pseudoparenchymatous, more or less circular, crustose thalli. Genetically the thalli are of two kinds, **haploid** and **diploid**. Morphologically the two kinds of thalli are alike. Occasionally the lower cells of the main upright filament bear rhizoid-like branches which provide additional anchorage. *Ectocarpus siliculosus* is cosmopolitan.

Cell structure (Fig. 17.2 B-C). The uninucleate cells are small, cylindrical and rectangular sometimes, barrel-shaped. The cell wall is thick and composed of three pectic-cellulose layers. Within the cell-wall is the protoplast. It is differentiated into the nucleus, cytoplasm and one or more chromatophores. The chromatophores are variable in form. According to the species there may be a number of small, disc-shaped chromatophores with a smooth margin (C) or a few ribbon-like chromatophores (B) golden brown in colour. The band-shaped chromatophores are irregular in outline. The chromatophore contains a naked projecting pyrenoid-like structure (D). According to Evans (1966) a new pyrenoid may develop from the old one by budding. A brown carotenoid pigment **fucoxanthin** is located in the plastids in addition to chlorohyll *a*, and *c*, carotenes and xanthophylls.

Growth (Fig. 17.4). It is apical in the prostrate system but the projecting system shows great diversity. Apical growth in the upright branches is considered exceptional. It has been reported in *E. lucifugus* by Oltmanns (1922). Setchell and Gardner recorded it in *E. chantransioides*. In most of the species of *Ectocarpus* (including *E. siliculosus*) the growth is intercalary and diffuse (B). It takes place by the division of any of the cells of the upright branches. In *E. granulosus* it is restricted to certain cells in the branch and is thus localised. Some species such as *E. paradoxus* and *E. irregularis* have as well-defined intercalary meristem situated at the base of the hair and is called the **trichothallic meristem** (A). The meristematic cells cut off new cells above and below. The former increases the length of the terminal hair and the latter give rise to the vegetative cells of the branch. This method of adding new cells is described as **trichothallic growth.**

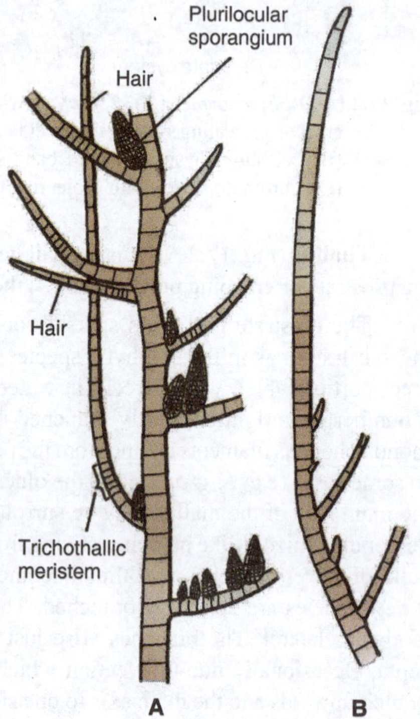

Fig. 17.4 (A-B). *Ectocarpales.* ***(A),*** *E. coniferus* showing trichothallic meristem, hairs and plurilocular sporangia; (B), *E. siliculosus* showing diffuse growth and branch terminating in a point and not hair (A after Beorgessen and B after Kuckuck).

Reproduction

It is both asexual and sexual but is confined to certain cells of the thallus.

1. Asexual Reproduction (Fig. 17.3). It takes place by means of biflagellate zoospores produced in two kinds of sporangia namely the multicellular or **plurilocular sporangia** and the one-celled **unilocular sporangia** borne on the diploid, sporophyte plant.

(*a*) Plurilocular Sporangia (Fig. 17.3). These are elongated cone-like multicellular structures which may be sessile or stalked. They are borne singly on the diploid or asexual plants (sporophytes) at the ends of the lateral branchlets.

(i) Development and structure of plurilocular Sporangium (Fig. 17.5). The terminal cell of the branchlet functions as the sporangial mother cell (A). It enlarges and contains numerous chromatophores. It then undergoes repeated transverse division to form a vertical row of 6-12

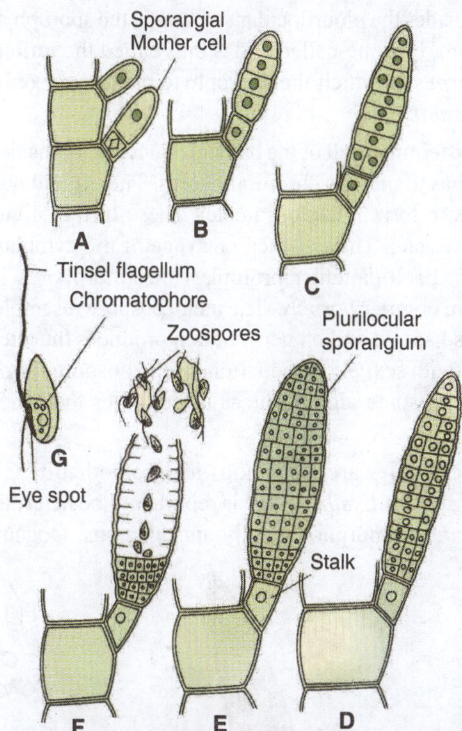

Fig. 17.5 (A-G). *Ectocarpales.* (A-D), stages in the development of plurilocular sporangium in *Ectocarpus*; (E), young plurilocular sporangium; (F), plurilocular sporangium liberating zoospores; (G), liberated zoospore (After Kuckuck).

cells (B). Subsequently all these cells undergo vertical division (C). The nuclear divisions involved are mitotic. As a result of repeated transverse and vertical divisions an elongated multicellular cone-like structure consisting of several hundred small cubical cells arranged in 20-40 transverse tiers is produced (D-E). It is the **plurilocular sporangium** (E). The protoplast of each compartment metamorphoses into a single, biflagellate **zoospore**. Since the nuclear division involved in the differentiation of spores is mitotic, the resultant zoospores (F) are **diploid** structures belonging to the category of mitospores. The mature sporangium dehisces usually by an apical aperture. The septa between the compartments disappear and the zoospores pass out though the apical aperture one by one in a slow stream (F). Those beneath the pore pass out in an irregular mass (Smith).

(ii) Structure of diploid zoospore. The liberated zoospore of *Ectocarpus* (G) is a biflagellate pearshaped structure usually with an eye-spot which is situated near the point of origin of the flagellae and is covered by the chloroplast membrane. The two flagella are inserted laterally and are of unequal lengths. The longer tinsel flagellum which bears a double row of flimmer hairs is directed forwards in motion. The shorter whiplash flagellum is directed backwards. At the hinder end of the zoospore is a brown chromatophore.

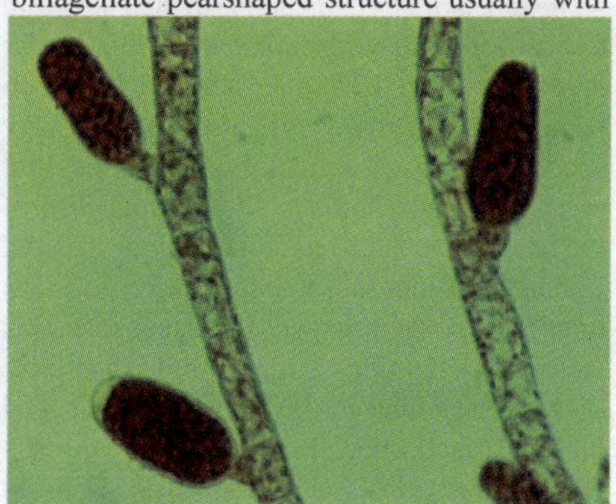

Plurilocular Sporangia of *Ectocarpus*

(iii) Germination of diploid zoospore. The liberated zoospores after swimming about for sometime, come to rest and settle on some solid substratum with the anterior end down. Each quiescent zoospore resorbs its flagella, rounds off and secretes a membrane around it. It then germinates by putting a tubular prolongation which subsequently forms the prostrate system of the diploid sporophyte. The diploid zoospores produced in plurilocular sporangia thus serve to reduplicate the sporophytic generation. They play no role in the phenomenon of alternation of generations.

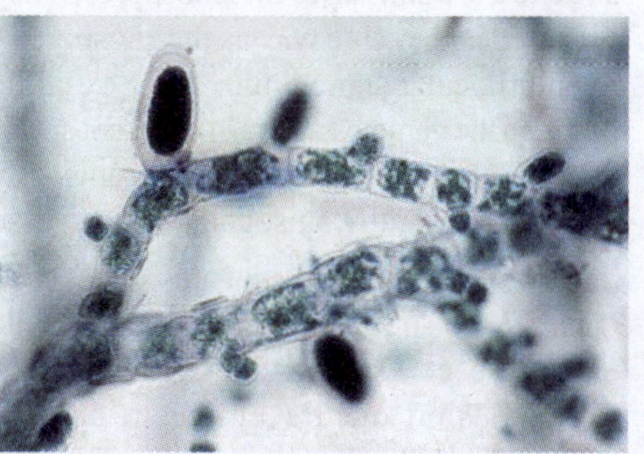

Ectocarpus Unilocular Sporangia

(*b*) Unilocular Sporangium (Fig. 17.3). Besides the plourilocular sporangia, the sporophytes of *E. siliculosus* produce another kind of sporangium. It is one-celled and is thus called the **unilocular sporangium**. There are other species of *Ectocarpus* in which the sporophyte plants bear only one type of sporangium, either *unilocular* or *plurilocular*.

Development of unilocular sporangium. The terminal cell of the branchlet increases considerably in size to take on a globose or ellipsoid form. It has numerous chromotaphores. The diploid nucleus of the unilocular sporangium undergoes meiosis to form 4 haploid nuclei. These further divide by mitosis repeatedly to produce 32 to 64 daughter nuclei. There is then cleavage of the cytoplasm to form as many uninucleate daughter protoplasts. Each daughter protoplast metamorphoses into a biflagellate haploid swarmer called a *meiozoospore* or *gonozoospore*. The meiozoospore resembles the diploid zoospore in every respect except that it is haploid and on germination produces the alternate haploid plant (gametophyte) which is concerned with sexual reproduction. Meiozoospore formation is considered a stage under sexual reproduction by some algologists as it multiplies the beneficial effects of a single act of fertilization.

2. Sexual Reproduction. Majority of the species are isogamous and homothallic. Goebel (1878) recorded isogamy in *E. globifer.* Another species *E. siliculosus* is reported to be heterothallic and exhibits physiological anisogamy. *E. secundus* is morphologically anisogamous. Oogamy, as a rule, is absent in this order. The gametes are produced in a large, elongated, conical sex organ partitioned into a large number of small cubical cells (Fig. 17.6 A). It is called the *plurilocular gametangium*. The gametangia in their structure and development resemble the plurilocular sporangia (Fig. 17.5 A-E) but are borne on haploid or gametophyte plants which in many species are physiologically of two kinds but morphologically similar. The terminal cell of the lateral branchlet gets inflated (Fig. 17.5 A). It then undergoes repeated transverse divisions to produce a vertical row of flat cells (B). This is followed by longitudinal and transverse divisions resulting in the formation of several hundred small cubical cells or chambers (C-E). These chambers are arranged in 24 to 40 transverse tiers (Fig. 17.6A). The protoplast of each chamber gives rise to one, sometimes, to two biflagellate pyriform gametes (C) which in structure are similar to the zoospores. They are liberated in the same way as the zoospores from the plurilocular sporangium (B). Sexual fusion between the gametes varies in different species. It may be isogamous, physiologically anisogamous or morphologically anisogamous.

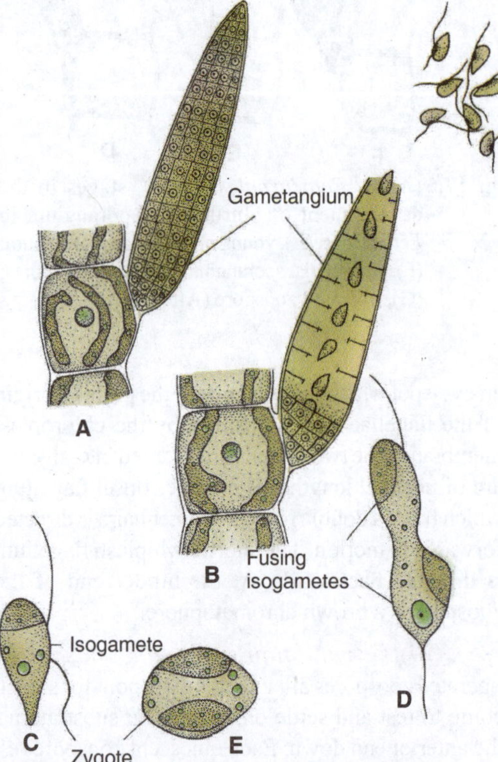

Fig. 17.6 (A-E). *Ectocarpales.* Isogamous sexual reproduction in *Ectocarpus sp.* (A), part of the filament bearing a young gametangium; (B), part of the filameut bearing a mature gametangium discharging gametes; (C), single isogamete; (D), fusion of isogametes; (E), zygote.

Goebel reported isogamy in *E. globifer* (Fig. 17.6). The fusing gametes are similar in every respect (D). They look alike and behave alike. Fusion occurs between isogametes coming from the same plant or even the same gametangium (D).

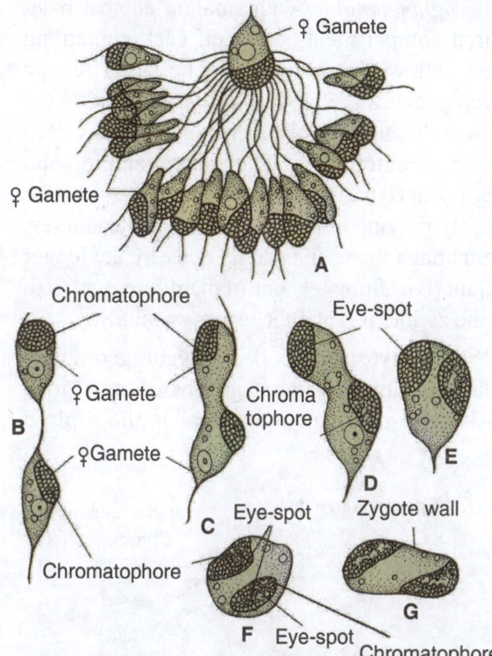

Fig. 17.7 (A-G). *Ectocarpales.* Physiologically anisogamous sexual reproduction in *Ectocarpus siliulosus.* (A), clump formation; (B-E), stages in gametic union; (F), naked zygote with 2 chloroplasts and 2 eye spots; (G), zygote at later stage (After berthold).

Berthold (1881) reported physiological anisogamy in *E. siliculosus* (Fig. 17.7) It is dioecious. Fusion in the dioecious species occurs between gametes from different plants. The fusing gametes are morphologically identical but different in their sexual behaviour. One is less active and often called the female gamete. It becomes passive and motionless after a short time. The more active male gametes cluster around and fix themselves to the body of the female gamete by the forwardly directed flagella (A). Soon the anchoring flagellum of one of these gametes contracts (B). Its body is brought into contact with that of the female gamete (C). Finally the bodies of the two gametes fuse, nucleus with nucleus and cytoplasm with cytoplasm, to form the zygote (G). The remaining active gametes swim away. The clus-tering of the active gametes around the passive female gamete is known as **clump formation**. Despite the morphological similarity between the two fusing gametes there exists **physiological anisogamy**. The less active ones represent the female gametes and the more active ones as the males.

Muller (1968) reported that the female gamete secretes a volatile substance. It has pleasant, sweet odour and this attracts the male gametes to the settled female gamete. This substance was later designated as *ectocarpin* by Machlis (1978).

Pronounced anisogamy is exhibited by *E. secundus* (Fig. 17.8). The fusing gametes in this species are of unequal size but both are motile. They are produced in two kinds of gametangia borne on the same plant (A). The smaller male or microgametes are produced in

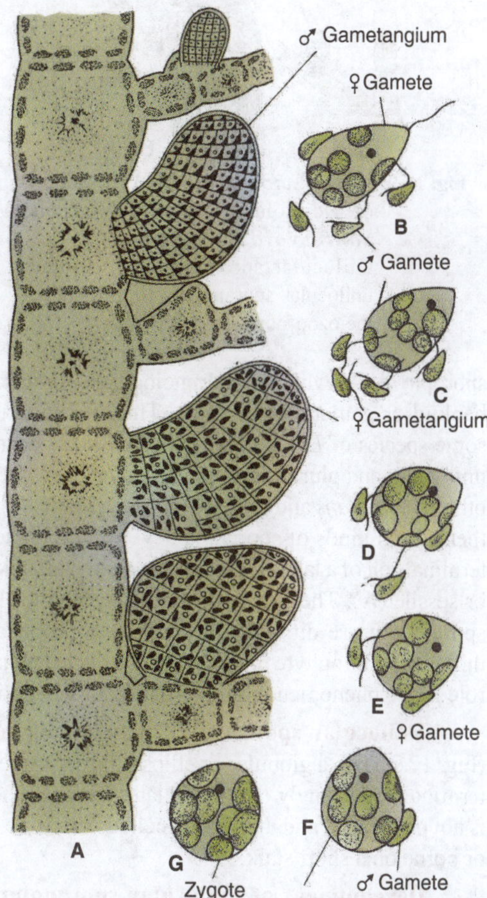

Fig. 17.8 (A-G). *Ectocarpales.* Anisogamous sexual reproduction in *Ectocarpus secundus.* (A), portion of filament bearing micro-and megaganetangia; (B-F), progressive stages in gametic union; (a), Young zygote (After Sauvagean).

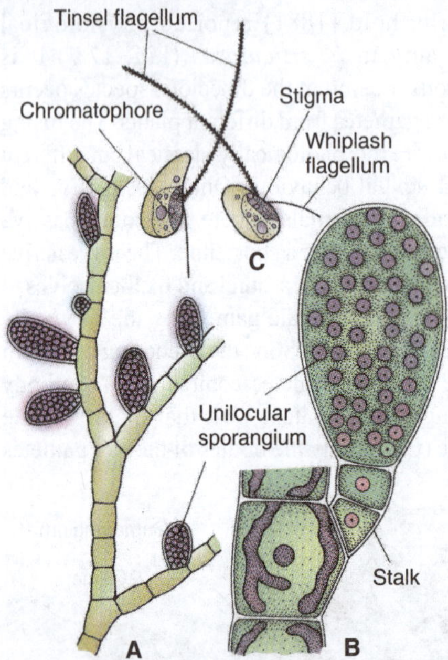

Fig. 17.9 (A-C). *Ectocarpales.* (A), portion of a diploid thread of *Ectocarpus confervoides* bearing sessile unilocular sporangia; (B), stalked unilocular sporangium; (C), liberated meiozoospores. (A after Taylor).

the *microgametangium* with smaller, cubical pale-coloured compartments or loculi each containing a small yellow chromatophore. The larger female or macrogametes are produced in *megagametangia* with larger loculi, each of which contains many deep brown chromatophores. The macrogamete soon comes to rest (B). The active microgametes seek the sluggishly moving or passive macrogamete and few of them attach themselves to its body by the longer flagellum (B). Ultimately one of them fuses with it to form the zygote (G) and the others swim away.

Sporophyte (Fig. 17.9). The zygote germinates without going into a resting stage. There is no zygotic meiosis. The resultant individual is the diploid sporophyte (A) which is morphologically similar to the haploid or gametophyte plants. The diploid thallus in some species of *Ectocarpus (E. siliculosus)* bears both unilocular and plurilocular sporangia on the same plant, others (*E. reptans* and *E. confervoides*) produce either of them. Both kinds of sporangia develop singly from the terminal cell of a lateral branchlet and may be stalked (B) or sessile (A). The spores produced in the plurilocular sporangium are diploid. They serve to propagate the diploid or sporophyte phase in the life cycle and play no role in the phenomenon of alternation of generations.

Unilocular sporangium or meiosporangium (Fig. 17.9). It is a globular or ellipsoidal structure borne terminally and singly on a small lateral branchlet (B). It is not partitioned and thus is one-celled. It may be sessile or borne on a short stalk.

Development of unilocular sporangium and differentiation of meiospores. (Fig. 17.10). During its development the terminal cell of the branchlet increases considerably in size (A). The young sporangium has a single conspicuous large, diploid nucleus and dense cytoplasm. Which completely fills it. The cytoplasm

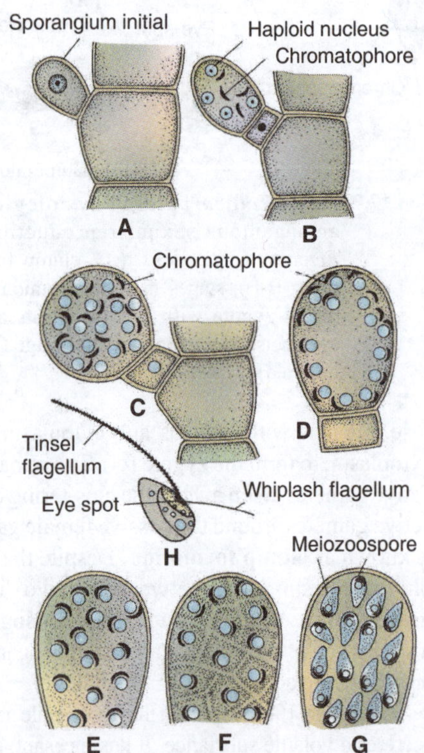

Fig. 17.10 (A-H). *Ectocarpus.* Development of unilocular sporangium and differentiation of meiozoospores. (A), sporangium initial; (B), 4-nucleate stage; (C-D), 8 and 16 nucleate stage; (E), later stage; (F), cleavage of cyto-plasmm; (G), formation of meiospores; (H), liberated meiozoospore (H, after Kucmkuck, A-G, diagramatic).

contains a number of parietal chromatophores. The diploid nucleus of the unilocular sporangium undergoes repeated division to form 32 to 64 daughter nuclei (Fig. 17.10 C-E). The first two divisions constitute **meiosis**. The cytoplasm thus becomes multinucleate. No walls are laid between the nuclei. At maturity the cytoplasm undergoes cleavage into uninucleate daughter protoplasts (F). Each daughter protoplast which contains a single chromatophore metamorphoses into a **meiospore** which is motile and is called a **meiozoospore**. Each meiozoospore is provided with two laterally inserted flagella and contains an eye-spot (H). The meiozoospores are small and pyriform in shape. The longer tinsel flagellum is directed forwards in motion and the shorter whiplash backwards. The meiospores escape all at one time in a gelatinous mass through a small apical aperture which is formed by the dissolution of the wall of the sporangium at the apex. On liberation they are enclosed in a thin vesicle which soon vanishes. The cell below the empty sporangium may grow into its cavity to form a new zoosporangium. The liberated meisopores disentangle their flagella and swim away in all directions.

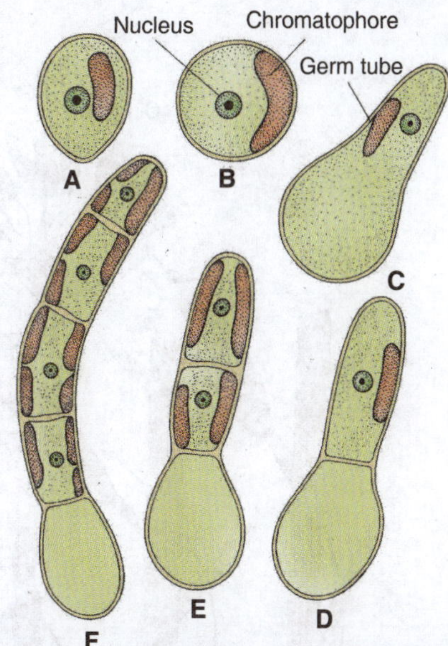

Fig. 17.11 (A-F). *Ectocarpus.* Stages in the germination of meiozoospores (diagrammatic).

Germination of meiozoospore (Fig. 17.11). After swimming for a while each meiospore comes to rest on some solid object. It then withdraws its flagella and secretes a membrane around it (A-B). In this condition it immediately germinates by putting out a short germ tube (C). The latter is then separated by a septum (D). The cell thus cut off divides and redivides (E-F) to produce the prostrate system from which arise the erect threads. This alternate plant in the life cycle is the **haploid** gametophyte or the **sexual** plant. In *E. siliculosus,* the asexual plants which produce the meiospores have sixteen chromosomes. The meiospores thus contain the haploid number eight.

Alternation of Generations (Fig. 17.13). The life history of *Ectocarpus* is of considerable scientific interest. In *Ectocarpus siliculosus* there occur two distinct plants in the life cycle-the diploid sporophyte and the haploid gametophyte. Morphologically both are similar (A-B). The sporophyte

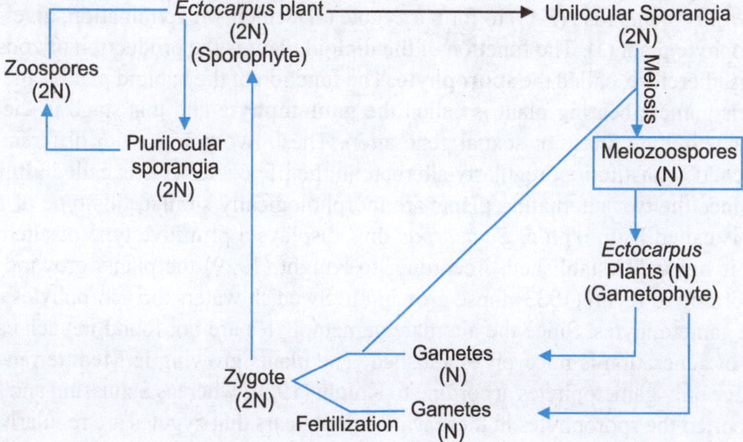

Fig. 17.12. Graphic representation of the life cycle of *Ectocarpus sp.*

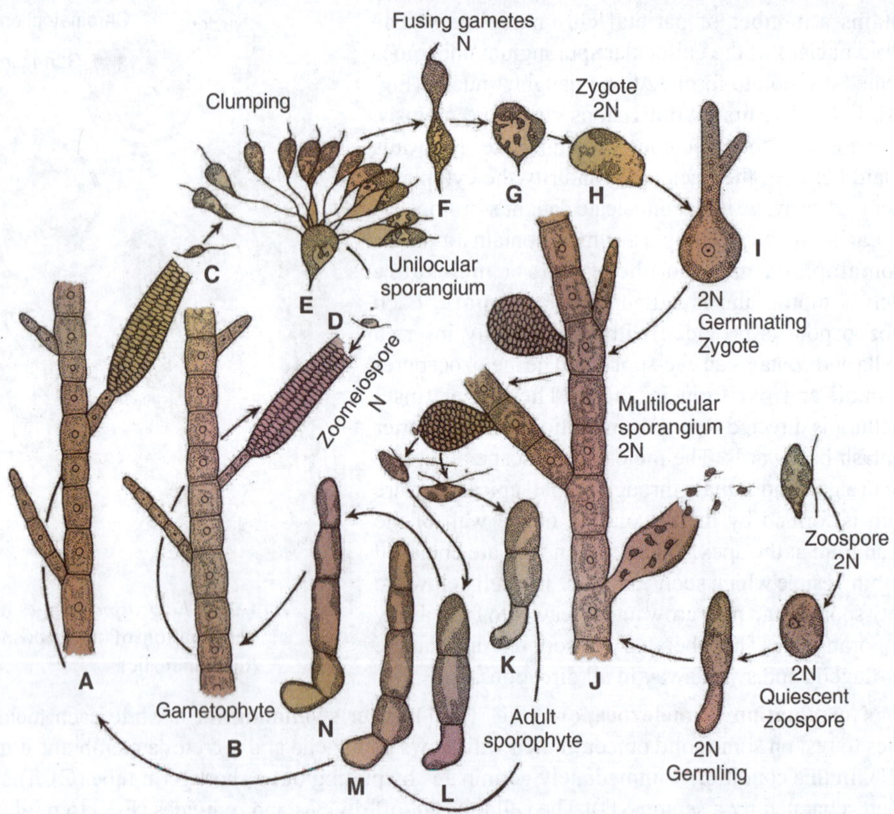

Fig. 17.13. *Ectocarpus sp*. Pictorial life cycle sketch.

plant has 16 chromosomes. It bears both **unilocular** and **plurilocular sporangia**. The former by meiosis produces haploid meiozoospores and the latter by mitosis forms diploid zoospores. The meiozoospores, on germination, give rise to the alternate haploid plants, the **gametophytes** with 8 chromosomes (A-B). The diploid zoospores develop into other sporophytes. The gametophyte plants bear multicellular reproductive structures known as the **plurilocular gametangia**. Each compartment of the plurilocular gametagium produces a single pyriform gamete (C-D). The gametes from two different *Ectocarpus* plants fuse (F-G) to form a zygote (H) which, on germination, gives rise to the diploid or sporophyte plant (J). The function of the diploid plant is the production of zoospores and meiospores. It is, therefore, called the **sporophyte.** The function of the haploid plant is the formation of gametes. This gamete bearing plant is called the **gametophyte** and this stage in the life cycle is called the gametophyte phase or sexual generation. These two plants, with different functions and different genetic constitution, regularly alternate in the life cycle. This is called **alternation of generations**. Since the two alternating plants are morphologically similar, this type of alternation of generations is called **isomorphic**. *Ectocarpus* thus displays a primitive type of alternation of generations. It is not well established. According to Knight (1929) the plants growing in British waters and according to Kylin (1933) those growing in Swedish waters are sporophytes and do not give rise to the gametophytes. Since the alternate gametophytes are not found in such cases hence the alternation of generation is not well established. The plants growing in Meditterranean waters near Naplses have only gametophytes according to Knight (1929) whereas Schussing and Kothbauer (1934) have reported the sporophytes in these waters. It appears that zygotes are regularly produced but rarely develop into sporophytes. In United States both gametophyte and sporophyte plants are found near Massachusettes.

Reduplication of Sporophyte. The sporophyte is reduplicated in two manners :

(*a*) By means of diploid zoospores produced in the plurilocular sporangia borne on the diploid plant. They serve to repeat and multiply the sporophyte generation but play no role in the alternation of generations.

(*b*) Sometimes by the fusion of meiozoospores produced in the unilocular sporangia. Knight (1922) reported that the haploid meiozoospores produced in the unilocular sporangia behave as gametes. They fuse to form a zygote which, on germination, gives rise to a diploid plant. Thus *Ectocarpus* provides an illustration of incomplete dissociation of zoosporic and gametic habits.

Reduplication of the Gametophyte. Gametophyte is reduplicated only by the parthenogenetic germination of the gametes.

Salient Features

1. The fine, tufted, brownish thallus is a **heterotrichous** filament consisting of well-defined erect and prostrate systems.
2. The aerial system consists of profusely or sometimes, sparsely branched filaments which generally grow upright in tufts, sometimes decumbent.
3. The prostrate system is irregularly and frequently branched. It serves as a holdfast. In some species the older portions of main axis may be surrounded by descending rhizoid like branches thus become *corticated*.
4. The filaments are generally **uniseriate** and usually consist of a single strand of cells.
5. The branching is always **lateral**. The branches in some species end in colourless **mucilage hairs**.
6. The cells composing the filament are short and cylindrical. The cell wall is thick and made up of three **pectic-cellulose** layers.
7. The cell protoplast is differentiated into cytoplasm, nucleus and small, golden brown disc or band-shaped chromatophores. The characteristic pigment **fucoxanthin** and others are located in the plastids.
8. The growth in the rhizoidal portion is **apical.** The upright portion exhibits **intercalary** growth and in a few species it is **trichothallic**.
9. Genetically the thalli of *Ectocarpus* are of two kinds, **haploid** and **diploid**. Morphologically they are alike.
10. The haploid thalli are concerned with **sexual reproduction** and diploid with asexual reproduction.
11. The diploid plants bear two kinds of sporangia, **unilocular** and **plurilocular**.
12. The unilocular sporangia are sessile or stalked, globose or ellipsoid. They give rise after **meiosis** to 32-64 **haploid zooids** or **swarmers** which are the **meiozoospores**. On germination they always produce the **alternate plant** which is haploid (sexual). The haploid zooids thus play a definite role in the phenomenon of **alternation of generations**.
13. The plurilocular sporangia which may be sessile or stalked are multicellular, each consisting of a number of small cells. The protoplast of each cell gives rise to a single **diploid zoospore**. No meiosis occurs at the time of their differentiation. The diploid zoospores serve to reduplicate the diploid plant. They play no role in the alternation of generations.
14. The zooids of both kinds, **meiozoospores** and diploid **zoospores** are biflagellate. The flagella are of unequal lengths and are inserted laterally. The longer tinsel flagellum is directed forwards in motion and the shorter whiplash backwards. Eyespot and chromatophores one or more are also present.

15. Sexual reproduction takes place by the formation of biflagellate gametes structurally similar to the zoospores.
16. The gametes are produced in **gametangia** similar in every respect to the plurilocular sporangia borne on the diploid plant.
17. The gametangia are developed on the **haploid** or **sexual** plants only.
18. In some species the mating gametes are similar (isogametes) and sexual reproduction is **isogamous,** others exhibit distinct anisogamy which may be physiological or morphological anisogamy.
19. The zygote is not a resting body. It germinates immediately to give rise to the diploid plant.
20. The haploid and diploid plants are morphologically alike. *Ectocarpus* thus exhibits an **isomorphic** type of **alternation of generations**.

Taxonomic Position

Division	:	**Phaeophyta** (Phaeophycophyta)
Class	:	**Isogeneratae**
Order	:	**Ectocarpales**
Family	:	**Ectocarpaceae**
Genus	:	*Ectocarpus*
Species	:	*siliculosus*

QUESTIONS

ESSAY TYPE

1. Describe the structure and mode of reproduction in *Vaucheria* or *Ectocarpus*. (*Allahabad, 1991; Awadh, 1998; Lucknow, 1998; Kumaon, 1999*)
2. Give a brief account of the life history of *Ectocarpus* and explain the alternation of generation of in this alga. (*Allahabad, 1993; Rohilkhand, 1998*)
3. With the help of labelled diagrams only, describe the life cycle of *Ectocarpus*. (*Allahabad, 1996; Awadh, 1997; Gorakhpur, 1994*)
4. Discuss alterantion of generations in *Ectocarpus* and mention where reduction division takes place. (*Allahabad, 1999*)
5. What is meant by alternation of generations ? Describe in detail with reference to an alga (*Ectocarpus*) of class Phaeophyceae. Give its systematic position upto order. (*Purvanchal 1998; Lucknow, 1996; Gorakhpur, 1991, 1993, 1995, 1998; Punjab, 1996*)
6. Describe habitat, vegetative structure and reproduction of *Ectocarpus.* (*Purvanchal, 1999*)
7. Give an account of salient features of Phaeophyceae and describe in brief the structure and reproduction in *Ectocarpus.* (*Awadh, 1996*)
8. Give an account of environmental effect on the life of *Ectocarpus.* (*Kanpur, 1996*)
9. Give an account of structure and reproduction in *Ectocarpus*. (*Rohilkhand, 1993*)
10. Write about habitat, thallus structure and different types of life cycles in *Ectocarpus.* (*Rohilkhand, 1996*)
11. Establish that *Ectocarpus* belongs to Phaeophyceae. (*Madras, 1995*)
12. Bring out the role played by sporangia and gametangia in the life-cycle of *Ectocarpus.* (*Trichy, 1995*)

13. Give a brief illustrated account of structure and reproduction of any brown alga which you have studied. *(Punjab, 1999)*

14. Describe the life history of *Ectocarpus*. *(Bundelkhand, 1994; Raipur, 1994)*

15. What are unilocular and plurilocular sporangia? Is there any difference between the zoospores produced by the two? Give reasons. *(Lucknow, 2001)*

16. Define Alternation of Generations. Discuss the phenomenon with reference to the life-cycle of *Ectocarpus*. *(Allahabad, 2003; Lucknow, 2001)*

17. Describe the reproduction in *Ectocarpus*. *(Kanpur, 2004)*

Short Answer Type

18. Write one important character for the pair which distinguishes its component genera.

(*a*) *Ectocarpus* and *Polysiphonia* *(Allahabad, 1992)*

(*b*) *Plurilocular* and *unilocular sporangia.* *(Purvanchal 2000; U.P. College, 1995; Rohilkhand, 1999; Bundelkhand, 1997)*

(*c*) *Ectocarpus* and *Batrachospermum* *(U.P. College, 1999)*

19. Give only brief answer :

(*i*) Meiosis in *Ectocarpus* *(Allahabad, 1995)*

(*ii*) Write the systematic position and important morphological features of *Ectocarpus.* *(Bundelkhand, 1993; Rohilkhand, 1991)*

20. Write short notes on :

(*a*) Plurilocular sporangia of *Ectocarpus* *(Allahabad, 1997; Awadh, 1995; Kanpur, 1994; Kumaon 1995, Rohilkhand, 1997)*

(*b*) Sporangia in *Ectocarpus* *(Allahabad, 2005)*

(*c*) Asexual reproduction in *Ectocarpus* *(Purvanchal 1997; U.P. College, 1998; Rohilkhand, 1992, Bundelkhand, 1995)*

(*d*) Sexual reproduction in *Ectocarpus* *(U.P. College, 1996; Agra, 1995)*

(*e*) Unilocular and Plurilocular sporangia *(Kanpur, 1997)*

(*f*) Structure of cell in *Ectocarpus* *(Agra, 1991)*

(*g*) Unilocular sporangium *(Allahabad, 2001; Madras, 1997)*

(*h*) *Ectocarpus*. *(Lucknow M.Sc., 2006)*

21. Draw labelled diagrams of the thallus and mention systematic position of *Ectocarpus.* *(Lucknow, 2007; Awadh, 1999)*

22. Why the alternation of generation in *Ectocarpus* is called isomorphic ? *(U.P. College, 1998)*

23. (*a*) Give the structure of chromatophore in *Ectocarpus.* *(U.P. College, 2000)*

(*b*) Give the classification of *Ectocarpus.* *(Kumaon, 1995, 1996; Rohilkhand, 1996, 2000)*

24. Draw well labelled diagrams of the following :

(*a*) Reproductive structures in *Ectocarpus.* *(Gorakhpur, 1996)*

(*b*) Structure of the thallus in *Ectocarpus*

(*c*) Cell structure in *Ectocarpus.* *(Bharathiar, 2000)*

(*d*) Life cycle of *Ectocarpus*. *(Kanpur, 2007)*

25. What are unilocular and plurilocular reproductive bodies and where do they occur ? What are they concerned with ?

26. Comment upon:

(*a*) Alternation of generation in *Ectocarpus.* *(Lucknow, 2002)*

(*b*) Isomorphic alternation of generation *(Lucknow, 2003, 2005)*

(*c*) *Ectocarpus* and *dictyotales.*

27. Write ten sentences on
    (*a*) Unilocular Sporangia of *Ectocarpus* (*Lucknow, 2005*)
    (*b*) Two characteristics of gametophytic and sporophytic generations of *Ectocarpus*. (*Allahabad, 2004*)
    (*c*) Plurilocular sporangia.

28. What do you understand by isomorphic alternation of generation? Briefly explain. (*Allahabad, 2006*)

29. What is the difference between unilocular and plurilocular sporangium? (*Kanpur, 2006*)

30. Give systematic position of *Ectocarpus*. (*Kanpur, 2005*)

OBJECTIVE TYPE

31. **Select the correct answer :**

(*i*) In *Ectocarpus* during fertilization the male and female gametes attach themselves to the female gamete by the anterior flagellum which is
(*a*) Long and whiplash
(*b*) Long and tinsel
(*c*) Short and whiplash
(*d*) Short and tinsel.

(*ii*) Thallus construction in *Ectocarpus* is
(*a*) Filamentous unbranched
(*b*) Heterotrichous
(*c*) Heterothallic
(*d*) Trichothallic.

(*iii*) In the motile reproductive bodies of *Ectocarpus* the anterior flagellum compared to the posterior one is
(*a*) Tinsel type and longer
(*b*) Whilplash type and shorter
(*c*) Tinsel type and shorter
(*d*) Whilplash type and longer.

(*iv*) Isomorphic alternation of generation is observed in
(*a*) *Hydrodictyon*
(*b*) *Batrachospermum*
(*c*) *Pinnularia*
(*d*) *Ectocarpus.*

(*v*) Plurilocular sporangia are characteristic of
(*a*) *Sargassum*
(*b*) *Dictyota*
(*c*) *Ectocarpus*
(*d*) *Ulothrix.*

(*vi*) In the life cycle of *Ectocarpus* meiosis occurs during
(*a*) Zoospore formation in plurilocular sporangia
(*b*) Zoospore formation in unilocular sporangia
(*c*) Gamete formation in the gametangium
(*d*) Germination of the zygote.

(*vii*) In the life cycle of *Ectocarpus* formation of mitospores is for
(*a*) Multiplication of the type of plants that produce them
(*b*) Formation of gametophyte
(*c*) Sexual reproduction
(*d*) Vegetative reproduction.

(*viii*) The unilocular sporangium is called so because it has
(*a*) Only one zoospore
(*b*) Haploid zoospores
(*c*) One meiotic division
(*d*) One chamber.

(*ix*) The zoospores of *Ectocarpus* are
(*a*) Haploid
(*b*) Diploid
(*c*) Both haploid and diploid
(*d*) Diploid in the sporophyte and haploid in the gametophyte.

(*x*) In *Ectocarpus* alternation of generation is said to be isomorphic because of the morphological similarties between the
(*a*) Male and female gametes
(*b*) Zoospores of unilocular and plurilocular sporangia
(*c*) Gametes and the zoospores
(*d*) Gametophyte and the sporophyte.

Dictyota sp.

CHAPTER 18

Order 2. DICTYOTALES

- Dictyota Dichotoma
- Alternation of Generations

The order includes marine forms in which the thallus is erect, flat and parenchymatous. Branching is regularly dichotomous. Apical growth is either by means of a single apical cell or by an apical meristem consisting of a mariginal row of cells (*Zonaria* and *Padina*). Sexual reproduction is oogamous sometimes anisogamous. The sex organs are borne in sori in the gametophyte plants which are typically unisexual. The sporophyte plant bears unilocular sporangia, each of which by meiosis produces usually four, sometimes eight, large, non-motile, naked meiospores. Both sporophytic and gametophytic plants are morphologically similar, thus alternation of generations is of isomorphic type.

The Dictyotales occur both in the temperate and tropical seas but attain their greatest development in the warmer oceans of the tropics and subtropics. The order comprises about 180 species which are allotted to 20 genera. The three best known genera are *Dictyota, Zonaria* and *Padina.* They are placed in the family Dictyotaceae.

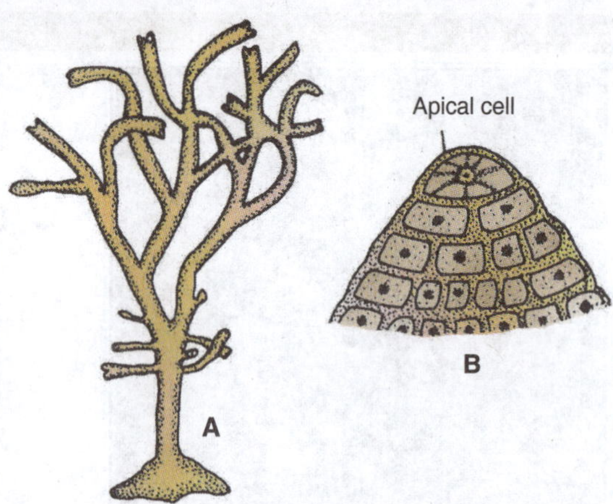

Fig. 18.1 (A-B). *Dictyota dichotoma.* (A) A plant showing habit; (B), Surface view of thallus apex.

Family Dictyotaceae. *Dictyota* is the type genus of the family with a flat, ribbon-like dichotomously branched thallus. It includes about 35 species. Of these 12 species have been reported from India by Misra (1966). *Padina* is another important genus of the family with a fan-shaped thallus in which lime occurs in the cell walls-a feature rare in the Phaeophyta. *Dictyota* is being discussed in detail here.

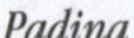

Padina

Dictyota bartrayresiana

Dictyota linearis

DICTYOTA DICHOTOMA

Dictyota dichotoma is a widely distributed, cosmopolitan *annual* commonly found in winter in the warmer waters of the tropics. It also occurs along the Indian coast, usually in the rocky tidepools submerged and attached to rocks with the help of discoid holdflast. In the temperatre regions it flourishes in summer.

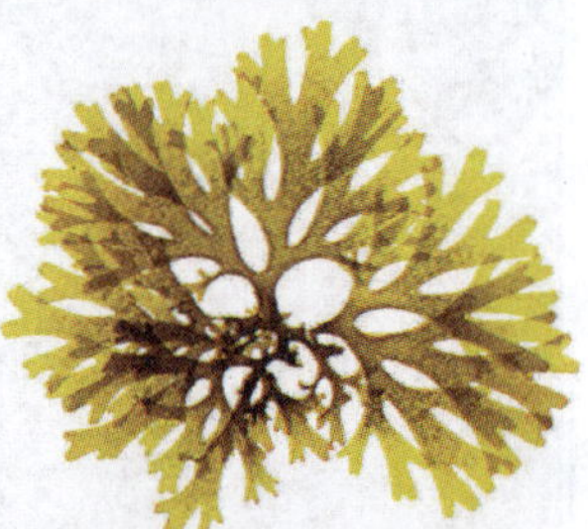

Dictyota dichotoma

Thallus (Fig. 18.2, A-B). *(a) External morphology.* The thallus of *D. dichotoma* is yellow brown in colour. It attains a height of about 30 cm. and shows differentiation into three parts namely, the **frond,** the **stipe** and the **holdfast.** Frond is the terminal upright, forked ribbon-shaped part of thallus (A). It shows perfect dichotomy and lacks a midrib. In texture it may be leathery or membranous. All the branches of the yellow brown frond lie in the same plane. Each flattened branch bears a single biconvex apical cell at its tip.

The middle, stalk-like more or less cylindrical or subcylindrical, unbranched region of the thallus (B) which gradually expands into the upper ribbon-like repeated by forked-frond is known as the **stipe** (B). The stipe at its lower-end is fixed to the substratum by means of a tuft of rhizoids constituting the **holdfast**. The stipe bears adventitious branches. Some of these may grow into horizontal stolons and serve as a means of vegetative propagation.

Dictyota attached by holdfast

Internal structure of the thallus (Fig. 18.2 H). The thallus in section is seen to consist of three layers of cells (H). There is an upper and a lower assimilatory or photosynthetic layer. Between these two is the median layer consisting of large, colourless medullary cells which contain fucosan vesicles and reserve food in the form of globules but lack chromatophores. There is a single large vacuole in the centre. The septa between the medullary cells are characterised by the presence of pits. This layer serves as a food storage region. The cells of the peripheral photosynthetic or epidermal layers are small and contain abundant chromatophores. They constitute the photosynthetic region of the thallus. Here and there groups of epidermal cells on both the surfaces of the thallus develop into colourless, multicellular mucilage hairs. The hairs are shed when the plant enters the reproductive phase.

Forked Ribbon shaped thallus of *Dictyota dichotoma*

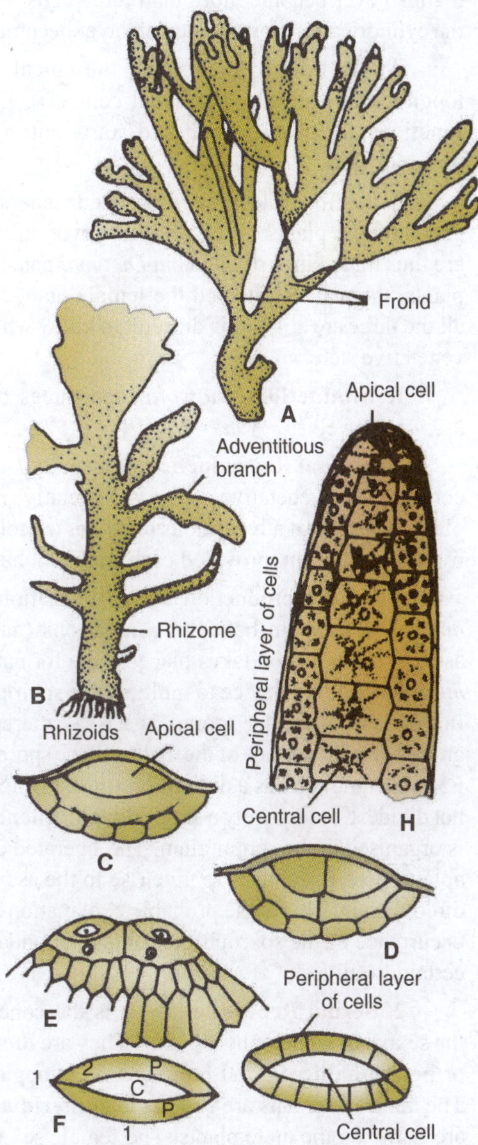

Fig. 18.2 (A-H). *Dictyota dichotoma.* (A), repeatedly forked upper portion of the erect frond showing habit; (B), basal cylindrical part of the erect frond; (C), surface view of thallus apex; (D), apical cell divided showing beginning of dichotomy; (E), segmentation during dichotomy; (F-G), seg-ments of thallus cross section showing sequence of segmentation (1-1, 2-2); (H), Median L. S. of thallus apex in the plane of the wide axis showing structure.

Growth (Fig. 18.2 C-H). The growth of the thallus is apical (H). It takes place by means of a single biconvex apical cell situated at the tip of each flattened branch of the frond (C). It appears elliptical when viewed from above. The apical cell cuts off a single series of segments at its posterior face. Each segment divides by a curved wall in a plane parallel to the thallus surface (F, 1-1). As a result two unequal segments are formed. The larger segment or the daughter cell divides by a second curved wall (F, 2-2). The second division is also parallel to the plane of flattening of the thallus. In this way two small peripheral or primary cortical cells are cut off, one towards either face of the thallus. Between them lies a single, large, central primary medullary cell (F,C). The primary cortical or peripheral cells divide by anticlinal walls (G). Meanwhile the large primary medullary or central cell also divides and redivides vertically in the same plane. Consequently the mature portions of the thallus never become more than three cells in thickness (H). In the cylindrical portion the median layer becomes 3-6 layers thick.

At the time of dichotomy the apical cell divides by a longitudinal wall into two equal cells (D). Each daughter cell functions as an apical cell and starts cutting segments at its posterior face.

In the life cycle of *Dictyota* occur separate sporophyte and gametophyte plants. The latter, however, are **dioecious**. There are thus three kinds of plants in *Dictyota* namely, the sporophyte plants, the male plants and the female plants. Morphologically, all the three are alike. It is difficult to know which is which in the vegetative state.

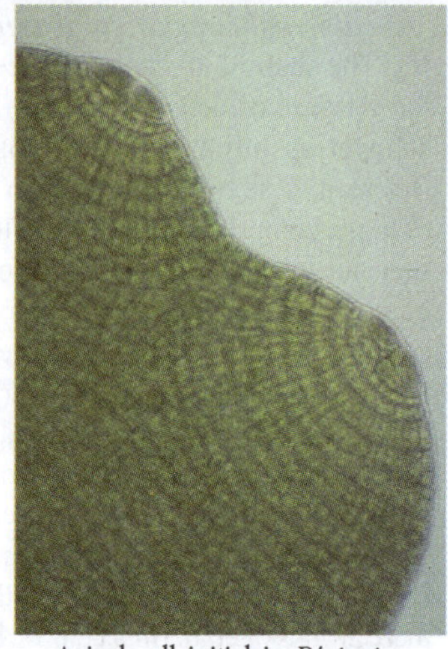

Apical cell initial in *Dictyota*

Reproduction. *Dictyota* reproduces by both methods, *asexual* as well as *sexual*.

1. Asexual Reproduction. Asexually *Dictyota* reproduces commonly by vegetative methods. Vegetative reproduction takes place by means of adventitious branches or stolons. It also occurs by **fragmentation**, provided each fragment has an apical cell.

Asexual reproduction by the formation of special kind of asexual spores (Mitospores) is *normally absent* in both the generations (gametophyte and sporophyte). Sometimes, however, asexual reproduction takes place by the formation of *mitospores*. It is a device to multiply the sporophytes. In this case unilocular sporangia arise scattered or in groups over the surface of the thallus of the sporophyte. Each sporangium has a diploid protoplast which does not divide. Consequently a single diploid **aplanospore** is organised in the sporangium. The liberated diploid aplanospore germinates to give rise to the asexual or diploid plant. This is the probable explanation for the occurrence of the sporophyte plants in abundance in certain localities.

2. Sexual Reproduction. It is the concern of the sexual or gametophyte plants. They are **dioecious** or **heterothallic**. Sexual reproduction is **oogamous**. The male sex organs are called the **antheridia**. They are borne on the male plants. The female sex organs are called the **oogonia**. They are produced by the female plants. Both the antheridia and oogonia are developed in definite projecting groups called **sori**. The female plants can be recognised from the males by the deep brown colour of the oogonial sori. The sori are developed in successive crops on both sides of the thallus.

The new ones appear between the scars left by the old ones. When the whole surface of the thallus has been used up the plant perishes.

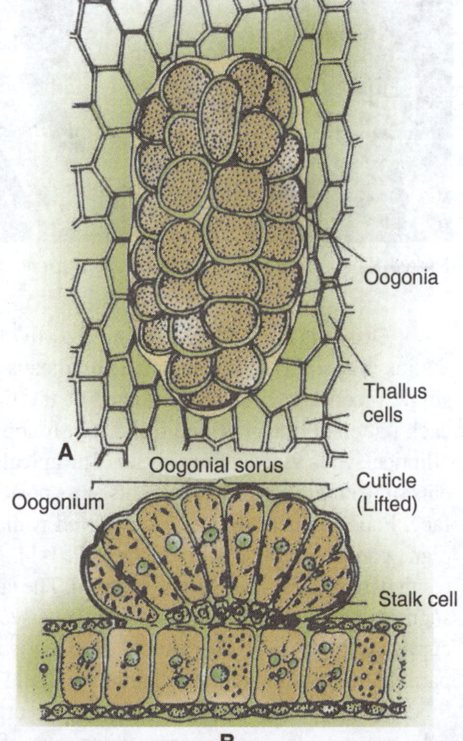

Fig. 18.3 (A-B). *Dictyota dichotoma.* (A) Surface view of oogonial sorus; (B), T..S. thallus passing through the oogonial sorus (A, after Thuret).

Oogonia (Fig. 18.3 A-B) The oogonial sori are deep brown in colour. There is either no involucre or a

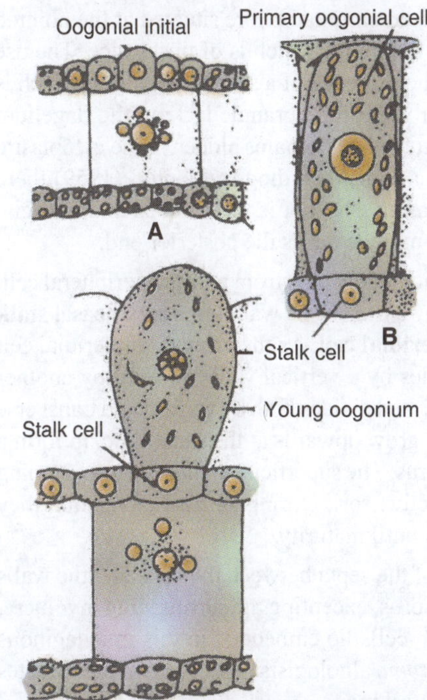

Fig. 18.4 (A-C). *Dictyota dichotoma.* Development of oogonium. (A), Oogonim initial; (B), Differentiation of stalk cell and primary oogonial cell; (C), Young oogonium.

rudimentary one around the oogonial sorus. The number of oogonia in a sorus varies from 25 to 50 (A).

(*a*) *Structure.* The mature oogonium is a large, sac-like structure supported on a small, basal stalk cell (Fig. 18.4 C). The body of the oogonium contains a single large brown uninucleate non-flagellate egg or ovum. Eventually the wall of oogonium gelatinise at the apex. The single ovum is liberated through the apical pore into the surrounding water and drifts passively (Fig. 18.6 A).

(*b*) *Development of oogonium* (Fig. 18.4). Each oogonium in a sorus arises from a single peripheral cell which functions as the **oogonial initial** (A). The latter elongates vertically and divides by a asymmetrical transverse division into a small, basal **stalk cell** and an upper, large **primary oogonial cell** (B). The former remains small and embedded in the peripheral layer. The primary oogonial cell enlarges considerably in size to form the **oogonium** (C). Its contents develop into a single egg or ovum.

Antheridia (Fig. 18.5). The antheridial sori look like white, shining spots on the surface of the male plants. The number of antheridia in a sorus varies from 100 to 300. The marginal vegetative cells of the antheridial sorus grow upwards to form a cup-shaped **involucre** around it. The male sorus is thus enclosed by an outer zone consisting of 3 or 4 rows of sterile, undivided cells which grow nearly to the same height as the antheridia (A).

(*a*) *Structure* (Fig. 18.5 B). The mature antheridium is a large colourless structure. It consists of a small basal **stalk cell** and a **body**. The body of antheridium is a multicellular structure composed of small, cubical cells called the locules. The locules are arranged in transverse tiers, there being 16 cells in each tier. There are 20 to 24 such tiers in height. The protoplast of each cell or locule gets metamorphosed into a single **sperm**. The

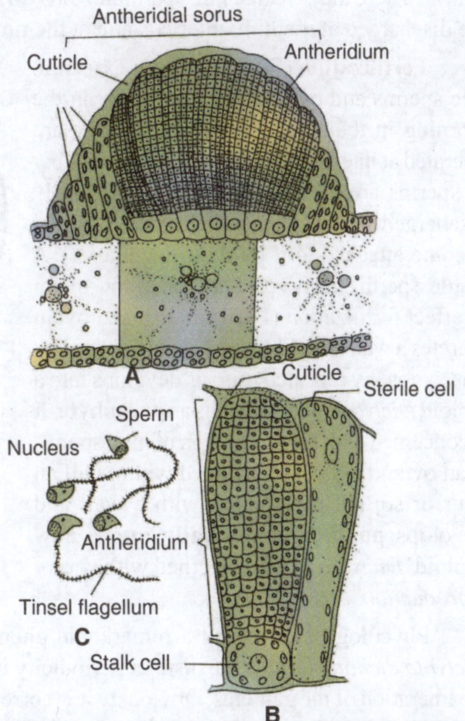

Fig. 18.5 (A-C). *Dictyota dichotoma.* (A), T.S. thallus passing through the antheridial sorus; (B), Single antheridium and the surrounding sterile cell; (C), Sperms.

sperm (C) is a small pyriform structure with a much reduced chromatophore situated at the pointed anterior end. There is a single forwardly directed lateral **flagellum** which is of tinsel type. The use of an electron microscope (Manton, 1959) has revealed the presence of a second flagellum which is rudimentary. There are two basal granules. The posterior part of the granule lacking the flagellum projects to the surface of the nucleus and forms a minute stub which remains hidden in the cytoplasm. The sperm has a distinct **nucleus** at the broader posterior end. According to Manton (1959) there is no eye spot. What Lloyd Williams (1905) interpreted as an eye spot is the reduced and vestigial chromatophore. The single flagellum is inserted laterally more towards the posterior end.

(*b*) *Development.* Each antheridium, like the oogonium, develops from a single peripheral cell, the **antheridial initial**. The latter elongates and divides by a transverse wall into a small basal **stalk cell** and an upper large primary spermatogenous or **antheridial cell**. As the primary antheridial cell enlarges in size it loses its chromatophores. It then divides by a vertical wall followed by another vertical wall at right angles to the first. The quadrate thus formed divides both by vertical and transverse walls. The marginal vegetative cells adjacent to the sorus grow upwards at the same time, to form a cup-shaped protective envelope or involucre around the sorus. The superficial thallus cells developing into sex organs remain in close apposition in their respective sori. During vertical elongation they gradually lift the cuticle which invests the respective sori until maturity.

(*c*) *Dehiscence.* At maturity there is gelatinisation of the septa between the cells and the walls of the mature antheridia. As a result the entire antheridial sorus, excepting the surrounding involucre, becomes a mass of mucilage. The minute rounded sperm cells lie embedded in this mucilaginous matrix. There they change into spermatozoids. In *D. dichotoma* algologists observed that the gametes are discharged at monthly intervals and at the time of the full moon.

Fertilization (Fig. 18.6). It is **external**. The sperms and ova are liberated early in the morning in the surrounding water. They are liberated at intervals of 14 days. A large number of sperms are attracted towards each freshly discharged passively, floating ovum (A) and become attached to it by their flagella (A). A single sperm, however, penetrates the ovum to effect fertilization (B). The fertilized ovum secretes a wall around it and becomes a zygote. The zygote by repeated mitosis develops into a diploid *Dictyota* plant which is a **sporophyte**. It is concerned with the production of meiospores. If an ovum fails to be fertilized within half an hour or so, it clothes itself with a wall and develops **parthenogenetically** into a new haploid *Dictyota* plant concerned with *sexual reproduction*.

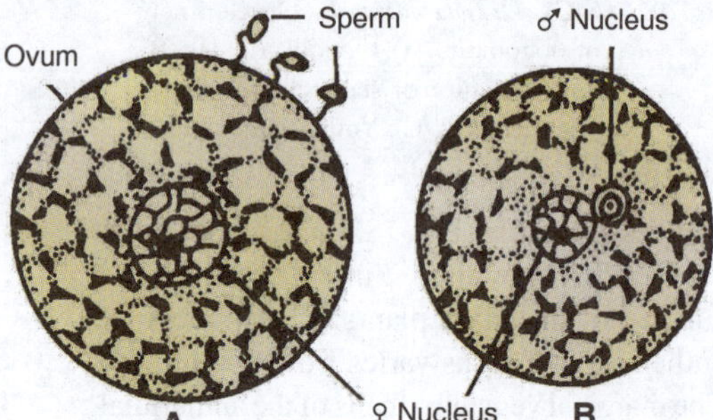

Fig. 18.6 (A-B). *Dictyota dichotoma.* Fertilization. (A), Freshly discharged spherical ovum surrounded by sperms; (B), Fertilized ovum with male and female nuclei lying side by side prior to fusion.

Phycologists observed a remarkable phenomenon in regard to the sexual reproduction of *Dictyota dichotoma.* It is the distinct periodicity in the development of sex organs on the thallus and the liberation of the gametes. Obviously it is correlated with the lunar month and the tidal cycle (neap and spring tides). According to Williams (1905) this periodicity is hereditary. He noticed that in the British waters the sex organs are produced during summer months from July onwards in fortnighly crops corresponding to the interval between two spring tides. The production of sex organs is about a week later for several days following the highest spring tide during neap tides period. The gametes are thus liberated during the next series of neap tides.

Germination of Zygote. The zygote germinates immediately without undergoing any resting period. The mitotic division of the diploid zygote nucleus is followed by the formation of a bipartition wall resulting in two daughter cells. One of these elongates to form the primary rhizoid consisting of a few cells. The sister daughter cell by repeated mitotic divisions and growth grows into an alternate diploid plant called the **sporophyte**. An apical cell is established very early in the development of the sporophyte.

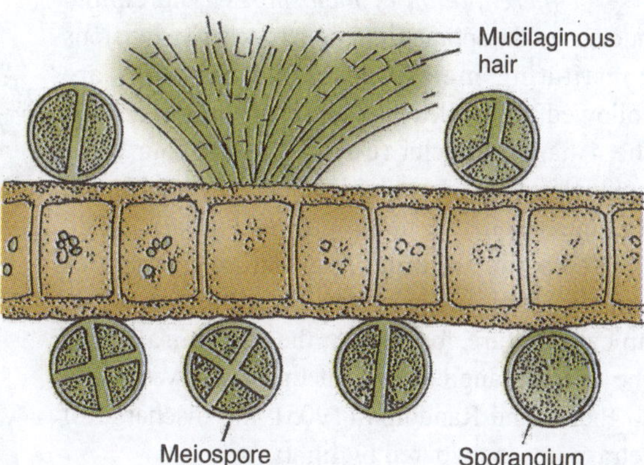

Fig. 18.7. *Dictyota.* T.S. thallus bearing tetrasporangia containing tetra-spores.

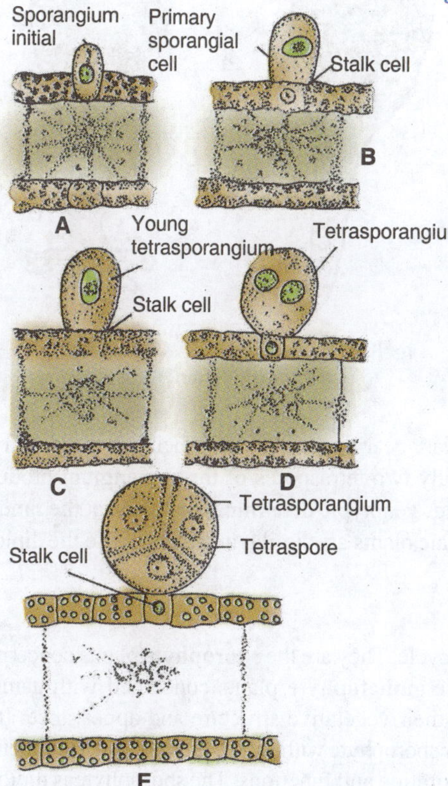

Fig. 18.8 (A-E). *Dictyota dichotoma.* Development of tetrasporangium. (A), peripheral cell functioning as sporangium initial; (B), sporangium initial divides to form a stalk cell and primary sporangial cell; (C), young sporangium with a nucleus at right angles to the surface of the thallus; (D), diploid nucleus undergoes first meiotic division; (E), next stage showing second meiotic division followed by cleavage of cytoplasm to form haploid tetraspores.

Sporophyte (Fig. 18.7). Morphologically the diploid thallus is exactly similar to the male and female thalli but instead of producing antheridia and oogonia it bears spherical sporangia. The production of sporangia shows no correlation with the tidal periodicity. They are borne either singly or in ill-defined groups (sori) not surrounded by involucres on the both the surfaces of the thallus. The sporangia project well above the surface of the thallus. The sporangium consists of a basal **stalk cell** and a **spherical body** which is unicellular. The stalk cell is embedded in the peripheral layer of the thallus. However, Heil (1924) reported that it projects into the sporangium like a columella and is said to help in dehiscence. The sporangium is filled with diploid protoplast.

Development of the sporangium (Fig. 18.8). The sporangium arises from a single peripheral cell of the thallus. It is the *sporangial initial* (A). It grows into a papilla-like outgrowth two or three times its original length. It divides by a transverse wall into a small basal **stalk** cell and an upper large cell (B). The basal stalk cell divides no further. Usually it remains embedded amidst the peripheral cells. In a few cases it elongates into a columella-like structure projecting into the sporangium at maturity. The upper large cell functions as the **primary sporangial** cell. The latter enlarges considerably in size and becomes spherical in form to form the tetrasporangium. It has a distinct large diploid nucleus (C).

Differentiation of meiospores. The diploid nucleus of the sporangium undergoes two divisions constituting **meioses.** Nuclear divisions are followed by the cleavage of the cytoplasm around the 4 haploid nuclei (e). As a result four large uninucleate naked, haploid meiospores are formed. The meiospores are non-motile and usually called the **tetrasporangium**. The mature tetraspores usually round off prior to liberation through an apical aperture formed by the gelatinisation of the tetrasporangium wall at the apex. According to Pierce and Randolph (1905), the discharge of tetraspores is favoured by light.

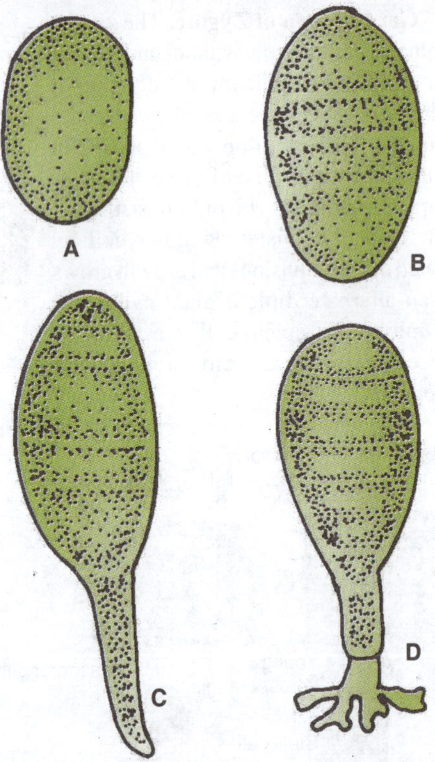

Fig. 18.9 (A-D). *Dictyota sp.* Germination of tetraspore (After Tthrust).

Germination of tetraspores (Fig. 18.9) The liberated tetraspore secretes a cellulose wall, around it (A) and then germinates without any resting period. Inoth (1936) reported that the first wall is at right angles to the incident ray. Of the two sister cells thus formed, the upper one functions as the apical cell of the embryo whereas the lower forms the rhizoid. The apical cell cuts off segments in the same manner as the apical cell of the thallus. During further development the apical region of the embryo flattens to form the ribbon-shaped thallus. From its basal portion arise numerous lateral branches which creep along the substratum. Some of these grow to form the rhizome portion of the thallus. The resultant plant differs from the parent plant bearing tetraspores in being haploid. It may be called the **sexual**, or gametophyte plant. Normally two tetraspores of the sporangium produce **male** plants and other two **female** plants. There is thus genotypic determination of sex at the time of meiotic division. Morphologically the male and female plants are similar and identical to the diploid or asexual plant.

Alternation of Generations (Fig. 18.10)

There occur two kinds of plants in a single life cycle. They are the **sporophyte** plants concerned with the spore production and the male and female **gametophyte** plants concerned with gamete production. The two kinds of plants are similar in their vegetative structure and appearance. It is difficult to identify a *Dictyota* plant as gametophyte or sporophyte without the presence of reproductive structures. Of course they differ in their genetic constitution and functions. The sporophyte is diploid, or $2N$ plant with 32 chromosomes. It produces the so-called **tetrasporangia**. The protoplast of the tetrasporangium produces usually four meiospores commonly called the **tetraspores**. Since the tetraspores are produced following a meiotic division they are haploid, each has 16 chromosomes. They germinate to produce the haploid, or the gametophytic plants. The latter bear sex organs arranged in definite **sori**. The sex organs give rise to the haploid gametes, the **sperms** and the **eggs**. The male and the female gametes fuse to form the diploid **zygote**. The zygote by mitotic divisions forms the diploid *Dictyota* plant.

The **zygote**, the **sporophyte** plant and the **tetrasporangia** are all diploid. They constitute the sporophyte generation or phase. The meiospores (tetraspores), sexual or the **gametophyte** plants,

antheridia, oogonia and the **sperms** and the **eggs** are all haploid. They constitute the **gametophyte** generation or phase.

These two generations occur regularly one after the other in a single life cycle of *Dictyota*. This is known as **alternation** of **generations**. Since the alternating diploid and haploid plants are similar in appearance, this type of alternation of generations is called **isomorphic**. The life cycle of *Dictyota* which is characterised by distinct alternation of generation with sporogenic meiosis is called **diplohaplontic**.

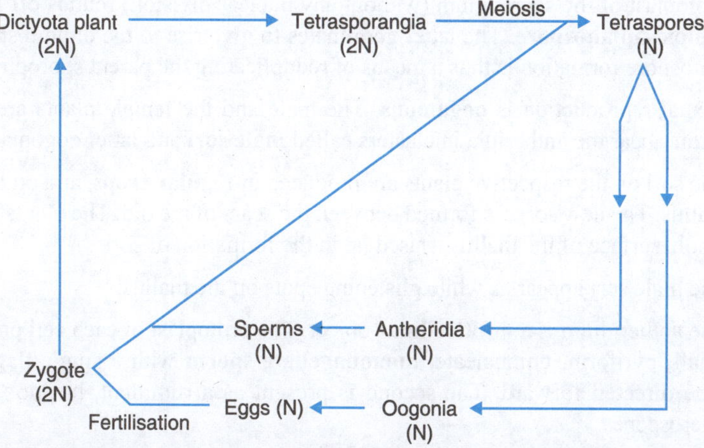

Fig. 18.10. Graphic representation of the life cycle of *Dictyota sp.*

Salient Features

1. The thallus is differentiated into (*i*) a prostrate irregularly shaped, rhizome-like or disc-shaped basal **holdfast** and (*ii*) upright **frond**.
2. Each frond has a lower, **cylindrical** protion which gradually flattens upwards into the **ribbon-like** portion which may be leathery or membranous.
3. The ribbon-like portion of the frond is repeatedly **forked** and shows perfect **dichotomy**.

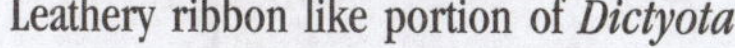

Leathery ribbon like portion of *Dictyota*

Fronds of *Dictyota* showing pefect Dichotomy

4. Viewed in transverse section the flattened frond consists of three layers of cells, a **median** layer of large, colourless storage cells banded on each side (above and below) by a layer of small **assimilatory** cells.
5. Here and there from the surface cells arise groups of colourless, muticellular, **mucilage hairs**. They are shed when the reproductive phase ushers.

6. The growth of the thallus takes place by means of a single lenticular **apical cell**.
7. The thalli are of two kinds, **diploid** sporophytes and **haploid** gametophytes.
8. Vegetative propagation of both the haploid and diploid thalli takes place by means of **adventitious branches** and **fragmentation**.
9. Normally there is no specialised asexual reproduction of either generation. Sometimes asexual reproduction takes place by the method of mitospore formation in which the entire protoplast of the sporangium (without any nuclear division) rounds off to form a single **diploid aplanospore**. The latter germinates to give rise to the diploid sporophyte plant. Mitospore formation is thus a means of reduplicating the parent sporophyte plant.
10. Sexual reproduction is **oogamous**. The male and the female plants are **separate**. The former bear the antheridia in clusters called **male sori** and latter oogonia in **female sori**.
11. The sori on the respective plants are produced in **regular crops** and on both sides of the thallus. The new sori are formed between the scars of the old. The plants perish when the whole surface of the thallus is used up in the formation of sori.
12. The male sori appear as white glistening spots on the thallus.
13. The antheridium is a **multicellular** object. The protoplast of each cell produces a single, small, pyriform, uninucleate, **monoflagellate sperm** with a single flagellum of tinsel type directed forward. The second is present as a rudiment, but not seen with light microscope.
14. The female sori appear as deep brown objects on the thallus.
15. Each oogonium is a **unicellular** object and produces a single **ovum** which is liberated through an apical pore.
16. Fertilization is thus **external**.
17. The zygote is not **resting spore**. It germinates at once to form a fertile **diploid** plant or the **sporophyte**.
18. The sporophyte reproduces by the formation of meiospores commonly called the **tetraspores.**
19. Four tetraspores are differentiated by meiosis from the diploid protoplast of the sporangium also known as the **tetrasporangium** borne on the diploid sporophyte plant.
20. The rliberated **tetraspore** secretes a wall around it and at once germinates into an alternate plant in the life cycle which is the haploid gametophyte plant.
21. Of the four tetraspores from each sporangium two give rise to the **male plants** and the other two to the **female plants**.
22. The diploid and the haploid plants are morphogically alike.
23. There is thus **isomorphic alternation of generations** in the life cycle which is **diplohoplantic**.

Taxonomic Position

Division	:	**Phaeophyta** (Phaeophycophyta)
Class	:	**Isogeneratae**
Order	:	**Dictyotales**
Family	:	**Dictyotaceae**
Genus	:	*Dictyota*
Species	:	*dichotoma*

QUESTIONS

ESSAY TYPE

1. Describe the habitat and structure of the thallus of *Dictyota*.
2. Describe sexual reproduction in *Dictyota*. (*Himachal Pradesh, 1992*)
3. Explain the phenomenon of alternation of generations in *Dictyota*. (*Himachal Pradesh, 1995*)
4. Give a diagrammatic representation of the life cycle of *Dictyota* illustrating the relative lengths of the haploid and diploid generations. (*Himachal Pradesh, 1996*)
5. Give significant features in the structure and methods of reproduction in *Dictyota*. (*Punjab University 1945*)
6. What are tetrasporangia ? Where are they produced and to what structures do they give rise to? What are the functions of the plants produced by the tetraspores?
7. Describe the morphology of thallus of *Dictyota* (both external and internal). Discuss the occurrence, structure, development and functions of tetrasporangia. (*Punjab, 1999*)
8. Give a comparative account of the structure and life cycle of
   (*a*) *Dictyota*
   (*b*) *Padina* (*Lucknow M.Sc., 2004*)

SHORT ANSWER TYPE

9. Diagrammatically give the life cycle of *Dictyota.* (*Gurunanak Deo, 1991*)
10. Draw well labelled diagrams of T.S. of *Dictyota* passing through male and female sori. (*Gurunanak Deo, 1992*)
11. Describe the internal structure of *Dictyota* thallus. (*Trichy, 1995*)
12. Write short notes on :
    (*a*) Oogonial sorus of *Dictyota* (*Punjab, 1993*)
    (*b*) Fertilisation in *Dictyota* (*Himachal Pradesh, 1991*)
    (*c*) Formation of thallus of *Dictyota* (*Himachal Pradesh, 1997*)
    (*d*) Sexual reproduction in *Dictyota*
    (*e*) Asexual reproduction in *Dictyota*
    (*f*) Internal structure of thallus of *Dictyota.*
13. The gametophytic thalli of *Dictyota indica* are often found missing on our shores. Explain its absence. (*Himachal Pradesh, 1990*)
14. Discuss the structure of tetrasporangia in *Dictyota.* (*Himachal Pradesh, 1991*)
15. Define tetraspore. (*Himachal Pradesh, 1993*)
16. Compare life cycles of *Dictyota* and *Fucus.*
17. Comment upon Ectocarpales and Dictyotales.

OBJECTIVE TYPE

18. **Select the correct answer :**
    (*i*) The antheridium in *Dictyota* is
    (*a*) Unicellular (*b*) Bicellular
    (*c*) Multicellular (*b*) Acellular
    (*ii*) Which of the following is exclusively marine ?
    (*a*) *Ulothrix* (*b*) *Spirogyra*
    (*c*) *Nostoc* (*d*) *Dictyota*
    (*iii*) The sperm in *Dictyota* is
    (*a*) Monoflagellate with tinsel type of flagella
    (*b*) Monoflagellate with whiplash type of flagella
    (*c*) Biflagellate with tinsel and whiplash flagella
    (*d*) Aflagellate

(*iv*) In which of the following the thallus is differentiated into prostrate irregularly shaped disc shaped hold fast and upright fond ?

(*a*) *Ectocarpus* (*b*) *Sargassum*
(*c*) *Dictyota* (*d*) *Coleochaete*

(*v*) The process of sexual reproduction in *Dictyota* is

(*a*) Isogamous (*b*) Anisogamous
(*c*) Oogamous
(*d*) None of the above

(*vi*) In which of the following alga, the thallus is dioecious ?

(*a*) *Spirogyra* (*b*) *Ulothrix*
(*c*) *Dictyota* (*d*) *Ectocarpus*

(*vii*) Where does meiosis take place in *Dictyota* ?

(*a*) During the formation of gametes
(*b*) During the formation of tetraspores
(*c*) During the production of sperms
(*d*) During the germination of zygote

(*viii*) Tetraspores are characteristic of

(*a*) *Ectocarpus* (*b*) *Oedogonium*
(*c*) *Dictyota* (*d*) *Spirogyra*

(*ix*) In sporophytic thalli of *Dictyota* asexual reproduction takes place by

(*a*) Mitospores (*b*) Zoospores
(*c*) Akinetes (*d*) Heterocyst.

(*x*) Fertilization in *Dictyota* is

(*a*) Internal (*b*) External
(*c*) Both (*d*) None

CHAPTER 19

HETEROGENERATE : LAMINARIALES

Laminaria ochroleuca

- Family : Laminariaceae
  Genus : Laminaria
- Distribution
- Gertilization
- Germination of Zygote

General

The order includes algae that have attained a large size and are popularly called the **kelps**. They lead the other thallophytes in their morphological and anatomical complexity. In fact they are among the giants of the plant kingdom notable for their size and external elaboration of form. In form they range from the long, whiplike sporophyte of *Chorda* through the sea palm (*Postelsia* Fig. 16.3), sea fern (*Thallassiophyllum*) ending in the sea giants like *Macrocystis* (Fig. 16.1) which vary from 30 to 50 metres in length. The basic unit of structure, however, is the filament. The diplohaplontic life cycle exhibits heterologous type of alternation of generations. The sporophytes are the largest sea weeds whereas the gametophytes are dwarfed, microscopic and of filamentous growth. The latter bear the sex organs called **antheridia** and **oogonia** on separate plants. The sporophyte (Fig. 19.1) is nearly bilaterally symmetrical. It grows by means of an intercalary meristem and usually consists of one to several expanded leaf blades connected by a stalk (**stipe**) to a root-like holdfast (**hapteron**). The holdfast is firmly attached to some rock at the bottom. These parts superficially resemble the roots, stem and leaves. Many of the sporophytes show a tree-like form and present palm-like appearance (*Postelsia,* Fig. 16.2). There are about 100 species of kelps placed under 30 genera. The Laminariales are distinguished from the Fucales by their intercalary growth. The main region of cell division lies in the transition zone between the stipe and the lamina.

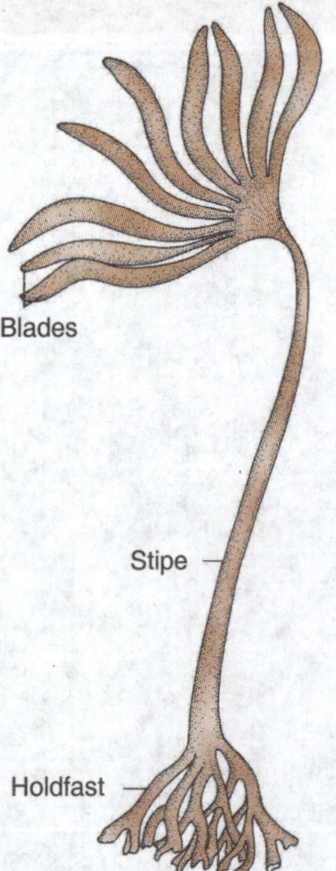

Fig. 19.1. *Laminaria.* Sporophyte with divided lamina.

Chorda

Postelsia

Macrocystis

The kelps generally grow attached to the rocks and are known as **lithophytes**. Epiphytes are uncommon. Some Laminariales are present on rock fragments which are not strong enough to anchor the adult thallus. As a result they get carried to the sea together with the rock pieces. Plants of *Laminaria cloustoni* sometimes bring ashore blocks of rock weighing several pounds. The sporophyte reproduces by the formation of unilocular sporangia. The sporangia form extensive sori with paraphyses on the blades.

Distribution

Laminariales are the inhabitants of the colder oceans of the Arctic and Antarctic zones. They reach their greatest size along the western coasts of North America. Some of them attain a length of over 100 feet. They occupy huge areas on the Pacific coast of North America and Canada, and grow on substrata that are free from permanent ice. They are also found in abundance in the temperate zones. Tropical and sub-tropical regions also have a few representatives of the Laminariales. Some species of *Laminaria* are also found in the adjacent parts of Asia. *Laminaria lejolisii* flourishes well in the Atlantic coast of Morocco. *L. pallida* is common in the southern hemisphere. The Pacific ocean also inhabits a great number of diverse forms of this order.

Classification

The order is divided into the following four families by Setchell :

1. *Chordaceae.* It includes the monotypic genus *Chorda.*
2. *Laminariaceae*. It is divided into four sub-families and includes various genera like *Laminaria, Phyllaria, Pleurophycus* etc.
3. *Lessoniaceae*. It is further split into three sub-families and includes *Lassonia, Postelisa, Macrocystis, Nerecystis,* etc.
4. *Alariaceae.* It is further divided into 3 sub-families and includes *Alaria, Underia* etc.

Laminaria.

Pleurophycus.

Nerecystis.

Alaria.

FAMILY : LAMINARIACEAE
GENUS : LAMINARIA

Distribution. The genus includes about 40 species. They are widely distributed in colder as well as temperate and tropical waters of the ocean. Some species are found in both the Northern and the Southern hemispheres. *L. flavicans* is found in New Zealand, *L. laminarioides* is common in Japan and the adjacent shores of Siberia. *L. Sinclarii* is found in the Pacific. *L. rodriaguezii* is the only species recorded in the Mediterranean. Several species are found in the Northern hemisphere. They occur in the sublittoral or in the lower littoral zone. Most of the species are perennial. A few are annuals (*L. ephemera*).

Morphology of the Sporophyte (Fig. 19.2). The thallus which is the diploid sporophyte may be six or more feet long. It is differentiated into a holdfast, stipe and blade or lamina.

1. Holdfast. It is an attaching organ variable in form. It may be a solid disc but usually it consists of root-like, branched (Fig. 19.1) or unbranched (Fig. 19.2) threads constituting the **hapteron**.

2. Stipe. It is the smooth, cylindrical or somewhat flattened, stem-like middle portion of the thallus. It arises from the basal holdfast and is unbranched, tough and pliable. It may be as long as 30-40 ft. and as thick as the forearm.

3. Blade or Lamina. It is the terminal expanded leaf-like portion borne terminally on the stipe. It has no mid-rib. The lamina is rather tough and leathery in texture and may reach a length of 2 metres or more. It may be simple and undivided into segments as in *L. saccharina* (Fig. 19.2). In *L. digitata* and *L. cloustoni* (Fig. 19.1) the blade is divided palmately into several segments in the upper part in the old specimens. It is undivided in the young plants. The latter resemble *L.*

Laminaria digitata.

Laminaria saccharina.

Laminaria setchelli.

Laminaria japonica

Blade
Stipe

Fig. 19.2. *Laminaria.* Sporophyte with undivided *Lamina.*

saccharina. The lamina is renewed at the base each year. The new blade appears at the apex of the stipe by the activity of the intercalary meristem. The old one is pushed in front and withers away or remains attached at the tip of the new blade (Fig. 19.3 C). Later on it disorganises. The plant may be perennial or annual. *L. cloustoni persists* for ten to twenty years. *L. ephemera* is an annual. In the annual species a new sporophyte is produced each spring.

Growth. The sporophyte grows by means of an **intercalary** meristem. It is situated at the junction of the stipe and blade. By repeated division of the meristematic cells the stipe and the lamina increase in length. In the perennial species the sporophyte lives from year to year. The blade, however, dies every year in autumn. The following spring a new blade is formed by the meristem at the base. The new lamina appears as the result of slight widening of the tip of the stipe (A). It is separated by a well-developed constriction from the previous year's blade (C-D). With the enlargement of the new blade the older one is reduced to a mere appendage and ultimately breaks off.

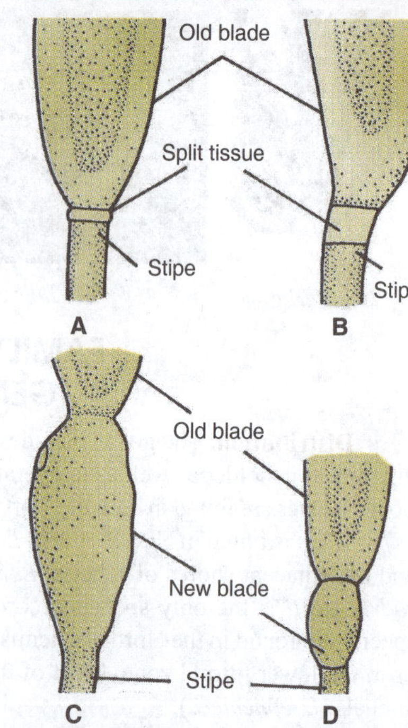

Fig. 19.3 (A-D). *Laminaria* sp. successive stages in blade-renewal (A and B after Setchell; C and D after Sau-vageau).

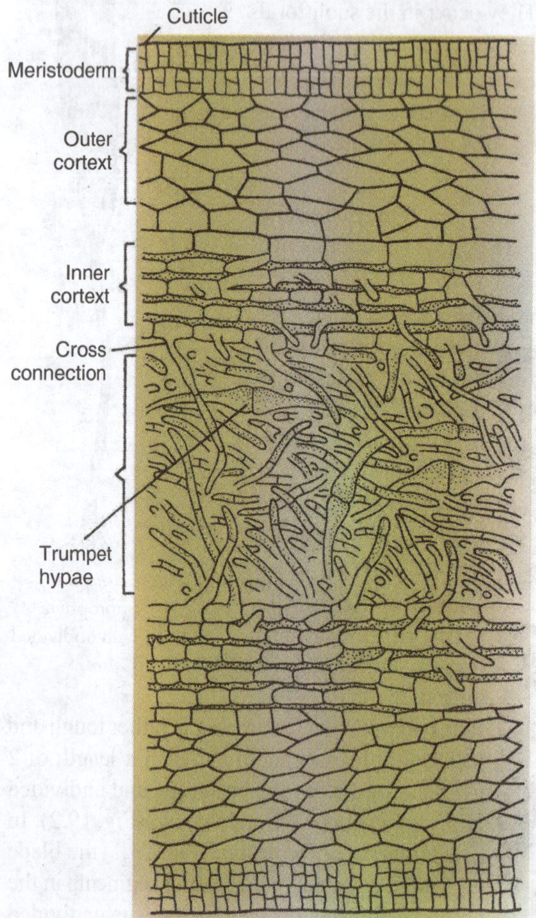

Fig. 19.4. *Laminaria.* L. S. Young stipe.

Anatomy of thallus

(Fig. 19.4). The thallus has a complex internal structure showing a greater degree of tissue differentiation than any other algal group. Viewed in a cross section the **stipe** shows 3 or 4 tissue regions namely, **meristoderm, cortex** (outer and inner) and **medulla** with cells quite distinct in size and shape.

(*a*) Meristoderm. It is usually a single surface layer of actively dividing (meristematic) cells which are cuboidal in form and have outer thickened preclinical walls. The thickening material forms a mucilaginous, non-cellular layer called the **cuticle**. Sometimes the meristoderm consists of two layers of small cells which contain pigmented plastids and retain the capacity to divide. It is thus both photosynthetic and meristematic. This superficial meristem contributes towards additional growth in girth in contrast to growth in length by an intercalary meristem. This results in lateral and longitudinal growth of the stipes and adds cells to the outer cortex (Sidemax and Scheirer, 1977).

(b) Cortex. It lies within the meristoderm and surrounds the central axis which constitutes the medulla. The cortex consists of large cells. It is wide (several cells deep) and appears as *outer* and *inner* cortex.

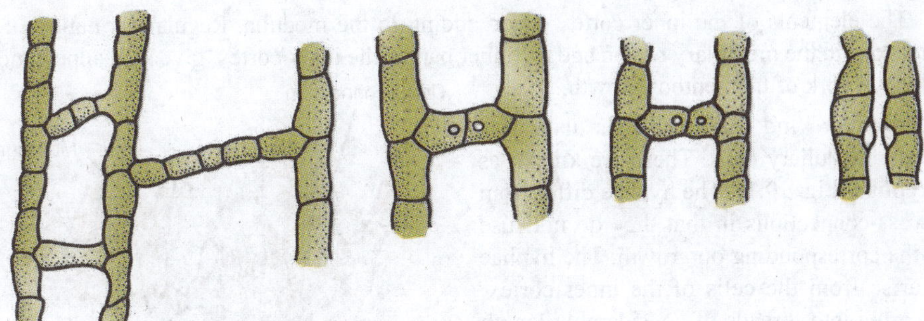

Fig. 19.5. Formation of cross-connections in *Laminaria* (from right)

(i) Outer Cortex. Immediately within the merstoderm is the several cells wide **outer cortex**. It is composed of elongated cells that are also broad. They are somewhat pale in colour. The cell walls are not thickened. The cells have pointed ends. They show horizontal septation. Each cell has a single large vacuole and a thin layer of peripheral cytoplasm but lacks dense vacuolate inclusions characteristic of meristoderm cells and have fewer chromatophores.

(ii) Inner Cortex. It consists of cells which are comparatively squarish at the ends. The cells of the inner cortex are elongated and show gelatinisation of the longitudinal walls. These cells have been seen to develop horizontal thread-like outgrowths or the cross-connections and the hyphae (Fig. 19.5). The cortex continuously increases in diameter at the periphery owing to the division and redivision of superficial cells. There is difference in the rate of cell division. Cells formed early in the growing season are larger than those formed late in the season. This results in the formation of concentric zones of light and dark cells constituting "growth ring". The latter resemble somewhat the annual rings in the secondary wood of a dicot stem.

(c) Medulla. It forms the central axis. It is composed of mass of elongated, anastomosing cells with pores on the walls between them. These cells are considered to function in transporting food as do the sieve tubes of the vascular plants.

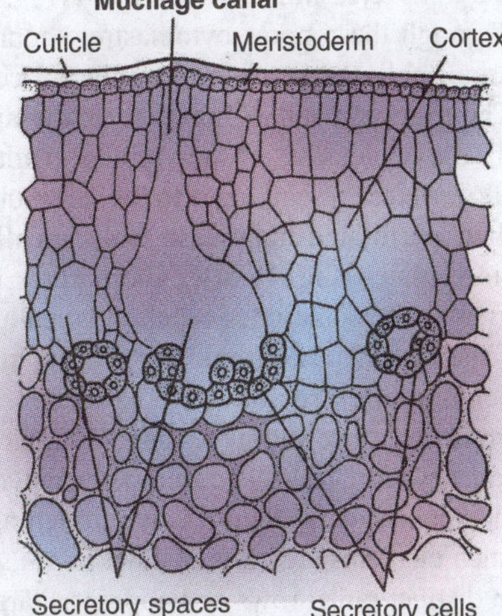

Fig. 19.6. *Laminaria sp.* Transverse section of stipe showing development of mucilage canals in the cortical region. (After Cuignard).

There is gradual transition of the cells of one region into the other. The cells of the limiting layer or the meristoderm often become palisade-like in shape and contain numerous chromatophores which lack pyrenoids. They form the principal seat of photosynthesis. The cells cut off towards the inner side, become elongated and broaden slightly to form the outer cortex. They also assume a spindle-like shape. The gradual thickening of the cells in the innermost layer of the outer cortex adds to the inner cortex. The longitudinal thickening is due to mucilage. The mucilaginous walls result in the separation of the cells of the inner cortex at certain points to form the **mucilage tracts** or **canals** (Fig. 19.9). Killain has reported the formation of cross-connections between these separated cells by the formation of protrusions on the longitudinal walls. These protrusions (Fig. 19.5) elongate and the two protrusions formed at the corresponding level between the two opposite cells grow towards each other until they meet. The protrusions are formed by the appearance of an oblique septum at the upper ends of the parent cells. Often the protrusions become septate and appear to be cellular structures.

The elements of the inner cortex go on adding to the medulla. Regular formation of cross connections in the medullary region and the inner part of the inner cortex, give it an appearance of a tangled network of filamentous growth.

Another kind of protrusions also arise from the medullary cells. These are known as the **hyphae** (Fig. 19.7). The hyphae differ from the cross-connections in that they do not fuse with the corresponding outgrowth. The hyphae also arise from the cells of the inner cortex. They grow into threads of considerable length and often show branching. They grow mainly in the radial direction across the medulla. Often they pursue a tortuous course in the medulla and it, therefore, becomes difficult to trace them to full length. Sometimes in the medullary region certain cells elongate and become considerably swollen at the septa. Such swollen cells are called the **trumpet cells** or **trumpet hyphae.** There are thus four different kinds of cells in the medulla, namely, (*i*) medullary cells derived from the inner cortex, (*ii*) cross-connections, (*iii*) hyphae, and (*iv*) trumpet hyphae (Fig. 19.7).

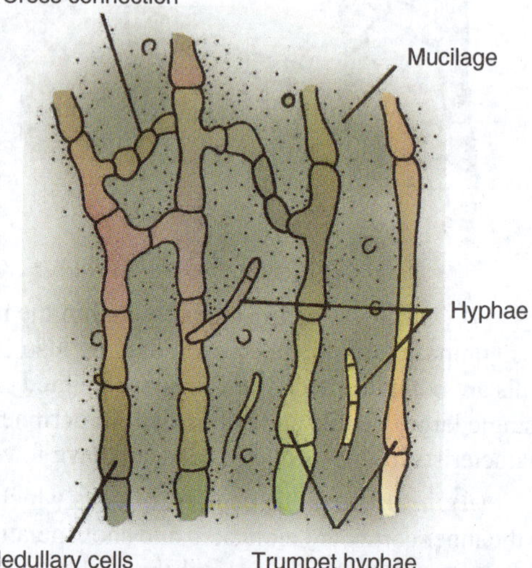

Fig. 19.7. *Macrocystis* sp. part of a medulla in longitudinal section showing four different kinds of constituent cells. (After Rosanthal).

The nature of the trumpet hyphae is a matter of dispute. Numerous pores have been noted in the transverse wall between them (Fig. 19.8B). Through these pores cytoplastmic connections are established between the adjacent cells (A). The trumpet hyphae thus structurally approach the sieve elements of the vascular plants. They show a close resemblance to the sieve tubes with their perforated transverse walls recalling the sieve plates (Fig. 19.8B). These pores in older hyphae are sometimes blocked by callus pads. This confirms their being sieve tubes in nature. Schmitz and Srivastava (1976) stated that the trumpet hyphae and sieve tubes are basically the same type of cells. The trumpet shape of these elements is due to their passive stretching during extension growth of the organ in which they occur. Such elements are quite distinct in *Macrocystis* and *Nereocystis* in inner cortex. They are arranged in distinct radial rows and appear as large thick-walled elements. Riggs regards trumpet hyphae of the Laminariales as the main conducting elements specially concerned with the transport of protein. The cortical cells are regarded by Wille as storage and mechanical elements. They furnish the chief mechanical elements in the young stipe. The older sporophytes have their mechanical elements located in the medulla and the innermost elements of inner cortex. The strengthening elements are responsible for the marked tensile strength of the stipe. In *Laminaria* the trumpet hyphae in the younger stages

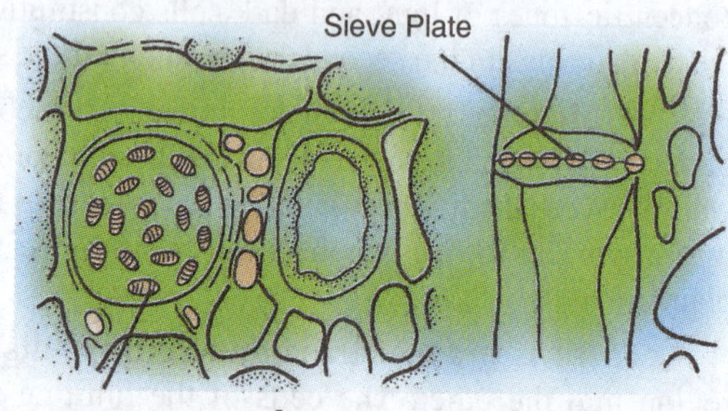

Fig. 19.8 (A-B). *Macrocystis* sp. Cells resembling sieve tubes. (A), cortical cell in transverse section showing pits; (B), in longitudinal section showing sieve plates (After Oltmanns).

become much attenuated and become extremely thick-walled and filled with callus. They are the chief elements of mechanical support to the stalk or the stipe.

Lamina or Blade. Anatomically, the blade resembles the stipe in many details. It shows four distinct regions namely upper epidermis, outer cortex, inner cortex and medulla. The outer cortex and the epidermal cells contain chromatophores and are the seat of photosynthetic activity. Mucilage canals are present in the cortex. Medulla occupies a rather narrow tangential zone and is composed of a network of cross-connections and medullary cells. They lose their regularity with the enlargement of blade. The medullary cells are dragged out of the longitudinal course. The cross connections are displaced and it becomes impossible to distinguish between these two kinds of structures in the nature blade.

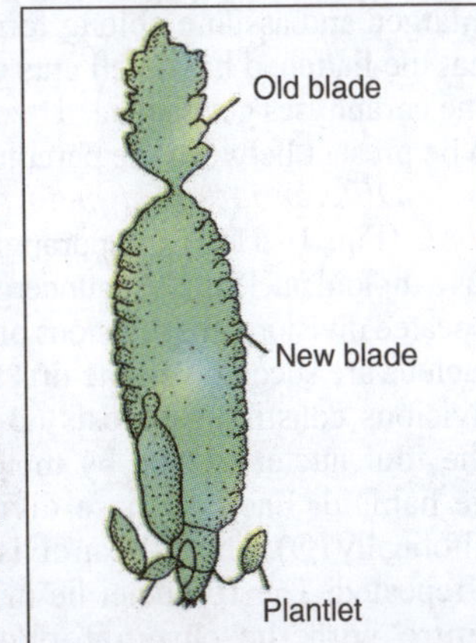

Fig. 19.9. *Laminaria.* Plant with stolons. (After Bamet).

Haptera. The rhizoidal outgrowths of the young plants are replaced at maturity by the broadened base of the stipe. The flattened base is fastened to the substratum by means of numerous rhizoid-like threads. It shows no internal differentiation except that the cells in the middle are somewhat elongated. These rhizoid-like processes arise in whorls and in regular succession from the higher levels of the stipe. These branches are called the haptera. In *L. cloustoni* they are formed in the period of active growth and appear in the beginning as small superficial outgrowths.

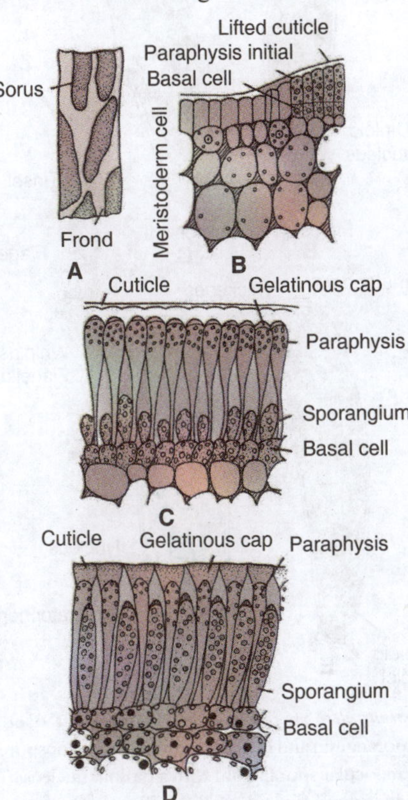

Fig. 19.10 (A-D). *Laminaria.* Reproduction by meiospores and succesive stages of development of sporangia. (A), part of a frond with sori; (B), Early development of sporangial sorus; (C) Latter stage in the development of sorus; (D), transverse section through a part of the nearly mature sorus. (A-C after Sauvageau and D after Smith).

Reproduction.

Asexual reproduction by means of asexual spores (mitospores) is absent in *Laminaria.* In *L. reodreguezu, L. longipes* and *L. sincelarii* vegetative propagation of the sporophyte takes place by means of stolons (Fig. 19.9). The horizontal stolons arise from the region of holdfast and turn up at their tips to grow into new plantlets.

The sporophyte nor-mally reproduces by means of **meiospores**.

1. Reproduction by Meiospores (Fig. 19.10). The meiospores are motile swarmers and thus may be called the **meiozoospores** or **gonozoospores**. They are produced within club-shaped **unilocular sporangia**, which are arranged in **sori** on both sides of the blade (A). The sporangia in the sorus occur interspersed with thick, sterile, protective hairs called the **paraphyses**. The paraphyses overtop the sporangia which lie perpendicularly to the surface of the blade. The sori are dark brown and irregularly arranged. They are produced on the older plants towards the end of the growing season. In *L. gyrata,* a Japanese species, the sori are arranged along the margins of the blade. These are suborbicular or transversely elongated. In *L. religiosa* the blade is depressed in the middle and elevated on either side. The depressed surface faces the light. The sori are arranged on the elevated part that is more or less in shade. There is an indication of the dorsiventrality of the blade in this species.

(*a*) Development of sporangia. The sporangia arise from the superficial (meristoderm) cells of the blade. These cells elongate and become palisade-like (B). The cuticle is also lifted along with the enlargement of the cells. Each enlarged cell divides transversely into an upper **paraphysis initial** and a **lower basal cell** (B). The paraphysis initial continues to lengthen and forms an elongated thread with a club-shaped upper portion. Meanwhile the basal cell broadens. As a result the paraphysis now occupies only a part of apex of the basal cell (C). The club-shaped upper ends of the paraphyses contain numerous chromatophores and fucosan vesicles and remain in contact with each other. Each paraphysis at its swollen tip develops a cap-like gelatinous thickening. The basal cells become enlarged and assume oblong forms. These oblong basal cells are the **sporangia**. The other view is that the flattened basal cell cuts off a sporangium at its distal end by the side of the paraphysis (C). The paraphyses get displaced to one side so that the developing sporangia from the basal cells appear to be present between the paraphysis (C).

Differentiation of meiozoospores (Fig. 19.11). The sporangium has a diploid nucleus (A). It undergoes repeated division. The divisions of the nucleus are successive. The first two divisions constitute **meiosis** (B-C). The four nuclei formed by meiosis are haploid. Each of these divides mitotically (D). The nuclear division is repeated. The 16 nuclei lie in the centre with the chromatophores occupying peripheral position (E). The nuclear division continues until there are 32, sometimes 64 haploid nuclei in the unilocular sporangium. The chromatophores also increase in number simultaneously. Later on they move to the periphery and become associated with the chromatophores. The cleavage of the protoplast follows and haploid, uninucleate bits of protoplast with one chromatophore each are formed (F). These uninucleate bits get metamorphosed into motile meiospores, called the **meiozoospores** or **gonozoospores**. Each meiozoospore has a single chromatophore occupying the base (G). It is doubtful whether there is an **eye spot** or not. It must be very small and thus difficult to make out. The two laterally inserted flagella are unequal in length. The front one is considerably longer and bears flimmers (hairs). It is thus of pantonematic or flimergeissel (tinsel) type. The hinder smaller flagellum is of whiplash type. The meiospores avoid light and move away from it. The wall of the sporangium near the apex becomes thickened. The sporangium dehisces at this point. Dehiscence is brought about by the pressure of the gelatinous paraphyses on the sporangia. The contents of the sporangium escape enclosed by a watery, gelatinous sheath that passes out between the paraphysis. After extrusion the

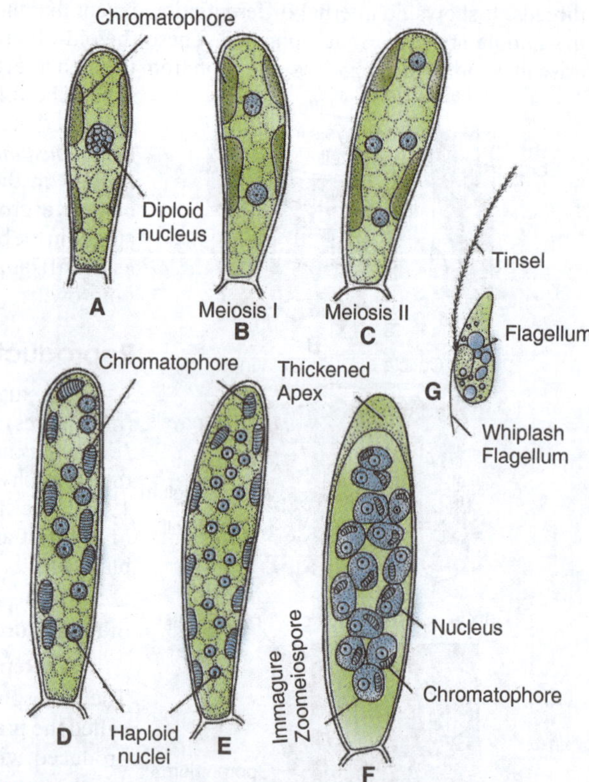

Fig. 19.11 (A-G). *Laminariales.* Stages in the development of the unilocular sporangium and differentiation of meiozoospores; (A), young unilocular sporangium with a diploid nucleus; (B-C), meiotic division of diploid nucleus; (D), 8-nucleate stage; (E), 16-nucleate; (F), differentiation of meiozoospores; (G), liberated meiozoospore (After Kylin).

sheath dissolves. The flagellate meiospores are thus liberated. They swim actively in all directions. In *Laminaria saccharina* out of the 32 meiospores, 16 are scheduled to form **male gametophytes** and 16 **female gametophytes**. This type of genotypic differnetiation of sex was recorded by Schreiber. Some investigators have reported two sizes of meiospores in one or two cases.

Germination of Meiozoospore (Fig. 19.12)

After a brief period of activity the meiozoospore settles down to germinate. It withdraws its flagella, becomes rounded and secretes a wall (A). The rounded cell formed from the meiozoospore is termed an **embryospore**. It puts out a slender tube or protuberance into which move bulk of the cell contents (B). The tube increases in length and its distal end enlarges to become bulbous (C). The germling appears to be dumb-bell-shaped at this stage. The contents of the tube migrate to the upper swollen or bulbous end (D) which becomes separated by the formation of a septum (E). This swollen part later produces the gametophyte which is mainly filamentous. The filamentous gametophytes are composed of elongated cells that contain numerous discoid chromatophores. They may be branched and show a tendency towards heterotrichous growth. The meiospores in *Laminaria* produce two kinds of gametophytes. The gametophytes in *Laminaria* are thus dioecous, filamentous and few-celled. They are totally different in morphology and much smaller in size from the highly organised sporophyte. The gametophytes vary in size and shape. The dwarf male gametophytes have smaller cells.

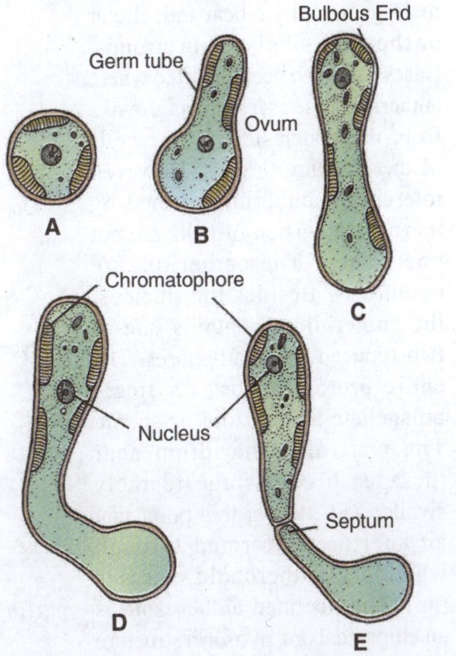

Fig. 19.12 (A-E). *Laminaria* Successive stages in the germination of meiozoospores. (A), Embryospore; (B), First stage in germination; (C-E), Later stages in germination.

2. Sexual Reproduction. It is **oogamous**. The male sex organs are called **antheridia**. They are borne on the **male** gametophyte. The female sex organs are called the **oogonia**. They are produced by the **female** gametophyte. It has been observed that the gametophytes of *Laminaria* produce sex organs only when the temperature of surrounding water is below 15°C. Perhaps this accounts for their abundance in the colder waters only. Vadas (1972), however, found that for *Nereocystis* light was the single critical factor in gametogenesis. Low light and short day lengths retarded sexual maturity. In the Laminarialean algae sex is genotypical determined in the gametophytes. In *Saccorhiza polyschides,* Evans (1965) observed a large X chromosome in the female gametophyte and a small Y chromosome in the male gametophyte. These were observed to pair during meiosis.

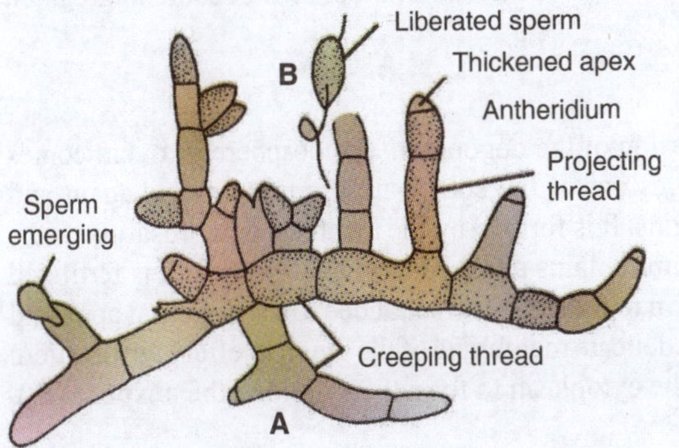

Fig. 19.13. *Laminaria.* Male gametophyte (After Sauvageau).

Male Gametophyte (Fig. 19.13). The male gametophyte is comparatively small. It consists of 3 cells at least. Normally it is filamentous. The filament is much branched and heterotrichous (A). The cells are small but more numerous with scanty chromatophores. They are pale in colour. The antheridia are formed at a relatively lower temperature. Higher temperatures (above 12-16°C) retard the formation of reproductive organs in *Laminaria*.

The male gametophyte degenerates after the gametes are shed. It is thus short-lived.

Antheridia. The terminal cells of the projecting threads of the male gametophyte bear unicellular **antheridia** singly or in groups. Cases have also been reported where antheridia arise as lateral outgrowths from the upper sides of the cells of the creeping threads. However, intercalary antheridia formed by transverse division of cells are not uncommon. The antheridia are colourless. Besides the nucleus, the antheridium contains one or two reduced chromatophores. The entire protoplast forms a single biflagellate antherozoid or sperm. The wall of antheridium near the apex becomes considerably swollen (S). It is at this point that an aperture is formed through which the antherozoid emerges (an). The liberated antherozoid is an ellipsoidal or a pyriform structure (B). It is furnished with two laterally inserted flagella. The eye spot is absent.

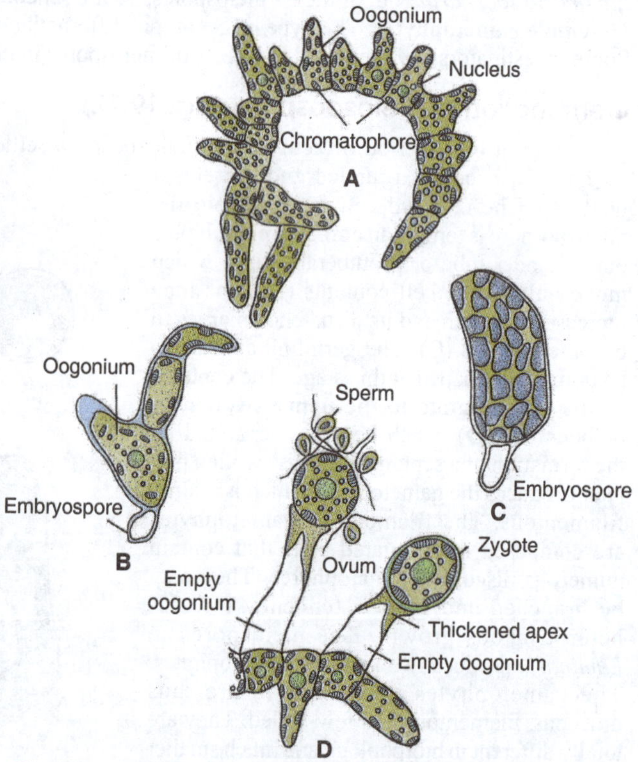

Fig. 19.14 (A-D). *Laminaria.* Female gametophyte. (A), female gemetophyte with oogonia; (B-C), reduced female gemetophytes; (D), Ovum extruded and seated on the thickened, cup-like platform surrounded by sperms F_1 (A after Sauvageau and B-C after Kylin).

Female Gametophyte (Fig. 19.14). It is scarcely branched (A) and generally reduced to a few cells (B). Sometimes it consists of only one cell (C), which may function as an oogonium. The cells composing it are large. Any cell (both terminal or intercalary) can function as an **oogonium** (A). The cells contain chromatophores. The cell destined to form an oogonium enlarges considerably and assumes a tubular or a pyriform shape. Its contents round off to form a single egg or **oosphere**. The oogonial wall at the distal end becomes thickened and is differentiated into three layers. The chromatophores become aggregated near the thickened part.

Fertilization

A slit develops in the thickened apical portion of the oogonium. The oosphere or ovum comes out through the break (F_1). As the ovum emerges out of the split wall, the latter meets again and forms a gelatinous cushion with thickened margins. It is formed by the localized gelatinisation of the oogonial wall (Bilasputra *et al*, 1971). The ovum remains attached to this platform and is fertilized in this condition. The liberated antherozoids swim to the ovum still attached to the oogonial apex and surround it (F_1). Only one is able to penetrate the delicate membrane of the ovum to effect fertilization. The male and the female nuclei fuse and so do the cytoplasm to form the **zygote** or the **oospore** (F_2).

Germination of Zygote (Fig. 19.15)

The zygote undergoes no resting period and starts germinating while still attached to the tip of oogonium (A). It is doubtful whether the attached condition represents anything more than physical adhesion. The zygote undergoes considerable elongation and the diploid nucleus divides mitotically. Transverse walls are laid in between the daughter nuclei. Each daughter cell divides one or more times. Finally a filament 6 to 10 cells in height is formed (S_1-S_2). Later, vertical divisions take place in the median and the upper parts of the filament (S_3). As a result a leaf-like expansion is formed (B). The lower cell remains undivided. It elongates to give rise to a rhizoid-like process which pierces through the empty oogonium (S_4). Further divisions in the transverse and the longitudinal planes form a large, leaf-like tissue consisting of hundreds of cells. The tissue is one layer of cells in thickness. Additional rhizoid-like processes of limited growth may appear from the lowermost cells of the leaf-like young sporophytes (C). The female gemetophyte dies after the development of 3 or 4 rhizoid-like processes. The leaf-like sporophyte by division, cell differentiation and growth becomes differentiated into parts characteristic of the adult sporophyte.

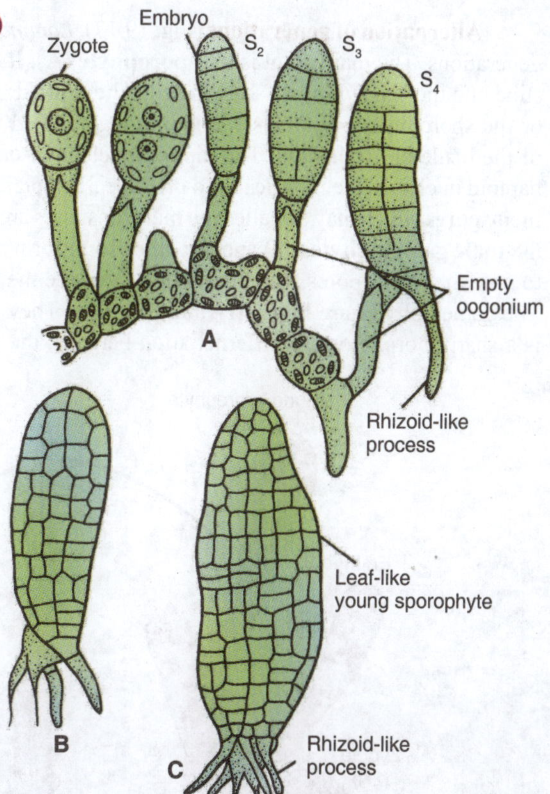

Fig. 19.15 (A-C). *Laminaria.* Germination of zygote and development of young sporophyte. (A), Part of female gemetophyte showing early stages of development of embryo sporophyte (S_1, S_2, S_3 and S_4) still seated and attached to the cup-like apical platform of empty oogonium; (B-C), Later stages in the development of sporo-phyte. (After Sauvageau).

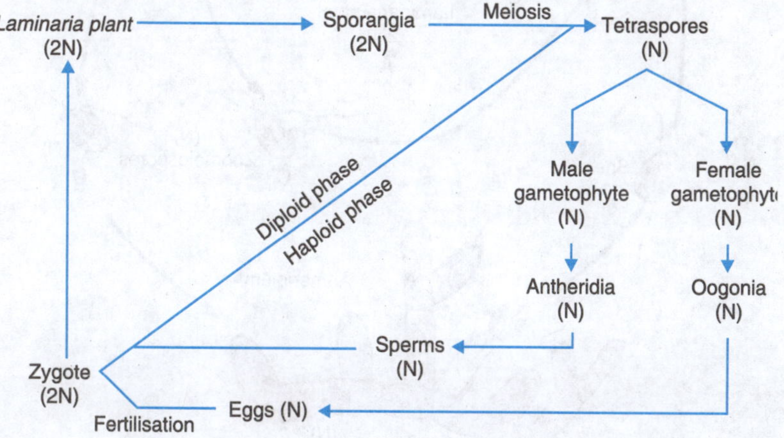

Fig. 19.16. Graphic representation of the life cycle of *Laminaria* sp.

Alternation of generations (Fig. 19.17). *Laminaria* exhibits an advanced type of alternation of generations. The main plant is the **sporophyte** (A). It is differentiated into the **holdfast,** cylindrical stipe and the broad **lamina**. It represents the diploid phase in the life cycle. The diploid individual or the sporophyte bears clusters of sporangia called the **sori** (B). They are present on both sides of the blade or the lamina. The diploid nucleus of each sporangium undergoes meiosis. The four haploid nuclei divide mitotically to produce 32 nuclei. Due to cleavage of the cytoplasm 32 haploid **meiospores** which may be called the meiozoospores are formed. Out of these, 16 germinate to produce the **male gametophytes** (F) and the other 16 to form the **female gametophytes** (G). It is important to note that meiospores, on germination, produce the gametophytes and not the sporophyte plants. The gametophytes are **heterotrichous** in habit. They are much smaller than the sporophyte. There is a sharp morphological differnetiation between the two. The male and the female gametophytes

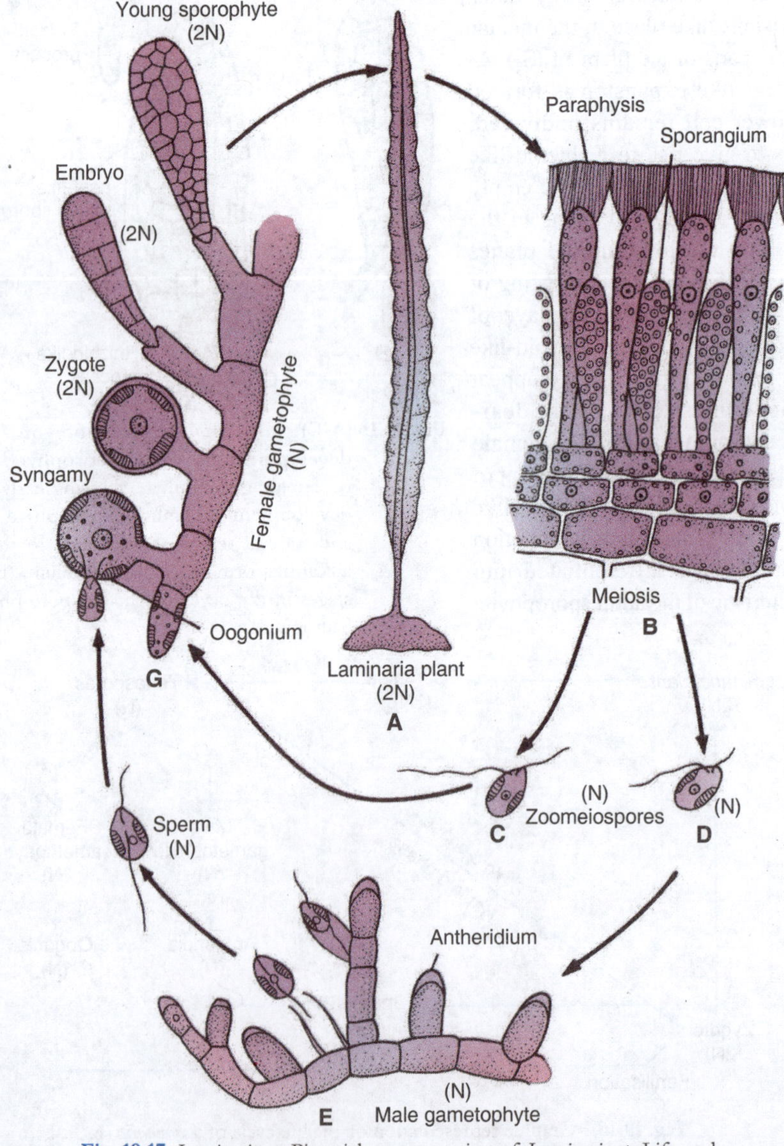

Fig. 19.17. *Laminaria sp.* Pictorial representation of *Laminaria sp.* Life cycle.

produce **antheridia** and **oogonia** respectively. These in turn produce the **sperms** (E) and the **eggs**. The gametophyte phase, therefore is represented by two distinct, male (E) and the female (G) filamentous individuals which differ from the sporophytes in morphological as well as cytological details—a feature in sharp contrast to the other classes of Algae.

The male and the female gametes fuse (G, *a*) to form the diploid zygote (G, *b*). The zygote on germination produces the diploid **sporophyte** and not the gametophyte. Two crucial points in the life cycle of *Laminaria* are the **meiotic division** of the diploid nucleus of the sporangium and the **fusion** of the gametes to form the zygote. At these points one generation switches on to the other generation. The large sporophytes regularly alternate with microscopically male and female gametophytes. This is called alternation of generations. Such an alternation of generations in which the two alternating individuals in the life cycle differ morphologically, physiologically and in genetic constitution is designated and **heteromorphic**. The life cycle of *Laminaria* which is characterised by distinct alternation of generations with sporogenic meiosis is called **diplohaplontic**.

Economic Importance

The sporophytes of some species are eaten as food. Kelps are a valuable source of iodine and potassium fertiliser. Iodine is still manufactured from *Laminaria, Ecklonia* and *Eisenia* in Japan. They are also the source of **algin** and **alginate** which are of so great importance in the production of plastics and artificial fibres.

Salient Features

1. The diploid thallus is large in size reaching upto 6 to 8 feet in length and differentiated into a **holdfast, stipe** and a **blade**.
2. The growing region lies in the transitional zone between the strip and the blade thus growth is intercalary.
3. The lamina is renewed at the base each year. The old blade is pushed away and dies.
4. The thallus sporophyte shows a high degree of morphological and anatomical differentiation.
5. The trumpet hyphae with their perforated cross walls recall the sieve elements of the vascular plants. The presence of cross-connections between the cells of the inner cortex and medulla is another anatomical feature unique to the algae.
6. The sporangia are arranged in distinct groups called **sori**. The mode of development of sporangia is unique.
7. Each sorus consists of paraphyses intermingled with unilocular sporangia. The diploid protoplast of each sporangium undergoes meiosis to produce 32 **meiozoospores**.
8. Out of these, 16 germinate to produce the **male** and the other 16 to produce the **female gametophytes**.
9. The gametophytes are **heterotrichous** in habit and are very much reduced as compared with the large sporophytes.
10. The male gametophytes are much branched and have smaller cells with scanty or no chromatophores.
11. The antheridia occur in terminal clusters on the erect filaments. Each antheridium produces a single biflagellate **antherozoid**. It is liberated through an apical pore formed by the thickening or gelatinization of the wall at the apex of the antheridium.
12. The female gametophytes are scarcely branched. They possess large cells filled with chromatophores. Any cell can function as an **oogonium**. The oogonial cell enlarges and becomes tubular or pyriform. Its contents round off to form a single **ovum** or **egg**.

13. The apical portion of the oogonial wall gets thickened and splits to allow the single ovum to escape. As the ovum is coming out the split wall again meets and forms a platform for the ovum to rest on.
14. The zygote starts germinating while still attached to the apex of the oogonium. The nucleus divides by simple mitosis. There is no meiosis.
15. The zygote does not undergo any period of rest.
16. There is heteromorphic alternation of generations in the life cycle which is **diplohaplontic**.

Taxonomic Position

Division	:	**Phaeophyta** (Phaeophycophyta)
Class	:	**Heterogeneratae**
Order	:	**Laminariales**
Family	:	**Laminariaceae**
Genus	:	*Laminaria*
Species	:	*laminariodes*

QUESTIONS

ESSAY TYPE

1. Give an illustrated account of the structure of the thallus and reproduction in *Laminaria.*
2. By means of labelled sketches only describe the life-cycle of *Laminaria.*
3. Give an illustrated account of the gametophyte phase in *Laminaria.*
4. What are the cross connections ? How are they formed ? Describe in detail.
5. Give an account of the structure of thallus in *Laminaria.* How are new blades formed ?
6. Write what you know about the formation and structure of the male organ in *Laminaria.*
7. Describe the formation of the female gametophyte and the structure of the female sex organ in *Laminaria.*

SHORT ANSWER TYPE

8. How is fertilization affected in *Laminaria*? Describe briefly the germination of zygote of *Laminaria.*
9. Describe briefly the anatomy of stipe in *Laminaria.*
10. What are the trumpet hyphae ? How are they comparable to phloem of higher plants ? Describe their probable functions.
11. Describe the germination of meiozoospores to form male gametophyte. *(Bharathiar, 2000)*
12. Write short notes on :
    (*a*) Sporophyte of *Laminaria*
    (*b*) Alternation of generations in *Laminaria*
    (*c*) Distinguishing characters of the order Laminariales
    (*d*) Fertilization in *Laminaria*
    (*e*) Male gametophyte of *Laminaria*
    (*f*) Female gametophyte of *Laminaria*
    (*g*) L .S. and T. S. of stipe
    (*h*) Meiozoospores.
13. Give a brief account of sexual reproduction in *Laminaria.*

OBJECTIVE TYPE

14. *Select the correct answer :*

(*i*) *Laminaria* is a

(*a*) Xerophyte (*b*) Hydrophyte

(*d*) Epiphyte (*d*) Lithophyte

(*ii*) The members of the order Laminariales are commonly known as

(*a*) Pond silk

(*b*) Net algae

(*c*) Kelps

(*d*) Sea anemone

(*iii*) In which of the following the adult plant is a sporophyte

(*a*) *Spirogyra*

(*b*) *Oedogonium*

(*c*) *Laminaria*

(*d*) *Coleochaete*

(*iv*) Where does meiosis take place in *Laminaria* ?

(*a*) In the sporangium

(*b*) During formation of mitospores or zygote

(*c*) During the formation of gametes

(*d*) During the formation of meiospores.

(*v*) Which of the following statements are true regarding the behaviour of 32 meiospores of *Laminaria saccharina*?

(*a*) All the 32 meiospores give rise to a monoecious gametophyte

(*b*) While 24 meiospores form the male gametophyte, only 8 meiospores form the female gametophyte.

(*c*) 16 meiospores form male gametophyte and 16 meiospores form the female gametophyte

(*d*) All the 32 meiospores form the male gametophyte

(*vi*) Algin or Alginate are produced from

(*a*) *Ectocarpus* (*b*) *Fucus*

(*c*) *Laminaria* (*d*) *Sargassum*

(*vii*) Kelps are valuable source of

(*a*) Nitrogen (*b*) Iodine

(*c*) Sodium (*d*) Copper

(*viii*) The trumpet hyphae are characteristic of

(*a*) *Polysiphonia* (*b*) *Sargassum*

(*c*) *Ectocarpus* (*d*) *Laminaria*

(*ix*) The alternation of germination in *Laminaria* is

(*a*) Isomorphic and diplohaplontic

(*b*) Heteromophic and diplohaplontic

(*c*) Isomorphic and haplontic

(*d*) Heteromorphic and haplontic

(*x*) Subclass heterogeneratae is so called because

(*a*) It has two different types of sporangia

(*b*) It has heteromorphic alternation of generations

(*c*) It has two different types of sex organs

(*d*) It has two types of zoospores.

Sargassum sp

CHAPTER 20

CLASS : CYCLOSPORAE Order : Fucales

- Family : Fucaceae
- Fucus
- Family : Sargassaceae
- Sargassum

General

The outstanding features of the class are :

1. The thallus is a diploid **sporophyte**.
2. The reproductive organs are lodged in special flask-shaped cavities of the thallus called the **conceptacles**.
3. The conceptacles may be scattered over the surface of thallus but more often confined to the swollen tips called the **receptacles**.
4. Sexual reproduction is always **oogamous**.
5. The unicellular sex organs, **antheridia** and **oogonia**, are borne on the diploid plant. Genetically they are thus diploid.
6. **Meiosis** takes place at the time of the differentiation of gametes. The gametophyte or haplophase is thus represented by the gametes only.

7. Unlike other brown algae, the sperm in the Fucales has a shorter anterior and longer posterior flagellum.
8. Asexual reproduction by the formation of asexual spores (mitospores) is entirely absent.
9. The life cycle consists of only one morphologically differentiated generation. It is the **sporophyte**. The gametophyte is reduced to merely the gametes. There is thus *complete absence of alternation of free-living, multicellular generations*.
10. The life cycle of Fucales which is characterised by a diploid adult (sporophyte) and gametanginal meiosis is called **diplontic**.

The class Cyclosporae includes a single order **Fucales** with about 40 genera and 350 species. Commonly the representatives of this order are called the **rock weeds** as they grow in shallow water attached along the rocky sea coasts in the intertidal zone. In size they are much smaller than the Laminariales. The order is divided into seven families *viz*. Fucaceae , Himanthaliaceae, Cystoseiraceae, Sargassaceae, Hormosiraceae, Durivlleaceae and Ascoseiraceae. Of these, Fucaceae and Sargassaceae are discussed here.

FAMILY : FUCACEAE

The thallus is usually flattened and dichotomously branched. The holdfast is disc-like. The important genera of the family are *Fucus, Pelvetia* and *Ascophyllum*. Of these *Fucus* is the best known. It includes a number of species.

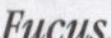

Fucus. *Pelvetia.* *Ascophyllum.*

FUCUS

Occurrence. *Fucus* is a cold sea water alga which is found attached to rocks along the shores. For this reason it is popularly called a **rockweed**. It occurs between the low and the high tide marks covering the rocks of the intertidal zone thickly. The plants are slippery on account of the copious mucilage spread on their surface and thus are treacherous to walk on. The genus comprises a number of species which usually occur attached to rocks by a holdfast in the intertidal zone where they often show a distinct horizontal zonation (bands). Each band consists of almost a pure stand of a member of the Fucales. The majority of species are inhabitants of the sea coasts of the North temperate and Arctic regions. They are left high and dry above the water line at low tide and covered at high tide. The two common species are *F. vesiculosus* (A) and *F. serratus* (B-C). The former is characterised by the swollen receptacles, paired air bladders and smooth margin of the frond, while the latter by the flattened receptacles and serrate margin (C). *F. spiralis* lacks air bladders. The receptacles are irregularly swollen and have a sterile border (D). All the species are common seaweeds. They are perennial, some of them living for as long as four years.

Organisation of the Thallus (Fig. 20.1. A)

Fucus has a thallus which is much smaller than that of *Laminaria*. It rarely exceeds 30 cm. in length. It is dark brown, leathery and slimy to touch. The frond is a repeatedly forked, ribbon-like structure with a distinct midrib. Morphologically the thallus is differentiated into three regions, namely, the holdfast, the stipe and the frond (A).

The **holdfast** or **hapteron** is a special attaching organ. In some species it grows into the crevices on the rocky coast and secretes glue-like mucilage which makes it very difficult to dislodge it. In others the holdfast is discoid. It secretes mucilage, but there is no penetration of the disc into the subtratum. Above the holdfast is a short cylindrical, stalk-like **stipe**. It is the lowermost region of the thallus from where all other portions except the midrib have been torn away by wave action and age. The stipe expands above into the broad frond and is continuous with its midrib. On either side of the midrib are the thinner wings. The flattened ribbon-like frond, which forms the upper major portion of the thallus, is repeatedly forked. It is leathery in texture and slippery. A prominent **midrib** traverses each thallus branch but never reaches the apex. The margin of the expanded wing-like region may be entire (*Fucus vesiculosus*) or serrate (*F. serratus*). At the tip of each growing branch is a narrow elongated depression or groove. At the bottom of this groove is situated a small **apical cell**. The apical groove is filled with mucilage. In *F. vesiculosus* and others there are small bladder-like expansion filled with air. These swollen vesicles are called the **air bladders**. They are scattered along the sides of the thallus in close association with the midrib in the regions of forking. The air bladders act as floats to lift the forked branches towards surface and thus keep the plant upright in water. They bouy

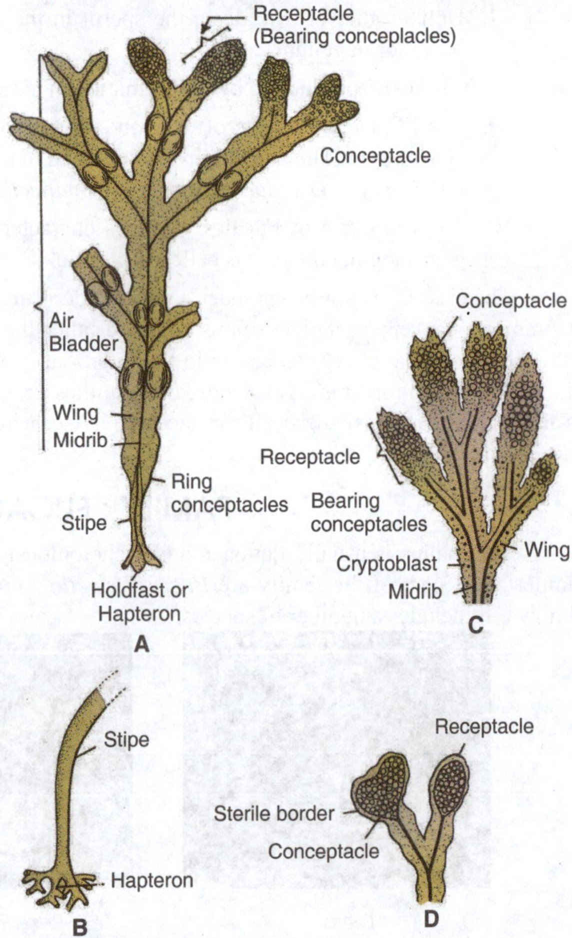

Fig. 20.1 (A-D). *Fucus* sp. (A), plant of *F. vesiculosus* showing habit; (B), holdfast and stipe of *F. serratus*; (C), fertile branch of *F. serratus* showing flattened receptacle; (D), fertile branch of *F. spiralis* bearing receptacles.

Fucus vesiculosus.

Fucus serratus.

Fucus spiralis.

up the plants during high tide. They arise by the accumulation of gas or air in the large intercellular spaces of the thallus. At low tide the plants are left exposed to drought above the water line. In this condition they are protected against excessive desiccation by the mucilaginous nature of the thallus.

Fucus plant (Fig. 20. 1 A) is diploid and the diploid number of chromosomes in one species is 64. When the plants enter the reproductive phase, the ends of the branches become swollen. The swollen ends are covered with small, scattered, pimple-like projections called the **papillae**. Each papilla has a minute pore at its tip which is called an **ostiole**. The pores appear like small scattered dots. Each pore leads into an approximately spherical cavity developed in the tissue of the thallus. These cavities are called the **conceptacles**. The conceptacles contain the sex organs. These fertile swollen portions of the thallus containing the conceptacles are called the **receptacles**. The receptacles, therefore, are the metamorphosed distal ends of the vegetative branches. Each receptacle has a small pore at its apex and is hollow. The mid-rib, which runs through the flattened branches, does not extend into the region of the inflated receptacle. The inflation is caused by the abundant production and accumulation of mucilage. In some species such as *F. serratus* (Fig. 20.1, C), the flattened lobes bear irregularly scattered small dots which mark the position of pits containing sterile hairs. These pits with hairs are called the *sterile* **conceptacles** or **cryptoblasts**.

The thallus of *Fucus*, as in *Laminaria,* is complex and exhibits differentiation of the plant body into parts of organs. Each organ has a special function to perform. The holdfast anchors the plant to the substratum. The tough stipe weathers the storm protecting the plants being torn away from its base by the turbulent wave action. The flattened frond carries on photosynthesis. The terminal receptacles constitute the localised reproductive regions.

Structure of the thallus (Fig. 20.2). Viewed in transverse section the flattened **frond** shows the distribution of tissues into the following three regions :

1. *Meristoderm* or *Palisade Layer*. It is the outermost part. It consists of a single layer of cells containing the plastids (B). The cells are columnar in shape and without any air spaces between them. They stand with their long axes perpendicular to the surface of the frond. This limiting surface layer of photosynthetic cells is called the **palisade layer** or **meristoderm** (B). The meristoderm cell has a central vacuole and a lining layer of cytoplasm. In the latter lies embedded the single nucleus and

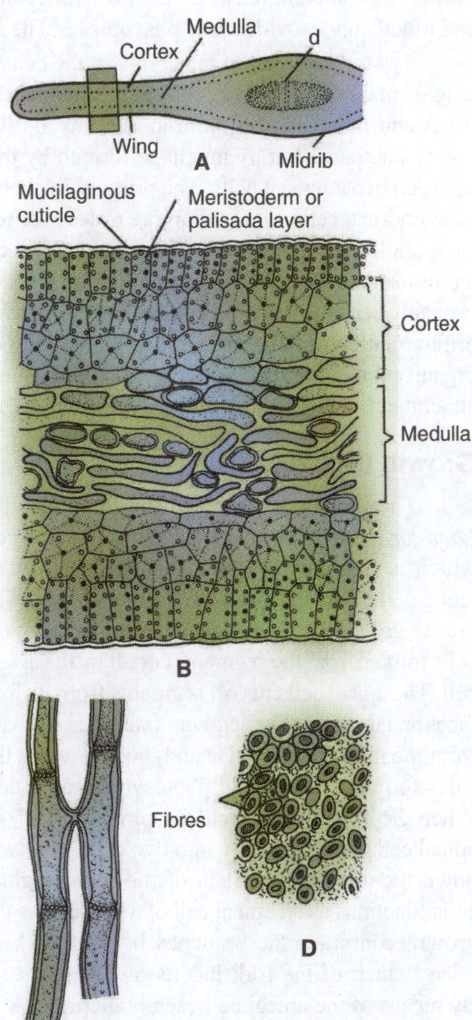

Fig. 20.2 (A-D). *Fucus*. Structure of frond. (A), T. S. frond, outline sketch showing one wing and midrib; (B), detailed structure of a part of the wing marked *b* in A above; (C), cells from the medullary zone showing pits in the transverse walls; (D), detailed structure of a part of midrib marked *d* in A above. It consists exclusively of fibres. (B after Oltmanns and C after Hansteen).

numerous plastids called **phaeoplast**. Besides photosynthesis, the cells of this layer are capable of active division. The meristoderm is covered on the outside by noncellular mucilaginous cuticle which protects the thallus from desiccation at low tide.

2. *Cortex.* Internal to the meristoderm is a broad zone of thin-walled polygonal, parenchymatous cells constituting the **cortex**. It is several cells deep. The outer cortical cells are comparatively large and well packed, have larger central vacuoles but are well supplied with comparatively fewer small, discoid phaeoplasts. The nucleus lies suspended by cytoplasmic strands in the central vacuole. Further in, the component cortical cells, become loosely organised, are more and more elongated and mucilaginous with fewer phaeoplasts. The inner cortex probably serves as the **storage region**.

3. *Medulla.* Within the cortex in the centre of the frond is the **medulla**. It consists of a loose, tangled mass of elongated cells called the **medullary hyphae**. The latter possess thick, mucilaginous walls and often pits on the transverse walls (C). Apparently the medullary cells appear isolated not by air spaces but by mucilage formed by the swelling of the middle layer of the common wall between two adjacent cells. The primary hyphae are thin-walled. The medullary hyphae form chains of cylindrical cells which run more or less vertically up and down the mid-rib and stipe. They run at right angles to the mid-rib and horizontally in the expanded membranous portions of the frond. In a cross-section they look elongated in the wings and appear as circles in the stipe and midrib. They probably serve as the food conducting elements. *Fucus* thus is an example of a genus which shows differentiation of structure with corresponding differentiation of function leading to the origin and organisation of distinct tissues. It has tissues specialised for photosynthesis, conduction, storage and attachment. The midrib consists exclusively of fibres (D).

Growth (Fig. 20.3)

The growth in length of the thallus unlike *Laminaria* takes place by means of **apical cell** which is four-sided. It lies at the bottom of an elongated pit at the tipt of each branch (A). The branch grows in length by the differentiation of the cells formed from the segments cut off by the apical cell. The apical cell cuts off segments from its four faces and the base. The segments cut off at the base form the medullary hyphae and those towards the sides form the cortex and photosynthetic tissues. When the branch has reached a certain length, its apical cell divides equally into two cells by a wall down the middle (B). Each of the two daughter cells functions as an apical cell of a branch. As the growth continues the branches become forked. Each branch of the fork has its own apical cell. By means of the apical cell each branch grows to its usual length. This type of branching is called **dichotomous**. In addition to the apical cell, other tissues of the thallus (meristoderm), as mentioned above, also have meristematic activity.

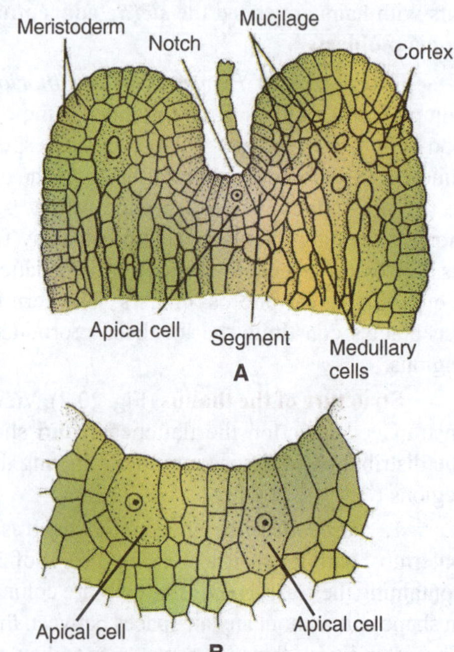

Fig. 20.3 (A-B). *Fucus vesiculosus.* (A), median longitudinal section of the apex of a young frond in the plane of the wide axis; (B), longitudinal section of apex of a forking young frond (After Oltmanns).

Reproduction. It takes place by vegetative and sexual methods only. Asexual reproduction by spores (Mitospores) is entirely absent. Vegetatively *Fucus* multiplies by the method of **fragmentation**. *Fucus* thallus is exposed to violent wave action.

Consequently it is often damaged by being dashed against rocks. The broken branches or fragments have a great power of regeneration. They become re-attached in salt marshes where they can easily get a foothold and grow as independent plants. Even young plants can grow out from detached branches of older plants.

Sexual Reproduction

It is the sole method of reproduction and is **oogamous type**.

(*a*) *Structure of a conceptacle* (Fig. 20.4 A-B). Sex organs are developed within roughly spherical cavities termed the **conceptacles**. Each conceptacle opens to the exterior by means of a pore called the **ostiole**. The conceptacles are aggregated at the tips of the branches. These fertile, terminal tracts of the thallus are called the **receptacles**. The branches bearing receptacles are shed annually. The receptacles is a swollen conical structure. Each conceptacle lies in the superficial part of the medullary zone of the receptacle (A). The cavity of the conceptacle is lined by a more or less compact tissue of two or three cells in thickness (B). It is known as the **fertile sheet**. From the superficial cells of the fertile layer arise many long, hair-like branched or unbranched filaments, the **paraphyses**. The paraphyses have barrel-shaped cells filled with scanty chromatophores. The septa between the cells are often oblique. The paraphyses may produce the sex organs or surround them. In addition, sterile, colourless, unbranched hairs arise from the upper portion of the lining layer

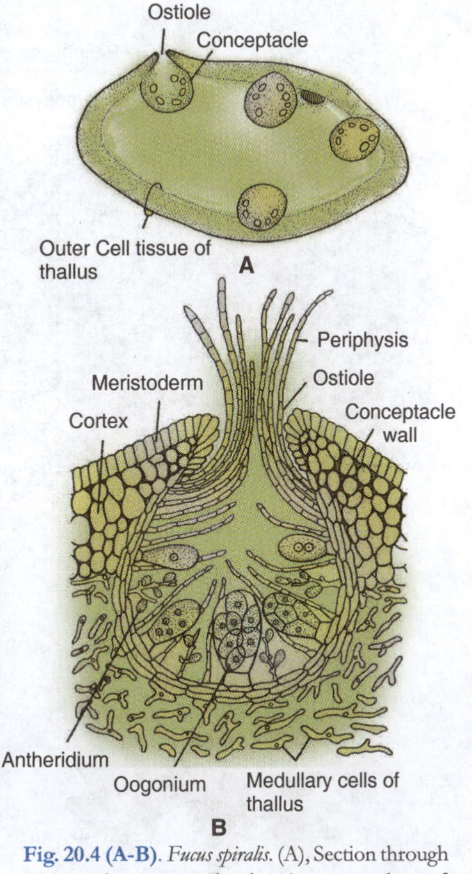

Fig. 20.4 (A-B). *Fucus spiralis.* (A), Section through the receptcale showing a number of conceptacles embedded in the superficial part of medulla; (B), Section through a bisexual conceptacle showing structure.

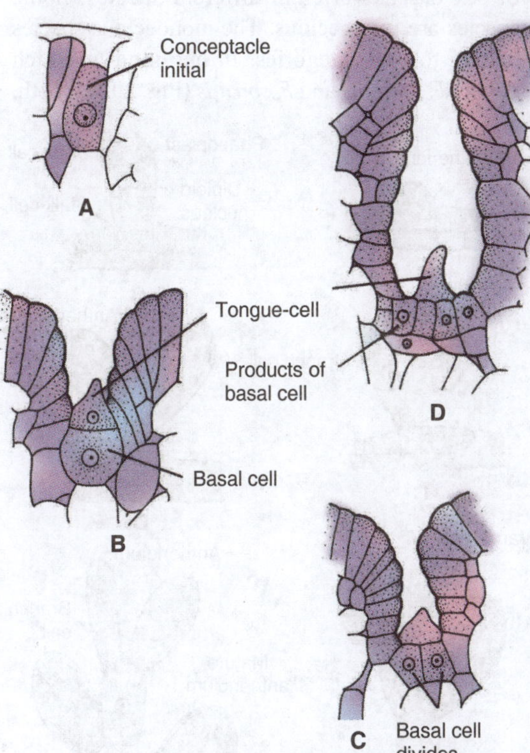

Fig. 20.5 (A-D). *Fucus serratus.* Early stages in the development of a conceptacle. (A), differentiation of a conceptacle initial; (B), division of initial cell into upper tongue cell and lower basal cell; (C), basal cell divides by a vertical wall into two; (D), Later stage showing the products of the basal cell forming the floor of conceptacle. (After Nienborg).

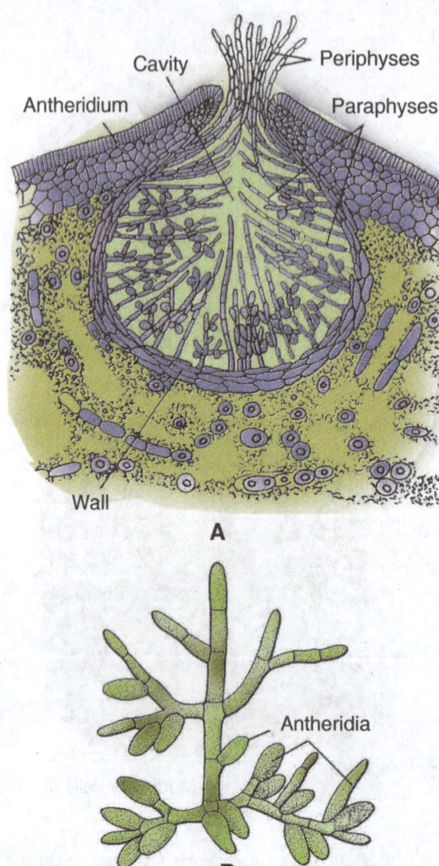

Fig. 20.6 (A-B). *Fucus sp.* (A), section of male conceptacle; (B), Fertile paraphysis bearing antheridia.

of the conceptacle. These are the **periphyses**. The are confined to a small area near the aperture (B). The periphyses in *F. spiralis* extend as a tuft through and project markedly beyond the ostiole. The cavity of the conceptacle is filled with mucilage secreted probably by the paraphyses.

(*b*) *Development of conceptacle* (Fig. 20.5). It is formed from a single superficial cell which still lies in the apical notch near the apical cell. It is called the **conceptacle initial** (A). The conceptacle initial increases in size and divides by cross wall into an upper **tongue cell** and a lower **basal cell** (B). The adjacent vegetative cells divide rapidly, grow up around the tongue and the basal cell divides repeatedly by vertical and cross walls to form the floors of the conceptacle (D). The sides are formed by the adjacent vegetative cells of the thallus.

(*c*) *Distribution of sex organs.* The distribution of sex organs varies in different species. Some species are **monoecious**. The monoecious species belong to two categories. In one kind to which belong *F. furcatus* and *F. spiralis* (Fig. 20.4B) both, the oogonia and the antheridia are developed in the same conceptacle. In the second category the oogonia and antheridia are produced in separate conceptacles on the same plant. Many species such as *F. serratus* (Fig 20.6 A) and *F. vesiculosus* are **dioecious**. They have male and female sex organs produced in separate conceptacles on different plants. In this case *Fucus* plants are either male or female. Meiosis in *Fucus* occurs during gametogenesis as in animals.

Antheridia. *Structure* (Fig. 20.6). The antheridia are small oval or club-shaped unicellular structure borne usually on the basal branches of the fertile richly branched paraphyses which arise from the surface cells forming the sides and floor of the conceptacle

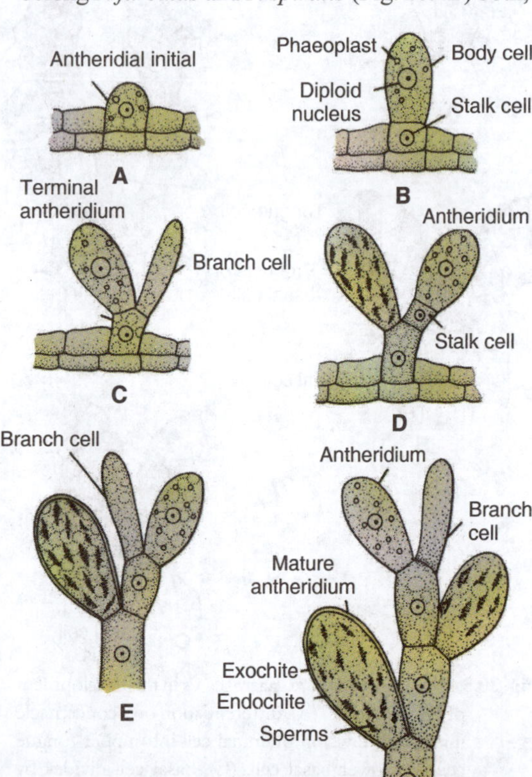

Fig. 20.7 (A-F). *Fucus sp.* Development of fertile paraphysis and antheridia. Explanation in the text.

(A). There are many antheridia in each conceptacle. The mature antheridium is an orange coloured oval or club-shaped sac with the wall differentiated into two layers. The outer layer is firm and called the **exochite**. The inner gelatinous layer is known as the **endochite** (Fig. 20.7 F). It has 64 sperm nuclei or sperms which are haploid structures. The young antheridium is unicellular and has a diploid protoplast.

Development of fertile hair and antheridia (Fig. 20.7). A surface cell of the conceptacle wall grows into a papilla-like outgrowth (A). It is the **antheridial initial**. The latter divides by a basal wall into a lower or basal, sterile **stalk cell** and an upper fertile **body cell** (B). The body cell subsequently enlarges to form the terminal antheridium. The stalk cell resumes activity and grows into a branch cell which shifts the terminal antheridium to one side (C). The branch cell divides again producing a second antheridium and a stalk cell (D). The latter again resumes growth pushing the second antheridium to one side (E). This process is repeated indefinitely (F). The result is a richly branched fertile paraphysis, the upper branches of which may remain sterile (Fig. 20.6 B).

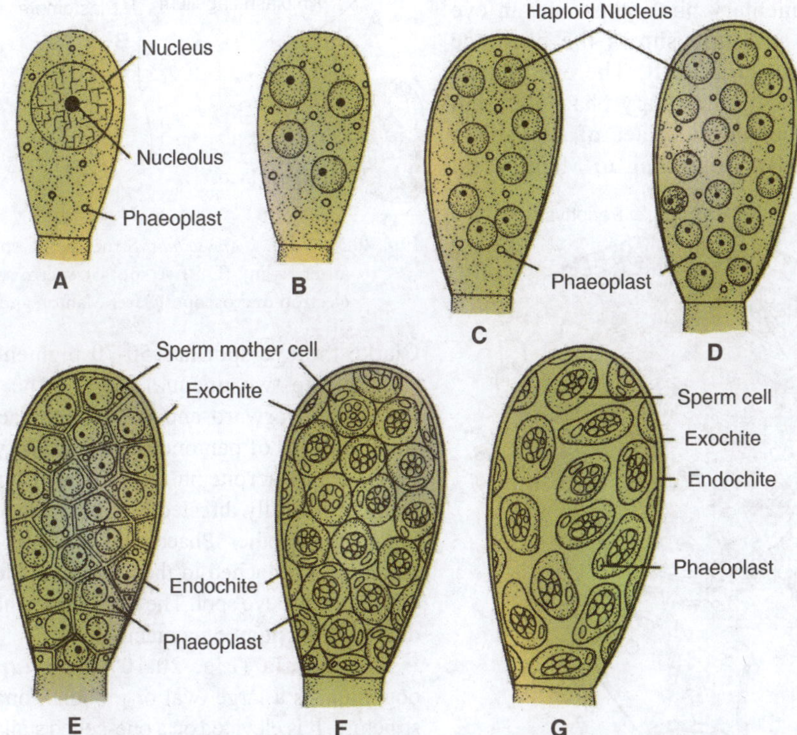

Fig. 20.8 (A-G). *Fucus* sp. Stages in the differentiation of sperms (spermatogenesis). (A), Young antheridium, with a diploid nucleus; (B), antheridium after meiosis containing 4 haploid nuclei; (C), 8- nucleate stage; (D), 32-nucleate stage; (E), Cleavage of protoplast at 32-nucleate stage to delimit sperm mother cells; (F), The sperm mother cells separate and round off; (G), Each sperm mother cell divides to form 2 sperm cells.

Differentiation of Sperms (spermatogenesis, Fig. 20.8). The young uninucleate antheridium has a large diploid nucleus with a conspicuous nucleolus (A). It is located in the centre and is surrounded by alveolar cytoplasm which, in addition, contains yellow-green chromatophores or phaeoplasts. The primary antheridial nucleus undergoes meiosis. Four haploid nuclei are produced (B). These divide by free mitotic nuclear divisions until there are 32-nuclei in an antheridium (D). The nuclear divisions are accompanied by the division of phaeoplasts which lose pigment at the 8-nucleate stage of antheridium. At the 32-nucleate stage, cytoplasm of an antheridium undergoes cleavage. Consequently

32 uninucleate daughter protoplasts separated by finely granular lamellae are formed (E). These are the **sperm mother cells**. Each sperm mother cell undergoes the last division to produce two **sperm** or **male cells** (F). This results in the formation of 64 male or sperm cells in an antheridium. The sperm cell besides the nucleus has a phaeoplast. It metamorphoses into a biflagellate sperm. The mature antheridium in which the antheridial wall is differentiated into two layers thus has a mass of 64 sperms. The sperms escape still enclosed in the inner layer (endochite).

Structure of Sperm. (Fig. 20.9).

The sperm is a minute, slightly elongated pear-shaped, laterally biflagellate cell with the anterior end pointed (A). It consists largely of nuclear material. Besides the massive nucleus it contains a single rudimentary phaeoplast and an eye spot. Fine structure shows the presence of mitochondria as well. The eye spot is attached to the rudimentary phaeoplast and is situated behind the place of attachment of flagella (B). According to Manton and Clarke the eye spot has 50-70 pigment facets (D). Of the two unequal flagella the shorter is directed forward and the longer backward. The former is of pantonematic (tinsel) type and the latter of acronematic (whiplash) type. The shorter forwardly directed tinsel type flagellum is in contrast to other Phaeophyceae. The posterior flagellum is attached to the surface of the sperm cell above the eye spot. The flagellar membrane is dilated at the point of contact.

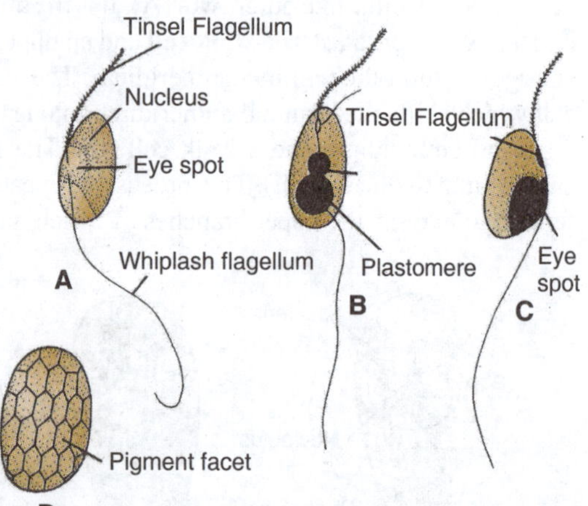

Fig. 20.9 (A-D). *Fucus spiralis*. Structure of sperm (A-C) (after Kylin); (D), Eye spot of sperm as seen with electron microscope. (After Manton and Clarke).

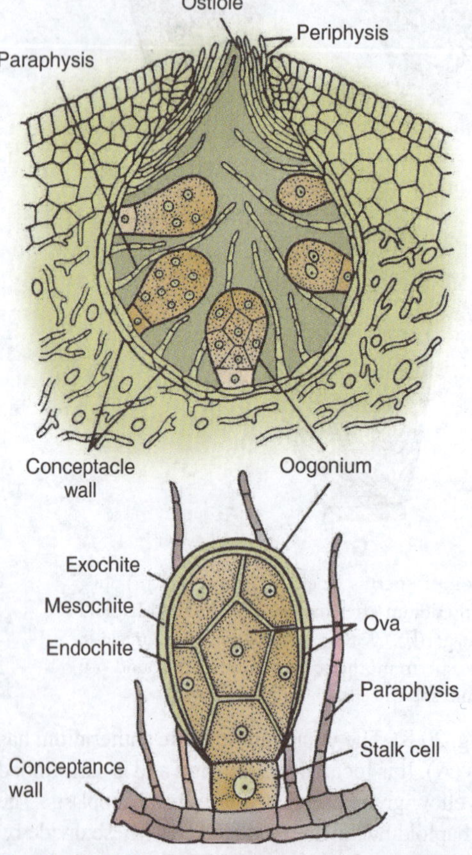

Fig. 20.10 (A-B). *Fucus*. (A), V.S. Female conceptacle; (B), A segment of conceptacle wall bearing a mature oogonium and a few paraphyses.

Oogonia (Fig. 20.10). *Structure*. The oogonium is a large oval or globular, one-celled structure. It is elevated on a one-celled stalk arising directly from the wall of the conceptacle (A). At maturity it contains eight eggs or ova. The oogonial wall is thick and differentiated into three layers, the outer **exochite**, the middle **mesochite**, and the innermost **endochite** (B). The exo- and endochite are thin. The mesochite is thick. There is a space between exochite and mesochite. In the region of the basal pit, where the stalk cell and the oogonium adjoin, the oogonial wall remains thin.

Development of oogonia (Fig. 20.11). The oogonia arise directly from the wall of the conceptacle. Any surface cell of the fertile sheet may function as an **oogonial initial.** The later grows up to form a **papilla** (A). A cross wall appears near the base of the papilla (B), and divides it transversely into a lower **stalk cell** and an upper **body cell** (C). Neither of them divides again. The stalk cell stands well above the conceptacle wall. The body cell forms the oogonium proper which grows into a large usually spherical structure. The young oogonium has a single diploid nucleus (C). The dense cytoplasm which surrounds it is devoid of vacuoles but contains numerous old drops.

Maturation of oogonium and differentiation of Ova (Oogenesis). During maturation of the oogonium the diploid nucleus undergoes three successive nuclear divisions. The first two constitute **meiosis** and occur soon one after the other. The resultant four haploid nuclei occur tetrahedrally and not in the same plane as shown in sketch E. After a pause, the third nuclear division occurs. It is mitotic. The oogonium at this stage contains eight nuclei (F). Cleavage of the oogonial cytoplasm forms eight

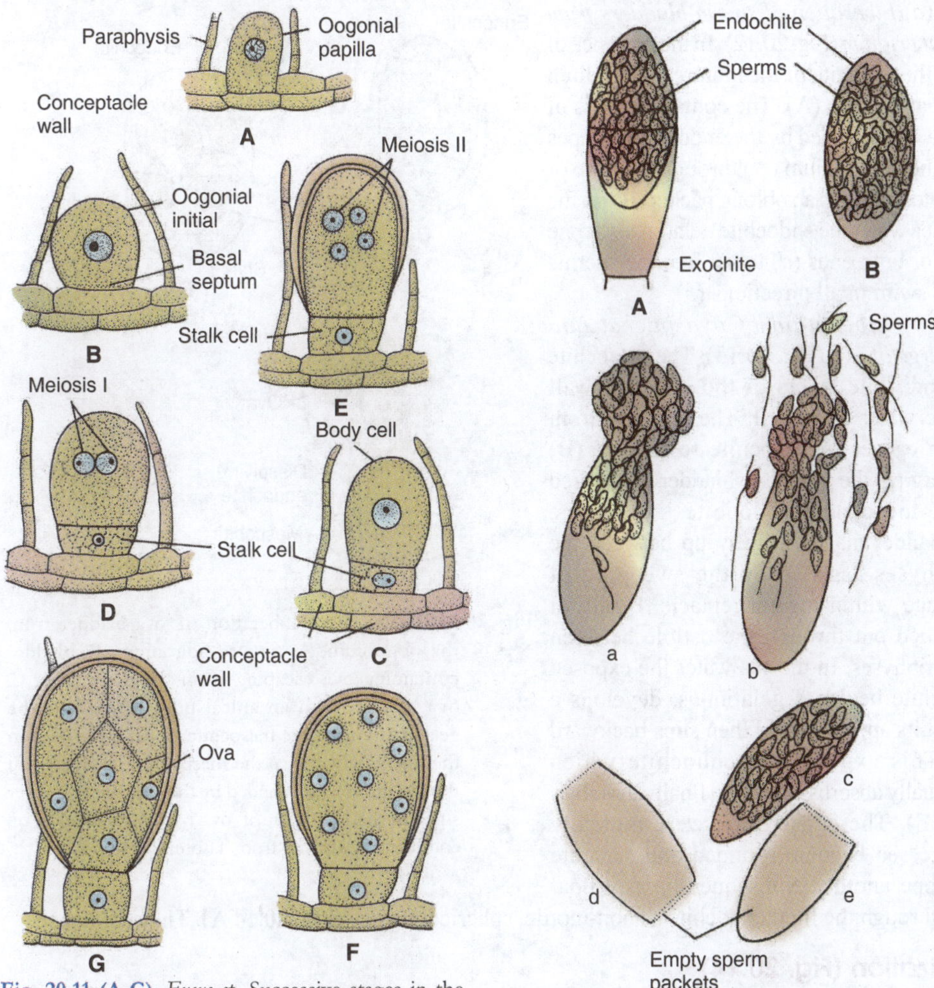

Fig. 20.11 (A-G). *Fucus sp.* Successive stages in the development of oogonium and differentiation of ova. Explanation in the text. Note that the sketch E and F are highly diagrammatic. The nuclei are shown as if arranged in the same plane. Actually they are arranged tetrahedrally and all are not seen in one section.

Fig. 20.12 (A-C). *Fucus sp.* Liberation of sperm bladder from an antheridium and escape of sperm from the packet. A, sperm bladder emerging from the antheridium; B, liberated sperm bladder with the endochite still enclosing the sperms; C, release of sperms from the sperm bladder. (C after Thuret).

uninucleate eggs (G). They are separated by fine membranes or septa connected with the endochite.

Liberation of sperm and Oogonial packets. The sperms and ova in *Fucus* escape from their respective sex organs in a mass still enclosed in the inner layer or layers of the wall. These packets of sex cells are called the **sperm** or **ova bladders**. They are released when the plants are refolded by the incoming tide.

(*a*) *Liberation of sperm bladders from the antheridium* (Fig. 20.12). In the presence of water the exochite of the mature antheridium bursts at its apex (A). The contained mass of sperms till enclosed by the endochite escapes from the antheridium (A) through the pore of the conceptacle as an oblong packet (B). In the open sea water the endochite gelatinizes at one (*a*, *b*) or both ends (*d*) to set free the sperms which swim in all directions (*b*).

(*b*) *Liberation of ova packet and discharge of ova* (Fig. 20.13). The mesochite and endochite layers of the oogonial wall imbibe water and swell. The pressure from within causes the exochite to rupture (B) discharging the ova in a bladder composed of the meso- and endochite layers. The ova bladder makes its way up between the paraphyses assisted by the sweeping of mucilage within the conceptacle. Finally it is pushed out through the ostiole between the periphyses. In the sea water the exposed mesochite becomes gelatinous, develops a pore at its apex (C), and then slips backward (D). This exposes the endochite which continually absorbs water and finally dissolves in it (E). The eggs which were mutually compressed become rounded and separate from one another. Subsequently they float away through the fluid endochite as non-motile, spherical bodies (Fig. 20.14 A). They are naked.

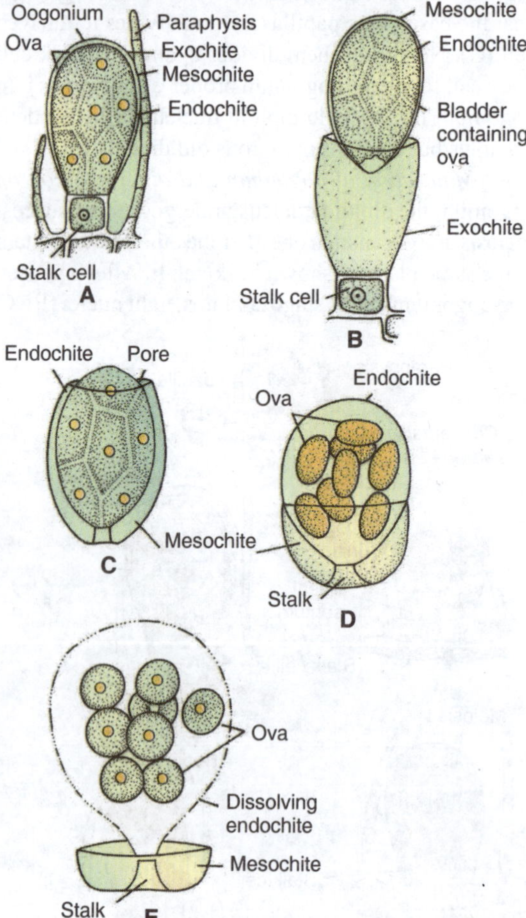

Fig. 20.13 (A-E). *Fucus.* Liberation of ova bladder from the oogonium. A, mature oogonium; B, bladder containing ova escaping from the oogonium; C, ova bladder with an apical pore formed by the gelatinisation of the mesochite; D, an early stage in the liberation of ova, the mesochite rolls back and the ova are still surrounded by the endochite; E, later stage in the liberation of ova by the dissolution of endochite. (Adapted from Thuret).

Fertilization (Fig. 20.14)

The release of gametes in *Fucus* is correlated with the tidal cycle. When at low tide the plants are left exposed, the tissues of the fruiting frond shrink due to desiccation. The contraction exerts pressure on the mucilage (within the conceptacles). Consequently the packets of ova and sperms embedded in the mucilage are forced out through the ostioles. They are picked up by the flood tide. The walls of the packets dissolve to liberate the ova (A) and the sperms (B) in the surrounding water at high tide. Pollock (1970) observed that a period of about 20 minutes was required for the release of ova and sperms from the freed oogonial and sperm packets. *Fucus* is peculiar in discharging its eggs into the water. Fertilization in *Fucus* thus takes place in the open sea water. It is described as **external**. The eggs contain a lot of reserve food material and are many times larger in diameter than the sperms. The sperms according to Thuret (1854) are attracted towards and cluster around the non-motile eggs floating passively in water. The source of attraction is a substance of unknown composition secreted by the mature ovum. Muller (1972 B) reported that the attractant of sperms is a volatile substance released by the ovum. Muller and Jaenicke (1973) designated this sex attractant *fucoserratin.* Furthermore, to air and light, the sperms react negatively but positively to gravity. The heavy eggs as they tend to sink to the bottom are followed by the sperms. Soon each egg is surrounded by numerous, tiny sperms which get attached to the surface of the egg by one flagellum whereas the second remains free, lashing in water (C). The combined lashing action of the flagella of the attached sperms causes the non-motile eggs to rotate or spin in water. This may continue until a single sperm penetrates the egg (E) and fuses with it (F). The other sperms disperse because the secretion of attractive material ceases. The fertilized ovum secretes a wall around it and becomes a zygote. The diploid zygote settles down on a rock. The mucilaginous walls of the zygote attach it to the substratum. By repeated mitotic cell division the zygote develops into a young diploid *Fucus* plant.

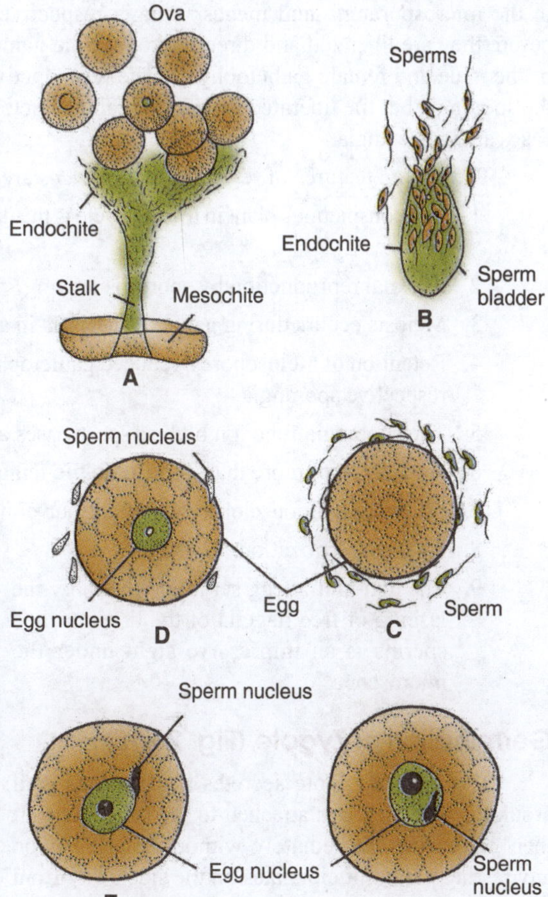

Fig. 20.14 (A-E). *Fucus*. Liberation of ova and sperms and stages in fertilization. (A), liberation of ova from the ova packet; (B), liberation of sperms; (C), ovum surrounded by sperms; (D), sperm nucleus penetrating into egg; (E), sperm nucleus and egg nucleus lie side by side; F, fusion of sperm and egg nuclei.

Latest Interpretation of Sex Organs of Fucus. There is increasing tendency among the modern algologists to consider the young diploid antheridium as the **microsporangium** and the diploid oogonium as the **megasporangium**. They liken these sporangia to the unilocular sporangia of *Ectocarpus.* The diploid protoplast of the microsporangium undergoes meiosis to form four **microspores**. The microspores do not germinate to form any vegetative tissue representing the male gametophyte. On the other hand each microspore divides four or more times to produce from 16-64 sperms. The microsporangium containing the sperms now becomes an **antheridium**. The diploid protoplast of the megasporangium undergoes meiosis to form four large haploid **megaspores.** The megaspores do not germinate to form the vegetative tissue representing the female gametophyte. On the other hand each megaspore divides once to produce directly a total of eight eggs. The megasporangium containing the eggs now becomes the oogonium. On the basis of this view *Fucus* sporophyte is heterosporous. It produces two kinds of spores, small **microspores** and large **megaspores** in the microsporangia and megasporangia respectively. The micro- and megaspores germinate before they are liberated and directly produce the gametes (sperms and eggs) without intervention of the male and female gametophytes. The reproductive cells in *Fucus* thus originate as spores in the sporangia but the liberated products are the gametes. The sporangia thus function both as spore cases and gametangia.

The unique features of reproduction in *Fucus* are :

1. The conspicuous plant in the life cycle is the diploid sporophyte which produces the haploid gametes.
2. Asexual reproduction by spores is lacking.
3. Meiosis occurs during gametogenesis as in animals.
4. Retention of meiosopores, reduced gametophytes and eventually the gametes within their respective sporangia.
5. The sporangia function both as spore cases and gametangia (sex organs).
6. Production of more than one egg in the female gametangium.
7. Liberation of non-motile eggs in the surrounding water.
8. External fertilization.
9. The non-motile egg set into rotation by the lashing of free flagella of the tiny attached sperms is an impressive sight under the microscope.

Germination of Zygote (Fig. 20.15)

The diploid zygote secretes a gelatinous wall around it by which it is attached to the substratum. It then germinates immediately without entering upon any resting period. It elongates on the side away from light (A) and then divides by a cross wall (B). The lower daughter cell divides transversely and its pointed end grows into the first rhizoid which attaches the young sporophyte to any solid object in water (C). The attachment is affected by the mucilaginous wall. There is no penetration. The upper cell by repeated division builds the club-shaped body of the young sporophyte. The first division is vertical. More rhizoids in the meantime may grow out at the base and one or two hairs at the apex (F). The adjacent cells surrounding the hairs separate from one another to form an apical depression. At the bottom of this depression is established the **apical cell**. The subsequent growth of the young sporophyte (H) takes place by the activity of the apical cell which is three-sided at first but becomes four-sided later on.

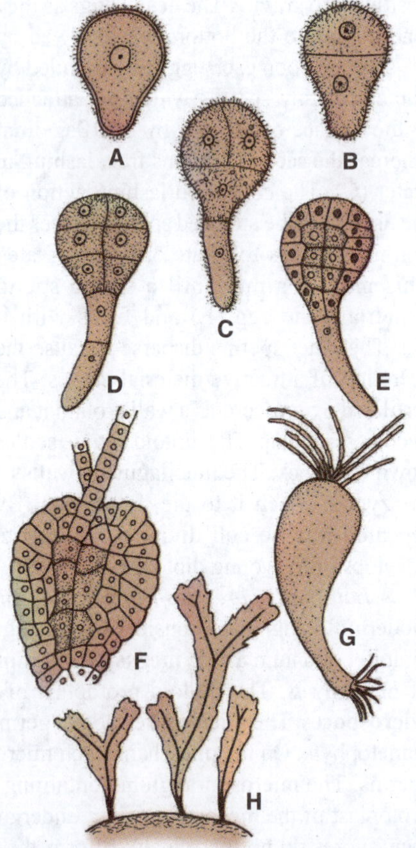

Fig. 20.15 (A-H). *Fucus sp.* (A-G), Stages in the germination of the zygote; (C-F, after Nienberg; G, after Oltmanns); (H), Young sporophytes.

Alternation of Generations (Fig. 20.16). The *Fucus* plant (A) is diploid with 64 chromosomes. It is the sporophyte plant even though it bears the male and the female sex organs. The young oogonium and the antheridium nuclei contain 64 chromosomes.

They are diploid structures equivalent to the *sporangia* usually borne on the sporophyte plant. Meiosis occurs at the time of the formation of gametes (sperms and eggs). The eggs and the sperms, therefore, contain the haploid or gametic number of chromosomes which is 32. They are the first and the last structures that represent the gametophyte phase in the life cycle of *Fucus* plant. There is no distinctly organised gametophyte plant. The gametophore thus starts with the differentiation of gametes in the sex organs. It ends with fertilization which restores the diploid number of chromosomes in the resultant zygote.

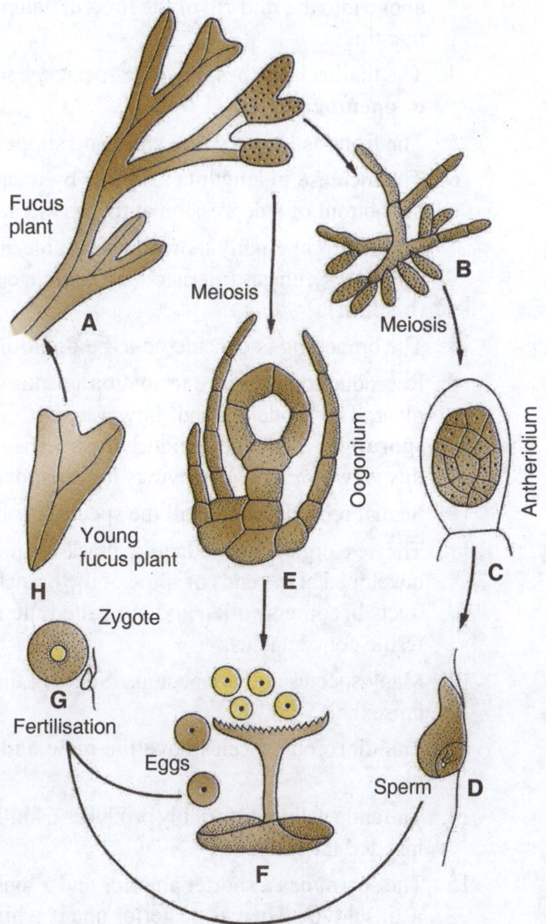

Fig. 20.16 (A-H). Diagrammatic representation of the life cycle of *Fucus*.

Fucus, therefore, is unique in the fact that it represents the end of one line of evolutionary development (gametophyte stage) in the water plants. Even the terrestrial angiosperms lag behind *Fucus* in attaining so high a degree of elimination of the gametophyte. Some phycologists, however, consider the antheridia and oogonia of *Fucus* equivalent to microsporangia and megasporangia which directly proceed to the formation of gametes without any intervening vegetative gametophyte stage.

The zygote in *Fucus* is the first sporophyte structure. In *Fucus*, therefore, there is only an alternation of chromosome number from diploid to haploid and back to diploid state. This is referred to as cytological alternation of generations. The corresponding alternation of multicellular plants with different functions is lacking. There is only one multicellular generation which is the sporophyte. It alternates with the unicellular gametophyte as represented by the gametes. There is thus the *morphological* (externally recognizable) alternation of generations. Some algologists hold a different view. They consider that there is not only a cytological alternation of generations in *Fucus* but morphological one as well. The sporophyte generation no doubt is multicellular and parasitic. It is reduced to merely the gametes. To support their viewpoint they point to the Angiosperms for an analogous situation. This interpretation will form the basis for placing Fucales in the Heterogeneratae.

Salient Features

1. The increase in size and complexity of thallus structure resulting in the differentiation of tissues is worthy of note.
2. The dark brown or olive green thallus is **diploid**. It attains a considerable size and is differentiated into a disc-like **holdfast, stipe** and **frond**.
3. The stipe is usually a short stalk attached to the substratum by the holdfast and continues

above into the mid-rib of the forked, flattened frond. The frond has either entire or serrate margin.

4. The thallus branches, in some species, bear swollen vesicles known as the **air bladders** or **pneumatocysts**.
5. The frond is leathery in texture and slippery.
6. The increase in length takes place by means of a single four-sided apical cell situated at the bottom of a depression at the tip of each branch.
7. The thallus internally shows considerable differentiation of tissues. It has tissues specialised for photosynthesis (meristoderm), storage (cortex), conduction (medulla) and attachment (holdfast).
8. The branching is considered to be **dichotomous**. Some hold it to be **monopodial**.
9. Reproduction by spore formation is entirely lacking–a remarkable feature in the diploid plant. The modern trend, however, is to consider the diploid sex organs as **unilocular sporangia**. The latter produce the microspores and megaspores by meiosis. According to this view *Fucus* sporophyte is **heterosporous**.
10. Sexual reproduction in all the species is oogamous.
11. The sex organs are lodged in flask-shaped cavities called the **conceptacles** which are developed at the ends of some of the branches of the mature plants. These fertile, terminal tracts become swollen and are called the **receptacles**. The receptacles are studded with fertile conceptacles.
12. Many species are monoecious. Some of them bear hermaphrodite conceptacles and other unisexual.
13. The dioecious species have the male and the female sex organs occurring on separate plants.
14. The antheridium invariably produces 64 biflagellate sperms, each with two unequal flagella inserted laterally.
15. The sperm has a shorter anterior and a longer posterior flagellum. The anterior flagellum is tinsel type whereas posterior one is whiplash type.
16. Oogonium develops 8 eggs which are shed in the surrounding water.

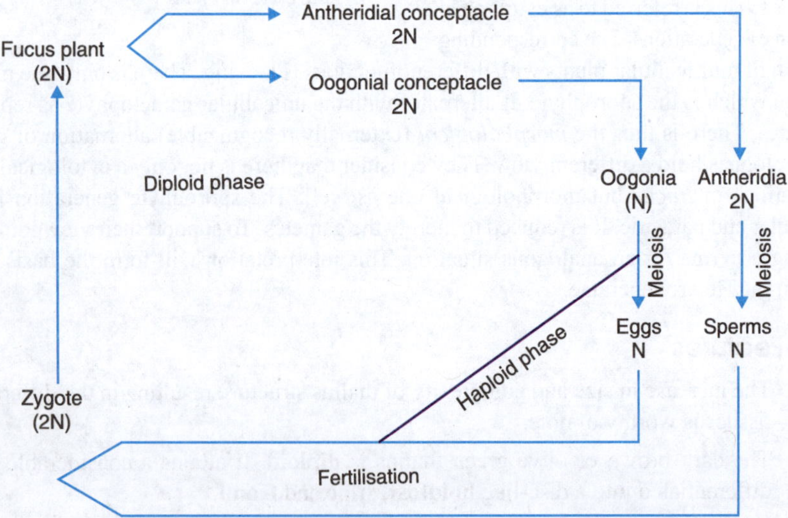

Fig. 20.17 . Graphic representation of the life cycle of *Fucus*.

17. Fertilization is thus external.
18. Meiosis takes place at the time of differentiation of the gametes–a feature in sharp contrast to the majority of the green algae in which zygotic meiosis takes place.
19. The diploid zygote undergoes no resting period–another feature in contrast to the delayed germination of the zygote of green algae.
20. The zygote germinates immediately to give rise to a diploid thallus whereas in the green algae the parent plant is usually haploid.
21. *Fucus* displays only *cytological alternation of generations*. There is no morphological alternation of generations because there is only one morphological generation in its life cycle. It is the diploid or sporophytic generation. The alternate sexual generation or gametophytic generation is completely suppressed and is solely represented by the gametes.

Adaptation of Fucus to its peculiar habitat (intertidal zone). *Fucus* grows in a habitat where it is subject to the buffeting action of violent sea waves and exposed to drying effect of the atmosphere for half the time. It is well adapted to this habitat. A unicellular or a free-floating alga cannot live in such a habitat. It will be swept away. *Fucus* resists the sweeping action of the waves by attaching itself firmly to the substratum. It withstands the tearing action by its slimy nature, leathery texture and non-brittle (rubbery) toughness of its stipe and repeatedly dichotomously branched frond. It escapes desiccation when exposed to the atmosphere by making good use of the water absorbing and moisture retaining capacities of the mucilage secreted over its surface. The air bladders, which some species possess along the midrib region and near the tips, give buoyancy when the plant is submerged and thus assist in floatation and spreading of the thallus. The presence of fucoxanthin, as the characteristic pigment, is considered as a chromatic adoption. It assists in the absorption of blue light waves which penetrate into water more effectively than the yellow and red waves.

Taxonomic Position

Division	:	**Phaeophyta** (Phaeophycophyta)
Class	:	**Cyclosporae**
Order	:	**Fucales**
Family	:	**Fucaceae**
Genus	:	*Fucus*
Species	:	*vesiculosus*

FAMILY : SARGASSACEAE

Some members of the family are bilaterally symmetrical but the great majority are radially symmetrical. They are characterised by their complex morphology among the Fucales and are distinguished by certain special features of their lateral branch systems. The branching is **monopodial**. The branches appear to arise in the axil of the leaf-like expansion of which there are one or two at the base of each branch. Growth in length is carried on by a three-sided apical cell situated at the bottom of a deep, funnel-shaped depression. The family includes five genera. Of these *Sargassam* is widely known and is the most abundant brown seaweed in India (Chauhan and Krishnamurti, 1971). It is chiefly confined to the tropical seas and has 150 species. Of these, fourteen species were reported from India by Greville. *Sargassam swartzii* (Turn.) is one of the more common species at port Okha and occurs as a sublittoral species.

SARGASSUM

Distribution. This marine alga is popularly called the **gulf weed**. Most of the species occur in tropical or temperate zones attached along the rocky shores. They are specially abundant in the warmer seas. Some species occur along the Indian coast (East and West) and that of the Australian seas and also in Japan. The common Indian species are *Sargassum tennerimum, S. carpophyllum, S. duplication, S. plagiophyllum, S. ilicifolium, S. christifolium, S. wightii, S. myriocystum* and *S. cinerium.* A few species are found in the Floridian sea and the West Indies and some in the British waters. *S. linifolium* is met with in the Mediterranean. *S. natans* occurs in huge floating masses in the North Atlantic ocean around the Gulf of Mexico. It perpetuates itself by fragmentation. The area in which it occurs is called the Sargasso sea. It lies off the African coast between 20° and 35° latitude north.

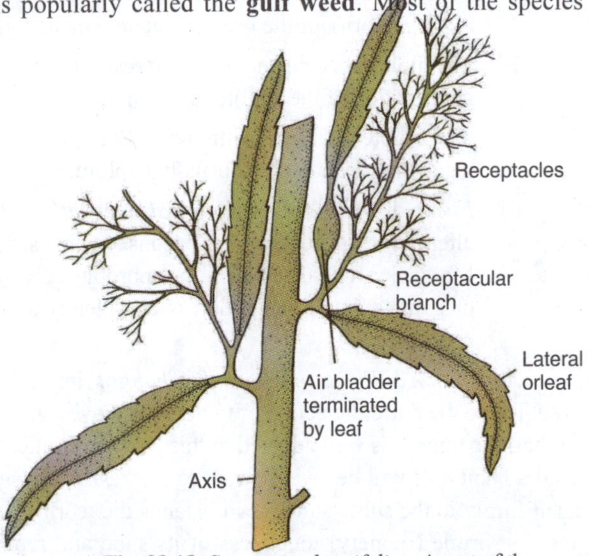

Fig. 20.18. *Sargassum longifolium.*A part of the thallus showing laterals and receptacular branches.

Sargassum polyphyllum. *Sargassum hystrix.* *Sargassum echinocarpum.*

Sargassum wightii. *Sargassum muticum.* *Sargassum vulgare.*

Sargassum natans.

Habit. The *Sargassum* plant is macroscopic and more or less bushy in habit. Apparently it looks like a small angiospermic plant. The plant body is radially symmetrical (*S. patens*). Some species such as *S. vulgare* and *S. filipendula* are found attached to the substratum. Free floating unattached forms like *S. natans* and *S. baciferum* are also fairly common. *S. natans* occurs in huge floating masses. It has no attaching disc or holdfast and is sterile. It multiplies exclusively by fragmentation of the thallus. The other species of *Sargassum* found in the Sargasso sea is *S. hystrix*.

Organisation of the thallus (Fig. 20.19). *Sargassum* plant is a diploid **sporophyte**. The main or central axis (*ma*) in some species (*S. filipendula*), is 30 cm. or more in length. It bears richly branched **primary laterals.** The phyllotaxy is 2/5. Each primary lateral or branch (*lb*) is of unlimited growth. It is also called the **long shoot**. The long shoot bears numerous **secondary laterals** of limited growth. Of the latter the basal one (*bl*) is usually the most conspicuous. It is shaped like a leaf. It subtends more or less clearly the remainder of the branch system. These so-called leaves are in some cases very narrow and drawn out with a serrate margin. In *S. patens* these leaf-like expansions have a conspicuous mid-rib. The mid-rib is absent in the secondary laterals of *S. enerve* (*bl*). The secondary laterals are flattened in a vertical plane. In the lower part of long shoots these so-called lower leaves or laterals of limited growth bear **axillary rudiments** (*abr*). In the upper region of the adult plant they subtend apparently considered axillary **branch systems** (Fig. 20.20, *ax*). The basal member or members of axillary branch system are swollen into stalked and rounded structures called the **air bladders** (Fig. 20.20, *ab*). The subsequent members of the axillary branch system may be cylindrical or flattened. They are called the **receptacles** (Fig. 20.20, *rc*). The receptacles are studded with fertile, flask-shaped **conceptacles**. The receptacles are produced when the plants become fertile. In *Sargassum* the receptacles are specialised branches and not metamorphosed vegetative branches. In some species the so-called leaf subtending the branch system may be modified into an air bladder. Such air bladders are homologous to the laterals of limited growth of "leaves". They may (*S. longifolium*) or may not be terminated by the leafy growth (Fig. 20.20). The air bladders contain the same gases as the atmosphere. Hence they act as floats keeping the plant upright in water. In the unattached species they help the plant to float. Colla

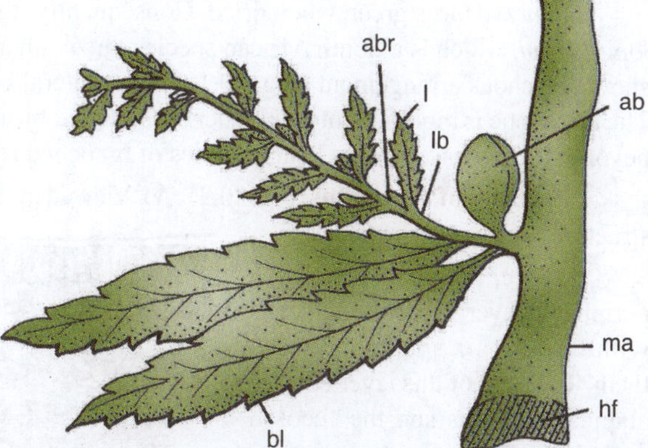

Fig. 20.19. Basal portion of the plant of *Sargassum enerve*., *ab*, air blad-der; *abr*, axillary branch; *bl* basal lateral; *l*, lateral branch, *ma* main axis; *hf*, holdfast.

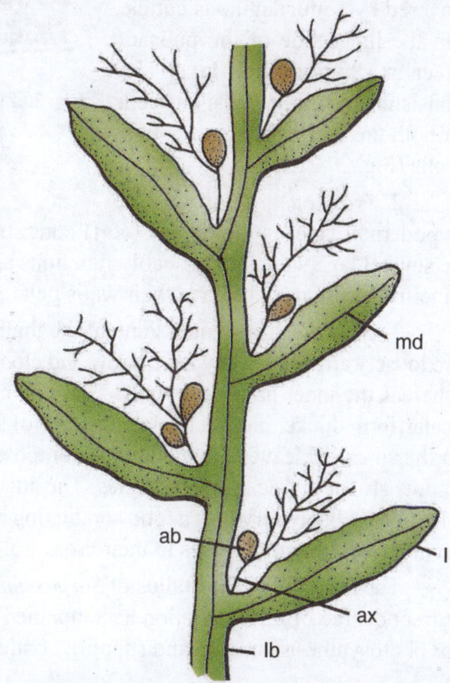

Fig. 20.20. *Sargassum peron.* Lateral branch (*lb*) bearing lateral (*l*); with axillary branches (*ax*); air bladder (*ab*); *md*, midrib; *rc*, receptacles bearing con-ceptacles.

and Willis believe them to be respiratory in function while others look upon them as a mechanism for decreasing specific gravity. They arise as surface outgrowths accompanied by increase in the thickness of the cortex.

Sargassum plant is anchored to the substratum by means of an attaching disc (Fig. 20.19 *h f*). It has an irregular outline and more or less rough appearance. In addition numerous stolon-like structures arise from the base of the main axis. They provide additional anchorage.

S. enerve turns green when dried. Consequently it is often used as a piece of decoration. In *S. longifolium*, which is a South African species, the main axis is flattened. The laterals arising from it show distichous arrangement (Fig. 20.18). Each lateral bears two basal leaves alternately arranged. The upper one is modified into a bladder surmounted by a leafy growth (Fig. 20.18). The branch axis beyond the two basal leaves bears a series of branched receptacles.

Structure of the thallus (Fig. 20.21, A). Viewed in cross section the thallus reveals the following structure :

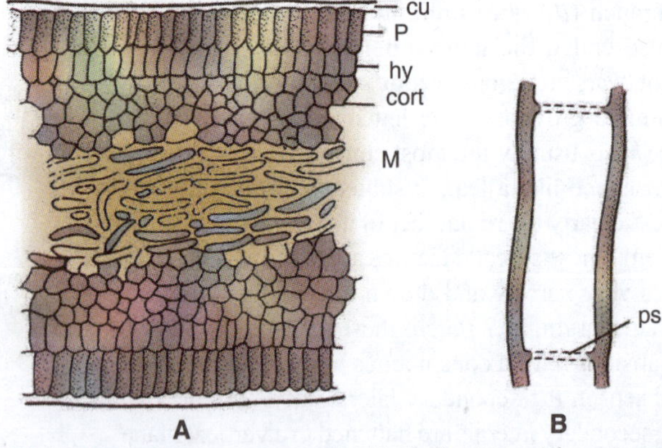

Fig. 20.21 (A-B). *Sargassum sp.* (A), T.S. Thallus (*cu*, cuticle; *cort*, cortext; *hy*, hypodermis; *M*, medulla; *p*, palisade layer); (B), Single cortical cell showing perforated transverse septa (*ps*).

1. *Meristoderm*. It is a peripheral layer of columnar cells without any air spaces between them. The cells of this layer contain abundant plastids and the fucosan granules. This **palisade layer** (*p*) of cells constitutes the *meristoderm*. It is concerned with photosynthesis. On the outside the palisade cells are covered by a mucilaginous cuticle. On the inner side of the palisade layer is a **hypodermal layer** (*hy*) consisting of similar cells. The cells of both the layers have pits in their walls (Fig. 20.21 B).

2. *Cortex*. Within the hypodermal layer is the **cortex** (cort) consisting of thick-walled polygonal cells. The cortex may be several layers thick. It probably functions as the storage region and plays a mechanical role. The innermost cortical cells have their walls gelatinised and merge imperfectly into the **central zone**.

3. *Medulla*. The central zone of the thallus is called the **medulla** (M). The cells in this region are loosely arranged. They are narrow and elongated. The outer cells of the medulla have thick walls whereas the inner have thin walls. There is, however, no gelatinized material. Hanstein described scalariform thickenings on the lateral walls of the medullary cells. The medullary hyphae are absent. In the so-called leaves the medulla is composed of thin-walled cells only. The medullary region is almost absent in free floating species. The attaching organs contain hyphae in the medullary region. Medulla probably serves as a food conducting region. According to Hanstein the cells of all the zones are inter-connected by pores in their cross walls (Fig. 20.18 B). Others, however, deny this.

The anatomy of the thallus of *Sargassum* reveals that the vegetative cells constituting it show a great degree of differentiation accompanied by division of labour. It contains tissues specialised for photosynthesis, storage and support, conduction and attachment.

Reproduction

It takes place by vegetative and sexual methods only. Asexual reproduction by the formation of spores is absent.

1. *Vegetative Reproduction. Sargassum* reproduces vegetatively by fragmentation. The older parts die and decay separating the younger parts. The latter continue to grow and finally develop into full-fledged *Sargassum* plant. *S. Natans* and *S. hystrix* multiply exclusively by this method.

2. *Sexual Reproduction*. It is the usual method of reproduction and is **oogamous**. The sex organs are produced in flask-shaped cavities called the **conceptacles** borne on the sporophyte plant.

The conceptacles are aggregated in the upper region of the axillary branch systems. These may be cylindrical or flattened and are called the **receptacles.**

(*a*) *Mature conceptacle*. It is a depression or cavity (Fig. 20.23B) in the thallus. It lies in the superficial part of medulla of the receptacle. The conceptacle has a wall of its own. The wall consists of a few layers of compactly arranged flat cells rich in chromatophores.

The conceptacle wall separates the cavity from the loose tissue of the medulla. The greater part of the wall bears many branched hair-like filaments, the **paraphyses.** The latter consist of barrel-shaped cells with scanty chromatophores. The septa between the cells are often oblique. Each conceeptacle opens to the exterior by means of an aperture termed the **ostiole.** Near the aperture are present the colourless hair. They are called the **periphyses.** Between the paraphyses and the colourless hair near the ostiole are present hairs with abundant chromatophores.

(*b*) *Development of conceptacle* (Fig. 20.22, *A-G*) It is formed from a single superficial, flask-shaped **initial cell** (*A, ic*). It is distinguishable from others by its larger size, slower rate of division and large nucleus. The conceptacle initial divides by a curved septum into two unequal daughter cells (*C*). The upper daughter cell is called the **tongue cell** (*C, tc*) and the lower **basal cell** (*C*, bc). In the meanwhile the cells surrounding the conceptacle initial divide in such a manner that it sinks down. The tongue cell remains separate from the adjoining tissue. The basal cell divides repeatedly by vertical walls to form the wall of the conceptacle except its upper narrow portion. The latter is formed from the adjoining tissues. The tongue cell merely elongates and in some cases divides to form a row of few cells (*G, tc*). Eventually the tongue cells disappear.

Sex Organs. The distribution of sex organs varies in different species. Some species are **monoecious** and others **dioecious**. The conceptacles are always unisexual. In the monoecious species the male and the female sex organs are developed in separate conceptacles borne on the same plant. In

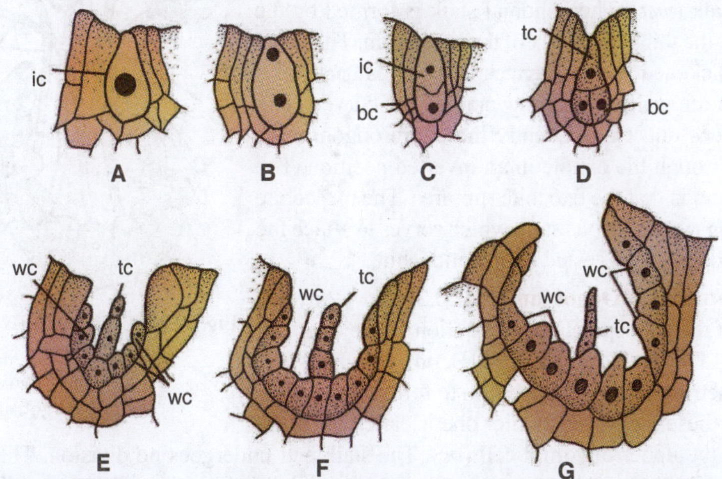

Fig. 20.22 (A-G). Development of conceptacle of *Sargassum* sp. (A-D), Earlier stages. (A), Flask shaped initial (*i c*); (B), Nuclear division; (C), Formation of basal cell (*b c*) and tongue cell (*t c*); (D), Basal cell divided into two by a vertical wall; (E-G). Later stages in development of the conceptacle.

the dioecious species the male and female con-ceptacles are borne on different plants. The dioecious plants are thus either **male** or **female**. The male and female plants are to some extent distinguishable. The male receptacles are **smooth** and the female **spinous**. In *S. horneri* the male receptacles are called the **antheridium** and the female **oogonium**.

Oogonium (Fig. 20.23, B). The oogonia are sessile and borne directly on the wall of the conceptacle. They are, as a matter of fact, embedded in the wall. The young oogonium is filled with dense cytoplasm. It has a single diploid nucleus and contains plenty of oil drops. During maturation the diploid nucleus undergoes three successive divisions. The first two constitute **meiosis**. As a result, eight haploid daughter nuclei are formed. The cytoplasm undergoes no cleavage. Towards maturity the protoplast of the oogonium rounds off to form a single egg or ovum. The young ovum is eight nucleate. Out of these eight nuclei one enlarges. The other seven are, at first arranged along the periphery. They finally degenerate during fertilization. Some investigators hold that the nuclear divisions take place when the oogonium is young and enclosed in the conceptacle. Rao and others hold that the nuclear divisions occur when the oogonia are discharged from the conceptacle and remain outside the ostiole attached to the conceptacle wall. At maturity the oogonial wall thickens and gets differentiated into three layers. These are the outer thin **exochite**, the middle thick **mesochite** and the innermost **endochite**.

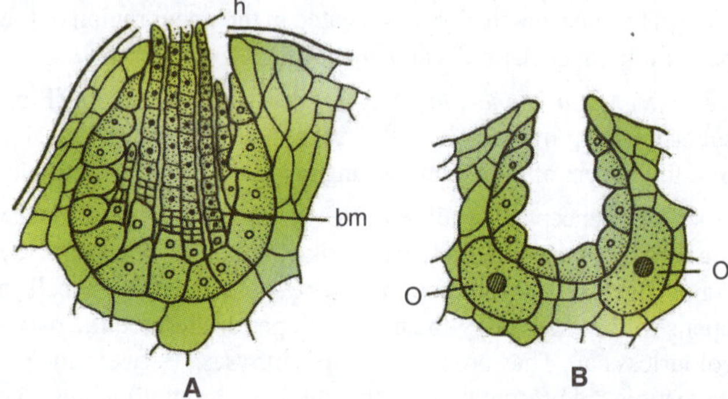

Fig. 20.23 (A-B). (A), Vertical section of young cryptoblast. *bm*, basal meristem; *h*, hair; (B), V.S. of young female conceptacle with two embedded oogonia (*o*).

The mature oogonia (Fig. 20.24) are discharged through the ostiole. Even after emergence each oogonium remains attached to the conceptacle wall by means of a long gelatinous stalk (*mst*). The oogonial stalk is formed by the elongation of the thickened apex of the oogonium. The distal end of the thickened apex elongates and curves back till it fixes itself to the wall of the conceptacle. The curved stalk then straightens out. Consequently the entire oogonium is pushed out through the ostiole in an inverted position. The other viewpoint is that the exochite ruptures. The mesochite forms the long mucilaginous stalk which serves to attach the discharged oogonium invested by the endochite.

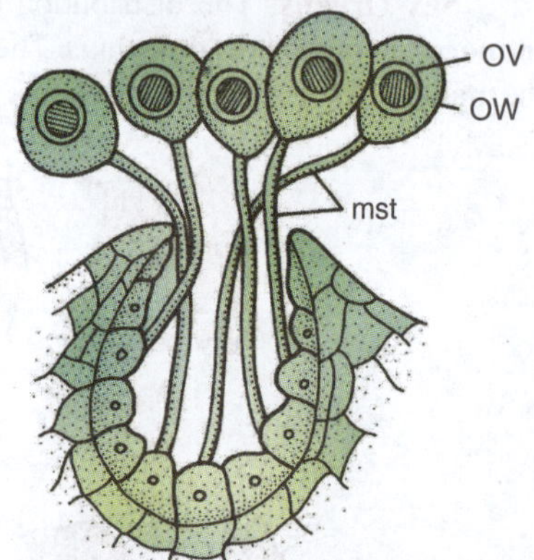

Fig. 20.24. *Sargassum sp.* Extruded oogonia attached to the conceptacle wall by long gelatinous stalks (*mst*); *ov*, ovum; *ow*, oogonial wall.

Development of Oogonium (Fig 20.25, A-F). Any cell in the wall of the conceptacle may function as the oogonial initial (A, *oi*). The latter grows in size (B, *oc*). It then divides by an asymmetrical transverse division to form two unequal cells (C). Of these the lower smaller one is called the **stalk cell** (*sc*) and the upper **oogonial cell** (*oc*). The stalk cell undergoes no division. This accounts for the embedded nature of the oogonium. The oogonial cell increases in size. It is filled with dense cytoplasm and is devoid of vacuoles. It has plenty of oil drops. Its single nucleus undergoes three successive divisions to form eight daughter nuclei (D). Of these the first and the second division follow rapidly one upon the other and the third after a short period of rest. The first two nuclear divisions

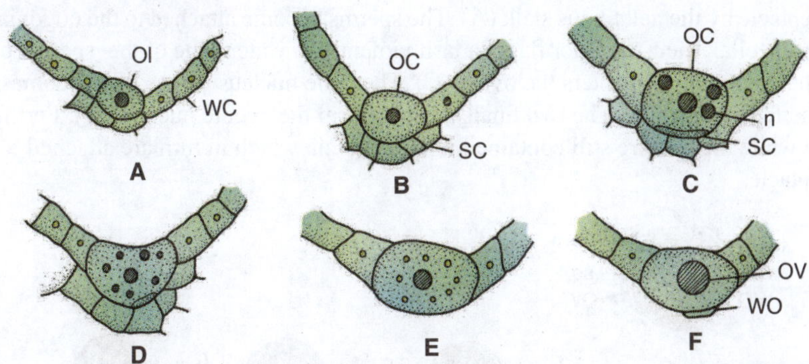

Fig. 20.25 (A-F). Development of Oogonium in *Sargassum sp.* (A), Very early stage; *oi*, oogonial initial; *wc*, conceptacle wall; (B), and (C), Formation of stalk cell (*sc*) and oogonial cell (*oc*) and the division of the nucleus (*n*) into four daughter-nuclei; (D), 8-nucleate stage; (E), One large functional nucleus and 7 degenerating nuclei; (F), Oogonium

constitute **meiosis**. In *Sargassum*, reduction division thus takes place during gamete formation. The nuclear division in the oogonium is not followed by cleavage of the cytoplasm. the protoplast of the oogonium withdraws from the oogonial wall and rounds off to form a single ovum. It has 8-nuclei. One of these migrates to the centre. The other seven nuclei become arranged along the periphery. There they gradually degenerate (E). The central one that survives enlarges to become the functional haploid female nucleus (F).

Antheridia. The structure and development of antheridia is nearly similar to that of *Fucus* (Fig. 20.7). They occur in large numbers and are so crowded in a single conceptacle that observation is made difficult. They are borne terminally usually on the lower branches of the scantily branched paraphyses. The paraphyses arise from the wall of the male conceptacle. The mature antheridium is a small, ovoid structure. It contains 64 biflagellate male cells or sperms. When sperms reach maturity the antheridia separate from their respective stalk cells by the action of water currents. They emerge from the conceptacle through its ostiole and float on the surface of water. The wall fo the antheridium gelatinises and sperms are liberated. The liberated sperm is a pear-shaped structure with a pointed anterior end. It has two laterally inserted flagella of unequal length. The longer flagellum is directed backward and the shorter is directed forward. The greater part of the body of the sperms is occupied by the large nucleus. The flagella arise from the blepharoplasts. The reported eye spot is derived from the chromatophore which is vestigial.

Development. The first antheridium arises from the cell of the wall of the male conceptacle. It grows into a papilla-like outgrowth. A transverse wall appears dividing it into two cells. The lower cell functions as the wall cell of the conceptacle. The upper cell again divides into two. Of these the lower is called the **stalk cell** and the upper **antheridial cell**. The stalk cell grows into a branch pushing the first terminal antheridium to one side. The branch cell may again divide into the terminal antheridium and basal branch cell. Finally the stalk cell forms one or two-celled paraphysis.

Each antheridial initial contains a single prominent nucleus and a few chromatophores. As it increases in size the antheridial nucleus divides repeatedly to form 64 daughter nuclei. The first two divisions constitute **meiosis**. The subsequent ones are mitotic. At the 32-nucleate stage the cytoplasm of the antheridium becomes divided by finely granular lamellae. The latter become prominent at the 64-nucleate stage of the antheridium. Consequently the mature antheridium contains 64 male cells each with one chromatopore. Each male cell is a pear-shaped structure. It develops two flagella of unequal lengths. They are inserted laterally.

Fertilization (Fig. 20.26, A-D). The liberated sperms swim towards the female conceptacles. Several of them surround each oogonium which is outside the ostiole still attached to the wall of the female conceptacle by the gelatinous stalk (A). The sperms become attached to the oogonial wall by their anterior flagella. Their posterior flagella lash violently in water. One of the sperms penetrates the wall of the oogonium and enters the ovum (C). The male nucleus passes in and comes to lie by the side of the female nucleus. The two finally fuse to form the zygote nucleus (D). Fertilization is thus affected while the ova are still contained in the oogonia which in turn are attached to the wall of the conceptacle.

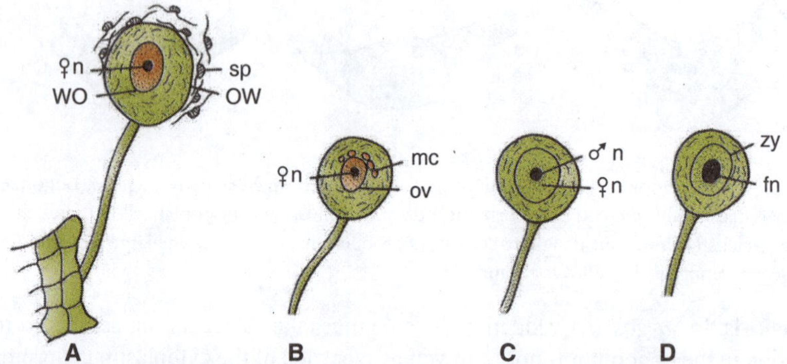

Fig. 20.26 (A-D). Fertilization in *Sargassum sp.* (A), Sperm (*sp*) surrounding the oogonium; (B), Sperms having penetrated the oogonial wall; (C), Male and female nuclei lying side by side; (D), Act of fertilization accomplished; *fn*, fusion nuclei; *mc*, male cell; *sp*, sperm; *ov*, ovum; *zy*, zygote; *ow*, oogonal wall; *fn*, female nucleus.

Germination of Zygote (Fig. 20.27). Germination of zygote starts while it is still enclosed in the oogonium and the latter is still attached to the conceptacle wall. Eventually the oogonial wall gelatinises. The zygote is liberated. The liberated zygote rests on some solid object. There it divides by a transverse wall into an upper and lower cell (A). The lower cell gives rise to the fibrous rhizoid like processes and the upper forms the erect diploid plant.

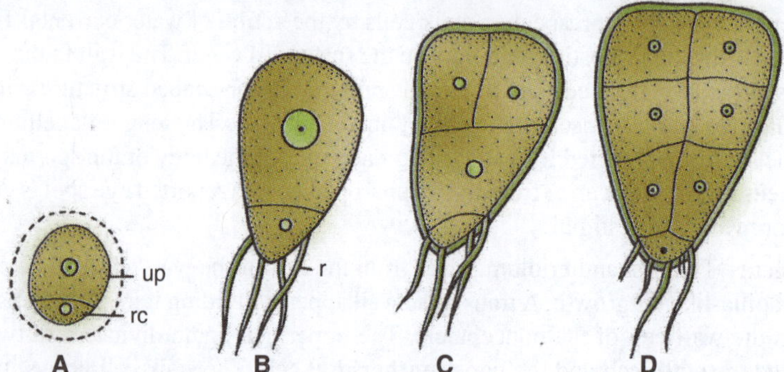

Fig. 20.27 (A-D). Germination of zygote in *Sargassum sp. up*, upper cell; *rc*, rhizoid cell; *r*, rhizoid.

Sexual reproduction in *Sargassum* differs from that of *Fucus* in the following respects :

1. The conceptacle are always unisexual.
2. The oogonium is sessile and remains embedded in the wall of the conceptacle.
3. The oognoium contains only a single oosphere or ovum.
4. The oogonium after liberation remains attached to the conceptacle wall by a long gelatinous wall.
5. The ovum is 8-nucleate at first, but 7 of these nuclei degenerate leaving behind only one nucleus at the time of fertilization.

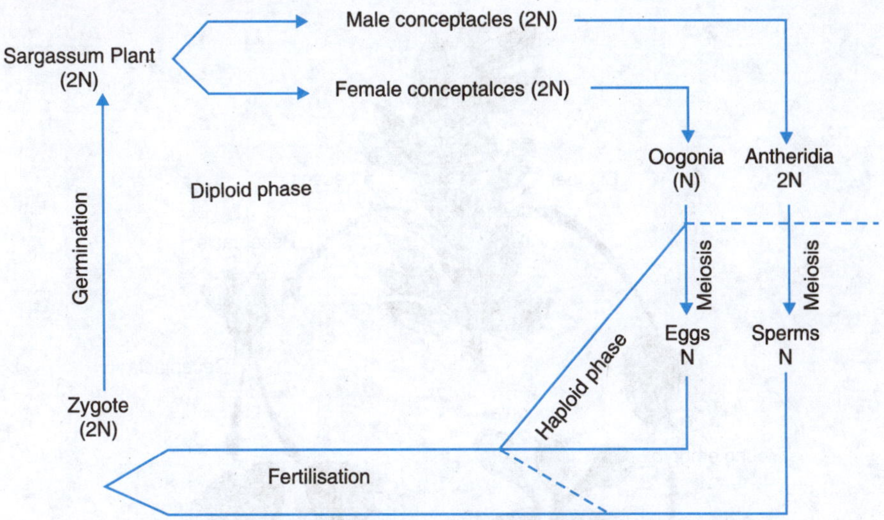

Fig. 20.28. *Sargassum sp.* Graphic representation of the life cycle.

6. The ovum or egg is fertilised while it is still contained in the oogonium which remains attached to the wall of the conceptacle. The fertilization is thus internal.
7. The zygote starts germinating while still attached to the parent by the gelatinous stalk.

Cryptoblasts. Besides the fertile conceptacle there are found in certain species of *Sargassum* sterile flask-shaped structures called the **cryptoblasts** (Fig. 20.23, A). They are borne on the primary branches. They open to the outside by openings called **cryptoblamata**. The cryptoblasts do not contain any sex organs. They develop in the same way as the conceptacles and have the same structure. From the centre of the floor of the cryptoblast arises a group of colourless unbranched hairs. Each hair has a definite basal meristem (Fig. 20.23, *bm*). These hairs project in tufts through the cryptoblamata. They fall off with age. Only the basal portions persist in the old cryptoblasts.

Intermingled with the colourless hairs are found coloured ones. In the fertile conceptacles sterile hairs are only present near the aperture. The cavity of the cryptoblasts is filled with mucilage secreted by the hairs. It is held that plants exposed to extreme conditions of the environment have the cryptoblasts developed in abundance.

Similar structure and development of the cryptoblasts and the conceptacles indicate that the two organs are homologous. This is confirmed by the fact that some people have reported the occurrence of abortive antheridia in cryptoblasts in certain cases.

Alternation of Generations (Figs. 20.28, 20.29)

Sargassum has a typical cyclosporian life cycle as in *Fucus*. There occurs only a single morphologically organised individual. It is the diploid **sporophyte** which is elaborate in structure and bushy in habit. It bears **diploid** sex organs (**antheridia** and **oogonia**) in flask-shaped depressions of the diploid thallus called the **conceptacles**. The thalli may be monoecious or dioecious but the conceptacles are unisexual. **Meiosis** takes place at the time of differentiation of the gametes from the diploid protoplasts of the sex organs. The gametes are thus **haploid**. The haploid male gamete fuses with the haploid female gamete to form the zygote in which the **diploid** condition is again resorted. The sexual or gametophyte generations in the life cycle has been completely suppressed. There is no morphologically organised gametophyte plant. The extremely reduced gametophyte is solely represented by the gametes only. From this we conclude that *Sargassum* exhibits no alternation of morphologically organised generations. There is, however, a **cytological** alternation of generation from the diploid condition to the haploid at the time of gamete formation (meiosis) and the back to

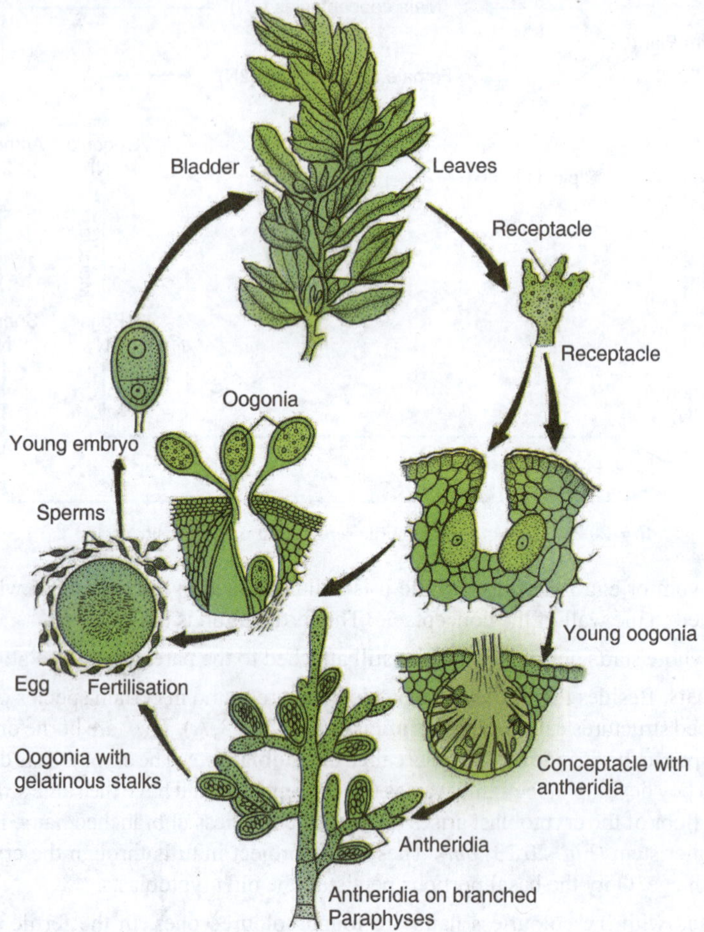

Fig. 20.29. *Sargassum sp.* Diagrammatic representation of the life-cycle. (Meiosis or reduction division takes place during gametogenesis).

the diploid condition at the time of zygote formation (fertilization). The two critical points in the life cycle thus are **meiosis** and **fertilization**.

Salient Features

1. The much branched, perennial diploid thallus apparently presents the appearance of a small, bushy angiosperm with its stem-like stipe, leaf blades and berry-like air bladders.
2. It is either cylindrical with the primary laterals arranged spirally and thus radially symmetrical or flattened with distichous arrangement of the laterals. In the latter case the thallus is bilaterally symmetrical.
3. The main axis is usually attached to the substratum by means of a well-developed holdfast with fibrous processes.
4. A few species occur in huge, free floating masses and multiply exclusively by the method of fragmentation.
5. The thallus is mainly composed of the much branched primary, long laterals.
6. The primary, long laterals generally bear the leafy shoot, secondary laterals of which the basal one is usually conspicuous and apparently subtends the branch system.

7. The primary and secondary laterals are sterile. The secondary laterals subtend axillary branch systems in the upper region of the thallus.
8. The lower members of the axillary branch system may be reduced to **air-bladders**. The subsequent ones are fertile and may be cylindrical to flattened.
9. These cylindrical, fertile tracts are called the **receptacles**. They are studded with conceptacles.
10. Growth in length takes place by means of a three sided apical cell situated in a funnel-shaped depression.
11. Asexual reproduction by the formation of spores is entirely absent—a feature unusual for the diploid, asexual or sporophyte plant.
12. It reproduces entirely by **fragmentation** and **sexual reproduction**.
13. Sexual reproduction is oogamous. The sex organs as in *Fucus*, are lodged in conceptacles which are confined to the receptacles.
14. Some species are **monoecious** and others **dioecious**.
15. The conceptacles are always unisexual.
16. The oogonia and antheridia are diploid structures—another unusual feature.
17. Meiosis takes place at the time of differentiation of the gametes.
18. The ripe oogonium has only one oosphere.
19. The oogonium is generally embedded in the wall of the conceptacle.
20. The antheridium produces 64 small, biflagellate sperms. The two flagella are of unequal length and inserted laterally.
21. Fertilization takes place when the oogonium is discharged but remains just outside the ostiole still attached to the wall of the conceptacle by means of a long, mucilaginous stalk.
22. The zygote germinates without undergoing any resting period and forms the diploid thallus.

Economic Importance. *Sargassum* is a source of alginates and alginic acid. According to Chauhan and Krishnamurthy (1971), *S. swartzii* yields a fairly high amount of algin of high viscosity.

Taxonomic Position

Division	:	**Phaeophyta** (Phaeophycophyta)
Class	:	**Cyclospore**
Order	:	**Fucales**
Family	:	**Sargassaceae**
Genus	:	*Sargassum*
Species	:	*cinereum*

Sargassum cinereum.

Origin and Affinities of brown algae

The affinities of this group are still obscure. From the absence of motile members in the vegetative state it appears that the brown algae are neither related to nor have evolved from any other algal group. In the organisation of the thallus, heterotrichous filament and methods of reproduction the browns begin where the Chlorophyta end. This suggests that probably the brown algae have arisen from the heterotrichous members of green algae such as the Chaetophorales and represent an entirely independent line of evolution which ended blindly. The basic objection against this view is the different pigment constitution of the green and brown algae. The other view is that both the green and brown algae originated long ago from some common ancestor along two parallel lines of development. This common ancestor is suggested to be a flagellate that has since become extent.

The occurrence of motile reproductive cells (zoospore and sperms) and their constant structure in the brown algae lend support to the view that brown algae may have had a flagellate ancestry. These flagellate progenitors may have been very much like the present-day unicellular Chrysophyceae or Xanthophyceae. Somewhat similar pigmentation, presence of flimmer hairs on one of the two unequal flagella and absence of starch as reserve food further indicate that either the browns have originated directly from such unicellular forms or have arisen at an early stage from a common stock and then evolved independently. On the basis of this view the original browns must have been unicells. Such forms among the brown algae, however, are no longer extant. Perhaps these unicell progenitors disappeared long ago as have the other connecting transitional links when the salt contents of the primeval oceans increased.

The brown unicell progenitor exploited the particular kind of pigmentation and metabolism which proved satisfactory in the marine environments. It gradually attained a fairly high degree of specialization of the vegetative plant body which is the present-day browns is generally differentiated into the holdfast, stipe and the blade with floats. In fact such a plant body in its vegetative organisation is not far removed from a primitive vascular plant. However, it lacks xylem for conduction and mechanical support. It has no cuticle for protection against excessive transpiration. It lacks the protective jacket layer of sterile cells around the multicellular sex organs. Because of the above mentioned handicaps the browns failed to give rise to any higher group of plants.

The brown algae posses distinctive common characters such as the structure of the motile cells, pigmentation of chromatophores and products of metabolism. These suggest that they all had a common ancestor. However it is very difficult to establish phylogenetic relationships between the three classes (Isogeneratae, Heterogeneratae and Cyclosporeae) with three different kinds of life histories. The widely accepted view is that two diverging lines of development were established early in the evolution of brown algae, (*i*) one with homologous alternation of generations and (*ii*) the second with heterologous. In both these lines of development there was progressive evolution of sexual reproduction from isogamy to oogamy. It was accompanied by increasing elaboration of thallus structure. This view fails to explain the origin of Cyclosporae in which there is no free-living haploid generation. It is restricted to the gametes only. Kniep (1928) and Svedelius (1921 and 1928) suggested that the Cyclosporeae evolved by a gradual reduction of the gametophyte to a single-celled structure. This is supported by the occurrence of single-celled female gametophytes in certain Laminariales. According to this view, Cyclosporeae will be considered to be highly specialised forms occupying the top of the heteromorphic series.

Smith (1938) put forth a different hypothesis. He postulated that the Cyclospore are evolved by dropping off the multicellular gametophyte and not by its gradual reduction to a single-celled structure. This was made possible by the zoospores of unilocular sporangia functioning directly as gametes instead of germinating to give rise to the gametophytes. This viewpoint is supported by the fact that in rare cases the zoospores from unilocular sporangia function actually as gametes. From the electron microscope investigations of the sperms of *Fucus* and several species of Cysteseiraceae Manton and her co-workers (1957 and 1959), concluded that *Fucus* is a highly specialised form and is not primitive.

QUESTIONS

ESSAY TYPE

SARGASSUM

1. Give an account of structure and reproduction of *Sargassum*. (*Awadh, 1998*)
2. Enumerate distinguishing features of phaeophyceae. Describe male and female conceptacles of *Sargassum* with suitable diagrams. (*Awadh, 1993*)
3. With the help of suitable diagrams, describe the life cycle of *Sargassum*. (*Kanpur, 1997; Gorakhpur, 1990; Bundelkhand, 1998*)
4. Give an account of habitat and sexual reproduction in *Sargassum*. (*Kanpur, 1999*)
5. (*a*) Describe the sexual reproduction of *Sargassum*. (*Agra, 1993*)
   (*b*) Give a detailed account of sexual reproduction in *Sargassum*. (*Rohilkhand, 2005*)
6. Give a brief account of structure and distribution of *Sargassum*. (*Agra, 1995*)
7. Give an account of the structure of the thallus and mode of reproduction in *Sargassum*. (*Poorvanchal, 2007*)
8. Describe the post fertilisation changes in the life history of *Sargassum*. (*Kanpur M.Sc., 2002; Bhagalpur, 1990*)
9. Describe in detail the nuclear changes and alternation of generations in *Sargassum*.
10. Give diagrammatic representation of the life cycle of *Sargassum* illustrating the relative lengths of the haploid and diploid generations.
11. State the salient features of the lie cycle of *Sargassum*.
12. Describe sexual reproduction with the help of self-explanatory sketches, in *Sargassum*. Point out its differences from *Fucus*. (*Vir Kumar Singh, 1993*)
13. *Sargassum* exemplifies a reduced life-cycle. Explain. (*Madras, 1997*)
14. Describe the habitat, thallus structure and reproductive methods in *Sargassum*. Compare its habitat with that of *Gracilaria*. (*Manomaniam Sundarnar, 1995*)
15. Give a self-explanatory diagram of *Sargassum* plant and its graphic life cycle. (*Bhagalpur, 1993*)
16. Discuss the reproduction in *Sargassum*. (*Ranchi, 1991*)

FUCUS

17. Describe the structure and reproduction in *Fucus*. (*Allahabad, 1995; Magadh, 1996*)
18. Give an account of the occurrence, thallus structure and reproduction in *Fucus*. (*Allahabad, 1998*)
19. Give an account of the habit and habitat of *Fucus* and explain how its form and structure is adapted to the environmental conditions.
20. Give an account of the life history of *Fucus* and discuss the question of alternation of generations in the genus.
21. Give a diagrammatic account of the life cycle of *Fucus* illustrating the relative length of the haploid and diploid generations.
22. Write an illustrated account of the sexual reproduction in *Fucus*. Describe sexual reproduction in *Fucus*. (*Magadh, 1990, 1994*)
23. Give an account of the habit, structure of thallus and mode of reproduction in *Fucus*.
24. Write an account of the life history of *Fucus*. (*Vikram Singh, 1994, 98; Bihar, 1991*)
25. Give a brief account of the life history of *Fucus*. (*Vir Kumar Singh, 1990; Magadh, 1990*)

26. Give an account of vegetative structures of *Fucus* and *Sargassum*. (*Vir Kumar Singh, 1997*)

27. Give illustrated account of the thallus structure of *Fucus*. (*Vir Kumar Singh, 1998*)

28. Describe in detail the thallus structure and method of sexual reproduction in *Sargassum*. What is the importance of alternation of generation in its life cycle? *Poorvanchal, 2003*)

29. Write a detailed note on mode of reproduction and isomorphic alternation of generation in *Sargassum*? (*Kanpur M.Sc., 2004*)

30. Discuss the range of sexual reproduction in different genera of phaeophyceae studied by you. (*Kanput M.Sc., 2005*)

SHORT ANSWER TYPE

31. Write short notes on

(*a*) Male conceptacle of *Sargassum*. (*Awadh, 1991, 1996; Kanpur, 1995*)

(*b*) Sexual reproduction in *Sargassum*. (*Awadh, 1994*)

(*c*) Thallus structure of *Sargassum*. (*Awadh, 1997; Agra, 1991*)

(*d*) Conceptacles of *Sargassum*. (*Awadh, 1998; Purvanchal, 1996, 1999; Lucknow, 1999; Gorakhpur, 1991, 1995; Rohilkhand MSc, 2006;*)

(*e*) Structure of conceptacle. (*U.P. College, 1995*)

(*f*) Sex organs of *Sargassum*. (*Lucknow, 2000; Bundelkhand, 1997*)

(*g*) Male and female conceptacles in *Sargassum*. (*Kumaon, 2000*)

(*h*) Cryptoblamata.

(*i*) Structure and development of sex organs in *Fucus*. (*Agra, 1947*)

(*j*) Habit and habitat of *Fucus*. (*Agra, 1959*)

(*k*) Conceptacles in *Fucus* and *Sargassum*. (*Vir Kumar Singh, 1998*)

(*l*) Conceptacles and Receptacles. (*Meerut, 1998*)

(*m*) Vegetative thallus of *Fucus*. (*Bhagalpur, 1991*)

(*n*) Cryptoblasts. (*Allahabad, 2001*)

(*o*) Conceptacles of *Fucus* (*Allahabad, 2002, 2006*)

(*p*) Growth in *Fucus*. (*Allahabad, 2004, 2005*)

32. Where does reduction division occur in *Sargassum*, *Volvox* and *Ectocarpus*? (*Awadh, 1991*)

33. What are the functions of air bladders in *Sargassum*? (*Bundelkhand, 1989*)

34. Draw labelled diagrams of the thallus and give the most important characteristic and classification of *Sargassum*. (*Awadh, 1992, 99; U.P. College, 1998*)

35. With the help of labelled diagrams, show the structure of *Sargassum*. (*Awadh, 1994*)

36. Draw well labelled diagrams of the following :

(*a*) V.S. male conceptacle of *Sargassum*. (*Purvanchal, 1997; Lucknow, 1992, 1999*)

(*b*) V.S. female conceptacle of *Sargassum*/mature female conceptacle. (*Purvanchal, 2000; Lucknow, 1996; Gorakhpur, 1996; Kanpur, 2006*)

(*c*) Male and female conceptacles of *Sargassum*. (*Gorakhpur, 1992; Kanpur, 2003*)

(*d*) Conceptacles of *Sargassum*. (*Lucknow, 1993*)

(*e*) Male conceptacle of *Sargassum*. (*Lucknow, 1994*)

(*f*) External morphology of *Sargassum* thallus. (*Gorakhpur, 1994; Poorvanchal, 2006*) (*g*) Diagrammatic life cycle of *Sargassum*/ *Fucus*.

(*h*) Diagrammatic life cycle of *Ectocarpus*. (*Kanpur, 2005*)

37. Distinguish between :

(*a*) *Sargassum* and *Chara*. (*U.P. College, 1999*)

(*b*) *Sargassum* and *Fucus*.

(*c*) *Ectocarpus* and *Sargassum*.

OBJECTIVE TYPE

38. Select the correct answer :

(*i*) The receptacles are found in

(*a*) *Ectocarpus* (*b*) *Sargassum*
(*c*) *Polysiphonia* (*d*) *Nemalion*

(*ii*) The life cycle of *Sargassum* is

(*a*) Haplontic
(*b*) Diplontic
(*c*) Haplo diplobiontic
(*d*) None of the above

(*iii*) Which of the following is typically a marine alga?

(*a*) *Ulothrix*
(*b*) *Spirogyra*
(*c*) *Chlamydomonas*
(*d*) *Sargassum*

(*iv*) The alga commonly called *Rockweed* is

(*a*) *Ectocarpus* (*b*) *Fucus*
(*c*) *Sargassum* (*d*) *Laminaria*

(*v*) *Fucus* plant is

(*a*) Haploid (*b*) Diploid
(*c*) Triploid (*d*) Tetraploid

(*vi*) Reduction division in *Fucus* takes place during

(*a*) Germination of zygote
(*b*) Growth of thallus
(*c*) Formation of sex organs
(*d*) Gametogenesis

(*vii*) The spermatozoid in *Fucus* is a

(*a*) Minute, slightly elongated pear shaped structure
(*b*) Large terminally uniflagellate structure
(*c*) Minute ovoid biflagellate cells
(*d*) Large pear-shaped terminally biflagellate cell

(*viii*) The mature oogonium in *Fucus* contains

(*a*) One egg (*b*) Two eggs
(*c*) Four eggs (*d*) Eight eggs

(*ix*) Which of the following is called the Gulf weed?

(*a*) *Ectocarpus* (*b*) *Sargassum*
(*c*) *Fucus* (*d*) *Nemalion*

(*x*) In *Sargassum* growth of the thallus is by the activity of a

(*a*) A three sided apical cell
(*b*) A four sided apical cell
(*c*) An intercalary meristem
(*d*) All cells of the thallus

(*xi*) In *Sargassum* the part of the *oogonium* that forms the pseudostalk is called

(*a*) Exochite (*b*) Mesochite
(*c*) Endochite (*d*) Stalk cell

(*xii*) In which of the following alga the *oogonia* are raised on gelatinous stalks out of the conceptales?

(*a*) *Ectocarpus*
(*b*) *Sargassum*
(*c*) *Fucus*
(*d*) *Polysiphonia*

(*xiii*) Sterile flask-shaped cavities present in the leaves of *Sargassum* are called

(*a*) Conceptacles
(*b*) Cryptoblasts
(*c*) Cryptoblamata
(*d*) Cryptostomata

(*xiv*) The photosynthetic region in the thallus of *Sargassum* is

(*a*) Epidermis (*b*) Cortex
(*c*) Medulla (*d*) Cuticle

(*xv*) The product of photosynthesis is *Sargassum* is
(*a*) Starch (*b*) Protein
(*c*) Manitol (*d*) Glycerol

(*xvi*) *Sargassum* conceptacles occur
(*a*) In the reciptacles
(*b*) In the holdfast region
(*c*) On the stem
(*d*) On the leaf

(*xvii*) In the eight-celled nucleated oogonium of *Sargassum*, the number of diploid nuclei that develop into eggs are
(*a*) Only one
(*b*) Only four
(*c*) Less than four
(*d*) All the eight nuclei

(*xviii*) In *Sargassum* the fertilization of the female gametes takes place when the egg is
(*a*) In the oogonium
(*b*) In water
(*c*) Outside the conceptacle still attached to the plant
(*d*) Inside the conceptacle still attached to the plant

(*xix*) The name of gulfweed is
(*a*) *Fucus*
(*b*) *Sargassum*
(*c*) *Polysiphonia*
(*d*) *Ectocarpus*

Polysiphonia sp.

CHAPTER 21

RHODOPHYTA (Rhodophycophyta) — Red Algae

- Distribution and General Features
- Thallus
- Cell Structure
- Glass : Rhodophyceae
- Subclassss : Bangioideae (Bangiophycidae)
- Genus : Porphyra

Distribution and General Features

The members of this division are popularly called the **red algae**. They are perhaps the most showy of all the seaweeds and are predominantly marine. They occur in all the oceans but are a few in number in the polar seas. They reach their greatest development in the deeper and warmer tropical and subtropical waters. Some 5250 species and 831 genera of red algae are known. There are, however, many fresh water forms (about 200 species) many of which occur in cold, fast flowing waters in the tropics *e.g. Batrachospermum, Lemanea, Thorea,* and *Hildenbrandia* while others *e.g. Compsopogen* and *Asterocystis,* grow in stagnant water bodies. The marine forms are recognised by their bright pink colour caused by the biliprotein pigments

R-phycoerythrin and R-phycocyanin. Other shades such as red, purple and violet are by no means rare. The fresh water forms, however are, of bluish green colour. The majority of the marine forms occur from low tide marks to great depths as 100 metres beneath the surface of the sea. They occur attached to rocks or other solid substratum by holdfasts. Many species grow on other red algae, brown algae or marine green algae either as epiphytes or parasites.

The **epiphytic** species have normal thallus and pigmentation whereas parasitic species show a great reduction in their form and pigment constitution. Moreover **parasitic** species exhibit considerable specificity to their hosts, *e.g. Polysiphonia lanosa* (*host-Ascophyllum nodosum*), *P. violacea* (*host-Fucus vesiculosus*) , *Ceratocolax hartzii* (*host-Phyllophora membranifolia*), *Ceramium condicola* (*host-Codium fragile*), *Gelidiocolax suhriae* (*host-Gelidium*), *Pterocladiophila hemispherica (host-Pterocladia lucida)* and *Polysiphonia fastigata* is known as a *semiparasite* on *Ascophyllum nodosum,* a brown algae. Some **epiphytic** forms growing on other red algae include *Choreocolax, Acrochaetium* and *Rhodochorton* whereas *Colaconema* occurs as an endophyte within the thalli of other algal members. Many species are **epizoic** and grow on sponges, hydroids and molluscs. A few, however, grow in the littoral zone in rock pools which do not dry out at low tide.

Thallus of red algae in general is more delicate than the browns. They are thus poorly adapted to the turbulent habitat of the brown algae. Instead they flourish and occur in abundance at great depths where other plants do not occur in abundance, why? The sunlight as it penetrates water, portions of spectrum such as red, orange, yellow and green light rays which are of short wave lengths are filtered out. Only the blue and violet rays of high wavelength remain and penetrate to great depths. The green pigment chlorophyll cannot trap these light rays of high wavelength and the green plants are thus unable to carry on photosynthesis at these depths. On the other hand the red pigment **r-phycoerythrin** and a blue pigment **r-phycocyanin** which are characteristic pigments of all the red algae, can utilize wavelengths of light (blue and violet rays) not absorbed by chlorophyll. The light energy so trapped is then transferred to chlorophyll *a* for use in photosynthesis. This enables red algae to grow at greater depths than other plants. Some of them grow as much as 300 feet beneath the surface of water. The red algae owe their bright pink colour to these two biliprotein pigments which occur in their chromatophores. In addition to these two phycochromoproteins the other pigments are chlorophyll *a, d,* carotenes and xanthophylls. One of the xanthophylls is *lutein.* Chlorophyll *d* is a minor component in the red algae. In fact it has not been detected in Bangiophycidae. The red phycoerythrin and blue phycocyanin are chemically different from the similarly named pigments of the Cyanophyceae.

Thallus

Motile forms are unknown. In fact they are the only eukaryotic algae which produce no motile stage. Even the reproductive cells are non-flagellated. Unicellular and colonial forms are rare. *Porphyridium* is the common unicellular red alga. With the exception of two genera, all the red algae have a multicellular macroscopic thallus which is very diverse in form. It may be filamentous (*Goniotrichum*), ribbon-like or plate-like parenchymatous (*Porphyra*). The filamentous thallus is the most beautiful of all the algae. It is mostly branched and tufted. The simple filament which is so commonly found in the fresh water green algae, is met with very rarely. The apparently simple filament of red algae, in reality, consists of a system of a varying number of parallel filaments. These in some forms may have an additional covering of corticating filaments. In fact the thallus of red algae is built on the following two architectural plans :-

1. *Uniaxial.* The thallus in this case consists of an axial filament with or without cortication. The axial filament bears richly branched laterals. *Acrochaetium* is the example of this category. The laterals in some (*Grimelia*) may be organised to form a uniaxial pseudoparenchymatous thallus.

2. *Multiaxial.* The thallus consists of a tuft of axial filaments each bearing laterals which radiate out to the margin. The central filaments together with the branches form a pseudoparenchymatous thallus.

The filamentous forms have beautiful plant bodies with feathery branches more lacier and delicate than the robust brown algae. In fact *Delesseria* and *Plocamium* are a real delight to behold. The thalli usually are attached to rocks or solid substratum in water by holdfasts. Rarely do they occur in a free-floating state. The stony thalli of some calcareous red algae such as *Corallina* and *Lithothamnion* are heavily impregnated with lime and form the coral reefs.

Except the Bangioideae the growth of the thallus is **apical**. The only exceptions in the Florideae are Corallinaceae and Delesseriaceae in which it is intercalary. Branching in red algae is chiefly **monopodial**. In fact, on the basis of this uniformity in morphology (monopodial branching) and reproduction, some algologists suggest monophyletic origin of the red algae.

Cell Structure

The cells of red algae show a typical eukaryotic structure. The cell wall consists of two layers. The outer is made up of pectic materials together with polysulphate esters and the inner of cellulose. The cellulose microfibrils are scattered through a granular matrix. Mostly there is an outer coating of mucilage but in many genera the outer pectin layer is impregnated with calcium carbonate.

The cell protoplast, except in a few red algae, has a tonoplast bound **central vacuole**. The cytoplasm in them forms a thin layer next to the cell wall. In the orders Bangiales, Nemalionales, Cryptonemales and Gigartinales there is usually a single **nucleus** which is flat when parietal in position but spherical if located in the centre of the cell. It has a well-defined nuclear envelope, a conspicuous nucleolus and divides by mitosis. In other orders of red algae the cells tend to be multinucleate. In *Griffthsia* the vegetative cells may contain several thousand nuclei (3000 to 4000) in each cell. Between the cell wall and the cytoplasm is present the **cell membrane**. The cytoplasm contains one or more chromatophores. Usually the primitive red algae (Bangiales and some Nemalionales) have a single, large stellate (lobed) chromatophore in the centre of the cell (Fig. 21.3). It has a centrally located dense proteinaceous body often termed the **pyrenoid**. The so-called pyrenoid lacks a starch sheath typical of the pyrenoids of green algae. It is thus described as naked. The advanced members of red algae normally have more than one chromatophores in each cell (*Polysiphonia*). They are parietal in position and are devoid of pyrenoids. In form they are usually discoid. Parietal plate-like or band-shaped (*Furcellaria*) chromatophores have also been recorded. Though diverse in its gross form, the chromatophore of red algae shows remarkable uniformity in its fine structure. It is a double membrane bound structure containing a homogeneous matrix, the **stroma**. The stroma is traversed by a number of widely separated single thylakoids or discs which are not organised into bands (Fig. 23.3). The photosynthetic pigments which are located in the chromatophores include chlorophyll *a*, chlorophyll *d*, *a*-and -carotene, xanthophylls (lutein, teraxanthin, zeaxanthin, violaxanthin) and biliproteins (r-phycoerythrin and r-phycocyanin). Phycoerythrin is typically the characteristic and predominant accessory pigment of red algae. It often completely marks the presence of green chlorophyll *a* and is thus responsible for imparting red colouration to the species growing in deep waters or in shaded situations. Bright pink colour of the thallus is caused by predominance of both the biliprotein pigments (r-phycorythrin and r-phycocyanin). Other shades such as purple, violet brownish, yellow, greenish and the like often exhibited in the species of intertidal distribution are the result of photodestruction of phycoerythrin. Crosett *et al* (1965) termed this capability of the algal population to change the proportion of their pigments in response to the different intensities of incident light as *Chromatic adaptation*.

Electron microscopic investigations have revealed the presence of dictyosomes, endoplasmic reticulum and mitochondria (Fig. 21.3). Most of the ribosomes lie free in the cytoplasm. The reserve food occurs in the form of small grains of a peculiar solid carbohydrate called **Floridian starch**. It is a glycogen-like polysaccharide which turns yellow or brown with iodine. The floridian granules occur scattered in the cell cytoplasm outside the chromatophore (Fig. 21.3). Aggregates of phycobiliproteins in the form of spherical granules termed **phycobilisomes** are located on the outer surface of the thylakoids.

Bouck (1962) reported the occurrence of spherical or pearshaped gland cells in *Lomentoria baileyana*. (Fig. 21.1). They are found in the inner part of the thallus and have denser contents than the adjacent vegetative cells. Their function is not known at present.

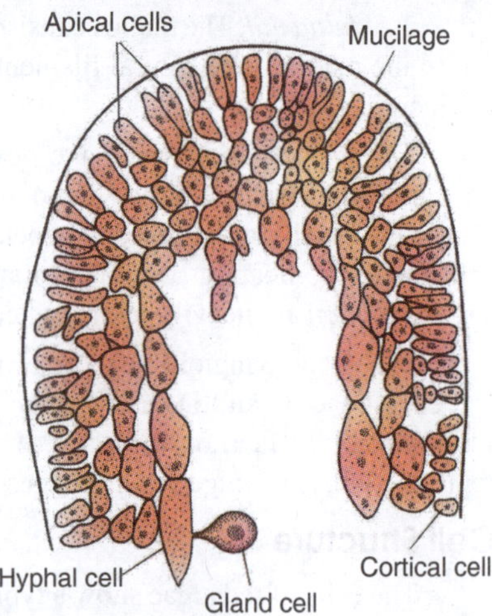

Fig. 21.1. *Rhodophyceae*. L.S. through the apical region of the thallus of *Lomentaria baileyana* showing gland cell.

The higher orders of red algae are unique in having **pits** on the cell walls of adjacent cells. These pits assist in maintaining protoplasmic continuity between these cells. According to Ramus (1971), a pit connection is neither a "pit" nor an intercellular connection between the adjacent cells. It is a lens-shaped plug held within a septal aperture. It consists of a circular septal aperture and a lens-shaped plug fitted into the aperture. On each side of the plug is membrane-bounded plug cap.

Porphyra. *Porphyridium.* *Goniotrichum.* *Arcochaetium.*

Delesseria. *Lithothamnion.* *Griffithsia.* *Lomentoria.*

Reproduction

Reproduction takes place both by asexual and sexual means. The spores are non-motile and of diverse types such as monospores, neutral spores, carpospores or meiospores (bispores, tetraspores or polyspores).

Sexual reproduction (Fig. 21.2) is complex and of a highly specialised type by which the red algae are distinguished from other algae. Some species are monoecious and others dioecious. In the latter case the male and female plants are often identical. The most distinguishing feature is the nature of the male gamete. It is non-flagellated and produced singly in the male sex organ called the **spermatangium** (A). The non-motile male gametes are called the **spermatia**. They are carried passively by the action of water currents and lodged against the terminal narrow elongated receptive part (trichogyne) of the female sex organ called the **carpogonium** (B). The carpogonium in *Nemalion*, one of the simplest red alga, consists of a swollen basal part containing the female nucleus and an elongated terminal hair-like structure called the **trichogyne**. During fertilization the spermatium discharges its contents into the latter. The spermatium nucleus migrates downwards to fuse with the egg nucleus in the basal swollen part of the carpogonium. Soon after fertilisation the trichogyne shrivels (C). The zygote nucleus undergoes meiosis. The haploid daughter nuclei pass through a series of mitotic divisions to yield a mass of short haploid filaments called the **gonimoblasts**. The terminal cells of these filaments mature into **carposporangia** each containing a carpospore. The carpospore germinates directly into a new plant. The entire structure consisting of carpogonium, loosely organised gonimoblasts and carposporangia collectively forms the **cystocarp**. Carpogonium in the advanced members of red algae is produced terminally on a special lateral branch called the **procarp** or **carpogonial filament**. The latter consists of a few cells and is developed by the division of a pericentral parent mother cell called the **supporting cell**. Associated with the procarp is a special cell known as the **auxiliary cell**. The zygote does not grow directly into the sporophyte plant but it forms directly or indirectly a number of spores called the **carpospores**. The carpospore in the higher forms is diploid and on germination produces a sporophyte plant. It is called the **tetrasporophyte**.

The division Rhodophyta includes a single class, the **Rhodophyceae**.

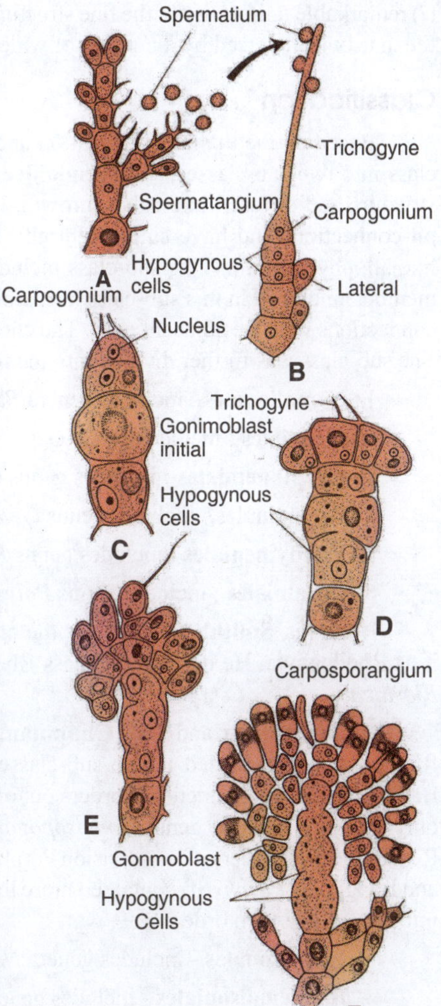

Fig. 21.2 (A-F). *Rhodophyceae*. (A) Sexual reproduction in *Nemalion multifidum*. (A), Spermatangial branch; (B), branch bearing carpogonium; (C), fertilized carpogonium; (D), development of gonimoblast initials (E), later stage of D; (F), mature gonimoblast filament bearing carposporangia.

CLASS : RHODOPHYCEAE

The chief characteristic features of the class are : (*a*) complete absence of motile stages in the life cycle, (*b*) presence in the cells of phycobilin pigments r-phycoerythrin, r-phycocyanin, together with chlorophyll *d* and xanthophyll teraxanthin in addition to the usual pigments chlorophyll *a*, and carotenes and the customary xanthophylls lutein, zeaxanthin and others, (*c*) unique sexual reproductive system which is oogamous, highly complex and depends on the passive dispersal of the male gamete (spermatium) to the trichogyne of the female sex organ called the **carpogonium**, (*d*) occurrence of distinct **post-fertilization changes** not found in any other algal division, (*e*) accumulation of reserve carbohydrate food in the form of grains of different sizes (in the cell-cytoplasm outside the chromatophore) of a peculiar solid carbohydrate known as **floridean starch** and galactoside floridosides, (*f*) presence of polysulphate esters of carbohydrate in addition to cellulose and pectin in the cell wall, (*g*) another unique feature of multicellular red algae is the occurrence of modified areas known as **pits** on the cell walls which have pit connections with the adjacent cells (*h*) occurrence of **gland cells** in the thallus of certain members of the group, and (*i*) remarkable uniformity of the fine structure of a double membrane bound chromatophore in which the matrix is traversed by a number of widely separated, single discs or thylakoids.

Classification

Based on the thallus organisation and post fertilization changes, **Fritsch** (1935) divided the class into two sub-classes: (I) **Bangioideae** : which included primitive forms with simple thallus structure and diffused intercalary growth. Usually the cells of the members of this sub-class lack pit-connections and have single centrally located stellate chromatophores which have pigments like allophycocyanin. This sub-class includes only a single order **Bangiales;** (II) **Florideae** : The members included in this sub-class have uni- or multiaxial thalli with distinct apical growth and pit connections with the adjacent cells. The chromatophores are discoid and several in number per cell. The sub-class was further divided into the following six orders :

1. **Nemalionales** : includes genera *Batrachospermum, Chantransia, Thorea.*
2. **Gelidiales** : includes genus *Gelidium*
3. **Cryptonemiales** : includes genus *Cryptonemia*
4. **Gigartinales** : includes genus *Gigartina*
5. **Rhodymeniales** : includes genus *Rhodymenea*
6. **Ceramiales** : includes genus *Polysiphonia*

However, **Smith** (1955) raised the class Rhodophyceae to the rank of Division and named it as Rhodophyta. He divided the class Rhodophyceae into two sub-classes viz. *Bangioideae* and *Florideae.*

V.J. Chapman and **D.J. Chapman**, (1962) placed the class Rhodophyceae into Phylum Rhodophyta and divided it into sub-classes **Bangiophycidae** and **Florideophycidae**. Sub-class Bangiophycidae contained five orders out of which, two orders viz. Compsopoganales included the only single fresh water genus *Compsopogon* and Rhodochaetales has the sole genus *Rhodochaete.* Remaining three orders of sub-division Porphyridinales (*Porphyridium*), Goniotrichales (*Asterocystis*) and Bangiales (*Porphyra*) contained more than one genus. Sub-class Floridiophycideae was divided into following seven orders :

1. **Nemalionales** - includes genera *Nemalion, Batrachospermum*
2. **Bonnemaisoniales** - includes genera *Asparagopsis, Scinaia.*
3. **Gelidiales** - includes genera *Gelidium, Pterocladia*
4. **Cryptonemiales** - includes genera *Dudresnaya, Hildenbrandia, Lithothamnion*
5. **Gigartinales** - includes genera *Gracilaria, Furcellarea, Gigartina*
6. **Rhodymeniales** - includes genera *Rhodymenia, Lomentaria*
7. **Ceramiales** - includes genera *Polysiphonia, Callithamnion.*

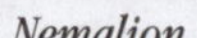

Nemalion.

Batrachospermum.

Asparagopsis.

Scinaia. *Gelidium.* *Pterocladia.* *Dudresnaya.*

Gelidium. *Pterocladia.* *Furcilaria.* *Rhodymenia.*

Round (1973) followed the scheme proposed by Chapman and Chapman (1962) and placed the class Rhodophyceae into Phylum Rhodophyta which was further divided into two sub-classes, **Bangiophycidae** and **Florideophycedae**. According to him sub-class Bangiophycidae contained a single order Bangiales whereas sub-class Florideophycidae included six orders *viz.* Nemalionales, Gelidiales, Cryptonemiales, Gigartinales, Rhodymeniales and Ceramiales. According to leading phycologists sub-class Bangioideae includes a greater proportion of fresh water forms than the sub-class Florideae.

Comparison between sub-class Bangioideae and sub-class Florideae.

Sub-class : Bangioideae	Sub-class : Florideae
1. Members may be fresh water, marine or terrestrial.	1. Members are mostly marine.
2. Thallus simple unicellular or multicellular.	2. Thallus relatively complex, uniaxial or polyaxial.
3. Cells always uninucleate.	3. Cells multinucleate the number of nuclei variable and may some times range between 3000-4000.
4. Growth through diffused intercalary cells.	4. Growth through a single apical cell.
5. Each cell has a single centrally located stellate chromatophore.	5. Each cell has many discoid chromatophores, present along the periphery.
6. Chromatophore has a single pyrenoid.	6. Pyrenoids absent in chromatophores
7. Pit connections absent.	7. Pit connections present between the adjacent cells.
8. Motile cells absent.	8. Motile cells absent.
9. Zygote divides meiotically to form haploid carpospores which on germination give rise to alternate dissimilar vegetative phase.	9. The zygote may divide to from haploid or diploid gonimoblast filament, the terminal cell of which develops into a carposporangium having a single carpospore.

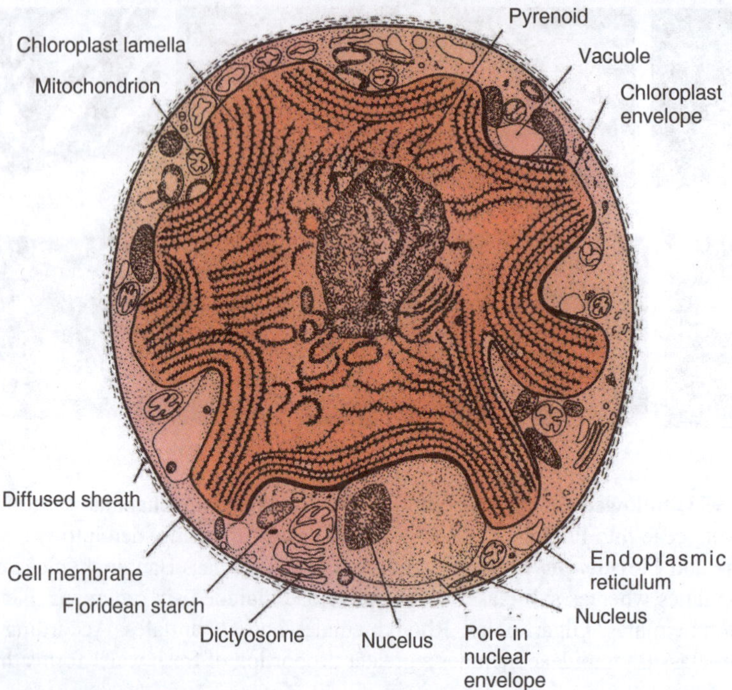

Fig. 21.3. *Rhodophyceae*. Diagrammatic representation of fine structure of a unicellular red alga *Porphyridium*.

SUBCLASS : BANGIOIDEAE (BANGIOPHYCIDAE)

General Characters. The sub-class also receives the name **Protoflorideae**. It includes forms with a simple thallus which may be unicellular (*Porphyridium*), filamentous (*Gonitrichum*) or sheet-like parenchymatous (*Bangia* and *Porphyra*). The filament may be simple or branched. The cells are uninucleate and in most genera contain, a single, axile stellate chromatophore with a central pyrenoid. Growth in the multicellular forms is diffuse (intercalary). Usually there are no pit connections between the adjacent cells. Recently these have been observed in one genus.

Cell structure. The fine structure of the *unicell of Porphyridium* has been investigated by some workers (Fig. 21.3). The cell envelope has been reported to consist of a diffuse **sheath** having a fibrillar appearance. Internal to it is the **cell membrane**. Further in is the granular cytoplasm in which lie embedded all the structures characteristic of an eucaryotic cell. There is the large, axile, stellate chromatophore (**rhodoplast**). It has a prominent centrally placed pyrenoid without a starch sheath.

The irregularly shaped rhodoplast with extensions in various directions consists of a granular matrix bounded by a double membraned envelope. The matrix is traversed by photosynthetic lamellae or thylakoids which are more numerous at the periphery especially the extensions. The thylakoids run parallel to one another and have a uniform space in between. Small granules are located on the outer side of each lamella. These are termed the **phycobilosomes**. They contain either phycoerythrin or phycocyanin. The phycobilosomes containing phycoerythrin are spherical and those containing phycocyanin are discoid. The single nucleus of the unicell is eccentric in position. It is bounded by a double membraned nuclear envelope which has pores. The nucleus is located immediately adjacent to the rhodoplast. The existence of mitochondria in the cytoplasm has been demonstrated by Spur, Dougharty and Jones (1964) and Gnatt and Conti (1965). The mitochondrion is a double membraned sac with tubular cristae arising from the inner membrane. The dictyosomes and endoplasmic reticulum in the form of tubules and vesicles run parallel to the cell membrane. Most of the ribosomes lie free in the cytoplasm. Floridean starch grains of various sizes occur randomly scattered in the granular cytoplasm. Vacuoles, each bounded by a single membrane, the **tonoplast** are found in the cytoplasm between the cell organelles and the starch grains. The cell in the filamentous Bangioideae has several parietal disc-shaped chromatophores which are devoid of pyrenoids.

Asexual reproduction in the unicellular forms takes place by the vegetative method of cell division called **fission.** The parent cell simply divides into two daughter cells. In the multicellular forms asexual reproduction takes place by means of asexual spores (mitospores) of two types, **monospores** and **neutral spores**. The monospores are produced singly in special cells smaller than the vegetative cells called **monosporangia** (*Kyliniella, Rhodochaete*). The protoplast of the monosporangium forms a single **monospore**. The neutral spores are formed by direct transformation of vegetative cells into spores (*Porphyra, Bangia*).

Sexual reproduction. It is oogamous. The male sex organ is called a **spermatangium**. The spermatangia develop directly from the vegetative cells by repeated division until 64 or 128 small cells called spermatangia are formed. The walls of spermatangia gelatinise, each liberating a small uninucleate non-motile protoplast which functions as a male gamete called **spermatium**. The female sex organ is called a **carpogonium**. It is a unicellular vegetative cell which lacks a trichogyne or has a short one. The non-motile spermatium is carried passively by water currents to and fuse with the egg nucleus in the carpogonium to form a zygote. The latter divides directly into a number of **carpospores**. Meiosis is said to occur at the time of carpospore formation. The liberated carpospore germinates to form a new thallus.

Classification of Bangioideae. The sub-class comprises 15 genera and about 70 species. All are placed in a single order, **Bangiales**. On the basis of thallus structure the order Bangiales is split into three families, which include the unicellular, filamentous and parenchymatous forms, respectively. Skuja (1939) suggested the rank of an order for each of these families. *Porphyra* is usually taken as an example of the order Bangiales.

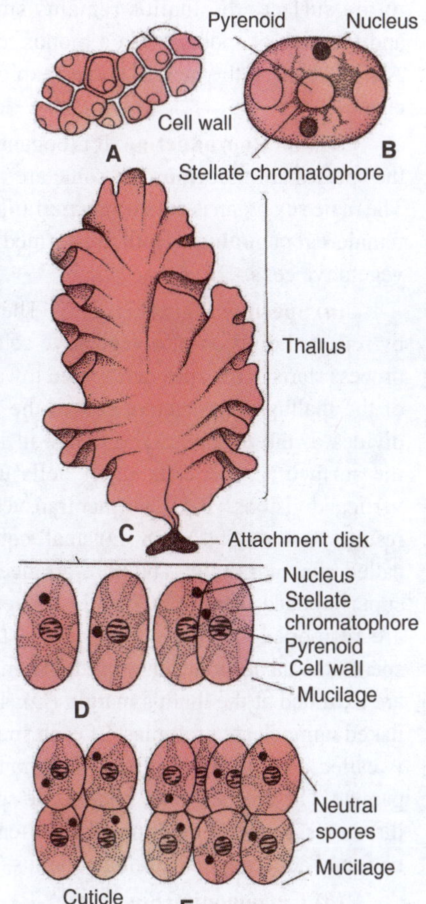

Fig. 21.4 (A-E). *Bangiales.* (A) Part of a stratum showing habit of *Porphyridium cruentum*; (B), An old unicell of the latter showing structure; (C), Thallus of *Porphyra* with an attachment disc; (D), T.S. thallus showing structure; (E), T.S. thallus of C showing neutral spores (A-B after Geitler and D after Ishikawa).

GENUS : PORPHYRA

Distribution. It is a common marine red alga found in the intertidal zone on the rocky seashores. It occurs on both the coasts, Atlantic and Pacific of North America. Some species grow attached to rocks (lithophytes) and others are epiphytic.

Thallus (Fig. 21.4 C). The plant body is a plate-like or sheet-like parenchymatous blade apparently resembling that of a green marine alga *Ulva*. The margin of the blade ranges from, smooth wavy to greatly convoluted. The thallus is often unbranched and attached to the substratum by a small, basal disc or cushion-like holdfast. Depending on the species the blade-like thallus is one to two cells thick. Frequently it is one cell layer thick.

Cell structure (Fig. 21.4 D). The cells are elongated. They are cubical or ellipsoidal in form and lie embedded perpendicular to the thallus surface in the tough gelatinous matrix derived from the cell walls. The outer walls of the cells are strongly thickened and covered with **cuticle**. The cells are uninucleate. Each cell has s single, large, axile stellate chromatophore with a centrally located pyrenoid. In certain species there are two chromatophores in each cell. The nucleus lies adjacent to the chromatophores but between the adjacent cells. The growth is intercalary. It takes place by the division of cells at intervals irrespective of their position.

Asexual Reproduction (Fig. 21.4 E). It takes place by means of asexual spores formed by direct transformation of vegetative cells. These are known as the **neutral spores**. The process of conversion mostly commences at the thallus apex. The spores are formed by the anticlinal division of vegetative cells. Since the division is in a plane perpendicular to the surface, the thallus remains single layered and the neutral spores lie in a monostromatic layer. After liberation the neutral spores exhibit amoeboid changes.

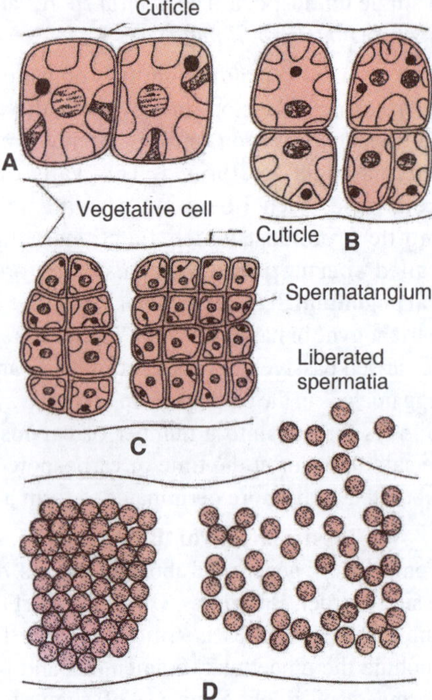

Fig. 21.5 (A-D). *Porphyra.* Development of spermatangia. (A-C), division of spermatangia. (A-C), division of vegetative cell to form sperma-tangia; (D), liberation of spermatia (A-C after Ishikawa; D after Newton).

Sexual Reproduction. It is oogamous. Most of the species are **dioecious** but some are **monoecious**. The male sex organ is called **spermatangium** and the female **carpogonium**. Both are formed from single vegetative cells.

(*a*) Spermatangia (Fig. 21.5). They are formed by repeated division of a vegetative cell (A-C). The process starts at the apex and spread towards the base of the thallus. A vegetative cell of the male thallus divides in three planes. The first wall is parallel to the surface (A). The daughter cells undergo two vertical divisions. (B-C). Further transverse divisions result in a mass of 32-128 (D) small colourless cells called **spermatangia**. They are arranged in several superimposed tiers of 4,8 or 16. When the plants are reflooded by the incoming tide the walls of spermatangia become gelatinised. The male elements are extruded at the thallus margin (D). The liberated naked uninucleate protoplast of each spermatangium is called a **spermatangium**. It is a nonmotile male gamete. On liberation the non-motile spermatia are carried passively by the water currents in all directions. Some of them may be drifted to and lodged against carpogonia. Several spermatia may be found adhering to the female thallus (Fig. 21.6 B).

(*b*) Carpogonia (Fig. 21.6 A-B). The carpogonium is formed by slight modification of an ordinary vegetative cell of the female thallus (A). It increases in size. The swollen cell undergoes no division. Its protoplast functions directly as an egg. In certain species (*P. perforata*) the ellipsoid **carpogonium** extends to the thallus surface at one or both the ends by giving out a small papillate outgrowth which is considered a rudimentary **trichogyne** (A). In species which lack the protuberance or trichogyne (*P. umbilicus*), the spermatium itself puts forth a narrow process containing a thin stream of cytoplasm (B). It establishes a connection between the spermatium and the carpogonium.

The spermatial nucleus migrates through it into the carpogonium to fuse with the egg nucleus.

(c) Formation of carpospores (Fig. 21.6 C). Immediately after fertilization the zygote nucleus undergoes meiosis. The four haploid nuclei divide mitotically to form 8, 16 or 32 nuclei. Cleavage of the protoplast at each nuclear division leads to the formation of a group of 8-32 uninucleate haploid meiospores known as the **carpospores**. The mature naked, nonflagellate carpospores are released by the breakdown of the surrounding cell walls when the female thallus is submerged by the incoming tide. In monoecious species the male areas are segregated from the female areas. The former are yellowish white or white and the latter purple. The carpospore has a stellate chromatophore.

Germination of carpospores. The liberated naked carpospore becomes amoeboid. It exhibits slow amoeboid movements of 2 or 3 days. It then comes to rest, becomes spherical, secretes a cell wall and germinates to produce a branched filamentous structure which in some species (*P. umbilicalis*) resembles an alga known as *Conchocelis rosea*. The *Conchocelis* stage produces monospores whose function is unknown. They may serve to multiply the conchocelis stage or germinate to form the plate-like *Porphyra* thallus.

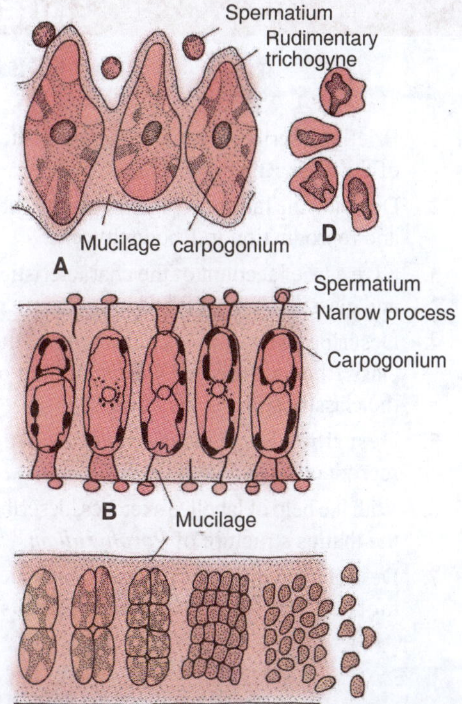

Fig. 21.6 (A-D). *Porphyra.* (A), V.S. thallus of *P. tenera* with carpogonia; (B), V.S. Female thallus of *P. umbilicus* with spermatia adhering to it; (C), stages in the development of carpospores; (D), carpospores in the amoeboid condition (A based on Kunieda, B after Dangeard, C and D after Smith).

Taxonomic Position:

Division	:	**Rhodophyta** (Rhodophycophyta)
Class	:	**Rhodophyceae**
Sub-class	:	**Bangiodeae** (Protofloridae)
Order	:	**Bangiales**
Family	:	**Bangiaceae**
Genus	:	*Porphyra*
Species	:	*umbilicus*

Genus Porphyridium (Fig. 21.4 A-B). It is a unicellular, mainly terrestrial red alga. The cells occur massed together forming a brightly coloured reddish mucilaginous layer on damp soil and moist wall (A). The unicell is globose in form. It has a single large, axile, stellate chromatophore with a pyrenoid located at the centre (B). The single nucleus lies adjacent to the chromatophore. The unicells are surrounded by a gelatinous matrix. The alga multiplies by the method of fission. *P. cruentum* is the commonly known species of this genus.

Fig. 21.3 is the diagrammatic representation of fine structure of unicellular red alga *Porphyridium*.

QUESTIONS

ESSAY TYPE

1. Briefly describe the main characteristics of the class Rhodophyceae.
2. Describe the range of vegetative structure and reproduction in Rhodophyceae.
3. Give a brief account of the characteristics and classification of Rhodophyceae.
4. Describe the general characters of the class Rhodophyceae. Give an outline of the classification of the class.
5. Describe the thallus structure and reproduction in *Porphyra*.
6. With the help of labelled sketches, describe the thallus structure of *Porphyridium*.
7. Describe the different types of life cycles met within Rhodophyceae. Discuss their interrelationships and evolutionary tendencies. (*Rohilkhand M.Sc., 1989, 1990, 1995, 1998*)
8. Discuss in detail the reproduction in Rhodophyceae. (*Rohilkhand, 1991*)
9. Discus in brief the main characters of Myxophyceae, Phaeophyceae and Rhodophyceae. (*Meerut, 1997*)
10. Describe the main distinguishing features of Chlorophyceae, Phaeophyceae, Rhodophyceae and Cyanophyceae. (*Rohilkhand, 1998*)
11. Describe the cell structure in Rhodophyceae.
12. Describe the reproduction in Rhodophyceae.

SHORT ANSWER TYPE

13. Describe life cycle patterns in Rhodophyceae. (*Lucknow M.Sc., 2008*)
14. Describe different life cycle patterns in Rhodophyceae. Discuss their interrelationships thinks and evolutionary tendencies. (*Rohilkhand M.Sc., 2006*)
15. Discuss the following:
    (*a*) Why are Red Algae considered primitive amongst eukaryotic algae? (*Lucknow M.Sc., 2005*)
    (*b*) Post fertilization changes in Florideae. (*Lucknow M.Sc., 2005*)
16. Write short notes on the following :
    (*a*) Spores in Red Algae. (*Rohilkhand, 1993*)
    (*b*) Carpospores (*Kanpur, 1995; Awadh, 1991*)
    (*c*) Pigments in Rhodophyceae. (*Rohilkhand, 1995*)
    (*d*) Gonimoblast filaments. (*Kumaon, 1995; Gorakhpur, 1990*)
    (*e*) Carpogonium. (*Allahabad, 1995*)
    (*f*) Carposporophytes. (*Lucknow, 2007*)
    (*g*) Cystocarp. (*Bundelkhand, 1994*)
    (*h*) Distinguishing features of Phaeophyceae and Rhodophyceae. (*Gorakhpur, 1996*)
    (*i*) Carposporophyte of Red Alga. (*Punjab, 1999; Kanpur M.Sc., 2007*)
17. Give the salient features of class Rhodophyceae. Why the members of the class are known as Red algae? (*Kumaon, 1995*)
18. Differentiate between
    (*a*) Rhodophyceae and Cyanophyceae (*Lucknow, 1998*)
    (*b*) Rhodophyceae and Chlorophyceae
    (*c*) Rhodophyceae and Phaeophyceae
    (*d*) Rhodophyceae and Xanthophyceae (*Lucknow, 2005*)
19. Write economic importance of Red algae. (*Punjab, 1992*)
20. Write explanatory notes on
    (*a*) *Porphyra*
    (*b*) *Porphyridium*

(*c*) Rhodophyceae
(*d*) Alternation of generations in Rhodophyceae.

21. Give distinguishing features of class Rhodophyceae. (*Kanpur, 2005*)

22. What is the nature of reserve food of Rhodophyceae class? (*Kanpur M.Sc., 2002*)

23. Name male gametes of Red algae. (*Kanpur M.Sc., 2002*)

24. Discuss in brief affinities of Rhodophyceae. (*Kanpur M.Sc., 2005*)

25. Give a brief account of characteristic features of Rhodophyceae and its different orders studied by you. (*Lucknow M.Sc., 2002*)

26. Write briefly on distriguishing features of Rhodophyceae. (*Allahabad, 2005*)

OBJECTIVE TYPE

27. Select the correct answer :

(*i*) Reserve food material in Rhodophyceae is
(*a*) Starch
(*b*) Floridean starch
(*c*) Cyanophycean starch
(*d*) Laminarin

(*ii*) Members of class Rhodophyceae are commonly called
(*a*) Green algae
(*b*) Blue algae
(*c*) Red algae
(*d*) Blue-green algae

(*iii*) Red colour of the red algae is due to the pigment named
(*a*) Chlorophyll
(*b*) Phycoerythrin
(*c*) Phycocyanin
(*d*) All of the above

(*iv*) In Rhodophyceae non-motile male gametes are called
(*a*) Spermatozoids
(*b*) Anthrozoids
(*c*) Zoospores
(*d*) Spermatia

(*v*) Algae belonging to Rhodophyceae are
(*a*) Only marine
(*b*) Only fresh water
(*c*) Both marine as well as fresh water
(*d*) Fresh water and terrestrial

(*vi*) Cystocarp consists of
(*a*) Oogonium
(*b*) Carpogonium only
(*c*) Carpogonium and gonimoblast filaments
(*d*) Carpogonium, loosely arranged gonimoblasts and carposporangia

(*vii*) In higher red algae, Carpospores are usually
(*a*) Haploid
(*b*) Diploid
(*c*) Neither haploid nor diploid
(*d*) May be haploid or tetraploid

(*viii*) In *Porphyridium* cell, the axial stellate chromatophore contains
(*a*) No pyrenoid
(*b*) A single pyrenoid
(*c*) Two pyrenoids
(*d*) Many pyrenoids

(*ix*) Which of the following is a unicellular red alga?
(*a*) *Chlamydomonas*
(*b*) *Hydrodictyon*
(*c*) *Porphyridium*
(*d*) *Polysiphonia*

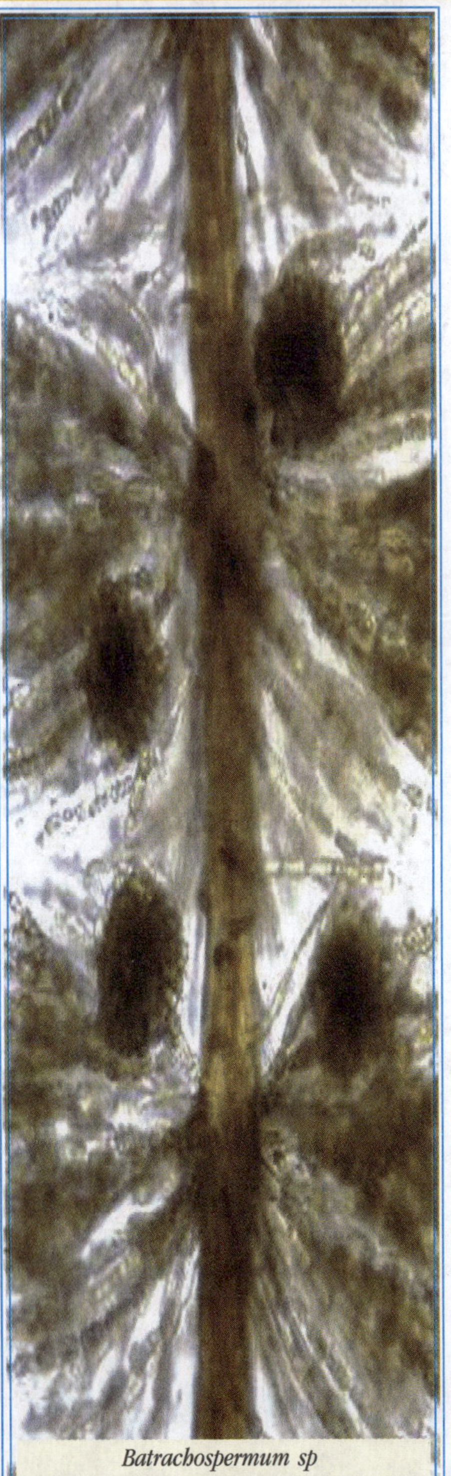
Batrachospermum sp

CHAPTER 22

SUB-CLASS : FLORIDEAE (Florideophycideae) Order : Nemalionales

- Order : Nemalionales
- Family : Batrachospermaceae
- Batrachospermum

The sub-class Florideae is characterised by the following features :

1. The thallus is fundamentally filamentous and even in elaborate forms it consists of coalescing filaments with a **uniaxial** or **multiaxial** type of construction.
2. Growth of all filaments is apical.
3. There are perforations between the adjacent cells. These are known as the **pit connections**.
4. There is complete absence of **motile** cells in the life cycle.
5. Sexual reproduction is **oogamous** and the sex organs exhibit complexity of form and structure.

6. Carposporangia, unlike the Bangioideae, are borne on definite filamentous structures called the **gonimoblast** filaments.
7. Zygote nucleus may or may not divide reductionally. Reduction occurs in *Batrachospermum* and does not occur in *Polysiphonia*.
8. Most of the members are met with in the sea, a few fresh water forms also occur.

This sub-class includes six orders. These are Nemalionales, Gelidiales, Cryptonemiales, Gigarteniales, Rhodomeniales and Ceramiales. Of these we shall study structure and reproduction of the two orders, Nemalionales and Ceramiales.

Nemalion.

Rhodochorton.

Bonnemaisonia.

MEMBERS OF ORDER MEMALONIALES.

ORDER : NEMALIONALES

The characteristic features of the order are as follows :

1. Usually marine but a few forms like *Batrachospermum, Lamania, Thorea* and *Tuomeya* are fresh water forms.
2. The plant body may be uniaxial or multi-axial. The filament may be heterotrichous.
3. The number of chromatophores varies from a single large to several small discoid or rod-shaped structure in each cell.
4. There is no diploid plant in the life cycle as the reduction division usually takes place immediately after fertilisation.
5. The life cycle is **haplobiontic**. Morphologically it may be diphasic or triphasic.
6. Carposporangia are haploid and bear haploid carpospores.
7. Asexual reproduction is by monospores produced singly within monosporangia.
8. Carpogonial branch is simple and arises from a vegetative cell.
9. Auxiliary cell is absent.
10. Presence of a juvenile stage or the **chantransia** stage from which the main plant arises as a lateral outgrowth.

The order includes eight families *viz.* Acrochaetiaceae, Batrachospermaceae, Lemaneaceae, Naceariaceae, Bonnemaisoniaceae, Thoreaceae, Nelmenthocladiaceae and Chaetangiaceae. Out of these a representative of the family Batrachospermaceae will be dealt with in detail.

FAMILY : BATRACHOSPERMACEAE

The filamentous thallus is of uniaxial construction. The axial filament is often corticated. The cortical threads arise from the basal cells of the laterals. The haploid carpospores are borne singly and terminally on the gonimoblast filaments. They germinate to give rise to the *Chantransia* stage which multiplies by means of monospores. The *Batrachospermum* plant arises as a lateral outgrowth from the Chantransia stage. The family includes both marine and fresh water forms. Of the five fresh water genera included in this family, *Batrachospermum* and *Lemanea* are commonly known. The former is taken as a type. It comprises about 40 species.

BATRACHOSPERMUM

Occurrence. It is one of the few, inland fresh water red algae commonly found in the tropical, subtropical and temperate regions. It likes cool, well aerated, clean fresh water and prefers shady situations. Hence it occurs in slow flowing streams, water falls, margins of lakes, in springs or the outflow from them but usually under shade attached to rocks. Some species are found in rapidly flowing brooks and rivers. *Batrachospermum moniliforme* is the common example. It occurs attached to stones and sticks in the rapid flowing water in the tropics, subtropics and temperate regions. Majority of the species are annuals but a few such as *B. vagum* are perennial. This species has been reported to form an interesting association with the snail *Planorbis planorbis* L. This association was reported by Iltis, H. in 1913.

Organisation of thallus (Fig. 22.1). *Batrachospermum* thallus is bluish or olive green or violet green in colour when the plants are growing in shallow water. In deep water the specimens are of dark violet or reddish colour. The variation in thallus colour is the result of differences in light intensity to which the plants are exposed to. The *Batrachospermum* plant or thallus occurs in a spawn-like mass covered with soft, thick mucilage. The adult thalli or plants, which are of uniaxial construction in its simplest form, may attain a length of 20 cm or even more. It is considered to be a heterotrichous form in which the prostrate system is completely reduced. The branched projecting system thus floats freely

Batrachospermum growing in shallow water.

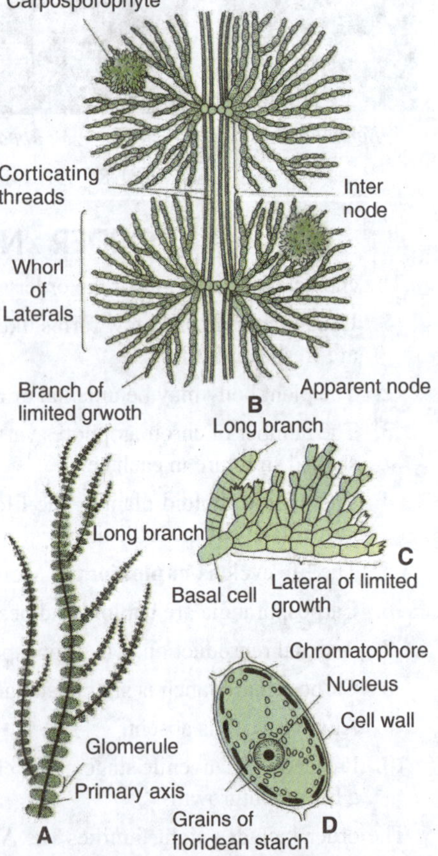

Fig. 22.1 (A-D). *Batrachospermum.* (A), part of thallus presenting bead-like appearances; (B), the same under higher magnification showing two glomerules each consisting of a cluster of laterals of limited growth at the node; (C), origin of a long branch from a basal cell of the lateral; (D), cell showing structure. (A-C after Sirodot, D, after Schmitz).

in water. In a few cases it is attached to the substratum by rhizoids arising from the basal cells of the primary axis. The primary of main axis of the monosiphonous thallus is corticated and consists of a number of elongated cylindrical cells placed end to end in a single row. The terminal apical cell is hemispherical. The primary axis is freely branched and is of unlimited growth. The branches or laterals it bears are of two kinds namely (*i*) branches of limited growth or the primary laterals and (*ii*) the long branches or branches of unlimited growth.

(i) Primary laterals or branches of limited growth. These arise in clusters or whorls from the upper ends of the cells just below the septa. They usually consist of moniliform cells arranged like chains of beads. Each cluster or whorl consists of several (usually 4 of 6) laterals. All the laterals in a whorl are of about the same length and are densely branched. Consequently the laterals of a single whorl form a spherical or a globose mass consisting of a cluster of branches. Many of these remain sterile and some bear the sex organs. The spherical or globose cluster of laterals is called a **glomerule**. The glomerules may be close to or remote from one another. This depends upon the length of the cells constituting the central filament. This peculiar arrangement of laterals in small globose masses on a central axis gives the whole thallus a beaded appearance of branching chain of beads to the naked eye (A) and axial filament appears differentiated into nodal and internodal regions very much like the green alga *Draparnaldiopsis*.

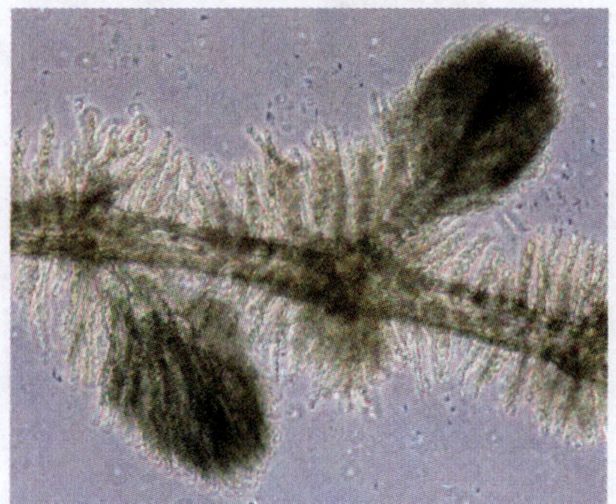

Batrachospermum dense branches enveloped in soft mucilage.

(ii) Long branches. In addition to the branches of limited growth, the central filament also bears long branches or branches of **unlimited** growth here and there (A). The long branch arises singly from the basal cell of one of the laterals in a whorl (C) and consists of cells much longer than those of the dwarf laterals. The long branches like the primary axis bear the branch whorls at intervals and present a beaded appearance to the naked eye and thus appears differentiated into nodal and internodal regions. Each bead or glomerule consists of a cluster of densely branched primary laterals of limited growth. The entire thallus consisting of an axial filament with whorls of dense branches is usually enveloped in copious soft mucilage.

Cell structure. (Fig. 22.2). The filament consists of a large number of elongated cylindrical cells placed end to end in a single file. Pits in the septa between the cells are lacking. Each cell contains either a diffuse chromatophore or a number of disc-like chromatophores which are parietal in position. Each chromatophore has a single pyrenoid and contains the pigments chlorophyll *a* and *d*, r-phycocyanin and r-phycoerythrin. The latter two pigments mask the green colour of the chlorophyll. The cell is uninucleate and contains grains of floridean starch scattered in the cytoplasm which at its

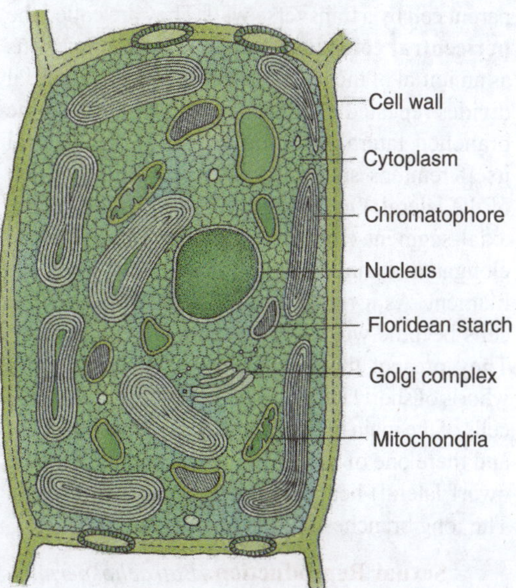

Fig. 22.2. *Batrachospermum.* Structure of the cell as seen under electron nucroscope.

periphery is limited by the plasma membrane. The cell wall consists of two layers. The outer layer is composed of pectic materials and the inner of cellulose which is made up of microfibrils.

Cortication (Figs. 22.1 B and 22.3). The basal cells of the laterals grow into narrow threads consisting of short cells. These run down along the axial filament to form an envelope or a sheath around it. It is called the **cortex**. The cortex completely covers the primary or central filament and conceals it from view.

Growth. The growth of axial filament and the long branches is apical. It takes place by means of an **apical cell** which is dome-shaped (Fig. 22.3). The apical cell divides by transverse wall and cuts off a series of segments parallel to the posterior face. The segments thus cut off rapidly increase in length and breadth but usually remain broader in the septa. In the older portion of the main filament they elongate considerably to become large cylindrical cells.

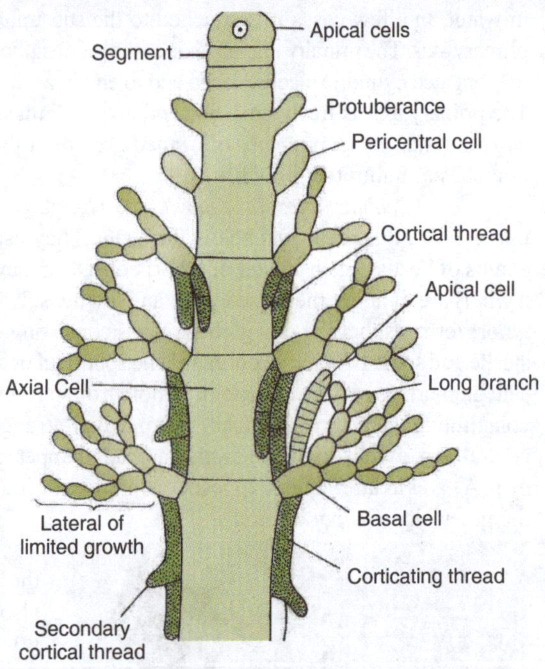

Fig. 22.3. *Batrachospermum.* Apical part of thallus showing segmentation of the apical cell, cortication and origin of branches (Diagram-matic).

Origin of Branches (Fig. 22.3). Branching is **monopodial**. A little distance behind the apex each axial segment, as it increases in size, grows into 4-6 small lateral outgrowths at its upper end.

These outgrowths are then separated from the parent cell by a transverse wall. They are called the **pericentral cells**. Each pericentral cell functions as an initial of the lateral branch. The lateral initial divides repeatedly and actively to give rise to the branched lateral of limited growth. The initial itself remains small and is called the basal cell of the lateral. On the other hand the parent or the axial segment elongates enormously to form an elongated, cylindrical cell constituting the main filament. As a result groups of small pericentral cells become widely separated from one another. They present the appearance of nodes bearing whorls of short laterals. The long cylindrical central cells of the main filament look like internodes. Here and there one of the nodal cells (basal cells of the dwarf lateral) bears (Fig. 21.6 C) a long branch. The long branches thus arise singly.

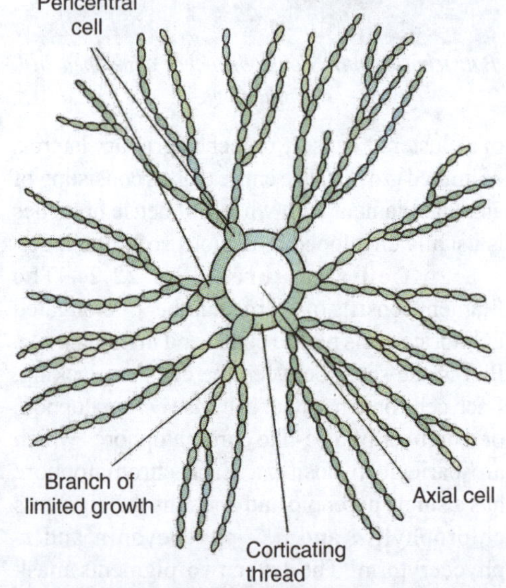

Fig. 22.4. *Batrachospermum.* End view of pericentral cells bearing the nodal cluster of laterals of limited growth (Diagrammatic).

Sexual Reproduction. *Batrachospermum* reproduces sexually. Sexual reproduction is of an advanced **oogamous** type. Some species bear both kinds of sex organs on the same plant (**monoecious**) and others on different individuals (**dioecious**). The male sex organs are termed **antheridia** or **spermatangia** and the female **carpogonia**. They arise on some of the specialized dwarf branches in the whorl, others remain sterile.

Carpogonia. (*a*) *Position and Structure* (Fig. 22.5). Near the base of a primary lateral arises a 3-5 celled special branch of secondary order. It is the female or carpogonial branch (B). The terminal cell of the latter is the **carpogonium**. It is a flask-shaped structure consisting of two parts. There is the basal, dilated portion which contains the egg nucleus and a more or less drawn out narrow, distal part known as the **trichogyne** (C). The trichogyne is demarcated by a median constriction from the basal, swollen portion of the carpogonium. It is usually meant to receive the male cell. This receptive structure varies in shape in different species. It may be cylindrical, club-shaped or spatula-shaped according to the species. The basal swollen egg cell contains dense cytoplasmic contents which later become laden with food reserves. At the early stages of development it may also contain chromatophores which start degenerating afterwards. The single female or egg nucleus is located at the centre. There is no ovum free from the carpogonial wall. The cytoplasm in the region of the trichogyne is colourless and usually without a nucleus. It may be that the nucleus degenerates in the earlier stages of development of the trichogyne. From the vegetative cells below the carpogonium arise short, erect, sterile filaments which grow up and around the carpogonium enclosing it partially.

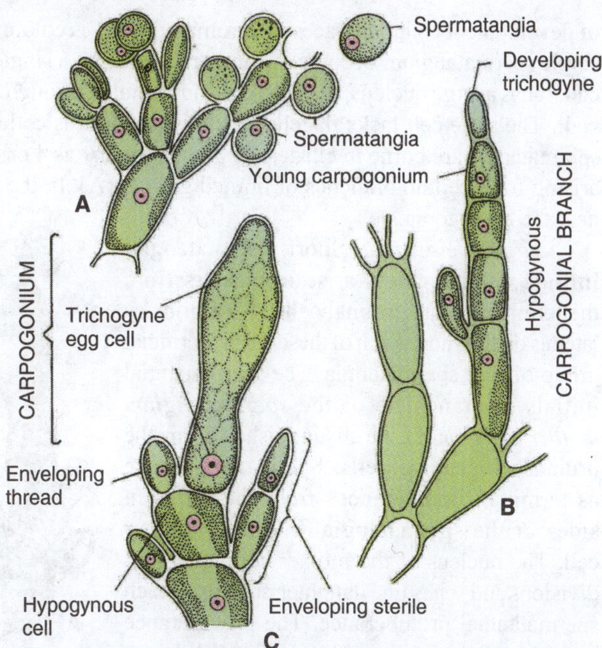

Fig. 22.5 (A-C). *Batrachospermum*. Sex organs. (A), antheridial branch bearing spermatangia in clusters, the terminal spermatangium has discharged the spermatium; (B), carpogonial branch with a terminal young carpogonium; (C), mature carpogonium ready for fertilization (After Kylin).

(*b*) *Development of Procarp* (Fig. 22.5 B). Any cell usually the basal one of a primary lateral of limited growth cuts off a segment. It is the **procarp initial** (B). The latter divides by transverse divisions 3 or 4 times to form a 3 or 4-celled offshoot which is usually known as the **carpogonial** branch or **procarp.** The terminal cell of the carpogonial branch enlarges in size and assumes a flask-shaped appearance to become the carpogonium. The swollen, nucleate basal part of the carpogonium is called the **egg cell** and an elongated enucleate neck, the **trichogyne**. Davis recorded the presence of a nucleus in the trichogyne but this has been denied by majority of the recent workers. May be it possesses a nucleus which degenerates in the early stages.

Towards maturity the female nucleus travels towards the base. The cytoplasm and the female nucleus lying in the swollen egg cell constitute the **female gamete** or the **egg.** There is no contraction of the female protoplast in the earlier or later stages of development. The cell next to the carpogonium (that lies at its base) is termed the **hypobasal** or **hypogynous cell** (C).

Spermatangia

(*a*) *Structure and Position* (Fig. 22.5A). The spermatangium when ripe is a single-celled, colourless, spherical structure. The wall is fairly thick. The uninucleate protoplast which at early stages of development contains traces of chromatophores, is colourless at maturity. The entire contents of a single spermatangium become metamporphosed into a single non-motile, colourless, spherical male cell. It has a large nucleus, but is naked and liberated through a narrow, apical slit in the spermatangial wall. The male cell lacks flagella. For this reason it is called a **spermatium** and not a sperm. The spermatangia are borne in clusters of 2 or 3 as many as 4 or 5 at the tips of short ultimate branchlets arising from certain branches of limited growth (A). In the monoecious species they usually occur near the carpogonia.

(*b*) *Development.* Short, spermatangial **initials** or **branchlets** arise in clusters from the terminal or subterminal cells of the primary laterals of the whorl. Each of these bears a terminal group of four spermatangia. The spermatangial initials or branchlets or the *spermatangium mother cells* cannot be distinguished from the ordinary vegetative cells. Spermatangia arise as terminal protuberances from the different sides of the spermatangial initial or mother cell. The nucleus of the mother cell undergoes divisions and sends one daughter nucleus to each spermatangial protuberance. The protuberance becomes separated from the mother cell by an annular ingrowth of the wall and becomes the **spermatangium**.

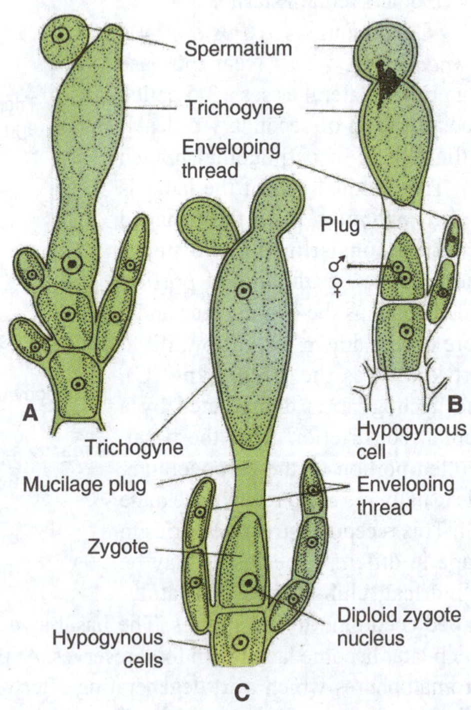

Fig. 22.6 (A-C). *Batrachospermum.* Successive stages in fertilization. A, mature carpogonium with the spermatium attached to the trichogyne; B, later stage with the spermatium nucleus (*o*) having migrated to basal part of the carpogonium to fuse with the female nucleus; C, still later stage in which the male and female nuclei have fused to form a diploid zygote nucleus (After Kylin).

Fertilization (Fig. 22.6). The liberated spermatium floats on the surface of water and is carried passively by water currents to the tip of the trichogyne. A large number of spermatia fail to reach the trichogyne and degenerate. Their formation in larger numbers, however, ensures contact with the trichogyne. The tip of the trichogyne becomes mucilaginous and the spermatium sticks to it without any difficulty (A). The adjoining wall between the spermatium and the trichogyne dissolves. The male nucleus enters through the opening, moves down the trichogyne and finally joins its female partner (B). It has been recorded that in *Batrachospermum* the nucleus of the spermatium undergoes division into two daughter nuclei. One of these fuses with the female nucleus to form the diploid fusion nucleus. The other remains in the trichogyne. After fertilization the trichogyne is separated from the basal part of the carpogonium containing the zygote by **plug** of mucilage (C). Finally it shrivels away. The zygote undergoes no resting period.

Post-fertilization Changes (Fig. 22.7). The post-fertilization changes begin immediately after fertilization with the division of the diploid zygote nucleus (A). Lying at the base of the carpogonium it divides twice giving rise to four nuclei. The division is **meiotic**. The resultant four nuclei are thus haploid. Many small, lateral protuberances arise from the side walls of the basal part of the carpogonium. The haploid daughter nuclei continue to divide mitotically and pass out into these protuberances one each. The protuberances grow and are separated by a wall from the carpogonium. These outgrowths

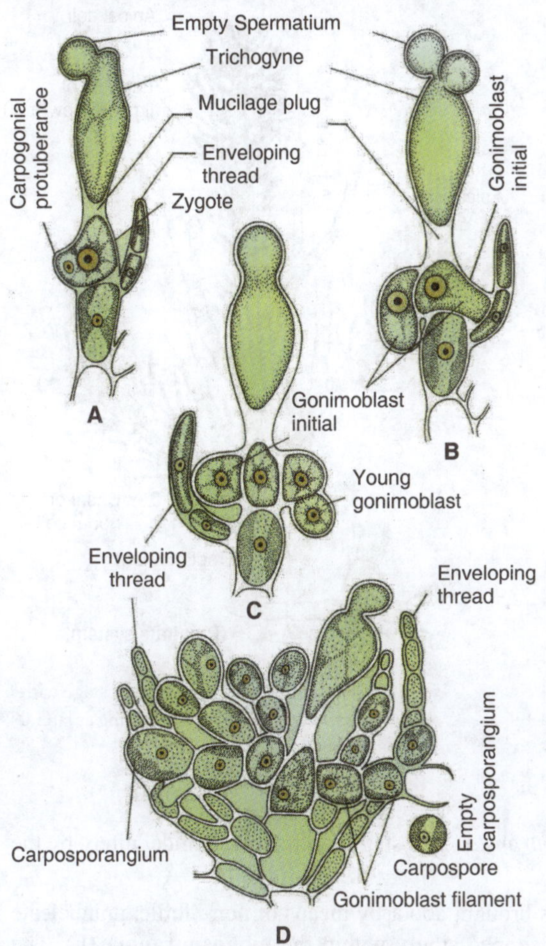

Fig. 22.7 (A-D). *Batrachospermum.* Post-fertilization changes resulting in the development of carposporophyte or cystocarp. (A), fertilized carpogonium having given out a lateral protuberance from the wall at its base; (B), first gonimoblast initial cut off and second forming; (C), later stage of development of gonimoblasts, first gonimoblast initial has formed a young filament and the second has been separated; (D), carposporophyte or cystocarp consisting of gonimoblast filaments, each ending in a carposporangium containing a carpospore (After Kylin).

are known as the **gonimoblast initials** (B). Several gonimoblast initials may be formed. The gonimoblast initials undergo repeated division (C) and form a mass of branches or unbranched filaments (D). These filaments are called the **gonimoblast filaments**. They are usually short, compact and branched so as to form a cluster. The terminal cell of each gonimoblast filament which contains dense cytoplasm and food reserves becomes enlarged to function as a **carposporangium** (D). The contents of the carposporangium develop into a single, naked, nonmotile spherical **carpospore**. The carpospore is a haploid structure. It contains a single nucleus and a chromatophore. While all these changes are taking place the sterile cells below the carpogonium give rise to numerous lateral, sterile threads. These threads enclose the gonimoblast filaments and the carpogonium to form a sheath around them. The entire structure consisting of the enveloping threads, the gonimoblast filaments bearing the carposporangia together with the carpogonium constitute a characteristic fruit body known as the **cystocarp**. Some call it the **carposporophyte**. It remains parasitic on the parent gametophyte. The carpospores are discharged as spherical, non-motile, naked structures. The empty walls of terminal carposporangia remain behind.

Carposporophyte in Batrachospermum.

Germination of Carpospores (Fig. 22.8). The liberated carpospore comes to rest and secrets a wall around it. It then germinates by putting out a process on one side (A). The contents of the carpospore migrate into the process or the protuberance. The latter becomes cut off by a cross wall. The process undergoes transverse septation and forms a creeping filament. The creeping filament later on branches and rebranches to form a prostrate system (E). From the latter arise the projecting or the erect filaments. The erect filaments also branch (F). This filamentous, protonema-like heterotrichous structure is called the **chantransia stage** (F). It is the juvenile stage in the life cycle. It is microscopic

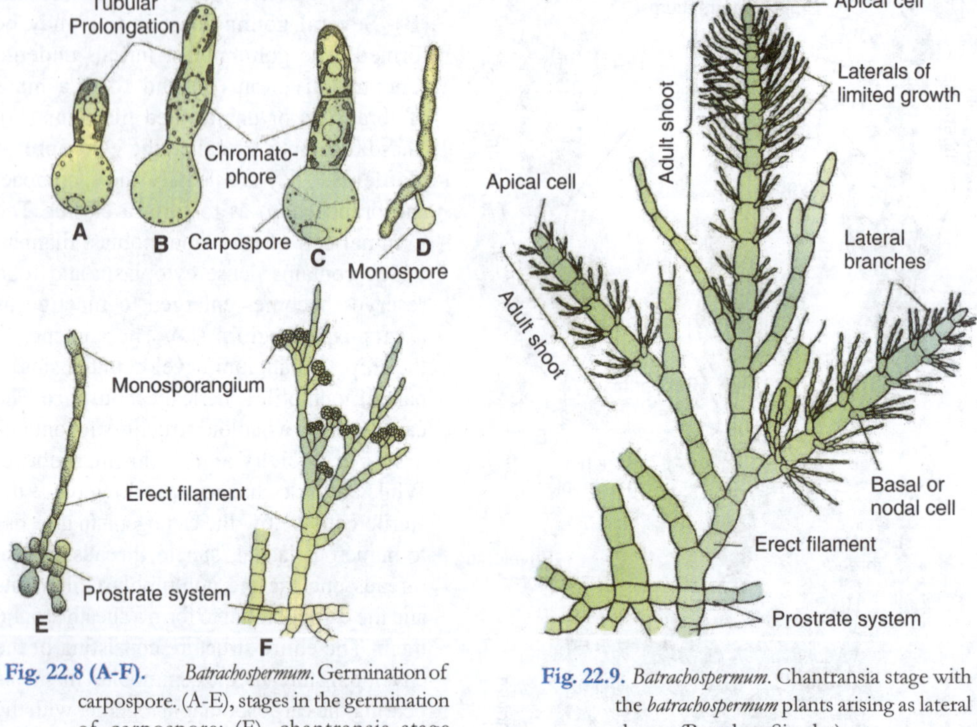

Fig. 22.8 (A-F). *Batrachospermum.* Germination of carpospore. (A-E), stages in the germination of carpospore; (F), chantransia stage reproducing asexually (A-E, after Geitler; F, after Sirodot).

Fig. 22.9. *Batrachospermum.* Chantransia stage with the *batrachospermum* plants arising as lateral shoots (Based on Sirodot).

and looks very much like an alga called *Chantransia*. This stage can reproduce indefinitely by the formation of **monospores.**

Asexual Reproduction (fig. 22.9). It is brought about by means of non-motile, uninucleate spores called the **monospores.** They are developed singly within the **monosporangia** (F). The monosporangia are nothing but the swollen terminal cells of the short, lateral branches of the erect filaments of the chantransia stage (F). The contents of each monosporangium become metamorphosed into a uninucleate monospore. The monosporangia have rarely been noticed to develop on the main filament of *Batrachospermum*. They frequently develop on the chantransia stage and are spherical or oblong in shape. Each monospore germinates to produce the chantransia stage. The monospores thus serve to perpetuate the **chantransia** stage. *Batrachospermum* plant arises as a lateral outgrowth from the erect filaments of *chantransia* stage (Fig. 22.9). The latter is often compared to the protonema of mosses. According to this view the erect free floating *Batrachospermum* plant and the juvenile chantransia stage constitute the two systems of a heterotrichous form.

Alternation of generations (Fig. 22.10 and 22.11). *Batrachospermum* plant is a free-living gametophyte. It produces male and female gametes which fuse to form the zygote. The zygote is the only diploid structure in the life cycle. It is transitory because reduction division occurs at the time of development of the gonimoblast initials. The gonimoblast filaments, the carposporangia and the carpospores are all **haploid** and constitute the *carposporophyte phase* in the life cycle. The chantransia stage formed by the germination of the carpospore is also haploid. From the haploid chantransia stage arises as an outgrowth the main haploid *Batrachospermum* plant.

There are thus three haploid somatic phases or generations in the life of *Batrachospermum*. These are :

1. The first is the *Batrachospermum* plant (A). It represents the free living parent gametophyte concerned with the production of gametes. It is independent.
2. The second is the haploid, dependent *carposporophyte* or *cystocarpic* stage representing the second gametophyte phase concerned with the production of haploid carpospores (H). It is parasitic on the parent gametophyte.
3. The third haploid phase or generation is the *chantransia stage* (M). It is free living and is formed by the germination of the haploid carpospore (I-L). The *Batrachospermum* plant arises as a lateral outgrowth from this juvenile stage (M).

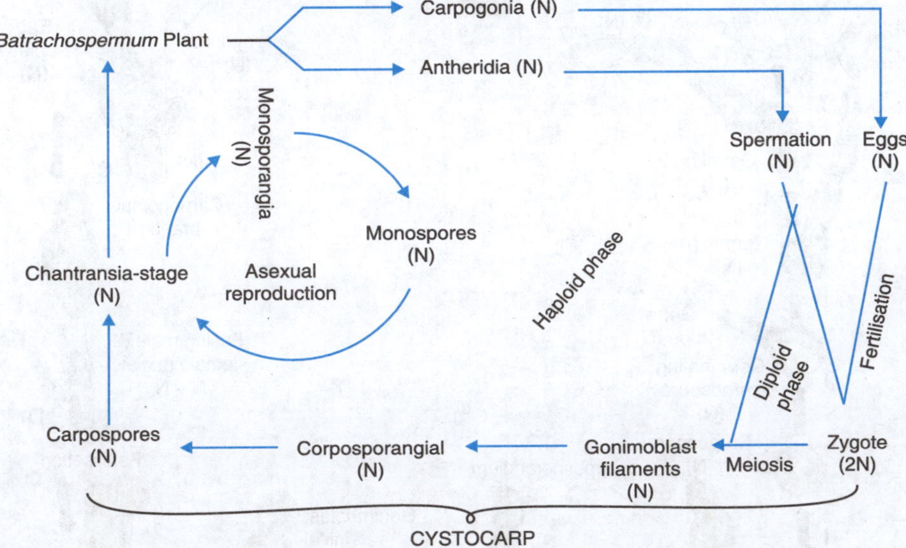

Fig. 22.10. Graphic representation of the life cycle of *Batrachospermum*.

The life cycle of *Batrachospermum* thus involves alternation of a succession (3 in number) of haploid somatic generations with a transitory diploid phase as represented by the zygote. It means between one zygote and the next there occur 3 haploid somatic generations one after the other. Cytologically, such a life cycle is called **haplobiontic** because the three alternating individuals are of one kind (gametophyte). Morphologically it is described as **triphasic** or **trigenic**. It is also termed as **haplohaplohaplontic**.

Taxonomic Position

Division	:	**Rhodophyta** (Rhodophycophyta)
Class	:	**Rohodophyceae**
Sub-class	:	**Florideae**
Order	:	**Nemalionales**
Family	:	**Batrachospermaceae**
Genus	:	*Batrachospermum*
Species	:	*moniliforme*

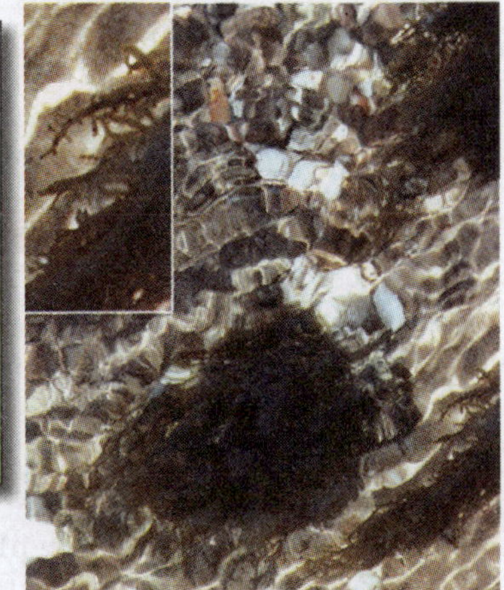

Batrachospermum moniliforme.

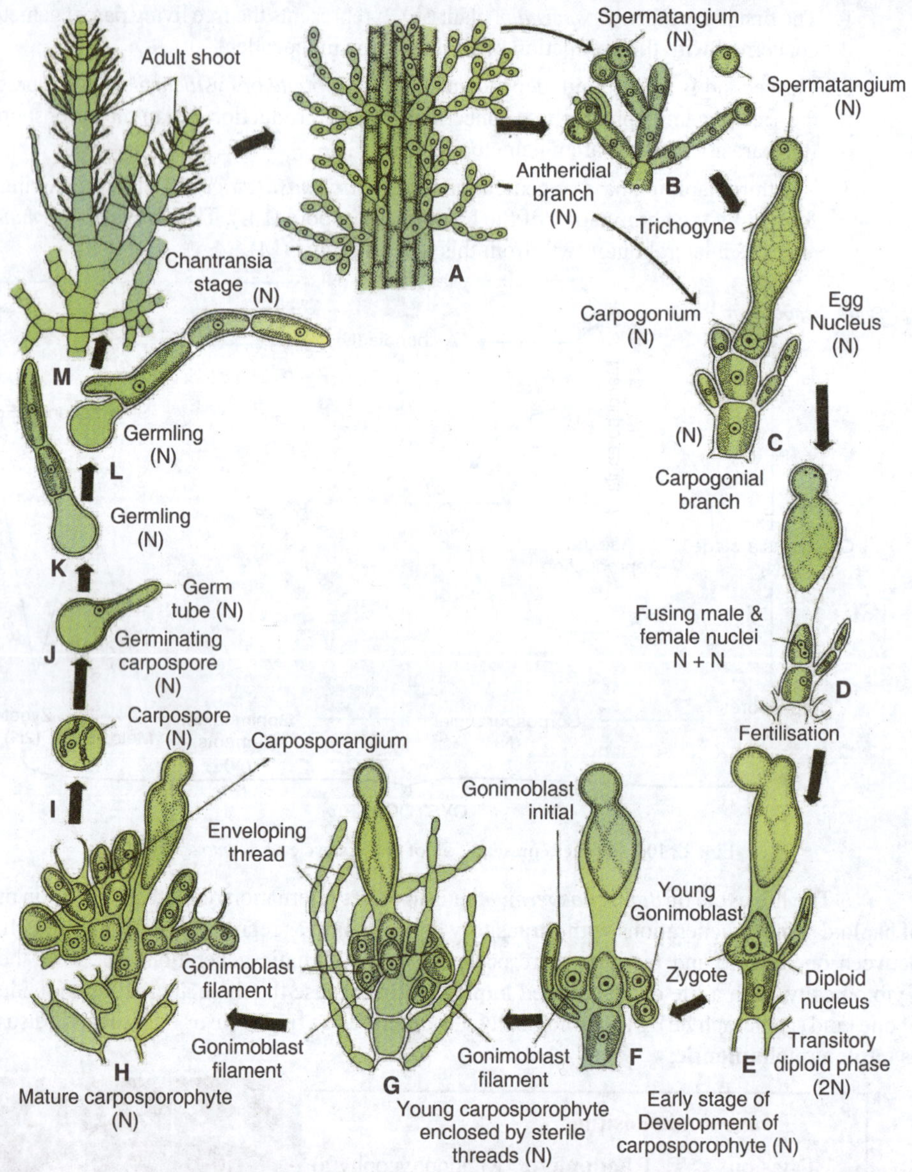

Fig. 22.11 (A-M). *Batrachospermum.* Pictorial sexual life cycle.

QUESTIONS

ESSAY TYPE

1. Describe the thallus and reproduction in *Batrachospermum*. Mention its features of significance.

   (Vir Kumar Singh Univ., 1996)

2. Describe the structure of thallus and development of cystocarp in *Batrachospermum*.

3. Give an account of the life history of *Batrachospermum*.

   (U.P. College, 1999; Allahabad, 1991, 1997; Lucknow, 1997; Vinoba Bhave, 1997; Magadh, 1994; Vir Kumar Singh Univ., 1995; Bihar, 1992; Lucknow M.Sc. 2004)

4. Give an illustrated account of post fertilization changes in the life cycle of a fresh water red alga studied by you. *(Purvanchal, 1997; H.P. 1994)*
5. Describe the mode of sexual reproduction in *Batrachospermum*. *(Gorakhpur, 1992, 1999; Vir Kumar Singh Univ., 1997; Allahabad, 2001)*
6. Give an account of the thallus structure and life cycle of *Batrachospermum*. *(Gorakhpur, 1997)*
7. Compare life cycle patterns of *Batrachospermum* and *Polysiphonia*. *(Allahabad, 1999)*
8. Compare the process of sexual reproduction in *Ectocarpus* with that of *Batrachospermum*.
9. Bring out clearly with a series of sketches and brief explanation the differences between the modes of reproduction of *Dictyota* and *Batrachospermum*.
10. What is chantransia stage? How does it help in multiplication and production of plant body in *Batrachospermum*? Illustrate the development of cystocarp in *Batrachospermum*. *(Punjab, 1997)*
11. Describe the structure of sex organs and process of cystocarp formation in *Batrachospermum*. *(Punjab, 1999)*
12. How would you illustrate that *Batrachospermum* exhibits an example of advanced oogamy, while maintaining simple plant body? *(Punjab, 1991)*
13. Give detailed account of development of cystocarp in *Batrachospermum*. *(Himachal Pradesh, 1993)*
14. Compare the life history of *Sargassum* with that of *Batrachospermum*. *(Bharathiar, 2000)*
15. Describe the structure reproduction and life cycle in *Batrachospermum*. *(Bihar, 1990; Allahabad, 2005)*
16. Give an illustrated accounts of sexual reproduction in *Batrachospermum* and give its systematic position. *(Poorvanchal, 2005)*

SHORT ANSWER TYPE

17. **Write short notes on :**
    (*i*) Chantransia stage *(U.P. College, 1995; Gorakhpur, 1994, 1998; Allahabad, 1993, 1995; Punjab, 1996, 1997, 1998; Nagpur, 1996; Meerut, 1991, 1993, 1997; Lucknow M.Sc., 2003; Poorvanchal, 2007)*
    (*ii*) Carposporophyte or cystocarp in *Batrachospermum*. *(U.P. College, 1996, 1998; Purvanchal, 1995, 1999; Gorakhpur, 1993; Allahabad, 1995, 1996, 1997; Lucknow 1999; Kumaon, 1997; Punjab, 1996)*
    (*iii*) Structure of Chromatophore in *Batrachospermum*. *(U.P. College, 2000)*
    (*iv*) Post fertilization in *Batrachospermum*. *(Purvanchal, 1998, 2000; Meerut, 1989)*
    (*v*) Gonimoblast filaments. *(Purvanchal, 1990; Kumaon, 1995)*
    (*vi*) Carpogonium. *(Gorakhpur, 1996; Awadh, 1998; Allahabad, 1992)*
    (*vii*) Carpospores. *(Awadh, 1991)*
    (*viii*) Sexual reproduction in *Batrachospermum*. *(Awadh, 1995)*
    (*ix*) Cystocarp formation in *Batrachospermum*. *(Punjab, 1999)*
    (*x*) Branching in *Batrachospermum*. *(Himachal Pradesh, 1990)*
    (*xi*) Carpospore in *Batrachospermum*. *(Vir Kumar Singh Univ., 1990)*
    (*xii*) Somatic phase of *Batrachospermum*. *(Bhagalpur, 1993)*
18. (*a*) Where does meiosis occur in the life cycle of *Batrachospermum*? *(Gorakhpur, 1990, 1996; Lucknow, 1998)*

(*b*) Draw a labelled diagram showing structure of cystocarp. (*Kerala, 2001*)

19. Name the reserve food materials found in the cells of red algae. (*Gorakhpur, 1990*)

20. Draw neat and labelled diagrams of

(*a*) Cystocarp of *Batrachospermum*. (*Gorakhpur, 1991, 1995, 1998*)

(*b*) Thallus of *Batrachospermum*. (*Awadh, 1997; Lucknow, 1995, 2001*)

(*c*) Sex organs and chantransia stage of *Batrachospermum*. (*Himachal Pradesh, 1992*)

(*d*) Structure of mature carposporophyte of *Batrachospermum*. (*Poorwanchal, 2003*)

21. Give name and systematic position of a freshwater red alga. (*Gorakhpur, 1996, 1998; Awadh, 1993, 1998; Allahabad, 1993; Kumaon, 1996*)

22. Give the most important characteristic of *Batrachospermum*. (*Awadh, 1992*)

23. Name an alga which produces tetraspores. (*Allahabad, 1992*)

24. Throw light on the significance and formation of chantransia stage. (*Himachal Pradesh, 1990*)

25. Throw light on the development of cystocarp in *Batrachospermum*.

26. Describe chantransia stage in *Batrachospermum*. (*Himachal Pradesh, 1994*)

27. Name an alga which shows Chantransia stage. (*Allahabad, 2002*)

28. Differentiate between:

(*a*) Asexual reproduction in *Batrachospermum* and *Ectocarpus*. (*Allahabad, 2004*)

(*b*) Post fertilisation stages of *Batrachospermum* and *Polysphonia*. (*Allahabad, 2004*)

29. Where does *Batrachospermum* occur? Give a note on its habit and habitat.

OBJECTIVE TYPE

30. Fill in the blanks :

(*i*) Dwarf branches of *Batrachospermum* form a thick clusters of branches at node called the .... (*Punjab, 1999*)

(*ii*) In *Batrachospermum*, the gonimoblast filaments, carpospores and carpogonium constitute the .... (*Punjab, 1998*)

(*iii*) The terminal cell of gonimoblast filament in *Batrachospermum* functions as ....

(*iv*) The red colour in red algae is due to the pigment named ....

(*v*) *Batrachospermum* belongs to the order ....

(*vi*) Chantransia stage is characteristic of the red alga ....

(*vii*) The fresh water red alga you have studied is ....

(*viii*) The life cycle of *Batrachospermum* is ....

(*ix*) Adult plant of the red alga *Batrachospermum* is ....

(*x*) Carpospores after germination produce Protonema like stage that resembles the alga............... .

31. Select the correct answer :

(*i*) *Batrachospermum* is

(*a*) Red alga of sea

(*b*) Red alga of fresh water

(*c*) Green alga of fresh water

(*d*) Brown alga of sea.

(*ii*) The adult plant of red alga *Batrachospermum* is

(*a*) Haploid

(*b*) Diploid

(*c*) Triploid

(*d*) Tetraploid

(*iii*) Chantransia stage is the characteristic stage of the alga

(*a*) *Polysiphonia*

(*b*) *Sargassum*

(*c*) *Batrachospermum*

(*d*) *Porphyra*

(*iv*) The branching in *Batrachospermum* is

(*a*) Monopozdial

(*b*) Sympodial

(*c*) Both of the above

(*d*) None of the above

(*v*) Spermatia in *Batrachospermum* is

(*a*) Uniflagellate

(*b*) Apically biflagellate

(*c*) Laterally biflagellate

(*d*) Aflagellate and non-motile

(*vi*) *Batrachospermum* belongs to the order

(*a*) Ceramiales

(*b*) Nemalionales

(*c*) Fucales

(*d*) Volvocales

(*vii*) Asexual reproduction in *Batrachospermum* takes place by

(*a*) Zoospores

(*b*) Chlamydospores

(*c*) Oospores

(*d*) Monospores

(*viii*) The alternation of generation in *Batrachospermum* is

(*a*) Haplontic

(*b*) Haplo-diplobiontic

(*c*) Haplohaplohaplontic

(*d*) Diplobiontic

(*ix*) Chantransia stage is produced as a result of germination of

(*a*) Oospores

(*b*) Zoospores

(*c*) Zygote

(*d*) Carpospores

(*x*) The thallus of *Batrachospermum* consists of

(*a*) Only primary laterals

(*b*) Only long branches

(*c*) Both of the above

(*d*) None of the above

Batrachospermum sp

CHAPTER 23

ORDER : Ceramiales

- Family : Rhodomelaceae
- Family : Polysiphonia
- Economic Importance of Rhodophyceae

The members of this order are characterised by : a plan of growth by a primary filament with an apical cell; the primary or axial filament becomes corticated and thus **polysiphonous; diplobiontic** life cycle with isomorphic alternation of generations; a definite, 4-celled **procarp; auxiliary** cell formed directly from the supporting cell after fertilization; fusion of the carpogonium and auxiliary cell, and occurrence of usually tetrahedral **tetraspores**. The order includes about 900 species which are placed under nearly 160 genera. It comprises four families which are *Ceramiaceae, Delesseriaceae, Rhodomelaceae* and *Dasyaceae*. Out of these Rhodomelaceae is considered here.

FAMILY : RHODOMELACEAE

The family includes about 100 genera. All the genera are characterised by the **polysiphonous** structure of the thallus and formation of two different types of laterals on the thallus. The carpogonial branch is four-celled. Following fertilization the supporting cell cuts off auxiliary cell towards its apex. The auxiliary cell directly functions as the gonimoblast initial. The cystocarp becomes enclosed in a prominent pericarp with a definite aperture. Falkenberg (1901) divided the family into two sub-families, *Polysiphonieae* and *Lophothalieae*. We take *Polysiphonia*, which is the commonest and the best known genus, as an example of this family. It is included in the sub-family *Polysiphonieae* which is characterised by the radial construction of the thallus.

POLYSIPHONIA

(*Sub-family : Polysiphonieae*)

Distribution. *Polysiphonia* is exclusively marine and is formed both in tropical and temperate seas. It usually grows in abundance in the intertidal belt and sub-littoral region. It includes about 200 species (Womersley 1979). Most of them are found in the littoral zone in tidal marshes, brackish estuaries and tide pools frequently growing as epiphytes (e.g. *P. ferulacea* and *P. urceolate*) on rock weeds. They are commonly met with in the British seas and also in abundance along with Atlantic coast of the American continent. *Polysiphonia* is less common along the Pacific coast. Some sixteen species have been recorded from India. Of these *Polysiphonia platycarpa* is common on wooden wharves in the Kumari harbour. *P. variegata* occurs along the Karachi coast in Pakistan and is also found in Okha port in India. *P. ferulacea* grows epiphytically of *Gellidium pusillum* and shells of barnacles in rock pools near the lower water mark.

Habitat. Most of the species occur in quiet waters. Some, such as *P. nigrescens*, are found in rough waters. *P. variegata* inhabits polluted water near estuaries. It is frequently found on the roots of mangrooves. *P. urceolata* is epiphytic on *Laminaria*. Since this species occurs on parts of *Laminaria* where there is accumulation of organic matter, Tobler considered it as an indication of saprophytism. *P. fastigiata* is found on the fronds of a brown sea weed *Ascophyllum nodosum* and occasionally on *Fucus*. It too destroys some of the host cells near the point of attachment. It is thus considered to have an endophytic attachment. Some botanists regard it to be a semiparasite and an example of an alga which may be on the process of becoming a parasite. This view is supported by the fact that it can still live an independent life but the plants living apart from the host are smaller and less healthy. Some species grow as annuals and some as perennials.

Organisation and structure of the thallus (Fig. 23.1). *Polysiphonia* has a filamentous thallus which is generally brownish red to purplish red in colour. The filaments branch and rebranch several times giving the plant body a beautiful, delicate, feathery appearance. The thallus is attached to the substratum in water by a holdfast. Many species of *Polysiphonia* such as *P. urceolata* and *P. nigrescens* form thick tufts or dense bushes on the substratum (Fig. 23.1). The filamentous thallus in most of the species, if not all, is of heterotrichous form. It consists of an erect or projecting system of branched filaments and a filamentous prostrate system (Fig. 23.2).

Fig. 23.1. *Polysiphonia sp.* showing habit.

1. Basal Prostrate system. (Fig. 23.2 A). The filaments constituting this part of the thallus creep over the substratum and are anchored to it by means of thick-walled, elongated unicellular rhizoids arising from the peripheral cells facing the substratum. The distal ends of rhizoids expand to form flat, irregularly lobed attachment pads or discs which are usually only developed in contact with the substrate. The creeping filaments lack trichoblasts and function as a means of perennation.

2. Upright or vertical filaments. These arise from the creeping filaments and may attain height of about 25 cm or even more. Some of them may be formed by the upturned tips of the creeping axes.

There are other species (*P. elongata* and *P. violacea*) in which the creeping or prostrate system is absent. The thallus simply consists of the upright axis anchored to the substratum by unseptate rhizoids arising from the lower peripheral cells of the erect axis. The rhizoids may combine to form a massive disc at their distal ends (Fig. 23.2B).

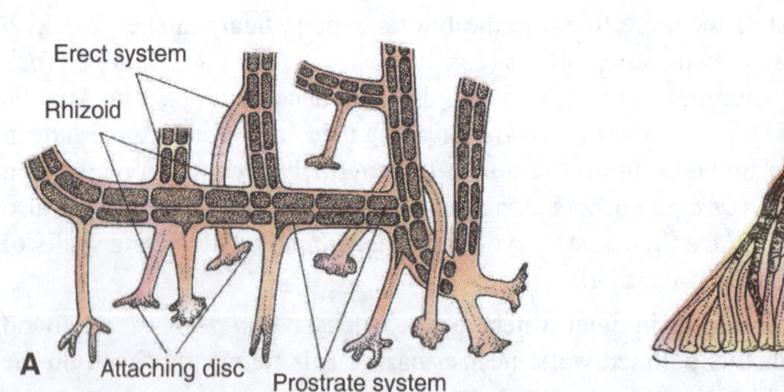

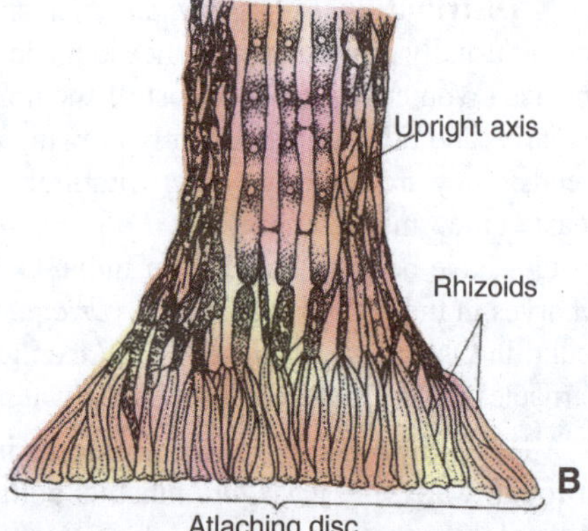

Fig. 23.2 (A-B). *Polysiphonia.* (A) Lower portion of thallus of *P. nigresceus* showing the prostrate part with its attachment system and some erect filaments arising from it; (B), lower portion of the upright axis of *P. violacea* with a massive attaching disc at its distal end (After Battin).

Structure (Fig. 23.3 A-C). The aerial or upright portion of the thallus is feathery in appearance (Fig. 23.1). It consists of branched filaments which branch and rebranch several times giving the thallus a beautiful feathery appearance (Fig. 23.1). The branching is usually lateral and the branches

Polysiphonia denudata.

Polysiphonia nigrescens.

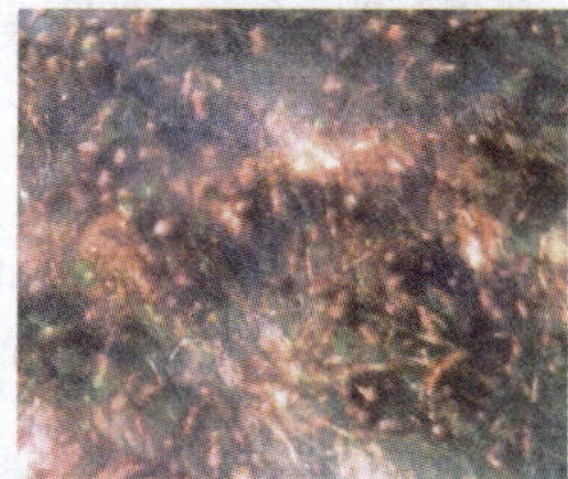

Polysiphonia tuticoriensis.

Polysiphonia violacea.

Polysiphonia dotyi.

Polysiphonia fucoides.

are of two kinds, long and short. The short branches are of limited growth and are known as the **trichoblasts**. The trichoblasts are colourless, hairlike and forked. They are borne on the long, erect branches of unlimited growth. Closer examination, however, reveals that the main filament and the long branches, each consists of a system of parallel filaments. These are called the **siphons** (A). There is one axial filament termed the axial or **central siphon** surrounded by a jacket of a variable number (4-20 or even more) of peripheral filaments called the **pericentral siphons**. Such a thallus is called **polysiphonous**. This algal genus thus derives its name from the polysiphonous nature of its thallus (*A*). The number of peripheral or pericentral siphons is generally constant for a given species. The thallus of *Polysiphonia* in surface view under low power of the microscope is thus seen to consist of several tiers of cells placed end to end in the direction of the length of the axis. The cells of the central siphon may be comparatively large and elongated and usually remain uninucleate. They are called the **axial or central cells**. The cells of the peripheral siphons are called the **pericentral cells**. Because of nearly the same length of the cells in the central and peripheral filaments and the regular manner in which the tufts of cells are arranged the main axis appears to be differentiated into nodes and internodes. The cell walls are gelatinous in character. The whole filament is thus clothed externally by a common firm gelatinous investment usually infested with diatoms and minute epiphytic filaments. In *P. spiralis* the central siphon is surrounded by four pericentral siphons. The latter do not run parallel to the central siphon but are spirally twisted around it. The apical region of the thallus consists only of the central siphon (Fig. 23.3 C-D). In the ecorticated species the pericentral cells remain undivided.

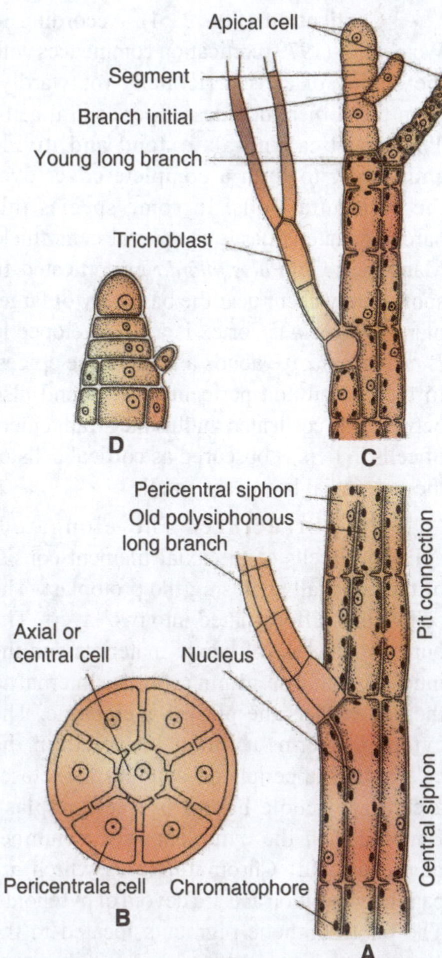

Fig. 23.3 (A-D). *Polysiphonia.* (A).L. S. through mature portion of the thallus; (B), cross section of thallus showing pit connections between the adjacent pericentral cells; (C), apical portion of the thallus in surface view; (D), apex showing separation of pericentral cells from the central cell.

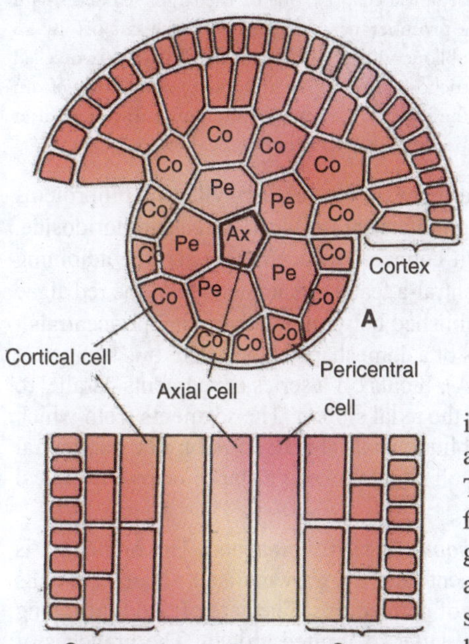

Fig. 23.4 (A-B). *Rhodomelaceae.* Diagrams illustrating cortex formation. (A), Transverse section; (B), Longitudinal section (After Falkenburg)

Cortication (Fig. 23.4). According to Womersley (1979) cortication commences with the cutting of cells periclinally (outwardly) from the corners or sides of pericentral cells. These cortical initials extend and divide anticlinally to form a complete cover over the pericentral cells. In some species this parenchymatous tissue is several cells thick. Many species of *Polysiphonia* are corticated. In some it may occur near the base only of larger plants (*P. mollis*). Cortex is well developed in *P. crassiusula*. It extends almost to the apices. In the central and pericentral cells and also between the corticated thallus the arrangement of cells in tiers is obscured as cortical cells of the superficial layers are small.

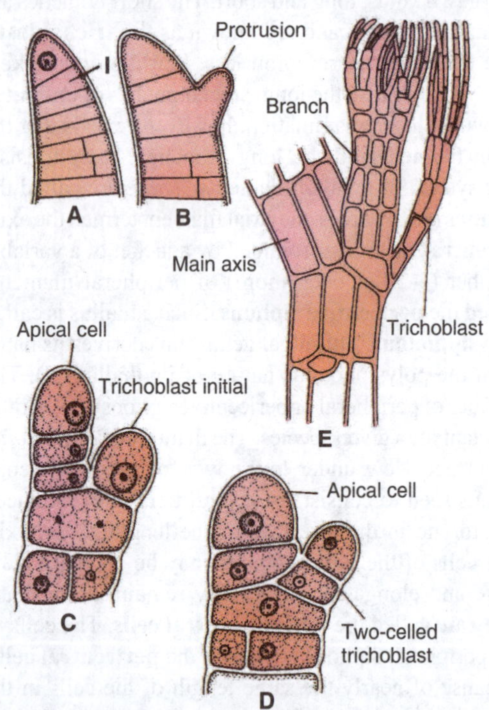

Fig. 23.5 (A-E). *Polysiphonia.* (A), showing segmentation of apical cell; (B), one of the larger segments with a protuberance; (C), protuberance cut off by an oblique wall as a trichoblast initial; (D), two celled tricho-blast; (E), polysiphonous branch arising from the base of the trichoblast (A after Taylor, d, after Berthold and E, after Boergesson).

Cell Structure. The elongated, cylindrical cells of the axial filament consist of the cell wall enclosing the protoplast. The cell wall is differentiated into two layers. The outer is made up of pectic materials and the inner of cellulose in microfibrils. Internal to the cell wall is the plasma membrane. The cytoplasm forms a lining layer within the cell wall. The peripheral cytoplasm encloses a central vacuole bound by the tonoplast. Embedded in the cytoplasm are a number of red discoid Chromatophores which are parietal in position and are devoid of pyrenoids. The photosynthetic pigments located in the chromatophores are chlorophyll a, chlorophyll d, and carotene, a few xanthophylls, and biliproteins (r-phycoerythrin and r-phycocyanin). The reserve food products are floridean starch and floridoside. The single nucleus lies in the peripheral cytoplasm. Pit connections occur between the adjoining central cells and between the central cells and the pericentral-a feature characteristic of the red algae (Fig. 23.3 B). Later secondary connections may be established between the overlying pericentrals.

Growth (fig. 23.5 A). The thallus grows by means of a domeshaped apical cell. It is situated at the extreme tip of the naked part of the central siphon (A). It cuts off a series of segments parallel to its flat posterior face. These segments elongate to form the axial siphon. The segments from which laterals (branches) originate are larger and cut off by oblique walls (Fig. 23.5 A.1). The pericentral cells are cut off by the vertical division of the central cell situated some distance below the apical cell (Fig. 23.5 D).

Branching. The filamentous thallus of *Polysiphonia* is richly branched. The branching is usually lateral. The branches arise from the axial filament near the growing apex (about 2-5 cells removed from the apical cell) before the differentiation of pericentrals. They are of two kinds, long and short. The former are of unlimited growth and the latter of limited growth. The branches of limited growth are called the **trichoblasts** (Fig. 23.5, E). The trichoblast is simple and consists of a single row of cells with pit connections between the successive cells. It is of limited growth and forked. The trichoblasts are spirally arranged on the main filament. They are usually deciduous and are shed annually in the perennial species before winter. The basal cell of the lost trichoblast, however, remains buried between the pericentral cells as a "**scar cell**". The trichoblasts reappear in spring.

The trichoblasts bear the sex organs. In some species the trichoblasts are restricted to certain regions of the plant. In others they may be distributed over a considerable part of the thallus.

The long branches are of unlimited growth and are polysiphonous (Fig. 23.5 E) and may bear trichoblasts arranged in a spiral manner. The growth of the long branch is by means of the activity of an apical cell.

Origin of trichoblasts (Fig. 23.5 B-D). An axial segment two to five segments below the apical cell grows into a protrusion (B) which is cut off as a small cell by a diagonal wall formed at its base (C). It is the **trichoblast initial**. It divides repeatedly to form a dichotomously branched (usually with 2-4 furcahons) gradually tapering hair-like multicellular trichoblast (E). The cells in the trichoblast are arranged uniseriately in a linear file. The trichoblast is thus **monosiphonous**. The component cells are uninucleate, colourless and contain leucoplasts. After cutting of a trichoblast initial, the cells of the axial filament divide in a vertical plane to cut off an encircling layer of pericentral cells (E).

Origin of long branches (Fig. 23.5 E). In many species such as *P. violacea, P. decipiens* the long branches arise from the trichoblasts almost at the same time as the parent trichoblast. The basal cell of the few-celled trichoblast puts out a small lateral outgrowth. It is later cut off from the basal cell by an oblique septum and functions as an **apical cell** of the long branch. The apical cell cuts off a series of segments arranged in a single file constituting the central siphon of the long branch. These central cells later constitute the pericentral siphons surrounding the central siphon of the branch. The long branch continues growing in the same way as the main axis by the activity of the apical cell. As it continues to grow the trichoblast is displaced to one side so that the branch appears to arise from the axil of the trichoblast. In reality it is a lateral growth of the trichoblast. The long branches may also bear the trichoblasts.

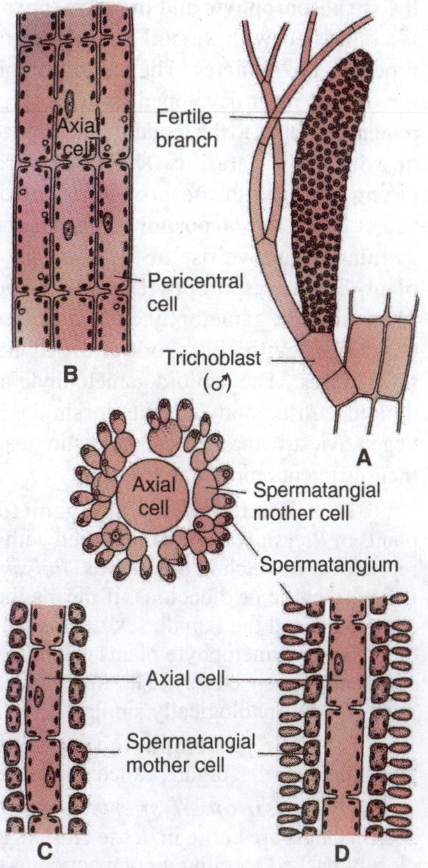

Fig. 23.6 (A-E). *Polysiphonia.* Male sex organs. (A) male trichoblast with the fertile branch seen in surface view; (B-D), stages in the development of spermatangia; (B), optical section of fertile branch with encircling pericentral cells; (C), pericentral cells divide to form spermatangium mother cells; (D), spermatangial mother cell bearing spermatangia; (E), cross section of fertile (spermatangial) branch with developing spermatangia.

The long branches in *P. nigresceus, P. elongata* and in some other species arise direct from the main axis in a spiral manner near the growing apex before the differentiation of the pericentral cells (Fig. 23.3 C). The initial of the long branch arises as a small lateral outgrowth from an axial segment, 2-5 cells away from the apical cell. The lateral protuberance is then cut off by an oblique wall from the parent cell and becomes the **apical cell** of the long branch. The apical cell cuts off a series of segments parallel to its flat face producing a file of axial cells constituting the central filament. The cells of the central filament cut off a peripheral layer of pericentral cells after having cut off a trichoblast initial.

Reproduction

In the life cycle of *Polysiphonia* occur three separate kinds of plants. These are the **gametophyte**, the **carposporophyte** and the **tetrasporophyte**. The gametophyte is a free living haploid plant. It is concerned with **sexual reproduction** and produces the **gametes**. The carposporophyte is the diploid plant developed from the zygote. It remains attached to the female gametophyte plant on which it is parasitic. The carposporophyte is concerned with the production of diploid spores called the **carpospores**. Each carpospore germinated to give rise to the tertrasporophyte plant. The tetrasporophyte is an independent plant like the gametophyte. It is sexless and concerned with the production of haploid **tetraspores**. The haploid gametophyte and the diploid tetrasporophyte plants are similar in their vegetative structure but can be distinguished by their different reproductive organs.

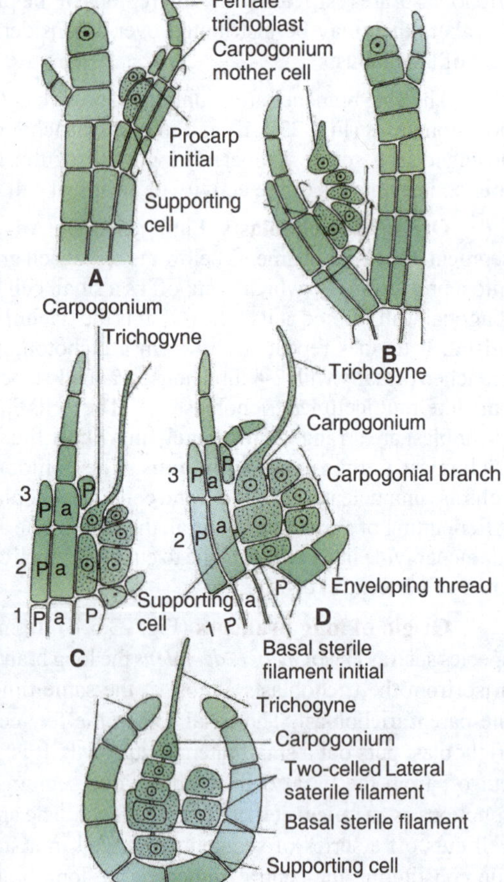

Fig. 23.7 (A-E). *Polysiphonia.* Successive stages in the development of procarp. (A-D), longitudinal sections of female trichoblast; (E), trangenial section of D showing structure of mature carpogonium ready for fertilisation.

1. Gametophytes. The gametophyte plants of *Polysiphonia* are concerned with sexual reproduction which is **oogamous**. *Polysiphonia* is heterothallic or dioecious. It means the male sex organs and the female sex organs are borne on different gametophyte plants called the **male** and **female** respectively. The male and female plants are morphologically similar.

(*a*) **Male gametophyte** (Fig. 23.6A). It bears the male sex organs called **spermatangia**.

(*i*) *Position of spermatangia.* The spermatangia are borne in dense clusters and are closely packed forming a compact cone-shaped structure on short monosiphonous branches near the apices of the male plants. These branches are called the **male trichoblasts** (Fig. 23.6, A). The male trichoblast usually consists of two basal cells constituting the stalk. The stalk usually forks into two branches. In some species such as *P. lanosa* both the branches may become fertile. In others one branch becomes fertile and the other remains sterile. The latter may develop into a repeatedly forked sterile axis (A). The fertile branch remains unbranched but is many cells in length.

(*ii*) *Development of spermatangium*. (Fig. 23.6 A-D). Excepting the two lowermost cells of the fertile branch of the male trichoblast, all others cut off a variable number of encircling pericentral cells (B). The fertile region of the male trichoblasts thus becomes polysiphonous. According to Grubb (1925) each pericentral cell divides one or more times to produce the spermatangial mother cells which are arranged in the form of a compact layer external to the central siphon (C). Each spermatangial mother cell in turn abstricts along its free surface two to four spermatangia (D) according to the species. This results in a compact, cone-shaped structure consisting of clusters of spermatangia. New spermatangia develop as the old ones are cast away or there may be proliferation of a new one within the empty wall of the discharged spermatangium.

(*iii*) *Spermatangial structure*. The spermatangia are spherical or oblong, unicellular structures. They appear as white or pale coloured patches or dots. At maturity the spermatangium has a large nucleus and a colourless cytoplasm. The spermatangial wall is thick and differentiated usually into three layers. These are an *outer* deeply staining layer, a *middle* gelatinous layer and an *inner* highly refractive layer. The uninucleate protoplast of the spermatangium produces a single male cell called the **spermatium**. It is a unicellular, spherical, non-motile structure. The spermatium is liberated through a narrow apical slit in the elastic spermatangial wall. The liberated spermatia are colourless and so numerous that they cover the central filament. Being non-motile they are transported by the sea water to the female sex organ.

(*b*) Female gametophyte. The female thallus of *Polysiphonia* bears the female sex organs called the **carpogonia**.

(*i*) *Position and structure* (Fig. 23.7 D-E). The female sex organ or the **carpogonium** is somewhat a flask-shaped structure. It consists of a basal swollen portion produced at its unattached end into a relatively long, drawn out tubular structure, the **trichogyne** (D). The female gamete is the uninucleate protoplast within the swollen portion. The trichogyne simply functions as a receptive organ. The carpogonium is situated at the summit of a short lateral curved branch consisting of 4 cells. It is called the **carpogonial** branch or **filament**. The basal cell of the carpogonial branch is termed the **supporting cell**. It cuts off two **sterile filament initials**, one towards its base and the second laterally. The basal sterile filament initial remains undivided (D) but the lateral sterile filament initial immediately divides to form a two-celled **lateral sterile filament**. This is the structure of the carpogonium at the time of fertilization (E). The carpogonial branch ending in the carpogonium is borne on a greatly reduced female trichoblast originating from the central siphon of the female plant near its apex. The pericentral cells of the female trichoblast adjacent to the supporting cell grow into outgrowths known as the **enveloping threads** (D).

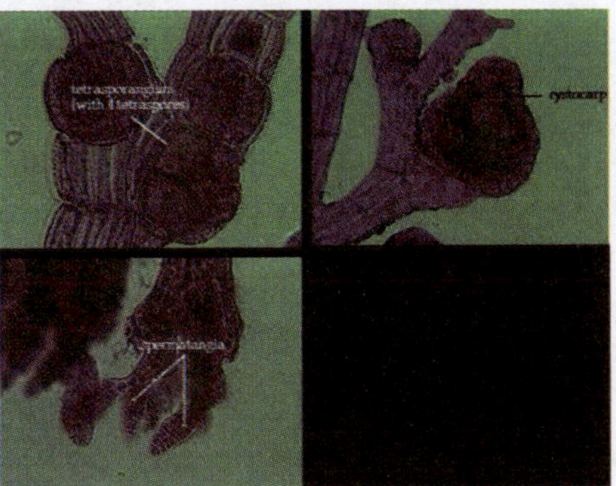

Spermatangia cystocarp and tetrasporangia in Polysiphonia

(*ii*) *Development of the female sexual apparatus* (Fig. 23.7 A-E). The female trichoblast initial arises from the central siphon three or four cells behind the apical cell (A). It divides repeatedly to form a reduced female trichoblast 5 to 7 cells in height (A). The three lowermost cells of the female trichoblast divide vertically to form an ensheathing layer of pericentral cells (*p*) which are arranged in three tiers one above the other (1-3). One of the pericentral cells in the middle tier on the adaxial side becomes the **supporting cell**. The latter cuts off a small initial cell at its free end known as the procarp initial (A). The initial cell divides and redivides to form a small, curved four-celled branch (B). It is the **carpogonial filament** or **procarp.** The end cell of the carpogonial branch functions as the **carpogonium mother cell**. It gets modified into the carpogonium consisting of a long, slender distal portion the **trichogyne** and a basal, swollen portion (C). In the meanwhile the supporting cell cuts off two sterile cells, one towards its base (D) called the **basal sterile filament initial** and another on one side called the **lateral sterile filament initial**. According to

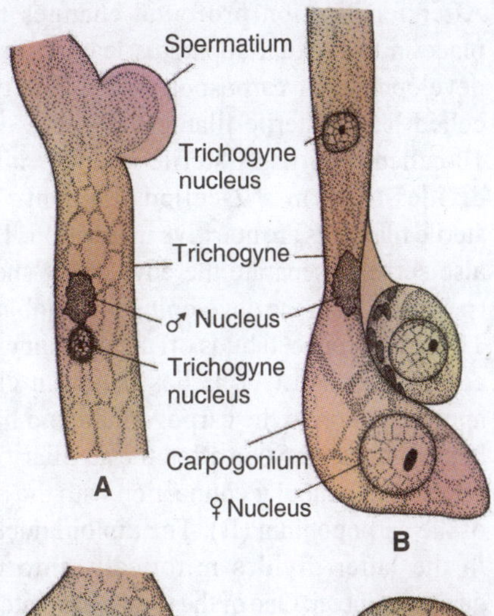

Fig. 23.8 (A-D). *Polysiphonia.* Successive stages in the migration of the male nucleus down the trichogyne and its fusion with female nucleus in the carpogonium (After Yamanouchi).

Kylin (1923) the latter divides immediately to form a two-celled **lateral sterile filament** (E). At this stage the carpogonium is ready for fertilization. The pericentral cells adjacent to the supporting cell grow out into outgrowths which after fertilization develop into an envelope or sheath around the fruit body. The sheath is known as the **Pericarp**.

Fertilization (Fig. 23.8). The liberated spermatia are carried passively by the currents of sea water. As they come in the vicinity of the carpogonium one of them adheres to the trichogyne. At the point of contact the wall between the two dissolves (A). The male nucleus then enters the trichogyne and moves down (B). It reaches the swollen basal portion of the carpogonium. Here it lies by the side of the female nucleus (C) with which it finally fuses to accomplish fertilization (D). The fusion or the diploid nucleus now lies at the base of the carpogonium.

Post-fertilization changes (Fig. 23.9). After fertilization profound changes take place in the sexual apparatus leading to the development of carposporophyte. The two-celled lateral sterile filament becomes 4 to 10-celled. The basal sterile filament initial divides to form a 2-celled filament. The sterile filaments are nutritive in function. They also serve to separate the enveloping sheath (pericarp) from the developing gonimoblasts. The supporting cell buds off an **auxiliary cell** at its upper end (A). It lies below in close approximation to the carpogonium and has a haploid nucleus. Soon after, the auxiliary cell establishes a tubular connection with the base of the carpogonium (B). The diploid nucleus in the latter divides mitotically into two daughter nuclei. One of these migrates into the auxiliary cell through the tubular connection. After the migration of the diploid nucleus into the auxiliary cell, the carpogonial branch gradually begins to shrivel. The outgrowths of pericentral cells adjacent to the supporting cells start producing an envelope around the developing carposporohyte. The haploid nucleus of the auxiliary cell degenerates at this stage. It now comes to possess only the migrated diploid nucleus. The latter divides

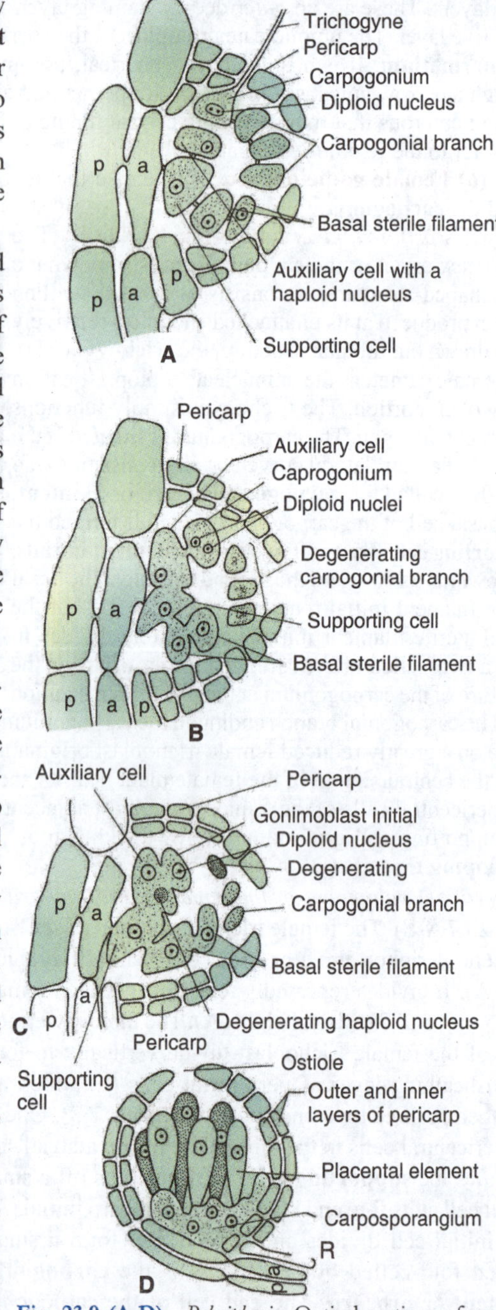

Fig. 23.9 (A-D). *Polysiphonia.* Optical sections of procarp at different stages showing postfertilization changes leading to the development of carposporophyte. (A), differentiation of auxiliary cell; (B), establishment of a tubular connection between the auxiliary cells and carpogonium; (C), migration of one diploid nucleus into the auxiliary cell; (D), mature cystocarp surrounded by a two layered pericarp. (After Kylin).

mitotically into two daughter nuclei. One of these remains in the auxiliary cell and the other migrates into a small lateral outgrowth arising from the upper side of the auxiliary cell (C). It is the **gonimoblast initial**. The latter is separated from its parent cell by a septum. The gonimoblast initial grows into a number of short threads, the **gonimoblast filaments** which form a densely compacted mass. The cells of the gonimoblast filaments have each a diploid nucleus. The end cells of the gonimoblast filaments develop into elongate, pear-shaped **carposporangia**. The carposporangium has a diploid nucleus. The diploid protoplast of the carposporangium develops into a single, diploid **carpospore**.

With the development of the gonimoblasts, there is gradual fusion of the supporting cell with the already fused auxiliary carpogonium (Fig. 23.9 B). The fusion finally extends up to and involves the cells of the sterile filaments and axial cell bearing the supporting cell (Fig. 23.9 C). The resultant large, irregularly shaped structure is called the **placental element** (Fig. 23.9 D). The latter provides nourishment to the growing carposporophyte. Apparently the gonimoblast filaments appear to arise from the placental element. As these structures develop the carpogonial filament gradually degenerates. It takes no part in the formation of the placental element. Meanwhile the pericentral cells of the female trichoblast adjacent to the supporting cell have grown out into outgrowths. These outgrowths finally develop into an urn-shaped envelope or a sheath around the developing fructification. The sheath is called the **pericarp**. It consists of two layers and has a wide aperture, the **ostiole** at its distal end (Fig. 23.10, A). The inner layer of the pericarp is formed directly by the division of the cells of enveloping threads arising from the pericentral cells adjacent to the supporting cell and adaxial pericentral of the basal segment. The entire structure consisting of the placental element, gonimoblast filaments bearing the carposporangia and the surrounding sheath or pericarp is called the **cystocarp**. It is partly a haploid and partly a diploid structure. The shape of the cystocarp varies from subspherical to ovoid.

Carposporophyte (Fig. 23.10 A-C). The diploid portion of the cystocarp constitutes the **cystocarpic plant** or the **carposporophyte**. It consists of the zygote and the gonimoblasts bearing the carposporangia. It is the second individual in the life cycle of *Polysiphonia*. It remains attached and thus parasitic on the female thallus of *Polysiphonia*. It produces the diploid carpospores. The carposporophyte is surrounded and protected by a two-layered pericarp open at the top (A). The latter is a haploid structure. At maturity the pericarp contains the diploid carpospores which float out through the opening or ostiole (A) and are carried by the water currents.

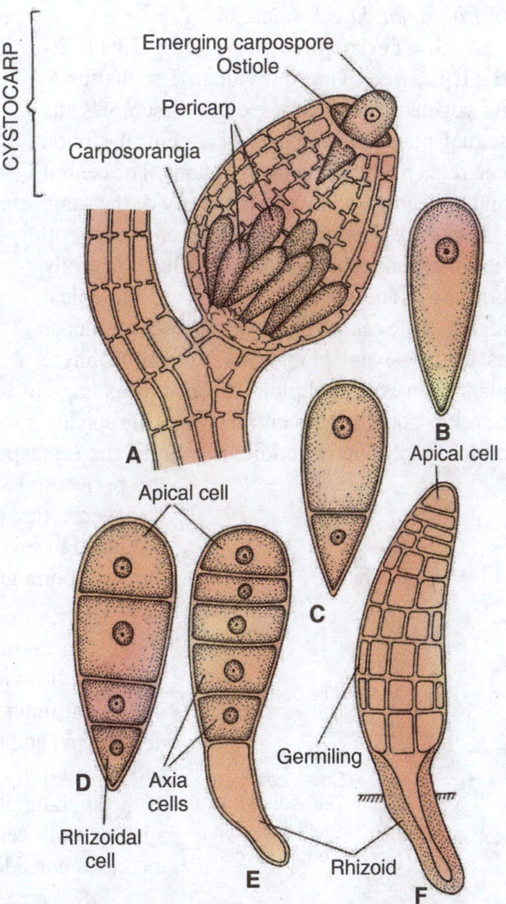

Fig. 23.10 (A-F). *Polysiphonia.* (A) Portion of a branch bearing a mature cystocarp in surface view; (B), a liberated carpospore; (C-D), early stages in the germination of carpospore; (E), Later stage in carpospore germination; (F), a tertrasporophyte germling.

Germination of carpospores (Fig. 23.10 C-F). On coming in contact with a solid object the naked diploid carpospore secretes a wall around it and attaches itself to the substratum (*B*). It then divides by repeated mitotic divisions to form a diploid *Polysiphonia* plant. The first division in the carpospore is asymmetrical and transverse. It divides it into a smaller lower cell and a larger upper cell (C). Each of these again divides transversely forming a four-celled small filament (D). The basal cell of the filament is called the **rhizoidal cell**. It is colourless, elongated and expanded into a disc at its distal end. The end cell of the filament is more or less dome-shaped. It functions as the **apical cell** cutting off segments parallel to its flat posterior face (C). These segments elongate and become the axial cells constituting the central siphon. The lower axial cells divide by vertical walls to cut off the pericentral cells (F). In this way a full-fledged tertrasporophyte individual of *Polysiphonia* is formed.

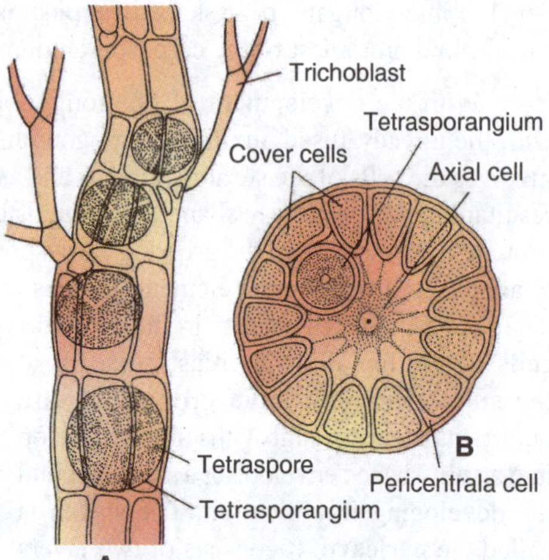

3. Tetrasporophyte (Fig. 23.11, A-B). It is a free living individual. The thallus in its vegetative structure exactly resembles the sexual plants. It consists of a central siphon encircled by the pericentral siphons. The central and pericentral cells are practically of the same size and are arranged from end to end in several tiers one above the other. The thallus is laterally branched. The branches arise from the apical region of the central filament in the same manner as in the sexual plants. The tetrasporophyte plant, however, is diploid and produces haploid **tetraspores** within sac-like structures. The latter develop from the pericentral cells in the apical region of the polysiphonous branches. These spherical diploid reproductive bodies are called the **tetrasporangia**. The tetrasporangia are developed from the pericentral cells. Only one of the pericentral cells in each transverse tier produces a tetrasporangium. Usually they are produced in several successive tiers. The fertile branches bearing the tetrasporangia become swollen and twisted.

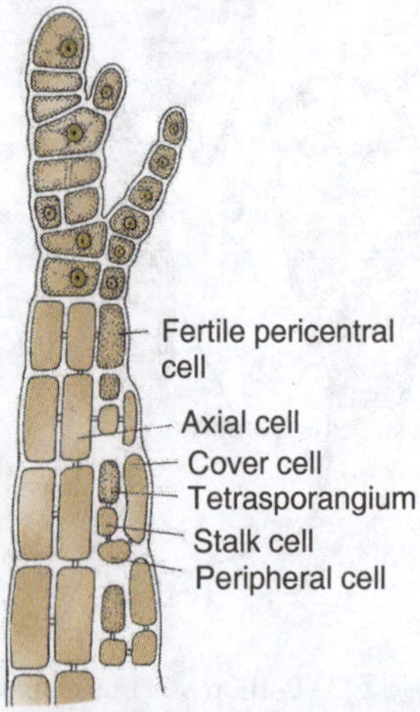

Fig. 23.11 (A-B). *Polysiphonia.* (A), Part of tetrasporaphyte thallus containing mature tetrasporangia in surface view; (B), transverse section of tetrasporphyte thallus passing through a young tetrasporangium (A, after Boergessen and B, after Falkenburg).

(i) Development of tetrasporangium (Fig. 23.12). The fertile pericentral cell is smaller in size than the other cells in the same tier. It divides by a vertical wall into two halves, **outer** and **inner**. The outer half may again divide to form two **cover cells**. (*P. nigrescens*) and in some species (*P. violacea*) a small **peripheral cell** in addition. The inner half of the fertile pericentral functions as the sporangial mother cell. It divides by a transverse wall into a lower **stalk cell** and an upper **tetrasporangium cell**. The latter increases considerably in size. It has a diploid nucleus.

Fig. 23.12. *Polysiphonia* sp. Optical section of apical region of tetrasporophyte thallus showing development of tetrasporangia.

(ii) Differentiation of Tetraspores. The diploid nucleus of the tetrasporangium divides by **meiosis**. The four resultant haploid daughter nuclei arrange themselves in a tetrahedral manner. This is followed by a deepening constriction (cleavage) of the cytoplasm of the tetrasporangium from theperiphery towards the centre. The

resultant four uninucleate meiospores are tetrahedrally arranged (A). They are usually naked and are called the **tetraspores.** Each haploid tetraspore contains twenty chromosomes.

(*iii*) *Liberation of Tetraspores*. When the tetraspores reach maturity, the sporangial wall ruptures and the two cover cells spread apart longitudinally. The tetraspores escape.

(*iv*) *Germination of Tetraspores*. The liberated tetraspore germinates in the same way as the carpospore. It gives rise to the haploid or the gametophyte plant concerned with sexual reproduction. In this way the life cycle continues. Two of the tetraspores of a tetrasporangium give rise to the male plants and the other two to the female plants.

Alternation of Generations (Fig. 23.13)

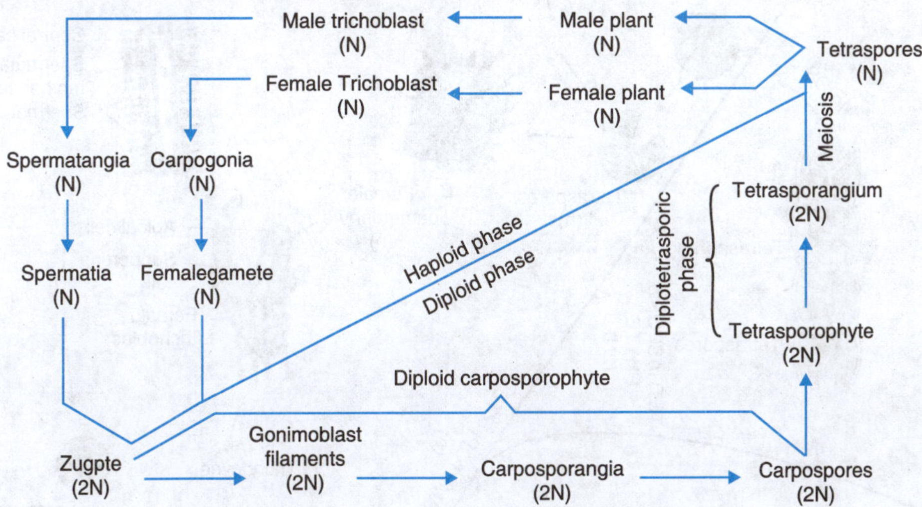

Fig. 23.13. Graphic representation of the life cycle of *Polysiphonia sp.*

In the single life cycle of *Polysiphonia* there occur three different individuals each representing a distinct phase. These are :-

1. The haploid male and the female plants which bear the sex organs. They represent the **sexual phase** or the **gametophyte generation**. It is characterised by *n* number of chromosomes, which is 20. It begins in the life cycle with the formation of haploid tetraspores and ends with fertilization. It is independent and free-living.
2. The diploid **carposporophyte** or the **cystocarpic plant**. It consists of the gonimoblast filaments, the carposporangia and the carpospores. The carposporophyte represents the **first diploid** or **sporophyte generation**. It is characterised by 2 *n* number of chromosomes which is 40. It starts with fertilization and ends with the formation of diploid carpospores. It is attached and dependent for its nutrition on the female plant.
3. The diploid **tetrasporophyte** or the **tetrasporic plant**. It represents the **second diploid phase** or the **sporophyte generation** in the life cycle. It has 40 chromosomes in the nuclei of its cells. It consists of the tetrasporophyte plant and the tetrasporangia. It begins with the germination of the carpospores and terminates with the formation of the haploid tetraspores. It is an independent, free living, diploid phase. The haploid tetraspores germinate to give rise to the new haploid or gametophyte plants.

Thus life cycle of *Polysiphonia* with two separate diploid generations, and one haploid can be termed morphologically triphasic or trigenetic. Cytologically, however, it is **diplobiontic** because it is

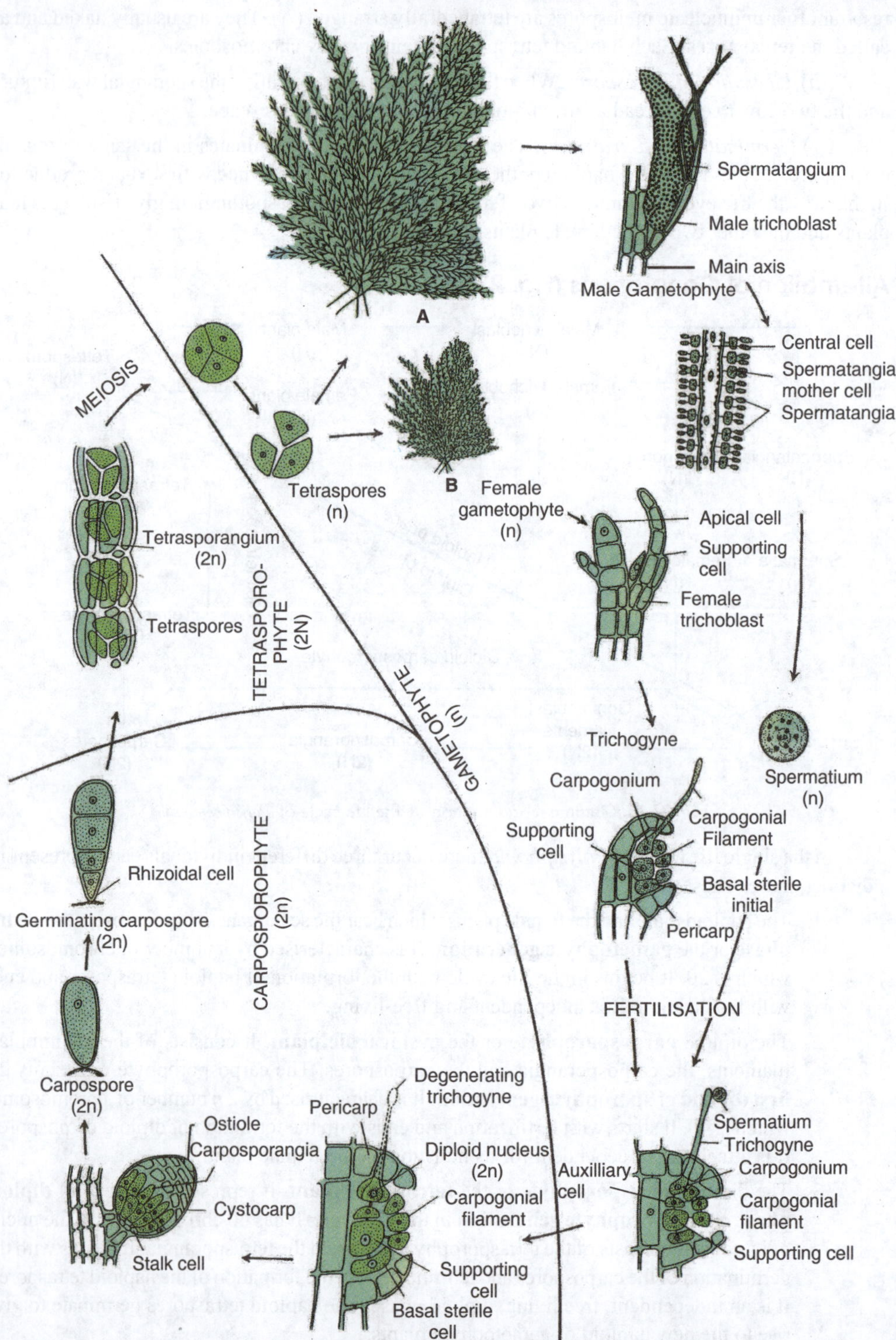

Fig. 23.14. Pictorial representation of the life cycle of *Polysiphonia sp.*

characterised by two kind of individuals, haploid and diploid. A life cycle with three phases in which one is haploid and two diploid is called **diplodiplohaplontic** or **diplobiontic** in contradistinction to the **haplohaplohaplontic** or haplobiontic life cycle of *Batrachospermum.*

In the diplobiontic life cycle of *Polysiphonia* there is regular alternation of one gametophyte phase with two sporophyte generation and those of the independent gametophyte phase with two sporophyte phases. The plants of one independent sporophyte generation and those of the independent gametophyte generation are similar in their morphological characters. This type of alternation of generations is called **isomorphic**.

Salient features

1. The polysiphonous nature of the thallus which consists of an axial siphon surrounded by a number of pericentral siphons.
2. The plant body consisting of the decumbent portion and the upright portion.
3. The feathery, upright portion consisting of a much branched system of relatively large branches bearing numerous small branches called the **trichoblasts**. The latter bear the sex organs.
4. Growth in length by means of a single, dome-shaped apical cell.
5. Occurrence of **pit connections** between the adjoining central cells and between the central cells and the pericentral cells, another characteristic feature of the red algae.
6. Occurrence of three kinds of plants – gametophytes, carposporophyte and tetrasporophyte.
7. Sexual reproduction is **oogamous**. The male sex organs are called the **spermatangia** and the female **carpogonia**.
8. The spermatangia are developed in large numbers on the male trichoblast. The male gamete or the spermatium is non-motile.
9. The carpogonium is a flask-shaped structure consisting of the basal swollen portion and the distal tubular portion-the **trichogyne**.
10. The carpogonium is situated at the summit of a short, lateral, curved branch consisting of three or four cells. It is the **carpogonium filament** or **procarp**. The latter arises from one of the pericentral cells of the reduced female trichoblast.
11. The diploid nucleus formed by fertilization lies at the basal swollen portion of the carpogonium. The latter communicates by means of a tubular connection with the **auxiliary cell** cut off from the supporting cell.
12. The migration of one of the diploid nuclei into the auxiliary cell.
13. The development of **gonimoblast** filaments form the auxiliary cell.
14. The formation of a diploid carpospore in the modified terminal cell of the gonimoblast filament now called the **carposrangium.**
15. The development of the carposporophyte bearing the **carpospores**.
16. The formation of a fruit body or fructification called the **cystocarp** consisting of an urn-shaped sheath containing the carpospores.
17. Development of the **tetrasporophyte** by the germination of the diploid carpospore. The tetrasporophyte is concerned with production of meiospores called tetraspores.
18. Formation of 4 tetraspores in each tetrasporangium borne on the tetrasporophyte.

19. Meiosis occurs at the time of the differentiation of tetraspores which are thus haploid.
20. Germination of haploid tetraspores to give rise to the **gametophyte** or **sexual** plants concerned with sexual reproduction.
21. Distinct alternation of generations consisting of morphologically alike sexual and asexual plants.

Taxonomic Position

Division	:	**Rhodophyta** (Rhodophycophyta)
Class	:	**Rhodophyceae**
Sub-class	:	**Florideae**
Order	:	**Ceramiales**
Family	:	**Rhodomelaceae**
Subfamily	:	**Polysiphonieae**
Genus	:	*Polysiphonia*
Species	:	*platycarpa*

Affinities of red Algae

The red algae are evidently a specialized class. Its affinities with other classes of Algae are uncertain. They have reached a marked degree of complexity of the thallus and have developed their own system of sexual reproduction quite different in the structure of the sex organs and sex cells from those in the greens and browns. The red colour, lack of motile cells, elaborate sexual apparatus and unparalleled post-fertilization changes sharply set them off from all other algae including the blue-greens which have no type of gametes or sexual reproduction. In spite of this an attempt has been made by some algologists to seek their affinities with the green algae. The pseudoparenchymatous investment of the cystocarp recalls in both form and formation that of the oogonium of the green alga *Coleochaete*. Pringsheim and Lambert reported that the meiozoospores of *Coleochaete* do not give rise directly to the thallus plant. They produce a number of asexual generations before the sexual plant is produced. The same type of life cycle is exhibited by a red alga *Batrachospermum.* A series of analogies were also observed in the process of sexual reproduction in *Coleochaete* and certain Nemalionales. However the trichogyne of red algae confirms this viewpoint. The trichogyne in *Coleochaete* is merely an outgrowth of the oogonium. The absence of chlorophyll *b* in red algae is another hurdle against the origin of red algae from green algae like *Coleochaete*. Furthermore, the photosynthetic thylakoids comprising the plastids are stacked into distinct bands in green algae whereas in the red algae they occur singly widely separated and are not pressed into bands.

Some algologists suggest a link with the Cyanophyceae on biochemical grounds. The red algae share the following features with the Cyanophyceae :

1. *Lack of motile structures*. This feature places the red algae apart from every other algal group except the Cyanophyceae.
2. *Similar pigmentation*. The water soluble phycobilins, phycocyanin and phycoerythrin pigments are present in both but no chlorophyll *b*.
3. *Storage of starch variants as reserve food in both.* In the blue-green algae it is cyanophycean starch and in red algae it is floridean starch.
4. *Presence of protoplasmic connections through pits between the adjacent cells* in certain genera of both.
5. Occurrence of unicells and false branching in certain blue-green and primitive Protoflorideae.

6. *Similar organization of photosynthetic lamellae*. They occur singly, widely separated and are not pressed into bands in both the red and blue-green algae.

The above-mentioned similarities have led some biologists to suggest a remote relationship between these two groups of Algae. The suggestion is that either both the algal groups have arisen from some non-flagellated common ancestor or the red algae arose from some simple cyanophycean ancestor by way of the primitive Protoflorideae in which unicells and falsely branched filaments occur.

Despite the existence of a few unicellular forms in the Protoflorideae, the origin of Rhodophyceae from the Cyanophyceae still presents insurmountable difficulties because the unicellular Protoflorideae are generally assumed as reduced rather than primitive. Moreover, the Protoflorideae although less elaborate than the Euflorideae, the two are generally considered to represent divergent lines of evolution from a common ancestral stock. Furthermore, the highly specialized thallus of the red algae and their complicated reproductive apparatus differ markedly from that of the blue-green algae and other algal groups. In fact the blue-greens lack any type of sexual reproduction. Besides, the lack of a true nucleus and clearly differentiated (membrane bound) chromatophores separate the blue-greens from red algae. In the face of all these differences the probability of a direct affinity between the two classes of algae is very small. The possibility, however, is that both could have evolved in the remote past from a common stock such as a nonflagellated ancestor possessing biliproteins. If we assume that the present-day red algae did arise from the unicellular ancestor, the latter are unknown.

Sachs (1875) suggested that the sex organs and fruit body of red algae have a striking similarity to those of the Ascomycetes. These are :

1. The *cystocarp* of the red algae is comparable to the *ascocarp* of Ascomycetes.
2. Presence of *trichogyne* on the female sex organ of both.
3. Occurrence of non-motile, male sex cells, the *spermatia* in both.
4. Resemblance of the gonimoblast filaments of the red algae to the ascogenous hyphae of the Ascomycetes.

In addition to the above, lack of motile cells in the life cycle, filamentous thallus with apical growth and a central pore in the transverse septa are other features which the red algae share with the Ascomycetes.

These similarities prompted Sachs and some other phylogeneticists to hold that the red algae may have given rise to the Ascomycetes by the loss of photosynthetic pigments coupled by a change in the mode of nutrition to parasitism or saprophytism. The adoption of land habit by the Ascomycetes resulted in the development of ascus as the spore producing structure and ascospores as the type of spores replacing tetrasporangia and naked tetraspores respectively. This hypothesis is mainly based on the studies of red algae growing in the deeper recesses of the ocean where light is minimum and thus the stimulus to take over to another mode of existence seems to be present. The opponents of this view consider these similarities nothing more than coincidental or consider them as case of parallel development.

Economic Importance of Rhodophyceae

The red algae furnish food for sea animals. Some species are fed to cattle. They have also been regularly used as human food. Some of them such as *Rhodymenia* is known as Dulse in Scotland, dillis in Ireland and Sol in Iceland and is used as human food. Lever and Amanori are popular Japanese dishes which are prepared from *Porphyra*, Irishmoss (*Chondrus*) and *Kanten* are the important examples of red algae which serve as food plants. They are eaten extensively by the people of Asiatic countries and rocky coasts of Europe and N. America.

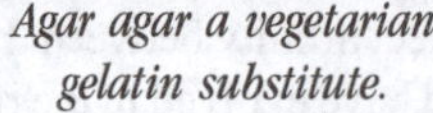
Agar agar a vegetarian gelatin substitute.

Agar agar used in jellies.

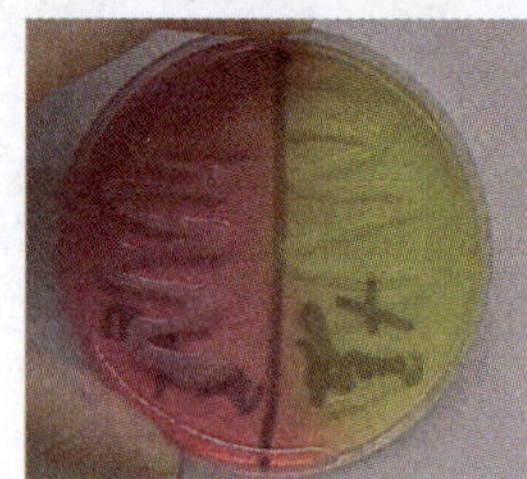
Agar agar used to culture micro organisms

Agar agar used to solidify pandan custard.

Porphyra (Fig. 21.3 C) is an important genus used as food by the Japanese and Chinese Americans. It forms an important ingredient in soups and is cooked as a flavouring with meat. In Japan it is considered a great delicacy and is raised in submarine gardens made of bundles of bamboos lowered in sheltered covers. *Rhodymenia palmata*, commonly known as Dulse, is boiled in milk in Ireland and then eaten. It is also used as a mendicament and eaten in dried form and in soups.

Polysaccharides (geloses) which are important agents of medicinal and industrial use are obtained from the cell walls of many red algae. The industrial uses of red algae are many. They are used as a substitute for animal gelatin in the manufactured puddings. The products obtained from many red algae are used in making shoe polish, cosmetics, glue, jellies, in pie fillings, in preserves, candies, tooth paste, as a sizing agent, as a filter, and as a stabilizer in ice cream. An extract of *Chondrus crispus* (Irish moss Fig. 24.15 B) which is known as **carrageenin** finds many uses. It is used in the preparation of chocolate milk and various pharamaceuticals including laxatives and cosmetics.

The striking use of red algae is in the manufacture of **agar** and agar was first extracted from *Eucheuma*. It is gelatinous, bleached material which is extracted from certain red algae like species of *Gelidium* and *Gracilaria*, *Petrochladia, Acanthopeltis,* and *Chondrus*. *Rhodymenia* is one of the most used red algae for agar extraction. In India agar agar is obtained from the species of *Gelidium, Gelidiella, Gigartine, Gracilaria, Gymnogangyrum* and *Hypnea*. Agar is sold in dry form. It can be dissolved in hot water and causes the water to yell "on cooling". Agar is used in medicine as a laxative, in clarification in liquors and also as a culture medium for the growth of bacteria and fungi. Its use in the bacteriological laboratories is indispensable for no suitable substitute for agar is known. It is added to culture media upon which microorganisms are grown. It causes the media to solidify. It is also used in baking, canning, cosmetics and medicinal purposes.

Calcareous red algae (*Lithothamnion, Lithophyllum, Porolithon* and *Goniolithion*) play an important role in the formation of so-called coral reefs.

QUESTIONS

ESSAY TYPE

1. Describe the life history of *Polysiphonia*.

   (*Kanpur, 1995; Awadh, 1999; Bundelkhand, 1999; Bharathiar, 2000, Lucknow, 2004*)
2. Illustrate the phenomenon of alternation of germination in the life history of *Polysiphonia* and compare it with that of the Archegoniatae.

   OR

   Explain the alternation of generations with reference to *Polysiphonia*.

   (*Bharathiar, 2000*)
3. Give an account of life history of *Polysiphonia* with necessary diagrams and point out the special features in it.

   (*Kanpur, 1998*)
4. Describe the reproduction of *Polysiphonia*. (*Bihar, 1991*)
5. Describe the process of cystocarp formation in *Polysiphonia*.

   (*Lucknow, 1994; Bharathiar, 2000*)
6. Where is *Polysiphonia* found? Describe morphology of the thallus and explain formation of tetrasporophyte.

   (*Lucknow, 1995*)
7. Give a detailed account of post fertilization changes in *Polysiphonia* till the formation of cystocarp.

   (*Lucknow, 1996; Agra, 1956; Vir Kumar Singh Univ., 1991; Vinoba Bhave Univ., 1998; Bhagalpur, 1991*)
8. Compare life cycle patterns of *Batrachospermum* and *Polysiphonia*.

   (*Allahabad, 1999*)
9. Define alternation of generation. Explain it with the help of life cycle of *Polysiphonia*.

   (*Awadh, 1993; Madras, 1995; Bundelkhand, 1995*)
10. Explain the changes that take place after male and female gametes fuse in *Polysiphonia*. Illustrate your answer with labelled diagrams. (*Agra, 1992*)
11. Give a diagrammatic presentation of *Polysiphonia* life cycle.

    (*Agra, 1995; Meerut, 1998*)
12. Describe briefly the illustrated life history of *Polysiphonia*. (*Kumaon, 1998*)
13. Give an illustrated account of structure and reproduction in *Oedogonium* or *Polysiphonia*.

    (*Kumaon, 2000; Ravi Shanker, 1991*)
14. Describe the reproduction in red alga studied by you. (*Rohilkhand, 1994*)
15. With labelled diagrams, describe the structure of thallus and sexual reproduction in *Polysiphonia*.

    (*Rohilkhand, 1997; Mahatma Gandhi Univ., 1998*)
16. Compare life history of *Polysiphonia* with that of a fern.
17. Give a diagrammatic representation of the life cycle of *Polysiphonia* illustrating the relative lengths of haploid and diploid phases.
18. Illustrate the phenomenon of homologous alternation of generation from the life history of *Polysiphonia*.
19. (*a*) Give a comparative account of the mode of reproduction of *Chara* and *Polysiphonia* and discuss the evolutionary tendencies in them.

    (*b*) Describe sexual reproduction in *Polysiphonia*. (*Kerala, 2001*)
20. Describe the triphasic diblobiontic type of life cycle of an alga studied by you. Illustrate your answer with suitable labelled diagrams.

    (*Allahabad, 2002, 2006*)
21. Describe post fertilization changes leading to cystocarp formation in *Rhodomeniales* and *Ceramiales*.(*Rohilkhand MSc, 2005*)

SHORT ANSWER TYPE

22. Identify the following by giving one important character and give their systematic position
    (a) *Polysiphonia*
    (*Kanpur, 1996, 1999; Awadh, 1992; Rohilkhand, 2000; Bundelkhand, 1993, 1998*)
    (b) *Porphyridium*
23. Describe the female reproductive structure of *Polysiphonia.* (*Kanpur, 2005, 2006*)
24. Compare:
    (a) *Ceramiales* with *Nemalionales*
    (b) *Polysiphonia* with *Batrachospermum*
25. Draw well labelled sketches of the following :
    (a) Cystocarp of *Polysiphonia* (*Lucknow, 1992, 1993*)
    (b) Thallus structure of *Polysiphonia* (*Awadh, 1992*)
    (c) Filament of *Polysiphonia* with male sex organs. (*Meerut, 1989*)
    (n) Trichoblasts in *Polysiphonia.* (*Allahabad, 2005; Rohilkhand M.Sc., 2007*)
26. Write short notes on :
    (a) Cystocarp of *Polysiphonia* (*Lucknow, 1991, 1996, 1997, 2000, 2001; Awadh, 1995; Agra, 1994, 1996; Osmania, 1991, 1992; Meerut, 1993; Bundelkhand, 1994, 1997; Vir Kumar Singh Univ., 1986; Raipur, 1994*)
    (b) Sexual reproduction in *Polysiphonia* (*Allahabad, 1998; Meerut, 1991*)
    (c) Reproduction in *Polysiphonia* (*Awadh, 1997*)
    (d) Female reproductive organs in *Polysiphonia* (*Agra, 1991*)
    (e) Post fertilisation changes in *Polysiphonia* (*Kanpur, 1999; Rohilkhand, 2005*)
    (f) Sex organs in *Polysiphonia* (*Rohilkhand, 1992*)
    (g) Structure of thallus in *Polysiphonia* (*Rohilkhand, 1996*)
    (h) *Polysiphonia* (*Osmania, 1992, 1993*)
    (i) Tetrasporophyte in *Polysiphonia* (*Andhra, 1991; Meerut, 1997; Lucknow, 2003; Allahabad, 2006*)
    (j) Alternation of generation in *Polysiphonia* (*Ravi Shankar Univ., 1995*)
    (k) Development of cystocarp in *Polysiphonia* (*Meerut, 1997*)
    (l) Cystocarp (*Bangalore, 2001; Allahabad, 2003; Rohilkhand M.Sc., 2007*)
    (m) Cystocarp formation in *Polysiphonia.* (*Lucknow, 2002*)
    (n) Trichoblasts in *Polysiphonia* (*Allahabad 2005; Rohilkhand, M.Sc., 2007*)
27. Name the type of branching found in trichoblasts of *Polysiphonia* (*Rohilkhand, 1992*)
28. Name the semi parasitic species of *Polysiphonia* (*Rohilkhand, 1993*)
29. What type of thallus is seen in *Polysiphonia*? (*Nagarjuna, 1994*)
30. Explain the tetrasporophyte in *Polysiphonia* (*Bharathiar, 1995*)
31. Differentiate between
    (a) Carpospore and tetraspore. (*Bundelkhand, 1994, 1997*)
    (b) *Polysiphonia* and *Fucus*. (*Allahabad, 2003*)
32. Where does reduction division takes place in *Polysiphonia* ?
33. Where would you find *Polysiphonia* in India?
34. Describe the corposporophyte of *Polysiphonia*. (*Kanpur, 2007; Rohilkhand, 2007*)

35. Write short notes on tetrasporophyte and development of carposporophyte. (*Kanpur, 2004*)

36. Differentiate between cystocarp and spermocarp. (*Lucknow, 2006, 2007*)

37. Name an alga which produces tetraspores. (*Allahabad, 2001*)

38. What are the different spores produced in the life cycle of *Polysiphonia*? In which generation are they produced? (*Rohilkhand, 2004*)

OBJECTIVE TYPE

39. Fill in the blanks :

(*i*) Heterotrichy means having .... and .... branches.

(*ii*) Tetraspores in *Polysiphonia* germinate to produce ....

(*iii*) The alga which yields agar agar is ....

(*iv*) Gonimoblast filaments in *Polysiphonia* is ....

(*v*) Semiparasitic species of *Polysiphonia* is *Polysiphonia* ....

(*vi*) The alga which produces tetraspores is ....

(*vii*) The alternation of generation in *Polysiphonia* is............... .

(*viii*) Reduction division in *Polysiphonia* takes place during the formation of ....

(*ix*) The adult plant of *Polysiphonia* is ....

(*x*) In *Polysiphonia* short branches of limited growth are also called ....

40. Select the correct answer :

(*i*) Which of the following alga is marine?
(*a*) *Vaucheria*
(*b*) *Polysiphonia*
(*c*) *Chlamydomonas*
(*d*) *Spirogyra*

(*ii*) Heterotrichy means having
(*a*) Long and short branches
(*b*) Prostrate and erect branches
(*c*) Branches modified into leaves and air bladders
(*d*) Rhizoidal and photosynthetic branches

(*iii*) Agar Agar is obtained from
(*a*) *Ectocarpus*
(*b*) *Fucus*
(*c*) *Polysiphonia*
(*d*) *Gelidium*

(*iv*) The tetraspore germinates to produce
(*a*) Tetrasporophyte
(*b*) sporophyte
(*c*) Carposporophyte
(*d*) gametophyte

(*v*) The Red Sea gets its name because of the presence in its water red coloured algae belonging to class
(*a*) Rhodophyceae
(*b*) Phaeophyceae
(*c*) Chlorophyceae
(*d*) Cyanophyceae

(*vi*) The male gamete in *Polysiphonia* is called
(*a*) Spermatium
(*b*) Spermatozoid
(*c*) androspore
(*d*) Zoospore

(*vii*) Cystocarp of *Polysiphonia* is
(*a*) Gametophyte
(*b*) Sporophyte
(*c*) Saprophyte
(*d*) lithophyte

(*viii*) In *Polysiphonia* the life cycle is
(*a*) Haplontic
(*b*) haplo-diplontic

(*c*) Haplo-haplobiontic

(*d*) Haplodiplodiplontic

(*ix*) The sterile sheath of the carposphorophyte in *Polysiphonia* is called

(*a*) Epicarp

(*b*) Procarp

(*c*) Pericarp

(*d*) Cystocarp

(*x*) Reduction division in *Polysiphonia* takes place during the

(*a*) Formation of carpospores

(*b*) Formation of tetraspores

(*c*) Germination of carpospores

(*d*) germination of tetraspores

(*x*) The urn. shaped structure in *Polysiphonia* is

(*a*) Procarp (*b*) Pericarp

(*c*) Archicarp (*d*) Cystocarp

(*xi*) Diploid structure is:

(*a*) Tetraspore

(*b*) Synzoospore

(*c*) Carpospore

(*d*) Hyponospore

Batrachospermum sp.

CHAPTER 24

ALGAE — GENERAL TOPICS

- Evolution of Thallus in Green Algae
- Pigment Constitution in Algae
- Fine Structure of Algal Plastids
- Origin and Evolution of Sex in Algae
- Reproduction in Algae
- Modes of Perennation in Algae
- Economic Importance of Algae
- Culture of Algae
- Water Bloom
- Fossil algae

It is extremely difficult to define the group Algae as they exhibit enormous range in vegetative structure, reproduction and the life history. The study of a large number of plants belonging to this group baffles our imagination and it becomes extremely difficult to draw any concrete conclusion. The group includes the simplest organisms which are represented by a single motile or non-motile cell. It performs all the vital functions of life. The group also includes the most complex types of thalli that bear resemblance to the higher plants. Processes of reproduction are, nonetheless, presenting complexities. On one extreme there are certain algae that reproduce by the simplest and most primitive methods of reproduction whereas on the other extreme there are certain algae that show highly advanced modes of reproduction. There are enormous variations in the life cycles of various algae and they present difficulties in drawing any safe conclusion.

Algae, as defined by Fritsch, "are the holophytic organisms that have failed to reach the higher levels of differentiation characteristic of the Archegoniates". Even this definition does not completely define the term. It has almost become impossible to define the term algae in a perfect manner. Some write of algae as the most primitive thallophytes whereas on the other hand they admit of the higher type of organisation of plant body and complex methods of reproduction met with in a large number of plants belonging to this group. Certain algae show a morphological elaboration and anatomical differentiation that is quite equivalent to or even exceeds that found in many Bryophytes. Complex sex organs of *Chara* possess a cellular wall that takes no part in the formation of the reproductive cells, although another interpretation has been given by Goebel. Presence of trumpet hyphae with perforated walls in the stipe of Laminariales and their heterologous life cycles, clearly speak of their advancement. Besides, there are certain algae (blue-green algae) which have no sexual apparatus and their life cycles show no alternation of generations. So much so their cells have a primitive structure in not possessing a definite nucleus. Such examples speak of their primitiveness. Any generalised statement of algae being primitive plants is not appropriate.

The wide comparisons between simplicity and complexity in the various plants of algal organisation are suggestive of their representing several lines of evolutionary descent; and if this is true, algae are not a natural group of plants. They are, as a matter of fact, an artificial assemblage of plants that according to Gibbs, D, "can more appropriately be spoken of as plants of algal organisation". In the following pages an attempt has been made to discuss some of the general topics which are (*i*) Evolution of thallus in algae, (*ii*) Pigment constitution in algae, (*iii*) Fine structure of algal plastids (*iv*) Origin and evolution of sex in algae, (*v*) Reproduction in algae, (*vi*) Modes of perennation in algae, (*vii*) Life histories of algae, (*viii*) Economic importance of algae, (*ix*) Culture of algae (*x*) Water blooms, (*xi*) Fossil algae, (*xii*) Algae and water pollutions, (*xiii*) Biofertilizers, (*xiv*) Bioluminiscence, (*xv*) Colourless Algae, (*xvi*) Toxic Algae and (*xvii*) Algal viruses or cyanophages.

(*i*) EVOLUTION OF THALLUS IN GREEN ALGAE.

Thallus in green algae presents a considerable variation in the degree of complexity of form. It ranges from motile unicells (*Chlamydomonas*), motile colonies (*Volvox*), non-motile unicells (*Chlorella*), non-motile colonies (*Pediastrum*), unbranched filament, branched filament, heterotrichous filament (*Stigeoclonium*) and even thallus type. This wealth of examples of various body types in green algae suggests a probable sequence in the evolution of multicellular organisms from unicellular ones by gradual progressive elaboration. The motile unicell type of thallus as that of *Chlamydomonas* is considered to be the most primitive. There is every possibility that from such a primitive motile unicell type of thallus arose three main evolutionary lines which resulted in the evolution of present-day more complicated forms (Blackman and West).

(A) Volvocine line (Fig. 4.34). In one of these evolutionary lines there has been a progressive evolution of the motile unicell in the direction of the motile colony probably by adherence of daughter cells of unicelled flagellates, after division, in a mucous envelope. It is called the **volvocine line** of development of the thallus of algae. Motility of vegetative cell is retained in this line. The series runs from the simple motile cell of *Chlamydomonas* through colonies of *Gonium* (4-celled), *Pandorina* (16-celled) and *Eudorina* (32, 64 or 128-celled). The series ultimately culminates in the large highly complex, motile spherical colony of *Volvox* consisting of thousands of *Chlamydomonas* like cells. If the number of cells is relatively few, the colony is disc or cup-shaped but when the number of component cells is large, the colony is a motile, hollow sphere. The development of motile colony has not proceeded beyond the *Volvox* stage. The chief weakness of this evolutionary scheme is the retention of motility by the whole colony. This presents mechanical limitations. The hollow spherical thallus as that of *Volvox* with a diameter beyond 1mm. would become physically unstable. The co-ordination between the component cells on the colony would become impossible. These factors stood in the way of further progress of the volvocine line beyond the *Volvox* stage. Thus, the volvocine line ends as a blind alley in evolution.

(B) Chlorococcine line (Fig. 5.20). In the second main line of thallus development known as the **chlorococcine line**, motility is confined only to the reproductive cells and that to in some genera. The dominant phase is a non-motile unicell. This *coccoid* habit evolved from the primitive unicellular *Chlamydomonas* like ancestor through the loss of flagella. The simplest chlorococcoid forms evolved along two lines, **azoosporic** and **zoosporic**. The former multiply by autospores (*Chlorella*) and the latter by zoospores (*Chlorococcum*). Apparently the chlorococcine line proceeded in more than one direction. In any case the loss of motility of the vegetative cell is the chief characteristic. In one case the chlorococcoid unicell exhibits a strong colonial tendency which is manifested by assemblage of non-motile cells into motionless colonies of different shapes. The series starts with *Scenedesmus*. It is an assemblage of 4 to 8 ellipsoidal or fusiform unicells arranged transversely in one row or two alternating rows. Next is the free floating flat, plate-like colony of *Pediastrum* consisting of 4-128 polygonal cells. The series runs through globular colonies of a variable number of cells and culminates in *Hydrodictyon* in which the colony has the form of an elongated sac-shaped net consisting of several cylindrical cells.

The other characteristic of chlorococcoid unicell is the lack of vegetative cell division. There may be division of the nucleus but no cytokinesis. The protoplast divides only at the time of reproduction. This tendency of the chlorococcoid unicell to become multinucleate for a short while before reproduction has led to the evolution of a coenocytic (siphonaceous) thallus which reaches its climax in the order Siphonales. A very early stage in the development of a siphoneous body is represented by *Characium apiculatum*. The unicell is elongated and has a tendency to become attached to the substratum. The single nucleus undergoes division prior to zoospore formation. The unicell thus becomes multinucleate for a short while before it enters the reproductive phase. Next step is illustrated by *Protosiphon* which is permanently multinucleate. The bladder like aerial thallus tapers to an elongated, colourless underground rhizoidal portion which anchors it to the substratum. Many algologists look upon this alga as a border line genus between the *Chlorococcales* and the *Siphonales*. It illustrates how the Siphonales may have evolved from the Chlorococcales. Some consider the *Siphonocladiales* as transitional stage leading to the evolution of the *Siphonales*. There are some who consider the Siphonocladiales as an off shoot of the order Siphonales derived by incomplete formation of cross walls. The coenocytic siphoneous tendency also enjoyed a limited success because this plan of thallus construction has its limitations. From the account given above it is evident that the chlorococcine line of thallus development is **polyphyletic**. Gupta and Nair (1962) even suggested the origin of filamentous habit (*Ulotrichales*) from the palmelloid Chlorococcales such as *Gloeotoenium* in which the cells are sometimes arranged in a linear series.

(*C*) Tetrasporine line (Fig. 24.1). Under certain conditions, the primitive motile *Chlamydomonas* like cell loses motility temporarily. It then divides into two. The daughter cells do not develop flagella but remain loosely aggregated within the gelatinised wall of the parent cell. The non-motile daughter cells further divide and redivide so that eventually numerous non-motile cells become associated within the gelatinous mass to form a temporary association. This phase of *Chlamydomonas* is called *Palmella stage* because of its apparent resemblance to another green alga *Palmella*–a member of the order Tetrasporales. This order mostly includes unicellular members occurring in an immobile palmelloid condition. The development of such a stage in which the non-motile condition is dominant and vegetative cell divisions are prominent from motile unicellular green algae like *Chlamydomonas* is called the **tetrasporine line** of development. This evolutionary line derives its name from one of the simplest members of the Tetrasporales called *Tetraspora*. In cell structure *Tetraspora* displays relationship to the Volvocales but lack of cell motility indicates that it may have been the progenitor of the tetrasporine line of evolution of the thallus of green algae.

Significance of Tetrasporine line of evolution of thallus. According to one view put forth by Blackman (1900) and supported by West (1916) and Smith (1955), the filamentous Ulotrichales and Ulvales with an expanded (foliaceous) thallus have evolved from the tetrasporaceous algae (Tetrasporales, palmelloid forms). To support his viewpoint Blackman cited an example of a palmelloid form *Palmodictyon varians* in which the cells in the colony are, sometimes, arranged in linear series. This provided him a clue to hypothesize that the restriction of cell division to a single plane (transverse) may lead through forms like *Radiofilum, Geminella interrupta* and *G. minor* to the evolution of a simple filament of *Ulothrix* which consists of a single row of cells firmly connected with one another. The simple filament is considered the forerunner of various branched types.

(a) Branched filament. It is the next step in increasing complexity of thallus structure. The branch arises as a lateral outgrowth when occasional cell in a filament divides in a second plane. If the outgrowth undergoes transverse septation like the main thread, the filament becomes branched. This kind of modification of the simple filament is very frequent. The common examples of green algae with thallus consisting of a branched filament are *Cladophora, Pithophora* and *Bulbochaete*.

(b) Heterotrichous filament. Branching of filament becomes more and more complex. The climax is reached in the heterotrichous filament. It is the most advanced and elaborate type of plant body in the green algae and is found in the Chaetophorales. In the evolutionary transition from the filament, a portion of the filamentous algal thallus developed vertically while the rest which constituted the basal part remained prostrate. The thallus in this case is thus divided into two parts, **prostrate system** and **projecting system**. In *Stigeoclonium* both the systems are well developed. Fritsch (1945) opined that the heterotrichous habit is the forerunner of the origin of land habit in plants. He visualized *Fritschiella tuberosa* as the possible precursor of land plants. In this green alga, the prostrate system develops rhizoids which anchor the thallus to the mud. In the erect system the cells divide in two or three planes to result in parenchymatisation of the thallus which is specialised for photosynthesis.

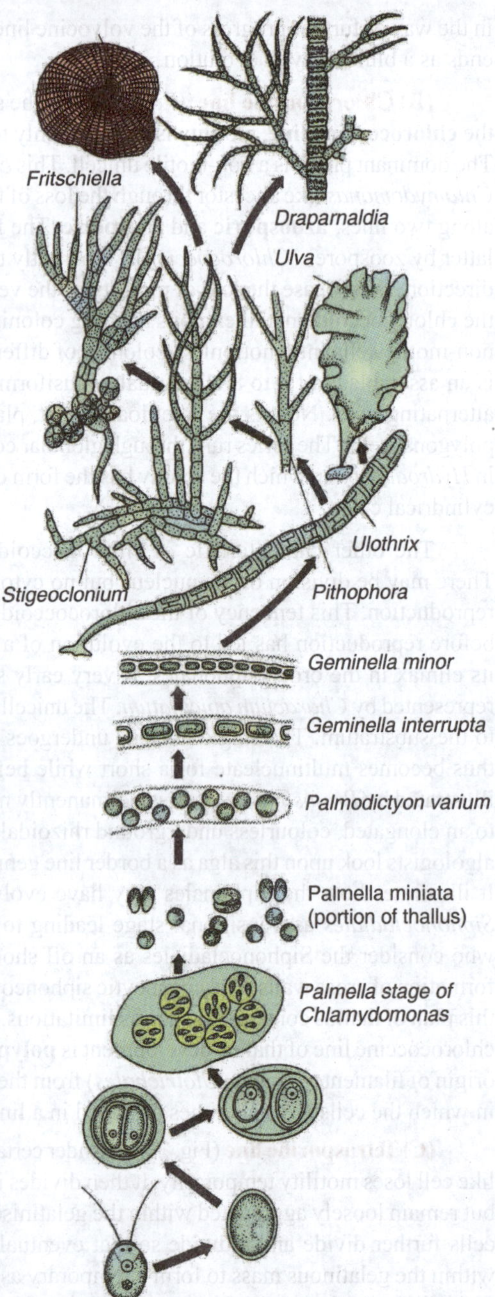

Fig. 24.1. Tertrasporine line of develop-ment of thallus in green algae.

The heterotrichous filament has undergone a variety of modifications by the suppression of the erect system and elaboration of the prostrate system or *vice versa*. The discoid thallus of *Coleochaete* is an example of the former. There is a view that a form similar to *Coleochaete* might have been the progenitors of the liverworts.

In *Draparnaldia* and *Draparnaldiopsis* (green algae) there is complete elimination of the prostrate system and corresponding elaboration of the erect projecting system.

(c) Foliaceous Thallus. Another tendency in the tetrasporine line has been an early departure from the filamentous condition to the leaf-like or foliaceous thallus. The division of the cells of the simple filament in transverse as well as longitudinal planes results in the flat, leaf-like thallus of *Ulva*. The foliaceous thallus in some forms consists of 2 or 3 layers in thickness. It is due to few divisions occurring in a plane parallel to the flattened surface. Such a parenchymatous thallus is tubular in *Enteromorpha*.

(ii) PIGMENT CONSTITUTION IN ALGAE.

Of the external features, colour of the algal thallus is most impressive and distinctive. It may be green, yellow, red, brown, and other shades of these. These striking differences of colour in thalli of Algae form one of the basic criteria for a preliminary classification (Table I). The colour of the thallus is due to the presence of definite chemical compounds called the **pigments.** Each pigment has its own characteristic colour. The particular colour that a thallus has may be due to the predominance of one in a combination of several others. Each algal division has its own particular combination of pigments and a characteristic colour. In all, there are four different kinds of pigments found in the Algae. These are five **chlorophylls**, 20 **xanthophylls**, five **carotenes** and **phycobilins**. Usually the algal pigments are located in the plastids. The latter may be ribbon, band or cup-like in form or a parietal plate. In some they are lens-shaped, disc or network like or an axial band, a stellate or an oval body, a lobed disc or a parietal ring. The Cyanophyta lack plastids, the pigments are located in the lamellae.

The chlorophylls. There are five known chlorophylls, namely, chlorophyll *–a, –b, –c, –d* and *–e*. They show different spectral properties. Of these chlorophyll-*a* occurs in all Algae. Chlorophyll-*b* is found in Chlorophyta, Euglenophyta and Charophyta. Chlorophyll-*c* occurs in Bacillariophyta, Pyrrophyta and Phaeophyta. Chlorophyll-*d* occurs only in red algae. Chlorophyll-*e* has been found in the Xanthophyta (*Tribonema,* and zoospores of *Vaucheria*). The green colour of the thallus in green algae is due to the presence of two closely related pigments **chlorophyll-*a*** and **chlorophyll-*b***. They are the dominant pigments in Chlorophyta, Charophyta and Euglenophyta. Commonly we call these two pigments simply as **chlorophyll**. In the cell the pigments are contained in a specialised part or parts of the cell protoplast called the **plastids**. The plastids containing both chlorophyll-*a* and *b* are called the **chloroplasts** and those which lack chlorophyll *b* and have carotenoids in excess over chlorophyll are usually called **chromatophores**. In the green algae *chlorophyll-a* and *chlorophyll-b* make up about two-thirds of the total pigment in the chloroplast. Associated with these are the yellow **carotenes** and yellow to orange brown **xanthophylls**. There are two carotene pigments namely *a*-carotene and carotene and eight or nine xanthophylls. The chlorophylls, carotenes and xanthophylls are present in different proportions. This accounts for the various external colours of thalli in the Chlorophyta.

Chlorophyll-a and *chlorophyll-b* are microcrystalline solids. The empirical formula for chlorophyll-*a* is $C_{55}H_{72}O_5N_4Mg$ and for chlorophyll-*b* is $C_{55}H_{70}O_6N_4Mg$. *Chlorophyll-a* and *-b* are not soluble in water even under prolonged boiling. They are fat soluble and dissolve in either, benzol, acetone, chloroform, carbon bisulphide and ethyl alcohol. *Chlorophyll-a* gives a bluish green coloration if dissolved in the above solvents whereas *Chlorophyll-b* gives a pure yellowish green coloration. The Chloropohylls absorb blue and red rays and are the import photosynthetic pigments.

TABLE I : Chief algal division and their principal characters

	DIVISION	*Green pigments*	*Other pigments*	*Stored food*	*Nature of cell wall*	*No. of flagella and their insertion*	*Sexuality*	*Cell structure*
1.	Chlorophyta	Chlorophyll *a, -b*	Carotene and xanthophylls	Starch	Cellulose, pectin rarely hemicellulose	2 or 4, equal anterior, whiplash	Isogamy to oogamy	Eukaryotic
2.	Charophyta	Chlorophyll *a, -b*	*y*-Carotene lycopene and xanthophylls	Starch	Cellulose, pectin	2, equal anterior whiplash	Oogamy	"
3.	Euglenophyta	Chlorophyll *a, -b*	Carotene and xanthophylls antheraxanthin	Paramy-lum, Fats	Absent, periplastic	1, rarely 2 or 3, anterior, tinsel	Rarely sexual (isogamous)	gullet present
4.	Chrysophyta	Chlorophyll *a*	*B*-carotenes, xanthophylls	Leucosin, fats, chry-solaminarin	Cellulose absent, or equal.	1 or 2, unequal or anterior, both whiplash or one whiplash and one tinsel	Rarely sexual (isogamous)	Eukaryotic
5.	Xanthophyta	Chlorophyll *a, e*	Carotene and xanthophylls	Oil fat, leucosin	Pectin, cellulose often absent	2, unequal, ante-rior one tinsel and one whiplash	Isogamy, anisogamy rare if present, oogamous in *Vaucheria*	"
6.	Bacillariophyta (Diatoms)	Chlorophyll *a, c*	Carotene, *e*-carotene, xantho-phylls (fucoxanthin)	Fats, leucosin	Sillicified cell wall in two halves, outer layer of hydrated silica and inner a pectin membrane	1, anterior (only in male gamete of some), tinsel	Isogamous, anisogamous and oogamous	"
7.	Phaeophyta	Chlorophyll *a, c*	Carotene, xanth-ophylls (fucoxanthin)	Laminarin, mannitol	Cellulose, with alginic acid and fucocinic acid	Only in reproduc-tive cells, 2 un-equal, lateral whiplash and tinsel	Isogamy to Oogamy	"
8.	Rhodophyta	Chlorophyll *a, d*	*r*-phycoerythin r-ph-ycocyanin, *a*-and Carotene, xantho-phylls (lutein and taraxanthin)	Floridean starch	Cellulose, and pectic material together with polysulphate esters	Absent	Advanced and complex (oogamous)	"
9.	Cyanophyta	Chlorophyll *a*	Blue *c*-phycocyanin, red *c*-phycoerythrin, Carotene, myxo-xanthin and myxo-xanthophyll	Myxophy-cean starch and cyano-phycin	Pectin hemicellulose and mucopeptide	Lacking	Absent	Prokaryotic

The chlorophyll pigments are formed from a substance called **protochlorophyll** which is synthesised in the dark. It is generally changed to chlorophyll only in the light.

The carotenoids. It is a group of yellow, orange, red, and brown pigments. About sixty different carotenoids have been reported in plants. They are placed under two categories, the orange yellow **carotenes** and yellow or brown **xanthophylls** or **carotenols**. The carotenoids are protective pigments functioning as screens to light. In addition they pass on energy of light waves to chlorophyll. They absorb blue and green light waves. There is a suggestion that the carotenoids function as co-enzymes in the photosynthetic process.

(*a*) Carotenes. The carotenes are linear unsaturated hydrocarbons represented by a chemical formula $C_{40}H_{56}$. There are five carotenes so far known, namely, carotene *-a, -B, -e, -y* and lycopene. Of these *B*-carotene is present in most algae. It is replaced by cartone-*a* in the Siphonales (Caulerpales), Cryptophyta and to a lesser extent in red algae. The Charophyta also lack *B*-carotene. It is replaced by *y*-carotene and lycopene. The Bacillariophyta possess both *B* - and *e*-carotene. Carotenes are fat-soluble pigments. They are insoluble in aqueous solution but are soluble in lipid solvents such as ethyl alcohol, chloroform and carbon bisulphide and absorb blue and green light waves.

The resting cells of some of the Algae (terrestrial species of *Trentepohilia*) contain carotenoid pigments outside the chloroplasts. The eyes-pots of zoospores and other motile stages, also contain carotenoid pigments. The haematochrome, or the yellow and the red coloured substances are according to some scientists (Zopf and Wisselingh) represented by two or more carotenoid pigments. The haematochrome in *Sphaerella* appears only when the surrounding medium lacks nitrogen (Pringsheim). The same fact has been observed in some of the Chlorophyceae growing in boggy places or the moorlands. Such algae are yellow in colour due to the fact that the moorlands usually lack nitrogen contents.

(*b*) Xanthophylls. They are yellow or brown pigment represented by a molecular formula $C_{40}H_{56}O_2$. They are closely related to the carotenes but contain oxygen in addition to carbon and hydrogen. Both carotenes and xanthophylls are not watersoluble but are soluble in chloroform. There are many xanthophylls so far known in algae. The chief among them are *Lutein, Siphonein, Zeaxanthin, Siphonoxanthin, Astaxanthin, Neoxanthin, Cryptoxanthin, Violaxanthin, Lycopene, Diatoxanthin, Diadinoxanthin, Flavoxanthin, Myxoxanthin, Dinoxanthin, Taraxanthin, Antheraxanthin, Oscilloxanthin, Myxoxanthophyll, Flavacin, Peridinin, Neofucoxanthin-A and -B and Fucoxanthin.* Of these **Fucoxanthin** is a characteristic pigment of the Phaeophyta imparting distinctive brown or olive brown coloration to the thallus. The diatoms too contain several brownish xanthophylls such as fucoxanthin, diatoxanthin and a few others. Peridinin is found only in the Pyrrophyta. Myxoxanthin and a few others are reported only in the Cyanophyta. Taraxanthin is characteristic of red algae and antheraxanthin of Euglenophyta. Like carotenes the xanthophylls absorb blue and green light rays and are fat soluble. Fucoxanthin is considered a supplementary light absorbing pigment in Phaeophyta and Bacillariophyeceae. An impure solution of fucoxanthin can be extracted from the dead specimens in water and a chemical formula, $C_{40}H_{60}O_6$ has been given to it by Willstatter and Page. *Ectocarpus, Dictyota* and *Laminaria* contain a good amount of fucoxanthin in their chromatophores. In *Fucus* the amount of this pigment is comparatively less.

The Phycobilins. It is another group of pigments comprising the tetrapyrrolic compounds joined to globulin-proteins. So far seven phycobilin pigments both blue and red have been enlisted. They are *Phycoerythrin -r, -c, -x, -b* and *Phycocyanin -r* and *-c*. The seventh is *Allophycocyanin*. The phycobilins are water-soluble pigments found in the red and blue green algae. Of the seven phycobilins, r-phycoerythrin and r-phycocyanin are common. The former absorbs blue, green and sometimes yellow rays whereas the latter absorbs green, and sometimes yellow rays whereas the latter absorbs green, red and yellow. Both the phycoerythrin and phycocyanin are closely related chemically. They are colloidal proteinaceous pigments which can be extracted from the cell in hot water. During extraction procedure of phycobilins the two components (pigment and protein moiety) cannot be separated. To indicate the existence of the pigment protein complex the name of the phycobilin pigments was changed to **biliproteins**. Both the phycoerythrin and phycocyanin pigments are found in the red algae and blue-green algae. Those of the former are of *r*-type and those of blue-green algae *c*-type. Chlorophyll content in them is quite low. In some forms small quantities of chlorophyll-*d* has been reported. Carotene and xanthophyll pigments are also present. Red algae occurring in colder regions show higher content of chlorophyll-*a*. Phycoerythrin crystallises in hexagonal prisms whereas crystals of phycocyanin are rhombic plates. The blue pigment allophycocyanin is present in the blue-green algae only. It is absent in the red algae. The deep red colour exhibited by some of the deep water Florideae is due to the higher concentration of phycoerythrin in the chromatophores. Phycocyanin is either lacking or present in negligible amounts. Some fresh water Florideae like *Batrachospermum* contain a large quantity of phycocyanin and small quantity of phycoerythrin. Their colour is blue-green. Red coloration is evident only after death. *Batrachospermum vagum* is believed to lack phycoerythrin. Phycobilins differ from chlorophylls and carotenoids in being water soluble.

Chlorophyll-*a* is of prime importance in photosynthesis. The accessory pigments function only indirectly. The wavelengths of light which are not absorbed by chlorophyll are absorbed by Phycocyanin and Phycoerythrin. The light energy trapped by the latter two pigments is then transferred to chlorophyll-*a* which utilizes it in food synthesis.

(*iii*) FINE STRUCTURE OF ALGAL PLASTIDS.

The plastids are the pigments bearing double membrane bound organelles found in the plant cells. Like those of the higher plants, the algal plastids consist of a system of membranes suspended in the surrounding granular matrix called **stroma** (Table II). The membrane units are called **lamellae**. Menke (1962) termed them **thylakoids**. Sagar and Palde (1957) coined the term **discs** for them. The disposition of the thylakoids in the plastid varies in the different algal groups. Consequently the algal plastids show diversity of structure which is often distinctive and thus useful in the primary classification of algae into various groups (divisions). In the algal plastids the thylakoids are long and commonly traverse the whole plastid. They occur singly or in multiple layers forming stacks (=bands). Except some of the green algae the stacks are not regular and well defined. The thylakoids in the algal plastids are only of one kind and restricted to the stack itself. They are not closely packed or fused as in the granum of the chloroplasts of higher plants. The thylakoids probably contain all the chlorophyll pigments of the plastid. The algal plastids are usually divided into two categories, **chloroplasts** and **chromatophores**. The chloroplasts contain both chlorophyll *a* and *b* whereas chromatophores possess only chlorophyll *a* but not *b*. On the basis of this distinction the plastids of three algal divisions, Chlorophyta, Charophyta and Euglenophyta are termed *chloroplasts* whereas those of Xanthophyta, Chrysophyta, Bacillariophyta, Pyrrophyta, Cryptophyta, Phaeophyta, and Rhodophyta are known as *Chromatophores*. The algal plastids differ from the plastids of other plants in that they frequently possess

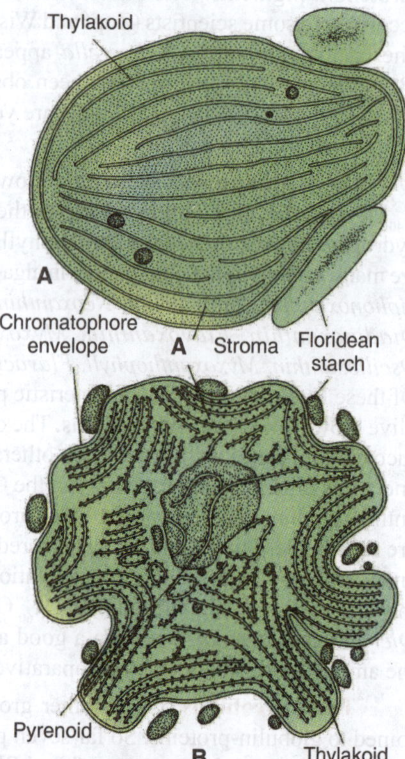

Fig. 24.2. *Rhodophyta.* Diagram showing fine structure of a chromato phore, (A), chromatophore of a red filamentous alga with single, free, widely separated discs and no pyrenoid; (B), stellate chromatophore of *Porphyridium* with a pyrenoid.

pyrenoids. The thylakoids occur singly or in multiple layers but are not found closely fused as in the **granum**. The algae, therefore, have plastids without clearly defined grana. The Cyanophyta lack membrane bound plastids. In them the photosynthetic apparatus comprises numerous, free, widely separated single thylakoids found scattered in the chromatoplasm. Below we discuss the structure of the plastids of different algal groups in order of increasing complexity.

(*a*) Rhodophyta (Fig. 24.2). Like the plastids of other eukaryotic algae, the chromatophores of red algae are limited by a narrow, double membrane (**Chromatophore envelope**) enclosing a dense granular matrix, the **stroma**. Embedded in the matrix are number of free, single thylakoids separated by wide uniform space in between. The presence of single discs which are not staked is a primitive feature unique to the reds only among the eukaryotic algae. On the basis of the simple structure of the chromatophore, the red algae may be considered the most primitive though morphologically they are as complex and advanced as the green algae. The thylakoids run roughly horizontally and extend the full length of the chromatophore (A). They may have small granules on the outer surface. Except the lower red algae (Bangiodeae, B) the chromatophores are devoid of pyrenoids. Even when present the pyrenoid has no starch sheath and is not considered equivalent in function to the pyrenoids of green algae. The pyrenoid has a dense matrix and is traversed by a number of single thylakoids. The grains of floridean starch are found scattered in the cytoplasm outside the chromatophore envelopes.

(*b*) Cryptophyta (Fig. 24.3). The chromatophore of cryptomonads (*Rhodomonas*) is limited by its own double membrane envelope. Within it is the dense matrix or stroma containing dense granules and lipid globules. Suspended in the matrix are the thylakoids stacked into bands. Each band consists of two loosely, sometimes closely, appressed thylakoids. Rarely there are 3 or 4 thylakoids per band. The 2-disc bands run along the length of the chromatophore. Many of them may extend nearly the entire length. At many places they are widely separated from each other and at others they run parallel to each other. External to the chromatophore envelope is the second double-membraned envelope. It is termed the **Chromatophore endoplasmic reticulum**. Gibbs (1962) reported that the chromatophore endoplasmic reticulum is continuous with the double membrane nuclear envelope. Starch ensheathed pyrenoids embedded in the chromatophore have been reported in some species but in some others they have been reported to lie free in the cytoplasm. Lamellar structures are absent in the pyrenoids. The starch grains occur outside the chromatophore and lie closely against its envelope within the chromatophore endoplasmic reticulum.

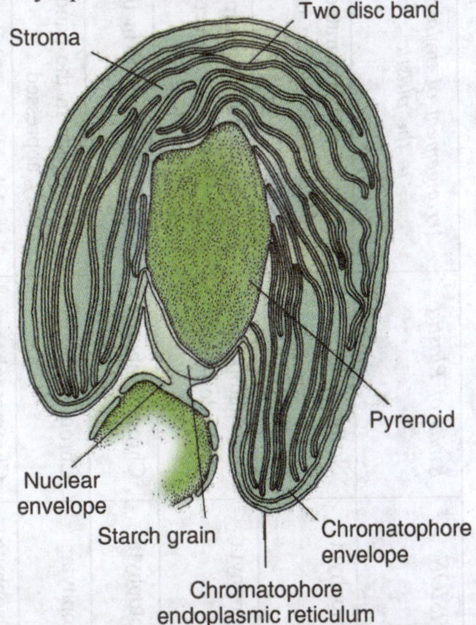

Fig. 24.3. *Cryptophyta*. Diagram showing fine structure of a chromatophore of *Rhodomonas*.

(*c*) Chrysophyta (Fig. 24.4). (includes Xanthophyceae, Chrysophyceae and Bacillariophyceae). The chromatophore has its limiting double membrane envelope. External to it is another double membrane envelope. The matrix contains small, dense granules and larger lipoid globules. Embedded in the matrix are numerous thylakoids which are pressed into a number of bands extending the whole length of the chromatophore. Usually there are 3 closely appressed thylakoids in a band which is much compressed. Individual discs are less wide than in the other algal groups. The discs in the band may separate at the ends of the chromatophore. A thylakoid in the band may sometimes be shorter than the others. There is one band in the chromatophore, the constituent thylakoids of which appear to be folded around the other bands. This band is called the "**girdle lamella**". Pyrenoids are commonly found in the chromatophores of

TABLE II : Plastid Composition of Algae

	DIVISION	*Kinds of Plastid*	*Arrangement of thylakoid in the plastic*	*Pyrenoid present or absent; with or without starch sheath*	*Kind and location of reserve food*
1.	Cyanophyta	No plastids	Single, widely separated thylakoids found free in the chromatoplasm.	Absent	True starch absent, Cyanophycean starch found in the Chromatoplasm.
2.	Rhodophyta	Chromatophores or rhodoplasts	Thylakoids single, simple, widely separated, never stacked into bands	Pyrenoids found in the plastid of a few species without a starch sheath	Floridean starch grains found scattered in the cytoplasm outside Chromatophore envelope
3.	Cryptophyta	Chromatophores	Thylakoids in twos in the band	Pyrenoid with a starch sheath	True starch outside the chromatophore envelope
4.	Pyrrophyta	Chromatophores	3-thylakoids in the band closely appressed	Pyrenoid with a starch sheath	Starch as large sheets in the cytoplasm outside the Pyrenoid
5.	Chrysophyta	Chromatophores	3-thylakoids in each band closely appressed; except Bacillariophyceae, girdle lamellae normaly present	Pyrenoids naked	Oil, leucosin, and chrysolaminarin found in the cytoplasm
6.	Phaecophyta	Chromatophores	3-thylakoids in the band closely associated with a space in between girdle lamella normally	Pyrenoid protruded (stalked) and naked	Manitol, laminarin and oil formed in pyrenoids, diffuses into the vacuoles in the cytoplasm present
7.	Euglenophyta	Chloroplasts	2-6 thylakoids in a stack usually 3, closely appressed	Pyrenoid within the chloroplasts; naked	Paramylen grains outside but usually close to the Chloroplast envelope and against the pyrenoid
8.	Chlorophyta	Chloroplasts	Usually 2-20 discs in grana like stacks chloroplast envelope	Pyrenoid with a starch sheath lying within the	True starch in the form of grains within the chloroplast envelope
9.	Charophyta	Chloroplasts	In *Nitella* all the 40-100 thylakoids are stacked in a single band	Pyrenoid absent	True starch

The latest trend among the algologists is to designate the algal plastids as chloroplasts irrespective of the Chlorophyll pigments that they contain.

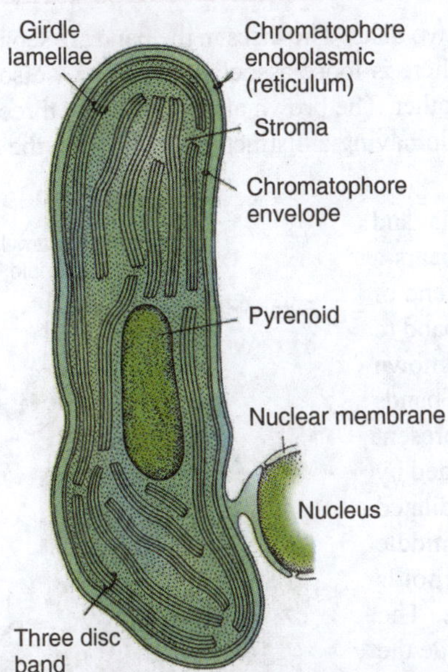

Fig. 24.4. *Chrysophyta.* Diagram showing fine structure of a chromatophore.

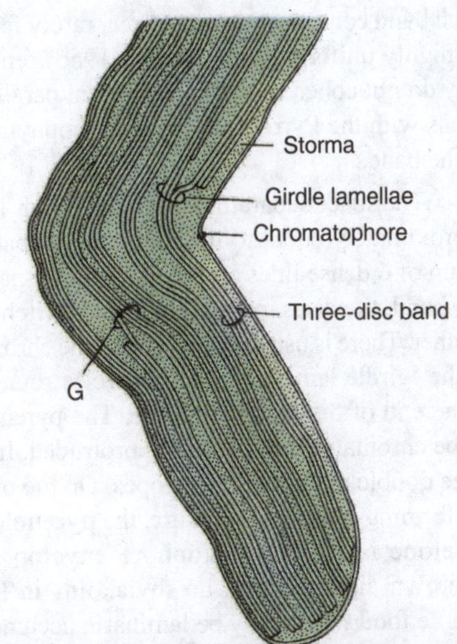

Fig. 24.5. *Pyrrophyta.* Diagram showing fine structure of a chromatophore of *Amphidinium*.

Chrysophyceae, Bacillariophyceae and some Xanthophyceae. They are traversed by bands continuous with the bands of the chromatophore. The number of thylakoids in the pyrenoid bands is generally reduced.

(*d*) Pyrrophyta (Fig. 24.5). The chromatophore structure of the Dinoflagellates (*Amphidinium*) is similar to that of Chrysophyta. The band is composed of 2-4 thylakoids but usually appressed discs run through the chromatophore matrix Dodge, (1971). The bands lie close to and more or less parallel to each other right across the chromatophore. According to Gibbs (1962) the 3-disc bands may occasionally split into two bands. The daughter bands may add one or two discs to it. It this way each daughter band comes to possess 3 discs in the band. The chromatophore has its own limiting double membrane tight envelope. There is little chromatophore matrix between the bands. It contains dense granule and lipid globules. The pyrenoid which is situated within the chromatophore has a few 3-disc bands passing through it (*Exuvialla*). Starch sheath is often present around the pyrenoid (*Amphidinium carteri,* Gibbs, 1962 *a*). Starch grains are deposited in the cytoplasm but closely appressed to it.

(*e*) Phaeophyta (Fig. 24.6). External to the limiting double membrane chromatophore envelope there is a second double membrane envelope called the **Chromatophore endoplasmic reticulum** which is continuous with the nuclear membrane. The granular matrix within the chromatophore envelope contains dense granules, lipid globules and thylakoids. Bouck (1965) observed that the thylakoids are stacked into bands.

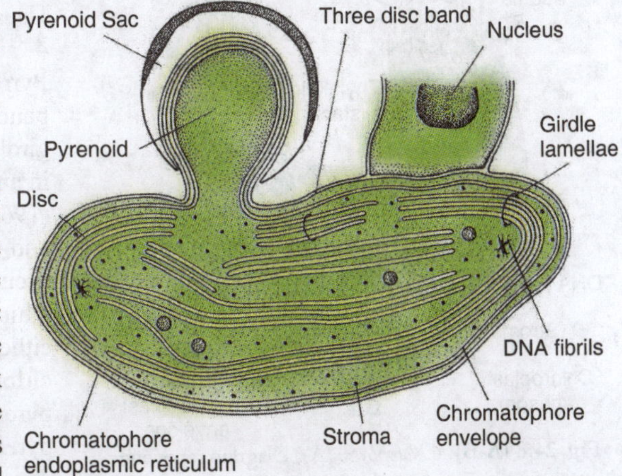

Fig. 24.6. *Phaeophyta.* Diagram showing fine structure of a chromatophore of a brown alga.

Each band comprises usually three, rarely four or even two discs. The discs in the band are separated by highly uniform space. Evans (1966) reported that there is loose association of 3 or 4 discs but they do not cohere into bands and run parallel to each other. The brown algae thus share three-disc bands with the Pyrrophyta and Chrysophyta but differ in having a distinct space between the discs of the bands.

The space separating the 3 thylakoids is very regular and approximately equal to the locule. Each band thus appears a group of 6 dense lines alternating with 5 spaces. At the end of the band, the discs may separate and switch from one band to another. There is usually one band in the chromatophore known as the 'girdle lamella' which encircles round the other bands at the end of the chromatophore. The pyrenoid when present in the chromatophore is usually protruded. It is surrounded by three double membrane envelopes. On the outside is a dilated sac forming a cap like structure, the **pyrenoid sac**. The middle envelope is the chromatophore envelope. The pyrenoids of brown algae contain no thylakoids in their matrix. The reserve food which may be laminarin accumulates outside the chromatophore near the pyrenoid.

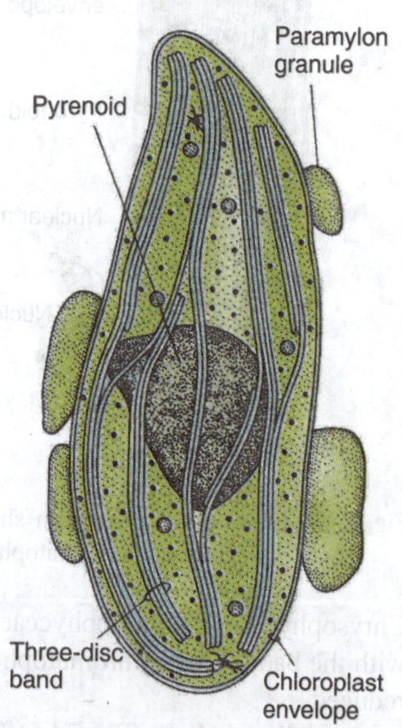

Fig. 24.7. *Euglenophyta.* Diagram showing fine structure of a chloroplast of *Euglena.*

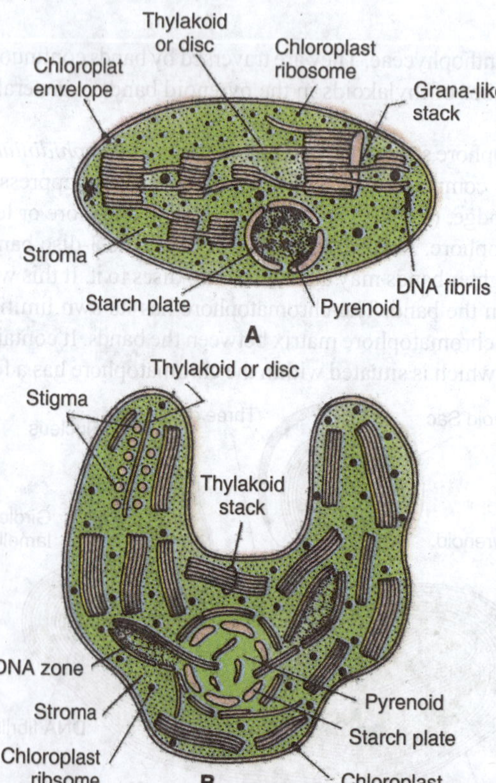

Fig. 24.8 (A-B). *Chlorophyta.* (A), Diagram showing fine structure of chloroplast of a green alga with grana-like stacks of discs. (B), Cup shaped chloroplasts of *Chlamy- domonas.*

(f) Euglenophyta (Fig. 24.7). The granular matrix within the limiting narrow double membrane chloroplast envelope of Euglenids is traversed by 10 to 45 bands (stacks), each consisting of 2-5 closely appressed discs. Usually there are three discs in a band but 2-, 4- and 5-disc bands are also found. As in the Pyrrophyta and Chrysophyta the discs in the band are, at least, partly fused. There is no girdle lamella. Each disc in the band terminates in an enlarged rim in *Euglena gracilis* Gibbs, (1962). The matrix contains small granules and large lipid globules. The reserve food accumulates in the form of paramylon grains which lie outside the chloroplast envelope either free in the cytoplasm or close to the chloroplast envelope usually against the pyrenoid but this is not always the case. The pyrenoid within the chloroplast is a region of dense matrix material without starch sheath. It is traversed by widely spaced bands of two appressed discs. According to Dodge (1968) the chloroplast envelope in Euglenophyceae comprises three membranes.

(g) Chlorophyta (Fig. 24.8). The chloroplast of green algae has the most complex structure among the eukaryotic algae. A tight double membrane envelope limits the chloroplast. The matrix within is often full of lamellar structure and is thus greatly reduced. The organisation of chloroplast lamellae exhibits much variety. The number of thylakoids (discs) in the stack varies within a wide range. Usually there are 2-20 discs in a band or stack but the chloroplast of *Nitella* may have all the 40-100 thylakoids in a single stack. The thylakoids in the stacks associate and dissociate in various ways giving a grana-like appearance. The stacks are well defined with discs closely appressed together. Some of the bands extend for a considerable distance. The discs in the stack may often shift from one stack to another. However in some green algae the discs appear to be randomly arranged with little suggestion of regular stacks. The discs come together and separate in an irregular manner. The chloroplast matrix contains starch grains besides the ribosomes and osmiophilic globules. In some green algae the pyrenoid is a region of dense matrix material in the chloroplast. It contains fibrils which are tight packed. In some green algae the pyrenoid lacs discs and in others it is traversed by single discs which may be continuous with the chloroplast bands (*Spirogyra* and *Cosmarium lundelii*). The pyrenoid has around it a sheath of starch grains which separates it from the chloroplast matrix.

(iv) ORIGIN AND EVOLUTION OF SEX IN ALGAE

There are two primary methods of reproduction among plants, **asexual** and **sexual**. In asexual reproduction no sex is involved. There is no union of elements of two cells. This type of reproduction is brought about by the formation of special cells set apart for this purpose. Each of these cells is capable of developing into a plant like the parent on separation. Such special cells are called the **asexual spores**. Sexual reproduction, on the other hand, involves the union of two special cells called the **gametes**. The two gametes fuse to form the zygote which develops into a new individual. Normally the gametes are incapable of growth alone.

(a) Origin of sex. It involves the origin of gametes. The most primitive algae, the Cyanophyceae reproduce only by the asexual methods of reproduction. *Protococcus* reproduces entirely by fission. Sexual reproduction is entirely absent in their case. Other algal forms whether unicellular or multicellular reproduce both by asexual as well as sexual modes of reproduction. With a few exceptions sexual or gametic reproduction has not replaced asexual reproduction, but has been added as a supplementary method. It is evident that originally the gametes were derived from the motile asexual spores of zoospores which had become too small and thus weak to form a new plant alone. Except for size the gametes are much like the zoospores.

The view is supported by the fact that in algae like *Chlamydomonas, Ulothrix* and many others with isogamous mode of sexual reproduction, the motile asexual spores (zoospores) and the gametes intergrade. In forms like *Chlamydomonas debaryanum,* the *zoospores* and the *gametes* are identical in every respect. They are similar in form, structure and mode of development. The chief morphological difference between them is their size. The zoospores are larger and the gametes are smaller. This difference in size depends upon the number of divisions that occur at the time of their differentiation. Where divisions stop early zoospores which are comparatively larger in size are formed. They contain the minimum level of factors essential to enable them to grow into new individuals alone. When division is long continued it results in the formation of smaller, swarmers resembling zoospores. They are incapable of giving rise to new plants alone. These come to be known as the **gametes** or **sexual cells**.

In *Ulothrix zonata* three kinds of zoospores have been reported :

(*a*) Quadriflagellated macrozoospores formed in small numbers per cell;

(*b*) Quadriflagellated microzoospores formed in large number per cell; and

(*c*) Biflagellated microzoospores formed in still large numbers per cell.

The gametes in *Ulothrix* are also biflagellated and are produced in large numbers. These gametes closely resemble the zoospores but are smaller than the macrozoospores and quadriflagellated microzoospores in size, and have two flagella instead of four. In structure, both the gametes and zoospores are similar and produced in the same manner. The difference in size and number of flagella between the gametes and zoospores of *Ulothorix* has been bridged over by the reported occurrence of biflagellated microzoospores in *U. zonata.* So the principal difference between the gametes and zoospores, as mentioned above, lies only in the number of divisions occurring at the time of their differentiation. The similarity of gametes to zoospores suggests that sex (union of two cells) has originated by the accidental fusion of zoospores which had become too small to form a new plant alone. The fusion probably took place in response to certain unfavourable environmental conditions. Since this accidental pairing proved of advantage to the plant in increased vitality and vigour, the method was maintained and finally became fixed.

This view is supported by the fact that sexual reproduction in the lower organisms is always associated with conditions unfavourable for growth. So long conditions for growth are exceptionally favourable they multiply by vegetative means. Under favourable conditions of growth they propagate by spore formation. Sexual reproduction is resorted to only under conditions unfavourable for growth. This is due to the fact that in enfeebled conditions caused by starvation, the plants produce small swarmers which along are unable to produce new plants. In pairing together they combine their resources and the resultant fusion cell or the zygote regains the power of producing a new individual directly. The fact that the zygote normally functions as a thick-walled resting spore in green algae suggests that sexual reproduction in green algae originated as an additional method to enable them to tide over a period unfavourable for growth. With the return of favourable conditions for growth the zygote in many green algae germinates to form asexual spores. It is merely a response to the favourable conditions.

The very fact that under certain conditions the gametes behave like spores and develop into new plants without fusion lends additional support to the view that the gametes are modified spores.

Evolution of Sex

Originally when sexual reproduction had become established both the gametes were flagellated and of the same size. Morphologically as well as physiologically they were identical. This condition is retained by some of the green and brown algae such as *Chlamydomonas debaryanum, Cladophora* and *Ectocarpus*. In all these forms sexual reproduction is seen in its simplest form. It consists in the fusion of similar, motile and naked gametes. They are called the **isogametes**. They fuse in water. The isogametes are indistinguishable as to sex and are produced in ordinary vegetative (undifferentiated) cells. This kind of sexual reproduction which consists in the fusion of morphologically and physiologically similar gametes is called **isogamy**. It is the most primitive method of sexual reproduction. It marks the transition from asexual reproduction to sexual reproduction.

There are some forms among the algae which represent all the steps through which plants passed in the evolution of sexual differentiation. In some green algae the fusing gametes are externally alike but dissimilar in their behaviour. In *Ulothrix,* for example, the gametes are similar but fusion takes place only between two gametes produced in distinct filaments. It means that the gametes though similar externally yet show functional disparity. They do not have male and female characteristic but positive and negative strains. It is because of this that the gametes of one filament fuse with those from another. The gametes of *Spirogyra* are alike but they are also distinguishable by their degree of motility. The gametes of one filament pass over and combine with the gametes in the other filament. This shows that there must be some difference (not structural) probably biochemical between the two filaments of a conjugating pair. Had there been no such difference there would have been random cross over of cell protoplasts between the two filaments. The active motile gamete may represent a primitive type of male gamete and the passive, non-motile gamete, a primitive type of female gamete. The difference in behaviour between the fusing gametes which are otherwise similar is a step towards anisogamy, a more advanced form of sexual reproduction than isogamy. It is called functional or **physiological anisogamy**. Anisogamy is more marked when the fusing gametes are of different sizes. They are called the **anisogametes** and reproduction as **anisogamy.** The best examples are some species of *Chlamydomonas (C. braunii)* and *Pandorina*. The anisogametes are produced in vegetative cells called the **gametangia**. The gametangia may be of different sizes. The smaller ones are called the **male gametangia** and the larger as **female gametangia**. The gametes produced by the male and the female gametangia are different in size. The smaller is regarded as **male** and the larger **female**. Both are flagellated.

Still more progressive elaboration of the sex organ is found in *Chlamydomonas coccifera* and *Eudorina*. These develop distinct sex organs, the male being called **antheridia** and the female **oogonia**. They are specially modified vegetative cells. The gametes produced within them are recognisable as **male** and **female**. They are dissimilar in size and shape. The smaller male gamete is active and may be called the **sperm**. The larger gamete is immobile and may be called the **egg.** This type of sexual reproduction shows close approach to **oogamy**. It may be regarded as well marked heterogamy or primitive type of oogamy. In a few species exhibiting heterogamy or primitive type of oogamy the gametes produced in these sex organs are shed (*Fucus*). They fuse externally in water.

The next stage in the evolution of sex is seen in forms like *Oedogonium* which exhibit more advanced type of sexuality. The male gamete is small and flagellated. It is called the **sperm**. It is produced in the single-celled male sex organ called the **antheridium**. The female gamete or egg is larger, non-motile and produced singly. It is retained within the **oogonium**. The antheridia and the oogonia are not specially modified vegetative cells but special sex organs differentiated from the vegetative cells of the thallus. Gametic union no longer takes place in water. The sperms swim to the egg retained within the oogonium to bring about fertilization.

The more advanced condition of oogamy is found in *Vaucheria* and *Chara*. In these forms the elaboration and differentiation of sex organs reaches its culminating point in the green algae. The gametes (sperms and eggs) in these forms are borne in special sex organs (the antheridia and oogonia). These are entirely distinct from the vegetative cells and are not developed from them. They arise as special reproductive branches.

From the above description it is clear that the evolution of sex, which involves the differentiation of gametes, was accompanied by differentiation of sex organs. During differentiation of gametes the male gamete retained motility and small size. The female gamete sacrificed its motility for the increased size and nutritive capacity. This adaptive feature of the female gamete or the egg to store excess food is of advantage to the species. It gives a good start to the organism in life which is of great survival value to the species. Besides, in forms inhabiting moving water isogamy which involves the fusion of similar, motile gametes, must be extremely wasteful. Both the gametes have to expend energy to meet and both being equally active none is able to store much food for the offspring.

The evolution of sexual differentiation in the algae has taken place along several independent lines and not in any one phylogenetic line. One of these is *Volvox* series starting with *Gonium,* passing through *Pandorina, Eudorina* and finally culminating in *Volvox*. In some species of *Volvox*, antheridia and oogonia develop on the same colony and others in different colonies. In the latter case a colony has either antheridia or oogonia but not both. In such species there is distinction into male and female colonies. Sexual reproduction in these forms has evolved to a point where there is sex differentiation.

(V) REPRODUCTION IN ALGAE

Reproduction actually means the production of young ones like the parent. The chief object of reproduction, therefore, is the perpetuation of the species and consequently the increase in number of the individuals of a species. There are two main types of reproduction, ***asexual*** and ***sexual***.

A. Asexual Reproduction. It consists in the separation from the parent of a highly specialised cell or a group of cells which directly develops into a new individual resembling the parent. Asexual reproduction is uniparental and takes place in a variety of ways. The most important ones are detailed below :-

1. Spore Formation. The spores are reproductive units specialised for asexual reproduction. Each unit can grow into a new organism by itself. It involves no sex and there is only one parent. The spores are usually one-celled structures and are formed by most of the Algae. They are produced either within ordinary vegetative cells (*Chlamydomonas*, *Ulothrix*) or in specially modified cells (*Oedogonium, Vaucheria, Ectocarpus*) called **sporangia**. Each sporangium may produce a single, large spore or more than one small spores. The spores may be motile or non-motile. They are able to grow into new individuals without fusion. When motile they are called the **zoospores**. The non-motile spores are called the **aplanospores**. The motile spores may also be known as the **planospores**. According to Klebs (1896) the factors which influence zoospore formation are: (1) changes in the concentration of culture medium, (2) darkness or a reduction in light intensity, (3) changes in temperature, (4) transfer to still from moving water and (5) transfer to water in terrestrial algae. Observations of Klebs were confirmed by subsequent workers. In addition it has been suggested by them that in certain algae endogenous, lunar or seasonal rhythm may also be involved.

(*a*) Zoospores. Formation of zoospores is the most characteristic method of asexual reproduction in most of the green algae. The zoospores are tiny, naked masses of protoplast furnished with fine protoplasmic threads, the flagella by means of which they can swim in water. In green algae the flagella are of whiplash type. They are usually of equal length and may be two (*Chlamydomonas*) or four (*Ulothrix*) per zoospore. They are inserted at the anterior end. In *Oedogonium* the zoospore has a circle of many whiplash flagella inserted at the base of a colourless beak at the anterior end. The zoospores of *Vaucheria* have numerous flagella distributed all over the surface of the body in pairs, one pair opposite each nucleus. The flagella in the pair are reported to be of unequal lengths (heterokont) but both are of whiplash type. In other yellow green algae the zoospores are furnished with two unequal flagella inserted at the anterior end. The longer one is of tinsel type and the shorter whiplash type. The zoospores may be produced singly or more than one per zoosporangium. They are liberated either through a small aperture in the wall or by transverse splitting of the wall of the zoosporangium. The liberated zoospores swim about actively in water for some time. After a short period of activity each comes to rest. It settles down on some solid object and secretes a wall around it. The flagella are resorbed. The eye spot and the vacuoles disappear. The zoospore in this condition grows into a new individual by cell division and cell enlargement.

Behaviour of the zoospores in *Pediastrum* needs special mention. They do not germinate or divide to form a colony. The zoospores formed by a single cell orientate themselves in a single plane and become apposed to form a colony much like the parent. This feature is of notable importance and is not met with in any other class of the Algae.

The shape and the structure of the zoospores vary with the individuals in the same class or in the different classes. Usually they are ovoid or spherical in shape. Each zoospore has a distinct nucleus and a chloroplast. Eye spot and vacuoles may or may not be present.

The zoospores of brown algae differ from those of green algae in having two unequal flagella inserted laterally on a pear or spindle shaped body. Of these the longer flagellum is of tinsel type. It is directed forward in motion. The shorter whiplash flagellum is directed backwards. In certain brown algae like *Ectocarpus* there are two kinds of zoospores. Of these one type is produced within unilocular sporangia borne on the diploid plants and the other within plurilocular sporangia. Zoospores produced within unilocular sporangia are differentiated by meiosis. They are haploid called the gonozoospores or meiozoospores. On germination they produce the alternate haploid sexual plant. Zoospores produced within plurilocular sporangia are usually diploid. They are differentiated by mitosis from the diploid contents of the sporangium. These belong to the category of mitospores. They serve to reduplicate the sporophytic stage on which they are borne. Zoospores are absent in all the red and blue-green algae a feature quite unusual for the aquatic algae. Multiplication by zoospore formation is the most effective and rapid means of asexual reproduction so long as the environmental conditions remain favourable.

(*b*) Aplanospores. They are non-motile and constitute a normal means of asexual reproduction in the terrestrial species of the yellow green algae (*Vaucheria*). In some green algae they are produced under certain unusual conditions (*Ulothrix* and *Microspora*). In aplanospore formation the protoplast of the cell may round off or divide into a number of non-motile, round bodies, each of which secretes a wall around it. Each aplanospore, on germination, develops into a new individual. The aplanospores are regarded as zoospores that have failed to develop flagella. Hence, they are considered analogous and not homologous to endospores or exospores of blue-green algae. The walls of the aplanospores are thin as compared with the other non-motile spores. The aplanospores produced in *Chlorella* have the same distinctive shape as the mother cell before liberation. They are called **autospores**.

(*c*) Hypnospores. Under certain more distressing environmental conditions the aplanospores in green algae secrete thicker walls around them. They are called the *hypnospores*. The hypnospores germinate into new plants with the return of conditions favourable for vegetative growth. In *Chlamydomonas nivalis* the walls of the hypnospores become red due to the presence of *haematochrome* and represent the phenomenon known as the **red snow**.

(*d*) Akinetes. They are resting cells which mainly serve as means of perennation rather than multiplication. In the formation of an akinete the entire protoplast of the vegetative cell rounds off and the original parent cell wall, which also serves as the spore wall, becomes considerably thickened. It is highly resistant. The akinetes contain abundant reserve food and are formed next to a heterocyst or at the end of a trichome in the blue-green algae. In some they are intercalary in position. In green algae they are formed in any cell and are frequently found in *Pithophora* and *Cladophora*. With the return of favourable conditions the akinetes develop into new filaments.

In certain Cyanophyceae (*Dermocarpa pacifica*) non-motile endospores are produced by division of the protoplast in three planes. As a result, a number of small usually spherical, daughter protoplasts are formed. Each endospore has a wall of its own distinct from the parent cell wall. In a few others such as *Chamesiphon* the cell wall ruptures at the distal end of the cell. The **exospores** are successively cut off at the exposed distal end of the protoplast.

(*e*) Cysts and Statospores. Many flagellate algae and some filamentous forms produce cysts under unfavourable conditions and these cysts act as resting stages. Cysts are commonly formed in *Euglena*. In this alga the cyst are composed of an unidentified carbohydrate. In *Protosiphon* also cysts are formed under dry periods. Since cysts in this alga are formed of multinucleate portions of protoplasts, these are also called as **coenocysts**.

Statospores are thick walled spores formed in diatoms. These spores may be smooth walled or the walls may be variously ornamented.

(*f*) Neutral Spores. In some alga, the protoplast of vegetative cells directly functions as spores and these are called **Neutral spores.** In *Asterocystis* the neutral spores are liberated by the rupture of gelatinous sheath of the filament, after which it secretes a wall, settles on the substratum and grows into a new filament.

In *Ectocarpus*, plurilocular sporangia produce diploid spores which are also called as **neutral** spores. Reproduction by neutral spores may be seasonal or may take place throughout the season.

(*g*) Monospores. Spores produced singly in sporangia are called as **Monospores** and the sporangia as **Monosporangia**. These are generally formed in members of Rhadophyceae and Phaeophyceae. Monospores are haploid and naked and formed when the entire protoplast of a uninucleate sporangium gets transformed into a spore. These are liberated by the rupture of the sporangial wall.

(*h*) Carpospores. Spores produced directly in the carpogonium after the fertilization of the egg or indirectly from the cells of the filament following fertilization are called as **carpospores**. These carpospores are common in *Polysiphonia, Batrachospermum* and other red algae.

(*i*) Hormospores. Multicellular spore like bodies having thick cell walls formed in some cyanophycean algae are called as **Hormospores**. These are somewhat similar to hormogonia but are surrounded by a stratified wall. Hormospores are usually formed in series. *Westiella* is an alga in which hormospores are usually formed.

(*j*) Nannocytes. In the members of Chroococcales, the contents of the cells repeatedly divide to produce very small endospore like structures. These do not enlarge enough to be formed as endospores. Geitler termed them as **Nannocytes** but some phycologists still regard them as endospores. *Macroycystis*, *Gloicocapsa* and several other genera of the order produce Nannocytes.

(*k*) Planococci. In some *Cyanophyta*, single-celled hormogones which show creeping movements are known as Planococci. Examples are *Demosiphon* and *Stauromatonema.*

(*l*) Pseudohormocysts are cluster of cells formed terminally on erect branches of *Westiellopsis* - a blue green alga. These are formed after repeated transverse and longitudinal divisions of terminal cells and the contents escape as *gonidia*.

(*m*) Tetraspores. In brown algae non-motile spores known as **tetraspores** are produced within specialized cells known as the **tetrasporangia**. Tetraspores are produced on the diploid plant and are haploid. The diploid nucleus of the tetrasporangium undergoes reduction division to form four tetraspores. They germinate to give rise to the alternate haploid or the sexual plants. They are commonly found in the Dictyotales. The tetraspores belong to the category of sexual spores known as the gonospores or meiospores.

In red algae too tetraspores are produced by the sporophyte plant. They germinate to give rise to the haploid or the sexual plant, *e.g., Polysiphonia*. In *Batrachospermu,m* non-motile monospores are produced singly within the monosporangia borne usually on the chantransia stage.

In some red algae, the tetrasporophyte instead of producing four tetraspores in each tetrasporangia may produce only two spores. Such spores are known as **Bispores** and the sporangia as bisporangia. Bispores may be uninucleate or binucleate. Examples are *Callithananion* and *Amphroa.*

When the tetrasporophyte produces sporangia which formed many spores, such spores are known as **Polyspores**. Examples are *Gastroclonium*.

In some algae, clustered masses of spores are also produced and these are termed as **Paraspores**. The paraspores give rise to triploid plants as in *Plumeria elegans* and *Seirospora occidentalis*.

2. Fission. It is the simplest and the chief method of vegetative reproduction in many one-celled green and blue-green algae. It is simple cell division of a unicellular alga into two new daughter cells. Each daughter cell retains half of the original parental membrane and forms anew the missing half. These daughter cells separate and carry on independent existence. In this way, two one-celled individuals result from a single parent.

3. Fragmentation. It is the breaking away of a few or many celled segments of a filament. Such bits of living cells are called the **fragments**. The fragment is able to grow into a new filament by cell division. It is the commonest method of vegetative reproduction in the filamentous green algae (*Spirogyra* and *Zygnema*) when conditions for vegetative growth are extremely favourable. The non-filamentous colonial forms in blue-green algae multiply by the splitting of the parent colony.

The filamentous blue-green algae form 'specialized fragments' called the **hormogones**. The hormogones are delimited in the trichomes of some forms (*Nostoc*) at the heterocysts, and in others (*Oscillatoria*) they are formed by the development of biconcave gelatinous separation discs, at short distances, in the trichome. The hormogones may move a short distance by spontaneous movements before they develop into new individuals by cell division.

4. By the formation of Adventitious Branches. In *Chara* certain nodal cells have been reported to produce naked adventitious branches that can give rise to new plants independently. In *Dictyota* adventitious branches arise from the lower cylindrical part of the thallus.

5. By the formation of Tubers. Rhizoids of *Chara* or its buried nodes form tuber-like growths that become filled with food material. The detached tuber germinates to produce the *Chara* plant. Such tuber-like structures have also been reported in *Cladophora*. *Chara* also multiplies vegetatively by the formation of characteristic starch amylums.

6. Budding. Bold (1933) reported vegetative propagation in *Protosiphon* by the formation of buds. The vesicle proliferates a budlike outgrowth which is subsequently cut off by transverse septum. The bud grows into a new *Protosiphon* plant.

Fission, fragmentation, budding, formation of adventitious branches and tubers constitute means of **vegetative propagation**. In vegetative reproduction the parent cell wall is not discarded.

B. Sexual Reproduction. It involves the fusion of two specialised reproductive cells called the **Sex cells** or gametes. Fusion may occur between two gametes from the same plant (*monoecious*) or from different plants (*dioecious*). The process of fusion is called **fertilization** and the product of fusion is called the **zygote**. The sex cells are always haploid or monoploid. Reduction division occurs either at (*Fucus*) or much before differentiation of the gametes. In the majority of the green algae it takes place prior to the germination of the zygote. The haploid gametes cannot grow into new plants alone but the zygote does. It is always a diploid structure. It is obvious, therefore, that further development of gametes depends upon the sexual stimulus provided by the act of fertilization. Sexual reproduction is absent in the blue-green algae. In general it is of two main types, **isogamous** and **heterogamous**.

(*a*) Isogamy. It is the simplest and the most primitive type of sexual reproduction. It is usually found among the lower or simpler forms. It consists in the fusion of morphologically identical gametes. They are similar in size and structure and are called the **isogametes**. The isogametes are usually naked and motile. They behave and look alike. They are produced in ordinary vegetative cells. The algae exhibiting **isogamous** type of sexual reproduction are usually monoecious. Many, however, are dioecious. In the latter case fusion takes place only between the gametes coming from different individuals (filaments) of the same species. *Ulothrix* is dioecious. The fusing gametes are alike in shape, size and activity but show functional disparity. They do not exhibit male and female characters but definitely belong to two opposite strains usually designated as **plus** (positive) and **minus** (negative) strains. The two kinds of gametes are produced on different individuals. Fusion takes place only when gametes of opposite kind meet.

In *Spirogyra* the gametes are morphologically alike but dissimilar in their behaviour. One of the fusing gametes is more active than the other. Difference in behaviour is a step towards **anisogamy**. The gametes in *Ectocarpus siliculosus* (brown alga) are identical but at the time of fusion one gamete becomes passive and is surrounded by a large number of other gametes (male). The passive gamete may be designated as female. This mode of fusion is called *clump formation*.

In *Chlamydomonas monoica* one of the gametes is more passive. It receives into its protoplast the cytoplasmic contents of other.

It has been suggested that in isogamous sexual reproduction each strain secretes its own specific type of chemotactic substance. The isogametes may not differ morphologically but they differ in their specific chemical secretions which they produce. Copulation occurs only when gametes with different secretions meet. The act of fusion in isogamy is called **conjugation** and the product of fusion, the **zygospore**.

(b) Heterogamy. It is a more advanced type of sexual reproduction as compared with isogamy. It includes the fusion of dissimilar gametes. The fusing gametes differ in size, structure and physiology. The act of fusion in this type of sexual reproduction is called **fertilization** and the product of fusion is termed the **zygote**. Heterogamous sexual reproduction is of two types, **anisogamous** and **oogamous**.

(i) Anisogamy. It involves the fusion of gametes which are dissimilar and are produced in modified vegetative cells called the **gametangia**. The larger, passive gamete is called the **female**. In *Chlamydomonas braunii* the female gamete is large and less motile. It soon comes to rest. It is sought by the smaller, active male gamete. In various groups of algae anisogamy leads over quite gradually to oogamy.

(ii) Oogamy. It is the highest and the most advanced stage of sexual reproduction. It is generally characteristic of the filamentous forms of algae. The gametes are produced in special sexual organs which are very sharply differentiated from the ordinary vegetative cells. The female sex organ is enlarged. It is called the **oogonium**. With the exception of *Fucus* and *Sargassum* (Fucales) it produces a single ovum. The male sex organ is smaller in size and is called the **antheridium**. It produces one or more male cells or gametes furnished with flagella, exception being the red algae. The male cells are very active in their movements and are called the **sperm**. The large ovum has plenty of food reserves. The sperms are destitute of food reserves. The ovum with the exception of Fucales is retained within the oogonium. The sperm gains entrance into the oogonium through an aperture in its wall. The zygote formed as a result of the fusion of the two gametes secretes a thick wall around it in the green algae and enters upon a period of rest. It contains a lot of food reserves. In the majority of marine algae the zygote at once develops into another individual. There is no resting period.

Red algae exhibit oogamous sexual reproduction of a very specialised kind. The male gametes are non-motile and are called **spermatia**. The spermatia are spherical and are produced singly within the antheridia. They are carried to the female sex organ or the carpogonium by the water currents. They adhere to the **trichogyne** a tubular prolongation of the carpogonium. The intervening wall between the two disappears. The nucleus of the spermatium wanders down to meet the female nucleus lying in the lower dilated portion of the carpogonium. Associated with the sex organs in the red algae are certain accessory structures. The zygote develops into a diploid cystocarpic plant which produces the carpospores. Exception is *Batrachospermum* where the zygote nucleus undergoes reduction division and produces haploid carpospores.

Advantages of sexual reproduction. It prevents the degeneration of the race which may result from continuous asexual reproduction for an indefinite period. Sexual reproduction gives variety to offspring through new combination of genes at the time of random union of gametes. Even offspring of the same parents differ. Moreover they are produced in larger numbers than can survive. Through natural selection the stronger survive and the weaker perish. Thus the race improves. Without variety evolutionary process cannot continue.

Disadvantages. (*i*) *Depends on chance.* Sexual process involves the fusion of two sex cells or gametes. For this the gametes meet but very often they do not. In this case much of the reproductive effort of the parent organisms will go to waste.

(*ii*) *Involves locomotion.* In order to meet either the gametes must be motile or be brought together by the movements of the parents.

(*iii*) *Requires aquatic medium.* Out of necessity the gametes must be naked because the protective coats around them would hinder fusion. The unprotected gametes will dry quickly in air. Sexual reproduction thus invariably requires an aquatic medium.

Conditions favouring sexual reproduction

(*i*) Light is considered to act as a determining influence as bright light stimulates the production of gametes.

(*ii*) Nitrogen deficiency, in general, induces gamete formation.

(*iii*) Temperature and hydrogen ion concentration (pH) are also involved in the initiation of the sexual process.

(*iv*) Deficiency of nutritive materials also initiates sexual reproduction.

Klebs suggested that sexual reproduction was usually associated with high light intensity and a reduction in the concentration of mineral salts. According to Coleman (1962) development of sexuality is largely controlled by nutritional factors. An alga can form gametes only when it attains a certain physiological state.

(*vi*) MODES OF PERENNATION IN ALGAE.

Perennation includes the methods employed by plants to pass through periods unfavourable for vegetative growth. It is as a matter of fact a temporary sleep. It includes the cessation of all normal activities of an organism. Under stressing circumstances when life processes cannot run smoothly, the living organism develops some stage in the life cycle which is protected by a thick resistant wall. It is also furnished with sufficient reserve food. The protoplasm of such a stage remains dormant by reducing its metabolic activities to bare minimum.

The marine algae live in conditions where variation in external conditions of life is rare. Some of them, which grow above low tide mark and are thus exposed to periodic desication, protect themselves by secreting copious mucilage. It baths and protects their bodies. The fresh water and sub-aerial algae are, however, exposed to the dangers of desication and extremes of temperature. They, therefore, develop certain devices which enable them to carry through the period of severe conditions into the next growing season. These have been dealt with in detail under the heading "Asexual reproduction in algae." Hence only a passing reference will be made to them.

(*a*) Akinete. It is a specially modified vegetative cell covered by a thick, resistant wall. It contains a lot of reserve food material and little water. It functions as a means of perennation enabling the plant to tide over periods unfavourable for growth. The akinetes are developed in many blue-green algae (*Nostoc*) and green algae (*Cladophora* and *Pithophora*). The whole filament dies leaving behind the akinetes. The latter survive and develop into new plants with the return of normal condition of life.

(*b*) Secretion of mucilaginous sheath. Many terrestrial blue-green algae possess a gelatinous sheath external to the cell wall. In some forms it is fairly thick. The gelatinous sheath coupled with low osmotic pressure of the cell, absence of vacuoles and gel-like consistency and viscosity of the protoplast account for the great resistance of the blue-green algae to desiccation.

(*c*) Zygospore. The formation of zygospore constitutes the normal method of perennation in green algae. Towards the end of the growing season, the green algae resort to sexual reproduction. It results in the formation of a structure called the **zygospore**. The latter secretes a many layered thick and resistant wall around it and goes into the resting stage. The zygospore remains dormant until the next growing season In all the green algae zygospore is the chief means of carrying the plant through the adverse conditions into the next growing season.

(*d*) Palmella stage. Under adverse conditions in some green algae such as *Chlamydomonas*, the contents of the vegetative cell divide to form the new daughter cells. These increase in number by repeated cell division. All the cells remain clustered together inside the mucilage envelope formed by the gelatinisation of the parent cell walls of the successive generation. This is the *Palmella stage*.

It is of short duration and represents a milder method of permeation.

(e) Hypnospores. The aplanospores in some algae (*Protosiphon*) under more distressing conditions secrete thick walls around them and become *hypnospores*. These thick-walled resting spores are able to remain in a quiescent condition for some time during winter, period of drought and thus function as perennation structures. With the onset of conditions favourable for growth each hypnospores germinates to give rise to a new plant.

(f) Hormospores. Under unfavourable conditions of growth the hormogonia produced by some of the filamentous Cyanophyceae (*Westiella*) secretes thick walls and function as **hormospores**. They remain *in situ*. With the advent of favourable conditions the hormospores creep out of their thick envelopes to grow into new individual.

(g) Tubers. In certain species of *Cladophora* the rhizoidal cells divide to form a number of cells which become filled with reserve food materials. Their walls become thickened. These are produced on the subterranean part of the plant and can withstand extremes of temperature and drought. With the return of normal conditions each tuber germinates to give rise to a new plant.

(vii) LIFE HISTORIES OF ALGAE.

The sequence of events through which an organism passes from the zygote to the zygote of the next generation constitute its life cycle. Broadly speaking at least five main types of life cycles can be distinguished among the algae reproducing sexually. They are described next:

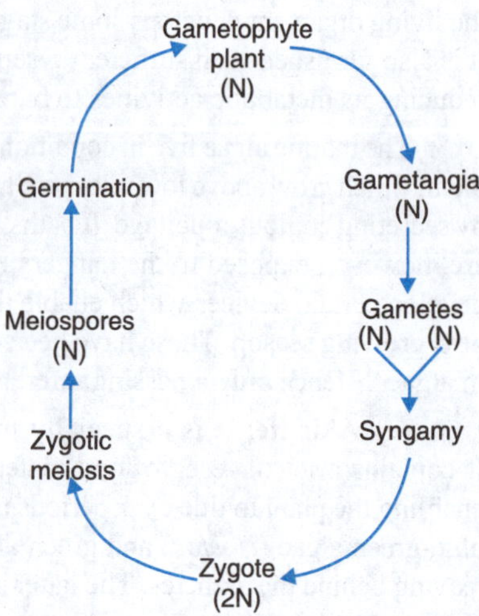

Fig. 24.9. Graphic sketch of a haplontic life cycle.

(a) Haplonic Life Cycle (Fig. 24.9). Sexual reproduction always involves a cyclic alternation between a haploid and a diploid condition. In the majority of green algae, Charophyta and *Bangia* in the red algae, there is a single thalloid vegetative individual in the life cycle. It is haploid. As it bears the gametes it is called the **gametophyte** plant. The latter multiplies by vegetative means and also by the formation of mitospores during the growing season. The mitospores, on germination, produce the gametophyte plants. They thus serve as additional means of multiplying the **gametophyte generation** which produces them and play no role in the phenomena of **alternation of generations.** Towards the end of the growing season the gametophyte plant produces gametes in vegetative cells called gametangia. These gametes fuse to form a **zygospore**. The zygote at the time of germination undergoes **zygotic** or **initial meiosis** producing 4 haploid spores called the **meiospores**. Each meiospore germinates to produce the gametophyte plant. The diploid condition in these algae is thus confined only to the zygospore itself. Although it gives rise to haploid spores but it can hardly be regarded as a **sporophyte**. In these plants, therefore, there is alternation between a prolonged, haploid vegetative gametophyte plant and a single-celled diploid zygospore with meiosis occurring on germination of the zygote. This cannot be considered as true alternation of generations. There is, however, alternation of chromosome numbers from haploid to diploid and back to haploid condition but no corresponding alternation of plants with different functions. Such a life-cycle is called **haplontic**. It is a primitive type of life cycle which is characterised by **zygotic meiosis** and haploid adults. Examples are *Spirogyra, Ulothrix, Chlamydomonas* and *Oedogonium*. Fritisch (1935, 1942*b*) and Stebbins (1960) considered it to be a primitive type of life cycle but according to Chaodifaud (1960) it has evolved from a diplohaplontic life cycle by reduction and loss of the diploid or sporophyte generation.

(b) **Diplontic Life Cycle** (Fig. 24.10). This type of life cycle is exhibited in many diatoms (Bacillariophyta), some of the Siphonales, Siphonocladiales and Dasycladiales among the green algae and Fucales (*Fucus* and *Sargassum*) in the brown algae. The vegetative thallus or plant constituting the dominant phase in the life-cycle, is **diploid**. It is the sporophyte plant but it bears the sex organs which produce the gametes. Meiosis or reduction division occurs at the time of differentiation of the gametes in the sex organs. This is known as **gametogenic meiosis**. In these organisms the haploid condition is limited to the gametes alone. The gametophyte phase in such forms is thus extremely reduced. It is a brief phase represented by a few haploid cells, the **gametes** only. At the time of gametic fusion the diploid condition is re-established. The zygote by equational mitosis develops into the sporophyte plant. Here again the life cycle includes one vegetative adult. It is the **sporophyte** plant. The sporophyte plant in the life cycle alternates with a few haploid cells, the **gametes**. There is, therefore, no true alternation of generation in the life cycle. Such a life cycle is called **diplontic**. It is characterised by **gametogenic meiosis** and a **diploid adult (sporophyte)**. The gametophyte is represented by the gametes, the only haploid cells in the life cycle.

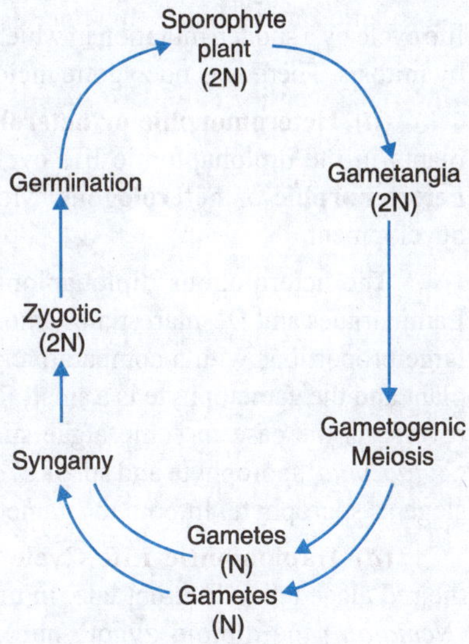

Fig. 24.10. Graphic sketch of a diplontic life cycle.

Many phycologists think that a diplontic life cycle has evolved from a diplohaplontic cycle by reduction of the gametophyte but in the Bacillariophytes the diploid vegetative plant is believed to have arisen by the elaboration of the diploid zygote of the primitive haplontic cycle.

(c) **Diplohaplontic Life Cycle** (Fig. 24.11.). The third type of life cycle is typical of the Ulvales and Cladophorales in the green algae and *Ectocarpus* and *Dictyota* in brown algae. There is alternation of two distinct vegetative individuals having not only different chromosome numbers but different functions as well. One of these individuals is the haploid **gametophyte**. It is concerned with sexual reproduction. The other is diploid **sporophyte**. It produces sexual spores (**meiospores**). This type of life cycle which consists in the alternation of two vegetative individuals, the gametophyte and the sporophyte with sporogenic meiosis is called **diplohaplontic**. It is a widespread type of life cycle and is of two categories:

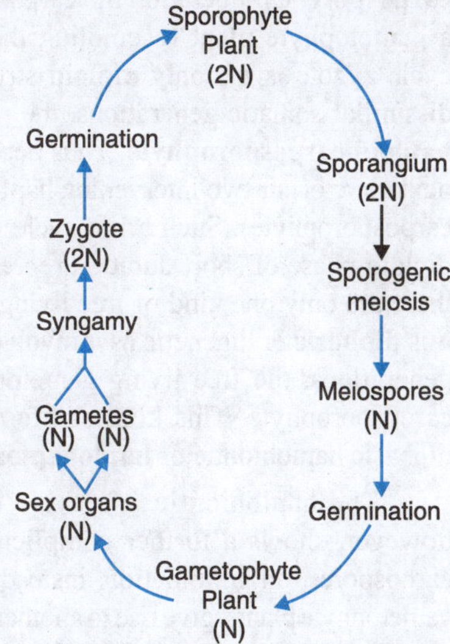

Fig. 24.11. Graphic sketch of a diplohaplontic life cycle.

(i) **Isomorphic or homologous diplohaplontic.** When the alternating vegetative individuals (the gametophyte and the sporophyte) are morphologically similar the life cycle is called **Isomorphic** or **homologous**.

Meiosis in such a life cycle is delayed so that the zygote yields a multicellular diploid plant. Meiosis takes place in the zoosporangium of Ulvalves and Cladophorales, unicellular sporangium in the Ectocarpales and the tetrasporangium in the Dictyotales and red algae. It is known as **sporogenic meiosis**. Isomorphic type of diplohaplontic life cycle is considered to have evolved from a haplontic life cycle by a sudden mutation in which the zygote directly developed into a new vegetative individual by mitosis. There was no zygotic meiosis.

(*ii*) Heteromorphic or heterologous diplohaplontic. When the two alternating vegetative plants in the diplohaplontic life cycle are morphologically dissimilar or unlike it is known as **heteromorphic** or **heterologous**. Most frequently the sporophyte has undergone most elaborate development.

The heterologous diplohaplontic life cycle with a dominant sporophyte is found in the Laminariales and Desmarestiales among the brown algae. There is an alternation of a sporophyte of large proportions with a comparatively small gametophyte. The sporophyte is a huge multicellular plant and the gametophyte is a small few celled structure. The reverse is the case in some algae such as *Cutlaria* with its '*Aglaozonia*' sporophyte and sp. of *Urospora* with a *Codiolum* stage as sporophyte. In both the gametophyte is dominant.

(*d*) Haplobiontic Life Cycle (Fig. 24.12). Some of the red algae tell a different tale. In many primitive red algae (*Nemaloin*) the diploid zygote nucleus undergoes meiosis and puts out a number of short, branched filaments which form a small somatic individual or plantlet. It is the **haploid carposporophyte**. It remains attached to the carpogonium and thus is parasitic on the parent gametophyte. The terminal cells of the carposporophyte function as **carposporangia**. The protoplast of each carposporangium forms a **haploid carpospore**. On liberation the carpospore germinates to form a gametophyte plant resembling the parent. In such a life cycle zygote is the only **diploid** structure but there are two dissimilar somatic generations, the parent **gametophyte**, the **haploid carposporophyte**. Thus between one zygote and the next there occur two intervening haploid somatic generations (free living gametophyte and parasitic carposporophyte). Such a life cycle involves alternation of successive haploid generations with a diploid phase of short duration represented only by the zygote. Cytologically it is haplobiontic as there are only one kind of free living individuals (gametophyte) in the life cycle. Morphologically it is **diphasic** or digenetic as it involves the alternation of two successive haploid dissimilar somatic generations, the free living gametophyte and the parasitic caropsporophyte. This kind of life cycle is also known as diphasic haplobiontic or **haplohaplontic**.

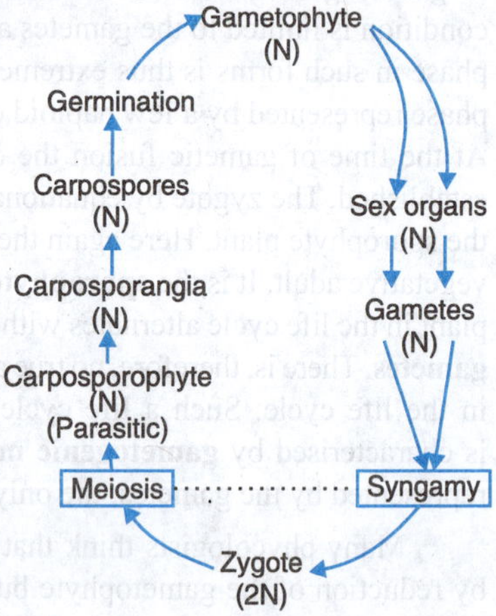

Fig. 24.12. Graphic sketch of a diphasic (haplohaplontic) haplobiontic life cycle of a red alga.

The haplobiontic life cycle of *Batrachospermum*, however, shows a further complication. In it the haploid carpospores, on germination, instead of forming the normal gametophyte plants, give rise to another juvenile form called the chantransia stage. It is capable of reproducing itself asexually by monospores. From the chantransia stage eventually arises the normal gametophyte plant as a lateral outgrowth. There is thus alternation of three successive dissimilar, somatic haploid generations (carposporophyte, chantransia stage and parent gametophyte) with a diploid phase of short duration represented by the zygote. The haplobiontic life cycle of *Batrachospermum* is therefore **triphasic**. Morphologically it is **haplohaplontic** (Fig. 24.13).

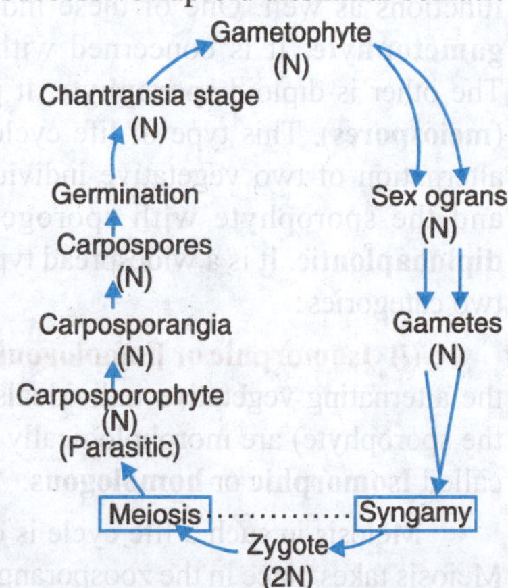

Fig. 24.13. Graphic sketch of a triphasic (haplohaplohaplontic) haplobiontic life cycle of a red alga.

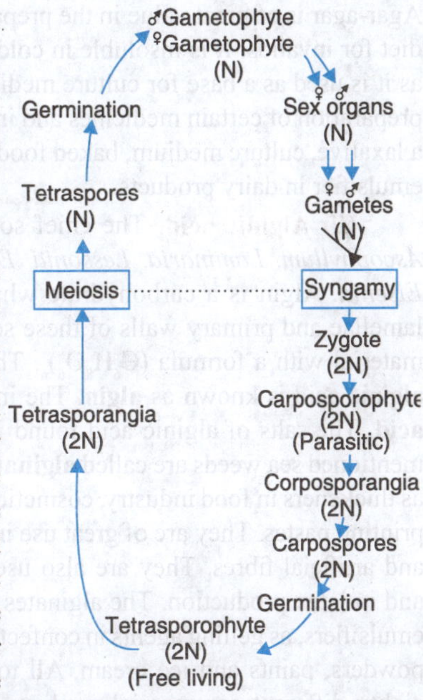

Fig. 24.14. Graphic sketch of a triphasic (diplodiplohaplontic) diplobiontic life cycle of a red alga.

(*h*) Diplobiontic life cycle (Fig. 24.14). In many other Florideae the diploid zygote divides mitotically to form a small, compact plantlet, the **diploid carposporophyte**. It remains attached to the gametophyte plant and produces diploid **carposporangia.** The protoplast of the carposporangium forms a single **diploid carpospre**. On liberation the latter germinates to form a free-living, diploid **tetrasporophyte**. The adult tetrasporophyte bears **tetrasporangia**. The diploid nucleus of the tetrasporangium undergoes meiosis to form four haploid **tetraspores**. The tetraspore, on germination, produces a free-living gametophyte plant. Such a life cycle which is characterised by two kinds of individuals, haploid as well as diploid is called **diplobiontic**. In the higher members of Florideae it usually consists of one gametophyte somatic phase and two sporophyte somatic phases (zygote together with carposporophyte and tetrasporophyte) and is thus **triphasic** or **diplodiplohaplontic**. In a triphasic diplobiontic life cycle, the two free-living generations may be morphologically similar (**isomorphic**) as in *Polysiphonia* or dissimilar (**heteromorphic**) as in certain Nemalionales. The diplobiontic heteromorphic life cycle may have an elaborate free-living gametophyte and a simple free-living sporophyte or *vice versa*.

According to Fritsch (1942 *b*) the diplobiontic types have evolved from the haplobiontic by sudden mutation. Drew (1954) suggested that the carposporophyte has arisen by progressive developments of the attached zygote.

(*viii*) ECONOMIC IMPORTANCE OF ALGAE

Exploration of the seas, which are full of algae, has brought home to man the usefulness of algal flora. The algae are of importance in the fields of agriculture and industry. In addition they are used as food, fodder and as manure.

Role in Industry. The algae are useful as the source of many commercial products. The four major products derived commercially from them are *agar-agar, alginic acid carrageenin*, and *diatomite*. The industrial utilisation of Algae, particularly the sea weeds, dates back to many hundreds of years.

(*i*) Agar-agar. It is a mucilage produced by certain red algae and stored along with cellulose in the cell walls. The main sources of agar-agar in Japan are the thalli of *Gelidium, Gracilaria* and *Gigartina*. Until 1939 Japan was the largest producer of *agar*. The other red algae used for this purpose are *Camphylaephora, Eucheuma, Hypnea, Ahnfeltia* and *Furcellaria*. Agar is a gelatinous, clear, nitrogen free extract from the above mentioned genera of the red algae. The extract is a gel containing galactose and a sulphate. Its melting point is between 90 and 100°F. At lower temperature it changes into a solid.

Agar agar sources and uses.

Agar-agar is of great value in the preparation of foodstuffs and is particularly used in the articles of diet for invalids. It is insoluble in cold but soluble in hot water. It is almost a necessity to research as it is used as a base for culture media for bacteria, fungi, algae and other tissues. It is used in the preparation of certain medicines and in cosmetics, leather and textile industries. It is largely used as a laxative, culture medium, baked foods, meat industry and as an emulsifier in dairy products.

Alginate acts as stabilizer for Icecreams.

***(ii)* Alginic acid.** The chief sources of alginic acid are *Ascophyllum, Laminaria, Lessonia, Ecklonia, Macrocystis* and *Eisenia*. Algin is a carbohydrate which occurs in the middle lamellae and primary walls of these sea weeds. It is a colloidal material with a formula $(C_6H_8O_6)_n$. The soluble calcium salt of alginic acid is known as **algin**. The insoluble extract is **alginic acid**. The salts of alginic acid found in the cell walls of above mentioned sea weeds are called **alginates**. The alginates are used as thickeners in food industry, cosmetic, and in textile industry as printing pastes. They are of great use in the production of plastic and artificial fibres. They are also used in the rubber industry, and in latex production. The alginates are also of great value as emulsifiers, as gelling agents in confectionary, dental impression, powders, paints and ice cream. All told algin is used in about eighty different commercial products. Because of its wide use the Kelp beds in shallow waters are harvested at regular intervals by special cutting and collecting machines.

***(iii)* Carrageenin** (caragheen). The chief source is a red alga *Chondrus crispus* (*Irish moss*) and to lesser extent *Gigartina*. It is a cell wall polysaccharide and is a mucilaginous extract. Carrageenin is used in food, textile, pharmaceutical, leather and brewing industries. It is also used to stabilize emulsions and as remedy for cough. Its use as a component of tooth pastes, deodorants, cosmetic and paints is no less important.

***(iv)* Diatomite.** It is a rock like deposit formed from the indestructible, siliceous frustules (cell walls) of the past diatoms that had collected over many millions of years on the floors of seas to form oceanic sediment. The latter extends hundreds and thousands of feet in depth in some localities. These diatomaceous deposits are mined in several parts of the world to obtain the diatomaceous earth which is put to several commercial uses. It is a whitish substance, firm but usually soft and light. It is highly porous and insoluble. It has abrasive qualities and is chemically inert. It is fire proof and highly absorbent. Because of these properties it is used in industry and commerce as a filter for oils and other solutions (sugar industry) and for clearing solvents. It is used in insulation of refrigerators, boilers, hot and cold pipes, hollow tile bricks for construction of constant temperature rooms, sound proof rooms, in packing corrosive chemical liquids, in polishes for metals, insulation for pipes and furnaces, in the manufacture of dynamite, a constituent of some tooth powders, bleaching powders, a reinforcing agent in both concrete and rubber. It is also used as a base on automobile and silver polishes. Its use in wine and paper industries is no less important.

Kelp algae.

(*v*) Japan produces about 100 tons of iodine annually from Kelps (brown sea weeds). The chief genera employed for the purpose are *Laminaria, Ecklonia, Fucus* and *Eisenia*. The kelps are also a source of soda and potash. The red algae *Rhodomela* and *Polysiphonia* are the source of bromine.

(*vi*) Glue manufacturing is another important algal industry in Japan. Red alga *Gleopeltis furcata* is used for this purpose. This glue is known as *funori*. It is used for sizing paper and cloth. It is also used as an adhesive.

Role in Agriculture. Soil is a living mass. Apart from the soil particles it harbours bacteria fungi, algae and other micro-organisms. The algae occupy a volume of three times that of the bacteria. Of the soil algae the blue greens are the most important. They are capable of fixing atmospheric nitrogen in their bodies. Upon liberation, nitrogen in usable form increases soil fertility and improves the growth of crop plants. P.K De (1939) has furnished a conclusive proof that the blue-green algae are the chief agents for nitrogen fixation in rice fields. They increase the fertility of the soil by fixing atmospheric nitrogen. Some of the important nitrogen fixing blue-green algae are *Oscillatoria princeps, O. formosa,* species of *Anabaena, Sprirulina, Nostoc* and *Cylindrospermum*. The practical application of these algae as fertilizers is the seeding of rice fields with nitrogen fixing species to increase the nitrogen content of the soil.

Another important use of the blue-green algae is in the reclamation of barren, alkaline soils R.N. Singh (1961). Such soils have been reclaimed and brought to a productive state by inducing a proper growth of certain Cyanophyceae. Successive crops of these neutralize the alkalinity and increase fertility of the soil.

The algae are also used as fertilizers particularly the sea weeds. The agricultural utilisation of sea weeds as manure is chiefly confined to the farm land near the coastal regions. Large brown and red algae are used as organic fertilizers. They are richer in potassium but poor in nitrogen and phosphorous than the farm manure. The sea weed may be used direct and ploughed. The direct application of sea weeds is in vogue close to the sea in some countries such as France, Ireland and Sri Lanka, for earth vegetables such as potatoes, and turnips, and also for coffee plantations. In Japan they are used in the rice fields whereas in China for groundnuts and sweet potatoes. In India *Turbinaria* is used as a fertilizer around palm trees. In Rajasthan *Anabaena* and *Spirulina* which are produced in enormous quantities in the Sambhar salt lake are employed as green manure by the local farmers. The sea weeds are also used as compost. Sometimes they are burnt and ashes scattered over the farm lands.

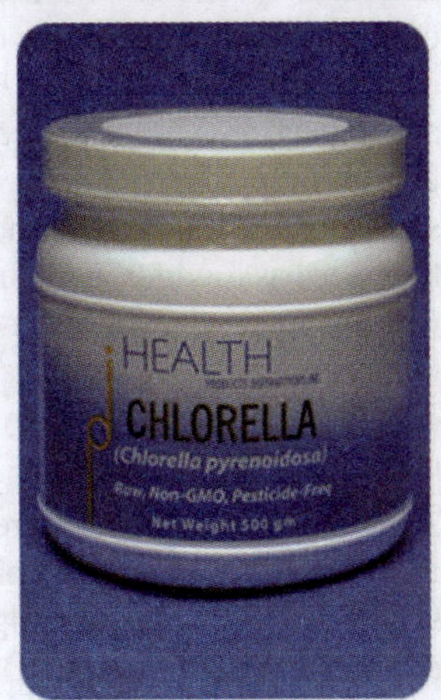

Chlorella pyrenoidosa tablets used as food supplement in space flights.

The sea weeds are processed into a sea weed meal for transport inland. Concentrated liquid extracts of sea weeds are sold as liquid fertilizers and also an insecticides. The sea weed products aid in binding sandy soils. They also help to breakdown clayey soils and thus promote a good crum structure.

In some countries there is a practice to grind up *Lithothamnion* and *Lichophyllum* and use it in place of lime. Similar use is made of species of *Chara* which are encrusted with calcium carbonate.

The algae such as *Chlorella* also help to aerate water by removing carbon dioxide and restoring supply of oxygen in photosynthesis. This is important to the fish. Their growth is encouraged in the sewage disposal plants. They take up nitrates and phosphates for their metabolism and liberate oxygen in photosynthesis which help the aerobic bacteria to decompose raw sewage.

Algae and space travel. The biologists have given a serious thought to construct a biological system in which there would be a complete recycling of the biological materials. This biological system will be used during a space flight trip lasting over 30 days. The spaceman will need a device to get rid of carbon dioxide and other body wastes and will also require sources of oxygen and food.

It has been considered beneficial to use microscopic, unicellular, algae such as *Chlorella pyrenoidosa* and *Synechococcus* as a possible food source in anticipated space flight. It multiples rapidly and thus will synthesise a rich harvest of food utlizing carbon dioxide and liberating sufficient oxygen as a by product for use. As a source of nitrogen for protein synthesis it will assist in the decomposition of human wastes such as urine and faeces.

Algae as Food. The algae are important as a source of food of the fishes, aquatic amphibia, mammals and other animals. Man's dependence on fish and other aquatic animals to supplement his diet is a well known fact. Indirectly, therefore, the algae are of great value to man. Some biologists, however, suggest that algae might be raised by man as a source of food. Miller (1972) believed that the algae can play an important role in the production of protein for human food. In fact in some coastal parts of the world they are used directly as human food. In the Pacific islands and orient the red, brown and green algae form a regular portion of diet. Over a hundred species appear on the diet list. In this respect *Spirogyra* and *Oedogonium* are valuable genera in India and *Ulva* (Fig.7.1 A) in Europe. The former two are dried, put up in small packets and sold on the market in India to be made into soup. The colonies of *Nostoc* (Fig. 2.21) are boiled and used as food in Brazil. In the Pacific islands raw algae (sp. of red, and brown algae) are chopped and added to other dishes. The young stipes of *Laminaria* (Fig.19.1) and sporophylls of *Alaria (E)* are also eaten. The laminarian food from stipes is called *Kombu* and from *Alaria* as *sarumen* in Japan. *Durvillea* and *Ulva* are dried, salted and sold. *Ulva lactuca* was formerly used in salad and soups in Scotland. *Porphyra* (laver) (Fig. 21.3 C) is considered a tasteful dish in England and is a common item of diet in Korea, Japan and China. It is rich in vitamins B and C. A red alga *Rhodymenia palmata* (Fig. 24.15 C) is used as food and also as a salty confection named *dulse*. Large quantities of *Durvillea antartica* and some species of *Ulva* are consumed in Chile. The prolific users of sea weeds as food are the coastal people in China and Japan. The commonest species used are *Porphyra tenera (Amanori), Laminaria,* and a green alga *Monostroma*. *Undaria* and *Sargassum* are used for this purpose. *Caulerpa rocemosa* is cultivated in Philippines as a source of food. In China and Japan some sea weeds are regularly harvested as food for man. In Japan many of them are cultivated in bamboo frames in water to meet their demand. It has been estimated that near about 25% of the daily diet of Japan consists of sea weeds. The total food production by marine algae is estimated to be eight times that by the land plants.

Algae as food.

The algae are considered rich in proteins, fats and vitamins A,B, C and E. The vitamins A and D which are commercially obtained from the livers of shark and similar fish originally come from synthesis by the plankton algae particularly which form the food of the fish. Diatom *Nitzschia* is rich in vitamin A. Vitamin B is found in *Ulva, Enteromorpha, Laminaria, Alaria valida, Porphyra, Nereocystis* and *Chondrus crispus. Ulva, Enteromorpha, Alaria valida* and *Undaria* also contain vitamin C. Dulse contains half as much vitamin C as oranges. Fucoids and *Porphyra* are even richer.

Use of Algae as Fodder. The sea weeds are used as a fodder for animals with beneficial effects. Norway, France, U.S.A. Denmark, and New Zealand are the countries where it is a common practice to use sea weeds as fodder for the cattle. In Great Britain, France, Scandinavia and Pacific coasts of U.S.A, kelps are chopped for sheep and chickens. Some countries have developed small industries for processing the weeds such as *Ascophyllum, Fucus* and *Laminaria* into suitable cattle feed. The processed food is given to cattle, poultry and pigs. Egg-yolks of fowls fed on chopped sea weeds have increased iodine and carotene content. The egg laying capacity of the poultry increases. *Macrocystis sp.* is also used being rich in vitamins A and E. A red alga *Rhodymenia* is used as a cattle fodder in France. The milk-yielding capacity of the cattle is enhanced when dried sea weeds such as *Pelvetia* forms an ingredient in cow feed. Similarly increased butter and fat content of milk is reported from cattle whose diet is supplemented with sea weed meal. In China *Sargassum* is used as fodder.

Role of Algae in Medicine. Many sea weeds contain a high percentage of iodine content and thus are employed in the preparation of various goiter medicines or are administered directly as a powdered weed. *Laminaria sp.* such as *L. japonica* and *L. religiosa* have a high value of iodine. Among the green algae *Codium intricatum* contains a considerable quantity of iodine. Red algae such as *Gelidium* and *Grateloupia* contain a medium amount.

A patient suffering from prolapsed stomach is fed on a diet containing dried agar-agar and asked to drink a lot of water. The stomach gets distended and regains its normal condition. A few algae are a source of **antibiotics** which inhibit the growth of other bacteria. **Chlorellin** from *Chlorella* is one such antibiotic. It has also been reported that extracts of *Cladophora* and *Lyngbya* possess antiviral properties and kill strains of certain bacteria (*Pseudomonas* and *Mycobacterium*). The Charales have been claimed to possess larvicidal properties. Caballero (1919) considered Charales useful in the destruction of mosquito larvae.

Fig. 24.15 (A-E). Some sea weeds of economic use. (A), *Gigartina;* (B), *Chondrus crispus;* (C), *Rhodymenia;* (D), *Gelidium amansi;* (E); *Alaria.*

Agar-agar is an important algal product used in the manufacture of pills and ointments. Besides it forms a base for many kinds of medicines used as **laxatives.**

Carrageenin extract, which is another product of algal origin, acts as a **blood coagulant**. Alginic acid stops bleeding.

An effective **vermifuge** is obtained from the extracts of *Digenea, Codium, Alsidium* and *Durvillea.* Many algae are used in the treatment of lung, kidney and bladder ailments by unani hakims.

Carrageenin – a product of algal origin.

Role of Algae in the origin of petroleum and fuel gas. The fuels such as petroleum and natural gas have their origin in the organic matter in the marine environment. Minute algae constituting the plankton trapped the sun's energy during photosynthesis. It was transferred to the marine animals as they fed upon them. The organic compounds derived from the dead bodies of plants and animals constituting the plankton gradually accumulated at the bottom of the ocean and were buried in the course of time by sedentary action. In the environment free from oxygen these compounds were decomposed and converted into oil and gas. The natural gas associated with oil is largely methane which is produced by the action of methane producing bacteria upon organic compounds.

Algae as biofuel.

Algae and sewage disposal. Sewage is the foul domestic and industrial liquid waste which is deficient in oxygen but abounds in dissolved and suspended organic and inorganic materials. It harbours microorganisms of decay and decomposition. The use of small green algae such as *Chlamydomonas, Chlorella* and *Euglena* in large, shallow tanks of effluent (sewage oxidation ponds) has proved a rapid, cheap and effective means of converting the dangerous and expensive waste into an odourless and valuable fertiliser. These tanks promote growth of the algae which flourish at the expense of the mineral nutrients present in the sewage. The rapid photosynthetic activity of the algae produces abundant oxygen which is used by the micro-organisms responsible for decomposition of remaining organic matter in the sewage.

Algae and water supplies. The algae are of negative value as well. In the rainy season and spring, the blue-green algae, some green algae, some golden brown algae and the diatoms become so abundant that the water in the ponds, lakes and reservoirs becomes cloudy and assumes a yellowish or greenish tinge. Sometimes a floating yellowish green scum may develop on the surface of water. Popularly we call these manifestations of algal growth as **water blooms**. Such concentrations of algae impart unpleasant odour, oily and fishy taste to the drinking water. Some of the blue green algae (*Microcysis, Aphanizomenon* and others) produce toxic protein products which are poisonous to fish, cattle, sheep and other domesticated animals.

Harmful algal blooms.

(*ix*) CULTURE OF ALGAE

Algae need be cultured in the laboratory for conducting various studies like studies on cytology and genetics, morphological details, physiology and nutrition and ecology of algae. However, it is very difficult to grow pure cultures of algae in the laboratory since the ideal conditions for culturing and maintaining the algal cultures in the laboratory are rather not possible for a long time. Over the years it has been conclusively proved that algal cultures are important in ascertaining the correct identity of the algae besides studying the physiology and biochemistry of algae.

Algal cultures.

Pringsheim (1964) defines cultures as populations which can be maintained indefinitely by subcultures. It was probably Famintzin (1871) who emphasised the utility of inorganic salts for obtaining algal cultures. Later Bouilhac (1897) in addition to inorganic salts used organic compounds also. Initially gelatin was used as solidifying agent but Tischutkin (1897) used agar-agar in place of gelatin. Pringsheim (1912) was the first to obtain pure cultures of several members of Conjugales.

Generally algae may be cultured on liquid media, solid media and both solid and liquid media. Some of the media used for algal cultures are Knop's medium, Pringsheim's medium, Czurda's medium, Chu 10 medium, modified Chu 10 medium, Bold's basal medium and Ott's artificial sea water medium. These are primary liquid media but can be converted to solid medium by adding agar-agar at the rate of 2% per litre of the medium. Since algae are photoautotrophs, they require the presence of light period alternating with dark periods. However, *Chlorella* requires the presence of continuous light. The pH of the culture medium should be around 7.5 as most algae prefer alkaline pH. Deionised water should be used in the preparation of the culture medium. Cultures should be maintained at a temperature between 20°C and 30°C. Generally bacteria contaminate the algal cultures. To obtain bacteria less cultures, antibiotics like aureomycin is used. Before inoculation treatment of algal suspensions with 0.01% Mercuric chloride has been found useful.

Almost all the algae including marine algae have been grown in pure cultures. Blue green algae are the easiest to be grown followed by green algae.

(X) WATER BLOOM

Water blooms or Blooms are formed by the growth of a solitary alga, rarely by a few algae. When the algae are abundant in a pond or lake as to give it a distinct colour to the entire body, such an algal growth is called as **water blooms**. The colour of the algae determines the colour of the water blooms. The algae involved in the development of water blooms are generally planktons.

Algal blooms.

Water blooms are generally formed by the rapid growth and multiplication of algae which usually belong to class Cyanophyceae. A few members of Chlorophyceae, Chrysophyceae, Euglenophyceae and Pyrophyceae are also known to form water blooms. Water blooms are formed in fresh waters, inland salt lakes and in the sea.

Water blooms may be temporary or permanent. Generally a number of temple ponds and a few lakes and big ponds have permanent water blooms. The water blooms are usually thick during winters and summers while during the rainy season the blooms become thin or get diluted because of sudden splashes of rain drops that dilute the blooms.

In India, the water blooms in temple ponds, lakes etc. are formed of *Microcystic aeruginosa* and *M. flos aquae*. Water bloom forming algae of class Cyanophyceae include species of *Anabaenopsis, Spirulina, Oscillatoria, Arthrospira* etc., those of Chlorophyceae are *Chlamydomonas, Pandorina, Euderina, Gonium, Volvox, Cosmarium, Closterium* etc. *Euglena* of Euglenophyceae, *Synura* of Chrysophyceae, *Peridinium, Ceratium* and *Gymnodinium* of Pyrrophyceae. In salt takes of Sambhar, where pH ranges between 14 and 17, blooms are formed of *Anabaenopsis*. In the esturine regions of rivers in South India Colonies of *Anaebaenopsis arnoldi v. indica* form the blooms. In the India seas, *Trichodesmium erythraeum* and *T. thiebantic* are the bloom forming algae.

Although bloom forming factors are yet to be understood, it is generally believed that optimum conditions necessary for sound algal growth must be present in water bodies.

Water blooms formed of Cyanophycean algae are good nitrogen fixers and therefore can be used as fertilisers. Water blooms also provide food for immense flocks of flamingos and other migratory birds. However, water blooms may be harmful because they are indirectly responsible for fish mortality due to depletion of oxygen. Sometimes they are responsible for unpleasant taste to drinking water and foul smell. These also liberate deleterious substances to aquatic animals.

Water blooms can be controlled by application of copper sulphate or Chlorine. Cyanophages have also been used to control blooms besides algicides.

(*xi*) FOSSIL ALGAE

Since almost all algae are soft and delicate, their fossilisation should not be expected. Yet a sizable number of algal fossils have been discovered although fossil algae are imperfectly preserved. Very few algae have been well preserved and have been identified since identification of algae is difficult. Algal fossils have also been imperfectly explored. According to Champman and Champman (1981) the preservation of algae, particulary in the lower geologically state either has not occurred or has not been very good to be identified. Only those algae that deposit lime or silica have been preserved most successfully and therefore these constitute the groups of fossil algae about wihich we know. Even at the present time fossilisation of algae is taking place.

Algae have been preserved as impressions, carbon fillows, molds, casts and petrifications. Fossils of algae have been reported from the precambrian period to the recent times. Largest number of algal fossils belong to Cretaceous period although algal fossils have also been reported from Palaeozoic, Triassic, Jurassic and Tertiary periods.

TABLE

Sl.	*Class of algae*	*Period in which they appeared*
1.	Cyanophyceae	
	(a) Unicellular	Early precambrian
	(b) Multicellular	
	Oscillatoriaceae	Middle precambrian
2.	Rhodophyceae	Late precambrian
	Chlorophyceae	Late precambrian
	(Chlorococcaces)	Late precambrian
	Phaeophyceae	Late precambrian
	Heterocystous blue green	Late precambrian
3.	Rhodophyceae	Cambrian
4.	Chlorophyceae	Ordovician
5.	Charophyceae	Silurian
6.	Bacillariophyceae	Jurassic
7.	Chrysophyceae (?)	Jurassic
8.	Xanthophyceae	Cretaceous
9.	Chrysophyceae	Cretaceous

The oldest fossil alga is reported to be a cyanophycean alga — *Archaeosphaeroides barbertonensis*. Blue green algal fossils have been reported from early precambrian era. These fossils are mostly impressions which are globular or thread-like and unicelluar (?). In the middle precambrian period diversity of fossils increased as coccoid and filamentous blue green algae belonging to Oscillatoriaceae appeared and in late precambrian period the algae which appeared were heterocystous prokaryote and eukaryote. The eukaryote belong to class Chlorophyceae, Phaeophyceae and Rhodophyceae. The following table summarises the algal classes that appeared in different geological period:

Archaeosphaeroides barbertonensis oldest fossil alga

A large number of algal fossils have been reported from India as well. These have been reported from Punjab, Bihar, Madhya Pradesh, Rajasthan, Kuch, Kashmir, Kerala, Andaman & Nicobar, Pondichery, Kathiawar, Chhatisgarh, Vindhyans, Indore etc. A list of Indian algal fossils is given below:

LIST OF ALGAL FOSSILS

S.No.	*Class*	*Name of the algal fossil*
1.	Cyanophyceae	*Aphanocapsa*
		Synechocystis
		Scytonema
		Palaeonostoc
		Archaeonema
2.	Chlorophyceae	*Dissocladella*
		Dictylospora
		Neomeris
		Piania
		Botryococcus
		Palaeochyla
		Oedogonium
		Ulothrix
		Spirogyrites
3.	Charophyceae	*Chara (Gyrogites)*
		Nitellites
4.	Bacillariophyceae	*Fossil diatoms*
5.	Rhodophyceae	*Archaeolithothamnion*
		Lithothamnion
		Melobsia
		Distichoplax
		Amphira
		Corallina
		Petrophyton
		Solenopora
		Lithophyllum

The study of algal fossils has assumed importance since it was revealed that algal fossils may be indicators of oil deposits. Uses of the study of algal fossils are as follows :

(*i*) Algal fossils may indicate the presence of oil deposits.

(*ii*) Algal fossils have been used to understand the past environment and ecological conditions.

(*iii*) Algal fossils also act as guide fossils since Ordovician of Paleozoic era is also called the Age of Rhodophyta" due to dominance of Rhodophycean algae.

(*iv*) Algal fossils are important because they form the limestones. Bricks made of algal limestones are very light and hence used for the construction of high rise buildings.

(*v*) Diatomaceous earth formed of fossil diatoms are used for various purposes. Alfred Nobel is reported to have used the diatomaceous earth to absorb nitroglycerine from which the dynamite is made. Diatomaceous earth is also used for insulation of boilers and earth furnaces. In sugar refining process and brewing industry processed diatomaceous earth is used mainly as filtration aid. It is also used as filters for paints, varnishes and paper products. Diatomite is also used for sprinkling to reduce secondary explosions in coal mines.

(*xii*) ALGAE AND WATER POLLUTION

Rapid industrialisation and urbanization associated with population explosion has increased the extent of water pollution many fold. Water bodies have been used as sink for the disposal of sewage and industrial wastes since long. This has caused an increase in the extent of water pollution and a decrease in natural regeneration capacities of the water bodies. Unwise use of agrochemicals specially Nitrogen and Phosphate fertilizers has enriched the permanent water bodies with nutrients, which facilitates the growth of diverse aquatic plants specially algae. Sometimes massive growth of bloom forming algae may take place due to increased nutrient status of permanent water bodies which deteriorates the water quality of these bodies. Certain algae flourish well in organically polluted water and ultimately cause Eutrophication of ponds, pools and lakes as a result of continuous increase of the organic matter. Besides some algal genera serve as indicators of water pollution. These have been used as "biological markers" of the extent of water pollution or specific nature of water pollutant. Kolkwitz and Marsson (1908) were probably the first to describe the composition of algal communities of rivers in relation with its extent of pollution and identified three main zones of polluted river which favour the growth of diverse species of algae. These zones are :

Algae polluting water.

(*a*) Polysaprobic Zone which is rich in the compounds of high molecular weight and is biodegradable in nature. It is oxygen defficient zone and favours the growth of *Euglena, Oscillatoria* etc.

(*b*) Mesosaprobic Zone which lies in between polysaprobic zone and oligosaprobic zone and is characterized by having less decomposition of organic matter as compared to polysaprobic zone. This segment is not much oxygen defficient and there is moderate availability of oxygen. Few species of *Oscillatoria, Ulothrix* and *Phermidium* usually have better growth in this zone.

(*c*) Oligosaprobic Zone usually have very little amount of biodegradable organic matter, is rich in dissolved oxygen and favours the growth of a number of algal genera such as *Cladophora, Calothrix, Batrachospermum, Spirogyra* etc.

In subsequent years several Phycologists got interested in the field and a lot of work has been done regarding the growth of algal flora in relation with severity of water pollution. Some of the important workers in the field include Palmer (1969), Patrick (1973), Rai and Kumar (1976), Prasad and Saxena (1980), Prasad and Singh (1980, 1982), Sinha et. al. (1988, 89), Shukla (1991) Shukla and Anjum (1994), Tripathi & Shukla (1994), Singh and Sinha (1994). The above mentioned studies concluded that there is definite correlation between the growth of algal genera and the extent of water pollution. The algal species also give an idea about the nature of water pollutant and their source. Following is the list of some algal genera and species which are specifically related with the nature of pollutants and their possible sources :

A. Algae of domestic sewage (organically polluted water)

Domestic sewage wastes which are carried in open passages gets accumulated in low lying areas and exhibits abundant growth of *Oscillatoria rubescens*, *O. amphibia*, *O. boryana*, *O. mougeolia* and *O.okenii*, *Euglena acus*, *E. charkowiensis* and *E.gracilis* and *Spirulina major* wherever a pool of domestic sewage is formed *Euglena* becomes dominant along with few diatoms whereas *Oscillatoria sp.* dominates in small open passages. In the perennial open passages of domestic (residential) sewage the species of *Schizomeris* and *Stigeoclonium* flourish well.

In India the common algal species found in organically polluted water bodies include, *Euglena viridis, Nitzschia palaea, Oscillatoria lemosa, O. tennis, Arthrospira jenneri, Stigeoclonium tenue, Phacus longicaudey, P. pleuronectus, Anabaena constrieta, Chlorella vulgares, Scenedesmus bijuga, S. dimesphus, Pandorina morum, Cyclotella menenghiniana, Navicula muralis, N. pygmea, N. cryptocephala, Gomphonema parvulum, G. sphaeropleurum, Nitzschia acicularis* and *Synedra ulna.*

B. Algal species as indicators of industrial wastes :

Following algal associations indicate the specific nature of pollutants of diverse industries:

Ulothrix zonata

(*i*) **Indicators of paper mill wastes** are *Ulothrix zonata, Scenedesmus bijuga, Pandorina morum, Oscillatoria splendida, Cymbella ventricosa, Amphora ovales, Navicula cryptocephala, N. radiosa, Surirella ovata, S. ovata var. Salina* and *Synedra ulna.*

(*ii*) **Indicators of phenolic compounds** include *Acanthes affinis, Ceratoneis arcus, Cocconeis placentula, Cyclotella kuetzengii, Cymbella naviculiformis, Fragillaria virescens; Gomphonema parvulum, Navicula cryptocephala, Nitschia palea, Pinnularia borealis* and *Synedra ulna.*

(*iii*) **Indicators of oil wastes** may be *Amphora ovalis, Diatoma vulgares, Gomphonema herculaneum, Navicula radiosa, Melosera varians, Synedra acus* and *S. ulna.*

(*iv*) **Indicators of distillary wastes** are *Chlorogonium gracillima, Chlorobranchis gracillima* and *Chlamydobotrys sp.*

(*v*) **Indicators of copper as wastes** include *Calothrix braunii, Scenedesmus obliquus, Chlorococcum botryoides, Stigeoclonium tenue, Navicula viridula, Cymbella ventricosa, C. naviculiformis, Nitzschia palea* and *Acanthes affinis.*

(*vi*) **Indicators of iron as wastes** are generally *Stauroneis phoenicenteron, Surirella linearis, S.delicatissima,* and *Trachelomonas hispida.*

(*vii*) **Indicators of chromium wastes** may be *Euglena acus, E.viridis, E.sociabilis, E. oxyuris, Navicula atomus, N. cuspidata, Tetrospora lubrica, Stigeoclonium tenue* and *Nitzschia sp.*

(*viii*) **Algal associations indicating high acidity or discharge of acidic wastes** generally consist *of Cryptomonas erosa, Chromulina ovalis, Euglena mutabiles, E. viridis, E. stellata, Ulothrix zonata* and *Lepocinclis ovum*.

C. Diatoms as indicators of water pollution

Patrick (1956) opined that probably diatoms are more suitable markers of water pollution than other groups of algae because different or specific species are found in different severity of water pollution. Moreover diatoms need no special preservatives because of their silicified walls and there are large number of species due to which statistical analysis of the data is more convenient. Later on the studies of Palmer (1969), Rana and Kumar (1971), Patrick (1973), Marathe and Nandkar (1976), Prasad and Singh (1982) also indicated that the abundance of species and presence of diverse taxa of diatoms in the polluted water furnish more dependable results on the extent of water pollution. According to them diatoms like *Amphora veneta, Cyclotella meneghiniana, Gomphonema parvulum, Melosira crenulata, M. granulata, Navicula accomoda, N. cryptocephala, Nitzschia diducta, N. palea* are indicators of water pollution specially the sewage wastes contamination. Besides the pollution indicating genera of diatoms, there are certain genera and species which prefer to grow in clean water such as *Cocconeis placentula, Cymbella affinis, Gomphonema micropus* and *Nitzschia amphibia* and serve as the indicators of clean waters. Moreover, the presence of a single species of alga does not serve as the reliable indicator of water pollution or its severity. For this purpose, the algal community growing in polluted water has to be taken into account, because many genera of algae have some species growing in polluted waters whereas other species prefer clean water for their growth. The studies of Fjerdingstad (1950) confirmed that in *Ulothrix zonata* one of its variety grows in polluted water while another variety prefers clean water. Later on in the studies of Prasad and Saxena (1980) it was reported that there are some species of blue green which are abundant in polluted water e.g. *Oscillatoria formosa, O. nigra, O. curviceps, O. amphibia, O. tenuis, O. limosa, Lyngbya cryptovaginata* and *L. shackletoni* whereas other species of above genera e.g. *Oscillatoria acutissima, O. martinii Lyngbya hieronymusii, L. borgerti* and *L. spirulinoides* prefer unpolluted water for their growth. It may thus be concluded that algal species may serve as indicators of the water quality of water bodies. Polluted waters with abundant algal growth may make the water unfit for human consumption.

(*xiii*) BIOFERTILIZERS

The idea of biofertilizers originated due to the high cost of chemical fertilizers and high level of industrialisation. Moreover, manufacture of nitrogenous fertilizers consume high energy generally generated from fossil fuels like oil and coal which are already depleting in developing countries like India. In addition indiscriminate use of chemical fertilizers has not only put an additional burden on conventional sources of energy but has also disturbed the aquatic as well as soil ecosystems.

Nitrogen fixing biofertilizer

Since most of the nutrient supply of the soil is dependent upon the activity of the micro-organisms and plenty of N_2 present in atmosphere can not be used by the plants directly. The same is true with sulphur and phosphorus, present in soil in complex forms which have to be mineralized by microbes in order to make them available to the plants for utilization. Therefore, it became necessary to develop fertilizers which can be used directly the microorganisms were tested as organic manures and these were called Biofertilizers.

Algal cultivation to produce biofertilizers

The term 'biofertilizer' denotes all the "nutrient input of biological origin for plant growth" where 'biological origin' denotes microbial process synthesizing complex compounds and their further release into outer medium which are again used by the plants, thus, biofertilizers are also called "Microbial inoculants" (Subba Rao, 1982)

Cyanobacteria and bacteria are widly used as biofertilizers as they are known to fix atmospheric nitrogen.

A: Cyanobacteria: Algal biofertilizers

Several blue green algae (BGA) *e.g. Anabaena, Nostoc, Aulosira, Cylindrospermum, Plectonema, Tolypothrix, Gloeotrichia* are good nitrogen fixers and their role in fixation of atmospheric nitrogen in the paddy fields has been realised much earlier (Singh, 1961). In suitable conditions cyanobacteria multiply rapidly and fix atmospheric N_2 and release it into the surroundings. Unlike N_2 fixing bacteria, cyanobacteria can photosynthesize and do not need external supply of carbon compounds. The nitrogenous compounds released in the growing medium include aminoacids, proteins and growth promoting substances (Sewart, 1970).

Cyanobacteria *e.g. Anabaena Azolla* are known to fix atmospheric N_2 in symbiotic association with a water fern *Azolla*. It contains the alga as an endophyte in the leaf cavity and the significance of it has been realised in Vietnam, China, Philippines, India and Bangladesh. The species which are known to fix nitrogen in association with cyanobacteria include *A. caroliniana*, *A. filiculoides*, *A. mexicana*, *A. microphylla*, *A. nilotica*, *A. pinnata* and *A. nibra*.

A. pinnata is a commonly occuring species in India and noteworthy work has been done in Central Rice Research Institute, Cuttack regarding mass cultivation of *Azolla* and its use as biofertilizer (Algalisation, Venkatraman 1961) has been on rise in states of in Tamilnadu, Uttar Pradesh, Andhra Pradesh, Karnataka, Maharastra and Haryana in India.

B. Bacterial Biofertilizers

Several free living and symbiotic bacteria e.g. *Azospirillum, Clostridium, Beijerinkia, Klebsiella, Rhodospirillum*, *Azotobacter*, *Rhizobium*, *Bacillus* etc. are known to fix atmospheric N_2 and add to the fertility of soil.

The legume *Rhizobium* relationship is well known symbiotic association so far as atmospheric N_2 fixation is concerned. *Rhizobium* occures as free living microbe in the soil but does not fix N_2 unless and until it establishes symbiotic relationship with leguminous plants. The method which are used to increase the number of such N_2 fixing bacteria in soil is known as bacterization. It is a technique of seed dressing with bacteria. It has been observed that bacteria (*e.g. Azotobacter*, *Bacillus*, *Rhizobium*) can successfully establish symbiotic association with the roots of host plants and fix N_2. Bacterial fertilizers 'Azotobakterin' and 'Phosphobacterin' containing cells of *Azotobacter chroococcum* and *Bacillus megaterium var phosphaticum* respectively have been used in USSR and East Europian countries and are known to increase the crop yield about 10-20% (Cooper 1959).

Bacterization of seeds in different countries like England, USSR, Romania, Poland, Bulgaria, Hungary and India noteworthy increase in crop yield such as wheat, barley, maize, sugarbeet, cabbage and potato has been reported.

Bacteria are also known to secrete growth substances and secondary metabolites in rhizosphere (Subba Rao, 1982 and Dwivedi et al, 1989) which help in seed germination and plant growth.

Moreover, free living bacteria *Azotobacter* in association with *Azospirillum* and symbiotic bacteria *Rhizobium* associated with phosphate solubeilizing bacteria *Bacillus megaterium, B. polymyxa and Pseudomonas striata* are successfully used to increase crop yield.

C. Mycorrhiza; As biofertilizer

Mycorrhizae are a group of fungi which include a number of types based on the different structures formed inside or outside the roots of plants. Mycorrhizae develop symbiosis with the roots of specific plants facing nutrient scarcity especially nitrogen and phosphorus. The fungal symbiont gets shelter and food from the plant while the plant gets an array of benefits like better uptake of phosphorus, salinity and drought tolerance, heavy metal toxicity and increased plant growth and development.

Uses of Biofertilizers: Advantages of biofertilizers over chemical fertilizers:

1. Biofertilizers are economical and free from pollution hazards.
2. Cyanobacteria secrete growth promoting substances e.g. IAA, IBA, NAA, aminoacids, proteins, vitamins which inhance growth and development of plants.
3. Cyanobacteria grow and multiply rapidly and add sufficient amount of organic matter in the soil.
4. Cyanobacteria can be used to reclaim saline or alkaline soils as they are capable to grow under wide pH range.
5. Blue green algae are of immense economic value as their use in paddy fields can increase the crop yield by 10-45% and about 40-50 kg nitrogen per hectare is left over in the soil which can be used by other crops.
6. Bacterial biofertilizers (*Azotobacter* and *Azospirillum*) secrete antibiotics beside N_2 fixation which help in crop protection.
7. Rhizobial biofertilizers can fix 50-150 kg N/ha/annum which reduces the use of chemical nitrogenous fertilizers considerably.
8. *Azolla* in association with cyanobacteria (*Anabaena Azollae*) fix atmospheric nitrogen and increases organic matter and fertility in the soil and enhances the tolerance against heavy metals.
9. Mycorrhizal biofertilizers play important role in selective absorption of immobile and mobile elements to the plants, also increase area of absorption of roots, and increases tolerance regarding adverse edaphic factors and pathogenecity.
10. The physio chemical properties have been observed to be improved by the use of biofertilizers like texture, soil moisture, pH, organic content and macro and micro nutrients.

(*xiv*) **BIOLUMINESCENCE: (*Greek, bios* = living; *Latin, lumen* = light)** *i.e.*

Bioluminiscence

Bioluminescence is the phenomenon of "production and emission of light by living organisms as a result of chemical reaction during which chemical energy is converted to light energy".

Bioluminescence is a form of luminescence or "cold light". In this, less than 20% of the light generates thermal radiations. About 90% of deep sea marine life produces bioluminescence. Most of the light emitted belongs to blue and green light of the spectrum. Most forms of bioluminescence are lighter and only exist at night following circadian rhythm. It should not be confused with the phenomenon of light like phosphorescence, fluorescence or refraction of light.

Organisms showing bioluminescence: Every organism produces some form of bioluminescence within the electromagnetic spectrum, but most is neither visible nor noticeable to the naked eye. Bioluminescence of every organism is unique in wavelength, duration, timing and regularity of flashes. A list of organisms to emit visible bioluminescence has been given below:

(*i*) Non-marine organisms	1. Certain Arthropods Fireflies, glow worms (e.g., rail road worms), certain mycetophilid flies, certain centipedes, certain millipedes 2. Some Annelids 3. Mushrooms. Jack O'Lantern mushroom, Honey mushroom, *Panellus stipticus*,
(*ii*) Fishes	Cookie cutter shark, Marine hatchetfish, anglerfish, Flashlight fish, Pinecone fish, Porichthys, Gulper Eel.
(iii) Marine Invertebrates	Many cnidarians like sea pens, corals, *Aequorea- victoria* (a jellyfish); Ctenophores or "comb jellies"; certain echino-derms; certain clams certain nudibranchs; certain crustaceans like ostracods krill; certain octopuses (like Balitaenidae); certain squids like order Teuthida, colossal squid, mastigoteuthidae, sapiolidae, fireflies squid.
(*iv*) Planktons and Microbes	Dinoflagellates Vibrionaceae (e.g. *Vibrio fischeri Vibrio harveyi Vibrio phosphoreum*)

Algae showing bioluminescence: Bioluminescence is caused by many dinoflagellates, a single- celled mesokaryotic algae. They figure hundreds per litre in sea water.

Bioluminiscence by algae in a bloom

The first dinoflagellate discovered showing bioluminescence was ***Noctiluca*** (**nox = night; licere = to shine**). It is very familier dinoflagellate because of its large size (≈ 2 mm in diameter) and luminescent capabilities. It produces tomato soup in the coastal regions. *Noctiluca* is monotypic represented by a single species *N. miliaris*.

The dinoflagellate *Lingulodinium polyedrum* show brilliant glowing red tides at a beach in Carlsbad California in 2005.

The other algae showing bioluminescence are *Gonyaulux, Pyrocystis, and Pyrodinium.*

Correlation between bioluminescence and Circadian rhythm: *Circadian rhythm* (*Latin, Circa = around and diem/dies = day*) literally meaning "about a day" (**Franz Halberg**). It is roughly 24 hour cycle in physiological processes of living beings including plants, animals and micro-organisms. It is regulated by endogenous factors.

Sweeney and Hastings describes two types of bioluminescence in *G. polyedra,* one showing luminescence intensity and other showing luminiscence capacity. They studied them under two condition, induced and spontaneous luminescence conditions:

(*i*) When cells are vigorously agitated, they give off flash of light which is at peak in mid of dark period and at lowest in the mid of light period, in case of plants growing under normal light dark cycle (LD cycle).

(*ii*) When these cultures are transferred to DD cycle (Continuous dark), it can be observed that endogenous circadian rhythm is indeed present, the cycle of the peaks of bioluminescence continued for several days but amplitude gradually decreases. A similar fact was observed in LL (continued light) where peaks of light emission was not so great as under LD cycle but circadian rhythm continued indefinitely.

Mc Murray and Hastings (1972): Studied spontaneous luminescence. They found that amount of light produced gradually increases from the begining of dark period and reaching a peak at mid night.

Reasons of photoemission/bioluminescence

1. Earlier it was thought that some crystalline inclusions called "**Scintillons**" were the source of photoemission. Later these particles were found in both luminescent and non-luminescent dinoflagellates.

2. Schmitter (1971) while working on *Gonyaulax polyedra* reported the presence of polyvescicular bodies from the peripheral cytoplasm. These bodies exhibited luminescence even in purified preparations from extracts of algae.

3. Soli (1966), Swift and Tsylor (1967), and Biggley and others suggest that the substrate enzyme complex of *Luciferinluciferase* which is characteristic of all bioluminescent organisms and also found in bioluminescent dinoflagellates is the same and is responsible for the phenomenon of bioluminescence among certain dinoflagellates. These enzymes may be present in those bodies, that cause luminescence.

Utility of Bioluminescence for the organisms:

The utility of bioluminescence for the organisms follows:

(*a*) Comouflage: The cookiecutter sharks uses bioluminescence for camouflage. A small patch on its under belly remain dark and appears as a small fish to large predators like tuna and mackerel fishes. When these fishes try to consume the 'small fish', they are bitten by these sharks.

(*b*) Attraction

(*i*) Dinoflagellates have interesting twist on this mechanism when a predator of plankton is sensed through motion in the water, the dinoflagellate luminesces. This in turn attracts even larger predators which will consume the would be predator of the dinoflagellates.

(*ii*) Bioluminescence is also utilised for attraction of mates e.g. fireflies utilize periodic flashing in their abdomen to attract the mates in the mating season.

(*iii*) The honey mushroom attract insect using bioluminescence to disseminate the fungus spores into the environment.

(*c*) **Repulsion:** Certain squid and small crustaceans utilize bioluminescence for repelling their predators.

(*d*) **Communication:** Bioluminescence is thought to play a direct role in communication between bacteria. It promotes the symbiotic induction of bacteria into host species and may play a role in colony aggregation.

(xv) COLOURLESS ALGAE AND THEIR POSSIBLE PIGMENTED RELATIVES

The numerous colourless motile and non-motile forms of algae belong to different classes. Some colourless forms show close resemblences with the living forms while others are very much different and hardly have any resemblence with the living forms. The feature usually taken to trace the affinities are cell morphology, method of reproduction, resting stages, food reserves, occurrence of eye spot etc.

The following is the list of colurless forms and their nearest pigmented relatives:

Class	Pigmented form	Non-pigmented form
A; Cyanophyceae	1. *Oscillatoria*	*Vitreoscilla, Beggiatoa Oscillospira*
	2 *Microcoleus*	**Thioploca**
	3. *Spirulina*	*Spirulina albida, Thiospirillopsis*
	4. *Microchaete*	*Thiochaete*
	5. *Lyngbya* (?)	*Leucothrix, Thiothrix*
	6. *Synechococcus*	*Achromatium*
B; Chlorophyceaec	1. *Dunaliella*	*Hyaliella*
	2. *Tetrachloris*	*Polytomella*
	3. *Chlamydomonas*	*Polytoma, Tusselia*
	4. *Chlorogonium*	*Hyalogonium*
	5. *Sphaereliopsis*	*Sphaerellopsis hyalina*
	6. *Carteria*	*Telrablepharis*
	7. *Brachiomonas*	*Hyalobrachion*
	8. *Solenochloris*	*Furcilla*
	9. *Coccomonas*	*Chlamydoblepharis*
	10. *Phacolus*	*Phacollus pallidus*
	11. *Tetraspora*	*Cloeochaete*
	12. *Chlorella*	*Prototheca*
C; Chrysophyceae	1. *Ochromonas*	*Monas*
	2. *Chromulina*	*Oikomonas*
	3. *Mallomonas*	*Mallom*
	4. *Synochromonas pallida*	*Anthophyra vegetans*
	5. *Chrysodendron ramosum*	*Dendromonas virgaria Monadodendron distans*
	6. *Epiphyxis and Stylophyxis*	*Stokesiella lepteca*
	7. *Chrysarchnion insidians*	*Leukochrysis*
	8. *Eleutheropyxis*	*Leukopyxis asymmetrica*
	9. *Kybotion eremita*	*Peatytheca micropora*
	10. *Lagynion*	*Heterolagynion*
	11. *Kehyrion*	*Donatomonas cylindrica*
	12. *Stylochromonas minuta*	*Monosiga*

D; Bacillariophyceae	*Nitzschia*	*Nitzschia putrida*
E; Cryptophyceae	1. *Cryptomonas* and *Chroomonas*	*Chilomonas paramecium* *Cyathomonas truncata* *Phyllomitus Kathablepharis*
F; Euglenophyceae	1. *Eutreptiella*	*Distigma*
	2. *Euglena*	*Astasia, Cyclidiopsis, Khawkinea.*
	3. *Lepocinelis*	*Gyropaigne*
	4. *Phacus*	*Phacus triqueter*
	5. *Trachelomonas*	*Trachelomonas reticulata*
G; Pyrrhophyta	1. *Amphidinium, Gymnodinium etc.*	*Amphidinium elenkinii* *Gymnodinium helicocoster* *G. blax* *G.. colymbeticum* *Glenodinium daxi*

Loss of pigments in colourless algae is supposed to be accompanied by loss of plastids. The colourless members of Chrysophyceae appear to have originated in two ways:

1. In *Rhizochrysis* the chromatophores do not always divide at the same rate as the protoplasts with the result that some of the individuals receive no plastids at all.
2. According to **Pascher** the colourless forms more usually arise by progressive reduction of chromatophores (e.g Chlamydomonadaceae).

Terenetz have also reported the degeneration of the plastids in few forms of *Euglena gracilis*. Other aspects of apochlorosis have not been fully understood even now

Colourless species must be capable of utilising organic compounds available in the environment because reduction in the number of plastids may lead to starvation of the species.

(*xvi*) TOXIC ALGAE

Toxic algae are those which cause sickness or death to aquatic livestock particularly animals. Almost all aquatic algae in growing medium secrete variety of extracellular substances which may be beneficial or harmful/toxic to aquatic flora and fauna. Besides, the massive growth of algae, particularly Blue green algae (BGA) form water blooms which adversly affect the aquatic animals. Bloom forming algae causes severe oxygen depletion in the aquatic bodies which results in heavy fish mortality particulary in cloudy days.

Several cases of death of animals have been reported from different parts of world due to algal poisoning. Studies in the algal toxicity in South Africa has clearly indicated that the deaths of thousands of cattles sheeps and other animals was due to blooms of *Microcystis toxica* and *Anabaena flos-aquae*. The toxic substances secreted by *Gymnodinium veneficum* is known to be toxic to aquatic animals along the coasts of California and Washington.

Algal toxins are also known to be toxic to human beings. They cause gastrointestinal disorders and severe illness or even death (Tisdale 1931, Veldee 1931). The death of human beings in Japan was due to the consumption of fishes fed on *Gonyaulax*, a dinoflagellate. Gastrointestinal disorders in human beings are also known to be caused by *Prototheca protoricensis, Oscillatoria intestenii. Anabaena circularis, Aphanomezon and Microcystis* (Schwimmer et at, 1964). While *Gymnodinium brevis* is reported to cause respiratory tract irritation, burning of eyes and throat infection, *Anabaena* circularis and *Lyngbya majuscula* are reported to cause skin disorders and allergic conjuctivitis for swimmers in swimming pools.

A list of toxic algae has been given in the Table:

Class	*Toxic algae*
1. **Cyanophyceae**	*Microcystis aeruginosa, M. toxica.*
	Aphanothece nidulans, A. cyanea.
	Anabaena flos aquae, A. circularis, A. limnetica
	A. schevemetievi, A. nadsonii, A. inaqualis
	Gloeotrichia echinulata.
	Lyngbya bergei.
	Nodularia spumigena.
2. **Dinophyceae**	*Gonyaulax catenella, G. monilata, G. tamerensis.*
	Gymnodinium brevis, G. flavium, G. mikimotoi
	G. veneficum.
	Pyrodinium phoneux
	Prymnesium parvum.
3. **Chlorophyceae**	*Scenedesmus sp.*
	Chlorella sp.

Chemical Nature of Algal toxins:

Most of the algal toxins are water soluble complex compounds.

1. The toxins of *Microcystis* is non volatile acidic compound having much absorptive qualities. It is a neurotoxin composed of polypeptides of aminoacids.
2. The toxin of *Gonyaulax* is an alkaloid and acid stable non-volatite compound. It can be stored for several months without losing its toxicity.
3. The toxin of *Gymnodinium veneficum* is acid labile, water soluble and ether insoluble. It is a neurotoxin and exibit high toxicity.
4. *Prymnesium parvum* secretes toxin of non dialysable nature. It is both thermo labile and acid labile.

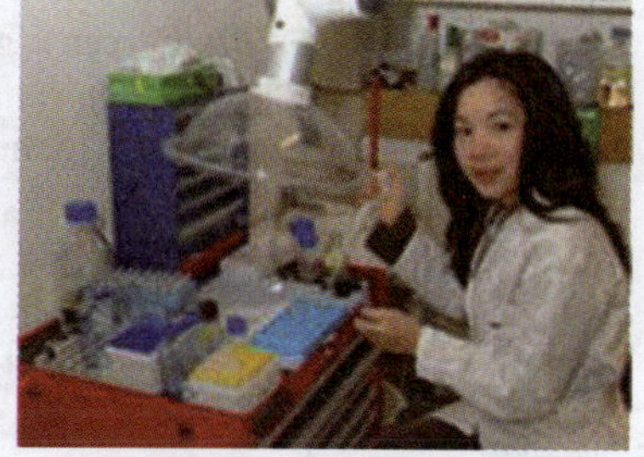

Analysis of algal toxins.

Effects of Algal toxicity: Generally the effects of algal toxicity vary and has been summarised below:

1. The feeding of *Microcystis toxica* results in loss of weight, weakness, enlargement of liver and conjestion within the spleen.
2. *Gymnodinium brevis* cause respiratony tract irritation, burning of eyes and throat in human beings.
3. *Anabaena sp.* and *Lyngbya majuscula* cause skin disorders and conjuctivities for swimmers in swimming pools.
4. *Oscillatoria intestenii etc.* cause gastrointestinal disorders in human beings.
5. *Gonyaulax sp.* toxicity may cause even death of human beings.
6. Algal toxically affect all the aquatic biota adversely and may cause even death of flora and fauna of the aquatic system.

However, the algal toxicity depends upon the existence of toxic and non-toxic strains and environmental conditions. It is surprising that *Microcystis aeruginosa* is common in temple ponds in India but no toxicity has been reported. *Anabaena flos-aquae* is also common in Indian water bodies.

Light intensity has also marked effect on the production of toxins of *Anabaena flos-aque* (Peary and Gorham, 1966). The endotoxins of *Microcystis* can exibit its impact only after the death and decay of the cells. Though heavy fish mortality in water bodies having water blooms of *Microcystis* have been reported but it appears that the main cause of fish mortality in such cases is due to depletion of oxygen instead of algal toxins.

(*xvii*) ALGAL VIRUSES OR CYANOPHAGES (Greek, Phages = to eat)

"Viruses attacking Blue-Green Algae"

"Cyanophages are those viruses which attack and eat the members of Cyanophyceae."

Krauss (1961) was probably the first to suggest disappearence of blue green algae may be due to algal viruses. **Lewin (1960)** reported a phage which attack on colourless blue green algae, *Spirochaeta rosea.* For the first time **Saffermann** and **Moris (1963)** isolated a phage attacking a photosynthetic blue green algae and since then several algologists became interested and isolated various cyanophages.

These agents are now known to attack a wide range of Blue-green algae. These are called as **Phycoviruses** (Schneider, T.O. Diener and R.S. Safferman, 1964), **Algophages** (Goriushyn V.A, and S.M. Chaplinskyaya), **Blue-green algal viruses** (Padam and Shel 1971;) and Cyanophages (Luftig. R. and R. Haselkorn, 1968, Padan and Shilo, 1971). The latter designation seems most appropriate because of their close analogies to the bacteriophages.

(*a*) **Nomenclature:** Like the plant viruses the cyanophages have been named according to their known hosts. It should be done by taking first letters of generic names of host followed by arabic number which designate serological sub group; the different isolates of similar types were marked by suffix consisting of first letter of the provinance.

e.g.

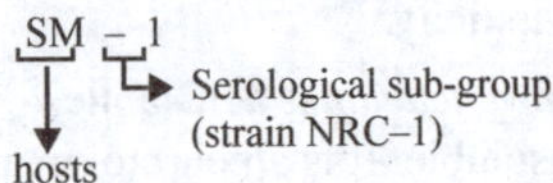

S = *Synechoccus elongatus*

M = *Microcystis aeroginosa*

Some of the well known cyanophages are:

1. LPP-group affect 3 genera of filamentous blue green algae that do not form heterocyst; *Lyngbya, Phormidium, and Plectonema.* It is divided into two distinct serological groups LPP-1 and LPP-2.

2. *C*-1 Affects *cylindrospermum*

3. *TAuHN*-1 affects the host *Tolypothrix, Aulosira, Haplosiphon,* and *Nostoc.*

4. A-1, the long tailed virus infects *Anabaena variabilis.*

5. *AS*-1, *Anacystic nidulans and Synecoccus cedroum.*

6. N-1 attacks the filamentous *Nostoc muscorum.*

(*b*) **Composition:** The Cyanophages are composed of nucleic acid and proteins. The nucleic acid is linear, double stranded DNA. The length of the nucleic acid ranges from 13.2nm to 24.3nm. The protein forms capsid, generally the molecular weight of major head proteins varies between

13000 and 39000 while the major tail protein has a molecular weight of 80,000. No internal proteins were detected. The sum of molecular weights accounts for about 35% of DNA.

(c) Structure: The structure of Cyanophages is based on the shape and the bonding characteristics of the capsid proteins. The morphology of LPP-1 has been most extensively studied. **Lufting and Haselkorn** who proposed a model of LPP-1 suggested that LPP-1 is head tail like Cyanophage similar to bacteriophages.

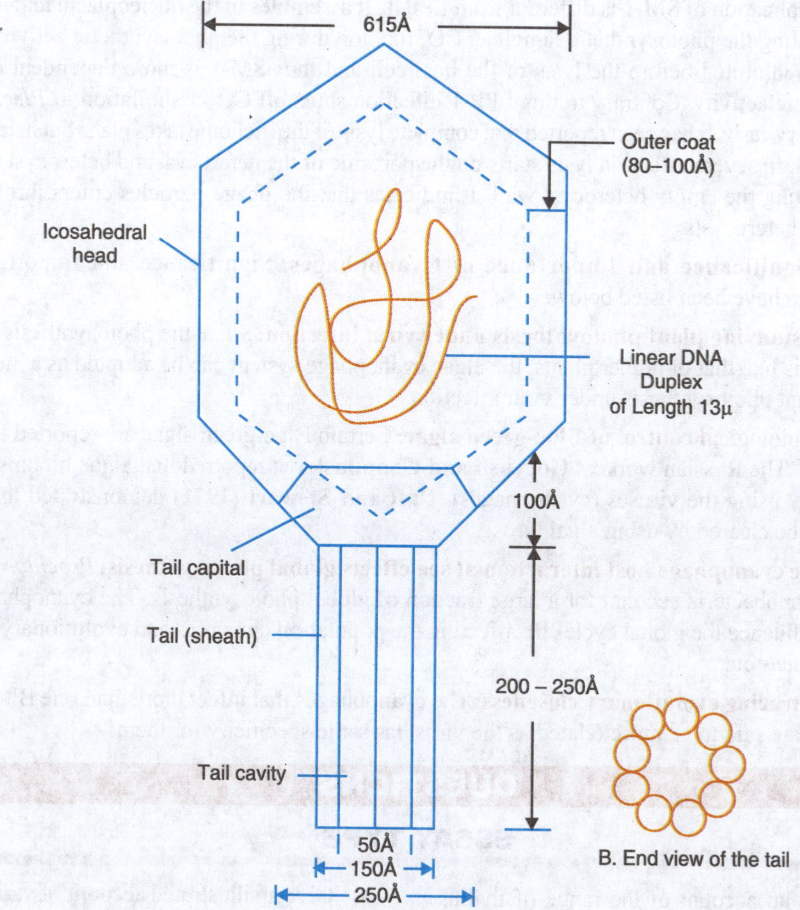

Fig. 24.16. Structure of Cyanophage LPP-1 as proposed by Lufig and Haselkorn.

The viral head capsid is a polyhedron, appearing hexagonal in projection. The edge to edge distance of the viral head is 600 ± 20*A*°. Recently the detailed structure of head has been studied by **Adolph and Haselkorn** who suggested an icosahedral form. The head encloses linear DNA duplex of 13μ length. The entire head is covered with double membranous coat. A short tail is attached to one of the vertices of the head. It measures 200 ± 25*A*° in length and 150*A*° in width. The tail is a hollow cylinder with sixfold radial symmetry made up of several (2 – 4) rings of six subunits.

(d) Growth cycle of Cyanophages: Smith *et al.* suggested that the tail of LPP-1 phage first gets penetrated into the vegetative cell of the host as in the case of bacteriophage. The first sign of infection of algal virus is pushing of photosynthetic lamellae to one side and in early stage of infection, many particles, probably α-particles are observed between the photosynthetic lamellae. At later stage, the virogenic stroma is full of developing virus particles. After infection of DNA into the host cell, algal Viral DNA is formed in the nucleoplasm and the DNA moves into the space between the photosynthetic lamellae where helices begin to form. Ultimately these helices move into the virogenic stroma where the final assembly of the viral particles takes place. The protein coat is now built around the DNA helices and they regain their characteristic shape.

LPP-1 is reported to break down half of the host DNA and incorporate the breakdown products into LPP-1 DNA. Finally, the lysis of host cell takes place and phages are released.

The replication of SM-1 is different from LPP-1. It assembles in the nucleoplasm and is rarely seen penetrating the photosynthetic lamellae. CO_2 fixation during the photosynthetic activity is not significantly inhibited before the lysis of the host cell and thus SM-1 is more dependent on host photosynthetic activity. Contrary to this LPP-1 infection shuts off CO_2 assimilation in *Plectonema boryanum* very early. It has been reported that complete lysis of the trichome takes place but heterocysts are not lysed. In severe infection lysis starts on the poreside of the heterocyst and heterocyst content is lysed leaving the empty heterocyst wall. It indicates that the phage particles enters through the pores of the heterocysts.

(*e*) Significance and Importance of Cyanophages: Significance and Importance of Cyanophages have been listed below:

1. In studying plant photoynthesis under viral infection: Since the photosynthesis of blue green algae is like that of higher plants, the algae cyanophage system can be adopted as a model for studying plant photosynthesis under viral infection.

2. In biological control of Blue-green algae: Certain Blue-green algae are reported to cause algal bloom. The Russian workers **Goryush and Chaplinskaya** reported that algae blooms can be destroyed by using the viruses (cyanophages). **Daft and Stewart (1971)** demonstrated that algal blooms can be cleared by using algal viruses.

3. The cyanophage host interaction at sea effects global photosynthesis: *Prochlorococcus* group of cyanobacteria account for a large fraction of global photosynthesis. The cyanophages are known to influence the global cycles by affecting the population dynamics and evolutionary path of *Prochlorococus* etc.

4. In tracing evolutionary closeness: The cyanophages that infect more than one Blue-green algae hosts are said to be more related as the virus has same specificity for them.

QUESTIONS

ESSAY TYPE

1. Give an account of the range of thallus organisation in algae studied by you.
   (Meerut, 1994, 1999, 2000; Kanpur M.Sc. 2002; Poorvanchal, 2007; Kanpur, 2003; Rohilkhand, 2006)
2. Give an account of the origin and evolution of sex in algae. *(Kerala, 2001)*
3. Give an account of the asexual reproduction in algae studied by you. Give suitable examples.
   (Meerut, 1995; U.P. College, 1996; Lucknow, 2000; Allahabad, 2003, 2006)
4. Give an illustrated account of various life cycle patterns found in Algae.
   (Meerut, 1998; Kumaon, 1995, 1999; Punjab, 1992)
5. Describe various modes of reproduction in Algae.
   (Rohilkhand, 1992; Kumaon, 1998)
6. (*a*) Write an essay on economic importance of Algae.
   (Rohilkhand, 2000; Kumaon, 1996, 1998; Kanpur, 1997, 2006; Agra, 1996; Allahabad, 1998; Punjab, 1996, 1997, 1998; Trichy, 1995; Himachal Pradesh, 1998; Bundelkhand, 1994, 1998; Kanpur M.Sc. 2006; Rohilkhand, 2006)

(*b*) Describe in detail the economic importance of Algae. (*Kanpur, 2007*)

7. (*a*) Give an account of various types of chloroplasts in green algae. (*Kumaon, 1995*)

(*b*) Give an account of asexual reproduction in Chlorophyceae. (*Kerala, 2001*)

8. Give a brief account of various types of habitats occupied by algae. (*Kumaon, 1995*)

9. Comment upon the evolution of thallus in algae. (*Kumaon, 1996*)

10. Describe briefly with suitable examples and illustrations the alternation of generation in Algae. (*Kumaon, 1997*)

11. What are Algae? With the help of suitable examples and illustrations trace the origin and evolution of sexuality in Algae. (*Kumaon, 2000*)

12. Describe the various modes of perennation in Fresh water algae studied by you. (*Kanpur, 1995; Awadh, 1992*)

13. Write an account of origin and development of sex in algae. (*Agra, 1955*)

14. Describe the methods of sexual reproduction in Algae giving suitable examples. (*Lucknow, 1992*)

15. Give a detailed account of algal pigments and their significance in algal classification. (*Lucknow, 1993; Allahabad, 1996; Himachal Pradesh, 1999*)

16. Write about thallus organisation in algae. Also describe the importance of heterotrichous habit in algae. (*Rohilkhand M.Sc., 2007*)

17. Illustrate different types of life cycles in algae studied by you. (*Kanpur, 2004*)

18. With the suitable diagrams, illustrate asexual reproduction in algae. (*Kanpur, 2004*)

19. Give a detailed account of the benefits we obtain from algae. (*M.S. Univ., 1995*)

20. Explain the evolution of thallus in algae giving suitable examples from the genera you have studied. (*Bundelkhand, 1995*)

21. Discuss the economic importance of algae. (*Gurunanak Deo, 1992*)

22. Write a short essay on the industrial uses of algae. (*Ravi Shankar Univ., 1992*)

23. Trace the origin and evolution of sexuality in green algae. Illustrate your answer with suitable sketches and examples.

24. (*a*) Write what you know about various types of life cycles in algae.

(*b*) Describe the different life cycle patterns in algae. (*Rohilkhand, 2006*)

25. Discuss

(*a*) Sexual reproduction in Algae (*Lucknow, 2003*)

(*b*) Reserve food storage in Algae (*Lucknow, 2003*)

26. Justify the statement that heterotrichoses habit of filamentous algae show advanced stage of thallus. (*Kanpur M.Sc., 2002*)

27. Write an essay on evolution of sex in algae. (*Kanpur M.Sc., 2004*)

28. Describe the evolution of sex in algae and reproductive diversity in different groups of algae. (*Lucknow M.Sc., 2004*)

29. Describe/Write briefly on:

(*a*) Range of thallus structure in Algae (*Lucknow, 2006*)

(*b*) Names of some products of industrial importance obtained from algae. (*Poorvanchal, 2007*)

(*c*) Alternation of generations (*Lucknow, 2007*)

(*d*) Reserve food products in algae (*Lucknow, 2007*)

(*e*) Reproductive diversity in algae (*Lucknow M.Sc., 2006*)

(*f*) Symbiotic algae (*Lucknow M.Sc., 2006*)

(*g*) Protein rich agae (*Lucknow M.Sc., 2006*)

(*h*) Sites of meiosis in algae (*Lucknow M.Sc., 2006*)

(*i*) Crude culture of algae (*Lucknow M.Sc., 2005*)

(*j*) Role of macro and micro nutrients in algal culture (*Poorvanchal, 2006*)

SHORT ANSWER TYPE

30. What do you understand by isomorphic alternation of generation? Give two examples. *(Lucknow, 2004)*

31. Which algae is rich in protein? *(Kanpur M.Sc., 2002)*

32. Give a brief account of algal blooms. *(Kanpur M.Sc., 2004; Lucknow M.Sc., 2006)*

33. Name an algae which is used as food. *(Allahabad, 2001)*

34. Name three economically important algae. *(Allahabad, 2006)*

35. How algae is useful as food ? *(Poorvanchal, 2005)*

36. Write detailed/short notes on the following:

(*i*) Algae as food or edible algae. *(Meerut, 1994; Kumaon, 1979; Lucknow, 2007)*

(*ii*) Pigments in Algae *(Meerut, 1994, 1995; Allahabad, 1992; Gorakhpur, 1996, 1997; Awadh, 1991, 1994, 1995; U.P. College, 1996, 1998; Bundelkhand, 1998; Lucknow, M.Sc., 2003; Rohilkhand, Adse 2007)*

(*iii*) Algae in agriculture/use of algae in agriculture. *(Meerut, 1995; Kumaon, 2000; Poorvanchal, 1998)*

(*iv*) Plastids in Algae *(Kanpur, 1994; Awadh, 1993; Bundelkhand, 1999)*

(*v*) Alternation of generation *(Kanpur, 1994, 1996; Awadh, 1997)*

(*vi*) Flagella in Algae. *(Kanpur, 1995; Awadh, 1991, 93, 98; Kanpur M.Sc., 2002)*

(*vii*) Algal bloom or Water bloom *(Kanpur, 1996; Punjab, 1997; Lucknow, 2003, 2007; Rohilkhand, 2007)*

(*viii*) Importance of reserve food materials in the classification of algae. *(Allahabad, 1994)*

(*ix*) Use of algae in industry. *(Poorvanchal, 1999)*

(*x*) Useful algae. *(Rohilkhand, 2007)*

(*xi*) Algae as nitrogenous fertilizers *(Bundelkhand, 1995)*

(*xii*) Bloom formation *(Lucknow, 2005)*

(*xiii*) Algae and water pollution *(Kanpur M.Sc., 2006)*

(*xiv*) Reserve food of algae *(Lucknow M.Sc., 2005)*

(*xv*) Algal indicators of pollution *(Lucknow M.Sc., 2005)*

(*xvi*) Structure of Algal chloroplasts *(Lucknow M.Sc., 2004)*

(*xvii*) *Spirulina* *(Lucknow M.Sc., 2004)*

(*xviii*) Modes of perennation in algae *(Rohilkhand M. Sc., 2007, 2006)*

(*xix*) Role of algae in maintenance of aquatic ecosystem. *(Rohilkhand, 2007)*

37. Describe the role of algae in industry and as manure. *(Agra, 1994)*

38. Differentiate between :

(*a*) Zoospore and Oospore *(Poorvanchal, 1998)*

(*b*) Aplanospore and Akinete *(M.D.S. Univ., 1998)*

(*c*) Aplanospore and Oospores *(Poorvanchal, 1998)*

(*d*) Aplanospore and Carpospore *(U.P. College, 1995)*

(*e*) Parasitic algae and Epiphytic algae *(U.P. College, 1997)*

(*f*) Isomorphic and Heteromorphic alternation of generations

(*g*) Zoospore and synzoospore

(*h*) Hypnospore and Akinetes *(Kerala, 2001)*

(*i*) Coenobium and Colony (*Lucknow, 2001*)

(*j*) Cystocarp and Spermocarp (*Lucknow, 2001*)

(*k*) Zygospore and oospore. (*Lucknow, 2007*)

39. Describe the role of algae in agriculture with special reference to (*i*) Nitrogen fixers (*ii*) Soil reclamation and (*iii*) Fertilisers and manures.

40. (*a*) Name four industrial products obtained from algae and mention the class to which they belong. (*Himachal Pradesh, 1993*)

(*b*) Name any three economically important algae. (*Lucknow, 2004*)

41. What are the harmful effects of algae? (*M.S. Univ., 1995; Gurunanak Dev., 1991*)

42. Write an explanatory note about algae as biofertilisers. (*Bhagalpur, 1990*)

43. Give the salient features of alternation of generation. (*Gurnnanak Deo, 1991*)

44. Name an alga which is used in reclamation of alkaline usar soils. (*Kerala, 2001*)

45. Comment on the following :

(*i*) Economic importance of Algae. (*Meerut, 1996; Rohilkhand, 1997; Lucknow, 1997, 1999, 2000, 2004, 2006; Allahabad, 1996, 2002; Gorakhpur, 1990, 1993, 1998; Awadh, 1993, 1994; U.P. College, 1999; Bundelkhand, 1997; Poorvanchal, 2005*)

(*ii*) Range of vegetative structures (thallus organisation) in Algae. (*Rohilkhand, 1993, 1997; Gorakhpur, 1997; Punjab, 1993; Madras, 1997*)

(*iii*) Shapes of chloroplasts in Algae. (*Rohilkhand, 1994; Lucknow, 1995*)

(*iv*) Biofertilisers or Algae as fertilisers/ Algal. Biofertilisers (*Rohilkhand, 1995; Agra, 1991; Lucknow M.Sc., 2004*)

(*v*) Evolution of sex in Algae. (*Agra, 1993*)

(*vi*) Haplobiontic life cycle. (*Awadh, 1996; Punjab, 1991*)

(*vii*) Vegetative reproduction in algae (*Awadh, 1997*)

(*viii*) (*a*) Diplohaplontic life cycle with examples and graphic life cycle (*Poorvachal, 2004*)

(*b*) Asexual reproduction in algae. (*Himachal Pradesh, 1999; Lucknow, 2004*)

(*c*) Diplontic life cycle (*Poorvachal, 2005*)

(*ix*) Role of algae in medicines (*Punjab, 1991*)

(*x*) Importance of algae as food and in industry (*Bundelkhand, 1996*)

(*xi*) Origin of sex in algae.

(*xii*) Evolution of thallus in green algae

(*xiii*) Fossil algae (*Kanpur M.Sc, 2006; Lucknow M.Sc., 2003*)

(*xiv*) Culture of algae

(*xv*) Chloroplasts in algae (*Lucknow, 2001*)

(*xvi*) Algae in relation to water pollution (*Lucknow, 2002, 2007*)

(*xvii*) Algae in medicine (*Rohilkhand M.Sc., 2007*)

(*xviii*) Algal as biofertilisers (*Kanpur M.Sc., 2005; Lucknow M.Sc., 2004; Poorvanchal, 2006*)

(*xix*) Toxic algae. (*Lucknow M.Sc., 2004*)

OBJECTIVE TYPE

46. Select the correct answer :

(*i*) Auxospore formation takes place in
(*a*) Desmids
(*b*) Diatoms
(*c*) Green algae
(*d*) Red algae

(*ii*) A motile flagellated cell that reproduces asexually is called as
(*a*) Sperm
(*b*) Zoospore
(*c*) Oospore
(*d*) Zygospore

(*iii*) Which of the following part of an algal cell is alive?
(*a*) Cell wall
(*b*) Chloroplast
(*c*) Vacuole
(*d*) Starch

(*iv*) The Red sea gets its name because of the presence in its water red coloured algae belonging to class
(*a*) Rhodophyceae
(*b*) Chlorophyceae
(*c*) Phaeophyceae
(*d*) Myxophyceae

(*v*) Fusion of mature individuals which act as gametes is called
(*a*) Isogamy
(*b*) Conjugation
(*c*) Hologamy
(*d*) Autogamy.

(*vi*) Chloprophyll-e is present in
(*a*) Chlorophyceae
(*b*) Xanthophyceae
(*c*) Phaophyceae
(*d*) Rhodophyceae

(*vii*) Reproduction through hormogonia is found in
(*a*) *Spirogyra*
(*b*) *Oscillatoria*
(*c*) *Ectocarpus*
(*d*) *Sargassum*

(*viii*) In Cyanophyceae the site of nitrogen fixation is
(*a*) Akinetes
(*b*) Incipient nucleus
(*c*) Heterocysts
(*d*) Plasma membrane

(*ix*) Alginic acid derivatives are extracted from.
(*a*) *Cladophora*
(*b*) *Zygnema*
(*c*) *Laminaria*
(*d*) *Acetabularia*

(*x*) Red rust disease of tea is caused by
(*a*) *Noctiluca*
(*b*) *Cephaleuros*
(*c*) *Polysiphonia*
(*d*) *Batrachospermum*

(*xi*) Diplobloutic triphasic life cycle is found in
(*a*) *Nostoc*
(*b*) *Ectocarpus*
(*c*) *Polysiphonia*
(*d*) *Chlamydomonas*

(*xii*) Diploid structure is
(*a*) Tetraspore (*b*) Synzoospore
(*c*) *Carpospore* (*d*) *Hypnospore*

SELECTED BIBLIOGRAPHY

Allen, M.M. 1968. Photosynthetic membrane system in *Anacystis nidulans*. *J. Bact.* **96**(3); 836-841.

– 1968. Ultrastructure of cell wall and cell division of unicellular blue-green algae. *J. Bact.* **96** (3) : 842-852.

Allen. N.S. 1974. Endoplasmic filaments generate motive force for rotational streaming in *Nitella. J. Cell Biol.* 63 : 270-287.

– 1980. Cytoplasmic streaming and transport in the characean alga *Nitella Can. J. Bot.* **58** (7) : 786-796.

Allen, R.D. 1959. The moment of fertilization. *Scient. Am.*, **201 :** 124-134.

Ashraf, M., and M.B.E. Godward. 1980. Ultrastructure and chemistry of the zygospore wall of siprogyra. *Ann. Bot.* **46 :** 485-487.

Badour, S.S., C.K. Tan, L.A. Van caesele and P.K. Isaac. 1937. Observations on the morphology, reproduction and fine structure of *Chlamydomonas segnis* from Dalta Marsh, Manitoba. Can. Jr. Bot., **51 :** 67-72.

Bazin, M.J. 1968. Sexuality in a blue-green alga : genetic recombination in *Anacystis nidulans, Nature*, Lond. **218 :** 282-83.

Benerji, J.C. 1936. Studies on the Myxophyceae of Lower Bengal-I. *Jour. Ind. Bot. Soc.* **15 :** 285-302.

Bertagnolli, B.L. and M.J. Nandakavukaren. 1970. An ultrastructural study of pyrenoids from *Chlorella pyrenoidosa J.cell Sci.*, **7 :** 623-30.

Bharadwaj, Y. 1933. A new species of *Draparnoldiopsis* (*D. indica*) *New Phytol*., **32 :** 165-174.

Bharadwaj, Y. 1933. Contributions to our knowledge of Myxophyceae of India. *Ann. Bot.* **47 :** 117-143.

Birchen, R. and G. Kochert. 1979. Mitosis and Cytokinesis in androgonidia of Volvox carteri f. Weismannia. *Protoplasma* **100 :** 1-12.

Blank, R.B. Grobe and C.G. Arnold. 1978. Time sequence of nuclear and chloroplast *fusions* in zygote of *Chlamydomonas reinhardii*. *Planta* 138 (1) : 63-64.

Blatt, M.R., and W.R. Briggs. 1980. Blue light induced cortical fibre reticulation concomitant with chloroplast aggregation in alga *Vaucheria sessilis*. *Planta* **147 :** 365-362.

– N.K. Wessella and W.R. Briggs. 1980. Actin and cortical fibre reticulation in the siphonaecous alga *Vaucheria sessilis*. *Planta* 147 (4) : 363-375.

Boergesen, F. 1930. Some Indian Green and Brown algae especially from the shores of the Presidency of Bombay-I. *Jour. Ind. Bot. Soc.* **9 :** 151-174.

Boergesen, F. 1937. Contributions to a South Indian marine algal flora II. *Jour. Ind. Bot. Soc.* **16 :** 311-357.

Bold, H.C. 1933. The life history and cytology of *Protosiphon botryoides*. *Bull. Torrey Bot. club*, **60 :** 241-299.

– 1962. *The morphology of plants*. Second ed., Harper and Row, New York.

Bold, H.C. and M.J. Wynne. 1978. Introduction to the Algae-Structure and Reproduction. Prentice Hall of India Private Ltd., New Delhi.

Bonner, J.T. 1950. *Volvox*, a colony of cells. Scient. Am., **182 :** 52.

Bonsor, S.C., and Agona John L. 1973. The '*Palmeloid*' stage in a blue green alga, *Anabaena sp*. I Preliminary report. *Am. Jr. Bot.*, **60 :** 223-227.

Bouck, G.B. 1962. Chromatophore development, pits and other fine structure in the red alga *Lomentaria baileyana*. *J. Cell Biol.*, **12 :** 553-569.

– 1964. Fine structure in *Acetabularia* and its relationship to protoplasmic streaming in motile systems, Academic Press, New York.

– 1965. Fine structure and organelle associations in brown algae. *J. Cell Biol.*, **26 :** 523-537.

– 1969. Extracellular microtubules. The origin, structure and attachment of flagellar hairs in *Fucus* and *Ascophyllum* antherozoids. *J. cell Biol*. **40 :** 446-460.

Braten, T. 1971. The ultrastructure of fertilization and zygote formation in green alga *Ulva mutabilis*. *J. cell Sci* **9 :** 621-635.

– 1973. Autoradiographic evidence for the rapid disintegration of one chloroplast in zygote of the green alga *Ulva mutabilis*. J. cell Sci **12 :** 385-389.

Brody M., and A.E. Vatter, 1959. Observations on cellular structures of *Porphyridium cruentum*. *J. Biophysic and Biopchem. Cytol.*, **5 :** 289-294.

Brown, R.M.L. (Jr.) O.P.S.C. Johnson, and H.C. Bold. 1968. Electron and phase contrast microscopy of sexual reproduction in *Chlamydomonas moewusii*. *J. Phycol.*, **4 :** 100-120.

Burr, F.A., and M.D. Mc. Cracken, 1973. Existence of a surface layer on the sheath of *Volvox*. *Jour. Phycology* **3 :** 345-346.

Burger-Wiersman T., Veenhuis M., Korthals H.J., Van der wiel C.C. M. and Mur L.R. (1986). A new procaryote containing chlorophylls a and b. Nature 320: 262-264.

Catt, J.W., G.J. Hills, K. Robbers, 1976. A structural glycoprotein containing hydroxyproline isolated from the cell wall of *Chlamydomonas reinhardii*. Planta. **131 :** 165-171.

Cavalier Smith, T. 1976. Electron microscopy of zygote formation in *Chlamydomonas reinhardii*. *Protoplasma* **87 :** 297-309.

Chapman, D.J., and Chapman V.J. 1961. Life histories in the algae. *Ann. Bot. N.S.* **25 :** 547-561.

Chapman, V.J. 1962. The *Algae*. Macmillan, London.

– 1970. Seaweeds and their uses. 2nd. end. Methuen and Co. 11, New Fetter Lane, London.

Chauhan, V.D., and V. Krishnamurthy. 1971. Ecology and seasonal succession on *Sargassum swartzii*. Phykos, **10 :** 1-11.

– and F. Thivy. 1964. On the occurrence of some *Sargassum* species in Gujrat. *Phykos*, **3 :** 19-25.

Cocucci, A.E., and E.J. Caceres. 1976. The ultrastructure of the male gametogenesis in *Chara contrarea Var. Nitelloides* (Charyophyta). Phytomorphology **26 :** 5-16.

Copeland, J.J. 1936. Yellowstone Thermal Myxophyceae. *Ann. N.Y. Acad. Sci.* **36 :** 1-232.

Couch, J.N. 1932. Gametogenesis in *Vaucheria*. Bot. Gaz., **94 :** 272-296.

Cronshaw, J., A. Myers and R.D. Preston. 1958. A chemical and physical investigation of the cell walls of some marine algae. *Biochemet. Biophysica Act*, **27 :** 89-103.

Davis. J.S. 1964. Colony formation in *Pediastrum*. Bot. Gaz. **125 :** 129-131.

– 1962. Resting cells of *Pediastrum*. Am. J. Bot. **49 :** 478-481.

Dawes, C.J., F.M. Scott, and E. Bowler. 1961. A light and electron microscope survey of algal cell walls. I, Phaeophyta and Rhodophyta, Am. J. Bot. **48 :** 925-934.

Dawes, C.J. 1965. An ultrastructural study of *Spirogyra*. J. Phycol., **1 :** 121-127.

Desikachary, T.V. 1945. Germination of the heterocysts in two members of the Rivulariaceae. *Jour. Ind. Bot. Soc.* **25 :** 11-17.

Desikachary, T.V. 1969. Cyanophyta. I.C.A.R. New Delhi.

– and Sundaralingam, V.S. 1962. Affinities and interrelationships of the Characeae. Phycologia **2 :** 16-19.

De. P.K. 1938. Nitrogen fixation by the Blue-green algae. *Nature*. London.

Di Orio, J. and W.P. Mellington. 1978. Dictyosome formation during reproduction in Colchicine-Treated *Pediastrum boryanum. Protoplasma* **97 :** 329-338.

Dixton, P.S. 1963. Terminology and algal life histories with particular reference to the Rhodophyta. *Ann. Bot.* Lond. N.S., **27 :** 353-355.

Dodge , B.O. 1914. The morphological relationships of the Florideae and Ascomycetes. Bull. *Torrey Bot. Club* **41 :** 157-202.

Dodge, J.D. 1968. the fine structure of chloroplasts and pyrenoids, in some marine dinoflagellates. *J. cell sci.*, **3 :** 41.

– 1969. A review of the fine structure of algal eyespots. Br. Phycol. **J, 4 :** 199-210.

– 1973. The fine structure of algal cells. Academic Press, London.

Doriaswamy, S. 1940. On the morphology and cytology of *Eudorina indica*. Iyengar; *Jour. Ind. Bot. Soc.* **19 :** 113-149.

Drew, K.M. 1954. Studies in the Bangioideae. III, The life history of *Porphyra umbilicalis*. vas. laciniata. The conchocelis phase in culture. *Ann. Bot*. London, *N.S.* **18 :** 183-211.

Drew, K.M. 1954. The organization and inter-relationships of the carposporophytes of living florideae. *Phytomorphology* **4 :** 55-69.

– 1955. Life histories in the algae with special reference to the Chlorophyta, Phaeophyta and Rhodophyta. *Biot. Rev.*, **30 :** 343-390.

– 1956. Reproduction in Bangiophycideae. *Bot. Rev.*, **22 :** 553-611.

– 1926. The leaf of *Nitella opaca Ag.* and adventitious branch development from it. *Ann. Bot.*, **40 :** 321-348.

Echlin, P. and I. Morris. 1965. The relationship between the blue-green algae and bacteria. *Biol., Rev.* **40 :** 143-187.

Edwards, P. 1968. Life history of *Polysiphonia denudata* (Dillwyn) Kut, in culture *J. Phycol* **4 :** 35-37.

Erben, K. 1962. Sporulation in Physiology and Biochemistry of *Algae*, editor R.A. Lewin, Academic Press, London. pp. 701-710.

Evans, L.V. 1962. Cytological studies in the genus *Fucus*, Ann. Bot., N.S. **26 :** 345-360.

– 1966. Distribution of pyrenoids among some brown algae. *J. cell Sci.*, **1 :** 449-454.

– 1968. Chloroplast morphology and fine structure in British fucoids. *New Phytol.*, **67 :** 173.

Fay, P., W.D.P. Stewart, A.E. Walsby and G.E. Fogg. 1968. Is the heterocyst the site of nitrogen-fixation in blue-green algae? *Nature*, Lond., **220 :** 810-12.

Ferguson, J.M. 1932. On the mitotic division of *Draparnaldia glomerata Ann. Bot.* **46 :** 703-709.

Fischer, P.A., J. Dainty and M.T. Tyree. 1974. A quantitative investigation of symplasmic transport in *Chara corallina*. ultrastructure of the nodal complex and cell walls. *Can J. Bot.* **52 :** 1209-1214.

Floyd, G.L., K.D. Stewart, and K.R. Matox, 1971. Cytokinesis and plasmodesmata in *Ulothrix.J. Phycol.* **7 :** 306-309.

Fogg, G.E. 1970. Gas vacuoles and their ecological significance *First interna. Symp. on Taxonomy and Biology of blue-green algae*. Univ. Madras.

– and D.M. Collyer. 1954. The accummulation of fats as characteristic of certain classes of algae. *Proc. VIII Interna Bot. Congr.* **17 :** 28.

– 1951. Growth and heterocyst production in *Anabaena cylindrica*. III, The cytology of heterocysts. *Ann. Bot.* Lond. N.S., **15 :** 23-36.

Fraser, T.W., and B.E.S. Gunning 1969. the ultrastructure of plasmodesmata in filamentous green alga *Bullochaete hiloensis* (Nordst.) Tiffany *Planta*: **88 :** 244 -254.

Friedmann, I. 1966. Microscopy of algal chromatophores *Phycologia* **6 :** 29-36.

Fritsch, F.E. 1916. The algal ancestry of the higher plants. *New Phytol.* **15 :** 235-250.

– 1935. The structure and reproduction of Algae. Vols I and II Camb. Uni. Press.

– 1942 a. Studies in the comparative morphology of the algae. I Heterotrichy and Juvenile stages. *Ann. Bot.* **NS 6 :** 397-412.

– 1942 b. II. The algal life cycle, Ann Bot. **NS 6 :** 533-563.

– 1944. The present day classification of algae. *Bot. Rev.*, **10 :** 233-277.

– 1949. The lines of algal advance. *Biot Rev.*, **24 :** 94-124.

– 1951. The heterocyst-a botanical enigma. *Proc. Linn. Soc.*, London, **126 :** 194-211.

– 1942. The interrelationships of the Myxophyceae (Cyanophyceae), *New Phycol*, **41 :** 134 -148.

Fulton, A.B. 1978. Colonial development in *Pandorina morum* I structure and composition of the extra cellular matrix. Develop. Biol. **64 :** 224-235.

– 1974. Colony morphogencsis and formation of extracellular matrix. *Develop Biol.* **64** (2) **:** 236-251.

Gantt, E., and S.F. Conti. 1965. Ultrastructure of *Porphyridium cruentum*. *J. Cell Biol.*, **26 :** 365-381.

Geitler, L. 1935. Reproduction and life history in diatoms. *Bot. Rev.* **I :** 149-161.

– 1961. Spontaneous partial rotation and oscillation of the protoplasm in *Coleochaete* and other Chlorophyceae *Am. J. Bot.* **48 :** 738-741.

Gergis, M.S. 1971. The presence of microbodies in three strains of *Chlorella*. Planta (Berlin), **101 :** 1980-184.

Ghosh, S.C. and M.S. Randhawa. 1933. Aplanospore formation in *Vaucheria uncinata.* Kutz. *Curr Sci.* **2 (1) :** 15-16.

Gibbs, S.P. 1960. The fine structure of *Euglena gracilis* with special reference to the chloroplasts and pyrenoids. *J. Ultrastruct. Research* **4 :** 127-148.

– 1962. The ultrastructure of chloroplasts of algae. *J. ultrastruct.* Res., **7 :** 418.

– 1962a. The ultrastructure of pyrenoids of algae exlusive of green algae I. *J. ultrastruct. Res.*, **7 :** 247.

– 1962. Nuclear envelope chloroplast relationships in algae. *J. Cell Boil.*, **14 :** 433-444.

– 1962. The ultrastructure of pyrenoids of green algae. *J. ultrastruct. Res.* **7 :** 262.

– 1970. The comparative ultrastructure of the algal chloroplasts. *Ann. N.Y. Acad. Sci.*, **174 :** 454 - 473.

Godward, M.B.E. 1942. The life cycle of *Stigeoclonium amoenum* Kutz. *New phytol*, **41 :** 293-301.

– 1966. *Chromosomes of the algae*. London.

Gonzalves, Ella A., and G.R. Sonnad, 1961. The genus *Oedogonium* in Mysore state. *Jour. Bom. Nat. Hist. Soc.* **58 :** 715-723.

– and S.C. Jain 1970. Some Oedogoniaceae from Thana District. *Phykos*, **9 :** 1-16.

Govindji and P.K. Mahanty, 1970. Photochemical aspects of photosynthesis in Blue-green Algae. *First Internat. Symp. on taxonomy and Biology of Blue-green Algae*. Univ. Madras.

Graham, L.E. 1982. The occurrence, evolution and phylogenetic significance of parenchyma in *Coleochaete* Breb. (Chlorophyta). *Ann. J. Bot.* **69** (3) : 447-454.

Greenwood, A.D., I. Manton, and B. Clarke. 1957. Observations on the structure of zoospores of

Vaucheria. J. Exp. Bot.; **8 :** 71-86.

Greenwood, A.D., 1959. Observations on the structure of zoospore of *Vaucheria II. J. Exp. Bot.*, **10 :** 55-68.

Grobe, B., and C.G. Arnold. 1977. The behaviour of mitochondria in the zygote of *Chlamydomonas reinhardii. Protoplasma* **93 :** 357-361.

Gross, F. 1937. Life history of some marine plankton diatoms. *Phil. Trans. Roy Soc.* Lond. *Seb* **228 :** 1-48.

Grote, M. 1977. Studies in the copulation process of green alga *Spirogyra majuscula, Protoplasma* **91 :** 71-82.

Grubb, Violet M. 1925. Male organs of Florideae. *Jour. Linn. Soc. Bot.*, London, **47 :** 177-255.

Gupta, A.B., and G.U. Nair, 1962. Colony formation in *Gloeotaenium* and significance in the evolution of the filamentous habit. Bot. Gaz., **124 :** 144-146.

Gupta, A.B. and G.C. Shrivastava, 1963. Production of an antibacterial substance by *Hydrodicyton reticulum. Curr. Sci.*, 23 : 19-20.

Gwalik, S.R. and Mellington, W.F. 1969. Pattern formation and the fine structure of developing cell wall in colonies of *Pediastrum boryanum, Amer. J. Bot.*, **56 :** 1084-1093.

Haine, L.A., and J.S. Graigie. 1969. Studies on the algal cuticle. *J. Phycol.*, **5:** 89-102.

Halfen, L.A., and R.W. Castenholtz, 1970. Gliding in a Blue-green Alga ; a possible mechanism. *Nature*, **225 :** 1163-1165.

Hawkins, A.F., and G.F. Leadale, 1971. Zoospore structure and colony formation in *Pediastrum* spp. and *Hydrodicyton reticulatum. Ann. Bot.* **35 :** 201-211.

Hayama, T., T. Shimmen and M. Tazwa 1979. Participation of Ca^{2+} in cessation of cytoplasmic streaming induced by membrane excitation in Characeae internodal cells. *Protoplasma* **99 :** 305-321.

Heitz, E. 1960. Structure of the Chloroplast in *Chlamydomonas. Experientia*, **16 :** 265.

Hendey, N.I. 1959. The structure of the diatom cell wall as revealed by electron microscope. *J. Queckett Microsc. club.*, **5 :** 147-175.

Hill, J.C., and L. Michalis, 1968. An ultrastructural study of the vegetative cell division in *Oedogonium borisianum. J. Phycol*, **4 :** 261-271.

Hoare, D.S., S.L. Hoare and A.J. Smith, 1970. Heterothallic potentialities of the blue-green algae.
First Intern. Symp. on taxonomy and Biology of Blue-green Alga. Univ. Madras.

Hoffman, L.R. 1967. Observations on the fine structure of *Oedogonium* III Microtubular elements in the chloroplasts of *O. cardiacum. J. Phycol.*, **3 :** 212-221.

– and I. Manton. 1962. Observation on the fine structure of the zoospore of *Oedogonium cardiacum* with special reference to thc flagellar apparatus, *J. Exp. Bot.*, **13 :** 443-449.

Hunter, S.H., and L. Provasoli, 1964. Nutrition of Algae. *A. Rev. Pl. Physiol.*, **15 :** 37-56.

Huskey, R.J., and B.E. 1979. Genetic control of somatic cell differentiation in *Volvox. Develop. Biology,* **72** : 226-235.

Huskey, R.J. 1979. Mutants affecting cell orientation *Volvox carteri. Develop. Biology,* **72 :** 236-243.

Ingold, C.T. 1973. Cell arrangement in the coenobia of *Pediastrum. Ann. Bot.*, **37 :** 379-384.

Iyengar, M.O.P., 1925. *Hydrodictyon indicum*, a new species from Madras, *Jour. Indian Bot. Soc.*, **4:** 193-201.

– 1925. Notes on two new species of *Botrydium* from India. *Jour. Indian Bot. Soc.*, **4:** 193-201

– 1932. *Fritschiella*, a new member of the Chaetophoraceae. *New Phycol* **31 :** 329-335.

– 1933. On the formation of gametes in *Caulerpa. Jour. Indian Bot. Soc.*, **12:** 325.

Iyengar, M.O.P. 1933. A contributions to our knowledge of the colonial volvocales of south India. Jour. Linn. Soc. London. Bot. **49 :** 323-373.

– 1937. Fertilization in *Eudorina elegans* Eherenberg. Jour. Ind. Bot. Soc. **16 :** 111-118.

– 1944. On the reduction division and auxospore formation in *Cyclotella meneghiniana*, *Ibid*. **23 :** 125-152.

– 1958. Special type of lateral conjugation in *Spirogyra jogensis*. Jour. Ind. Bot. Soc. **37** (3) : 388

Jao, C.C. 1936. New Zygnemataceae collected in China. Amer. Jour Bot. **23.**

Jarosch, R, 1962. *In the Physiology and Biochemistry of algae* (ed. Lewin, R.A.). PP.573-581, Academic Press, New York.

JeejiBai, N. 1962. *Trentepohlia monilia* de Wilderman from Madras. *Phykos I*., (**2**) : 79-83.

Jones, J. 1930. An investigation into the bacterial associations of some Cyanophyceae with special reference to their nitrogen supply. Ann. Bot. **44:** 721.

Johnson, L.N. 1893. Observations on the Zoospores of *Draparnaldia. Bot. Gaz.*, **18 :** 294.

Jose, G., and Y.B.K. Chowdary. 1980. New records of *Trentepohlias* from India. *Mycologia* LXXII: 725-731.

Joseph. Thomas, K.A. David and A.R. Gopal Iyengar. 1970. On the biology of heterocyst in Blue-green algae. *First Internat. Symp. on Taxonomy and Biology of Blue-green algae*. Univ. Madras.

Joshi, M.C. 1957. On the occurrence of *Fritischiella tuberosa* Iyeng. In Pilani (Rajasthan). *Jour. Bomb. Nat. His. Soc.* **54 :** 970-973.

Joshi, H.V., and Krishamurthy 1972. The species of *Enteromorpha* from India. *Bot. Jour. Linn. Soc.*, **65**(**1**) **:** 119-127.

Joshi, R.D., J. Prakash and L.N. Dube, 1978. Brown Rust- A threat to grape cultivation. *Curr. Sci.* **47** (**14**): 516-517.

Kamiya, N. 1981. Physical and chemical bases of cytoplasmic streaming. *A Rev. Pl. Physiology*. **32 :** 205-236.

Kala, S.R. 1967. *Enteromorpha gujratensis*, a new species from Gujarat, India. *Phykos.*, **6 :** (12).

Kamitsubo, E. 1980. Cytoplasmic streaming in Characean cells-role of subcortical fibrils. Can. J. Bot. **58** (**7**) **:** 760-763.

Kellan, J.L. 1977. Inversion in *Volvox. J. Phycol.* **13 :** 373-378.

Kirk, J.T.O., and Tilney-Bassel. Richard, A.E. 1967. *The plastids*. W.H. Foreman and Co. London and San Francisco.

Klein, R.M., and A.A. Cronquist. 1967. A consideration of the evolutionary and taxonomic significance of some biochemical, micromorphological and physiological characters in the thallophytes. Q Rev Biot. **42 :** 105-296.

Knight, M. 1929. Studies in the Ectocarpaceae II. The life history and cytology of *Ectocarpus siliculosus* Dillw. Ibid., 56 **II** (5) : 307-332.

Kraml, M.M., M. Enders and N. Burkel. 1984. Kinetics of the dichroic key orientation of phytochrome during photoconversion in *Mougeotia planta* **161**(**3**): 216-222.

Kochert, G. 1975. Developmental mechanisms in *Volvox* reproduction. Develop. Biot (Supplement) **8 :** 55-80.

Kylin, H. 1948. On the nature of cell wall constituents of Algae. In *M.O.P. Iyengar Commemoration Volume*. pp. 97-99.

Kumar, H.D. 1962. Apparent genetic recombination in a blue-green alga. *Nature*, Lond., 196: 1121-1122.

– and H.N. Singh., 1976. *A Text Book of Algae*. East West Press, New Delhi.

Laloraya, V.K. and A.K. Mitra 1970. Fixation of elementary nitrogen by some blue-green algae.

First Internat. Symp. on Taxonomy and Biology of Blue-green Algae. Univ. Madras.

Lander, C.A. 1929. Oogenesis and fertilization in *Volvox. Bot. Gaz.*, **87 :** 431-416.

Lang, N. 1968. The fine structure of blue-green algae. *A Rev. Microbiol.*, **22:** 15-42.

– 1970. Cellular differentiation in blue-green algae. *First Internat. Symp. on Taxonomy and Biology of Blue-green Algae*. Univ. Madras.

Lang, N.J. 1963. Electron microscopy of the Volvocaceae and Astrephomenaceae. Ann. J. Bot. 50 : 280-300.

Lazaroff, N., and W. Vishniac. 1962. The participation of filament anastomosis in the developmental Cycle of *Nostoc muscorum*, a blue-green alga. *J. Jen. Microbiol*, **28 :** 203-210.

Lewin, R.A. 1962. *The Physiology and biochemistry of algae*. Academic Press, London.

Lewin, R.A. and Cheng L. 1989. *Prochloron*, a microbial Enigma chapman and Hall, New York/ London.

Lewin, R.A. (ed.) 1993. Origin of Plastids: Symbiogenesis, Prochlorophytes and the origins of chloroplasts. Chapman and Hall, New york/London.

Lewin, R.A. 2002. Prochlorophyta-a matter of class distinctions. Photosynthesis Research, **73**: 59-61.

Lewis, I.F. 1912. Alternation of generations in certain Florideae. *Bot. Gaz*. **53 :** 236-242.

Lund, J.W.G. 1959. Buoyancy in relation to the ecology of the fresh water phytoplankton *Brit. Phycol. Bull.*, **1 :** 1-17.

Manton, I. 1957. Observations with the electron micrscope on the internal structure of the zoospore of a brown alga. *J. Exp. Bot.*, **8 :** 294.

– 1959. Observations on the internal structure of the spermatozoid of *Dictyota. J. Exp. Bot.* **10 :** 448-461.

– 1965. Some phyletic implications of flagellar structure in plants. In *Advances in Botanical Research*. Vol. II. : 1-34 (ed. Preston, R.D.)

– 1966. Some possible significant structural relations between chloroplasts and other cell components. In *The Biochemistry of chloroplasts*. V. I : 23-47, (editor Goodwin, T.E.), Academic Press, London.

Manton, I. and B. Clarke, 1951. Electron microscope observations on the zoospores of *Pylaiella* and *Laminaria. J. Exp. Bot.*, **2 :** 242.

and B. Clarke, 1956. Observations on the fine structure of the male gamete of the marine centric diatom *Lithodesmium undulatum.* J.*R. Microsc. Soc.*, **85 :** 119-120.

Marchant, H.I. 1976. Actin in the green algae *Coleocheate* and *Mougetia*. Planta **131 :** 119-120.

– 1979. Microtubular determination of cell shape during colony formation by the algae *Pediastrum. Protoplasma,* **98 :** 1-14.

– 1977. Colony formation and inversion in the green alga *Eudorina ellegans*. *Protoplasma,* **93 :** 325-339.

Marathe, K.V. and P.V. Nandkar 1976. Drainage water algae. Indian J. Environ. Hlth. **18 :** 311– 334.

Marchant, H.J. and J.D. Pickett-heaps, J.D. 1978. Microtubules associated with plasma membrane isolated from protoplasts of green alga *Mougeotia*. Exp. Cell Res. 115:25.

– 1973. Mitosis and Cytokinesis in *Coleocheate scutata*. J. Phycol **9 :** 461-471.

Marchant, H.J. and J.D. Pickett-Heaps. 1974. The effect of colchicine on colony formation in Algae, *Hydrodictyon*, *Pediastrum* and *Sorastrum*. *Planta*, **116 :** 291-300.

– 1972 a. Ultrastructure and differentiation in *Hydrodictyon reticulatum*. III. Formation of the vegetative daughter net. *Aust. J. Biol. Sci.*, **25 :** 265-278.

– 1972 b. Ultrastructure and differentiation in *Hydrodictyon reticulatum*. IV. Conjugation of gametes and development of zygospores and azygospores. *Aust. J. Biol. Sci.*, **25 :** 279-291.

– 1972 C. Ultrastructure and differentiation in *Hydrodictyon reticulatum*. VI. formation of germ net. *Aust. J. Biol. Sci.*, **25 :** 1199-1213.

– 1971. Ultrastructure and differentiation in *Hydrodictyon reticulatum*. II. Formation of zooids within the coenobium. *Aust. J. Biol. Sci.*, **24 :** 471-489.

– 1970 I. Mitosis in coenobium of *Hydrodictyon reticulatum. Aust. J. Biol. Sci.*, **24 :** 471-489.

– 1970 I. Mitosis in coenobium of *Hydrodictyon reticulatum. Aust. J. Biol. Sci.*, **23 :** 1173-1186.

Mast, S.O. 1927. Structure and function of the eye spot in unicellular and colonial organisms. *Arch. Protistenk.*, **60 :** 197-220.

Matilsory M.B., and W.P. Jacobs, 1977. Regeneration in the coenocytic marine alga *Caulerpa* with respect to gravity. *Am. J. Bot.* **70 :** 635-638.

Mattox, K.R., and K.D. Stewart, 1977. Cell division in the scally green flagellate *Heteromastrix angulata* and its bearing on the origin of chlorophyceae. *Am. J. Bot.* **64 :** 931-945.

Maya Bahal, R.K. Shanta, and E.R.S. Talpasayi, 1973. Heterocyst development in *Anabaena amibigua* Rao: V. Effect of some growth regulators and inhibitors. *Phykos*, **12 :** (1-2) : 18.

Mcdonald, K.L. and J.D. Pickett-Heaps. 1976. Ultrastructure and differentiation in *Cladophora glomerata*. I cell division. *Ann. J. Bot. 63* : 592-601.

Mclanchlan, J. 1974. Effects of temperature and light on growth and development of embryos of *Fucus edentatus*, and *F. distichus Sp. distichus. Can. J. Bot.* **52 :** 943-951.

Misra, J.N. 1966. *Phaeophyceae in India* I.C.A.R., New Delhi.

Mitra, A.K. 1951. Certain new members of the Volvocales from Indian soils. Phytomorphology. **1** (**1-2**) : 58-64.

Mellington, W.F., and S.R. Gawlik 1970. Ultrastructure and initiation of wall pattern in *Pediastrum boryanum. Amer. J. Bot.* **57** (5) : 552-561.

Mellington, W.F. and S.R. Gawlik 1975. Cell shape and wall pattern in relation to cytoplasmic organisation in *Pediastrum simplex Amer. J. Bot.*, **62:** 824.

Miller, D.H., I.S. Melhman, D.T.A. Lamport and M. Miller. 1974. The chemical composition of the cell wall of *Chlamydomonas gymnogama* and the concept of a plant cell wall protein. *J. Cell. Biol*, **63 :** 420-429.

Moestrup, D. 1970. The fine structure of mature spermatozoids of *Chara corallina* with special reference to microtubules and scales. *Planta,* **93 :** 295-308.

Moner, J.C. 1955. Cell wall structure in *Pediastrum* as revealed by electron microscopy. *Ann. J. cell. Bot.* **42 :** 802-806.

Moris, I. 1968. *An Introduction to the Algae*. Second ed. Hutchinson and Co. London.

Mundie, J.R. 1929. Cytology and life history of *Vaucheria geminata*. Bot. Gaz. **87 :** 397-410.

Myers, J., and R.D. Preston. 1959. Electron microscope investigation into the structure of the Floridean Fat. *Ann. Bot.* **23 :** 258.

Nagai, R., and T. Hayama. 1979. Ultrastructure of the endoplasmic factor responsible for cytoplasmic streaming in *Chara* internodal cell *J. cell Sci.* **36 :** 121-136.

– and N. Kamiya. 1977. Differential treatment of *Chara* cells with cytocholasin B (CB) with special reference to its effect in cytoplasmic streaming *Exp. cell. Res.* **108 :** 231-237.

Neushul, M. 1970. a fine-etching study of the red alga *Porphyridium Ann. J. Bot.* **57 :** 1231.

–and A.L. Dahl. 1972. Ultrastructural studies of brown algal nuclei. *Ann. J. Bot.* **59 :** 401.

Ohashi, H. 1930. Cytological study of *Oedogonium*. Bot. Gaz. **90 :** 177 - 197.

Pal, B.P., B.C. Kundu, V.s. Sundaralingam and G.S. Venkataraman. 1962. *Charyophyta*. I.C.A.R., New Delhi.

Palmer, C.M. 1969. A composite rating of algal tolerating organic pollution. *J. Phycol.* **5:** 78-82.

Palmer, C.M. 1980. Algal and water pollution. Castle House, Publication Ltd. England.

Palvevitz, B.A. and P.K. Helper. 1975. Identification of actin in *situ* at the ectoplasm endoplasm interface of *Nitella J. cell. Biol.* **65 :** 29-38.

Pankratz, H.S., and C.C. Bowen. 1963. Cytology of blue-green Algae. I. The cells of *Symploca muscorum. Ann. Bot.* **50 :** 387-399.

Papenfuss, G.F. 1935a. Alternation of generations of Ectocarpus siliculosus Dillw. *Bot. Gaz*. **96 :** 411-416.

Papenfuss, G.F. 1935. Alternation of generation in *Ectocarpus Siliculosus*. Bot. Gaz. **96 :** 421-446.

– 1955. Classification of Algae. In *A century of progress in the natural sciences. Proc. Calif. Acad. Sci*. PP. 115-224, San Francisco.

– 1966. A review of the present system of classification of the Florideophycideae. *Phycologia* **5 :** 247-255.

Parker, B.C., and J. Huber. 1965. Translocation in *Macrocystis*. **II**. Fine structure of the sieve tubes. J. Phycol., 1 : 172-179.

Parker, B.C. 1964. The structure and chemical composition of cell walls of three chlorophycean algae. *Phycologia.*, **4**(2) **:** 63-74.

Patel, R.J. 1971. Cytotaxonomical studies of *Pithophora kewensis*. *Phykos*. **10** (1-2) **:** 18-28.

Patrick, R. 1948. Factors effecting the distribution of diatoms. *Bot. Rev.* **14 :** 473-524.

Phillips, R.W. 1896. On the development of the cystocarp in Rhodomelaecae. Ann. Bot. **10 :** 185-204

Phillipose, M.T. 1967. *Chlorococcales*. I.C.A.R., New Delhi.

– 1973. Use of algae especially diatoms in assessment of water quality : In. Biol. methods for the assessment of water quality. ASTM STP 528. amer. soc. for testing and materials. 76-95.

Pickett-Heaps, J.D. 1968. Ultrastructural and differentiation in *Chara fibrosa* IV. Spermatogenesis. *Anst. J. biol. Sci.* **28 :** 655-690.

– 1970. Some ultrastructure features of *Volvox* with particular reference to the phenomenon of inversion. *Planta* **90 :** 174-190.

– 1975. Green Algae, structure, reproduction and evolution in selected genera. Sinauer Associates, Inc, Sunderland, Mass.

– and H.J. Marchant. 1972. The phylogeny of the green algae : a new proposal, *Cytobios* **6 :** 255-264.

– K.L. Mc Donald, and D.H. Tippet. 1976. Cell division in a pennate diatom *Diatoma vulgare*. *Protoplasma* **86 :** 205-242.

Pinevich, A.V., Skulberg O.M., Matthys H.C.P., Schubert H., Willen E. Gavrilova D.V. and Velichko N. 1999. Characterization of a novel chlorophyll b containing *Prochlorothrix* species (Prochlorophyta) and its photosynthetic apparatus. Microbios 100:150–174.

Pocock, M.A. 1960. *Hydrodictyon*-a comparative biological study. J.s. Afr. Bot. **26 :** 167-419.

Prasad, B.N. and M. Saxena 1980. Ecological study of Blue-Green algae in River *Gomati. Indian J. Environ. Hlth*. **22**, (2), 151-168.

Prasad, B.N. and Y. Singh. 1982. On Diatoms as indicators of water pollution. *J. Indian, bot. Soc*. **61**, 326-336.

Prescott, G.W., 1969. *The Algae : A Review*. Thomas Nelson and Son, London.

Pringsheim, E.G. 1949. *Pure cultures of Algae*. Cambridge Univ. Press, Cambridge.

Rai, L.C. and H.D. Kumar, 1976. Systematic and ecological studies on algae of some habitats polluted with fertilizer factory effluent near Sahupuri, Varanasi, India. *Nova Hedwigia* **27 :** 805-811.

Ramanathan, K.R. 1964. *Ulotrichales*. I.C.A.R., New Delhi.

Rana, B.C., T. Gopal, and H.D. Kumar, 1971. Studies on the biological effects of Industrial wastes on the growth of algae. **13** : 138-143.

Randhawa, M.S. 1938. Observations on some Zygnematales from Northern India. Part I. *The Proceedings of the Indian Academy of Sciences* Vol. VIII. No. 3, Sec. B : (47-150).

Randhawa, M.S. 1959. *Zygnemaceae*. I.C.A.R., New Delhi.

Rao, V.N.R. 1970. Studies on *Cyclotella meneghiniana* Kutz. I. Sexual reproduction and auxopore formation. Proc Indian Acad. Sci., **72 B** : 281-287.

– 1971. Studies in *Cyclotella meneghiniana* Kutz. II. Induction of auxospore formation. *Phykos*, **10** (1-2) : 84-98.

Rattan, R.S. 1967. Notes on some Zygnemaceae from Punjab. *Phykos*, **6:** (1-2) : 95-99.

– 1971. Some new taxa of *Spirogyra* link from Punjab-*Phyks* **10** : (1-2) : 137-140.

Raven, P.H. 1970. A multiple origin for plastids and mitochondria. *Science*, Wash., **169** : 641-646.

Riemann, B.E.F., J.C. Lewin and V.E. Benjamin 1966. Studies on the biochemistry and fine structure of silica shell formation in diatoms II. The structure of the cell wall of *Navicula pelliculosa* (Breb) Hilse. *J. Phycol* **2** : 74-84.

Ris, H., and W. Plant. 1962. Ultrastructure of DNA containing areas in the chloroplast of *Chlamydomonas. J. cell Boil.*, **13** : 383.

Robards, A.W. 1970. *Electron microscopy and plant ultrastructure*. McGraw Hill book Co., New York.

Robberts, K. 1974. Crystalline glycoprotein cell walls of algae, their structure, composition and assembly. Philos. Trans. R. Soc. Lond. **268** : 129-146.

Rogalski, A.A., Jovertone, and M. Rudatt,. 1977. An ultrastructural and cyto-chemical investigation of the colonial green alga *Pediastrum tetras* during zoospore formation. *Protoplasma*, **91** : 93-106

Round, F.E. 1965. *The Biology of the Algae*. Edward Arnold Publishers, London.

Sagar, R. 1960. *Chlamydomonas. Science*, Wash., **132** : 460-464.

Saunder, H. 1931. Conjugation in *Spirogyra*. Ann. Bot. 44.

Schmitz, K., and M.L. Srivastava. 1976. The fine structure of sieve elements of *Nereocystis lutekeana. Ann. J. Bot.* **63** : 679-693.

– 1975. On the fine structure of sieve tubes and the physiology of assimilate transport in *Alaria* marginata. *Can. J. Bot.* **53** : 871-876.

– 1974b. Fine structure and development of sieve tubes on *Laminaria groenlandica. Cytobiologic* **10** : 66-87.

Schraudolf, H., and I. Frauenkron. 1979. Effects of Concanavalin on pattern formation in Hydrodictyoceae. *Protoplasma* **98** : 131-138.

Schultz, M.E., and F.R. Trainor. 1968. Production of male gametes and auxospores in the centric diatoms *Cyclotella meneghiana* and *C. cryptica*. J. *Phycol.* **4** : 85-88.

Seerwani, A.B. 1963. Occurrence of *Enteromorpha* at Khargone (M.P.). *Curr. Sci.* **32** : 182.

Shanta, R. Kale, Maya Bahal and E.R.S. Talpasayi : 1973. heterocyst development in *Anabaena ambigua* Rao. IV. Effect of some mineral salts. *Phykos*, 12 (1-2) : 11-17.

Shestakov. S.V. and N.T. Khyeen. 1970. Evidence for genetic transformation in a blue-green alga *Anacystis nidulans. Molec. gen. Genet.*, **107** : 372-375.

Shyam, R. 1980. On the life cycle, cytology and taxonomy of *Cladophora callicoma* from India *Ann. J. Bot.* 67 : 619-624.

Sideman, E.J., and D.C. Scheirer. 1977. Some fine structure observations on developing and mature sieve tube elements in the brown alga *Laminaria saccharina. Ann. J. Bot.* **64** : 649-657.

Silverberg, B.A., and T. Sawa. 1974. A cytochemical and ultrastructural study of the echinate cytoplasmic inclusion in *Nitella flexilis* (Characeae). *Can. J. Bot.*, 52 (1) : 159-165.

– 1974. An ultrastructure study of the pyrenoid in cultured cells of *Chlorella variegata-New Phytologist* **73 :** 143-146.

Singh, K.P. 1958. Some peculiarities in conjugation in *Zygnema terrestris* Randh. Sci. and Cult.24.

Singh, R.N. 1942. Reproduction in *Draparnaldiopsis indica* Bhardwaja. *New Phytol.* **41 :** 262-273.

– 1954. Comparative study of life cycle of two species (*S. farctum* and *S. amoenum*) of genus *Stigeoclonium. Rev. Algol.* (N.S.) (N.S.) **1** : 1.

– 1961. *Role of Blue-green algae in nitrogen economy of Indian Agriculture.* I.C.A.R., New Delhi.

– and K. Sinha. 1965., Genetic recombination in a blue-green alga *Cylindrospermum majus. Nature* Lond. **207** : 782.

– and P.K. Singh. 1967. Isolation of cyanophages from India. *Nature* (hand) **216** : 1020-1021.

Singh, V.P., and K. Trehan. 1973. Extracellular protein amino-acids of blue-green algae I. The production of extracellular amino acids by *Aulosira fertilissima* and *Anacystis nidulans. Phykos.* 12 (1-2) : 36.

– 1973, Extracellular protein amino acids of blue-green algae II. Effect of nitrate and amino acids on liberation of amino acids by *Aulosira fertilissima. Phykos*, 12 (1-2) : 42.

Singh, V.P., and K. Trehan. 1976. Recombination in Algae, *Recent Advances in Botany.* P.N. Mehra Commemoration Volume pp. 1-9.

Singh, V.P., A.K. Sinha and D.E. Reuben. 1994. Ecological studies on the algal flora of River *Sai* at Rai-Bareli. In Environment Management in developing countries vol. II : Water and its Management, Gyanodaya Prakashan Nainital. 406-410.

Singh, P.K. 1973. Occurrence and distribution of Cyanophages in ponds, sewage and rice fields. *Arch. Mikrobiol.* **89 :** 169-172.

Sinha, A.K. D.C. Pandey, A. Kumar and V.P. Singh 1989. Algal flora of River Ganga between Shuklaganj (unnao) and Kalakankar (Pratapgarh). *Advance in Applied Phycology* - II. 287-293.

Smith, G.M., and F.D. Kyiver. 1929. *Draparnaldiopsis*, a new member of the algal family Chaetophoraceae. *Trans. Amer. Microsc. Soc.*, **48 :** 196-203.

Smith, G.M. 1938. Nuclear Phases and alternation of generations in Chlorophyceae. Bot. Rev. **4** : 132-130.

Smith, G.M. 1950. *Freshwater algae of the United States.* Second edition, New York.

– 1951. *Manual of Phycology.* Chronica Botanica Publishing Co., Waltham, Mass.

– 1955. *Cryptogamic Botany*, Vol. I, Second edn. McGraw-Hill Book Co., New York.

Speer. H.L., W., Dougherty, and R.F. Jones. 1964. Studies on the fine structure of the red alga *Porphyridium cruentum. J. Ultrastruct. Res.*, II : 84.

Spessard, E.A. 1930. Fertilization in a living *Oedogonium.* Bot. Gaz. **89 :** 385-393.

Starr, R.C. 1968. Cellular differentiation in *Volvox Proc. Nat. Acad. Sci.* U.S.A., **59** : 1082-1088.

– 1970. Control of differentiation in *Volvox. Develop. Biol.* (Suppl.) **4 :** 59-100.

Stein, R. 1958. A morphological and genetic study of *Gonium pectorale. Ann. bot.* **45 :** 564-567.

Stewart, K.D., and K.R. Mattox. 1975. Comparative cytology, evolution and classification of green algae with some consideration of the origin of other organisms with chlorophyll *a* and *b. Bot. Rev.*, **41 :** 104-135.

– and G.L. Floyed. 1973. Mitosis, Cytokinesis and distribution of plasmodesmata and other cytological characteristics in the Ulotrichales, Ulvales and Chaetophorales-phylogenetic and other taxonomic considerations. *J. Phycol.* **9 :** 128.

Stewart, W.D.P. 1967. Nitrogen fixing plants. *Science.*

Sumons, E.B. 1906. A morphological study of *Sargassum philipendula.* Bot. Gaz. **41 :** 161-180.

Wash., 158 : 1426-1432.

– 1970. Algal fixation of atmospheric nitrogen. *Plant and Soil*, **32.** 355-586.

– 1970. Heterocysts of blue-green algae. *First Interna. Symp on Taxonomy and Biology of blue-green algae*. Univ. Madras.

Subrahmanyan, R. 1948. Somatic division, reduction division, auxospore formation and sex differentiation in *Navicula holophila, Iyengar Commemoration Volume*, pp. 239-266.

Sundaralingam, V.S. 1954. The developmental morphology of *Charazeylanica* Willd. J. Indian Bot. Soc., **33 :** 272-296.

– 1959. Comparative morphology of the Charophytes. *Proc. Symp. Algology*, I.C.A.R., New Delhi 78-84.

– 1962. Studies on Indian Charophytes. II Developmental morphology of three species of *Nitella.* Agardh. Phykos, I (2) : 61-75.

– 1965. Studies on Indian Charophytes, IV. Developmental morphology of three more species of *Nitella.* Agardh. *Phykos*, **4** (1) : 19-39.

Sundaralingam, V.S. 1948. The cytology and spermatogenesis in *Chara zeylanica. Iyengar Commemoration Volume* pp. 289-303.

Tahara, M. 1909. On the periodical liberation of oospores in *Sargassum*. Bot. Gaz. **23 :** 151-153.

Talpasayi, E.R.S., and S.K. Kale. 1967. Induction of heterocysts in the blue-green alga *Anabaena ambigua. Curr. Sci.*, **36 :** 318-319.

Tamiya, H. 1959. Role of algae as food. In *Proc. Symp. Algology* : 379-389, I.C.A.R., New Delhi.

Taylor, W.R. 1922. Recent studies of Phaeophyceae and their bearing on classification. Bot. Gaz. **74** : 431-441.

– 1936. Phaeophycean life histories in relation to classification. Bot. Rev. **3** : 554-563.

Thivy, F. 1959. Seaweed Utilization in India. In *Proc. Symp. Algology* : 345-365, I.C.A.R., New Delhi.

Trainor, F.R. 1958. Control of sexuality in *Chlamydomonas chlamydogana. Ame. J. Bot.* **45 :** 621-626.

– 1959. Comparative study of sexual reproduction in four species of *Chlamydomonas. Ame. J. Bot.*, **46 :** 65.

Turner, F.R. 1968. An ultrastructural study of plant spermatogenesis : Spermatogenesis in *Nitella. J. cell. Biol.* **37 :** 370-393.

Tyagi, V.V.S. 1973. Effect of some metabolic inhibitors on heterocyst formation in blue-green alga *Anabaena doliolum. Ann. Bot.* **37 :** 361-368.

– and A.S. Ahluwalia. 1978. Heterocyst formation in blue-green alga *Anabaena doliolum*. A study of some aspects of photoregulation. *Ann. Bot.* **42 :** 1333-1341.

Tyagi, V.V.S. 1975. The heterocysts of blue-green algae. *Biol Review* **50 :** 247-284.

Tyron, E.L., and J.S. Davis. 1976. A new asexual cycle in *Pediastrum simplex* strain. Bot. Gaz. 137(4) : 356-360.

Uedo, K. 1961. Structure of plant cells with special reference to lower plants VI. Structure of chloroplasts in algae. *Cytologia*, **26 :** 344-358.

Vasishta, P.C. 1960. On the structure and life history of *Rivularia mehrai* sp. Nov. *Punjab Univ. Res. Bull* (N.S.) 2 (3-4) : 237-244.

– 1962. Some observations on the life history of *Rivularia*. Joshi Sp. Nov. from Hoshiarpur, Punjab, India. *J. Indian Bot. Soc.*, Vol. XLI, No. **4 :** 516-523.

Venkataraman, G.s. 1961. *Vauchericeae*. I.C.A.R., New Delhi.

– 1973. *Algal Biofertilizers and Rice cultivation*. I.C.A.R., New Delhi.

– S.K. Goyal, B.D. Kaushik and Paromita Roy Choudhury, 1974. *Algae : Form and Function*

Today and tomorrow Printers and Publishers, New Delhi.

Viamontes, G.I., and Kirk, D.L. 1977. Cell shape changes and mechanism of inversion in *Volvox. J. cell. Biol.* **75** : 719-730.

Vidyavati and J. Nizam 1973. Conjugation studies in *Closterium acerosum* Ehren. *Phykos*, **12** (1-2) **:** 61-71.

– 1973. Cellular events of conjugation in *Cosmarium auriculatum* var. bogoriense. Phykos, 12 (1-2) **:** 72-78.

Walsby, A.E. 1980. The water relations of gas-vacuolate prokaryotes. Proc. R. Soc. Lond. **8208 :** 73.

West, G.S. 1916. Algae. Vol I. Bot. Handbooks. Cambridge, A treatise on the British fresh water Algae.

West, G.S., and O.E. Hood. 1911. The structure of the cell wall and the apical growth in the genus *Trentepohlia, New Phytol.* **10** : 241-248.

Westbrook, M.A. 1928. Contribution to the cytology of tetrasporic plants of *Rhodymelia palmata* (L) Grey and some other Florideae. Ann. Bot. **42 :** 149-172.

Wetmora, R.H., A.E. Demaggio, and G. Morel. 1963. A morphogenetic look at the alternation of generations. *J. Indian Bot. Soc.*, **42A** : 306-320.

Whitten, B.A., 1970. Biology of *Cladophora* in Fresh waters—a review paper. Water Research Pergamon Press, 4 : 457-476. Printed in Great Britain.

Wildman, R.B., and C.C. Bowen. 1970. Phycobilins in blue-green algae. *First Internat.Symp*. on *Taxonomy* and *Biology of Blue-green algae*. Univ. Madras.

Williamson, R.E. 1974. Actin in Alga. *Characorallina. Nature* **248 :** 801-802.

– 1975. Cytoplasmic streaming in *Chara.* A cell model activated by ATP and inhibited by cytochalasin *B.J. cell Sci.* **17 :** 665-668.

– Williamson, R.E. 1980. Actin in motility and other processes in plant cells *Can. J. Bot.* **58** (7) **:** 766-772.

Wolk, P. 1966. Evidence of a role of heterocysts in the sporulation of a blue-green alga. *Ame. J. Bot.* **53 :** 260-262.

Wolk, C.P. 1973. Physiology and cytological chemistry of blue-green algae. *Bacteriol. Rev.* **37 :** 32-101.

Womersley, H.B.S. 1979. Southern Australia species of *Polysiphonia grivelle* (Rhodophyta). *Aust. J. Bot.* **27 :** 439-528.

Wood, R.D. 1952. On characeae. Bot. Rev. **18** (5) **:** 317-393.

Wyatt, J.T., and J.K.G. silvey, 1969. Nitrogen fixation by *Gloeocapsa, Science* Wash., **165 :** 908-911.

Yamanouchi, S. 1906. The life history of *Polysiphonia violacea*. Bot. Gaz. **41 :** 425-433.

– 1906. The life history of *Polysiphonia violacea*. Ibid. **42 :** 401-449.

INDEX

A

B

C

D

E

H

I

K

L

P

Q

R

S

T

U